Signals and Communication Technology

More information about this series at http://www.springer.com/series/4748

Silvia Maria Alessio

Digital Signal Processing and Spectral Analysis for Scientists

Concepts and Applications

 Springer

Preface

This book is essentially a guided tour of the fundamental concepts in digital signal processing (DSP) and analysis, in which the guide is not a theoretician of the various algorithms involved, but a user of them; a scientist that early in her life started studying these techniques to apply them in a proper and fruitful way in her research work. Therefore the approach to the various topics is the approach of a user, more concerned about the way to best exploit the possibilities they offer, rather than their purely mathematical facets.

As a physicist, my research interests are in the field of paleoclimatology and analysis of meteo-climatic, oceanographic, and biomedical data. These interests led me, during the last three decades, to first study and then apply the techniques discussed in this book, as well as to teach many students how to use them correctly in their domain of interest. My personal experience with the vast literature available on the market often encouraged me to respond to unfulfilled curiosities with personal elaborations. For this reason, the present book is not a mere collection of the best treatments of the various methods found in the literature, but is rich in original insights and developments related to issues often neglected elsewhere, visualized in original and, hopefully, clarifying ways. All methods explained in the book have been tested in detail by the author analytically and numerically. The majority of the figures have been produced using the software written for these tests and simulations.

The present book evolved from course notes I developed over a period of about 15 years while teaching undergraduates and early graduates the basics of signal processing and spectral analysis. My students at the University of Torino, Italy, mostly had a background in physics and had been exposed to calculus, elements of analysis (e.g., Fourier series), basic linear algebra (vectors and matrices), and some elements of complex analysis, all of these topics being fairly standard in an undergraduate program in many scientific sectors. When the notes reached a sufficient maturity and completeness, the idea was born, of turning them into an easy-to-use introductory book for students and researchers, not only in physics but in all quantitative disciplines.

vii

In the past, DSP was taught, and related books were written, almost exclusively for people with a sound engineering background. I, however, deem that we should treat DSP in a similar way as how statistics is taught. Instead of claiming that every person who wants to know a little about statistics has to learn everything there is to know about statistics, many authors created special statistics courses and books for humanities majors, engineers, life sciences majors, and so on. Statistics is useful in many fields, and a helpful subset can be taught and used without previously studying all the background theories. DSP has these same characteristics: almost anyone can learn the basics of DSP and successfully apply these tools in their field, without being an engineer. As in statistics, there are trade-offs between the size of the "DSP toolbox" a person can put together and use skilfully, and the amount of study and background required. At one extreme, one can end up with just the step-by-step "recipe" for performing a certain task, without any need to understand what is behind it. At the other extreme, one can reach the point where they can select techniques from a wide array of possibilities, or even create their own, though this will require a very good background in mathematics.

This book falls some way in between these two extremes. It covers the basics of processing of monovariate discrete-time signals, digital filter design, and filtering implementation, as well as an in-depth discussion of classical and advanced methods of stationary and evolutionary spectral analysis. The books on these topics are uncountable, but often quite sectorial. For example, it is not easy to find a detailed treatment of the bases of signal processing and digital filters, coupled with a comprehensive review of spectral analysis methods. This book was meant to meet this need.

An effort has been made to make the text self-contained for readers with a good command of basic mathematical analysis, at the level taught in scientific bachelor programs, and to avoid the need for more advanced mathematical prior knowledge. Where some additional information is required, for instance about functions of complex variable, a brief review of the related notions has been integrated into the book, expressed in simple terms, separated from the main text and confined to an appendix, in order to make a first reading less cumbersome.

Mathematical rigor is not the emphasis of this book. I treated digital signal processing and spectral analysis as mathematics applied to a practical topic: not as mathematics for its own sake, but not as a set of applications without theoretical foundations either. I tried to be precise every time when mathematical issues are involved, but I also avoided going deep into mathematical facets, the knowledge of which does not add anything to the practical potential of a technique in data analysis. The book thus emphasizes results and their practical implications rather than calculations and formal demonstrations.

The approach is definitely practical: the book aims at making the reader aware of the purposes, the indications, the advantages, the drawbacks of the various tools, as well as the possible sources of misuse and consequent error. For example, the text shows how to filter a data record according to some desired specifications, and how to carry out a spectral analysis using the most suited method, depending on the characteristics of the series and on the spectral features to be detected. It was meant

to help the reader solve real problems. In order to illustrate the number and variety
of fields in which the methods described in the book may be useful to extract
information from data, several pages are devoted to real-world applications in
different scientific fields, from geophysics to medicine, macroeconomics, etc. The
book contains many figures illustrating in detail the various aspects of the methods,
to make understanding intuitive. These examples are based on synthetic data (noise
and sinusoids) and real measurements. A complete set of Matlab exercises requiring
no previous experience of programming is also proposed. However, no preliminary
tutorial on Matlab is given. It would not make any sense, when excellent books and
guides are available on this topic, including the Matlab User's Guide and Reference
Guide. Besides, only a very basic knowledge of Matlab is required to perform the
exercises, and all steps are described in detail.

The book is structured in such a way that some parts can be skipped without loss
of continuity. For example, if the reader is not interested in filters but only in
spectral analysis, they can pass directly from the first introductory part to the
chapters dealing with power spectrum estimation. However, even if digital filters
are traditionally employed by engineers, physicists, and mathematicians, they are
now being more and more extensively used in other fields such as economics,
sociology, life sciences, etc. On the other hand, spectral analysis is a fundamental
tool not only in physics (in particular in geophysics and in other sectors like
electromagnetism, optics, electronics, acoustics and so on) but also in economical,
socio-political, and biomedical research areas. Data analysis is becoming increas-
ingly needed in traditionally less quantitative disciplines, in response to the growing
amount of data now available to researchers: ranging from consumer preference and
sociological data offered by social networks to the huge flux of measurements
deriving from new high-throughput experiments in biological and medical research.

I, therefore, believe that the book can be useful not only to engineering and
physics students at an advanced undergraduate level, but also to students in geology, biology, economics, sociology, etc., because it contains all the main tools
useful to anybody who needs to process and analyze a sequence of measurements.
Also graduate-level students and researchers lacking an extensive knowledge of
advanced mathematics could profit by its approach, which, though being quantitative, has a discursive feel. Even if this work is mainly meant to be an upper
undergraduate university textbook, it can moreover be a useful tool of independent
study for further education and training in the industrial context.

I hope that with this book, digital signal processing can definitely spread out
from the domain of engineering to address the needs of all scientists and scholars.
I also hope that the techniques presented in this book will be useful to the readers in
their studies, as well as in their academic and professional lives, as they have been
in mine.

I welcome your feedback.

Torino   Silvia Maria Alessio
April 2015

Acknowledgments

First of all, I would like to thank my colleagues and friends Alfred R. Osborne, Renzo Richiardone, and Carla Taricco (Department of Physics, University of Torino, Italy) for encouraging me to write this book since I first mentioned what, at the time, was just an idea.

I am indebted to all the researchers who gave me permission to use the results of their studies as examples of real-world applications of the techniques presented in this book. In particular, I would like to thank warmly the following people for providing me with their data, and sometimes their software: Luís Aguiar-Conraria,[1] Michael Ghil,[2] Michele Greco,[3] Dmitri Kondrashov,[4] Lisa Sella,[5] Carla Taricco,[6] and Gianna Vivaldo.[7]

The data of seasonal climatic anomalies of sea surface temperature (SST) over the so-called NINO3 region of the Pacific Ocean used in this book was downloaded from the NOAA website. The data for another example involving an electrophysiological signal was downloaded from the DaISy (Database for the Identification of Systems), maintained by ESAT-Katholieke Universiteit Leuven, Leuven, Belgium. Thanks are due to all the researchers involved in the maintenance of these databases. The data for an example concerning spectral analysis of musical records are due to B. Porat.[8]

---

[1]NIPE and Departament of Economics, University of Minho, Braga and Guimarães, Portugal.

[2]Geosciences Department and Laboratoire de Météorologie Dynamique (CNRS and IPSL), École Normale Supérieure, Paris, France, and Department of Atmospheric & Oceanic Sciences and Institute of Geophysics & Planetary Physics, University of California, Los Angeles, CA, USA.

[3]Department of Matemathics, Physics and Informatics, University of Urbino, Italy.

[4]Department of Atmospheric and Oceanic Sciences and Institute of Geophysics and Planetary Physics, University of California, Los Angeles, CA, USA.

[5]Department of Economics, University of Torino, CNR-Ceris, Moncalieri, Torino, Italy.

[6]Department of Physics, University of Torino, Italy.

[7]IMT-Institute for Advanced Studies, Lucca, Italy.

[8]Department of Electrical Engineering, Technion-Israel Institute of Technology, Haifa, Israel.

I also would like to acknowledge the use of the Matlab-based wavelet software written by Christopher Torrence[9] and Gilbert P. Compo,[10] and of the SSA-MTM Toolkit distributed by the SSA-MTM Group.[11]

Thanks also to Sebastiano Acquaviva,[12] who managed to set aside some time in his very busy life to read the book and give me his precious opinion; to Gianna Vivaldo, who helped me with fine-tuning the chapter on singular spectrum analysis; to Roberto Giannino, who skillfully helped me with graphics issues; and to Richard D. Robinson, who found time in his round-the-world trip to proofread my work.

At last, some shameless lyricism.... While I was working to this book, my husband supported me in many ways: without him I would have starved before finishing. I would also like to thank a young physicist in love with neuroscience, Giulio Matteucci, for valuable discussions and suggestions. He happens to be my son too... so, thanks for preparing pizza for the whole family when the refrigerator was empty. Finally, thanks to my cat Misty, for the funny nonsense he (yes, he... a little misunderstanding between us when he was born...) typed so many times while he was walking on the keyboard of my computer, in the attempt to reach the tip of my nose and lay a feline kiss there. Laughter relieves fatigue!

Torino, April 2015   Silvia Maria Alessio

---

[9]Exelis, Boulder, CO, USA.

[10]CIRES, University of Colorado and Physical Sciences Division, NOAA ESRL, Boulder, CO, USA.

[11]Department of Atmospheric Sciences, University of California, Los Angeles, CA, USA.

[12]Encosys s.r.l.—Energy Control Systems, Italy.

Contents

Acronyms

AC	Autocorrelation/autocovariance
AR	Autoregressive
ARMA	Autoregressive moving-average
BK	Broomhead-King
BT	Blackman–Tukey
CESD	Cross-energy spectral density
c.l.	Confidence level
CPSD	Cross-power spectral density
CS	Compression score
CTFT	Continuous-time Fourier transform
DFS	Discrete Fourier series
DFT	Discrete Fourier transform
DOF	Degrees of freedom
DOG	Derivatives of Gaussian
DTFT	Discrete-time Fourier transform
DWT	Discrete wavelet transform
ESD	Energy spectral density
EOF	Empirical orthogonal function
FFT	Fast Fourier transform
FIR	Finite impulse response
Flop	Floating point operation
FWT	Fast wavelet transform
GLP	Generalized linear phase
IDFT	Inverse discrete Fourier transform
IDTFT	Inverse discrete-time Fourier transform
IDWT	Inverse discrete wavelet transform
LCCDE	Linear constant coefficients difference equation
LD	Levinson–Durbin
LMS	Least mean square
LP	Linear phase
LQE	Linear quadratic estimation

LS	Least squares
LTI	Linear time-invariant
MA	Moving average
MP	Maximal ripple
MSC	Magnitude square coherence
MSE	Mean squared error
MMSE	Minimum mean squared error
MUSIC	Multiple signal classification
IIR	Infinite impulse response
ON	Orthonormal
PAC	Partial AC
PR	Perfect reconstruction
PSD	Power spectral density
RC	Reconstructed component
RCSR	Real causal stable rational
RE	Retained energy (squared-norm recovery)
RLS	Recursive least square
RMSD	Root-mean-square deviation
SFG	Signal flow graph
SNR	Signal-to-noise ratio
SVD	Singular value decomposition
VG	Vautard–Ghil
WSS	Wide-sense stationarity, or wide-sense stationary
YW	Yule–Walker

Chapter 1
Introduction

1.1 Chapter Summary

Since this book is titled "Digital Signal Processing and Spectral Analysis for Scientists", first we should understand the meaning of the words contained in this title. While we do not need to explain what scientists and the Sciences are, we certainly need to define precisely the remaining words, and to do so we will re-arrange them in this order: Signal, Processing, Digital, Spectral Analysis. After doing so, we will go through a concise description of the historical background in which the techniques introduced in the book were developed. The structure of the book will then be presented. Finally, some possible further reading will be suggested.

1.2 The Meaning of the Book's Title

A *signal* is a formal description of a phenomenon evolving over time or space; by signal *processing* we mean any operation which modifies, analyzes or otherwise manipulates the information contained in a signal (Prandoni and Vetterli 2008). For example, consider atmospheric pressure at some site. We could think of measuring it at regular intervals—for example every 2.5 min—using a barometer and record the evolution of this variable over time: the resulting data set—a *sequence* of numbers with a proper measurement unit, e.g., hPa—represents an atmospheric pressure signal. Simple processing operations can then be carried out on this signal: for example, we can plot it versus time, we can compute the average pressure in a day, etc. From a conceptual point of view, it is essential to understand that signal processing operates on an abstract representation of a physical quantity and not on the quantity itself. In other words, we need a formal description of the signal in order to be able to operate on it. The description may just be a two-column table of values (the time and the value of the variable) as in the previous example, or a mathematical function, or a statistical description. Moreover, the type of abstract representation that

© Springer International Publishing Switzerland 2016
S.M. Alessio, *Digital Signal Processing and Spectral Analysis for Scientists*, Signals and Communication Technology,
DOI 10.1007/978-3-319-25468-5_1

we adopt for the physical phenomenon of interest determines the nature of the signal processing we will be able to perform on it. If, for instance, the signal is described by a mathematical function, then we will know all its past, present and future values with certainty; we will call it a deterministic signal and will treat it as such. If the signal has random characteristics, as in the case of physical measurements affected by random errors, then we could describe it in terms of some average values.

The adjective *digital* derives from "digitus", the Latin word for finger, and refers to the point of view on the real world according to which everything can ultimately be represented as an integer number—exactly as mankind started to do a long time ago, counting along the fingers of their hands. Digital signal processing is a branch of signal processing in which everything, including time, is abstractly represented by the set of natural numbers, regardless of the signal's origins (Prandoni and Vetterli 2008). In our former example, we decided to measure atmospheric pressure every 2.5 min, starting from some time origin. We make a first measurement, and say that the pressure value we get is relative to time $t = 0$. It is also our first pressure sample and we can assign it an adimensional time-index $n = 0$. Later, at $t = 2.5$ min, we collect another sample, and assign it to $n = 1$; and so on. Thus we have no difficulties with using only integer time values. If starting from discrete time n we want to go back to dimensional time, we will just multiply n by our *sampling interval* $T_s = 2.5$ min. We have obtained a *discrete-time signal* by sampling a physical quantity that conceptually exists at any instant of time, and is therefore a continuous signal as a function of continuous time, i.e., a *continuous-time signal*: we could also say it is an *analog signal*. We could use a recording machine capable of providing an analog representation of the physical phenomenon, for example a barograph. An aneroid barometer is an instrument for measuring pressure that uses a small, flexible metal box called an aneroid cell, which is made from an alloy of beryllium and copper. The evacuated capsule is prevented from collapsing by a strong spring. Small changes in external air pressure cause the cell to expand or contract. This expansion and contraction drives mechanical levers such that the tiny movements of the capsule are amplified and displayed on the face of the aneroid barometer. If the pointer in an aneroid barometer is replaced with a pen, then we have a barograph that produces a paper or foil chart called a barogram, recording the barometric pressure over time. The problem with these analog recordings is that they are not abstract signals but a conversion of a physical phenomenon into another physical phenomenon: we do not get sequences of numbers that can be processed. This is why we decided to sample our analog signal to turn it into a discrete-time signal.

What about the pressure values we get in our experiment? According to the measuring device we use, they will have a certain number of significant decimal places, and therefore a number of significant digits. In a proper way, we will be able to treat these values as integers. This is the concept of *discrete amplitude*. If a signal is both discrete-time and discrete-amplitude, then it is called a *digital signal*. Our example is useful for clarifying the meaning of several words with which we will soon become familiar, but depicts a particular case: the case in which a sequence of numbers is obtained by sampling a continuous variable of continuous time. Actually, there are signals that are intrinsically discrete-time, such as, for example, the record of the number of aircraft movements per hour in Fiumicino's airport in Rome—actually, this would be an intrinsically digital signal.

Now, if instead of sampling a physical quantity, we decided to sample some known mathematical function, in principle the number of digits would be unlimited, i.e., our signal would be discrete-time but *continuous-amplitude*, rather than digital. In this case time would not need to be discrete either. We would not need to perform any sampling, we could just treat that function of time using the standard tools for functions of continuous variables. However, if the processing we want to apply to our signal has to be done on a computer, a machine that uses finite-precision arithmetic, the continuous values of the signal will inevitably be converted into a vector—an indexed sequence—of discrete values.

It may be useful to add some words about the finite precision that our measurements inevitably have in any application.[1] If we model a phenomenon as an analytical function, not only is the independent variable (time) a continuous variable, but so is the function's value; a practical measurement, however, will never achieve an infinite precision. Consider our pressure example once more: we can use a certain type of barometer allowing us to record just the number hPa; we can use a better instrument and be able to record the tenth of hPa as well, but there is a limit beyond which we cannot go. Now, if we know that our measures have a fixed number of digits, the set of all possible measures is actually countable and we have effectively mapped our variable onto the set of integer numbers. This process is called *amplitude quantization*, or simply quantization, and it is the method that, together with sampling, leads to a fully digital signal. Moreover, recall that even if originally, during measurement operations, our signal had undergone no quantization effect, it would still undergo it when we processed it by a computer: a loss of precision with respect to the ideal continuous model is unavoidable in practice.

Due to its contingent nature, quantization is almost always ignored in the theory of signal processing and all derivations are carried out as if the signal samples were real numbers; therefore, in order to be precise, we should use the term "discrete-time" signal processing, and leave the word "digital" to the world of actual devices. Neglecting quantization will allow us to obtain very general results, but we must keep in mind that in practice, actual implementations will have to deal with the effects of quantization. Nevertheless, "digital signal processing" (DSP) is the most common expression to indicate the topics we are going to tackle.

In sampling a conceptually continuous-time variable, one issue arises concerning the rapidity of the variations we want to record. If the measured quantity varies rapidly, we can think of a device able to perform sampling with an adequately small

---

[1] In numerical analysis, "precision" is the resolution of the representation of the measured quantity, typically defined by the number of decimal digits; more precisely, resolution is the smallest change in the underlying physical quantity that produces a response in the measurement instrument. It must be noted that in the fields of science and statistics, the word "precision" traditionally has a very different meaning. It refers to repeated measurements of the same quantity and is related to reproducibility and repeatability. It is the degree to which repeated measurements under unchanged conditions provide the same results. On the other hand, the "accuracy" of a measurement is the degree of closeness of the measurement of a quantity to the true value of the quantity.

sampling interval, but we will never be able to match infinitely rapid variations: our measuring machine, however fast, still will never be able to take an infinite amount of samples in a finite time interval. Therefore, sampling is an operation that can be performed without loss of information only if the variations we want to record are not infinitely rapid, and we correctly choose the sampling time according to the characteristics of the phenomenon of interest. Once we have a set of measured values, the tools of calculus have a discrete-time equivalent which we can use directly, but the equivalence between the discrete and continuous representations only holds for signals which are sufficiently slow with respect to how fast we sample them: we need to make sure that the signal does not do crazy things between successive samples, and only if it is smooth and well behaved can we afford to have such sampling gaps. We will actually study the *sampling theorem*, that links the speed at which we need to periodically measure the signal to represent if faithfully to the maximum frequency contained in its *spectrum*.

But what is a spectrum? What is *spectral analysis*? All has to do with the Fourier idea of expressing a mathematical function of time as a function of frequency, through the transform that bears his name. For instance, the transform of a musical chord made up of pure notes without overtones, expressed as loudness as a function of time, is a mathematical representation of the amplitudes and phases of the individual notes that make it up. The function of time is often called the *time-domain representation* of the signal, and the corresponding function of frequency is called the *frequency-domain representation*. For discrete-time signals, Fourier tools exist that can be used in place of the continuous-time Fourier transform, and can provide a description of the frequency content of the signal. The main purpose of this description is the detection of any periodicities in the data.

Now, let us resume for the last time our atmospheric-pressure example. When we are through with our measurements, we will have a record with some duration, composed of a finite number of samples. We are interested in the characteristic of atmospheric pressure at a certain point of the Earth system, rather than in those of the particular sequence we measured. In fact, the ultimate subject of the investigation is the process that generates the measured values. For example, we typically want to get information about its spectral content, because this can help us to investigate external forcings that determine pressure variations, to establish the relation among the behavior of pressure and that of other atmospheric variables, etc. We must keep in mind that this process and the quantity we are considering may have, and usually do have, a temporal persistence that goes beyond the finite time limits of our measurements. From a theoretical point of view, this quantity is an infinitely-long signal, typically an analog signal, of which we can measure only a finite number of samples over a finite time span. Moreover, even if we measured an intrinsically deterministic quantity, the measure would be affected by random errors, i.e., *noise*, and the resulting digital signal should be interpreted as the superposition of a deterministic signal, and a random signal representing measurement errors. But most times, the process that generates the quantity is so complicated that we do not know it enough to be able to express it in deterministic terms, and anyway, even if we were able to do so, the deterministic description would be too complicated to be of practical use.

On the other hand, we saw that processing a signal requires a formal description of the signal—a *mathematical model* of the signal. A statistical-probabilistic model is then adopted, and the signal is treated as a *random signal*, the value of which is not exactly specified at a given past or current instant, and the future value of which is not predictable with certainty on the basis of past behavior. The issue then arises, about how to *estimate* the spectrum of atmospheric pressure at a certain site, from a single finite-length sequence of values it assumed in a particular experiment. This spectral estimation problem is addressed by proper techniques, collectively known as *statistical signal processing*.

1.3 Historical Background

The earliest recorded signals were digital. A classical example (mentioned, e.g., by Prandoni and Vetterli 2008) is provided by the Egyptian stele known as the Royal Annals of the Old Kingdom of Ancient Egypt, of which seven fragments have remained. This stele contained a list of the kings of Egypt from the First Dynasty to the early part of the Fifth Dynasty, and noted significant events in each year of their reigns, including measurements of the height of the annual Nile flood.

Digital representations of natural and social phenomena such as those depicted by the Royal Annals stele are adequate for a society in which quantitative problems are simple: counting sheep, days and so on. As soon as the interaction between man and his environment becomes more complex, the models used to interpret the world must evolve. Geometry, for example, was born of the necessity to measure and sub-divide land property. Splitting a certain quantity into parts implies difficulties with an integer-based world view. So, the idea of fractions, i.e., of rational numbers, was conceived, and proved helpful until another issue become evident in western civilization: in the 6th century BC, the Pythagorean School discovered that the diagonal and the side of a square are incommensurable, i.e., that $\sqrt{2}$ is not a simple fraction. The discovery of what we now call irrational numbers led to a more abstract model of the world, the one which today is called the continuum. Rooted in geometry and in the idea of the infinity of points in a segment, the continuum model assumes that time and space are an uninterrupted flow that can be divided arbitrarily many times into arbitrarily and infinitely small pieces. In signal processing language, we can say that the analog model of the world was born. The concepts of the infinitely big and the infinitely small contained in the analog model required almost two thousand years to be properly mastered (Prandoni and Vetterli 2008).

The digital and analog models of the world coexisted for several centuries, one as the tool of the trade for farmers, merchants etc., the other as an intellectual pursuit for mathematicians and astronomers. However, the increasing complexity of an expanding world pushed many minds towards science as a means to solve very tangible problems, and not only for describing the motion of the planets. Calculus, born in the 17th century from the work of Newton and Leibnitz, proved to be a powerful tool when applied to practical problems such as ballistics, ship routing, mechanical design and so on. Still, in order to apply calculus to any real-world signal, the latter has to be modeled as a function of a real continuous variable. As mentioned before, there are domains of research in which signals allow for such an analytical representation: astronomy, for example, and ballistics. If we were to go back to our pressure measurement example, however, we would need to model our measured quantity as a function of continuous time, which means that the value of pressure should be available at any given instant. A reasonable pressure function is not easy to obtain if all we can have is just a set of empirical measurements, even correctly spaced in time. So we need algorithmic procedures especially devised for discrete-time variables. These two tracks of scientific development, which ran parallel for centuries, have found their balance in digital signal processing that uses both continuous-time and discrete-time mathematical approaches.

Historically, modern digital signal processing started to consolidate in the 1950s, when mainframe computers became powerful enough to properly handle problems of practical interest. After World War II, computers based on the technology of vacuum tubes started being constructed. The first machine of this type was ENIAC (Elecronic Numerical Integrator and Computer) developed by Mauchly and Eckert at the University of Pennsylvania, USA, between 1943 and 1946. It was made of 18,000 electron tubes and 1500 relais; its electrical consuption was around 150 KW, its weight 30 tons, and occupied a room 30 m long. Another important machine was IAS, developed in Princeton, New Jersey, USA, by John Von Neumann. Von Neumann had worked to the ENIAC project; he conceived an innovative machine, able to execute instructions recorded in its memory. IAS started operating in 1952.

The foundations of digital signal processing were built in 1960–1965, and by the end of the 1970s the discipline had reached maturity: indeed, the major textbooks on the subject still in use today had already appeared by 1975. However, the theoretical bases of DSP had already been laid by Jean-Baptiste Joseph Fourier (1768–1830), who made important contributions to the study of trigonometric series, after earlier seminal studies by Leonhard Euler, Jean le Rond d'Alembert and Daniel Bernoulli. Fourier introduced trigonometric series to solve the equation for the conduction of heat in a metallic rod and published his initial results in 1807 ("Mémoire sur la propagation de la chaleur dans les corps solides"). In 1822 he published his fundamental work, "Théorie analytique de la chaleur". Later, major theoretical developments were achieved in the 1930 and 1940s by Nyquist and Shannon, among others, in the context of digital communication systems, and by the developers of the z-transform, notably Zadeh and Ragazzini in the West, and Tsypkin in the East.

The history of applied digital signal processing, at least in the electrical engineering world, began around the mid-1960s with the invention of the so-called fast Fourier transform (FFT). However, its rapid development started with the advent of microprocessors in the 1970s. Early DSP systems were designed mainly to replace existing analog circuitry and did little more than mimicking the operation of analog signal processing systems. It was gradually realized, however, that DSP had the potential for performing tasks that were impractical or even inconceivable by analog means.

An overview of the history and development of digital signal processing would be too long and tedious, also because the topic includes several main streams, while for our purposes we are interested only in a few of them. We may mention:

- numerical analysis. At the beginning, the development of discrete-time system theory was motivated by a search for numerical techniques to perform integration and interpolation, and to solve differential equations. When computers became available, the solution of differential equations modeling the behavior of physical systems was implemented by digital computers;
- communication theory. As digital computers became more advanced in their computational power, they were heavily used not only by the oil industry for geologic signal processing, but above all by the telecommunications industry for speech processing: as examples of major advancements in the field, we may think of the contributions by Shannon and Nyquist (Fig. 1.1), namely the Shannon theorem (Shannon 1949) and the Nyquist criteria for digital transmission (Nyquist 1928);
- z-transform theory. The basic idea of the complex series that we currently call z-transform was known to Pierre-Simon Laplace (1749–1827) at the beginning of the 19th century and was re-introduced in 1947 by Hurewicz as a means to solve linear constant coefficient difference equations (see Kanasewich 1981). In the Fifties, Ragazzini and Zadeh (1952) gave the z-transform its present name;
- frequency-domain techniques. During the Great Depression of the '30s, the U.S. Bureau of Standards retained its surplus employees and set them to developing a variety of mathematical tools. Perhaps the most useful of these was a technique to evaluate the Fourier transform from a number of discrete data points, using only multiplications and additions. This discrete Fourier transform (DFT) technique remained unused for a number of years before sampled-data control systems came into common usage around 1950. It was then realized that the DFT was ideal for use with digital computers and could be directly applied to analyzing the spectral features of these systems;
- fast Fourier transform (Cooley and Tukey 1965). This advancement was a giant leap and a sort of revolution in digital signal processing. Real-time processing became possible; and finally,
- digital filter theory.

Fig. 1.1 Claude Elwood Shannon (1916–2001), on the *left*, and Harry Theodor Nyquist (1889–1976), on the *right*

Fig. 1.2 Wilhelm Cauer
(1900–1945), on the *left*, and
James H. McClellan, on the
right

Analog filters were originally invented as hardware devices made of resistors, capacitors and inductors, for use in radio receivers and long-distance telephone systems. During the first couple of decades of the past century, analog filter design procedures were developed that were formalized in the 1930s by Paarmann (see Paarmann 2001). Outstanding contributions in the field were given by Butterworth and Cauer. What is known today as the Butterworth filter was published in 1930 (Butterworth 1930), while elliptic filters were invented by Cauer (1931) (Fig. 1.2, on the left); see also Norton (1937) and Weinberg (1962). Chebyshev filters, and inverse Chebyshev (also called Chebyshev Type II) filters, based upon Chebyshev polynomials (Chebyshev 1854; Chebyshev and Oeuvres 1899), were developed during the 1950s (Cauer 1931, 1939, 1958; Henderson and Kautz 1958; Storer 1957; Weinberg 1962; Stephenson 1985). As soon as digital computers became available, simulation of analog filters in software started being possible. It must be noted that the standard procedure for analog filter design was, and is, to first design a lowpass filter. Bandpass, highpass, or bandstop filters are then obtained by means of complex-variable transformations.

Concerning digital filters, we must distinguish between IIR and FIR digital filters. Indeed, depending on the length of the sequence that characterizes the filter in the time domain, i.e., its impulse response, digital filters can be divided into two categories: infinite-duration impulse response (IIR) filters and finite-duration impulse response (FIR) filters. For reasons that will become clear later in the book, IIR filters are sometimes also called recursive filters, while FIR are non-recursive. The history of IIR filters, at least as far as the classical approach to their design is concerned, is strictly related to the history of analog filters. Mathematical procedures were devised to transform these analog filters into digital ones, the most popular being the bilinear transform, also known as the Tustin method (Tustin 1947).

From the 1960s, the revolution of digital filtering began, and the majority of efforts were devoted to develop efficient, optimal approaches to design FIR digital filters. The most well-known design method was published by James McClellan (Fig. 1.2, on the right)[2] and Thomas Parks (Parks and McClellan 1972; McClellan

---

[2]Photo by Dicklyon (Own work) [CC BY-SA 3.0 (http://creativecommons.org/licenses/by-sa/3.0)], via Wikimedia Commons.

and Parks 2005), based on the Remez exchange algorithm. The latter was presented by Evgeny Yakovlevich Remez in the mid-1930s (Remez 1934a, b, c). The method, which became known as minimax or equiripple design method, involves the theory of optimal Chebyshev approximation. To explain what this means, we can add that approximation theory is a mathematical theory concerned with how functions can best be approximated with simpler functions. In particular, one can obtain polynomials very close to the optimal one by expanding the given function in terms of Chebyshev polynomials and then cutting off the expansion at the desired degree: this is Chebyshev approximation. The function that characterizes an FIR filter in the frequency domain, i.e., its frequency response, is a polynomial, so that FIR filter design is actually a problem of polynomial approximation.

Turning now to spectral analysis, the related historical background is very well described, for example, in Bloomfield (2000). The information provided here is mainly taken from Bloomfield's book. In 1772, Lagrange proposed a method based on the use of rational functions to identify periodic components in data series and used it to analyze the orbit of a comet (see Lagrange 1873). Like related methods, such as the one proposed by Prony (see Hamming 1987), the method was tedious to apply to any but the shortest series, and very sensitive to errors or other disturbances in the data. The first procedure to be feasible for a moderate number of data samples was introduced in 1847 by Buys-Ballot in a study on periodic variations in temperature (Buys-Ballot 1847; Whittaker and Robinson 1944): it was a tabular method of which a more sophisticated version was described by Stewart and Dodgson (1879). The method can be used to improve the estimate of an approximately known period and is quite cumbersome.

The Fourier analysis of a sequence may be carried out by a similar tabular technique described by Schuster (1897) (Fig. 1.3, on the left). This method was used in the second half of the 19th century to find periodic components with known periods in tidal data, meteorological series, and astronomical series, but still required tedious computations. Sir William Thomson, who later became Lord Kelvin, built an instrument called a harmonic analyzer for carrying out this analysis mechanically

Fig. 1.3 Franz Arthur Friedrich Schuster (1851–1934), on the *left*, and Maurice Stevenson Bartlett (1910–2002), on the *right*

(Thomson 1876, 1878) and claimed that it would reduce the time needed for an analysis by a factor of 10. Stokes (1879) pointed out that the harmonic analyzer also offered the possibility of determining unknown periods.

However, when Fourier analysis is used to search for periodic components of unknown periods in data series, the result may be very misleading: for instance, Knott (1897) used the harmonic analyzer on a series of Japanese earthquakes and found periods related to the lunar cycle, but Schuster (1898) showed that they were not statistically significant. Schuster (1898) further discussed Fourier analysis of data series and introduced the periodogram, a fundamental tool that is still in use. In subsequent papers published in 1900–1906 he applied the periodogram to the analysis of various sets of data, including the sunspot series.[3] In the early 1920s, other authors applied Schuster's method in different fields.

The statistical theory of random time series was developed in the 1920 and 1930s (see Wold 1954): the concept of spectrum was introduced in those years, though it had already been the subject of a note by Einstein (1914). This concept shifted the focus from the search for unknown periodicities to the study of the relative amplitude of all cyclicities in a data series.

In the 1940 and 1950s the attention of researchers concentrated on the problem of *estimating* the spectrum of a random series from a finite-length data record and reducing the exceedingly large variance that the periodogram shows. Daniell (1946) proposed a smoothed version of the periodogram as a suitable estimate, an idea that had already appeared in Einstein (1914). Bartlett (Fig. 1.3, on the right) introduced the idea of pseudo-ensemble average, i.e., the idea of dividing the series in shorter segments and then averaging the related periodograms (Bartlett 1948, 1950, 1955); Hamming and Tukey introduced the idea of spectral window and investigated the properties of windowed estimators (Hamming and Tukey 1949).

The following years saw a rapid development of the theory and practice of spectrum estimation: major contributions are due to Grenander and Rosenblatt (1953, 1956, 1957), Parzen (1957a, b), and Blackman and Tukey (1965). The Blackman-Tukey estimate is a way of computing the spectrum passing through the autocorrelation of the series, that quantitatively describes the relation between samples separated by a certain amount of time. This expansion of the field of spectral analysis was encouraged by the increased use of Fourier methods in several fields, mainly in electrical engineering, and by the parallel availability of computers to carry out the extensive computations required by these methods.

---

[3] Sunspots are temporary phenomena on the photosphere of the Sun that appear visibly as dark spots compared to surrounding regions. They are generated by strong magnetic fields that are created in the interior of the Sun and inhibit convective motions, thus forming areas of reduced surface temperature. Of all solar features, the sunspots are the most easily observed and were thus studied since the invention of the telescope by Galileo in 1609. This invention was based on previous work by Hans Lipperhey, a German spectacle maker, in 1608. Sunspots were first observed telescopically in late 1610 by the English astronomer Thomas Harriot and by the Frisian astronomers Johannes and David Fabricius, who published a description in 1611. The sunspot number series is one of the most popular indexes of solar activity and allowed for observing the well know 11-year period of Solar activity, known as the Schwabe cycle.

Fig. 1.4 James William
Cooley, on the *left*, and John
Wilder Tukey (1915–2000),
on the *right*

The next major advance was the introduction of the fast Fourier transform (FFT) by Cooley and Tukey (Fig. 1.4); this efficient algorithm (Cooley and Tukey 1965) allowed for significantly reducing the computational burden of practical spectral estimation.

Other advances in spectral estimation were published in the following years. The so-called parametric approach, for instance, is based on stochastic modeling of a time series, that after seminal work by Yule (Fig. 1.5, on the left) and Walker, dating back as early as the 1930s (Yule 1927; Walker 1931), developed around the 1970s. This approach is also closely related to linear prediction theory that can be traced back to the work of Kolmogorov (1941), who considered the problem of extrapolation of discrete time random processes. Other pioneers of this field are Levinson (1947), Wiener (1949), and Wiener and Masani (1957–1958); one of Levinson's contributions, namely, the Levinson-Durbin recursion, is still in wide use today. A particular approach to stochastic modeling and linear prediction, known as the autoregressive approach, was described in the 1951 thesis of Peter Whittle (Wittle 1951) and was later popularized by Box (Fig. 1.5, on the right) and Jenkins (Box and Jenkins 1970). Fundamental contributions to parametric spectral estimation were given by

Fig. 1.5 George Udny Yule
(1871–1951), on the *left*, and
George Edward Pelham Box
(1919–2013), on the *right*

Fig. 1.6 Michael Ghil, on
the *left*, and Stéphane Mallat,
on the *right*

Burg (1967, 1968), who introduced the maximum entropy method (MEM). Another
advanced, non-parametric spectral method, known as the multitaper method (MTM),
was developed in the 1980s (Thomson 1982, 1990a, b).

An outstanding contribution to arsenal of spectral techniques is represented by
the singular spectrum analysis (SSA) introduced by Robert Vautard, Michael Ghil
(Fig. 1.6, on the left) and collaborators (Vautard and Ghil 1989; Vautard et al. 1992;
Ghil et al. 2002) SSA is not, in a strict sense, a simple spectral analysis method: it is
a technique aimed at representing the signal as a linear combination of elementary,
data-adaptive variability modes. SSA does not provide a spectral estimate but rather
is a powerful de-noising filter, able to separate autocoherent features from random
features in a series.

As for evolutionary spectral analysis, it was born with Gabor (1946) and his short-
time Fourier transform (STFT). However, the present approach to the problem, i.e.,
wavelet analysis, is a relatively recent field. Wavelets have been developed independently in several fields (mathematics, quantum mechanics, engineering, geosys-
mology). The first ideas date back to 1909 and are due to Alfréd Haar, a Hungarian
mathematician, who introduced the first entity classifiable as a wavelet, which is
important mainly for historical and didactic reasons. Only many years after the first
papers on the subject, the connections among the various techniques that had been
proposed in different areas were recognized, and wavelet methods were unified into
a homogeneous corpus with a sound mathematical basis. This was accomplished
around the mid-Eighties by a French group composed of a geophysicist, Jean Morlet, a theoretical physicist, Alex Grossmann, and a mathematician, Yves Meyer.
Morlet proposed the wavelet transform as an alternative to STFT and applied it to
seismic data (Morlet et al. 1982). The original name given to the wavelet functions
was "wavelets of constant shape"; in French they were called *ondelettes*. Grossmann
noticed the similarity between the transform introduced by Morlet and the formalism
of coherent states in quantum mechanics. Grossmann, Morlet and their collaborators
studied the wavelet transform in detail and abbreviated the name of the new entities,
calling them simply *wavelets* (Grossmann and Morlet 1984; Grossmann et al. 1985).
So, the early work was in the 1980s by Morlet, Grossmann, Meyer, and others, but it
was the 1988 paper by the Belgian mathematician and physicist Ingrid Daubechies

(Daubechies 1988) that caught the attention of the larger applied mathematics communities in signal processing, statistics, and numerical analysis. Later, with more papers by Daubechies (see, e.g., Daubechies 1990, 1992) and by the French applied mathematician Stéphane Mallat (Fig. 1.6, on the right)—(see, e.g., Mallat 1989a, b, c, 2008)—the link was recognized between wavelet theory and many results obtained in the field of digital signal processing, such as those in *subband coding* (see, e.g., Crochiere et al. 1976; Vetterli and Kovačević 1995) and *pyramidal coding* (Burt and Adelson 1983). Ondelettes definitely became wavelets among the international scientific community.

Work by Donoho, Johnstone, Coifman, and others added theoretical reasons for why wavelet analysis is so versatile and powerful. Their results proved that wavelets have some inherent generic advantages and are near optimal for a wide class of problems (see, e.g., Donoho 1993), and that special wavelet systems can be created for particular signals and classes of signals. New methods for signal detection, denoising (e.g., Donoho 1995), and compression (e.g., Saito 1994; Guo 1997) emerged from the wavelet approach.

Finally, a few words about the Monte Carlo method, that will be mentioned several times in this book. The Monte Carlo method is a statistical trial-and-error technique for solving complex problems that are otherwise intractable using analytical deterministic techniques. Its essence is in generating random samples of data of known distribution to collect statistically valid results that would provide insight into the phenomenon or process being investigated. In other words, the method is a statistical testing approach to find approximate solutions to problems where exact mathematical treatment is too complex or time consuming. The Monte Carlo method was developed in the 1950s at Los Alamos (New Mexico, USA) during the famous Manhattan Project, by a group of researchers led by Nicholas Metropolis, and including John von Neumann and Stanisław Ulam. Actually, the idea for the method occurred to Stanisław Ulam, a Polish-American mathematician. The name of the method is down to Metropolis, who stated later: "I suggested an obvious name for the statistical method—a suggestion not unrelated to the fact that Stan [Ulam] had an uncle who would borrow money from relatives because he just had to go to Monte Carlo [as a center of gambling]" (see Metropolis 1987).

1.4 How to Read This Book

This book is structured into five Parts.

Part I introduces basic theoretical concepts. As explained above, discrete-time signals can be deterministic, i.e., described by a mathematical formula, or can have intrinsic random behavior, such as in the case of measurements repeated in time and affected by measurement errors. Part I approaches signals from a deterministic point of view. Chapter 2 deals with discrete-time signals and also with discrete-time systems, i.e., those algorithms that are used to process discrete-time signals. Linear time-invariant (LTI) systems are discussed, and the distinction between finite impulse response (FIR) and infinite impulse response (IIR) systems is introduced. In Chap. 3, the transforms used for discrete-time signals (the z-transform, the discrete-time Fourier transform, DTFT, the discrete Fourier series, DFS, and the discrete Fourier transform, DFT) are presented in the frame of their mutual relation. In Chap. 4, discrete time-signals are viewed as the result of sampling of continuous-time signals and the issues related to the sampling operation are discussed. In Chap. 5, the concept of correlation is introduced for deterministic discrete-time signals and the mathematical tools leading to their description in the frequency domain are given. This part should be studied by any reader who does not have a background in discrete-time signals and systems. The mathematical details can be overlooked in a first reading, but the fundamental concepts should be soundly acquired before proceeding further. In fact, this Part, and Chap. 3 in particular, opens the way from the time-domain descriptions of signals and systems to the corresponding frequency-domain description. Only thanks to this dual understanding can the reader be ready to tackle the concepts of filtering and spectral analysis.

Part II discusses digital filters. It starts with Chap. 6, presenting digital filter properties and filtering implementation methods and building the bases for the next two chapters: Chap. 7, which is about designing a FIR filter with some desired properties, and Chap. 8, which deals with IIR filter design. The Appendix to Chap. 8 covers topics in complex analysis connected to classical methods of IIR filter design from analog filters, such as trigonometric functions with complex arguments, elliptic integrals, Jacobi elliptic functions, and the rational elliptic function. This Part can be skipped if the reader is mainly interested in spectral analysis rather than in digital filtering.

Part III is devoted to random signals and to their spectral analysis. Chapter 9 introduces the statistical approach to signal analysis, presenting random variables, ensemble averages, stationary random processes, ergodicity, the power spectrum of a random signal and the cross-power spectrum of two random signals, as well as the issues that arise when the averages used to describe a random process must be estimated from a single, finite-length data record. Chapter 10 describes classical, Fourier-based spectral methods for stationary random signals; Chap. 11 deals with model-based stationary spectral methods, which are referred to as parametric methods. They rely on stochastic modeling of signals, which in turn is strictly related to the problem of linear prediction of signals. As such, a short account of these topics is also provided. Chapter 12 is devoted to Singular Spectrum Analysis (SSA), while Chap. 13 concerns the time-evolutionary spectral analysis of non-stationary signals: it briefly describes the related classical tool, the short-time Fourier transform (STFT), and then focuses on the continuous wavelet transform (CWT), which represents a more up-to-date approach to the problem.

Part IV is about the issues of signal decomposition, de-noising and compression. The main tool for these tasks is, nowadays, the discrete wavelet transform (DWT) that in Chap. 14 is properly placed into the general theoretical frame of signal expansion techniques. In Chap. 15, the DWT is exploited for devising schemes of de-noising and compression of signals. This is a somewhat specialized topic that some readers may want to skip.

Finally, Part V proposes a complete set of exercises that can be worked out with a computer in a Matlab environment. The reader thus can learn how to numerically perform the operations required by the digital signal processing and spectral analysis methods he has studied.

1.5 Further Reading

The most authoritative book on digital signal processing is Oppenheim and Schafer (2009). For background in signals and systems, the reader may refer to Oppenheim et al. (1996), or even to a very concise treatment, like the one offered by Hwei (2013). Other comprehensive books on digital signal processing include Porat (1996), Shenoi (2005), Proakis and Manolakis (2006), Mitra (2010). Digital filters are treated in detail in Parks and Burrus (1987) and in many other excellent books. For statistical digital signal processing and spectral analysis, Marple (1987), Kay (1993), Percival and Walden (1993), Stoica and Moses (2005), Hayes (2008) are very good references. Stochastic models and linear prediction are thoroughly described in Montgomery et al. (2008) and Vaidyanathan (2008). Finally, for wavelets two very good references are Qian (2001) and Misiti et al. (2007). We must also mention Meyer (1990), Daubechies (1992), Burrus et al. (1997), Vetterli and Kovačević (1995), Mallat (2008), Goswami and Chan (2011), which are among the most fundamental books on wavelets.

References

Bartlett, M.S.: Smoothing periodograms from time series with continuous spectra. Nature **161**, 686–687 (1948)

Bartlett, M.S.: Periodogram analysis and continuous spectra. Biometrika **37**, 1–16 (1950)

Bartlett, M.S.: An Introduction to Stochastic Processes, with Special Reference to Methods and Applications. Cambridge University Press, Cambridge (1955)

Blackman, R.B., Tukey, J.W.: The Measurement of Power Spectra from the Point of View of Communications Engineering. Dover, New York (1965)

Bloomfield, P.: Fourier Analysis of Time Series: An Introduction. Wiley, New York (2000)

Box, G.E.P., Jenkins, G.M.: Time Series Analysis: Forecasting and Control. Holden-Day, San Francisco (1970)

Burg, J.P.: Maximum entropy spectral analysis. In: 37th Annual International Meeting, Society of Exploration Geophysics, Oklahoma City, OK, USA (1967)

Burg, J.P.: A new analysis technique for time series data. In: Childers, D.G. (ed.) Modern Spectrum Analysis. NATO Advanced Study Institute of Signal Processing with Emphasis on Underwater Acoustics. IEEE Press, Enschede, The Netherlands (1968)

Burrus, C.S., Gopinath, R.A., Guo, H.: Introduction to Wavelets and Wavelet Transforms: A Primer. Prentice Hall, Upper Saddle River (1997)

Burt, P.J., Adelson, E.H.: The laplacian pyramid as a compact image code. IEEE Trans. Commun. **31**(4), 532–540 (1983)

Butterworth, S.: On the theory of filter amplifiers. Wirel. Eng. **7**, 536–541 (1930)

Buys-Ballot, C.H.: D: Les changéments periodiques de température. Kemint et Fills, Utrecht (1847)

Cauer, W.: Siebshaltungen. V.D.I. Verlag, G.m.b.H., Berlin (1931)

Cauer, W.: Ausgangsseitig Leerlaufende Filter. ENT **16**(6), 161–163 (1939)

Cauer, W.: Synthesis of Linear Communication Networks (translated from the German by Knausenberger, G.E., Warfield, J.N.). McGraw-Hill, New York (1958)

Chebyshev, P.L.: Théorie des mécanismes connus sous le nom de parallélogrammes. Mémoires des Savants étrangers présentées à l'Académie Impériale des Sciences de Saint-Pétersbourg **7**, 539–586 (1854)

Chebyshev, P.L., Oeuvres, I.: L'Académie Impériale des Sciences, Saint-Pétersbourg (1899). Reprinted by Chelsea (1962)

Cooley, J.W., Tukey, J.W.: An algorithm for the machine calculation of complex fourier series. Math. Comp. **19**, 297–301 (1965)

Crochiere, R.E., Weber, S.A., Flanagan, J.L.: Digital coding of speech in sub-bands. AT&T Tech. J. **55**(8), 1069–1085 (1976)

Daniell, P.J.: Discussion on the symposium on autocorrelation in time series. J. R. Stat. Soc. (Suppl.) **8**, 88–90 (1946)

Daubechies, I.: Orthonormal bases of compactly supported wavelets. Commun. Pure Appl. Math. **41**, 909–996 (1988)

Daubechies, I.: The wavelet transform, time-frequency localization and signal analysis. IEEE Trans. Inf. Theory **36**, 961–1005 (1990)

Daubechies, I.: Ten lectures on wavelets. In: CBMS-NSF Regional Conference Series in Applied Mathematics, Society for Industrial and Applied Mathematics (SIAM), Philadelphia, PA, USA (1992)

Donoho, D.L.: Unconditional bases are optimal bases for data compression and for statistical estimation. Appl. Comput. Harmonic Anal. **1**(1), 100–115 (1993)

Donoho, D.L.: De-Noising by soft-thresholding. IEEE Trans. Inf. Theory **41**(3), 613–627 (1995)

Einstein, A.: Method for the determination of the statistical values of observations concerning quantities subject to irregular fluctuations. Arch. Sci. Phys. Natur. **4**(37), 254–256 (1914)

Gabor, D.: Theory of communication. J. Inst. Electr. Eng. **93**, 429–457 (1946)

Ghil, M., Allen, M.R., Dettinger, M.D., Ide, K., Kondrashov, D., Mann, M.E., Robertson, A.W., Saunders, A., Tian, Y., Varadi, F., Yiou, P.: Advanced spectral methods for climatic time series. Rev. Geophys. **40**(1), 1.1–1.41 (2002)

Goswami, J.C., Chan, A.K.: Fundamentals of Wavelets: Theory, Algorithms, and Applications. Wiley, New York (2011)

Grenander, U., Rosenblatt, M.: Statistical spectral analysis of time series arising from stationary stochastic processes. Ann. Math. Stat. **24**, 537–558 (1953)

Grenander, U., Rosenblatt, M.: Some Problems in Estimating the Spectrum of a Time Series. In: Proceeding of Third Berkeley Symposium on Mathematics, Statistics and Probability. vol. 1, pp. 77–93. University of California Press, Oakland, CA, USA (1956)

Grenander, U., Rosenblatt, M.: Statistical Analysis of Stationary Time Series. Wiley, New York (1957)

Grossmann, A., Morlet, J.: Decomposition of hardy functions into square integrable wavelets of constant shape. SIAM J. Math. Anal. **15**, 723–736 (1984)

Grossmann, A., Morlet, J., Paul, T.: Transformations associated to square integrable group representations. J. Math. Phys. **26**, 2473–2479 (1985)

Hamming, R.W.: Numerical Methods for Scientists and Engineers. Dover Publications, New York (1987)

Guo, H.: Wavelets for approximate fourier transform and data compression. Ph.D. thesis, ECE Department, Rice University, Houston (1997)

Hamming, R.W., Tukey, J.W.: Measuring noise color. Unpublished Memorandum, Bell Telephone Laboratories (1949). Also in: The Collected Works of John W. Tukey, 1143–1153 (1985)

Hayes, M.H.: Statistical Digital Signal Processing and Modeling. Wiley, New York (2008)

Henderson, K.W., Kautz, W.H.: Transient responses of conventional filters. IRE Trans. Cir. Theory **CT-5**(4), 333–347 (1958)

Hwei, P.H.: Schaum's Outline of Signals and Systems. McGraw-Hill, New York (2013)

Kanasewich, E.R.: Time Sequence Analysis in Geophysics. Prentice-Hall, Englewood Cliffs (1981)

Kay, S.M.: Fundamentals of Statistical Signal Processing—Volume I: Estimation Theory. Prentice-Hall, Upper Saddle River (1993)

Knott, C.G.: On lunar periodicities in earthquake frequency. Proc. R. Soc. **60**, 457–466 (1897)

Kolmogorov, A.N.: Interpolation and extrapolation of stationary random sequences. Izv. Akad. Nauk SSSR Ser. Mat. **5**, 3–14 (1941)

Lagrange, J.L.: Recherches sur la manière de former de tables de planètes d'apres les seules observations. In: Oeuvres complètes de Lagrange, Tome **VI**, 507–627 (1873)

Levinson, N.: The Wiener RMS error criterion in filter design and prediction. J. Math. Phys. **25**, 261–278 (1947)

Mallat, S.G.: A theory for multiresolution signal decomposition : the wavelet representation. IEEE Trans. Pattern Anal. Mach. Intell. **11**, 674–693 (1989a)

Mallat, S.G.: Multiresolution approximation and wavelet orthonormal bases of L2. Trans. Am. Math. Soc. **315**, 69–87 (1989b)

Mallat, S.G.: Multifrequency channel decompositions of images and wavelet models. IEEE Trans. Acoust. Speech Signal Process. **37**, 2091–2110 (1989c)

Mallat, S.: A Wavelet Tour of Signal Processing. Academic Press, Waltham (2008)

Meyer, Y.: Wavelets and Operators. Cambridge University Press, Cambridge (1990)

Marple, S.L.: Digital Spectral Analysis. Prentice-Hall, Englewood Cliffs (1987)

McClellan, J.H., Parks, T.W.: A personal history of the Parks-McClellan algorithm. IEEE Proc. Signal Process. **22**, 82–86 (2005)

Metropolis, N.C.: The beginning of the Monte Carlo method, pp. 125–130, in: Stanislaw Ulam 1909–1984. Los Alamos Science, **15**, Special Issue, Report LA-UR-87-3600, Los Alamos Scientific Laboratory, Los Alamos, NM, USA (1987)

Misiti, M., Misiti, Y., Oppenheim, G., Poggi, J.-M. (eds.): Wavelets and their Applications. Wiley-ISTE, New York (2007)

Mitra, S.K.: Digital Signal Processing—A Computer-Based Approach. McGraw-Hill, New York (2010)

Montgomery, D.C., Jennings, C.L., Kulahci, M.: Introduction to Time Series Analysis and Forecasting. Wiley, New York (2008)

Morlet, J., Arens, G., Fourgeau, I., Giard, D.: Wave propagation and sampling theory. Geophysics **47**(2), 203–221 (1982)

Norton, E.L.: Constant resistance networks with applications to filter groups. Bell Syst. Tech. J. **16**, 178–193 (1937)

Nyquist, H.: Certain topics in telegraph transmission theory. Trans. AIEE, **47**, 617–644 (1928). Reprint as classic paper in: Proc. IEEE, **90**(2) (2002)

Oppenheim, A.V., Schafer, R.W.: Discrete-Time Signal Processing. Prentice Hall, Englewood Cliffs (2009)

Oppenheim, A.V., Willsky, A.S., Hamid, S.: Signals and Systems. Prentice Hall, Englewood Cliffs (1996)

Paarmann, L.D.: Design and Analysis of Analog Filters—A Signal Processing Perspective. Kluwer, Dordrecht (2001)

Parks, T.W., Burrus, C.S.: Digital Filter Design. Wiley, New York (1987)

Parks, T., McClellan, J.: Chebyshev approximation for nonrecursive digital filters with linear phase. IEEE Trans. Circ. Theory **19**(2), 189–194 (1972)

Parzen, E.: On consistent estimates of the spectrum of a stationary time series. Ann. Math. Stat. **28**(2), 329–348 (1957a)

Parzen, E.: On choosing an estimate of the spectral density function of a stationary time series. Ann. Math. Stat. **28**(4), 921–932 (1957b)

Percival, D.B., Walden, A.T.: Spectral Analysis for Physical Applications. Cambridge University Press, Cambridge (1993)

Porat, B.: A Course in Digital Signal Processing. Wiley, New York (1996)

Prandoni, P., Vetterli, M.: Signal Processing for Communications. EPFL Press, Lausanne (2008)

Proakis, J.G., Manolakis, D.G.: Digital Signal Processing—Principles, Algorithms and Applications. Prentice-Hall, Englewood Cliffs (2006)

Qian, S.: Introduction to Time-Frequency and Wavelet Transforms. Prentice Hall, Upper Saddle River (2001)

Ragazzini, J.R., Zadeh, L.A.: The analysis of sampled-data systems. Trans. Am. Inst. Elec. Eng. **71**(II), 225–234 (1952)

Remez, E.Y.: Sur la détermination des polynômes d'approximation de degré donné. Comm. Soc. Math. Kharkov **10**, 41–63 (1934a)

Remez, E.Y.: Sur un procédé convergent d'approximations successives pour déterminer les polynômes d'approximation. Compt. Rend. Acad. Sci. **198**, 2063–2065 (1934b)

Remez, E.Y.: Sur le calcul effectiv des polynômes d'approximation de Tschebyscheff. Compt. Rend. Acad. Sci. **199**, 337–340 (1934c)

Saito, N.: Simultaneous noise suppression and signal compression using a library of orthonormal bases and the minimum description length criterion. In: Foufoula-Georgiou, E., Kumar, P. (eds.) Wavelets in Geophysics. Academic Press, San Diego (1994)

Schuster, A.: On lunar and solar periodicities of earthquakes. Proc. R. Soc. **61**, 455–465 (1897)

Schuster, A.: On the investigation of hidden periodicities with application to a supposed 26 day period of meteorological phenomena. Terr. Magn. **3**, 13–41 (1898)

Shannon, C.E.: Communication in the presence of noise. Proc. Inst. Radio Eng. **37**(1), 10–21 (1949). Reprint as classic paper in: Proc. IEEE, **86**(2) (1998)

Shenoi, B.A.: Introduction to Digital Signal Processing and Filter Design. Wiley, New York (2005)

Stephenson, F.W.: RC Active Filter Design Handbook. Wiley, New York (1985)

Stewart, B., Dodgson, W.: Preliminary report to the committee on solar physics and on a method of detecting the unknown inequalities in a series of observations. Proc. R. Soc. **29**, 106–122 (1879)

Stoica, P., Moses, R.: Spectral Analysis of Signals. Prentice Hall, Upper Saddle River (2005)

Stokes, G.G.: Note on the paper by Stewart and Dodgson. Proc. R. Soc. **29**, 122–123 (1879)

Storer, J.E.: Passive Network Synthesis. McGraw-Hill, New York (1957)

Thomson, W.: On an instrument for calculating the integral of the product of two given functions. Proc. R. Soc. **24**, 266–268 (1876)

Thomson, W.: Harmonic analyzer. Proc. R. Soc. **27**, 371–373 (1878)

Thomson, D.J.: Spectrum estimation and harmonic analysis. Proc. IEEE **70**, 1055–1096 (1982)

Thomson, D.J.: Time series analysis of holocene climate data. Philos. Trans. R. Soc. London A **330**, 601–616 (1990a)

Thomson, D.J.: Quadratic-inverse spectrum estimates: applications to palaeoclimatology. Philos. Trans. R. Soc. London A **332**, 539–597 (1990b)

Tustin, A.: A method of analysing the behaviour of linear systems in terms of time series. J. Inst. Electr. Eng. IIA **94**(1), 130–142 (1947)

Vaidyanathan, P.P.: The Theory of Linear Prediction. Morgan & Claypool, San Rafael (2008)

Vautard, R., Ghil, M.: Singular spectrum analysis in nonlinear dynamics, with applications to paleoclimatic time series. Physica D **35**, 395–424 (1989)

Vautard, R., Yiou, P., Ghil, M.: Singular-spectrum analysis: a toolkit for short, noisy chaotic signals. Physica D **58**, 95–126 (1992)

Vetterli, M., Kovaĉević, J.: Wavelets and Subband Coding. Prentice Hall, Upper Saddle River (1995)

Walker, G.: On periodicity in series of related terms. Proc. R. Soc. London A **131**, 518–532 (1931)

Weinberg, L.: Network Analysis and Synthesis. McGraw-Hill, New York (1962)

Whittaker, E.T., Robinson, G.: The Calculus of Observations. Blackie and Sons, London (1944)

Wiener, N.: Extrapolation, Interpolation and Smoothing of Stationary Time Series. Wiley, New York (1949)

Wiener, N., Masani, P.: The prediction theory of multivariate stochastic processes. Part 1: Acta Math. **98**, 111–150 (1957); Part 2: Acta Math. **99**, 93–137 (1958)

Whittle, P.: Hypothesis Testing in Time Series Analysis. Almquist and Wiksell, Uppsala (1951)

Wold, H.O.: A: A Study on the Analysis of Stationary Time Series. Almqvist and Wiksell, Stockholm (1954)

Yule, G.U.: On a method for investigating periodicities in disturbed series with special reference to Wolf's sunspot numbers. Philos. Trans. R. Soc. London A **26**, 267–298 (1927)

Part I
Basic Theoretical Concepts

Chapter 2
Discrete-Time Signals and Systems

2.1 Chapter Summary

In this chapter we define more formally the concept of discrete-time signal. Discrete-time signals are, in general, infinite-length sequences of numerical values (Oppenheim and Schafer 2009) that may either arise from sampling of continuous-time signals, or be generated directly by inherently-discrete-time processes. They are also called sequences. Historically, discrete-time signals have often been introduced as the discretized version of continuous-time signals, i.e., as the sampled values of analog quantities, such as the voltage at the output of an analog circuit; accordingly, many of the derivations proceeded within the framework of an underlying continuous-time reality. However, the discretization of an analog signal is only one of the ways in which a discrete-time signal can be generated. Therefore, we will introduce discrete-time signals from an abstract and self-contained point of view, and will consider separately (in Chap. 4) their possible derivation from sampling of continuous-time signals.

Conceptually, a distinction must be made between deterministic signals and random signals. Deterministic signals have a univocal mathematical description, so that past and present signal values are known exactly and future signal values are perfectly predictable. Random signals do not allow for such a description and therefore the signal evolution cannot be exactly foreseen; their treatment requires statistical/probabilistic tools. The first two parts of this book deal with deterministic signals.

In this chapter, the fundamental concepts related to discrete-time signals (see, e.g., Haykin and Van Veen 2002; Proakis and Manolakis 2006; Shenoi 2007; Oppenheim and Schafer 2009; Mitra 2009) are presented, as well as the main features of those mathematical operators, called discrete-time systems, that are employed to process discrete-time signals. Constraints are normally imposed on discrete-time systems that make them suitable for digital signal processing. These constraints are: linearity, time invariance, stability and causality. The quantities used to univocally

© Springer International Publishing Switzerland 2016
S.M. Alessio, *Digital Signal Processing and Spectral Analysis for Scientists*, Signals and Communication Technology,
DOI 10.1007/978-3-319-25468-5_2

describe a system, i.e., the impulse response, the transfer function and the frequency response, are then defined, and a distinction based on the length of the impulse response is introduced. Discrete-time systems are thus classified as finite-impulse-response (FIR) or infinite-impulse-response (IIR). The input-output relation of discrete-time systems is finally presented, both in the form of an operation called linear convolution, and in the form of a recursive equation, referred to as the linear constant-coefficient difference equation (LCCDE).

2.2 Basic Definitions and Concepts

A *signal* can be defined as the set of values of a variable x, expressing the variation of a physical quantity as a function of time[1] t. If both x and t are continuous variables, we can mathematically represent such a function as $x = x(t)$. We call this function an *analog signal*. The term *continuous-time signal* can also be used, stressing the fact that time varies continuously.

The measurement or *sampling* of a continuous-time signal at a constant rate, i.e., at times separated by a constant increment T_s (periodic sampling; see Fig. 2.1), leads to a *sequence* of real (or complex) numbers that we call a *discrete-time signal*

$$x = x[n] = x(nT_s),$$

where the integer variable n is *discrete time*. Here, n is an adimensional variable. It measures the position of the value or *sample* $x[n]$ along the sequence, relative to a fixed origin $n = 0$, in number of samples. T_s is referred to as the *sampling interval*, *sampling period*, or *sampling (time) step*. We must always keep in mind that even if in practical settings discrete-time signals can arise from periodic sampling of continuous-time signals, a sequence may be generated directly from some discrete-time process.

As well as the independent variable being either continuous or discrete, the signal amplitude x may also be either continuous or discrete. *Digital signals* are those for which both time and amplitude are discrete: the signal amplitude is quantized in some way, usually depending on the nature of the measuring and/or recording process. In any case, processing a discrete-time signal with a computer, a machine that uses finite-precision arithmetic, transforms it into a digital signal. However, in this book the focus is on the discrete-time nature of sequences and only marginally considers the effects of signal amplitude quantization. In our discussions, x will always be a continuous variable.

---

[1]The independent variable might, of course, also be a spatial coordinate, however, here we will assume that we are only interested in *time* variations.

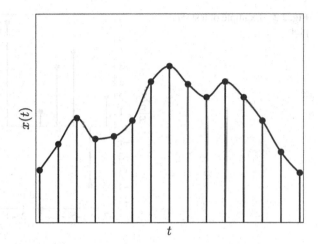

Fig. 2.1 An analog signal and its periodic sampling

The terms *processing* and *analysis* of a signal refer to the ensemble of all the techniques aimed at

- representing the signal,
- modifying it into another form that can be interpreted more easily,
- separating the content of information in a signal from "noise",
- extracting information that the signal contains about the state or the behavior of the physical system from which it derives.

The kind of signal processing and analysis that will be discussed in this book takes place in the dual domains of time and frequency and often involves applying a *discrete-time system* to the signal, i.e., a mathematical operator or a software algorithm that transforms it, mapping the input sequence into an output sequence,[2] according to the following scheme:

$$ input \quad \Rightarrow \quad \boxed{discrete\text{-}time\ system} \quad \Rightarrow \quad output. $$

---

[2]Signal processing systems are classified along the same lines as signals. We can thus have continuous-time systems for which both the input and the output are continuous-time signals, discrete-time systems for which both the input and the output are discrete-time signals, and digital systems, for which both the input and the output are digital signals. Strictly speaking, therefore, digital signal processing deals with the transformations of signals that are discrete in both amplitude and time. In this book, the word "digital" is used loosely as a synonym of "discrete-time", since we do not enter into details about the implications of amplitude quantization. In the same way, we will use the word "analog" as a synonym of "continuous-time".

Fig. 2.2 Example of a stem plot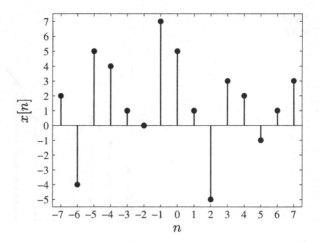

2.3   Discrete-Time Signals: Sequences

A real (or complex) discrete-time signal $x[n]$ exists only for integer values of n and is undefined elsewhere. In general, if we do not specify any particular limit values for n, by $x[n]$ we mean an infinitely-long sequence: $n = (-\infty, +\infty)$. This expression indicates the interval in which n varies.[3]

A discrete-time signal is typically represented graphically by a *stem plot*, like the one shown in Fig. 2.2.

We define the *energy* of the signal as

$$\mathscr{E} = \sum_{n=-\infty}^{+\infty} |x[n]|^2.$$

The signals for which $\mathscr{E}$ is finite are called *energy signals*.[4] However, $\mathscr{E}$ may be infinite as well. Bounded signals that do not have finite energy have, however, finite *average power*, defined as

---

[3]We use round and square brackets to indicate intervals, with the former meaning that the corresponding edge value is not included in the interval, and the latter meaning that the corresponding edge value is included in the interval. Intervals are referred to as *open* in the first case and *closed* in the second case. For example, $[-\pi, +\pi)$ indicates a range that includes the edge value $-\pi$ (square bracket) but *excludes* the edge value $+\pi$ (round bracket); this is an open interval; $[-\pi, +\pi]$ would be a closed interval. Whenever an edge value is $\pm\infty$, the interval will be open in the corresponding direction.

[4]Note that for real signals we could simply write energy as the integral of $x^2[n]$. However, here and in the rest of the book we will stick to the definition given above, holding for general complex-valued signals.

$$\mathscr{P} = \lim_{N \to \infty} \frac{1}{2N+1} \sum_{n=-N}^{+N} |x[n]|^2,$$

and are thus called *power signals*. If $\mathscr{E}$ is finite, obviously $\mathscr{P}$ is zero. Note that the expression "average power" is often simply shortened into "power".

A *periodic signal* is such that, for each n,

$$x[n+N] = x[n]$$

and its period is the minimum integer N for which the relation holds. A periodic signal has infinite energy over $n = (-\infty, +\infty)$, but its energy over a single period,

$$\mathscr{P} = \frac{1}{N} \sum_{n=0}^{N-1} |x[n]|^2,$$

is finite. The average power over $n = (-\infty, +\infty)$ is finite and equal to power calculated over a single period. Therefore periodic signals are power signals.

2.3.1 Basic Sequence Operations

Sequences can be manipulated in several basic ways. The product and sum of two sequences $x[n]$ and $y[n]$ are defined as the sample-by-sample product and sum, respectively. Multiplication of a sequence $x[n]$ by a number α is defined as the multiplication of each sample by α.

The following transformations of the independent variable n are considered:

1. $n \Rightarrow n - k$ with integer k: *translation* or *shift* (Fig. 2.3).

   $k > 0$: *time delay* (translation to the right) by k discrete-time steps;
   $k < 0$: *time advance* (translation to the left) by k time steps.

2. $n \Rightarrow -n$: *reflection* or *folding* of the signal around $n = 0$, i.e., around the origin of discrete times (Fig. 2.4).
3. $n \Rightarrow k - n$: folding associated with translation.

   - Since $k - n = -n - (-k)$, this is a folding operation followed by a translation by $-k$ time steps. But
   - since $k - n = -(n - k)$, this is also a translation by k steps, followed by a folding.

In Fig. 2.5, an example of folding and translation by $k = 2$ time steps is shown. The transformation can be achieved in two different ways. In Fig. 2.5a we see a sequence $x[n]$; Fig. 2.5b shows the same sequence folded around the origin, i.e.,

Fig. 2.3 Examples of
discrete-time signal
translations. **a** Original
signal; **b–c** shifted signals. In
order to make the
transformation easily visible,
empty circles and *dashed
lines* have been used for the
sample located originally at
$n = 0$. The signal $x[n - 3]$ is
delayed with respect to $x[n]$;
$x[n + 2]$ is advanced

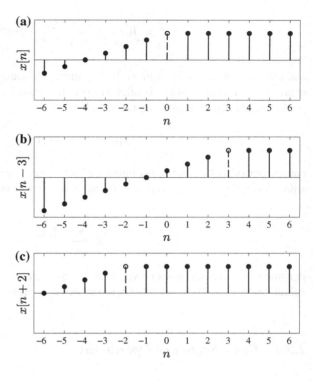

Fig. 2.4 Example of folding
around the origin of discrete
times. **a** A signal; **b** its
folded version. In order to
make the transformation
easily visible, *empty circles*
and *dotted, dash-dotted* and
dashed lines have been used
for some samples

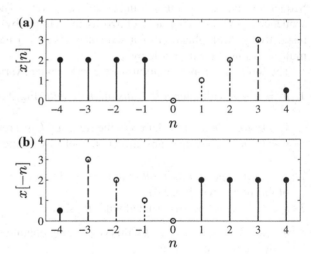

Fig. 2.5 Example of folding and translation by 2 time steps. **a** A sequence $x[n]$; **b** the same sequence folded around the origin, $x[-n]$; **c** the folded sequence translated to the left by 2 steps, $x[2-n]$; **d** the sequence $x[n-2]$ obtained by translating $x[n]$ to the right by 2 steps. If we now decided to take this last sequence and fold it, we would once again get $x[2-n]$. In order to make the transformations easily visible, *empty circles* and *dashed lines* have been used for some samples

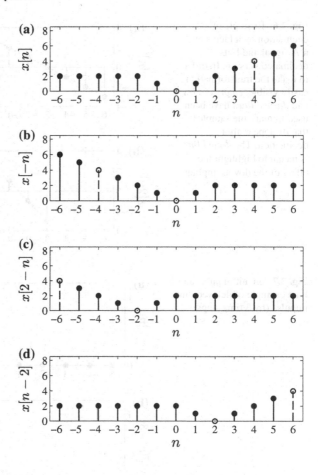

$x[-n]$; Fig. 2.5c shows the folded sequence shifted to the left by $k=2$ steps, that is, $x[2-n]$; Fig. 2.5d illustrates the sequence $x[n-2]$, obtained by shifting $x[n]$ to the right by two steps. If we now decided to take this last sequence and fold it, we would once again get $x[2-n]$.

4. $n \Rightarrow k_{dec}n$ with integer k_{dec}: *decimation* or *downsampling* of the signal.
   Fig. 2.6a shows a sequence, which is then downsampled by a factor $k_{dec} = 2$; Fig. 2.6b shows the decimated version.

2.3.2 Basic Sequences

When discussing the theory of discrete-time signals and systems, several basic sequences are particularly important.

Fig. 2.6 Example of decimation by a factor of 2. **a** A signal and **b** its decimated version. In order to make the transformation easily visible, *empty circles* and *dashed lines* have been used to mark the samples that disappear after decimation. The *dotted line* is meant to highlight the effect of the downsampling

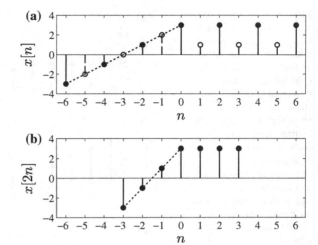

Fig. 2.7 **a** Unit impulse and **b** an example of delayed unit impulse (by 3 time steps)

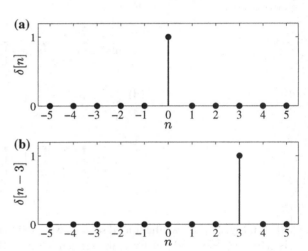

- *Unit impulse* or *unit sample* or *discrete-time* δ (Fig. 2.7a):

$$\delta[n] = \begin{cases} 1 & \text{for} \quad n = 0, \\ 0 & \text{elsewhere.} \end{cases}$$

- *Delayed unit impulse* (Fig. 2.7b):

$$\delta[n - k] = \begin{cases} 1 & \text{for} \quad n = k, \\ 0 & \text{elsewhere.} \end{cases}$$

Any discrete-time signal can be expressed as a sum of scaled and delayed impulses (see Sect. 3.8.5 in the appendix of Chap. 3):

$$x[n] = \sum_{k=-\infty}^{+\infty} x[k]\delta[n-k] = \sum_{k=-\infty}^{+\infty} x[n-k]\delta[k].$$

The term "scaled" here indicates multiplication of the delayed impulse by the factor $x[k]$. The third member of the equality derives from the second by substituting k with $n-k$.

- *Unit step* (Fig. 2.8a):

$$u[n] = \begin{cases} 1 & \text{for } n \geq 0, \\ 0 & \text{for } n < 0. \end{cases}$$

Fig. 2.8b shows a shifted version of the unit step.
The unit step is useful to define the so-called *causal sequences* —sequences for which $x[n] = 0$ for $n < 0$: indeed, given any signal $x[n]$, the signal $x[n]u[n]$ is causal. For example, $x[n] = a^n u[n]$, with a = constant, is a causal sequence.
The unit impulse and the unit step are related by

$$u[n] = \sum_{k=-\infty}^{n} \delta[k] = \sum_{k=0}^{\infty} \delta[n-k],$$
$$\delta[n] = u[n] - u[n-1].$$

- *Discrete-time sinusoid*:

we can derive its form from a continuous-time sinusoid. For example, let us take a sinusoidal function with unit amplitude, phase constant (initial phase) equal to

Fig. 2.8 a Unit step and **b** an example of anticipated unit step (by 2 discrete-time units)

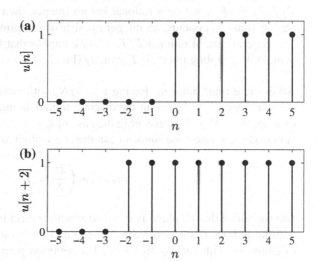

zero and analog frequency f, i.e., $x(t) = \sin(2\pi f t)$. Period is $T = 1/f$. By sampling $x(t)$ with a sampling interval T_s we obtain the sequence

$$x[n] = \sin(2\pi f n T_s).$$

If we now define an *adimensional frequency* or *normalized frequency*

$$\nu = f T_s = T_s/T,$$

measured in cycles/sample or (samples)$^{-1}$, we get

$$x[n] = \sin(2\pi \nu n) = \sin(\omega n),$$

where $\omega = 2\pi\nu$ is *discrete-time angular frequency*, an adimensional quantity measured in radians/sample.

A discrete-time sinusoid is periodic only if its ν is a rational number. To see this fact, we require that $x[n] = x[n+N]$, that is, $\sin[2\pi\nu(n+N)] = \sin(2\pi\nu n)$. This is true if, and only if, an integer k exists, such that $2\pi\nu N = 2\pi k$. However, this is only possible if

$$\nu = \frac{k}{N}, \qquad\qquad \omega = 2\pi\frac{k}{N},$$

that is, $1/\nu = T/T_s = N/k$, or

$$kT = NT_s.$$

The period of the discrete-time sinusoid is equal to the denominator of ν, after reducing ν, if necessary, to a ratio of numbers with no common divisor. If $T/T_s = N/k$ is not only rational but an integer, then a single period T of the analog sinusoid contains an integer number of sampling intervals T_s (Fig. 2.9a). The general case of rational $T/T_s = N/k$ implies that k periods T are needed to contain N sampling intervals T_s exactly (Fig. 2.9b).

All discrete-time sinusoids having $\nu = k/N$, with variable k and fixed N, share the same period N. Hence in the discrete-time domain, period is not the inverse of frequency: $N \neq 1/\nu$, except in the case of $k = 1$.

A periodic discrete-time sinusoid can thus be written as

$$x[n] = \sin\left(\frac{2\pi}{N}kn\right)$$

and we notice that its phase is symmetric with respect to k (frequency index) and n (time index). If we now keep N fixed and let k vary, thus considering a set of sinusoids with varying frequency but common period N in the time domain,

Fig. 2.9 Continuous-time sinusoid and corresponding discrete-time sinusoid: **a** the case of integer T/T_s and **b** the case of rational T/T_s. T is period in seconds, and T_s is the sampling interval in the same units

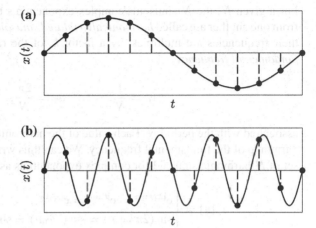

i.e., with respect to n, we see that periodicity with period N is present also in the frequency domain, i.e., with respect to k. This means periodicity in ν with period $N/N = 1$ and periodicity in ω with period $2\pi N/N = 2\pi$. The three variables k, ν and ω all express the concept of frequency, but the period changes its value when we express ourselves in terms of one variable or another: it is N in terms of k, 1 in terms of ν and 2π in terms of ω.

Two discrete-time sinusoids having their k values separated by a multiple of N, i.e., their angular frequencies separated by a multiple of 2π, are indistinguishable: for example, $\sin\left[(2\pi k + \omega)n\right] = \sin(\omega n)$. We thus understand that all possible distinct periodic discrete-time sinusoids with period N in k can be represented using only N values of k, for example $k = [0, N-1]$. This is analogous to the fact that in order to represent a periodic $x[n]$ with period N in n, we only need N values of n, for example $n = [0, N-1]$. We thus have only N possible distinct periodic discrete-time sinusoids with period N. Each of them is characterized by its *harmonic number k*.

Calculations often become easier if we represent real sines and cosines by combination of complex exponentials:

$$x[n] = e^{j\omega n} = \cos(\omega n) + j\sin(\omega n), \qquad j = \sqrt{-1}$$

with

$$\sin(\omega n) = \frac{1}{2j}(e^{j\omega n} - e^{-j\omega n}), \qquad \cos(\omega n) = \frac{1}{2}(e^{j\omega n} + e^{-j\omega n}).$$

The properties of complex exponential sequences are analogous to those of sinusoidal sequences. In the following discussion, the expressions "(discrete-time) sinusoid" and "(discrete-time) complex exponential" are considered as substantially equivalent and are used indifferently.

For a given N, the N sinusoids/complex exponentials that can be distinguished from one another are called *harmonically-related sinusoids/complex exponentials*: their frequencies are multiples, by a factor k, of the rational and non-negative *fundamental frequency*

$$\nu_0 = \frac{1}{N}, \qquad \omega_0 = \frac{2\pi}{N},$$

associated with the period N. Each value of the harmonic number k identifies one harmonic of the fundamental frequency. We can thus write the kth member of the set of N harmonic sinusoids or complex exponentials as

$$x_k[n] = \begin{cases} e^{j2\pi k\nu_0 n} = e^{jk\omega_0 n} = e^{j\frac{2\pi}{N}kn}, \\ \sin\left(2\pi k\nu_0 n\right) = \sin\left(k\omega_0 n\right) = \sin\left(\frac{2\pi}{N}kn\right). \end{cases}$$

Note that the choice $k = [1, N]$ would be perfectly reasonable as well, as any other choice: $k = [n_0, n_0 + N]$ with integer n_0. Nonetheless, the notation normally used is $k = [0, N - 1]$, including the frequency $\nu = 0$ in the harmonic set. Thus, n and k share the same range of values.

In summary, discrete-time sinusoids and complex exponentials with rational ν are:

• periodic with respect to n with period N, so that we only need to consider values $0 \le n \le N - 1$,
• periodic with respect to k with period N, so that we only need to consider values $0 \le k \le N - 1$,
• periodic with respect to ν with period 1, so that we only need to consider values $0 \le \nu < 1$,
• periodic with respect to ω with period 2π, so that we only need to consider values $0 \le \omega < 2\pi$.

An alternative and perfectly equivalent choice is $k = [-N/2, N/2 - 1]$, which is possible if N is even. This leads to working in the frequency intervals $[-\pi, \pi)$ in terms of ω and $[-0.5, 0.5)$ in terms of ν. From now on, we will assume that N is even unless stated otherwise, so that in our discussion k can be indifferently meant as varying in $[0, N - 1]$ or in $[-N/2, N/2 - 1]$. Angular frequency ω varies in the 2π-wide equivalent intervals $[0, 2\pi)$ or $[-\pi, +\pi)$; this 2π-wide range is referred to as the *principal interval* or the *fundamental interval*. As for ν, it varies in the 1-wide equivalent intervals $[0, 1)$ or $[-0.5, +0.5)$.

A remark is mandatory here. The above-mentioned periodicity implies that *any interval of length 2π can be used to express the variability range for ω*. For example, choosing $(-\pi, +\pi]$ instead of $[-\pi, +\pi)$ would be correct as well. It is just a matter of convention. Similar considerations hold for the other frequency variables k and ν.

2.3.3 Deterministic and Random Signals

Processing and analyzing a signal requires the assumption of a *model*—a mathematical description—of the signal. If a univocal description is possible using a mathematical formula or a well-defined rule, then the signal is called *deterministic*: at any instant, its past and present values are known and its future values are exactly predictable.

If we are not able to give such a description, or if such a description is too complicated to be useful, a different approach is adopted, in which the signal is seen as evolving in a way that cannot be exactly foreseen. In this case the signal must be treated by statistical tools (mean values, probability functions) and is classified as a *random signal*. Typically, all sequences deriving from periodically repeated measurements of physical quantities are affected by random measurement errors and thus belong to this category.

The distinction between deterministic and random signals is sometimes subtle. For example, even a signal constructed artificially as a sum of sinusoids may be treated as random if the phase constants are random numbers and some noise is superimposed to the sinusoidal signals. *Synthetic signals* of this kind are often used to test the behavior and the performances of algorithms in signal processing.

In the first two parts of this book we deal with deterministic signals, while in the third part we will discuss the statistical approach to random signals.

2.4 Linear Time-Invariant (LTI) Systems

A *system* is a univocal transformation, i.e., an operator T that maps an input sequence $x[n]$ into an output sequence $y[n]$, also called *response*:

$$input \quad x[n] \quad \Rightarrow \quad \boxed{operator \; T} \quad \Rightarrow \quad output \quad y[n].$$

In other words, a system is an algorithm that, given $x[n]$, determines $y[n]$:

$$y[n] = T\{x[n]\}.$$

In general, $y[n]$ for any $n = n_0$ depends on the values of $x[n]$ at all discrete times n. Such a property is expressed by saying that the most general system has *memory*.

Digital filters, which we will describe in Part II, are typical cases of systems. A very simple filter is, for example, the running average or moving average operator

$$y[n] = \frac{1}{M_1 + M_2 + 1} \sum_{k=-M_1}^{+M_2} x[n-k].$$

Classes of systems are then defined setting constraints on the properties of the transformation $T\{\cdot\}$. Let us examine these constraints.

1. *Linearity*:

$T\{\cdot\}$ is a linear transformation, which we will indicate by $L\{\cdot\}$, if

$$L\{ax_1[n] + bx_2[n]\} = aL\{x_1[n]\} + bL\{x_2[n]\},$$

where a and b are arbitrary constants.
Linearity implies

- additivity: $L\{x_1[n] + x_2[n]\} = L\{x_1[n]\} + L\{x_2[n]\}$;
- homogeneity: $L\{ax[n]\} = a\,L\{x[n]\}$.

As examples of linear systems we may list:

- the running average,
- the accumulator, $y[n] = \sum_{k=-\infty}^{n} x[k]$,
- the delayer, $y[n] = x[n - n_0]$, etc.

An example of nonlinear system is

- $y[n] = x^2[n]$ for a real input $x[n]$.

This system is also *memoryless* because $y[n]$ only depends on the value of the input $x[n]$ at the same discrete time n.

2. *Time invariance* or *translation invariance*:

if $y[n] = L\{x[n]\}$ implies $y[n - k] = L\{x[n - k]\}$ for any integer k, then the system is *linear time-invariant* (LTI).

Examples of LTI systems are:

- all the linear systems mentioned above.

An example of a non-LTI system is

- the downsampler, $y[n] = x[Mn]$ for any integer M.

To see this fact we define $x_1[n] = x[n - k]$ and consider the corresponding output $y_1[n]$: $y_1[n] = x_1[Mn] = x[Mn - k]$. But $y[n - k] = x[M(n - k)] = x[Mn - Mk] \neq y_1[n]$: if we send to the system the input $x[n]$ shifted by k time steps, as the output we do not simply get a shifted version of the output $y[n]$, unless we consider the trivial case $M = 1$.

3. *Causality*:

a system is causal if for any choice of $n = n_0$, the output $y[n_0]$ depends only on the values of the input at discrete times $n \leq n_0$: the system is non-anticipative, i.e., it does not anticipate future values of the input when producing the output. Examples of causal systems are the following:

- the delayer, $y[n] = x[n - n_0]$, is causal if $n_0 \geq 0$, i.e., if it produces a time delay, and non-causal if $n_0 < 0$, i.e., if it produces a time advance;
- the running average, $y[n] = 1/(M_1 + M_2 + 1) \sum_{k=-M_1}^{+M_2} x[n - k]$, is causal if $-M_1 \geq 0$ and $M_2 \geq 0$;
- the accumulator, $y[n] = \sum_{k=-\infty}^{n} x[k]$, is causal;
- the system $y[n] = x^2[n]$ is causal;
- the downsampler, $y[n] = x[Mn]$, is causal if $M > 1$. For example, for $M = 5$ we have $y[1] = x[5]$, $y[2] = x[10]$, etc.

4. *Stability*:

a system is stable in the *bounded-input, bounded-output* (BIBO) sense if, and only if, every bounded input sequence, i.e., every sequence whose absolute values are limited, produces a bounded output sequence: for all n,

$$|x[n]| \leq A < \infty \;\Rightarrow\; |y[n]| \leq BA < \infty,$$

where A and B are positive finite constants. Let us stress that system stability requires the BIBO property to hold for *any* bounded input.
Examples of stable systems include:

- the running average,
- the delayer,
- the downsampler,
- the system $y[n] = x^2[n]$.

An example of unstable system is

- the accumulator.

In order to verify that a given system does not satisfy the BIBO stability definition, we only need to find a particular bounded input for which the output is unbounded. In the case of the accumulator, if we set $x[n] = u[n]$, then we have

$$y[n] = \sum_{k=-\infty}^{n} u[k] = \begin{cases} 0 & \text{for } n < 0, \\ n + 1 & \text{for } n \geq 0. \end{cases}$$

Although this response is finite if n is finite, it is nevertheless unbounded, because there is no fixed positive finite value B such that $n + 1 \leq BA < \infty$ for all integer $n \in (-\infty, \infty)$. Therefore the accumulator is not a stable system according to the BIBO definition.

2.4.1 Impulse Response of an LTI System and Linear Convolution

Let us consider an LTI system and define its *impulse response* as

$$h[n] = L\{\delta[n]\},$$

assuming that it exists for the considered system. If we recall the equality $x[n] = \sum_{k=-\infty}^{+\infty} x[k]\delta[n-k]$, we can express the response of the LTI system to *any* input through $h[n]$. In fact, we can write

$$y[n] = L\{x[n]\} = L\left\{\sum_{k=-\infty}^{+\infty} x[k]\delta[n-k]\right\}.$$

But linearity allows for writing

$$y[n] = \sum_{k=-\infty}^{+\infty} x[k]L\{\delta[n-k]\}$$

and time invariance leads to

$$y[n] = \sum_{k=-\infty}^{+\infty} x[k]h[n-k] \equiv x[n] * h[n] = \sum_{k=-\infty}^{+\infty} x[n-k]h[k] \equiv h[n] * x[n].$$

This equation defines the *linear convolution*[5] of $x[n]$ and $h[n]$. Convolution is commutative, and also distributes over addition: $x[n] * (h_1[n] + h_2[n]) = x[n] * h_1[n] + x[n] * h_2[n]$. In the convolution sum, $y[n]$ appears to be influenced by all past, present and future values of the input $x[n]$, provided that $h[n]$ is not identically 0 for $n \in (-\infty, +\infty)$: therefore, in general, the system is non-causal. The impulse response sequence $h[n]$ completely characterizes the LTI system in the time domain.

If we linearly convolve two causal sequences $x[n]$ and $h[n]$ that are, though formally of infinite length, identically zero outside some finite interval of values of n: $x[n] \neq 0$ for $0 \leq n \leq N_1 - 1$, $h[n] \not\equiv 0$ for $0 \leq n \leq N_2 - 1$, then the convolution sum has finite summation limits and the effective duration of the convolution is finite: $y[n] \neq 0$ for $0 \leq n \leq N_1 + N_2 - 2$, implying an effective convolution length of $N_1 + N_2 - 1$. We will now justify this statement through an example.

---

[5]This definition of linear convolution is valid for general complex sequences.

2.4.2 An Example of Linear Convolution

Let us calculate the linear convolution of two causal sequences $x[n]$ and $h[n]$ that are not identically zero over the intervals $0 \leq n \leq N_1 - 1 = 12$ and $0 \leq n \leq N_2 - 1 = 4$, respectively. The sequences for this example are shown in Fig. 2.10a and b, respectively.

The procedure for obtaining $y[n] = \sum_{k=-\infty}^{+\infty} x[k]h[n-k]$ is the following:

1. we choose a value $n = n_0$ of discrete time for which we want to compute $y[n]$;
2. we fold $h[k]$ to get $h[-k]$;
3. if $n_0 > 0$ we shift $h[-k]$ to the right by n_0 samples; if $n_0 < 0$ we shift $h[-k]$ to the left by n_0 samples; we thus get $h[n_0 - k]$;
4. we multiply $x[k]$ by $h[n_0 - k]$, sample by sample, to obtain an intermediate product sequence $v_{n_0}[k] = x[k]h[n_0 - k]$;
5. we sum all the values of $v_{n_0}[k]$—only a finite number of these values will be different from zero—to obtain $y[n = n_0]$;
6. we repeat for any $n = n_0$—only a finite number of values of $y[n_0]$ will be different from zero.

In general, the sample $y[n_0]$ is zero when all values of $v_{n_0}[k]$ are zero: hence, finding the effective duration of $y[n]$ is equivalent to finding in which interval of n_0 values the intermediate sequence $v_{n_0}[k]$ assumes only zero values.

The procedure for performing linear convolution, in the case of the two sequences shown in Fig. 2.10, is illustrated in Fig. 2.11: the sequence $x[k]$ appears in Fig. 2.11a;

Fig. 2.10 Example of linear convolution: **a** the sequence $x[n]$ ($N_1 = 13$) and **b** the sequence $h[n]$ ($N_2 = 5$) to be convolved with $x[n]$; **c** the result of the linear convolution between $x[n]$ and $h[n]$. *Dashed lines* and *empty circles* mark the samples affected by edge effects, i.e., the transitory phases of the convolution (see text for details)

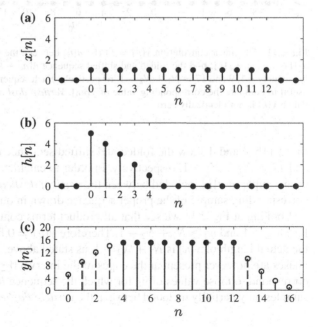

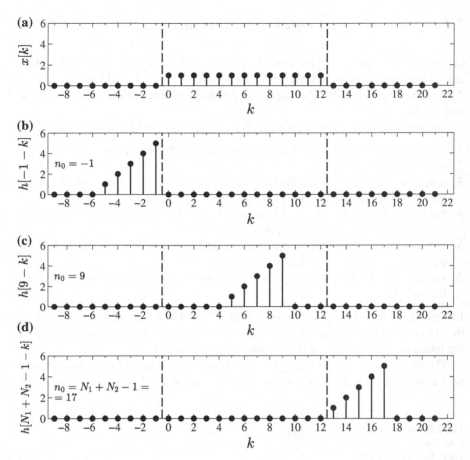

Fig. 2.11 The linear convolution $y[n] = x[n] * h[n]$ between the sequences shown in Fig. 2.10. **a** The sequence $x[k]$; **b–d** the folded and shifted sequence $h[n_0 - k]$ for three different values of n_0, namely -1, 9 and $17 = N_1 + N_2 - 1$, respectively. Each sequence in one of the panels b–d is useful to compute the corresponding value of $y[n_0]$. *Vertical dashed lines* enclose the k-range in which $x[k]$ is not identically zero

Fig. 2.11b, c and d show the folded and shifted sequence $h[n_0 - k]$ for $n_0 = -1, 9$ and $17 = N_1 + N_2 - 1$, respectively. In order to calculate $y[n_0]$, we must multiply each sample of $x[k]$ included between the *vertical dashed lines* in Fig. 2.11a by the corresponding sample of the proper sequence drawn in one of panels b–d.

Looking at Fig. 2.11 we see that all product terms contributing to $y[n_0]$ are zero for $n_0 \leq -1$ and $n_0 \geq N_1 + N_2 - 1$. Therefore $y[n] \neq 0$ for $0 \leq n \leq N_1 + N_2 - 2$: the actual length of $y[n]$ is $N_1 + N_2 - 1$, as stated above. Initial and final transitory phases are however present at the edges of the interval $0 \leq n \leq N_1 + N_2 - 2$. They correspond to those values of n_0 for which the sequence $h[n_0 - k]$ has its non-zero samples only partially included between the *vertical dashed lines* in Fig. 2.11, which enclose all non-zero $x[k]$ samples. The values of $n = n_0$ for which $y[n]$ is free from edge effects go from $n = n_0 = N_2 - 1$ to $n = n_0 = N_1 - 1$, and thus the length of the transitory-free linear convolution is $N_1 - N_2 + 1$.

Of course, the convolution of two finite-length sequences has effective summation limits that are finite. However, these limits depend on the $y[n]$ sample that is being computed, i.e., on the value of $n = n_0$. For a given value of n, the product terms included in $\sum_k$ will be, in general, different from zero when

$$0 \le k \le N_1 - 1,$$
$$0 \le n - k \le N_2 - 1 \quad \Rightarrow \quad -N_2 + 1 \le k - n \le 0 \quad \Rightarrow \quad n - N_2 + 1 \le k \le n.$$

We can summarize these two conditions writing

$$\max(0, n - N_2 + 1) \le k \le \min(N_1 - 1, n).$$

For two sequences with finite lengths N_1 and N_2 the convolution sum can thus be written as

$$y[n] = \sum_{k=\max(0,n-N_2+1)}^{\min(N_1-1,n)} x[k]h[n - k].$$

For instance, in Fig. 2.11c we can see that for $n = n_0 = 9$, with $N_1 = 13$ and $N_2 = 5$ we have the finite summation limits $\max(0, 9 - 5 + 1) = 5$ and $\min(13 - 1, 9) = 9$, so that $y[9] = \sum_{k=5}^{9} x[k]h[9 - k]$.

The sequence $y[n]$ resulting from the linear convolution between the sequences of Fig. 2.10a, b is visible in Fig. 2.10c. The total length is $N_1 + N_2 - 1 = 17$ including the transitory phases, and $N_1 - N_2 + 1 = 9$ without them. Further details on transitory phases in linear convolution will be provided in Sect. 16.4.1.1.

2.4.3 Interconnections of LTI Systems

Discrete-time systems, and LTI systems in particular, can be interconnected to form larger systems. There are two basic ways in which systems may be interconnected: in cascade (series) or in parallel. These interconnections are shown in Fig. 2.12.

- LTI systems *in cascade* (Fig. 2.12a):

$$y[n] = (x[n] * h_1[n]) * h_2[n].$$

The commutative property of convolution allows writing

$$y[n] = x[n] * (h_1[n]) * h_2[n]),$$

Fig. 2.12 Interconnections
of LTI systems: **a** in cascade
and **b** in parallel

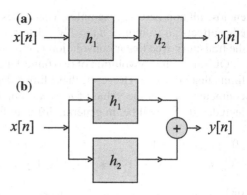

hence the equivalent system has an impulse response equal to the convolution of
the individual impulse responses:

$$x[n] \quad \Rightarrow \quad \boxed{h_1 * h_2} \quad \Rightarrow \quad y[n].$$

- LTI systems *in parallel* (Fig. 2.12b): since linear convolution distributes over addition,

$$y[n] = x[n] * h_1[n] + x[n] * h_2[n] = x[n] * (h_1[n] + h_2[n]),$$

hence the equivalent system has an impulse response equal to the sum of the
individual impulse responses:

$$x[n] \quad \Rightarrow \quad \boxed{h_1 + h_2} \quad \Rightarrow \quad y[n].$$

In general, cascade and parallel interconnections of systems can be used to construct larger and more complex systems. Conversely, it may be advantageous to take
a large system and break it down into smaller subsystems, for purposes of analysis
and implementation.

2.4.4 Effects of Stability and Causality Constraints on the Impulse Response of an LTI System

The impulse response of a stable and causal LTI system is a causal, absolutely
summable sequence.

This statement can be justified considering the consequences on $h[n]$ that derive
from requiring the LTI system to satisfy both stability and causality constraints.

1. *Stability*: requiring that if $|x[n]| \leq A < \infty$ for all n, then $|y[n]| \leq BA < \infty$ for all n, implies that $h[n]$ is *absolutely summable*, i.e.,

$$S = \sum_{k=-\infty}^{k=+\infty} |h[k]| < \infty.$$

The stability condition is sufficient to ensure that S is finite because if it is so, then the response $y[n]$ to a bounded input $x[n]$ is certainly bounded:

$$|y[n]| = \left| \sum_{k=-\infty}^{k=+\infty} h[k]x[n-k] \right| = \sum_{k=-\infty}^{k=+\infty} |h[k]||x[n-k]|,$$

and if $|x[n]| \leq A$ then

$$|y[n]| \leq A \sum_{k=-\infty}^{k=+\infty} |h[k]| = AS.$$

From this equation we also see that the constant B coincides with S and therefore does not depend on the input sequence $x[n]$, but only on $h[n]$, i.e., on the system. It can be shown that stability is also a necessary condition for $S < \infty$ because if $h[n]$ is not absolutely summable, a bounded input can be found, for which the output is unbounded.

A direct consequence of this fact is that in a stable LTI system, a finite-length input always produces a transient, i.e., finite-length, output.

2. *Causality*: for $y[n = n_0]$ to depend only on past input values $x[n \leq n_0]$, the condition $h[n] = 0$ for all $n < 0$ must hold, i.e., $h[n]$ must be a causal sequence.

This can be understood writing down the explicit expression for $y[n = n_0]$:

$$y[n_0] = \sum_{k=-\infty}^{k=+\infty} h[k]x[n_0 - k] =$$

$$= \sum_{k=-\infty}^{k=-1} h[k]x[n_0 - k] + \sum_{k=0}^{k=+\infty} h[k]x[n_0 - k] =$$

$$= \ldots h[-3]x[n_0 + 3] + h[-2]x[n_0 + 2] + h[-1]x[n_0 + 1] + h[0]x[n_0] +$$
$$+ h[1]x[n_0 - 1] + h[1]x[n_0 - 1] + h[2]x[n_0 - 2] + h[3]x[n_0 - 3]\ldots$$

from which it can be clearly seen that if future input values $(\ldots x[n_0 + 3], x[n_0 + 2]\ldots)$ must be absent from the expression of $y[n_0]$, the condition $h[n] = 0$ for all $n < 0$ must hold.

In causal systems, a causal input produces a causal output, since all $y[n < 0]$ are zero:

$$\ldots\ldots\ldots = \ldots\ldots\ldots$$
$$y[-1] = 0$$
$$y[0] = x[0]h[0]$$
$$y[1] = x[1]h[0] + x[0]h[1]$$
$$\ldots\ldots\ldots = \ldots\ldots\ldots$$

We may note that for a causal system, i.e., for a causal $h[n]$, the convolution sum becomes

$$y[n] = \sum_{k=0}^{+\infty} x[n-k]h[k] = \sum_{k=-\infty}^{n} x[k]h[n-k].$$

If we also impose causality of the input, then

$$y[n] = \sum_{k=0}^{n} x[n-k]h[k] = \sum_{k=0}^{n} x[k]h[n-k].$$

Therefore in a causal LTI system with a causal input the convolution sum giving $y[n]$ has finite limits $k = 0$ and $k = n$, even when $x[k]$ and $h[k]$ have infinite length. This does not mean that this sum can be calculated in practice: indeed, as n increases, it would require memorizing an unlimited number of samples of $x[k]$ and $h[k]$.

2.4.5 Finite (FIR) and Infinite (IIR) Impulse Response Systems

Now we will introduce the distinction between *finite impulse response* (FIR) and *infinite impulse response* (IIR) systems, through some examples.

- Running average system:

$$h[n] = \frac{1}{M_1 + M_2 + 1} \sum_{k=-M_1}^{+M_2} \delta[n-k] = \begin{cases} \frac{1}{M_1+M_2+1} & \text{for } -M_1 \le n \le M_2, \\ 0 & \text{elsewhere.} \end{cases}$$

This impulse response has *finite length* and we classify the system as an FIR system.

An FIR system is always stable, because S is certainly finite. Moreover, as we already know, the running average system is causal if $-M_1 \ge 0$ and $M_2 \ge 0$. In general, an FIR system may or may not be causal, but any non-causal FIR system can be made causal by cascading it with a proper delayer.

- Accumulator:

$$h[n] = \sum_{k=-\infty}^{n} \delta[k] = u[n] = \begin{cases} 1 & \text{for } n \geq 0, \\ 0 & \text{for } n < 0. \end{cases}$$

This impulse response has *infinite length* and we classify the system as an IIR system. The accumulator is causal, since $h[n] = 0$ for all $n < 0$. Since $S = \sum_{n=0}^{\infty} u[n]$ is infinite, the accumulator is unstable.
- However, an IIR system may be stable as well: for example, the system with impulse response given by

$$h[n] = a^n u[n],$$

with $|a| < 1$, has $S = \sum_{n=0}^{\infty} |a|^n = 1/(1 - |a|) < \infty$ and is therefore a stable IIR system.

2.4.6 *Linear Constant-Coefficient Difference Equation (LCCDE)*

An important subclass of LTI systems consists of those systems for which the input-output relation may be written not only in the form of a linear convolution, but also in the form of an Nth order *linear constant-coefficient difference equation* (LCCDE):

$$\sum_{k=0}^{N} a'_k y[n - k] = \sum_{r=0}^{M} b'_r x[n - r],$$

where a'_k and b'_r are $N + 1$ and $M + 1$ constant coefficients, respectively. Notice that actually this is not a single equation but a set of equations, one for each n, in which M and N in general can be finite or not. It can be shown that a system that satisfies an LCCDE is an LTI system. If the coefficients are not constant—i.e., if they vary with time n—the system is not an LTI system.

In literature, we can find this equation written in several ways, slightly different from one another:

1. $\sum_{k=0}^{N} a_k y[n - k] = \sum_{r=0}^{M} b_r x[n - r]$ with $a_0 \neq 1$ (as above);
2. in the same form, but with a_0 normalized to 1. This is the most frequent form. We will use the symbols a and b for this case, and the symbols a' and b' for the case with $a_0 \neq 1$;
3. $y[n] = \sum_{k=1}^{N} a_k y[n - k] + \sum_{r=0}^{M} b_r x[n - r]$.
   This is a form explicitly resolved with respect to $y[n]$, with a_0 normalized to 1 as in case (2) and coefficients a_k reversed in sign with respect to case (2). It is a *recurrence formula* by which the output at time n can be calculated from present/past values of the input and past values of the output.
4. $\sum_{k=0}^{N} a_k y[n - k] + \sum_{r=0}^{M} b_r x[n - r] = 0$,
   with coefficients b_r reversed in sign with respect to cases (1), (2) and (3), etc.

2.4.7 Examples of LCCDE

A general property of LTI systems is that we can write an unlimited number of different LCCDs to represent the input-output relation of a given LTI system. We illustrate this property through examples.

- Accumulator (a causal system):

  the definition we gave of this system, $y[n] = \sum_{k=-\infty}^{n} x[k]$, is in the form of a linear convolution. In fact, if we use the unit impulse $\delta[n]$ as the input, we see that the impulse response $h[n]$ of the accumulator is the unit step $u[n]$, hence $y[n] = \sum_{k=-\infty}^{n} x[k] = \sum_{k=-\infty}^{\infty} x[k]u[n-k] = x[n] * u[n]$. But since $y[n-1] = \sum_{k=-\infty}^{n-1} x[k]$, we can write

  $$y[n] = y[n-1] + x[n], \qquad\qquad y[n] - y[n-1] = x[n],$$

  that is an LCCDE with $N = 1$, $M = 0$, $a_0' = 1$, $a_1' = -1$, $b_0' = 1$. Thus, the accumulator is an LTI system that satisfies an LCCDE. This kind of LCCDE, in which the calculation of $y[n]$ involves the use of at least one past value of the output because $N \neq 0$, is said to be *recursive*.

- Running average in the case $M_1 = 0$, $M_2 > 0$ (a causal system):

  its definition, $y[n] = 1/(M_2 + 1) \sum_{k=0}^{M_2} x[n - k]$, actually is a non-recursive LCCDE with $N = 0$, $M = M_2$, $a_0' = 1$, and with $b_k' = 1/(M_2 + 1)$ for $0 \leq k \leq M_2$.

  However, the same system may be represented also by a recursive LCCDE, i.e., an LCCDE with $N \neq 0$:

  $$y[n] - y[n-1] = \frac{1}{M_2 + 1} \left\{ \sum_{k=0}^{M_2} x[n-k] - \sum_{l=0}^{M_2} x[n-l-1] \right\} =$$

  $$= \frac{1}{M_2 + 1} \{x[n] + x[n-1] + \ldots + x[n - M_2] - x[n-1] - \ldots - x[n - M_2 - 1]\} =$$

  $$= \frac{1}{M_2 + 1} \{x[n] - x[n - (M_2 + 1)]\},$$

  because of the cancellation of all intermediate terms. This is an LCCDE with $N = 1$, $M = M_2 + 1$, $a_0' = 1$, $a_1' = -1$, $b_0' = b_{M_2+1}' = 1/(M_2 + 1)$ and $b_k' = 0$ otherwise.

2.4.8   The Solutions of an LCCDE

We have stated that an unlimited number of different LCCDEs can be used to represent the input-output relation of a given LTI system. On the other hand, just as in the
case of linear differential equations with constant coefficients that are encountered
in the theory of analog systems, in the digital case an LCCDE of a given system
does not univocally identify the output $y[n]$ corresponding to a certain input $x[n]$.
An LCCDE does not express the input-output relation of the LTI system in a unique
way, unless we provide further information on the system by specifying proper initial conditions that $y[n]$ must satisfy. This happens since for any given $x[n]$, a whole
family of LCCDE solutions exists for $y[n]$.

Let us call $x_0[n]$ a particular input sequence for which the corresponding output
$y_0[n]$ satisfying the considered LCCDE has been found by some means. Then, any
sequence

$$y[n] = y_0[n] + y_h[n],$$

where $y_h[n]$ (homogeneous solution) is any solution of the homogeneous equation

$$\sum_{k=0}^{N} a'_k y[n-k] = 0,$$

will also satisfy the LCCDE with the same input. The sequence $y_h[n]$ is a member
of a family of solutions having the general form

$$y_h[n] = \sum_{m=1}^{N} A_m z_m^n,$$

where A_m indicates N coefficients—undetermined a priori—and z_m indicates N
complex numbers. These z_m are the roots, i.e., the zeros, of the *characteristic polynomial* $\sum_{k=0}^{N} a'_k z^{n-k}$. For the sake of simplicity, we will suppose that these roots are
all distinct, i.e., simple, with multiplicity equal to 1.

In order to see this fact we assume that the homogeneous solution has the form
$y_h[n] = z^n$ and we substitute it into the LCCDE:

$$\sum_{k=0}^{N} a'_k z^{n-k} = 0 = z^{n-N} \left(a'_0 z^N + a'_1 z^{N-1} + \ldots + a'_{N-1} z + a'_N \right).$$

The expression inside parentheses is the characteristic polynomial that must be set
to zero in order to satisfy the equation. Since its degree is N, the polynomial has
N roots that in general will be complex: z_m, with $m = [1, N]$. The most general
homogeneous solution is then

$$y_h[n] = A_1 z_1^n + A_2 z_2^n + \ldots + A_N z_N^n = \sum_{m=1}^{N} A_m z_m^n,$$

with coefficients A_m that must be uniquely determined according to the initial conditions specified for the system. We assumed that all roots of the characteristic polynomial are simple. The form of the terms associated with any multiple roots in $y_h[n]$ would be slightly different, but there are always N undetermined coefficients.

In conclusion, we can state that an LCCDE univocally represents the input-output relation of an LTI system for a given $x[n]$ only if we specify uniquely the homogeneous solution by imposing additional constraints, i.e., providing additional information in the form of auxiliary conditions. These auxiliary conditions usually consist of specified values of $y[n]$ at specific values of n, such as $y[-1]$, $y[-2]$, $\ldots y[-N]$ (initial conditions). Once this information has been provided, we must solve a set of N linear equations for the N undetermined coefficients A_m, so that $y_h[n]$ and then $y[n]$ can be explicitly calculated.

Alternatively, once $y[-1]$, $y[-2]$, $\ldots y[-N]$ have been specified, the subsequent values $y[0]$, $y[1]$, $y[2] \ldots$ can be iteratively generated by the recurrence formula

$$y[n] = -\sum_{k=1}^{N} \frac{a_k'}{a_0'} y[n-k] + \sum_{r=0}^{M} \frac{b_r'}{a_0'} x[n-r] = -\sum_{k=1}^{N} a_k y[n-k] + \sum_{r=0}^{M} b_r x[n-r],$$

where we set $a_k'/a_0' = a_k$, hence $a_0 = 1$, and $b_r'/a_0' = b_r$. To generate values of $y[n]$ for $n < -N$ we can rearrange the recurrence formula as

$$y[n-N] = -\sum_{k=0}^{N-1} \frac{a_k}{a_N} y[n-k] + \sum_{r=0}^{M} \frac{b_r}{a_N} x[n-r],$$

and set $n = -1, -2, \ldots$ so as to be able to recursively compute $y[-1-N]$, $y[-2-N]$, and so on.

Since we are interested in LTI systems, the auxiliary conditions must be consistent with the requirements of linearity and time invariance. In Chap. 3, where we will discuss the solution of an LCCDE using the z-transform, we will implicitly incorporate conditions of linearity and time invariance. As we will see in that discussion, even with these additional constraints, the solution to the LCCDE—and therefore the system—is not uniquely determined. In particular, there are generally both causal and non-causal LTI systems consistent with a given LCCDE. If we further prescribe causality, the solution becomes unique.

Causality is then the constraint chosen to get a unique solution. In this case the auxiliary conditions are often stated as *initial rest conditions*: the auxiliary information is that if $x[n] = 0$ for $n < n_0$, then the output $y[n]$ is constrained to be zero for $n < n_0$. Usually $n_0 = 0$ is chosen. This provides sufficient initial conditions to univocally generate $y[n]$ for $n \geq n_0$ through the recurrence equation.

The *order of an IIR system* is equal to the number N of past output values that must be memorized for use in the LCCDE.

In the case of FIR systems, we can always write the LCCDE in a non-recursive form with $N = 0$, i.e., $y[n] = \sum_{k=0}^{M} b_k x[n - k]$: no past values of the output are involved in the calculation of $y[n]$, and therefore no initial conditions are required. The coefficients b_k coincide with the samples of the impulse response:

$$h[n] = \sum_{k=0}^{M} b_k \delta[n - k] = \begin{cases} h[n] = b_n & \text{for } 0 \leq n \leq M, \\ h[n] = 0 & \text{elsewhere.} \end{cases}$$

The *order of an FIR system* is equal to the number M of past input samples that must be memorized to be used in the LCCDE in its "natural" non-recursive form.[6]

On the other hand, the LCCDE of an FIR system can always be put in recursive form, as we saw in the running average example. The advantages of one form over another in practical applications depend on considerations such as numerical accuracy, data storage, and the number of multiplications and additions required to compute each sample of the output. Conversely, an IIR system can be written in a non-recursive form, but then M goes to infinity, as shown in the example provided in the following subsection.

The *memory of an LTI system* is equal to the number of past input samples that must be memorized to be used in the convolution sum, i.e., in the non-recursive form of the LCCDE. Memory, expressed in number of samples, is thus finite and equal to the order for an FIR system, while it is infinite for an IIR system. The convolution sum thus appears as the way in which the memory of the system acts: for example, in the FIR case in which $y[n] = \sum_{m=0}^{M} x[n - m]h[m]$, the system forms the output at time n remembering $x[n]$ and M past samples of the input, $x[n - m]$ for $m = 1, M$, and combining them linearly with weights $h[m]$.

2.4.9   From the LCCDE to the Impulse Response: Examples

An LCCDE associated with proper initial rest conditions specifies an LTI causal system. Also the impulse response $h[n]$ completely specifies a given system. Then it must be possible to deduce $h[n]$ from the LCCDE and its associated initial conditions. We will now illustrate this fact through an example. This will also offer us the opportunity to verify once again that the same LTI system can be expressed both in recursive and non-recursive forms.

---

[6]Note that for order M, $h[n]$ in the above equation extends over $[0, M]$ and has $M + 1$ samples. This notation contrasts with the usual notation we adopted for finite-length sequences, according to which we would write $[0, M' - 1]$, i.e., M' samples, order $M' - 1$. We will have to be careful with this inconsistency in future discussion.

Let us consider a causal LTI system that exists and is unique,

$$y[n] - \frac{1}{2}y[n-1] = x[n] \qquad (N = 1, \ M = 0).$$

Let us choose $x[n] = \delta[n]$, so as to get $y[n] = h[n]$:

$$h[n] - \frac{1}{2}h[n-1] = \delta[n]$$

for all n. We selected a causal system, so we also know that $h[n] = 0$ for $n < 0$. In particular, in this case with $N = 1$, the initial condition is $h[-1] = 0$. Setting $n = 0, 1, 2, \ldots$ we get

$$h[0] - \frac{1}{2}h[-1] = \delta[0] = 1,$$

$$h[1] - \frac{1}{2}h[0] = \delta[1] = 0 = h[1] - \frac{1}{2} \quad \Rightarrow \quad h[1] = \frac{1}{2},$$

$$h[2] - \frac{1}{2}h[1] = \delta[2] = 0 = h[2] - \left(\frac{1}{2}\right)^2 \quad \Rightarrow \quad h[2] = \left(\frac{1}{2}\right)^2,$$

and so on. Generalizing to any n we can write

$$h[n] = \left(\frac{1}{2}\right)^n u[n].$$

Now let us consider the non-recursive system

$$y[n] = x[n] + \frac{1}{2}x[n-1] + \ldots + \left(\frac{1}{2}\right)^k x[n-k] + \ldots \qquad (N = 0, \ M = \infty).$$

Setting $x[n] = \delta[n]$ we get

$$h[n] = \delta[n] + \frac{1}{2}\delta[n-1] + \left(\frac{1}{2}\right)^2 \delta[n-2] + \ldots + \left(\frac{1}{2}\right)^k \delta[n-k] + \ldots =$$

$$= \begin{cases} \left(\frac{1}{2}\right)^n & \text{for } n \geq 0, \\ 0 & \text{for } n < 0, \end{cases}$$

so that we find again

$$h[n] = \left(\frac{1}{2}\right)^n u[n].$$

If the impulse response is the same, the two LTI systems are equivalent; they are IIR systems.

In summary, an LTI system satisfying an LCCDE

- is unique if it is causal;
  - systems in recursive form have at least one $a_k \neq 0$ with $k \neq 0$ (recall that $a_0 = 1$),
  - systems in non-recursive form have $a_k = 0$ for $k \neq 0$.

Notice that the distinction concerns the form in which the LCCDE is written and not the system itself. The system may always be put in any of the two forms;
- is classified according to the length of the impulse response $h[n]$:
  - FIR systems in non-recursive form have $N = 0$ and a finite M. For them,

$$h[k] = \begin{cases} b_k & \text{for } 0 \leq k \leq M, \\ 0 & \text{elsewhere.} \end{cases}$$

  An FIR system has
  · order equal to M,
  · finite memory equal to the order,
  · finite length $M + 1$ of the impulse response;
  - IIR systems in recursive form have $N \neq 0$; M and N are finite. Alternatively, IIR systems may be written in non-recursive form with infinite M.
  An IIR system has
  · order equal to N,
  · infinite memory,
  · infinite length of the impulse response.

2.4.10 Eigenvalues and Eigenfunctions of LTI Systems

LTI systems have *eigenfunctions* and *eigenvalues*, i.e., a type of signals exists, which can pass through an LTI system and emerge as a signal of the same kind, though in general it will be altered both in amplitude and phase.

To see this, let us take a complex number $z = re^{j\omega}$, where r is the (real) amplitude and ω is the angle in radians expressing the phase, and let us construct the discrete-time signal

$$x[n] = z^n = r^n e^{j\omega n}.$$

We now use this signal $x[n] = z^n$, with fixed z, as the input to an LTI system. The output will be

$$y[n] = \sum_{k=-\infty}^{+\infty} x[n-k]h[k] = \sum_{k=-\infty}^{+\infty} z^{n-k}h[k] = z^n \sum_{k=-\infty}^{+\infty} h[k]z^{-k},$$

that is,

$$y[n] = z^n H(z),$$

where we defined

$$H(z) = \sum_{k=-\infty}^{+\infty} h[k]z^{-k}.$$

$H(z)$ is, in this instance, a particular complex number, provided that the infinite sum converges for the considered value of z. Given a certain value of z, the expression written above represents the "recipe" to construct the output $y[n]$ corresponding to $x[n] = z^n$ at all discrete-time instants. The expression $y[n] = z^n H(z) = x[n]H(z)$ tells us that a signal of the form z^n is an eigenfunction of the LTI system, because the system response to an input of this form is a signal of the same z^n-type, though multiplied by the corresponding complex eigenvalue $H(z)$—a factor that can alter both the amplitude and the phase of the input signal.

We can now consider all the z-values for which the series defining $H(z)$ converges: z is now a complex variable, rather than a particular complex number. We can represent it in the *complex z-plane* (Fig. 2.13), i.e., a plane in which a Cartesian reference frame is established, having $\mathrm{Re}(z)$ on the abscissa and $\mathrm{Im}(z)$ on the ordinate. The *function $H(z)$* then defines the *z-transform* of $h[n]$. This is a particular case of a signal transform that will be discussed later. $H(z)$ is called the *transfer function* or the *system function* of the LTI system.

Let us now assume that $H(z)$ converges for values of z with unitary amplitude: $z = e^{j\omega}$. The values of z such that $|z| = r = 1$ fall on the *unit circle* in the z-plane (Fig. 2.13). On the unit circle, the eigenfunction is $z^n = e^{j\omega n}$: a complex exponential sequence with angular frequency ω. The transfer function $H(z)$ evaluated on the unit

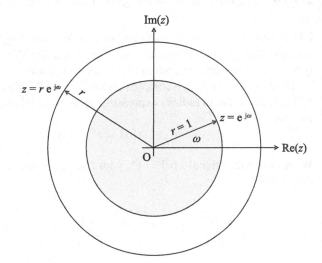

Fig. 2.13 The z-plane. The contour of the *gray-shaded* area represents the so-called unit circle, i.e., the locus of the points $z = e^{j\omega}$ having $|z| = 1$

circle, which will be indicated by $H(e^{j\omega})$, constitutes the *frequency response* of the LTI system:

$$H(e^{j\omega}) = \sum_{k=-\infty}^{+\infty} h[k]e^{-j\omega k}.$$

$H(e^{j\omega})$, the z-transform of $h[n]$ on the unit circle, is again a particular case of another kind of signal transform we will discuss later, namely it is the *discrete-time Fourier transform* (DFT) of the impulse response.[7] For a particular ω value, $H(e^{j\omega})$ is a complex number that describes the amplitude and phase variation that a discrete-time complex exponential signal with frequency ω undergoes when processed by an LTI system.

References

Haykin, S., Van Veen, B.: Signals and Systems. Wiley, New York (2002)
Mitra, S.K.: Digital Signal Processing. McGraw-Hill, Blacklick (2011)
Oppenheim, A.V., Schafer, R.W.: Discrete-Time Signal Processing. Prentice Hall, Englewood Cliffs (2009)
Proakis, J.G., Manolakis, D.G.: Digital Signal Processing. Prentice Hall, Englewood Cliffs (2006)
Shenoi, B.A.: Introduction to Digital Signal Processing and Filter Design. Wiley, New York (2007)

---

[7]The Fourier series and transform encountered in the theory of analog and digital signals are orthogonal signal expansions (a.k.a. representations). These expansions constitute a fairly standard topic in any undergraduate program in scientific sectors. However, a synthetic introduction to signal representations and to the related topic of vector spaces is provided in the appendix to the next chapter.

Chapter 3
Transforms of Discrete-Time Signals

3.1 Chapter Summary

Discrete-time signals allow for several invertible transforms that are fundamental analysis tools in digital signal processing. The first one is the z-transform: given a complex variable z, this transform is defined as an infinite series in the complex z-plane, which exists in the regions(s) of the plane where the series exhibits absolute convergence to an analytic function. Absolute convergence of the z-transform requires absolute summability of the corresponding infinite-length signal. Unit-amplitude z values identify the unit circle in the z-plane, on which z is a complex exponential signal characterized by its angular frequency ω that represents its phase angle. On the unit circle, provided that convergence exists there, the z-transform becomes a continuous function of frequency, called the discrete-time Fourier transform (DTFT) of the discrete-time signal. The DTFT representation can also be extended to sequences for which the z-transform does not exist, such as signals that are not absolutely summable but only square-summable, or periodic signals, like discrete-time periodic sinusoids and complex exponentials. Finally, if a discrete-time signal actually has finite length, it may be represented in the frequency domain by a finite number of DTFT samples, i.e., by a sequence obtained sampling its DTFT in a proper way: this sequence is the discrete Fourier transform (DFT) of the finite-length signal. The meaning and the properties of the DFT emerge clearly if this transform is introduced passing through the discrete Fourier series (DFS) of the signal's periodic extension. DFT can be efficiently computed via fast Fourier transform (FFT). Each inverse transform represents an expansion of the signal in a basis, which in the present case is an orthogonal basis. At the end of the chapter, an appendix provides an overview of the mathematical foundations of analog and discrete-time signal expansions.

© Springer International Publishing Switzerland 2016
S.M. Alessio, *Digital Signal Processing and Spectral
Analysis for Scientists*, Signals and Communication Technology,
DOI 10.1007/978-3-319-25468-5_3

3.2 z-Transform

The z-transform plays in discrete-time signal theory the same role that the Laplace transform (see Sect. 8.3.1) plays in analog signal theory: it allows replacing operations on real signals with operations on complex signals. This is convenient because in this way many results from the theory of functions of complex variable can be exploited for solving problems of digital signal processing. In other words, the z-transform brings the power of complex-variable functions' theory to bear on problems concerning discrete-time signals and systems.

The z-transform is a generalization of the discrete-time Fourier transform (DTFT) briefly introduced in Chap. 2, which is a key tool in representing and analyzing discrete-time signals and systems. A motivation for introducing the z-transform is that the DTFT does not converge for all sequences, and it is useful to have a generalization of the DTFT that encompasses a broader class of signals. Another motivation is that in analytical computations the z-transform notation is often more convenient than the Fourier transform notation.

We will now define the z-transform representation of a sequence and study how the properties of the signal are related to the properties of its z-transform.

For a generic sequence $x[n]$ that we will assume to be bounded, i.e., limited in amplitude ($|x[n]| < \infty$), the z-transform is defined as

$$X(z) = \sum_{n=-\infty}^{+\infty} x[n]z^{-n}.$$

This function exists for all values of z for which the series converges. The set of these values, which is a subset of the complex numbers, identifies a region of the z-plane called the *region of convergence*, often abbreviated as ROC. By definition, the ROC is the set of all z such that $\sum_{n=-\infty}^{+\infty} x[n]z^{-n}$ is *absolutely convergent*,[1] that is,

$$\sum_{n=-\infty}^{+\infty} \left| x[n]z^{-n} \right| < \infty.$$

To determine the shape of the ROC, it is convenient to express z in polar form, as $z = re^{j\theta}$, where r and θ are real, with $r \geq 0$ and $-\pi \leq \theta < \pi$. We then get

---

[1]Why do we base the convergence condition on a sum of absolute values, $|x[n]z^{-n}|$, while the z-transform is defined as a sum of complex numbers $x[n]z^{-n}$? The reason is that the value of an infinite sum can potentially depend on the order in which the elements of the series are added. In general, an infinite series $\sum_{n=1}^{\infty} a_n$ may converge (that is, yield a finite result) even when $\sum_{n=1}^{\infty} |a_n| = \infty$, but in such a case the finite value of $\sum_{n=1}^{\infty} a_n$ will vary if the order of the terms is changed. Such a series is said to be conditionally convergent. On the other hand, if the sum of absolute values is finite, the sum will be independent of the order of summation. Such a series is said to be absolutely convergent. In the z-transform we are summing the samples of a two-sided sequence and we do not want the sum to depend on the order of terms. The requirement of absolute convergence eliminates this issue and guarantees that the definition of the z-transform is unambiguous.

$$\sum_{n=-\infty}^{+\infty} \left| x[n]z^{-n} \right| = \sum_{n=-\infty}^{+\infty} |x[n]| \left| re^{j\theta} \right|^{-n} = \sum_{n=-\infty}^{+\infty} |x[n]| \, r^{-n}.$$

The convergence condition for a given bounded $x[n]$ depends only on r, and this suggests that the ROC will have annular shape centered in the origin of the z-plane.

The z-transform of a sequence is a Laurent series in the complex variable z (Churchill 1975). Therefore, the properties of Laurent series apply directly to the z-transform and allow establishing that if we define

$$R_- = \lim_{n \to +\infty} \left| \frac{x[n+1]}{x[n]} \right|, \qquad R_+ = \lim_{n \to -\infty} \left| \frac{x[n+1]}{x[n]} \right|,$$

then absolute convergence exists in

$$R_- < |z| < R_+,$$

provided that R_- does not exceed R_+. Thus, the ROC is a ring delimited by two circles with real, non negative radii R_- and R_+.

From the above statements we see that the determination of the ROC can substantially be reduced to the study of how $|x[n]|$ behaves when $n \to +\infty$ and $n \to -\infty$. For example, if we assume that the absolute value of $x[n]$ is bounded for $n \to \pm\infty$ by power laws of the type $|x[n]| \le M_- R_-^n$ for $n > 0$ and $|x[n]| \le M_+ R_+^n$ for $n < 0$, with $R_+ > R_-$, then R_- and R_+ are the radii delimiting the ring.[2] The lower radius R_- may go to zero and the upper radius R_+ may go to infinity: in other words, the ROC may extend outward to infinity and may extend inward to include the origin.[3]

Inside the ROC, $X(z)$ is an analytic function, i.e., a continuous function of z with derivatives of any order that are themselves continuous functions of z inside the ROC of $X(z)$. Examples of different ROCs are shown in Fig. 3.1.

Different sequences may share the same $X(z)$, but over different ROCs: when we specify $X(z)$ we must also specify the corresponding ROC, for $x[n]$ to be univocally determined. In other words, a certain analytic function $X(z)$ can exist in more than one region of the z-plane, and in each region the signal $x[n]$ for which the series sums up to $X(z)$ will be different.

---

[2] The ROC may also include the circle $|z| = R_-$ or the circle $|z| = R_-$ or both, but there is no general criterion for testing these possibilities.

[3] The *extended complex plane* is obtained by adding a single point $z = \infty$ to the conventional complex plane. The point $z = \infty$ has modulus larger than that of any other complex number and its phase is undefined. On the other hand, the point $z = 0$ has a modulus smaller than that of any other complex number and an undefined phase. The ROC may be extended to include the point $z = \infty$ if and only if the sequence is causal.

Fig. 3.1 Region of
convergence (ROC) of the
z-transform: **a** an example in
which $R_- < |z| < R_+$, and
b an example in which
$|z| > R_-$, and $R_+ \to \infty$

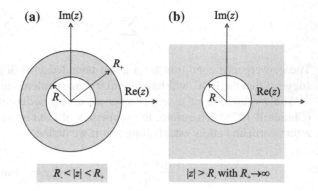

It must be noted that:

- the definition of $X(z)$ implies periodicity with respect to the phase of z, with period 2π;
- the convergence condition implies finite energy for $x_r[n] \equiv x[n]r^{-n}$, since

$$\sum_{n=-\infty}^{+\infty} |x_r[n]|^2 \le \left\{ \sum_{n=-\infty}^{+\infty} |x_r[n]| \right\}^2 .$$

3.2.1 *Examples of z-Transforms and Special Cases*

The following examples illustrate various points related to the convergence of the z-transform. We examine various kinds of sequences and discuss the characteristics of their ROC.

- Unit impulse:
  the z-transform of $x[n] = \delta[n]$ is obviously

$$X(z) = \sum_{n=-\infty}^{+\infty} \delta[n]z^{-n} = \delta[0]z^0 = 1$$

over the entire complex plane. The delayed impulse $x[n] = \delta[n-k]$ transforms into $X(z) = z^{-k}$ over the entire complex plane, with the exception of $z = 0$ if $k > 0$, or $z = \infty$ if $k < 0$.
- Unit step:
  the z-transform of $x[n] = u[n]$ is

$$X(z) = \frac{1}{1 - z^{-1}} = \frac{z}{z - 1}$$

over the region $|z| > 1$.

Another useful formula is that $x[n] = -u[-n-1]$ transforms into

$$X(z) = \frac{1}{1 - z^{-1}} = \frac{z}{z-1}$$

over the region $|z| < 1$.

- Let us consider the sequence

$$x[n] = \begin{cases} a^n & \text{for } n \geq 0, \\ 0 & \text{for } n < 0, \end{cases}$$

with $a = $ constant: a causal sequence that may be re-written as

$$x[n] = a^n u[n].$$

The z-transform is

$$X(z) = \sum_{n=0}^{+\infty} a^n z^{-n} = \sum_{n=0}^{+\infty} (az^{-1})^n = \frac{1}{1 - az^{-1}} = \frac{z}{z-a},$$

with the convergence condition $|az^{-1}| < 1$, i.e., $|z| > |a|$, from which we deduce that $R_- = |a|$ and $R_+ = \infty$ (Fig. 3.2a).

- Let us consider the sequence

$$x[n] = \begin{cases} -a^n & \text{for } n < 0, \\ 0 & \text{for } n \geq 0, \end{cases}$$

with $a = $ constant: a non-causal sequence that may be re-written as

$$x[n] = -a^n u[-n-1].$$

Fig. 3.2 Region of convergence for **a** the causal sequence $x[n] = a^n u[n]$, and **b** the non-causal sequence $x[n] = -a^n u[-n-1]$

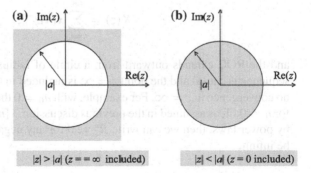

(**a**) Im(z)   (**b**) Im(z)

Re(z)   Re(z)

$|a|$ $|a|$

$|z| > |a|$ ($z == \infty$ included) $|z| < |a|$ ($z = 0$ included)

The z-transform is

$$X(z) = - \sum_{n=-\infty}^{+\infty} a^n u[-n-1] z^{-n} = - \sum_{n=-\infty}^{-1} (az^{-1})^n = - \sum_{n=1}^{+\infty} (za^{-1})^n =$$

$$= 1 - \sum_{n=0}^{+\infty} (za^{-1})^n = 1 - \frac{1}{1 - za^{-1}} = 1 - \frac{a}{a-z} = \frac{z}{z-a}.$$

The z-transform is the same as in the previous case but the ROC is different (Fig. 3.2b): the convergence condition is indeed $|za^{-1}| < 1$, i.e., $|z| < |a|$, from which we deduce that $R_- = 0$, and $R_+ = |a|$. The same z-transform with two different ROCs corresponds to two different sequences, and only one of them is causal.

- Periodic sequences, $x[n] = x[n+N]$ with $n \in (-\infty, +\infty)$:
  the z-transform does not exist.
  For example, let us consider the periodic sequence $x[n] = (-1)^n$. We may write

$$x[n] = x_1[n] + x_2[n] = (-1)^n u[n] + (-1)^n u[-n-1].$$

But from the previous examples we understand that

$$X_1(z) = \frac{z}{z+1} \quad \text{for} \quad |z| > 1 \quad \text{and} \quad X_2(z) = -\frac{z}{z+1} \quad \text{for} \quad |z| < 1,$$

so by linearity we have $X(z) = X_1(z) + X_2(z) = 0$.
What is the ROC in this case?
Since, with obvious notation, ROC_1 is $|z| > 1$ and ROC_2 is $|z| < 1$, and since the ROC must be given by the intersection of the two individual ROCs, i.e., $\text{ROC}_1 \cap \text{ROC}_2$, the ROC is empty: the z-transform series does not converge for any z for this signal. Indeed, $X(z)$ does not exist for any infinite-length periodic signal other than $x[n] = 0$. All one-sided (e.g., causal) periodic signals have a z-transform though.

- Right-sided sequences, $x[n] = 0$ for $n < n_0$, where n_0 is integer and finite:
  the z-transform is

$$X(z) = \sum_{n=n_0}^{+\infty} x[n] z^{-n},$$

and the ROC extends outward from a circle of radius R_- to ∞; if $n_0 \geq 0$ the sequence is causal and the point $z = \infty$ is included in the ROC; if $n_0 < 0$ there is no convergence in $z = \infty$. For example, with $n_0 = 0$ the signal is causal, $x[n] = 0$ for $n < 0$; if, as assumed in the previous discussion, $x[n]$ is bounded for $n \to \pm\infty$ by power laws, then we can write $R_+^n = 0$ for any negative n. Therefore R_+ must be infinite.

- Left-sided sequences, $x[n] = 0$ for $n > n_0$:
  the z-transform is

$$X(z) = \sum_{n=-\infty}^{n_0} x[n]z^{-n} = \sum_{n=-n_0}^{\infty} x[-n]z^n,$$

and the ROC extends inward from a circle of radius R_+ to the origin, with the exception of the point $z = 0$ if $n_0 > 0$. For example, with $n_0 = 0$, $x[n] = 0$ for $n > 0$ (anticausal signal); if $x[n]$ is bounded for $n \to \pm\infty$ by power laws, then $R_-^n = 0$ with positive n and therefore R_- must vanish.

- Two-sided sequences:
  this is the general case, in which the z-transform can be split into two parts:

$$X(z) = \sum_{n=-\infty}^{+\infty} x[n]z^{-n} = \sum_{n=-\infty}^{-1} x[n]z^{-n} + \sum_{n=0}^{+\infty} x[n]z^{-n},$$

where the first addend corresponds to a left-sided sequence with $n_0 = -1$, and the second addend corresponds to a right-sided sequence with $n_0 = 0$. As shown in Fig. 3.3, the first addend converges in $|z| < R_+$ with $z = 0$ included (Fig. 3.3a), while the second addend converges in $|z| > R_-$ with $z = \infty$ included (Fig. 3.3b). With $R_+ > R_-$, this implies convergence in $R_- < |z| < R_+$.

- Finite-duration sequences, $x[n] = 0$ for $n < n_1$ and $n > n_2$, where n_1 and n_2 are integer and finite:
  we have

$$X(z) = \sum_{n=n_1}^{n_2} x[n]z^{-n},$$

and $R_+ \to \infty$, $R_- \to 0$. Hence z can assume any value, with the exception of $z = 0$ when $n_2 > 0$ and $z = \infty$ when $n_1 < 0$. Convergence exists at least in $0 < |z| < \infty$, with the possible addition of $z = 0$ or $z = \infty$.

Fig. 3.3 Region of convergence for the z-transform of **a** a right-sided sequence, and **b** a left-sided sequence

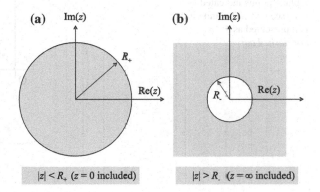

3.2.2   Rational z-Transforms

The z-transform is most useful when the infinite sum that defines it can be expressed in closed form, i.e., when it can actually be "summed" and expressed by a simple mathematical formula, as in the examples considered above.

The majority of the z-transforms encountered in practical cases are rational functions, i.e., ratios of polynomials. Moreover, the theory of rational approximation of functions (see, e.g., Powell 1981) states that we can always approximate a continuous function like the z-transform as closely as we want, under relatively large convergence constraints, by a rational function with sufficiently high numerator and the denominator degrees. The z-transforms of many important signals are rational functions; also, and more importantly, the transfer functions of LTI systems described by LCCDEs are rational functions.

Rational z-transforms are determined, up to a multiplicative constant, by their *zeros* (roots of the numerator polynomial) and by their *poles* (roots of the denominator polynomial).

Recall that a polynomial $P(z)$ with degree p has exactly p roots, including possible multiplicities: two or more roots can indeed be coincident. These roots are complex in general, but if the polynomial coefficients are real, the roots come in complex conjugate pairs: if λ is a root with multiplicity n, then λ^* is also a root. This is the case of the polynomials forming the $H(z)$ of an LTI system satisfying an LCCDE, because then the coefficients of the numerator and denominator polynomials derive directly from the LCCDE coefficients that are real and constant (Sect. 3.2.6). For rational z-transforms, the ROC is determined by the position of the poles: $X(z)$ goes to infinity at a pole and therefore does not converge at that point of the z-plane. So, the ROC cannot contain any pole. Figure 3.4 shows an example of a pole pattern in the complex plane. The unit circle is also shown as a *dot-dashed curve*. What are the possible ROCs of $X(z)$ in this example?

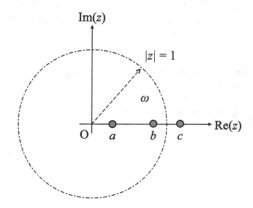

Fig. 3.4 An example in which a rational z-transform has three poles falling in the z-plane points indicated as a, b, c (see text). The unit circle is represented as a *dot-dashed curve*

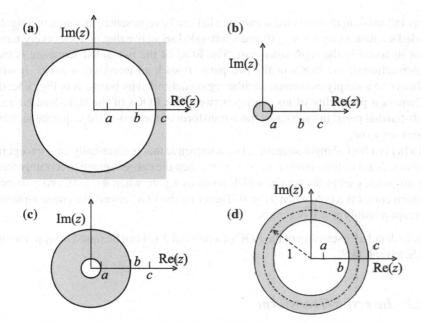

Fig. 3.5 The four different possibilities that exist for the region of convergence of a rational z-transform, when the poles are those shown in Fig. 3.4. Each ROC corresponds to a different sequence: **a** a right-sided sequence, **b** a left-sided sequence, **c** a two-sided sequence, and **d** another two-sided sequence, different from the previous one. The only sequence having a DTFT representation is the one corresponding to the ROC of panel d, because only in that case the ROC includes the unit circle (*dot-dashed curve*)

Given a pole pattern of a rational $H(z)$ like the one shown in Fig. 3.4, several possible choices for the ROC can exist, all of them satisfying the condition of containing no poles, and each choice is associated with a different sequence $x[n]$. In our example, there are only four possible choices for the ROC. They are shown in Fig. 3.5. Specifically, Fig. 3.5a corresponds to a right-sided sequence; Fig. 3.5b corresponds to a left-sided sequence; Fig. 3.5c, d correspond to two distinct two-sided sequences. Causality is the constraint that allows us to determine a unique solution: only the right-sided sequence in Fig. 3.5a is causal. If, and only if, the ROC contains the unit circle, the sequence has a DTFT representation. If we assume that the unit circle falls between the pole at $z = b$ and the pole at $z = c$, as in Fig. 3.4, then the only case in which the DTFT converges is the one in Fig. 3.5d.

To summarize, the following properties hold:

- if $x[n]$ is an infinite-length right-sided sequence, then the ROC extends outward from the outermost (i.e., largest magnitude) finite pole to (and possibly including) $z = \infty$;
- if $x[n]$ is an infinite-length left-sided sequence, then the ROC extends inward from the innermost (i.e., smallest magnitude) non-zero pole to (and possibly including) $z = 0$;

- any infinite-length two-sided sequence $x[n]$ can be represented as a sum of a right-sided sequence (say for $n \geq 0$) and a left-sided sequence that includes every term not included in the right-sided part. The ROC of the two-sided sequence is the intersection of the ROCs of the two parts. If such an intersection exists, it will always be a simply connected annular region delimited by poles, as in Fig. 3.5c, d. There is a possibility of no overlap between the ROCs of the right-handed and left-handed parts: in such cases the z-transform of the two-sided sequence simply does not exist;
- if $x[n]$ is a finite-length sequence, i.e., a sequence that is identically zero except in a finite discrete-time interval $n_1 < n \leq n_2$, then the entire z-plane is certainly free from poles, except the origin, which contains a pole when $n_2 > 0$, and $z = \infty$, which contains a pole when $n_1 < 0$. Therefore the ROC covers the entire z-plane, except possibly $z = 0$ or $z = \infty$.

A more detailed discussion of the ROC of a rational $X(z)$ can be found in Oppenheim and Schafer (2009).

3.2.3 Inverse z-Transform

One of the important roles of the z-transform is in the analysis of LTI systems. Often this analysis involves finding the z-transform of a sequence and, after some manipulation of its algebraic expression, finding the inverse z-transform. So, given the algebraic expression of $X(z)$ and its associated ROC, how can we calculate the corresponding $x[n]$?

Formally, we can use the Cauchy integral theorem to write (see, e.g., Oppenheim and Schafer 2009)

$$x[n] = \frac{1}{2\pi j} \oint_C X(z) z^{n-1} dz,$$

where C is a closed counterclockwise contour that encircles the origin and is contained inside the ROC. This integral can be evaluated using Cauchy's residue theorem (Churchill and Brown 1984).

If we focus on the case of practical interest in which the series that defines $X(z)$ converges to a rational function of z, then $X(z)$ has a finite number of poles and zeros. A technique of partial fraction expansion (or decomposition; see Sect. 3.2.8) can be adopted in this case. The aim is expressing $X(z)$ as a linear superposition of terms that can be easily inverted: in particular, simple terms whose inverse z-transform is known (see Oppenheim and Schafer (2009) for a list of some common z-transform pairs).

Though we will not enter into further details on this subject, it is worth mentioning that a rational z-transform can be expressed in the form of a ratio between polynomials containing powers of z^{-1}, as we did so far, or in the form of a ratio between polynomials containing powers of z. One form may be more convenient than the other as far as the inversion is concerned.

Causal sequences have a z-transform definition sum containing only negative powers of z, i.e., $X(z) = \sum_{n=0}^{+\infty} x[n]z^{-n}$. In this case, $X(z)$ is most conveniently given as a power series in z^{-1} with some coefficients c_n, and with the associated ROC. Then, the inversion becomes straightforward:

$$\sum_{n=-\infty}^{+\infty} x[n]z^{-n} = \sum_{n=-\infty}^{+\infty} c_n z^{-n}$$

leads directly to $x[n] \equiv c_n$. It is sometimes useful to expand an $X(z)$ given in closed form in series of powers of z^{-1}, so as to be able to perform this direct inversion. For example, let us take the non-rational z-transform $X(z) = \log(1 + az^{-1})$, with $|z| > |a|$. Using the series expansion for $\log(1 + \beta)$ valid for $|\beta| < 1$, which is

$$\log(1 + \beta) = \sum_{n=1}^{\infty} (-1)^{n+1} \frac{\beta^n}{n},$$

we can write

$$X(z) = \sum_{n=1}^{+\infty} \frac{(-1)^{n+1} a^n z^{-n}}{n}.$$

Comparison of this formula with the definition summation allows for direct inversion:

$$x[n] = \begin{cases} (-1)^{n+1} \frac{a^n}{n} & \text{for } n \geq 1, \\ 0 & \text{for } n \leq 0. \end{cases}$$

3.2.4 The z-Transform on the Unit Circle

When the z-transform converges on the unit circle (Fig. 2.13), there $X(z)$ represents the discrete-time Fourier transform (DTFT) of $x[n]$:

$$X(e^{j\omega}) = \sum_{n=-\infty}^{+\infty} x[n]e^{-j\omega n}.$$

Note that:

- $x[n]$ is a sequence, that is, a function of *discrete time*, while the DTFT is a function of the *continuous-frequency* variable ω;
- $X(e^{j\omega})$ is a periodic function with a period of 2π: indeed, since $e^{-j2\pi k} = 1$,

$$X[e^{j(\omega+2\pi k)}] = \sum_{n=-\infty}^{+\infty} x[n]e^{-j(\omega+2\pi k)n} = \sum_{n=-\infty}^{+\infty} x[n]e^{-j\omega n}e^{-j2\pi kn} = X(e^{j\omega});$$

- the expression of $X(e^{j\omega})$ can be seen as the Fourier series expansion of the periodic function $X(e^{j\omega})$. Frequency ω varies from $-\infty$ to $+\infty$ when we go through the unit circle an infinite number of times. This is referred to as the *wrapping* of the ω axis;
- $e^{j\omega n}$ and $e^{j(\omega+2\pi k)n}$ are indistinguishable, so $X(e^{j\omega})$ must be the same at these frequencies. For example, if we consider the frequency response $H(e^{j\omega})$ of an LTI system, the system must react in the same way to indistinguishable frequencies;
- when $x[n]$ derives from an analog signal $x(t)$ through periodic sampling with sampling interval T_s, we can relate the frequency variables of the discrete-time domain (ω and ν) to those of the continuous-time domain (f and Ω): $f = \Omega/(2\pi) = \omega/(2\pi T_s) = \nu/T_s$. The period of $X(e^{j\omega})$ is 2π when expressed in terms of ω, 1 in terms of ν, $1/T_s$ in terms of f, and $2\pi/T_s$ in terms of Ω.

3.2.5 Selected z-Transform Properties

We will now list some important properties of the z-transform. From this point on we will denote any reversible transform operation by the symmetric symbol $\Longleftrightarrow$; the context of the discussion clarifies what kind of transform is involved. So in this section $\Longleftrightarrow$ means "the right-hand term is the z-transform of the left-hand term and the left-hand term is the inverse z-transform of the right-hand term". For brevity, we omit symmetry properties.

1. *Linearity*:
   given two arbitrary constants a and b,

   $$ax_1[n] + bx_2[n] \Longleftrightarrow aX_1(z) + bX_2(z),$$

   and the ROC is at least the intersection of the individual ROCs of $X_1(z)$ and $X_2(z)$. If the linear combination is such that one or more zeros are introduced that cancel one or more poles, then the ROC may be larger.
2. *Folding or time reversal*:
   $$x[-n] \Longleftrightarrow X(z^{-1}).$$

   The ROC of $X(z^{-1})$ is such that if $X(z)$ converges in $R_- < |z| < R_+$, then the transform of the folded sequence converges in $1/R_+ < |z| < 1/R_-$.
3. *Time shifting, or translation, or "delay"*:

   $$x[n-m] \Longleftrightarrow z^{-m}X(z).$$

   The sequence $x[n-m]$ shares the same ROC as $x[n]$, with the exception of $z = 0$ e $z = \infty$. For $m > 0$, poles are introduced at the origin and zeros are introduced at infinity, and vice-versa for $m < 0$.

4. *Conjugation of a complex sequence*:

$$x^*[n] \Longleftrightarrow X^*(z^*),$$

with convergence in $R_- < |z| < R_+$.

5. *Multiplication by an exponential sequence a^n (modulation)*: given a real or complex constant a,

$$a^n x[n] \Longleftrightarrow X(z/a),$$

with ROC given by $|a|R_- < |z| < |a|R_+$, where R_- and R_+ delimit the ROC of $X(z)$. The positions of the poles and zeros are modified by a factor a:

- if a is real and non-zero, a compression (shrinking) or an expansion of the z-plane occurs for $|a| > $ and $|a| < 1$ respectively, i.e., both poles and zeros move along radial lines in the z-plane;
- if a is complex, with $|a| = 1$, so that $a = e^{j\theta}$, a rotation of the z-plane by an angle of θ occurs, i.e., both poles and zeros move along circles centered at the origin;
- if a is complex and $|a| \neq 1$, both a compression/expansion and a rotation of the z-plane occur.

An important special case is

$$(-1)^n x[n] \Longleftrightarrow X(-z).$$

Putting the folding, time-shifting and modulation properties together we can see that

$$(-1)^n x[m - n] \Longleftrightarrow (-z)^{-m} X(-z^{-1}).$$

These special cases will be useful in Chap. 14.

6. *Differentiation*:

$$nx[n] \Longleftrightarrow -z \frac{dX(z)}{dz},$$

with convergence in $R_- < |z| < R_+$, except the possible cancellation or introduction of the points $z = 0$ and $z = \infty$.

7. *Real and imaginary parts of a complex sequence*:

$$\text{Re}\{x[n]\} \Longleftrightarrow \frac{1}{2}[X(z) + X^*(z^*)] \quad \text{and} \quad \text{Im}\{x[n]\} \Longleftrightarrow \frac{1}{2j}[X(z) - X^*(z^*)],$$

with convergence at least in $R_- < |z| < R_+$.

8. *Convolution*:

$$y[n] = x_1[n] * x_2[x] \Longleftrightarrow Y(z) = X_1(z)X_2(z),$$

with convergence at least in the intersection of the ROCs of $X_1(z)$ and $X_2(z)$. If a pole of one of the individual z-transforms is canceled by a zero of the other, then the ROC of $Y(z)$ can be larger.

9. *Initial value theorem*:

   if $x[n]$ is causal, then $x[0] = \lim_{z\to\infty} X(z)$.

10. *Final value theorem*:

   if $X(z)$ exists for $|z| > r$ and $r < 1$, i.e., if the ROC includes the unit circle, then $x[n] \to 0$ as $n \to \infty$.

   If $X(z)$ is rational and is the z-transform of a causal sequence $x[n]$, then the ROC of $X(z)$ extends outward from a circle containing all its poles; in this case, the final value theorem implies that $x[n]$ vanishes for $n \to \infty$ if all poles are internal to the unit circle.

11. *Complex convolution theorem*:

   the z-transform of the product $w[n] = x[n]y[n]$ of two real sequences $x[n]$ and $y[n]$ takes the form of a *convolution integral*:

   $$w[n] = x[n]y[n] \Longleftrightarrow W(z) = \frac{1}{2\pi j} \oint_C X(z/v)Y(v)v^{-1}dv,$$

   where the counterclockwise contour C must fall inside the intersection of the individual ROCs of $X(z/v)$ and $Y(v)$. Alternatively we can write

   $$w[n] = x[n]y[n] \Longleftrightarrow W(z) = \frac{1}{2\pi j} \oint_{C'} X(v)Y(z/v)v^{-1}dv,$$

   where the counterclockwise contour C' must fall inside the intersection of the individual ROCs of $X(v)$ and $Y(z/v)$.

   If we choose a circular contour and express the variables in polar form, by setting $z = re^{j\omega}$ and $v = \rho e^{j\theta}$, we find

   $$W(re^{j\omega}) = \frac{1}{2\pi} \int_{-\pi}^{+\pi} X(\rho e^{j\theta})Y\left[\left(\frac{r}{\rho}\right)e^{j(\omega-\theta)}\right]d\theta =$$

   $$= \frac{1}{2\pi} \int_{-\pi}^{+\pi} X\left[\left(\frac{r}{\rho}\right)e^{j(\omega-\theta)}\right]Y(\rho e^{j\theta})d\theta,$$

   that formally is the convolution integral of $X(re^{j\omega})$ and $Y(re^{j\omega})$ as functions of ω, except for the fact that the integration limits are finite. Actually, this is an integral performed over a single period of $X(z)$ and $Y(z)$. This operation is referred to as the *continuous periodic convolution* of $X(z)$ with $Y(z)$. The ROC of $W(z)$ is given by $R_{x-}R_{y-} < |z| < R_{x+}R_{y+}$, with obvious notation.

12. *Parseval's theorem*[4]:

for generality, this time we explicitly consider two complex sequences $x[n]$ and $y[n]$. We can write

$$\sum_{n=-\infty}^{+\infty} x[n]y^*[n] = \frac{1}{2\pi j} \oint_C X(v)Y^*(1/v^*)v^{-1}dv.$$

The contour C must be contained inside the intersection of the individual ROCs of $X(v)$ and $Y^*(1/v^*)$. This property derives from the complex convolution theorem applied to $w[n] = x[n]y^*[n]$.

If $x[n] = y[n]$, Parseval's relation gives an expression for the energy[5] of $x[n]$ in terms of the z-transform:

$$\mathscr{E} = \sum_{n=-\infty}^{+\infty} |x[n]|^2 = \frac{1}{2\pi j} \oint_C X(v)X^*(1/v^*)v^{-1}dv.$$

On the unit circle, provided that convergence exists on it, this relation becomes

$$\mathscr{E} = \sum_{n=-\infty}^{+\infty} |x[n]|^2 = \frac{1}{2\pi} \int_{-\pi}^{+\pi} |X(e^{j\omega})|^2 d\omega.$$

This is Parseval's relation for the DTFT of $x[n]$, which will be resumed later.

3.2.6   Transfer Function of an LTI System

An LTI system satisfying an LCCDE with finite order has a rational transfer function $H(z)$, i.e., $H(z)$ is the ratio of two polynomials in z^{-1}. To demonstrate this, let us consider an LCCDE with order N,

$$y[n] + a_1 y[n-1] + \cdots + a_N y[n-N] = b_0 x[n] + b_1 x[n-1] + \cdots + b_M x[n-M],$$

and transform it according to the time-shift property of the z-transform: we obtain

$$Y(z) + a_1 Y(z)z^{-1} + \cdots + a_N Y(z)z^{-N} = b_0 X(z) + b_1 X(z)z^{-1} + \cdots + b_M X(z)z^{-M},$$

---

[4]The name of this relation originates from a 1799 theorem about series, formulated by Marc-Antoine Parseval, which was later applied to the Fourier series. Analogous relations for other transforms were then derived, like the one mentioned here for the z-transform.

[5]Parseval's relation written in this way is meaningful only for energy signals. It does not hold for power signals having a DTFT only in a generalized sense (see Sect. 3.3). For power signal we must reason in terms of average power and write Parseval's relation differently; this will be done in Chap. 5.

from which we get, with $a_0 = 1$,

$$Y(z)/X(z) = \frac{b_0 + b_1 z^{-1} + \cdots + b_M z^{-M}}{1 + a_1 z^{-1} + \cdots + a_N z^{-N}} = \frac{\sum_{r=0}^{M} b_r z^{-r}}{\sum_{k=0}^{N} a_k z^{-k}} = \frac{b(z)}{a(z)}.$$

But according to the convolution property of the z-transform,

$$y[n] = x[n] * h[n] \iff Y(z) = X(z)H(z),$$

so that

$$H(z) = Y(z)/X(z);$$

in conclusion,

$$H(z) = \frac{b_0 + b_1 z^{-1} + \cdots + b_M z^{-M}}{1 + a_1 z^{-1} + \cdots + a_N z^{-N}} = \frac{\sum_{r=0}^{M} b_r z^{-r}}{\sum_{k=0}^{N} a_k z^{-k}} = \frac{b(z)}{a(z)},$$

with $a_0 = 1$. The corresponding sequence $h[n]$ has, in general, infinite length.

We can summarize what we have learned about the transfer function of an LTI system as follows:

- $H(z)$ is the z-transform of the impulse response $h[n]$;
- $H(z)$ is the eigenvalue of the LTI system, corresponding to the eigenfunction z^n;
- $H(z)$ is the ratio of the z-transform of the output to the z-transform of the input, for any input $x[n]$.

A further comment on the relation $H(z) = Y(z)/X(z)$ is needed here. In order to derive this relation from the LCCDE, we only assumed

- that the system is LTI;
- that the ROCs of $X(z)$ and $Y(z)$ overlap at least partly, so that the ROC of $H(z)$ is not empty.

We did not assume the LTI system to be stable and/or causal: correspondingly, from the LCCDE we obtained $H(z)$ but not its ROC, and therefore the sequence $h[n]$ is not univocally defined. This is consistent with the fact that an LCCDE does not identify univocally the input-output relation of an LTI system, unless we associate proper initial conditions with it. As a consequence, given an LCCDE, different choices for the ROC of $H(z)$ may be available, all of them in the form of a ring delimited by poles but with no poles inside. Each choice leads to a different $h[n]$, as we already saw in the case of Fig. 3.5. However, if we constrain the system to be causal, $h[n]$ must be a right-sided sequence and therefore the ROC must extend outward from the outermost pole. If we further assume stability, the ROC must include the unit circle. The two requirements are not necessarily compatible: they are compatible if all $H(z)$ poles are contained inside the unit circle. This is easily seen comparing the stability condition for an LTI system with the condition defining the ROC of $H(z)$:

$$\sum_{k=-\infty}^{+\infty} |h[k]| < \infty,$$

$$\sum_{k=-\infty}^{+\infty} |h[k]||z^{-k}| < \infty.$$

Thus, if the ROC of $H(z)$ includes the points with $|z| = 1$, then the system is stable, and vice-versa.

For a stable and causal system,

- the ROC of the transfer function includes the unit circle and the part of the z-plane external to the unit circle, including $z = \infty$;
- all $H(z)$ poles are inside the unit circle; the zeros can be anywhere in the z-plane;
- the impulse response is a sequence $h[n]$ that tends to 0 for $n \to \infty$ (final value theorem);
- the values of $H(z)$ on the unit circle, i.e., $H(z)|_{z=e^{j\omega}} = H(e^{j\omega})$, represent the system's frequency response.

3.2.7 Output Sequence of an LTI System

Given the transfer function of an LTI system, $H(z)$, together with its ROC, and given an input sequence $x[n]$, we can, *in principle*, deduce the output sequence $y[n]$ in several ways.

1. Discrete-time domain

   - Convolution: we can invert $H(z)$ to get $h[n]$ and then compute $y[n] = x[n] * h[n]$.
   - LCCDE: from the expression of $H(z)$ we can deduce the LCCDE coefficients and then build the samples of $y[n]$.

2. z-transform domain

   - From $x[n]$ we can calculate $X(z)$. Then we form $Y(z) = H(z)X(z)$ and by inverse z-transform we obtain $y[n]$.

However, while an FIR system is easily actuated directly in the time domain via convolution, an operation that in the FIR case implies a finite number of additions and multiplications and the use of a finite number of memory locations on a computer, an IIR system is not. Therefore an IIR system will be actuated in the time domain via recursive LCCDE.

3.2.8 Zeros and Poles: Forms for Rational Transfer Functions

We now consider the rational transfer function of a stable and causal LTI system and discuss three more ways of expressing it. Recall that we assumed $a_0 = 1$.

- It can be shown that the transfer function,

$$H(z) = \frac{\sum_{m=0}^{M} b_m z^{-m}}{\sum_{k=0}^{N} a_k z^{-k}} = \frac{\sum_{m=0}^{M} b_m z^{-m}}{1 + \sum_{k=1}^{N} a_k z^{-k}},$$

can, provided that b_0 does not vanish, be factorized as

$$H(z) = A \frac{\prod_{m=1}^{M}(1 - w_m z^{-1})}{\prod_{k=1}^{N}(1 - p_k z^{-1})},$$

where w_m and p_k are the zeros and poles of $H(z)$, respectively, and the non-zero factor A is equal to b_0/a_0; but since we set $a_0 = 1$, we simply have $A = b_0$. In this expression, each factor in the numerator gives a zero of the transfer function at $z = w_r$ and a pole at $z = 0$, while each factor in the denominator gives a pole of the transfer function at $z = p_k$ and a zero at $z = 0$. As we stated in Sect. 3.2.2, the roots w_m of the numerator polynomial occur in complex-conjugate pairs, and the same is true for the roots p_k of the denominator polynomials.

If, on the contrary, $b_0 = b_1 = \cdots = b_{M_1} = 0$, i.e., if the b_m vanish up to some $m = M_1$, then the factorized formula for $H(z)$ is

$$H(z) = B z^{-(M_1+1)} \frac{\prod_{m=1}^{M-(M_1+1)}(1 - w_m z^{-1})}{\prod_{k=1}^{N}(1 - p_k z^{-1})},$$

where the factor B is equal to b_{M_1+1}/a_0.

- In addition to the formula given above, another factorized form of $H(z)$ exists, which is obtained starting from $H(z)$ expressed as a ratio of polynomials in z, rather than in z^{-1}—a transformation that is always possible. The procedure is the following.

Let b_{M-r} be the first numerator coefficient that is non-zero, with $r \leq M$; for example, $r = M$ if $b_0 \neq 0$. Then, with $a_0 = 1$, the following expressions can be written:

$$H(z) = \begin{cases} \dfrac{z^{N-M}(b_{M-r}z^r + \cdots + b_M)}{z^N + a_1 z^{N-1} + \cdots + a_N} & \text{for } N \geq M, \quad N - M \geq 0, \\[3ex] \dfrac{b_{M-r}z^r + \cdots + b_M}{z^{M-N}(z^N + a_1 z^{N-1} + \cdots + a_N)} & \text{for } N < M, \quad M - N > 0. \end{cases}$$

In the first case, the numerator degree is $N - M + r$ with $r \leq M$, hence the numerator degree is $\leq N$. The denominator degree is N. In the second case, the numerator degree is $r \le M$ and the denominator degree is M. Note that the exponent $M - N$ in the denominator is non-negative in this second case. In both cases, the numerator degree does not exceed the denominator degree (*proper fraction*), as is appropriate to a causal system. Indeed, positive powers of z are associated with time advances, while negative powers of z are associated with time delays. If $r = M$, i.e., $b_0 \ne 0$, the numerator and denominator degrees are equal (*exactly proper fraction*), and $y[n]$ depends on both past $(x[n - k])$ and present $(x[n])$ values of the input. If $b_0 = 0$, the numerator degree is smaller than the denominator degree (*strictly proper fraction*), and $y[n]$ only depends on past input values.

The previous expression is then properly factorized. For example, if $b_0 \ne 0$, $r = M$, then $H(z)$ can be written as

$$H(z) = b_0 z^{N-M} \frac{z^M + (b_1/b_0)z^{M-1} + \cdots + (b_M/b_0)}{z^N + a_1 z^{N-1} + \cdots + a_N},$$

and therefore we get

$$H(z) = b_0 z^{N-M} \frac{(z - w_1)(z - w_2) + \cdots + (z - w_M)}{(z - p_1)(z - p_2) + \cdots + (z - p_N)} = b_0 z^{N-M} \frac{\prod_{m=1}^{M}(z - w_m)}{\prod_{k=1}^{N}(z - p_k)}.$$

In this formula, the exponent $N - M$ can have any sign, or be zero as well. The factor b_0 is often indicated as the *gain*: $G = b_0/a_0 = b_0$ (recall that $a_0 = 1$). This form of $H(z)$ is referred to as the *zero-pole-gain form*.

From this expression we see that in the z-plane, at $z \ne 0$ and $z \ne \infty$, $H(z)$ has M zeros and N poles. It also has $|N - M|$ zeros (if $N \ge M$) or poles (if $N < M$) at $z = 0$. Therefore at finite z the transfer function $H(z)$ has N zeros and N poles. This allows for writing $H(z)$ as

$$H(z) = b_0 \frac{(z - w_1)(z - w_2) + \cdots + (z - w_L)}{(z - p_1)(z - p_2) + \cdots + (z - p_L)}, \qquad L = \max(N, M),$$

in which some w_k and p_k can be zero. Obviously, each pole at $z = 0$ corresponds to a zero at $z = \infty$, and each zero at $z = 0$ corresponds to a pole at $z = \infty$.

Table 3.1 summarizes what we have said about the number of poles and zeros of $H(z)$. The total number of poles and zeros indicated in Table 3.1 excludes those possibly existing at $z = \infty$. The number of zeros and poles present at $z = \infty$ is also indicated in Table 3.1, but in square parentheses, since these zeros and poles do not contribute to the total count.

Note that the numerator and the denominator polynomials might have one or more roots in common. If the numerator and denominator polynomials have no roots in common, then $H(z)$ is said to be *in minimal form*, meaning that the system cannot be described by an LCCDE having smaller values of N and M. Conversely, if the two polynomials do have a common factor, $H(z)$ is said to be in non-minimal form. A non-minimal $H(z)$ can be brought to a minimal form by canceling the

Table 3.1 Number of zeros and poles of $H(z)$

	$z \neq 0$, finite	$z = 0$	$[z = \infty]$	Total	$z = 0$	$[z = \infty]$	Total
		$N \geq M$			$N < M$		
Zeros	M	$N - M$		N		$[M - N]$	M
Poles	N		$[N - M]$	N	$M - N$		M

common factor of the numerator and denominator polynomials. In dealing with rational transfer functions, it is assumed that the function is given in minimal form, unless stated otherwise. FIR systems have $N = 0$, and therefore only have M zeros, besides a multiple pole at the origin.

From the previous discussion it is clear that a rational $H(z)$ is determined, up to a constant factor, by its poles and zeros, and as such, it can be described graphically by a diagram showing the positions of poles and zeros.

• Another form sometimes used for a rational $H(z)$ is the one that can be obtained by *partial fraction decomposition*. This form can be useful for the inversion of a rational $H(z)$ (Sect. 3.2.3).

In this case we start from a transfer function expressed as a ratio of polynomials in z^{-1}, and assume that the numerator has a degree M smaller than the degree N of the denominator. This is not restrictive, since if we have instead $M \geq N$, then we can preliminarily write

$$H(z) = \frac{b(z)}{a(z)} = c_0 + c_1 z^{-1} + \cdots + c_{M-N} z^{-(M-N)} + \cdots$$

$$\cdots + \frac{d_0 + d_1 z^{-1} + \cdots + d_{N-1} z^{-(N-1)}}{1 + a_1 z^{-1} + \cdots + a_{N-1} z^{-N}} =$$

$$= c(z) + \frac{d(z)}{a(z)},$$

and then work on the term $d(z)/a(z)$, for which the condition $M < N$ is certainly satisfied. The coefficients of the new polynomials $c(z)$ and $d(z)$ are obtained by equating the coefficients of the various powers of z^{-1} in the equation $c(z)a(z) + d(z) = b(z)$, and then solving the system thus obtained.

The term $c(z)$, provided it is present, can be inverted directly. As for the term $d(z)/a(z)$, assuming that all the poles of the LTI system are simple, i.e., no two of them are equal, it can be shown that it is possible to write

$$\frac{d(z)}{a(z)} = \frac{d_0 + d_1 z^{-1} + \cdots + d_{N-1} z^{-(N-1)}}{\prod_{i=1}^{N}(1 - p_i z^{-1})} = \sum_{k=1}^{N} \frac{A_k}{1 - p_k z^{-1}},$$

where p_k are the zeros of $a(z)$, that is, the poles of $H(z)$, and the A_k are coefficients called *residuals* of $d(z)/a(z)$ in the poles. The residuals A_k can be found multiplying the equation by each of the terms $1 - p_k z^{-1}$, and then substituting

$z = p_k$. This gives

$$A_k = \frac{d_0 p_k^{N-1} + d_1 p_k^{N-2} + \cdots + d_{N-1}}{\prod_{i \neq k}(p_k - p_i)},$$

in which the factors in the denominator are all different from zero, since the poles have been assumed to be all distinct.

In summary, the partial fraction decomposition of $H(z)$ for $M \geq N$ is

$$H(z) = c_0 + c_1 z^{-1} + \cdots + c_{M-N} z^{-(M-N)} + \sum_{k=1}^{N} \frac{A_k}{1 - p_k z^{-1}},$$

or

$$H(z) = \sum_{i=0}^{M-N} c_i z^{-i} + \sum_{k=1}^{N} \frac{A_k}{1 - p_k z^{-1}},$$

in the case of simple poles only. If $M < N$, there is no need for preliminary polynomial division, $d(z) \equiv b(z)$ and the first summation is absent. In this form, $H(z)$ is immediately invertible, since the inverse transforms of the terms involved are known. The impulse response of the system is then found considering that the terms $c_i z^{-i}$ correspond to terms of the form $c_i \delta[n-i]$, while the terms of the form $1/(1 - p_k z^{-1})$ correspond to $p_k^n u[n]$ or to $-p_k^n u[-n-1]$, depending on the ROC of H(z) (see Sect. 3.2.1). A causal LTI system will have

$$h[n] = \sum_{i=0}^{M-N} c_i \delta[n-i] + \sum_{k=1}^{N} A_k p_k^n u[n].$$

The partial fraction expansion formula can be generalized to the case of multiple poles. If r is a pole with multiplicity s_r, the corresponding term in the decomposition formula assumes the form

$$\frac{A_j}{1 - p_j z^{-1}} + \frac{A_{j+1}}{(1 - p_j z^{-1})^2} + \cdots + \frac{A_{j+s_r-1}}{(1 - p_j z^{-1})^{s_r}},$$

and can be easily inverted. In practice, LTI systems with multiple poles are rare in digital signal processing applications.

3.2.9 Inverse System

For a given LTI system with transfer function $H(z)$, the corresponding inverse system is defined to be the system with transfer function $H_i(z)$, such that if it is cascaded with $H(z)$, the overall effective transfer function is unity:

$$G(z) = H(z)H_i(z) = 1.$$

This implies that $H_i(z)$, provided it exists, is

$$H_i(z) = \frac{1}{H(z)}.$$

In the time domain, the equivalent condition is

$$g[n] = h[n] * h_i[n] = \delta[n].$$

The frequency response of the inverse system, if it exists, is

$$H_i(e^{j\omega}) = \frac{1}{H(e^{j\omega})}.$$

If $x[n]$ and $y[n]$ are the input and the output of the system with transfer function $H(z)$, respectively, then the system with transfer function $H_i(z)$, with $y[n]$ as the input, will give $x[n]$ as the output; but not all systems have an inverse.

For example, the ideal lowpass filter that will be presented in Sect. 6.1 (see Fig. 6.1a) does not have an inverse: indeed, there is no way to recover the frequency components of the input signal $x[n]$ that have been completely eliminated by the filtering operation. The same applies to the ideal highpass filters and to any ideal frequency-selective filter.

Many systems do have inverses, however. The class of systems with rational transfer functions provides the best example:

$$H(z) = \frac{b_0 \prod_{r=1}^{M}(1 - w_r z^{-1})}{a_0 \prod_{k=1}^{N}(1 - p_k z^{-1})}$$

gives immediately

$$H_i(z) = \frac{a_0 \prod_{k=1}^{N}(1 - p_k z^{-1})}{b_0 \prod_{r=1}^{M}(1 - w_r z^{-1})}.$$

If $H(z)$ is stable, $H_i(z)$ is not necessarily stable. Since the poles of $H_i(z)$ evidently are the zeros of $H(z)$ and vice-versa, an LTI system is stable and causal and also has a stable and causal inverse solely if not only its poles, but also its zeros lie inside the unit circle.

Note that if the original filter is FIR, i.e., if it has a polynomial transfer function, its inverse will be IIR, i.e., its transfer function will be a rational function. We can deduce that if in applications only the FIR class of filters is admitted for some reason, we just cannot invert a single filter while remaining inside the FIR class.

3.3 Discrete-Time Fourier Transform (DTFT)

Now we will focus on the representation of signals and system in the frequency domain.

Let us recall that the values of the complex variable z with unit amplitude, $z = e^{j\omega}$, define the complex exponential sequences $x[n] = e^{j\omega n}$, having the property of passing through an LTI system without changing their shape. The output is still a complex exponential sequence with the same frequency as the input, though with amplitude and phase determined by the system characteristics:

$$x[n] = e^{j\omega n} \Rightarrow y[n] = e^{j\omega n} H(e^{j\omega}),$$

where $e^{j\omega n}$ is the *eigenfunction* of the LTI system, corresponding to the *eigenvalue* $H(e^{j\omega})$. The set of values $H(e^{j\omega})$ for all possible ω, i.e., the function $H(e^{j\omega})$, is the *frequency response* of the LTI system, which coincides with the DTFT of $h[n]$,

$$H(e^{j\omega}) = H(z)|_{z=e^{j\omega}} = \sum_{n=-\infty}^{+\infty} h[n]e^{-j\omega n}.$$

The frequency response of an LTI system has the following properties:

- it exists if the system is stable;
- it describes, frequency by frequency, the amplitude and phase modulation operated by the system on complex exponential input signals.

If instead of the impulse response of an LTI system we consider any sequence $x[n]$, a similar summation gives the DTFT of $x[n]$,

$$X(e^{j\omega}) = \sum_{n=-\infty}^{+\infty} x[n]e^{-j\omega n}.$$

This definition is also referred to as the *analysis relation* of the DTFT.

$X(e^{j\omega})$ is a periodic complex function of ω, with period 2π, often referred to as the *spectrum* of $x[n]$. The amplitude of $X(e^{j\omega})$ is called *amplitude spectrum* of $x[n]$; the phase is called *phase spectrum*.

The DTFT of a sequence $x[n]$

- describes the composition of $x[n]$ in terms of periodic discrete-time complex exponentials with angular frequencies spanning the interval[6] $[-\pi, \pi)$;

---

[6]Recall from Chap. 2 that even if here choose a range of values for ω that spans $[-\pi, +\pi)$, *any interval of length* 2π can be used. For example, choosing $(-\pi, +\pi]$ instead of $[-\pi, +\pi)$ would be correct as well; we could also choose $[0, 2\pi)$, etc. Actually, whenever in our discussion we will consider a DTFT over the positive frequency half-axis only, we will write $\omega \in [0, \pi]$.

- exists if $x[n]$ is absolutely summable[7]: the condition

$$\sum_{n=-\infty}^{+\infty} |x[n]| < \infty$$

is sufficient to ensure uniform convergence of the DTFT definition sum to a continuous function of ω, since

$$|X(e^{j\omega})| = \left| \sum_{n=-\infty}^{+\infty} x[n]e^{-j\omega n} \right| \le \sum_{n=-\infty}^{+\infty} |x[n]|;$$

- is useful to describe the interaction between $x[n]$ and LTI systems, since complex exponentials are eigenfunctions of such systems;
- does not exist in the sense of uniform convergence if the ROC of $X(z)$ does not contain the unit circle.

Thus, signals and systems exist, for which the DTFT does not converge uniformly, but the z-transform exists in some domain not containing the unit circle. An absolutely summable sequence is certainly square-summable, i.e., it also has finite energy,

$$\mathscr{E} = \sum_{n=-\infty}^{+\infty} |x[n]|^2 \le \left\{ \sum_{n=-\infty}^{+\infty} |x[n]| \right\}^2,$$

but a finite-energy sequence is not necessarily absolutely summable. In order to be able to represent by DTFT not only absolutely summable discrete-time signals, but all energy signals, the condition of uniform convergence of the DTFT series is relaxed, and convergence in the *mean-square sense* is accepted.

The concept of mean-square convergence involves considering a truncated series $X_M(e^{j\omega})$, i.e., the sum of a finite number of series terms,

$$X_M(e^{j\omega}) = \sum_{n=-M}^{M} x[n]e^{-j\omega n},$$

and then evaluating this sum for increasing values of M. If the mean-square error between the truncated series and the function $X(e^{j\omega})$ decreases with increasing M, so that

$$\lim_{M \to \infty} \int_{-\pi}^{\pi} \left| X(e^{j\omega}) - X_M(e^{j\omega}) \right| d\omega = 0,$$

---

[7] Sequences that are absolutely summable are classified as ℓ^1-sequences; sequences that are square-summable are classified as ℓ^2-sequences. See the appendix to this chapter for these topics.

then convergence in the mean-square sense exists. The function $X(e^{j\omega})$ can be discontinuous in this case: a typical example is the frequency response of ideal frequency-selective filters, like the lowpass filter shown in Fig. 6.1a.

Note that for finite-energy signals that are not absolutely summable, the DTFT exists, but the z-transform does not exist: indeed, since $X(e^{j\omega})$ is discontinuous, it cannot be an analytic function. In this case, evidently we cannot think of the DTFT as the set of values assumed by the z-transform on the unit circle.

3.3.1 An Example of DTFT Converging in the Mean-Square Sense

Let us examine the frequency response of the ideal lowpass filter with cutoff frequency ω_c. Ideal frequency-selective filters will be discussed in Chap. 6; for the moment, see the shape of the ideal lowpass filter shown in Fig. 6.1a. The ideal lowpass filter is defined by

$$H(e^{j\omega}) = \begin{cases} 1 & \text{for } 0 \leq |\omega| \leq \omega_c, \\ 0 & \text{for } \omega_c < |\omega| \leq \pi. \end{cases}$$

This is a real function, that is, its phase is zero. So, if we hypothetically apply this filter to a sequence, we get no phase difference between the input and the output. The corresponding impulse response cab be computed as (see Sect. 3.3.3)

$$h[n] = \frac{1}{2\pi} \int_{+\omega_c}^{-\omega_c} e^{j\omega n} d\omega = \frac{\sin(\omega_c n)}{\pi n} = \frac{\omega_c}{\pi} \frac{\sin(\omega_c n)}{\omega_c n} = \frac{\omega_c}{\pi} \text{Sinc}(\omega_c n/\pi),$$

with $-\infty < n < +\infty$. Here Sinc$(\cdot)$ indicates the *normalized cardinal sine function* that in digital signal processing and information theory is commonly defined as[8]

$$\text{Sinc}(x) = \begin{cases} \frac{\sin(\pi x)}{\pi x} & \text{for } x \neq 0, \\ 1 & \text{for } x = 0. \end{cases}$$

The sequence $h[n]$ has infinite length and is non-causal; therefore system's the output cannot be calculated, either recursively or non-recursively. The ideal lowpass filter is not *computationally realizable*. This filter is useful in theory but is not applicable in

---

[8]In mathematics, the historical (unnormalized) Sinc function is defined as

$$\text{Sinc}(x) = \begin{cases} \frac{\sin(x)}{x} & \text{for } x \neq 0, \\ 1 & \text{for } x = 0. \end{cases}$$

In either case (unnormalized or normalized), the value at $x = 0$ is defined to be the limiting value Sinc$(0) = 1$.

practice. Later we will see that those causal approximations to frequency-selective filters that can be used in practical applications cannot have zero phase and cannot have discontinuities.

Moreover, the sequence $h[n]$ is not absolutely summable:

$$\sum_{n=-\infty}^{+\infty} \left| \frac{\sin(\omega_c n)}{\pi n} \right| = \infty,$$

and thus the DTFT of $h[n]$,

$$H(e^{j\omega}) = \sum_{n=-\infty}^{+\infty} \frac{\sin(\omega_c n)}{\pi n} e^{-j\omega n},$$

does not converge uniformly for all ω values, in connection to the fact that $H(e^{j\omega})$ is discontinuous at $\omega = \pm \omega_c$. However, convergence in the mean-square sense exists, because $h[n]$ is a square-summable sequence, i.e., has finite energy. Thus if we consider a finite number M of series terms,

$$H_M(e^{j\omega}) \equiv \sum_{n=-M}^{+M} \frac{\sin(\omega_c n)}{\pi n} e^{-j\omega n},$$

we have

$$\lim_{M \to \infty} \int_{-\pi}^{+\pi} |H(e^{j\omega}) - H_M(e^{j\omega})|^2 d\omega = 0.$$

This non-uniform convergence is characterized by the fact that in the vicinity of the discontinuity, the truncated series exhibits oscillatory behavior (*Gibbs phenomenon*; Fig. 3.6).

Thus, the ideal lowpass filter has a frequency response but does not have a transfer function. Also the other ideal frequency-selective filters, namely the ideal highpass filter, as well as the ideal bandpass and the ideal bandstop filters (see Fig. 6.1b, c and d, respectively), behave in a similar way. All these ideal filters allow for a frequency response description but do not have a transfer function.

3.3.2 Line Spectra

It is useful to have a DTFT representation for certain sequences that are neither absolutely summable nor square-summable. This is a further extension of the DTFT, based on the theory of generalized functions, also known as distributions (Lighthill 1958). A typical example of generalized function is the Dirac δ, an analog impulse of infinite height, zero width and unit area.

Fig. 3.6 The Gibbs phenomenon for a truncation of the frequency-response series at a finite number M of terms. **a–d**: $M = 1, 3, 7$ and 19, respectively

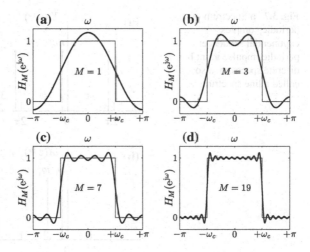

Let us consider the periodic complex exponential sequence $x_1[n] = e^{j\omega_0 n}$. This sequence is neither absolutely summable nor square-summable, and its z-transform does not exist. Nonetheless, it is possible and useful to define the DTFT of this sequence as the *periodic impulse train* shown in Fig. 3.7a,

$$X_1\left(e^{j\omega}\right) = \sum_{r=-\infty}^{\infty} 2\pi\delta(\omega - \omega_0 + 2\pi r),$$

where r is an integer number in $(-\infty, +\infty)$. The use of the periodic impulse train as a DTFT representation of the sequence $x_1[n]$ is justified primarily because formal substitution of the periodic impulse train into the inverse DTFT formula, given in the next subsection, leads to the correct result. Clearly, in this case the DTFT goes to infinity at $\omega = \omega_0 + 2\pi r$. Note that if $\omega_0 = 0$, $x_1[n] = 1$ for all n, and the DTFT reduces to $X_1\left(e^{j\omega}\right) = \sum_{r=-\infty}^{\infty} 2\pi\delta(\omega + 2\pi r)$.

Using the Dirac δ, the concept of DTFT can be extended rigorously to the class of sequences that can be expressed as a sum of discrete frequency components, such as

$$x[n] = \sum_k a_k e^{j\omega_k n}, \qquad -\infty < n < +\infty, \qquad 0 \le \omega_k \le \pi,$$

with the $\{a_k\}$ being some real constant coefficients. The expression

$$X(e^{j\omega}) = \sum_{r=-\infty}^{+\infty} \sum_k 2\pi a_k \delta(\omega - \omega_k + 2\pi r)$$

Fig. 3.7 a Spectrum of a
periodic complex
exponential sequence:
periodic impulse train; **b** a
discrete spectrum, also
called a line spectrum

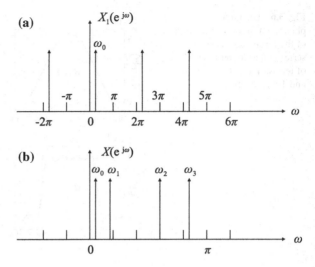

is a consistent DTFT representation of $x[n]$. Every Dirac δ in this expression represents a single spectral line at $\omega = \omega_k$. We can thus state that a linear combination of discrete-time complex exponential signals, which is called a *harmonic signal*, has a *discrete spectrum*, a.k.a. a *line spectrum* (Fig. 3.7b), as opposed to the *continuous spectrum* exhibited by signals that, being square-summable, allow for a conventional DTFT representation, at least in the mean square sense.

Another signal that is neither absolutely nor square-summable is the unit step sequence $u[n]$. It can be shown that this sequence can be represented by the following DTFT:

$$U\left(e^{j\omega}\right) = \frac{1}{1 - e^{-j\omega}} + \sum_{r=-\infty}^{\infty} \pi\delta(\omega + 2\pi r).$$

3.3.3 Inverse DTFT

Since $X(e^{j\omega})$ is a periodic function, it can be expanded in the form of a Fourier series:

$$X(e^{j\omega}) = \sum_{n=-\infty}^{+\infty} c_n e^{j\omega n},$$

where

$$c_n = \frac{1}{2\pi} \int_{-\pi}^{+\pi} X(e^{j\omega}) e^{-j\omega n} d\omega.$$

By direct comparison with the definition of DTFT, i.e., $X(e^{j\omega}) = \sum_{n=-\infty}^{+\infty} x[n]e^{-j\omega n}$, we deduce that $x[n] = c_{-n}$, hence

$$x[n] = \frac{1}{2\pi} \int_{-\pi}^{+\pi} X(e^{j\omega})e^{j\omega n} d\omega.$$

This is the definition of the inverse DTFT (IDFT), in which $x[n]$ appears as the superposition of an infinite number of elementary sinusoidal contributions with frequencies contained in $[-\pi, +\pi)$. Of course, the values of $X(e^{j\omega})$ at $-\pi$ and $+\pi$ coincide, due to the periodicity of $X(e^{j\omega})$. This formula is also called the *synthesis relation* of the DTFT, because it allows for synthesizing $x[n]$ from its $X(e^{j\omega})$ values in the frequency interval $[-\pi, +\pi)$.

The inverse DTFT represents an *expansion of the signal* in a basis, which in the present case is an orthogonal basis formed by sines and cosines, or complex exponentials. The reader interested in an overview of the mathematical foundations of analog and discrete-time signal expansions can refer to the appendix to this chapter.

About the DTFT, it may be noted that:

- the signal x[n] is a sequence, and correspondingly, the DTFT definition is a summation over discrete time n;
- the DTFT is a continuous function, and correspondingly, the IDTFT contains an integral in ω.

Moreover we can observe that

- in mathematical texts, the direct transform is most often written with $e^{+j\omega n}$, while
- in engineering texts, the direct transform is most often written with $e^{-j\omega n}$, and this is the convention normally adopted in digital signal processing—the one to which we adhere.

3.3.4 Selected DTFT Properties

The properties of the DTFT can be derived directly from those of the z-transform, provided that the z-transform exists and converges on the unit circle. However, the extensions of the DTFT concept examined above also share the same properties.

1. *Linearity*:
$$ax_1[n] + bx_2[n] \Longleftrightarrow aX_1(e^{j\omega}) + bX_2(e^{j\omega}),$$

   where a and b are arbitrary constants.
2. *Symmetry*:
   the symmetry properties of the DTFT are important because they simplify calculations. For briefness, here we neglect the general case of a complex sequence and focus on the symmetry properties of the DTFT of real sequences:

- $X(e^{j\omega}) = X^*(e^{-j\omega})$, i.e., the DTFT is a *conjugate-symmetric* function; we can also call it an *even function*;
- $\mathrm{Re}[X(e^{j\omega})]$ and $|X(e^{j\omega})|$ are even functions, while $\mathrm{Im}[X(e^{j\omega})]$ and $\arg[X(e^{j\omega})]$ are odd, or *conjugate-antisymmetric* functions; so $\mathrm{Re}[X(e^{j\omega})] = \mathrm{Re}[X(e^{-j\omega})]$, $\mathrm{Im}[X(e^{j\omega})] = -\mathrm{Im}[X(e^{-j\omega})]$, and so on.

  Then $\mathrm{Re}[X(e^{j\omega})]$ can be expanded as a series of cosines and $\mathrm{Im}[X(e^{j\omega})]$ can be expanded as a series of sines:

$$\mathrm{Re}[X(e^{j\omega})] = \sum_{n=-\infty}^{+\infty} x[n]\cos(\omega n), \qquad \mathrm{Im}[X(e^{j\omega})] = \sum_{n=-\infty}^{+\infty} x[n]\sin(\omega n);$$

- any real sequence can always be expressed as the sum of an even and an odd sequence: $x[n] = x_e[n] + x_o[n]$. The even part $x_e[n]$ transforms into the real part of $X(e^{j\omega})$ (i.e., into a purely real function), while the odd part $x_o[n]$ transforms into the imaginary part of $X(e^{j\omega})$ multiplied by j (i.e., into a purely imaginary function):

$$x_e[n] \Longleftrightarrow \mathrm{Re}[X(e^{j\omega})], \qquad x_o[n] \Longleftrightarrow j\,\mathrm{Im}[X(e^{j\omega})].$$

3. *Folding* or *time reversal*:

$$x[-n] \Longleftrightarrow X(e^{-j\omega}),$$

that for a real $x[n]$ coincides with $X^*(e^{j\omega})$.

4. *Delay*:

$$y[n] = x[n-m] \Longleftrightarrow Y(e^{j\omega}) = e^{-j\omega m} X(e^{j\omega}).$$

This property expresses the fact that a time shift does not alter the frequency composition of a signal: the transform of the delayed signal differs from the original one by just a phase factor.

5. *Conjugation*:

$$x^*[n] \Longleftrightarrow X^*(e^{-j\omega}).$$

If $x[n]$ is real, $x[n] = x^*[n] \Longleftrightarrow X^*(e^{-j\omega}) = X(e^{j\omega})$.
Moreover,

$$x^*[-n] \Longleftrightarrow X^*(e^{j\omega}).$$

6. *Differentiation*:

$$nx[n] \Longleftrightarrow j\frac{dX(e^{j\omega})}{dt}.$$

7. *Continuous periodic convolution*:

$$w[n] = x[n]y[n] \Longleftrightarrow W(e^{j\omega}) = \frac{1}{2\pi} \int_{-\pi}^{+\pi} X(e^{j\omega'})Y[e^{j(\omega-\omega')}]d\omega'.$$

8. *Convolution theorem*:

$$y[n] = x[n] * h[n] \Longleftrightarrow Y(e^{j\omega}) = X(e^{j\omega})H(e^{j\omega}).$$

9. *Modulation* or *frequency shift*:

$$e^{j\alpha n}x[n] \Longleftrightarrow X[e^{j(\omega-\alpha)}],$$

with α being some real constant. Note that if $\alpha = -\pi$, then $e^{-j\pi} = -1$ and $e^{j\alpha n} = (-1)^n$; therefore a particular case of modulation is

$$(-1)^n x[n] \Longleftrightarrow X[e^{j(\omega+\pi)}].$$

10. *Parseval's theorem* for the DTFT:

$$\sum_{n=-\infty}^{+\infty} x[n]y^*[n] = \frac{1}{2\pi} \int_{-\pi}^{+\pi} X(e^{j\omega})Y^*(e^{j\omega})d\omega,$$

and if $x[n] = y[n]$, then

$$\sum_{n=-\infty}^{+\infty} |x[n]|^2 = \frac{1}{2\pi} \int_{-\pi}^{+\pi} |X(e^{j\omega})|^2 d\omega.$$

This theorem expresses *energy conservation*: the energy of the signal in the time domain is equal to the energy of the signal in the frequency domain.[9]

11. *Transform of a real even sequence*:

$$x_e[n] = x_e[-n] \Longleftrightarrow X_e(e^{j\omega}) = x_e[0] + 2\sum_{n=1}^{+\infty} x_e[n]\cos(\omega n),$$

---

[9]If energy is finite, that is, if the signal is in some way transient, then this relation creates no infinity problems, and the quantity $|X(e^{j\omega})|^2$ appears as a frequency distribution of the energy of the signal: it suggests the notion of *energy spectrum* that will be discussed later. If energy is infinite, the notion of energy spectrum makes no sense. If the signal is a power signal, i.e., its energy is infinite *but* its power is finite (this could be the case of a deterministic periodic signal; see Chap. 2), another quantity may be defined in place of the energy spectrum. This quantity is the *power spectrum*, that we will actually study in detail in the case of random signals, rather than in the case of deterministic signals. In fact, random signals typically are power signals. For a more extended discussion of these topics in relation to deterministic signals see Chap. 5.

with

$$\arg[X_e(e^{j\omega})] = \begin{cases} 0 & \text{for } X_e(e^{j\omega}) > 0, \\ \pm\pi & \text{for } X_e(e^{j\omega}) < 0, \\ \text{undefined} & \text{for } X_e(e^{j\omega}) = 0. \end{cases}$$

Thus the DTFT of a real even sequence is a real even function of frequency.

12. *Transform of a real sequence with even symmetry around an integer $N_s \neq 0$* (Fig. 3.8a):

$$x[n] = x_e[n - N_s] \Longleftrightarrow X(e^{j\omega}) = e^{-j\omega N_s} X_e(e^{j\omega}).$$

Recalling that $X_e(e^{j\omega})$ is a real even function of ω, and using the delay property, we see that

- $X(e^{j\omega})$ and $X_e(e^{j\omega})$ have the same amplitude, and
- $\arg[X(e^{j\omega})] = \arg[X_e(e^{j\omega})] - N_s\omega$.

Thus, apart from $\pm\pi$ phase jumps caused by sign changes in $X_e(e^{j\omega})$, the phase of $X(e^{j\omega})$ is linear in ω. This property is important in digital filter design theory. Similar properties hold in the case of a half-integer N_s (Fig. 3.8b).

13. *Transform of a real odd sequence*:

$$x_o[n] = -x_o[-n] \Longleftrightarrow X_o(e^{j\omega}) = 2j \sum_{n=1}^{+\infty} x_o[n] \sin(\omega n),$$

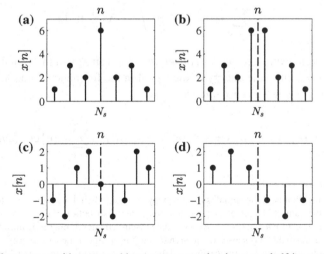

Fig. 3.8 Real sequences with even or odd symmetry around an integer or half-integer discrete-time value $N_s \neq 0$. **a** Even symmetry, integer N_s; **b** even symmetry, half-integer N_s; **c** odd symmetry, integer N_s; **d** odd symmetry, half-integer N_s

with

$$|X_o(e^{j\omega})| = 2\left|\sum_{n=1}^{+\infty} x_o[n]\sin(\omega n)\right|,$$

$$\arg[X_o(e^{j\omega})] = \begin{cases} \pi/2 & \text{for } X_o(e^{j\omega}) > 0, \\ \pi/2 \pm \pi & \text{for } X_o(e^{j\omega}) < 0, \\ \text{undefined} & \text{for } X_o(e^{j\omega}) = 0. \end{cases}$$

Thus the DTFT of a real odd sequence is an imaginary and odd function of frequency.

14. *Transform of a real sequence with odd symmetry around an integer $N_s \neq 0$* (Fig. 3.8c):

$$x[n] = x_o[n - N_s] \Longleftrightarrow X(e^{j\omega}) = e^{-j\omega N_s} X_o(e^{j\omega}).$$

Recalling that $X_o(e^{j\omega})$ is an imaginary and odd function of ω and using the delay property we see that

- $X(e^{j\omega})$ and $X_o(e^{j\omega})$ have the same amplitude, and
- $\arg[X(e^{j\omega})] = \arg[X_o(e^{j\omega})] - N_s\omega$.

Therefore, apart from $\pm\pi$ phase jumps that occur any time that $X_o(e^{j\omega})$ changes sign, the phase of $X(e^{j\omega})$ is again linear in ω. Similar properties hold in the case of a half-integer N_s (Fig. 3.8d).

3.3.5 The DTFT of a Finite-Length Causal Sequence

We will now introduce the *discrete Fourier transform* (DFT) in a qualitative way. Later we will follow a more rigorous approach to derive it properly. The DFT is the main tool for the analysis of discrete-time signals in the frequency domain.

Any infinite-length discrete-time signal can be reconstructed from the values of its $X(e^{j\omega})$ at all frequencies ω in the interval $[-\pi, +\pi)$, via inverse DTFT. However, if the sequence has finite length, i.e., if $x[n]$ is not identically zero only for $0 \leq n \leq N - 1$, the knowledge of a finite number of $X(e^{j\omega})$ values can be sufficient, provided that the set of ω values selected for sampling $X(e^{j\omega})$ are properly chosen.

Intuitively we can justify this statement as follows. The DTFT is a linear operation, and therefore N values of $X(e^{j\omega})$ taken at N frequency values ω_k, with $0 \leq k \leq N-1$, provide N linear equations in the N unknown values $x[n]$. The system has only one solution if the coefficient matrix is non-singular, i.e., if the determinant of the coefficient matrix does not vanish. Therefore, if the ω_k are chosen in such a way that this condition is satisfied, then the N samples of the signal $x[n]$ can be calculated univocally. The samples $X(e^{j\omega})|_{\omega=\omega_k} \equiv X[k]$ form the DFT of $x[n]$.

Fig. 3.9 Sampling of the DTFT on the unit circle: an example with $N = 12$

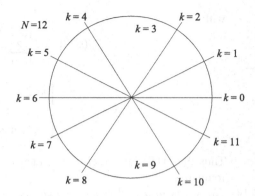

It turns out that the proper choice for the ω_k corresponds to subdividing the interval $[-\pi, +\pi)$ on the unit circle into N equal parts (Fig. 3.9):

$$\omega_k = \frac{2\pi k}{N}, \qquad 0 \le k \le N - 1.$$

The DFT can thus be seen as the result of periodic sampling of the DTFT at frequency intervals $\Delta\omega = 2\pi/N$ in $[-\pi, +\pi)$ or, equivalently, in $[0, 2\pi)$:

$$X[k] = X(e^{j\frac{2\pi}{N}k}) = \sum_{n=0}^{N-1} x[n]e^{-j\frac{2\pi}{N}kn}.$$

Note that $X[k]$ results from sampling a continuous function of ω, hence it is a sequence, as the signal $x[n]$.

More rigorously, deriving the DFT involves the following steps:

- taking a finite-length sequence;
- extending it periodically over the whole discrete-time axis;
- defining the *discrete Fourier series* (DFS) of the periodic sequence thus obtained. The DFS is a periodic sequence;
- defining the DFT as a single period extracted from the DFS.

The advantages of this procedure lie in the fact that

- it highlights the fundamental characteristic of periodicity inherent in the DFT and the related implications in practical applications;
- thanks to the introduction of the DFS, it allows to establish the formalism that is necessary to express the properties of the DFT;
- it provides the rationale for the choice of equally spaced frequencies for DTFT sampling.

3.4 Discrete Fourier Series (DFS)

We now undertake the path outlined above to derive the DFT in a more rigorous way. We start introducing the *periodic extension* of a causal finite-length sequence and its representation in the frequency domain via DFS.

Given a signal $x[n]$ with $0 \leq n \leq N-1$ (Fig. 3.10a), consider the sequence $x_p[n]$ obtained repeating $x[n]$ an infinite number of times along the n-axis, without any overlapping of adjacent $x[n]$ images. Obviously $x_p[n]$ is a periodic sequence with period equal to N, as illustrated in Fig. 3.10b.

The z-transform and the DTFT of $x_p[n]$ do not exist, but it is possible to represent $x_p[n]$ as a *discrete Fourier series* (DFS), i.e., as a superposition of N complex exponentials whose frequencies are harmonically related. Keeping in mind that in the discrete-time domain only N distinct angular frequencies exist in association with a period of N time steps, and that they are integer multiples of the fundamental frequency $2\pi/N$, the synthesis relation of the DFS, i.e., the inverse DFS (IDFS) can be written as

$$x_p[n] = \frac{1}{N} \sum_{k=0}^{N-1} X_p[k] e^{j\frac{2\pi}{N}nk}.$$

The coefficients $X_p[k]$ of this expansion, which constitute the DFS of $x_p[n]$, can be obtained exploiting the orthogonality of the complex exponential set: for any integer m,

$$\frac{1}{N} \sum_{n=0}^{N-1} e^{j\frac{2\pi}{N}(k-r)n} = \begin{cases} 1 & \text{for } k-r = mN, \\ 0 & \text{otherwise.} \end{cases}$$

Fig. 3.10 a A finite-length sequence $x[n]$ and **b** its periodic extension $x_p[n]$

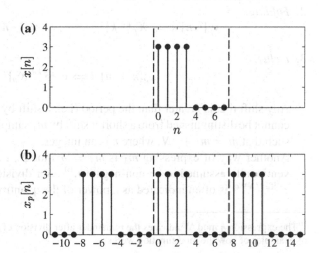

The analysis relation of the DFS can thus be written:

$$X_p[k] = \sum_{n=0}^{N-1} x_p[n] e^{-j\frac{2\pi}{N}kn}.$$

The DFS can be interpreted in two ways:

- as a finite-length complex sequence for $0 \leq k \leq N - 1$, or
- as a periodic complex sequence with period equal to N, for any integer k.

The second interpretation allows an exact duality between the time and frequency domains, in the sense that both the sequence $x_p[n]$ and its frequency representation are periodic. Therefore we will adopt this second point of view.

The sequence $X_p[k]$ can now be viewed as an infinite sequence of samples of the z-transform of $x[n]$, which certainly exists at least in $0 < |z| < \infty$, since $x[n]$ has finite length. The samples of $X(z)$ that form the DFS are taken treading the unit circle an infinite number of times and are equally spaced in angle around the origin of the z-plane.

3.4.1   Selected DFS Properties

We will now list the most important properties of the DFS.

1. *Linearity*:
$$ax_{p1}[n] + bx_{p2}[n] \Longleftrightarrow aX_{p1}[k] + bX_{p2}[k],$$

   where a and b are arbitrary constants.
2. *Folding*:
$$x_p[-n] \Longleftrightarrow X_p[-k], \qquad\qquad x_p^*[-n] \Longleftrightarrow X_p^*[k].$$

3. *Delay*:
$$x_p[n - m] \Longleftrightarrow e^{-j\frac{2\pi}{N}km} X_p[k].$$

Any shift that is greater than the period (i.e., a shift by m samples, with $m > N$) cannot be distinguished from a shorter shift by m_1 samples, with $0 \leq m_1 \leq N-1$, such that $m = m_1 + rN$, where r is an integer.

Another way of expressing m_1 is $m_1 = m \bmod N$, i.e., as the remainder, conventionally assumed to be non-negative,[10] after division of m by N. The factor $e^{-j(2\pi/N)km}$ is often indicated as a power of the quantity

---

[10]The expression $n \bmod N$ indicates the remainder after division of n by N. Let us make a couple of examples of how the mod function works:

- we want to compute $340 \bmod 60$. Now, 340 lies between 300 and 360, so $300 = 60 \times 5$ is the greatest multiple of 60 which is less than or equal to 340; we subtract 300 from 340 and get 40;

$$W_N = e^{j\frac{2\pi}{N}},$$

that is, $e^{-j(2\pi/N)km} = W_N^{km}$: so we can also write

$$x_p[n - m] \iff W_N^{km} X_p[k].$$

Note that for $m_1 = m \bmod N$ we have $W_N^{km} = W_N^{km_1}$.

4. *Modulation*:

$$W_N^{-ln} x_p[n] = e^{+j\frac{2\pi}{N}ln} x_p[n] \iff X_p[k - l].$$

The ambiguity of the shift in the time domain is thus also manifest in the frequency-domain representation.

5. *Conjugation*:

$$x_p^*[n] \iff X_p^*[-k].$$

6. *Duality*:

the DFS analysis and synthesis relations are similar. The DFS samples are simply given by the original periodic sequence order-reversed and multiplied by N:

$$x_p[n] \iff X_p[k] \qquad \rightsquigarrow \qquad X_p[n] \iff N x_p[-k].$$

This symmetry between the time and frequency domains is referred to as the duality of the DFS.

7. *Symmetry*:

for real signals, the real part and the amplitude of $X_p[k]$ are even sequences, while the imaginary part and the phase are odd sequences; the DFS is a *conjugate-symmetric* complex sequence, that is, $X_p[k] = X_p^*[-k]$.

8. *Periodic convolution*:

let us take two sequences $x_{p1}[n]$ and $x_{p2}[n]$ sharing the same period N. If we compute their DFSs, what sequence corresponds to $X_{p3}[k] = X_{p1}[k]X_{p2}[k]$? In other words, what is the inverse DFS of $X_{p3}[k]$? It can be demonstrated that

$$x_{p3}[n] = \sum_{m=0}^{N-1} x_{p1}[m]x_{p2}[n - m] \equiv x_{p1}[n] \otimes x_{p2}[n] = x_{p2}[n] \otimes x_{p1}[n].$$

The sequence $x_{p3}[n]$ is periodic with period N. The summation is extended to a single period of the two sequences $x_{p1}[n]$ and $x_{p2}[n]$, and the symbol $\otimes$ indicates the so-called *periodic convolution* of the two sequences.

In Fig. 3.11, an example of a periodic convolution is shown, for two sequences $x_{p1}[n]$ and $x_{p2}[n]$ having a period of $N = 8$ (Fig. 3.11a and c, respectively).

- we want to compute $-340 \bmod 60$. Now, -340 lies between -360 and -300, so $-360 = 60 \times (-6)$ is the greatest multiple of 60 less than or equal to -340; we subtract -360 from -340 and get 20.

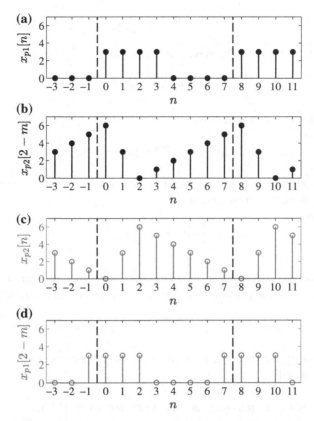

Fig. 3.11 An example of periodic convolution: the procedure for forming the sample at $n = 2$ of the sequence $x_{p3}[n] = x_{p1}[n] \otimes x_{p2}[n]$. The sequences have a period $N = 8$. **a** The original sequence $x_{p1}[n]$; **b** the sequence $x_{p2}[n]$, folded and shifted to the right by two samples; **c** the original sequence $x_{p2}[n]$; **d** the sequence $x_{p1}[n]$, folded and shifted to the right by two samples. The sample $x_{p3}[2]$ can be obtained multiplying sample by sample the two sequences drawn as *black lines* and *dots*, or the two sequences drawn as *gray lines* and *empty circles*; only the samples included between the *dashed lines* must be considered (a single period)

This figure illustrates the calculation of the sample $x_{p3}[2]$. The calculation can be done in two ways: if we combine the two sequences of Fig. 3.11a, b (*black lines* and *dots*) we have (see the samples included between the *dashed lines*) $x_{p3}[2] = 3 \times 6 + 3 \times 3 + 3 \times 1 = 30$; if we combine the two sequences of Fig. 3.11c, d (*gray lines* and *empty circles*) we get the same result, $x_{p3}[2] = 3 \times 3 + 6 \times 3 + 1 \times 3 = 30$. If, instead, we multiply the two sequences $x_{p1}[n]$ and $x_{p2}[n]$ sample by sample, what is the DFS of the product sequence? The answer is, if

$$x'_{p3}[n] = x_{p1}[n]x_{p2}[n],$$

then

$$X'_{p3}[k] = \frac{1}{N} \sum_{l=0}^{N-1} X_{p1}[l] X_{p2}[k-l] \equiv \frac{1}{N} X_{p1}[k] \otimes X_{p2}[k] = \frac{1}{N} X_{p2}[k] \otimes X_{p1}[k].$$

3.4.2 Sampling in the Frequency Domain and Aliasing in the Time Domain

We have seen that if $x_p[n]$ consists of non-overlapping, delayed repetitions of the finite-length sequence $x[n]$, with $n = [0, N-1]$, then $X_p[k]$ in the interval $0 \le k \le N-1$ consists of N samples of $X(z)$, taken along the unit circle, during a single revolution around the origin of the z-plane. Of course, these samples are also DTFT samples.

Now, suppose that we have an aperiodic sequence $x[n]$ with unspecified length, and we take an arbitrary number N of $X(e^{j\omega})$ values at frequencies that are integer multiples of $2\pi/N$. We then assume that these N samples form the DFS of a periodic sequence $x_p[n]$ and compute $x_p[n]$ through the DFS synthesis formula. What is the relation between $x_p[n]$ and $x[n]$? What happens if the actual length N' of $x[n]$ is smaller than N? What happens if it is larger? It can be demonstrated that

$$x_p[n] = \sum_{l=-\infty}^{+\infty} x[n + lN],$$

where l is an integer and where it is understood that $x[n] = 0$ for all values of n external to the interval $[0, N'-1]$. We thus see that $x_p[n]$ is formed joining an infinite number of repetitions of $x[n]$, with each repetition delayed by N steps in respect to the previous one, so that

- if $N' < N$, the repetitions are separated by zeros and remain distinct;
- if $N' = N$, the repetitions are juxtaposed; there are neither interposed zeros, nor overlapping tails;
- if $N' > N$, overlapping of the tails of contiguous repetitions occurs, and the shape of the original signal $x[n]$ becomes corrupted in the periodic extension $x_p[n]$: we would no longer be able to recover $x[n]$ from $x_p[n]$ by extracting a period. This phenomenon is referred to as *aliasing in the time domain*, or simply *time aliasing*.

Note that time aliasing can be avoided, choosing $N \ge N'$—i.e., sampling the DTFT of $x[n]$ with proper density—if, and only if, $x[n]$ has finite length.

Figure 3.12 shows, for a sequence $x[n]$ with length $N' = 9$ (Fig. 3.12a), two cases of $x_p[n]$, the first one free from time aliasing (Fig. 3.12b, where $N = 12$), and the second one contaminated by time aliasing (Fig. 3.12c, where $N = 7$).

Fig. 3.12 **a** A sequence $x[n]$ with length $N' = 9$ and two cases of periodic extension, $x_p[n]$: **b** a case which is free from time aliasing ($N = 12$) and **c** a case affected by time aliasing ($N = 7$). See text for details

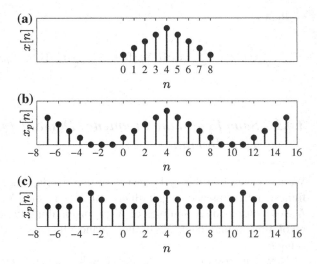

If $N' > N$, the information contained in the N samples of $X(\mathrm{e}^{\mathrm{j}\omega})$ turns out to be insufficient to reconstruct $x[n]$: the DTFT has been undersampled, and as a consequence, the possibility of reconstructing the original signal has been lost. If, and only if, $N' \le N$, the signal $x[n]$ can be recovered from $x_p[n]$ by extracting one period, which comprises N samples. In other words, $x[n]$ can be represented exactly by N transform samples taken along the unit circle with regular frequency spacing $\Delta\omega = 2\pi/N$, if, and only if, $N' \le N$.

When $N' \le N$ as required, the whole $X(z)$, and therefore the whole $X(\mathrm{e}^{\mathrm{j}\omega})$, can be reconstructed on the basis of those N samples: knowledge of $X(\mathrm{e}^{\mathrm{j}\omega})$ at all frequencies is not required if $x[n]$ has finite length. For the reconstruction of $X(z)$ from its samples, the following interpolation formula can be used:

$$X(z) = \frac{1 - z^{-N}}{N} \sum_{k=0}^{N-1} \frac{X_p[k]}{1 - \mathrm{e}^{\mathrm{j}\frac{2\pi}{N}} z^{-1}},$$

that for $z = \mathrm{e}^{\mathrm{j}\omega}$ provides the interpolation formula for $X(\mathrm{e}^{\mathrm{j}\omega})$:

$$X(\mathrm{e}^{\mathrm{j}\omega}) = \sum_{n=0}^{N-1} X_p[k]\varXi\left(\omega - \frac{2\pi}{N}k\right),$$

where

$$\varXi(\omega) = \mathrm{e}^{-j\omega\frac{N-1}{2}} \frac{\sin(N\frac{\omega}{2})}{N\sin(\frac{\omega}{2})} \equiv \mathrm{e}^{-j\omega\frac{N-1}{2}} D_N(\omega), \qquad \omega \in [-\pi, \pi).$$

Fig. 3.13 The Dirichlet
function for **a** $N = 5$ and
b $N = 6$

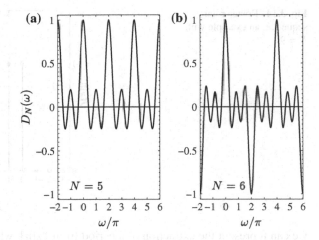

Here $D_N(\omega)$ indicates the *Dirichlet function*, or *periodic Sinc function*, shown in
Fig. 3.13a, b for an even and an odd value of N, respectively.[11] When N is odd, the
function is periodic with period 2π; when N is even, the function is periodic with
period 4π. For our purposes, the importance of these interpolation formulas is more
theoretical than practical.

3.5 Discrete Fourier Transform (DFT)

Consider a sequence $x[n]$ with length N—or even a shorter sequence augmented to
length N by adding zeros after the last sample. If the corresponding DTFT is sampled
at N equally spaced frequencies, then $x_p[n]$ is aliasing-free and we can write

$$x_p[n] = x[n \bmod N], \qquad n = (-\infty, +\infty).$$

Note that $n \bmod N$ varies in the interval $[0, N-1]$: $x[n \bmod N]$ comprises all the
existing samples of $x[n]$. If we introduce the *rectangular sequence* (Fig. 3.14),

$$R_N[n] = \begin{cases} 1 & \text{for } 0 \le n \le N-1, \\ 0 & \text{otherwise,} \end{cases}$$

---

[11]The precise definition of $D_N(\omega)$ is

$$D_N(\omega) = \begin{cases} \frac{\sin[N(\omega/2)]}{N\sin(\omega/2)} & \omega \ne 2\pi k, \qquad k = 0, \pm 1, \pm 2, \pm 3 \ldots, \\ (-1)^{k(N-1)} & \omega = 2\pi k, \qquad k = 0, \pm 1, \pm 2, \pm 3 \ldots. \end{cases}$$

This gives 1 at $\omega = 0$.

Fig. 3.14 Rectangular
sequence: an example with
$N = 9$

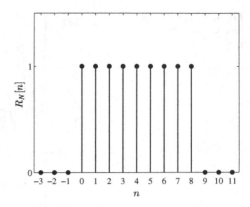

we can represent the extraction of a period from $x_p[n]$, which gives back $x[n]$, as

$$x[n] = x_p[n]R_N[n].$$

We can now define the DFT of $x[n]$ as the sequence $X[k]$ obtained extracting a period from $X_p[k]$:

$$X[k] = X_p[k]R_N[k].$$

In this way we preserve the exact duality between the time and frequency domains. The corresponding inverse relation is

$$X_p[k] = X[k \bmod N].$$

Thus, the definitions of DFT (analysis relation of the DFT) and IDFT (synthesis relation of the DFT) are respectively

$$X[k] = \begin{cases} \sum_{n=0}^{N-1} x[n]e^{-j\frac{2\pi}{N}kn} & \text{for } 0 \le k \le N-1, \\ 0 & \text{otherwise,} \end{cases}$$

and

$$x[n] = \begin{cases} \frac{1}{N}\sum_{k=0}^{N-1} X[k]e^{+j\frac{2\pi}{N}kn} & \text{for } 0 \le n \le N-1, \\ 0 & \text{otherwise.} \end{cases}$$

All summations in this formulas extend over $[0, N-1]$, and it must be understood that in this interval $x[n] \equiv x_p[n]$, $X[k] \equiv X_p[k]$.

It must be noted that:

1. the DFT is made of samples of $X(e^{j\omega})$ taken equally spaced in angle on the unit circle during a single revolution around the origin of the z-plane;

2. if $x[n]$ is computed by inverse DFT at values of n outside the interval $0 \leq n \leq N - 1$, the result is not zero but $x_p[n]$;
3. if $X[k]$ is computed by DFT at values of k outside the interval $0 \leq k \leq N - 1$, the result is not zero but $X_p[k]$. Therefore, the periodicity inherent in this signal representation is always present. In defining the DFT we simply recognize that we are only interested in $x[n]$ values for $0 \leq n \leq N - 1$, because actually $x[n]$ is identically zero outside that interval, and that we are only interested in $X[k]$ values for $0 \leq k \leq N - 1$, because these are the only values required to apply the inverse DFT and recover $x[n]$;
4. when $x[n]$ derives from periodic sampling of a continuous-time signal $x(t)$, i.e., when $x[n] = x(nT_s)$, the *analog spectrum* of $x(t)$, i.e., the *continuous-time Fourier transform* (CTFT) of $x(t)$, $f_k = k/(NT_s)$ (see, e.g., Bochner and Chandrasekharan 1949; Bracewell 2000),

$$X(f) = \int_{-\infty}^{+\infty} x(t)e^{-j2\pi ft}dt,$$

and the spectrum of the sequence $x[n]$, i.e., $X[k]$, express the same frequency content. We can see this qualitatively by writing, with $t = nT_s$, $f_k = k/(NT_s)$,

$$X(f_k) = \int_{-\infty}^{+\infty} x(t)e^{-j2\pi f_k t}dt \approx T_s \sum_{n=0}^{N-1} x[n]e^{-j2\pi f_k n} =$$

$$= T_s \sum_{0}^{N-1} x[n]e^{-j\frac{2\pi}{NT_s}knT_s} = T_s X[k].$$

Therefore

- up to a multiplicative factor T_s, the two spectra assume the same value at each harmonic number k, that is, at each member of the discrete frequency set $\{\omega_k = (2\pi/N)k\}$, $k = [0, N - 1]$, spanned by the DFT representation;
- the DFT samples do not depend on T_s, exactly as the sequence $x[n]$ does not contain any information about the sampling interval T_s adopted for sampling $x(t)$;
- T_s is the only dimensional parameter that relates the frequency-domain representation of a continuous-time signal to the frequency-domain representation of the corresponding discrete-time signal;

5. the rationale behind the subdivision of the interval $[-\pi, +\pi)$ into N equal parts is found in the DFS representation of a periodic sequence, which is based on the set of harmonically-related complex exponential sequences.

3.5.1 The Inverse DFT in Terms of the Direct DFT

DFT and IDFT are similar, so that the IDFT can be easily expressed in terms of the DFT. This *duality of the DFT* implies that a single software algorithm can implement both.

There are several commonly used methods to relate IDFT to DFT. A possible way is using conjugation:

$$
x[n] \equiv \text{IDFT}\,\{X[k]\} = \frac{1}{N}\sum_{k=0}^{N-1} X[k]\mathrm{e}^{+j\frac{2\pi}{N}kn} = \left(\sum_{k=0}^{N-1}\frac{X^*[k]}{N}\mathrm{e}^{-j\frac{2\pi}{N}kn}\right)^{\!*} =
$$

$$
= \left(\text{DFT}\left\{\frac{X^*[k]}{N}\right\}\right)^{\!*}.
$$

Thus if a software algorithm, given $x[n]$, provides $X[k]$, then we can use the same algorithm with $X^*[k]/N$ as the input to get $x^*[n]$, which coincides with $x[n]$ if $x[n]$ is real.

Another possible way consists in using sample-order reversal. We start from

$$
x[n] = \frac{1}{N}\sum_{k=0}^{N-1} X[k]\mathrm{e}^{+j\frac{2\pi}{N}kn} \quad \text{for}\ \ 0 \le n \le N-1,
$$

and substitute $N - n$ in place of n. We get

$$
x[N-n] = \frac{1}{N}\sum_{k=0}^{N-1} X[k]\mathrm{e}^{+j\frac{2\pi}{N}k(N-n)} = \frac{1}{N}\sum_{k=0}^{N-1} X[k]\mathrm{e}^{-j\frac{2\pi}{N}kn},
$$

or

$$
x[N-n] = \frac{1}{N}\text{DFT}\,\{X[k]\}.
$$

Now, as n goes from 0 to $N-1$, $N-n$ goes from N to 1. If we think of the underlying periodic extension of $x[n]$, we understand that in order to get $x[n]$ from $x[N-n]$ we simply need to take the first sample of $x[N-n]$ and bring it to the last position, and then reverse the sequence order. So we can use a DFT algorithm with $X[k]$ as the input to calculate $x[n]$, by re-arranging the output samples and dividing them by N.

3.5.2 Zero Padding

Let us take a real causal sequence $x[n]$ with finite length N. For instance, we can consider a sequence with even symmetry around the sample $n = (N-1)/2$. Note that symmetry would not be necessary here; this is just an example. In Fig. 3.15a, we

Fig. 3.15 **a** A real causal
sequence $x[n]$, with finite
length $N = 7$ and even
symmetry around the sample
$n = N_s = (N - 1)/2 = 3$,
and **b** its aliasing-free
periodic extension

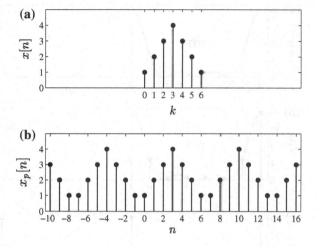

have $N = 7$, and the sequence is symmetric around the point $n = (N - 1)/2 = 3$.
Let us also generate its aliasing-free periodic extension $x_p[n]$ (Fig. 3.15b), in which
the repetitions of $x[n]$ are juxtaposed.

Now, consider the DTFT and DFT of $x[n]$, as well as the DFS of $x_p[n]$. In the
present example we expect $X(e^{j\omega})$ to be a real even function of frequency, up to a
linear phase factor that obviously disappears if we only look at $|X(e^{j\omega})|$. Even if for
brevity we did not mention this fact when listing the main properties of the DFS, the
frequency representation of the corresponding periodic sequence $x_p[n]$ behaves in
a similar way, and, as a consequence, the DFT of $x[n]$ also has this property. This,
however, is not the main point we wish to discuss here.

The main point is the following: if we are able to calculate $X(e^{j\omega})$ in an analytical
way, then we know its values over a continuous range of frequencies. So we can plot
it and observe its shape with any desired graphic resolution: ideally, as many details
as desired can be made visible. In practical applications, however, we often must
base our picture of $X(e^{j\omega})$ on the computation of its samples, $X[k]$. We might want
to use the interpolation formula given in Sect. 3.4.2 to improve the picture, but as we
shall see immediately, there is a better option. The magnitudes of $X(e^{j\omega})$, $X[k]$ and
$X_p[k]$ for the sequence of Fig. 3.15a are shown in Fig. 3.16a, b and c, respectively. We
assume, for the moment, that we are able to calculate $X(e^{j\omega})$ by analytical means. The
DFT magnitude (Fig. 3.16b) does not provide a detailed picture of the shape of the
underlying $|X(e^{j\omega})|$: $N = 7$ samples are enough to satisfy theoretical requirements
but inadequate to describe the true shape of $|X(e^{j\omega})|$. Indeed, the degree of accuracy
with which the samples of $|X[k]|$ describe the shape of $|X(e^{j\omega})|$ is dictated and
limited by the amount N of available $x[n]$ values. In order to improve the visual
quality of spectral graphs, very often the DFT is interpolated, i.e., the density of the
$X(e^{j\omega})$ frequency sampling is artificially increased, by adding a number of zeros at
the end of the sequence $x[n]$ before computing the DFT. This operation is referred
to as *zero padding*. In this way, more details of the underlying $|X(e^{j\omega})|$ become

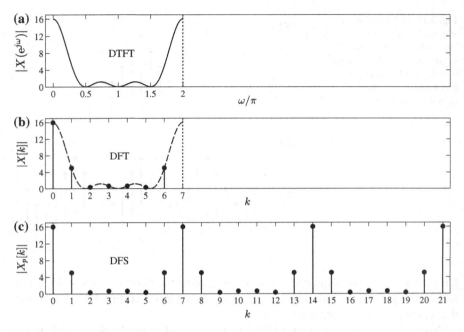

Fig. 3.16 Amplitude of **a** the DTFT and **b** the DFT of the sequence $x[n]$ shown in Fig. 3.15a; **c** amplitude of the DFS of the related periodic extension $x_p[n]$, shown in Fig. 3.15b. In panel b, the DTFT curve of panel a is reported as a *dashed line* to help visualization of the relation between the DTFT and the DFT. These plots extend over $\omega \in [0, 2\pi)$, corresponding to $k = [0, N - 1]$

visible. However, it must be kept in mind that this is only an artifice that does not add anything to the information we actually possess about the signal, because this information content resides in the N available samples of $x[n]$.

For example, given $x[n]$ with $n = [0, N - 1]$, we can form

$$x_z[n] = \begin{cases} x[n] & \text{for } n = [0, N - 1], \\ 0 & \text{for } n = [N, 2N - 1]. \end{cases}$$

The underlying DTFT remains the same, i.e., $X(e^{j\omega}) = X_z(e^{j\omega})$, but $x_{zp}[n] \neq x_p[n]$, hence also $X_{zp}[k] \neq X_p[k]$ and $X_z[k] \neq X[k]$:

$$X_z[k] = X_z(e^{j\omega})|_{\omega = \frac{2\pi}{2N}k} = X(e^{j\omega})|_{\omega = \frac{\pi}{N}k}$$

with $k = [0, 2N - 1]$. The frequency sampling operated on $X(e^{j\omega})$ is two times denser than before.

Zero padding is feasible with any number N_z of zeros: if N is substituted by $N + N_z$, the distance $\Delta\omega$ between any two adjacent transform samples, which originally was $\Delta\omega = 2\pi/N$, becomes $\Delta\omega' = 2\pi/(N + N_z)$. The improvement of the picture

Fig. 3.17 DFT magnitude for the sequence with $N = 7$ shown in Fig. 3.15a, after zero padding up to a length $N + N_z = 16$. The *dashed curve* depicts the DTFT magnitude

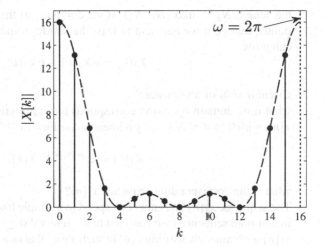

offered by the DFT can be appreciated in Fig. 3.17: in this case the sequence with $N = 7$ shown in Fig. 3.15a was augmented up to a length $N + N_z = 16$ by zero padding before calculating the DFT. The shape of the underlying DTFT is described much better than in Fig. 3.16b. If we arbitrarily increase N_z, thus reducing $\Delta\omega'$, we may be able to draw $|X(e^{j\omega})|$ as it appears in Fig. 3.16a, without any need for knowing its analytical expression. This artifice is perfectly acceptable, as far as we know that $N = 7$ sequence samples are enough to satisfy theoretical requirements and thus avoid aliasing.

Finally, let us mention that Fig. 3.16 also allows us to observe an even symmetry in the three frequency representations. For this example, an even sequence was chosen. We may wonder what the plot of $|X[k]|$ looks like when the sequence $x[n]$ has no symmetry properties. The answer can be found in the next subsection (Fig. 3.20): this symmetry is due to the reality of $x[n]$ and not to its even symmetry. The even symmetry of the sequence only has to do with its transform being real up to a linear phase factor. .

3.5.3 Selected DFT Properties

The properties of the DFT can be demonstrated considering the periodic extensions of the various sequences involved.

1. *Linearity*:
   if $x_1[n]$ has length N_1 and $x_2[n]$ has length N_2 and a and b are arbitrary constants, then

$$x_3[n] = ax_1[n] + bx_2[n]$$

has length $N_3 = \max(N_1, N_2)$. If we compute all three DFTs over at least N_3 points, that is, if we augment at least the shorter sequence by zero padding, we can write

$$X_3[k] = aX_1[k] + bX_2[k].$$

2. *Circular shift* of a sequence:
   what time-domain operation corresponds to multiplying the DFT of a sequence $x[n]$, with $0 \le n \le N - 1$, by a linear phase factor $e^{-j(2\pi/N)km}$? In other words: if

$$X_1[k] = e^{-j(2\pi/N)km} X[k],$$

   what is the corresponding sequence $x_1[n]$?
   The sequence $x_1[n]$ does not correspond to a simple linear time shift of $x[n]$, and in fact both sequences are confined to the interval $0 \le n \le N - 1$. The sequence $x_1[n]$ is obtained by shifting $x[n]$ in such a way that as a sample leaves the interval $[0, N - 1]$ at one end, it enters at the other end. This is equivalent to building the periodic extension of $x[n]$, shifting it linearly, and then extracting a period. This is exemplified in Fig. 3.18: we consider (Fig. 3.18a) a sequence $x[n]$, its periodic extension $x_p[n]$ (Fig. 3.18b), its periodic extension linearly shifted by m samples, with $m = -2$ (Fig. 3.18c), and finally the extraction of a period (Fig. 3.18d). We can observe that the sample that was originally located at $n = 0$ goes to the position $n = N - 2 = 4$.
   Now, let us informally imagine that $x[n]$ is wrapped around a *cylindrical surface*, along a circumference obtained cutting the cylinder by a plane perpendicular to its axis. Let us assume that this circumference is exactly N points long. This geometrical construction is illustrated in Fig. 3.19a, in which the sequence samples are represented by *arrows*. The sequence $x_p[n]$ can be obtained by repeatedly going along the circumference. The sequence $x_1[n]$ can be obtained by rotating the cylinder around its axis by a certain angle, with respect to the original position assumed as a reference.
   In Fig. 3.19a we see a 29-samples-long sequence, wrapped around a *black cylinder*. Note the *gray arrow*, representing the sample at $n = 0$. In Fig. 3.19b, the *black internal cylinder* reproduces the original configuration, which is assumed as a reference, while the *gray external cylinder* is rotated by $m = -2$ (i.e., by an angle $2\pi m/N$ in a clockwise direction) with respect to the reference position. The sequence, integral with the surface of the gray cylinder, appears circularly shifted by $m = -2$: the *gray arrow* that originally was located at $n = 0$ is now located at $n = N - 2 = 27$.
   Formally, the circularly shifted sequence can be expressed as

$$x_1[n] = \begin{cases} x[(n-m) \bmod N] R_N[n] & \text{for } 0 \le n \le N - 1, \\ 0 & \text{otherwise.} \end{cases}$$

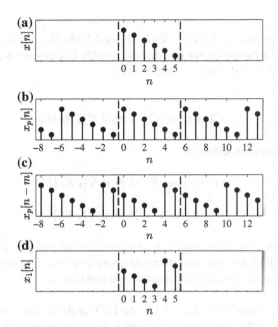

Fig. 3.18 Example of circular shift. **a** A sequence $x[n]$ with length $N = 6$; **b** its periodic extension; **c** the periodic extension, shifted by $m = -2$ samples; **d** a period extracted from the shifted periodic extension

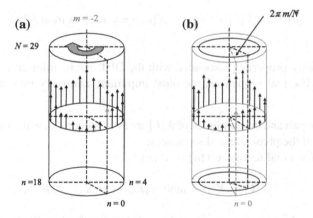

Fig. 3.19 Example of circular shift illustrated using a geometrical construction in which the sequence is wrapped along the circumference of a cylinder. **a** A sequence with $N = 29$ samples, wrapped around a *black cylinder*, with each sample represented by an *arrow*; the arrow representing the sample at $n = 0$ is highlighted in *gray*. The amount ($m = -2$) and the direction of the rotation to be applied in this example (see text) are also shown; the direction is clockwise, because m is negative. **b** The *gray external cylinder* is rotated by an angle $2\pi m/N$ in a clockwise direction with respect to the previous position, indicated by the *black internal cylinder*. The sequence, integral with the surface of the gray cylinder, is circularly shifted by $m = -2$, as witnessed by the *gray arrow*, originally located at $n = 0$ and now located at $n = N - 2 = 27$

3. *Modulation*:
   similarly, if we consider $X_1[k] = X[(k-l) \bmod N] R_N[k]$, that is, if we circularly shift the DFT by l samples, we find that the new DFT corresponds to the sequence $x_1[n] = e^{j(2\pi/N)ln} x[n]$, with $n = [0, N-1]$.

4. *Conjugation*:
$$x^*[n] \Longleftrightarrow X^*[-k \bmod N]_N R_N[k].$$

5. *Circular folding* of a sequence:

$$x[-n] \Longleftrightarrow X[-k \bmod N]_N R_N[k],$$

$$x^*[-n] \Longleftrightarrow X^*[k \bmod N]_N R_N[k].$$

   Circular folding, which corresponds to passing to the periodic extension of the sequence, folding the latter and then extracting a period, can be easily imagined with the help of the geometrical construction used above.

6. *Duality*:
   As we already remarked in Sect. 3.5.1, the DFT analysis and synthesis relations are similar. Usually, this duality property is formally expressed by stating (see Oppenheim and Schafer 2009) that $X[k]$ can be written as $Nx[k]$ index-reversed modulo N, i.e.,

$$x[n] \Longleftrightarrow X[k] \qquad \leadsto \qquad X[n] \Longleftrightarrow Nx[-k \bmod N] R_N[k].$$

7. *Symmetry*:
   the symmetry properties associated with the DFT can be inferred from those of the DFS. For real sequences, the most important symmetry properties are the following:

   - the real part and the amplitude of $X[k]$ are even sequences, while the imaginary part and the phase are odd sequences;
   - since for a real sequence $x[n] \equiv x^*[n]$,

$$X[k] = X^*[-k \bmod N] R_N[k] = X_p^*[-k] R_N[k].$$

   Concerning the second property, we may notice that for $k = [0, N-1]$ we have $k \bmod N = k$ and $-k \bmod N = N - k$. Therefore the following property holds:

$$X[k] = X^*[N-k] \quad \text{for} \quad k = [0, N-1].$$

   In terms of magnitudes this relation becomes

$$|X[k]| = |X[N-k]| \quad \text{for} \quad k = [0, N-1].$$

Fig. 3.20 Symmetry of the DFT amplitude for a real sequence. **a** A real sequence with $N = 8$; **b** its DFT; **c** its DFS; **d** the folded DFS (see text)

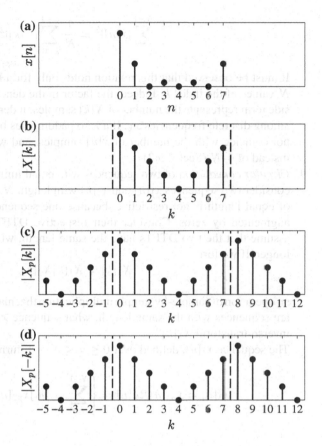

Since any spectral plot shows only the amplitude of the DFT, for a real sequence we can expect such a graph to exhibit symmetry of the samples around the point $k = N/2$, as in the example of Fig. 3.20, which shows a real sequence $x[n]$ without any symmetry (Fig. 3.20a) and the magnitude of its DFT (Fig. 3.20b). $|X_p[k]|$ and $|X_p[-k]|$ are shown in Fig. 3.20c and d, respectively. The plots in the last two panels are identical, due to the symmetry of $|X[k]|$ around $k = N/2$. Other important symmetry properties of the DFT of a real sequence $x[n]$ are listed below:

- $\text{Re}\{X[k]\} = \text{Re}\{X[-k \bmod N]\}R_N[k]$, since $\text{Re}\{X[k]\}$ is an even function;
- $\text{Im}\{X[k]\} = -\text{Im}\{X[-k \bmod N]\}R_N[k]$, since $\text{Im}\{X[k]\}$ is an odd function;
- $|X[k]| = |X[-k \bmod N]|R_N[k]$, since $|X[k]|$ is an even function;
- $\arg\{X[k]\} = -\arg\{X[-k \bmod N]\}R_N[k]$, since $\arg\{X[k]\}$ is an odd function.

8. *Parseval's theorem* for the DFT:
   starting from the definition of $X[k]$, the following relation can be demonstrated:

$$\sum_{n=0}^{N-1} |x[n]|^2 = \frac{1}{N} \sum_{k=0}^{N-1} |X[k]|^2.$$

It must be observed that this relation holds only for a DFT evaluated on a set of N values of the index k. Indeed, the factor in the denominator of the right-hand side term represents the number of $X[k]$ samples: it derives from the spacing $\Delta\omega$ among discrete frequencies ω_k. If a zero padding has been made, this value does not coincide with the number of $x[n]$ samples, and we must write $1(N + N_z)$ instead of $1/N$ (Sect. 3.5.2)

9. *Circular convolution* of two sequences with equal finite length:
   consider two sequences $x_1[n]$ and $x_2[n]$ with length N. Note that the assumption of equal length is not restrictive, because one sequence or both can always be augmented by zeros. Consider their respective DTFTs $X_1[k]$ and $X_2[k]$, and assume that the two DTFTs have the same length, which can be equal to N or longer. If we form

$$X_3[k] = X_1[k]X_2[k],$$

an operation of sample-by sample multiplication that makes sense since the factors are sequences with the same length, what sequence $x_3[n]$ is obtained when we inverse-transform $X_3[k]$?
The sequence $x_3[n]$, defined over $0 \le n \le N - 1$, turns out to be given by

$$x_3[n] = x_{p3}[n]R_N[n] = \left\{ \sum_{m=0}^{N-1} x_{p1}[m]x_{p2}[n-m] \right\} R_N[n],$$

or

$$x_3[n] = \sum_{m=0}^{N-1} x_1[m]\{x_2[(n-m)\bmod N]R_N[m]\} \equiv x_1[n] \circledast x_2[n].$$

The operation indicated by the symbol $\circledast$ is called *circular convolution* of the sequences $x_1[n]$ and $x_2[n]$ *over N points*.
For the time-frequency duality of the DFT, the following relation also holds:

$$x_1[n]x_2[n] \Longleftrightarrow \frac{1}{N}X_1[k] \circledast X_2[k]$$

over N points.
In circular convolution, the second sequence is circularly reversed (folded) and circularly shifted by n samples with respect to the first sequence: therefore, circular convolution is different from linear convolution. Circular convolution is really just periodic convolution, but if we visualize circular folding and shift using a cylindrical surface, we do not really need to actually build the periodic sequences involved. We do so in the next example.

3.5.4 Circular Convolution Versus Linear Convolution

We will now analyze in detail the difference between circular and linear convolution.

Given $x_1[n]$ with $0 \leq n \leq N_1 - 1$ and $x_2[n]$ with $0 \leq n \leq N_2 - 1$, linear convolution is defined as

$$x_1[n] * x_2[n] = x_3'[n] = \sum_{m=m_1}^{m_2} x_1[m] x_2[n - m],$$

where m_1 and m_2 are the n-dependent finite limits of the convolution sum that apply when we convolve two finite-length sequences: $m_1 = \max(0, n - N_2 + 1)$, $m_2 = \min(N_1 - 1, n)$. The length of $x_3'[n]$ is $N_1 + N_2 - 1$ (see Sect. 2.4.2). If $N_1 = N_2 = N$, linear convolution is $2N - 1$ samples-long, and therefore cannot coincide with the circular convolution over N points.

To make an example, we consider two rectangular sequences with $N = 6$, $x_1[n] \equiv x_2[n]$, and calculate both linear and circular convolution.

1. *Linear convolution*
   In Fig. 3.21 we can see:

   - the first rectangular sequence (Fig. 3.21a);
   - the second rectangular sequence folded around $n = 0$ (Fig. 3.21b), useful to compute $x_3'[0]$. Only one product term in the convolution sum is different from zero, so $x_3'[0] = 1 \times 1 = 1$;
   - the second rectangular sequence folded around $n = 1$ (Fig. 3.21c), useful to compute $x_3'[1]$. Two product terms are different from zero, so $x_3'[1] = 1 \times 1 + 1 \times 1 = 2$;
   - the second rectangular sequence folded around $n = 5$ (Fig. 3.21d), useful to compute $x_3'[5]$. Six product terms are different from zero, so $x_3'[5] = 1 \times 1 + 1 \times 1 + 1 \times 1 + 1 \times 1 + 1 \times 1 + 1 \times 1 = 6$;
   - the second rectangular sequence folded around $n = 6$ (Fig. 3.21e), useful to compute $x_3'[6]$. Five product terms are different from zero, so $x_3'[6] = 1 \times 1 + 1 \times 1 + 1 \times 1 + 1 \times 1 + 1 \times 1 = 5$, and so on.

   The resulting shape of $x_3'[n]$ appears as in Fig. 3.22; its length is $2N - 1 = 11$.
2. *Circular convolution over N points.*
   Let us indicate it by $x_3[n]$. We can imagine two coaxial cylinders, one for each sequence, with circumferences having the same length as the sequences. For example, in Fig. 3.23a we can see the position of the internal cylinder with respect to the external one that serves to calculate $x_3[2]$ (this time we are observing the cylinders from above; the numbers represent time-index values). To compute $x_3[2]$, we multiply each pair of corresponding $x_1[n]$, $x_2[n]$ samples and then sum the six products. The result is $x_3[2] = 1 \times 1 + 1 \times 1 + 1 \times 1 + 1 \times 1 + 1 \times 1 + 1 \times 1 = 6$. Hence $\{x_3[2] = 6\} \neq \{x_3'[2] = 3\}$. The other values of $x_3[n]$ are obtained in a similar way, after rotating the internal cylinder with respect to the external one.

Fig. 3.21 Linear
convolution
$x_3'[n] = x_1[n] * x_2[n]$
between two rectangular
sequences with length $N = 6$.
a The sequence $x_1[n]$;
b–d: $x_2[n]$ folded around
$n = 0, 1, 5,$ and 6,
respectively. The folded
sequences are useful for
computing samples of $x_3'[n]$
(see text)

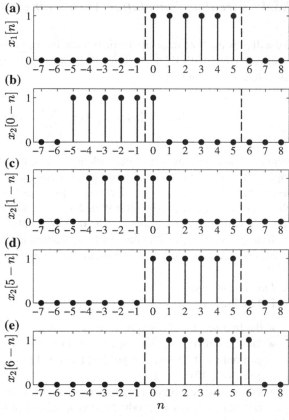

Fig. 3.22 The result of the
linear convolution between
two 6-samples-long
rectangular sequences

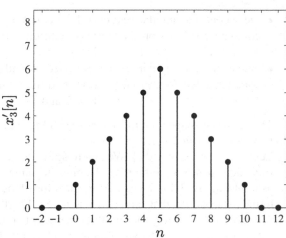

It is clear that the difference between the two types of convolution lies in the fact that in the circular convolution over $N = 6$ points, the tails of a sequence that leave the summation interval on one side enter the interval on the other side, thus corrupting circular convolution with respect to linear convolution. But linear convolution is what is wanted in DFT applications: it represents the theoretical concept of convolution, the one that we need to reproduce when using DFT as a frequency-domain signal representation tool. For example, if we want to apply an FIR filter to a sequence, we must implement linear convolution of the sequence with the impulse response of the filter, and if we want to do so operating in the frequency domain via DFT, we need a way to make circular convolution equivalent to linear convolution.

The issue here is *aliasing in the time domain*, as we can understand thinking of the corresponding frequency sampling. Indeed, if the result of linear convolution is $2N - 1$ samples-long, it must be represented by a DFT which is $2N - 1$ samples-long. This DFT can be seen as the sample-by-sample product of the DFTs of the individual factor sequences, if and only if the individual DFTs are computed over $2N - 1$ points each. Thus, we need to perform zero-padding before circular convolution. Linear and circular convolution give the same result if each sequence is augmented by $N - 1$ zeros, up to the length $2N - 1$ that is necessary to contain linear convolution.

Figure 3.23b illustrates the circular convolution of the two 6-points-long rectangular sequences, calculated in this way: $x_3''[n] = x_1[n] \circledast x_2[n]$ over $2N - 1 = 11$ points. In Fig. 3.23b, the circumferences of the cylinders have been increased to $2N - 1 = 11$ discrete-time units; the numbers represent index values. We can verify that in this case $x_3''[2] = 3 = x_3'[2]$, as required. Under these conditions we can compute linear convolution via IDFT of the product of two DFTs, and this is interesting because very efficient algorithms exist for calculating the DFT.

Given a sequence $x[n]$ with $n = [0, N - 1]$ and an impulse response $h[n]$ with $n = [0, N' - 1]$, in order to avoid aliasing when computing $x[n] * h[n]$ in the frequency domain via DFT

- $x[n]$ must be augmented by zero padding up to a length $L \geq N + N' - 1$;
- a DFT over L points then gives $X[k]$, $k = [0, L - 1]$;
- $h[n]$ must be augmented by zero padding up to the same length $L \geq N + N' - 1$;
- a DFT over L points then gives $H[k]$, with $k = [0, L - 1]$;
- the two transforms must be multiplied sample by sample: $Y[k] = X[k] \cdot H[k]$, $k = [0, L - 1]$;
- the inverse transform of $Y[k]$ finally provides $y[n]$, with $n = [0, L - 1]$. If $L > N + N' - 1$, at the end of the sequence $y[n]$ there will be $L - (N + N' - 1)$ zeros that can be eliminated.

We may finally notice that time aliasing in circular convolution can be avoided if, and only if, the sequences have finite length.

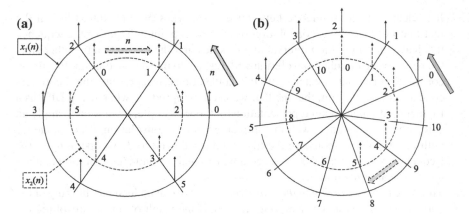

Fig. 3.23 **a** Two rectangular sequences $x_1[n]$ and $x_2[n]$ with $N = 6$, drawn as *arrows* on the lateral surfaces of two coaxial cylinders, as seen from above. The circumferences are equal to $N = 6$ discrete-time units; $x_1[n]$ lies on the external cylinder (*solid*) and $x_2[n]$ lies on the internal cylinder (*dashed*). The numbers represent time-index values. The mutual position of the cylinders is such that the sample at $n = 2$ of the circular convolution over $N = 6$ points can be obtained multiplying the corresponding solid and dashed samples, and then summing all products. **b** The same two $N = 6$ samples-long rectangular sequences, drawn on the $2N - 1 = 11$ points-long circumferences of two larger coaxial cylinders. The mutual position of the cylinders is such that the sample at $n = 2$ of the circular convolution over $N = 11$ points can be obtained multiplying the corresponding solid and dashed samples and then summing all products. See text for details

3.6   Fast Fourier Transform (FFT)

The *fast Fourier transform* (FFT) is an optimized algorithm to compute the DFT and its inverse transform rapidly. The invention of the FFT made the DFT practically applicable to the analysis of data sequences. The first important article on this subject was published by Cooley and Tukey (1965).

The computational complexity of the DFT is estimated in terms of the number of complex multiplications that must be performed; the additions are usually neglected. Since the quantity to be computed is

$$\sum_{n=0}^{N-1} x[n]e^{-j\frac{2\pi}{N}kn}, \qquad k = [0, N-1], \qquad n = [0, N-1],$$

N complex multiplications are needed for each n value, up to a total of N^2 complex multiplications, which are equivalent to $4N^2$ real multiplications. For high values of N, this can be cumbersome even with relatively powerful machines. Exploiting the symmetry and periodicity properties of complex exponentials, the efficiency of the DFT algorithm can be improved, by reducing its complexity. Such optimized way of computing the DFT is referred to as the FFT. Actually, the optimization can be attained in several ways, and many different varieties of FFT exist. As an example, we discuss the so-called *decimation-in-time FFT algorithm*.

Let us assume, for the sake of generality, that input values $x[n]$ are complex. We also assume that N is even: if it is actually odd, we just add one zero at the sequence's end. Then, we rearrange $x[n]$ in such a way that all the even-indexed samples come first, and are followed by the odd-indexed samples. In this way we can express the original N-points DFT as linear combination of two $N/2$-points DFTs, separately performed on even-indexed and odd-indexed samples. The linear combination requires N complex multiplications, so the complexity is reduced to $2(N/2)^2 + N$.

If N is equal to an integer power of 2, i.e., $N = 2^\nu$ with integer ν, the previous approach can be applied recursively until a linear combination of 2-points DFTs is obtained. Figure 3.24 shows the flow diagram of such an algorithm, for $N = 8 = 2^3$.

The computation is organized in $\nu = 3$ successive stages. In each stage, $N/2 = 4$ DFTs over 2 points must be computed. These basic building blocks are called "butterflies"; an example of a butterfly is shown in Fig. 3.25a. In the flow diagrams of Figs. 3.24 and 3.25a, transmission coefficients W_N^n, with $n = [0, N-1]$, appear in correspondence to the various branches. Let us recall that we defined $W_N = e^{-j(2\pi)/N}$. These coefficients are called *twiddle factors*, and their presence indicates that passing through that branch, a given numerical value is multiplied for the specified integer power of W_N.

In the mth stage, with $m = [1, \nu]$, a sequence of N values (that for $m = 1$ is the input sequence and in the other cases is the output of the previous stage) generates

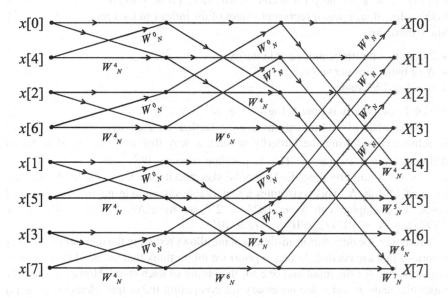

Fig. 3.24 Flow diagram for the computation of a DFT over $N = 8$ points, carried out via the decimation-in-time FFT algorithm

Fig. 3.25 **a** An example of DFT over 2 points (butterfly diagram); **b** a butterfly that requires two complex multiplications; **c** a simplified butterfly that requires only one complex multiplication

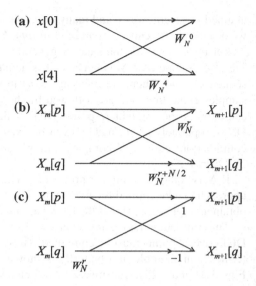

another sequence of N values. In the last stage ($m = \nu$) the output is the final result, i.e., the desired DFT. We can indicate the input sequence of the mth stage by $X_m[l]$, with $l = [0, N - 1]$ (but only if $m = 1$ this is the original sequence). The output sequence will be $X_{m+1}[l]$, with $l = [0, N - 1]$ (but only for $m = \nu$ this is the desired DFT). The generic butterfly of the mth stage is sketched in Fig. 3.25b. The exponent r in Fig. 3.25b is an integer that can assume values from 0 to $(N - 2)/2$, according to the value of m; p and q represent values of the index l in the range $l = [0, N - 1]$. Since there are

- 2 complex multiplications per butterfly,
- $N/2$ butterflies per stage,
- $\nu = \log_2(N)$ stages,

the DFT complexity is now $N \log_2(N) \ll N^2$.

The number of multiplications can be further reduced by a factor of 2: this is achieved modifying the butterfly in such a way that only one twiddle factor different from 1 is present. This is possible because $W_N^{N/2} = \left[e^{-j(2\pi)/N} \right]^{N/2} = e^{-j\pi} = -1$. Using the simplified butterfly, sketched in Fig. 3.25c, the final complexity is $(N/2) \log_2 N$. This constitutes a remarkable saving in terms of computational costs: for example, with $N = 1024 = 2^{10}$, we have $N^2 \approx 1.05 \times 10^6$, while $(N/2) \log_2 N = 1024/2 \times 10 \approx 5.12 \times 10^3$.

Moreover, the decimation-in-time scheme allows reducing the number of memory locations that are needed. In fact, a priori we might think that we need two memory registers, one for the input and one for the output of each stage. However, only the input elements p and q are necessary for computing the output elements p and q (Fig. 3.25a, b), so that *in-place computations* are possible: only one set of N memory locations is needed, because each output sequence of a given stage can cover the same memory locations previously reserved to the input, that is, the output of the preceding stage.

Other types of FFT algorithms exist, such as

- the decimation-in-frequency FFT, in which the DFT, rather than the input sequence, is progressively subdivided into shorter and shorter segments;
- algorithms devised to accommodate for values of N that are not integer powers of 2, but are the product of prime factors. Such algorithms are more computationally complex than those for $N = 2^\nu$. In most cases, it is advisable to augment the sequence up to the nearest $N = 2^\nu$ by zero padding;
- specific algorithms for real input sequences: the input can be arranged in the form of a complex sequence of length $N/2$, having the even-indexed samples of the input in its real part and the odd-indexed samples of the input in its imaginary part. Of course, FFT algorithms devised for complex input can be used also for real sequences, setting $\text{Im}\{x[n]\} = 0$, but this approach is much less efficient;
- at last, for a real sequence special algorithms are available, able to compute directly the coefficients of the so-called *discrete trigonometric expansion* of a real sequence, which will be briefly presented in the next section.

FFT is useful for a number of digital signal processing applications, such as

- performing convolution,
- estimating the autocorrelation/autocovariance of a sequence and the cross-correlation/covariance between two sequences (Part III),
- estimating the power spectrum of a discrete-time, finite-length random signal (Part III), etc.

Since we are interested in real sequences, we must always keep in mind that only half of the DFT contains essential information, while the rest is redundant and can be deduced from that half, exploiting symmetry properties: in fact, for real sequences we have $X[k] = X^*[N - k]$. Therefore, assuming without loss of generality that N is even, only the DFT samples with $k = [0, N/2]$ contain independent information.

3.7 Discrete Trigonometric Expansion

We merely mention here that an alternative way of expressing the DFT consists in using real sines and cosines instead of complex exponentials. Assuming that N is even (if it is odd we place a zero at the end), the *discrete trigonometric expansion* can be written as

$$x[n] = \frac{1}{N}\left\{a_0 + (-1)^n a_{N/2} + \sum_{k=1}^{N/2-1}\left[a_k \cos\left(\frac{2\pi}{N}kn\right) + b_k \sin\left(\frac{2\pi}{N}kn\right)\right]\right\},$$

with

$$a_0 = \sum_{n=0}^{N-1} x[n], \qquad a_{N/2} = \sum_{n=0}^{N-1} (-1)^n x[n],$$

$$a_k = 2 \sum_{n=0}^{N-1} x[n] \cos\left(\frac{2\pi}{N}kn\right) = 2\text{Re}\{X[k]\},$$

$$b_k = 2 \sum_{n=0}^{N-1} x[n] \sin\left(\frac{2\pi}{N}kn\right) = -2\text{Im}\{X[k]\}.$$

3.8 Appendix: Mathematical Foundations of Signal Representation

From the mathematical point of view, the Fourier-based signal representations discussed in the previous sections, as well as their analog counterparts, are signal representations, a.k.a. expansions, in an orthogonal basis established in a vector space.

In this appendix we present some basic concepts about *vector spaces* and the related operations and properties as the mathematical foundation for signal representation. Fourier series and transforms in both the continuous- and discrete-time domains are well-known examples of (orthogonal) signal expansions; however, also the wavelet transforms described in Chaps. 13 and 14 are signal expansions, and the synthetic notes on vector spaces and signal representations provided here will prove particularly useful when discussing them. For a more detailed treatment of these topics from the point of view of signal processing the reader can consult, for example, Wang (2013).

3.8.1 Vector Spaces

In our discussion, any signal, either a continuous-time one represented as a function $x(t)$, or a discrete one represented as a sequence $x[n]$, with a finite or infinite number of samples, will be considered as a vector in a *vector space*, which is just a generalization of the familiar concept of three-dimensional space. A vector space is a set V with two operations of addition and scalar multiplication defined for its members, referred to as *vectors*. We will denote vectors by bold letters.

- *Vector addition* maps any two vectors $x, y \in V$ to another vector $x + y \in V$ satisfying the following properties:

  - Commutativity: $x + y = y + x$.
  - Associativity: $x + (y + z) = (x + y) + z$.

- Existence of zero: there is a vector $\mathbf{0} \in V$ such that $\mathbf{0} + x = x + \mathbf{0} = x$.
- Existence of inverse: for any vector $x \in V$, there is another vector $-x \in V$ such that $x + (-x) = \mathbf{0}$.

- *Scalar multiplication* maps a vector $x \in V$ and a real or complex scalar a to another vector $ax \in V$ with the following properties:

  - $a(x + y) = ax + ay$;
  - $(a + b)x = ax + bx$, where b is another scalar;
  - $abx = a(bx)$;
  - $1x = x$.

We define a *function space* as a vector space whose elements are real- or complex-valued functions $x(t)$ of a real variable t, defined over either an infinite range, $-\infty < t < +\infty$, or a finite range, such as $0 \leq t < T$.

A function space that is particularly important in signal processing is $L^2(\mathbb{R})$. This is the space of all functions with a finite, well-defined integral of the square of the modulus of the function, i.e., *functions with finite energy* in the continuous-time domain. The symbol L signifies a *Lebesgue integral*,[12] the exponent 2 denotes the integral of the square of the modulus of the function, and $(\mathbb{R})$ states that the independent variable t is a number over the whole real line. If $x(t)$ is a member of that space, we will write $x(t) \in L^2(\mathbb{R})$.

The next most basic space is $L^1(\mathbb{R})$, which requires a finite integral of the modulus of the function. With these functions, one is allowed to interchange infinite summations and integrations, which is not necessarily true with $L^2(\mathbb{R})$ functions.

The concept of function space can be generalized to those functions satisfying $\int |x(t)|^p \, d\theta = K < \infty$, which are designated by $L^p(\mathbb{R})$. Although most of the definitions and derivations we may be interested in are in terms of signals belonging to $L^2(\mathbb{R})$, many of the results hold for larger classes of signals.

A more general class of signals than any $L^p(\mathbb{R})$ space contains what are called *distributions*. These are *generalized functions*, which are not defined by their having "values" but by the value of an "inner product" with an ordinary function (see the next subsection for the definition of inner product). An example of a distribution is the Dirac δ function. For an introductory study of the transforms we are interested in, such as Fourier and wavelet transforms, we can ignore the mathematical facets of signal space classes, use distributions as if they were functions, and assume Riemann integrals, but we should not forget that there are some cases in which these ideas are crucial.

In functional analysis, similarly to function spaces introduced for continuous-time functions, *sequence spaces* are defined for sequences. A sequence space is a vector

---

[12]Lebesgue integrals are somewhat more general than the basic Riemann integral. The value of a Lebesgue integral is not affected by values of the function over any countable set of values of its argument, or, more generally, a set of measure zero. For instance, a function defined as 1 on the rationals and 0 on the irrationals would have a zero Lebesgue integral. As a result of this, properties derived using Lebesgue integrals are sometimes said to be true "almost everywhere", meaning they may not be true over a set of measure zero.

space whose elements are sequences of real or complex numbers.[13] We may have $n = (-\infty, +\infty)$ or a finite range of n-values. Thus, $\ell^p(\mathbb{Z})$ is the space of p-power summable sequences; $\ell^2(\mathbb{Z})$ is the space of square-summable sequences; $\ell^1(\mathbb{Z})$ is the space of absolutely summable sequences, etc.

Here we will restrict our attentions to two typical vector spaces:

- the $\ell^2(\mathbb{Z})$ space of sequences $x[n]$ that are square-summable and therefore are discrete-time energy signals;
- the $L^2(\mathbb{R})$ space of functions $x(t)$ that are square-integrable and therefore are continuous-time energy signals.

Note that the term "vector" may be interpreted in two different ways. First, in the most general sense, it represents a member x of a vector space, e.g., a function $x(t) \in L^2(\mathbb{R})$ or a sequence $x[n] \in \ell^2(\mathbb{Z})$. Second, in a narrower sense, it can also represent a tuple of N elements, such as the N samples of a discrete-time signal, where N may go to infinity. From the context in the discussion, it should be clear what a "vector" represents; moreover, we will reserve bold symbols to vectors in the first, more general sense.

The sum of two subspaces $S_1 \subset V$ and $S_2 \subset V$ of a vector space V that are mutually exclusive, i.e., that do not intersect each other ($S_1 \cap S_2 = 0$), is called a direct sum, denoted by $S_1 \oplus S_2$. Moreover, if $S_1 \oplus S_2 = V$, then S_1 and S_2 form a direct-sum decomposition of the vector space V, and S_1 and S_2 are said to be complementary, in the sense that S_1 is the *complement* of S_2 in V, and S_2 is the *complement* of S_1 in V. The direct sum decomposition of V can be generalized to include multiple subspaces:

$$V = \oplus_{j=0}^{N-1} S_j = S_0 \oplus S_2 \oplus \cdots \oplus S_{N-1},$$

where all subspaces $S_j \in V$ are assumed to be mutually exclusive.

3.8.2 Inner Product Spaces

An inner product in a vector space V is a function that maps two vectors $x, y \in V$ to a real or complex scalar $s \equiv \langle x, y \rangle$ and satisfies the following conditions.

- Positive definiteness: $\langle x, x \rangle \geq 0$.
- Conjugate symmetry: $\langle x, y \rangle = \langle y, x \rangle^*$. If the vector space is real, the inner product becomes symmetric: $\langle x, y \rangle = \langle y, x \rangle$.
- Linearity in the first variable: $\langle ax + by, z \rangle = a \langle x, z \rangle + b \langle x, z \rangle$. It can be shown that linearity applies to the second variable only if the coefficients of the combina-

---

[13]Equivalently, a sequence space is a function space whose elements are functions from the field $\mathbb{Z}$ of integer numbers (values of n) to the field $\mathbb{R}$ of real or complex numbers (values of $x[n]$), exactly as a continuous-time-function space is a function space whose elements are functions from the field $\mathbb{R}$ (values of t) to $\mathbb{R}$ (values of $x(t)$).

tion are real: $a, b \in \mathbb{R}$. As a special case, when $b = 0$, we have $\langle a\mathbf{x}, \mathbf{y} \rangle = a \langle \mathbf{x}, \mathbf{y} \rangle$. A vector space with inner product defined is called an *inner product space*. All vector spaces in our discussion will be assumed to be inner product spaces, a.k.a. *Euclidean spaces*.

Some examples of the inner product are listed below:

- for two sequences of length N, the inner product is a summation,

$$\langle \mathbf{x}, \mathbf{y} \rangle = \langle x[n], y[n] \rangle = \mathbf{x} \cdot \mathbf{y} = \sum_{n=0}^{N-1} x[n] y^*[n];$$

- for functions $\mathbf{x} = x(t)$, $\mathbf{y} = y(t)$ the inner product is an integral,

$$\langle \mathbf{x}, \mathbf{y} \rangle = \langle x(t), y(t) \rangle = \int x(t) y^*(t) \mathrm{d}t,$$

where the integration limits depend on the particular functions being considered.

For real sequences or functions,

$$\langle x[n], y[n] \rangle = \sum x[n] y[n], \qquad \langle x(t), y(t) \rangle = \int x(t) y(t) \mathrm{d}t.$$

The concept of inner product is a concept of of essential importance,[14] based on which a whole set of other important concepts can be defined.

The *norm* (or length) of a vector $\mathbf{x} \in V$ is defined as

$$\|\mathbf{x}\| = \sqrt{\langle \mathbf{x}, \mathbf{x} \rangle}.$$

---

[14]In this part of the book we are dealing with deterministic signals. Starting from Chap. 9, our approach to signal representation and analysis will change: we will explicitly consider signals deriving from experimental measurements, which are better described in terms of random variables/random processes and of statistical/probabilistic arguments. The present discussion covers also the random case, except that in the case of random variables, the inner product should actually be defined through the so-called ensemble average operator,

$$\langle x(t), y(t) \rangle = \mathrm{E}\left[xy^* \right],$$

that is the "correlation" of the two random variables x and y (for zero-mean variables it is the "covariance"). This means (see Chap. 9) that the integral should include the probability density function related to each random variable. On the other hand, in the theory of wavelet representation of signals (Chaps. 13–15), the random nature of signals is not explicitly considered (see Sects. 14.2 and 14.3). Though keeping in mind that the statistical/probabilistic complication will come into play in Chap. 9, we can continue our discussion on inner-product spaces using the standard definition of inner product.

The norm is non-negative and it is zero if and only if $x = 0$. In particular, if the norm is 1, then the vector is said to be normalized and becomes a unit vector. Any vector can be normalized, dividing it by its norm.

The squared vector norm $\langle x, x \rangle = \|x\|^2$ represents the energy of the vector. For example, for a finite-length sequence $x[n]$ we have in $\ell^2(\mathbb{Z})$

$$\|x\| = \sqrt{\sum_{n=0}^{N-1} |x[n]|^2}.$$

This concept can be generalized to an infinite-length sequence, in which case the range of the summation will cover all integers $\mathbb{Z}$. This norm exists only if the summation converges to a finite value, i.e., if $x[n]$ is an energy signal. All such vectors are square-summable and form the vector space denoted by $\ell^2(\mathbb{Z})$. Similarly, the norm of a function vector $x = x(t)$ in $L^2(\mathbb{R})$ is defined as

$$\|x\| = \sqrt{\int |x(t)|^2},$$

where the lower and upper integral limits are two real numbers, which may tend to infinity, so that the integral spans the entire real axis. This norm exists only if the integral converges to a finite value, i.e., if $x(t)$ is an energy signal.

If the inner product of two vectors x and y is zero, they are orthogonal (perpendicular) to each other, denoted by $x \perp y$. Two subspaces $S_1 \subset V$ and $S_2 \subset V$ of an inner product space V are orthogonal, denoted by $S_1 \perp S_2$, if $x_1 \perp x_2$ for any $x_1 \in S_1$ and $x_2 \in S_2$. The *orthogonal complement* of a subspace $S_1 \subset V$ is the set of all vectors in V that are orthogonal to S_1. An inner product space V can be expressed as the direct sum of a finite or infinite number of mutually orthogonal subspaces:

$$V = \oplus_j S_j \quad \text{with} \quad S_l \perp S_m \quad \text{for all} \ l \neq m.$$

All of these definitions can be intuitively visualized in a three-dimensional space spanned by three perpendicular coordinates representing three mutually orthogonal subspaces. The orthogonal direct sum of these subspaces is the three-dimensional space, and the orthogonal complement of the subspace in the x direction is the two-dimensional yz-plane formed by coordinates y and z.

We now introduce the *distance between two vectors* x and y as the norm of their difference $x - y$. The distance is always non-negative and is symmetric, i.e., the distance between x and y is equal to the distance between y and x. When distance is defined between any two vectors in a vector space, the latter is called a *metric space*. In a metric space V, a vector sequence $\{x_1, x_2, x_3, \ldots\}$ is a called a Cauchy sequence if for any $\varepsilon > 0$ there exists an integer $L > 0$ such that for any $(l, m) > L$, the distance between x_l and x_m is smaller than ε.

A metric space V is a *complete metric space* if every Cauchy sequence $\{x_l\} \in V$ converges to $x \in V$, in the sense that the distance between x_l and x tends to 0 as l approaches infinity. In other words, for any $\varepsilon > 0$ there exists an $L > 0$ such that this distance becomes smaller than ε for $l > L$. Complete inner-product metric spaces are referred to as *Hilbert spaces*.

In what follows, to keep the discussion generic, the lower and upper limits of a summation or an integral may not be always explicitly specified, as the summation or integral may be finite or infinite, depending on each specific case.

3.8.3 Bases in Vector Spaces

In a vector space V, the subspace S of all linear combinations of a set of M vectors $b_k \in V$, with $k = [0, M - 1]$, is called the (linear) *span of the vectors*:

$$S = \text{Span}(b_k) = \sum_{k=0}^{M-1} c_k b_k,$$

where the $\{c_k\}$ are real or complex coefficients. A set of linearly independent vectors b_k that spans a vector space is called a *basis* of that space. Now, let $\{b_k\}$ be a set of linearly independent vectors in a Hilbert space H, and x an arbitrary vector that can be approximated in a finite-dimensional subspace with dimensionality M by

$$\hat{x}_M = \sum_{k=0}^{M-1} c_k b_k.$$

If the least-squares error of this approximation converges to zero when the dimensionality M of the space S tends to infinity:

$$\lim_{M \to \infty} \left\| x - \hat{x}_M \right\|^2 = 0,$$

then the approximation converges to the given vector:

$$\lim_{M \to \infty} \sum_{k=0}^{M-1} c_k b_k = x,$$

and the set of vectors $\{b_k\}$ is said to be a *complete basis*. If the vectors b_k are also orthogonal, they are called a *complete orthogonal basis*. If the orthogonal vectors have unit norm, we have a *complete orthonormal (ON) basis*.

The basis vectors are linearly independent, i.e., none of them can be represented as a linear combination of the other ones. The vectors of a complete basis are such that if we included any additional vector in the basis, the vectors of the set would no longer be linearly independent, and removing any of them would result in inability to represent certain vectors in the space. In other words, a complete basis is a minimum set of vectors capable of representing any vector in the space. Also, as any "rotation" of a given basis will result in a different basis,[15] we see that there are infinitely many bases that span the same space.

The concept of a finite space spanned by a basis composed of a finite number M of basis vectors can be generalized to a function space composed of all functions $x = x(t)$ defined over some interval $0 \leq t < T$, spanned by a set of countable but infinite basis vectors $\boldsymbol{b}_k = b_k(t)$, with $k = (-\infty, +\infty)$:

$$x(t) = \sum_{k=-\infty}^{+\infty} c_k \boldsymbol{b}_k = \sum_{k=-\infty}^{+\infty} c_k b_k(t).$$

The span of the basis vectors is[16] the subspace formed by all these linear combinations:

$$S = \text{Span}(\boldsymbol{b}_k) = \sum_{k=-\infty}^{\infty} c_k \boldsymbol{b}_k,$$

and conversely, $\{\boldsymbol{b}_k\}$ is the *spanning set* of S.

This idea can be further generalized to a function space composed of all functions defined over an infinite domain $-\infty < t < \infty$, spanned by a basis composed of a family of uncountably infinite functions $\boldsymbol{b}(f) = b(f, t)$, with $f = (-\infty, +\infty)$. Any function $x = x(t)$ in the space can be expressed as a linear combination of these basis functions, in the form of an integral:

$$x(t) = \int_{-\infty}^{+\infty} c(f) b(f, t) \mathrm{d}f = \int_{-\infty}^{+\infty} c(f) \boldsymbol{b}(f) \mathrm{d}f.$$

We see that the index k for the summation in the finite case is replaced by a continuous variable f for the integral, and the coefficient c_k is replaced by a continuous weighting function $c(f)$ for the uncountably infinite set of basis functions $\boldsymbol{b}(f) = b(f, t)$. An important issue is how to find the coefficients or the weighting function, given the vector x and the basis vectors $\boldsymbol{b}_k$ or $\boldsymbol{b}(f)$.

---

[15]Here we do not enter in details about how passing from one basis to another one can be seen as a rotation of axes; we leave this to the intuition of the reader.

[16]In several cases, the signal spaces encountered in wavelet theory are actually the *closure* of the space spanned by the basis set, meaning that the space contains not only all signals that can be expressed by a linear combination of the basis functions, but also the signals which are the limit of infinite expansions based on the considered (infinite) set. The closure of a space is usually denoted by an over-line, as in $\overline{\text{Span}}_k$, but we will neglect the over-line in our notation, for simplicity.

3.8.4 Signal Representation by Orthogonal Bases

Finding the coefficients or the weighting function can become straightforward if the basis is orthogonal. Let x and y be any two vectors in the Hilbert space H spanned by a complete orthogonal system $\{u_k\}$ satisfying

$$\langle u_l, u_m \rangle = \delta[l - m].$$

Note that in this case, $\{u_k\}$ is actually orthonormal. Then we have, in vector notation,

1. *series expansion*:

$$x = \sum_k c_k u_k = \sum_k \langle x, u_k \rangle u_k;$$

2. *Parseval's theorem*[17]:

$$\langle x, y \rangle = \sum_k \langle x, u_k \rangle \langle y, u_k \rangle^*,$$

$$\langle x, x \rangle = \|x\|^2 = \sum_k |\langle x, u_k \rangle|^2.$$

Here, the dimensionality of the space, i.e., the range of k values, is not specified to keep the discussion more general.

The results shown above can be generalized to a vector space spanned by a basis composed of a continuum of uncountable orthogonal basis vectors $u(f)$ satisfying

$$\langle u(f), u(f') \rangle = \delta(f - f').$$

Any vector x in the space can be expressed as

---

[17]Actually, the first equation above should be called Plancherel's theorem; the name "Parseval's theorem" should be reserved to the second equation, which is a particular case of Plancherel's theorem.

More precisely, Parseval's theorem refers to the result that the Fourier transform is *unitary*: loosely, this means that the sum (or integral) of the square of a function is equal to the sum (or integral) of the square of its transform. It originates from a 1799 theorem about series by Marc-Antoine Parseval (Parseval 1806), which was later applied to the Fourier series. It is also known as Rayleigh's energy theorem (Rayleigh 1889), or Rayleigh's identity, after John William Strutt, Lord Rayleigh. Plancherel's theorem (Plancherel 1910) is a more general result in harmonic analysis. It states that the integral of a function's squared modulus is equal to the integral of the squared modulus of its frequency spectrum. This result makes it possible to speak of Fourier transforms of quadratically-integrable functions and quadratically-summable sequences, rather than just of absolutely integrable functions and absolutely summable sequences. The unitarity of any kind of Fourier transform is often called Parseval's theorem in science and engineering fields, based on the above-mentioned earlier (but less general) result, but the most general form of this property should be called Plancherel's theorem.

$$x = \int c(f)u(f)\mathrm{d}f.$$

This equation represents a given vector in the space as a linear combination (an integral) of the basis function $u(f)$, weighted by $c(f)$. The weighting function $c(f)$ can be easily obtained exploiting the orthogonality of the basis. Indeed, taking the inner product with $u(f')$ on both sides of the equation we get

$$\langle x, u(f') \rangle = \left\langle \int c(f)u(f)\mathrm{d}f, u(f') \right\rangle =$$

$$= \int c(f)\langle u(f), u(f') \rangle \mathrm{d}f = \int c(f)\delta(f - f')\mathrm{d}f = c(f').$$

Therefore,

$$c(f) = \langle x, u(f) \rangle.$$

Parseval's theorem becomes, with obvious notation,

$$\|x\|^2 = \langle x, x \rangle = \int c(f)c^*(f)\mathrm{d}f = \langle c, c \rangle = \|c\|^2.$$

In the case of a countable set of vectors u_k, $\|x\|^2$ will still be equal to the norm of the coefficients, but since the coefficients depend on an index rather than on a continuous variable, $\|c\|^2$ will be a summation rather than an integral.

For example, let us consider the *continuous-time Fourier transform* (CTFT), which is defined (see, e.g., Bochner and Chandrasekharan 1949; Bracewell 2000) through the analysis and synthesis relations

$$X(f) = \int_{-\infty}^{+\infty} x(t)\mathrm{e}^{-\mathrm{j}ft}\mathrm{d}t,$$

$$x(t) = \int_{-\infty}^{+\infty} X(f)\mathrm{e}^{\mathrm{j}ft}\mathrm{d}f.$$

This is a signal representation of this type, the basis vectors $u(f)$ being sinusoidal functions of different frequencies f in $(-\infty, +\infty)$. On the other hand, the DFT of a sequence with finite length N is instead an expansion based on a finite number of basis vectors, namely, the set of complex exponentials $\mathrm{e}^{\mathrm{j}2\pi kn/N}$, etc.

3.8.5   Signal Representation by Standard Bases

As a special case of the class of orthogonal bases, we can consider the standard basis in the N-dimensional Euclidean space, in which a vector x representing N samples of a signal $x[n]$ can be expressed as a linear combination of *standard basis vectors*:

$$x = \sum_{n=0}^{N-1} x[n]e_n = Ix,$$

where I represents the identity operator in the form of a matrix having 1 as elements of the main diagonal and 0 as the other elements, and

$$e_1 = [1\,0\,0\,0\ldots]^T,$$
$$e_2 = [0\,1\,0\,0\ldots]^T,$$
$$e_3 = [0\,0\,1\,0\ldots]^T,$$

etc. The superscript T indicates the transpose of the vectors in square brackets: in fact, the vectors are normally written as column vectors.[18]

If we denote by $e[m, n]$ the mth element of e_n, then the mth sample $x[m]$ of x can be expressed as

$$x[m] = \sum_{n=0}^{N-1} x[n]e[m, n] = \sum_{n=0}^{N-1} x[n]\delta[m - n].$$

This result can be generalized to a vector space of infinite dimensions, spanned by e_n with $-\infty < n < \infty$, thus giving

$$x[m] = \sum_{n=-\infty}^{\infty} x[n]\delta[m - n].$$

We see that whenever a discrete-time signal is given in the form of a vector as a set of samples $x[m]$, each corresponding to a particular time instant, it is actually expressed in terms of the standard basis, which is implicitly used.

In particular, if we let $x[m] = \delta[m - m']$, the equation above becomes

$$\delta[m - m'] = \sum_{n=-\infty}^{\infty} \delta[n - m']\delta[n - m] = \langle \delta_{m'}, \delta_m \rangle,$$

indicating that the set of vectors $\delta_m = \delta[n - m]$ shifted by different amounts m of discrete time is indeed a set of orthonormal vectors that form a standard basis of a vector space, by which any vector $x = [\ldots, x[n], \ldots]^T$ in the space can be

---

[18]In the three-dimensional case, and thinking of the samples of $x[n]$, with $n = 0, 1, 2$ as the spatial coordinates x, y and z of a point with respect to a Cartesian coordinate reference system, this would be the familiar vector representation $xi + yj + zk$ that uses three standard basis unit vectors along each of the three mutually perpendicular axes. Thus we would have $e_1 = i, e_2 = j, e_3 = k$, the unit vectors indicating the direction of the axes being $[1\,0\,0]^T, [0\,1\,0]^T$ and $[0\,0\,1]^T$.

expanded as

$$x[n] = \sum_{m=-\infty}^{\infty} x[m]\delta[n-m] = \sum_{m=-\infty}^{\infty} x[m]\delta[m-n], \qquad n = 0, \pm 1, \pm 2, \ldots$$

It may seem only natural and reasonable to represent a signal vector by this standard basis, in terms of a set of time samples, but it is also possible, and sometime more beneficial, to decompose the signal into a set of components along some dimension other than time, which means to represent the signal vector by an orthogonal basis that can be obtained by rotating the standard basis.

The concept of representing a discrete time signal $x[n]$ by the standard basis can be extended to the representation of a continuous time signal $x(t)$ defined over some finite interval $0 \le t < T$ (see Wang 2013), and later can be further generalized to the interval $-\infty < t < \infty$, by using the Dirac delta. We will do exactly this when dealing with the sampling of continuous-time signals in Chap. 4. We can write the value of the signal at some $t = \tau$ as

$$x(\tau) = \int_{-\infty}^{\infty} x(t)\delta(t-\tau)dt = \int_{-\infty}^{\infty} x(t)\delta(\tau-t)dt.$$

If we let $x(\tau) = \delta(\tau-\tau')$, the above equation becomes, with $\delta(t-\tau) = \delta(\tau-t) \equiv \delta_\tau(t)$,

$$\delta(\tau-\tau') = \int_{-\infty}^{\infty} \delta(t-\tau')\delta(t-\tau)dt = \langle \delta_{\tau'}(t), \delta_\tau(t) \rangle,$$

indicating that the Dirac δ shifted by different amounts τ of time indeed represents a set of orthogonal functions that form a standard basis for a function space, by which any function $x(t)$ in the space can be expanded:

$$x(t) = \int_{-\infty}^{\infty} x(\tau)\delta(\tau-t)d\tau.$$

Again, it may seem natural to represent a continuous-time signal $x(t)$ by the corresponding standard basis, made of time impulses. However, this is not the only way or the best way to represent the signal. The time signal can also be represented by a basis other than the standard basis represented by the train of shifted Dirac δ, so that the signal is decomposed along some different dimension other than time. Such an alternative way of signal decomposition and representation may be desirable, as the signal can be more conveniently processed and analyzed, for any signal processing task. This is actually the fundamental reason why different orthogonal transforms are developed.

In summary, a signal vector can be represented under different bases that span the related space, and all these representations are equivalent in terms of the total energy (Parseval's theorem). However, these representations may differ drastically in terms of how different types of information contained in the signal are concentrated in different signal components, i.e., are carried by different coefficients or values of the weighting function. Sometimes certain advantages can be gained from one particular basis compared with another, depending on the specific application.

3.8.6 Frames and Biorthogonal Bases

So far we considered the representation of a signal vector $x \in H$ as some linear combination of an orthogonal basis u_k that spans the space:

$$x = \sum_k c_k u_k = \sum_k \langle x, u_k \rangle u_k,$$

with Parseval's theorem, written as $\|x\| = \|c\|$, indicating that x is represented by the expansion coefficients $\{c_k\}$ without any redundancy. However, sometimes it may be difficult or even impossible to identify a set of linearly independent and orthogonal basis vectors in the space. In such cases we could still consider representing a signal vector x by a set of vectors $\{f_k\}$ which may be linearly dependent and, therefore, do not form a basis of the space. A main issue is the redundancy that exists among such a set of non-independent vectors. Since it is now possible to find a set of coefficients $\{d_k\}$ so that $\sum_k d_k f_k = 0$, an immediate consequence is that the signal's representation is no longer unique:

$$x = \sum_k c_k f_k = \sum_k c_k f_k + \sum_k d_k f_k = \sum_k (c_k + d_k) f_k.$$

Moreover, Parseval's theorem no longer holds. The energy contained in the coefficients may be either higher or lower than the energy in the signal. In order to address this issue when using non-independent vectors for signal representation, the concept of *frame* is introduced.

For the expansion $x = \sum_k c_k f_k$ to be a reasonable representation of x in terms of a set of coefficients $c_k = \langle x, f_k \rangle$, first we must require the following relation to hold for any vectors $x, y \in H$:

$$\langle x, f_k \rangle = \langle y, f_k \rangle \quad \text{if and only if} \quad x = y.$$

This representation is also required to be stable in a twofold manner:

- *stable representation*: if the difference between two vectors is small, then the difference between the corresponding coefficients should also be small. So, if $\|x - y\|^2 \to 0$, then $\sum_k |\langle x, f_k \rangle - \langle y, f_k \rangle|^2 \to 0$, hence

$$\sum_k \left| \langle x, f_k \rangle - \langle y, f_k \rangle \right|^2 \le B \, \|x\|^2 \, ,$$

where $0 < B < \infty$ is a positive real constant. In particular, setting $y = 0$, so that $\langle y, f_k \rangle = 0$, we have

$$\sum_k \left| \langle x, f_k \rangle \right|^2 \le B \, \|x\|^2 \, ;$$

- *stable reconstruction*: if the difference between two sets of coefficients is small, then the difference between the corresponding *reconstructed vectors* (the vectors re-assembled using the coefficients and the basis vectors) should also be small. So, if $\sum_k \left| \langle x, f_k \rangle - \langle y, f_k \rangle \right|^2 \to 0$, then $\|x - y\|^2 \to 0$, hence

$$A \, \|x - y\|^2 \le \sum_k \left| \langle x, f_k \rangle - \langle y, f_k \rangle \right|^2 \, ,$$

where $0 < A < \infty$ is a positive real constant. Again, setting $y = 0$ and therefore $\langle y, f_k \rangle = 0$, we have

$$A \, \|x\|^2 \le \sum_k \left| \langle x, f_k \rangle \right|^2 \, .$$

Combining these two constraints we can say that a family of finite or infinite vectors $\{f_k\}$ in the Hilbert space H is a frame if there exist two real constants $0 < A \le B < \infty$ such that for any $x \in H$, the following inequality holds:

$$A \, \|x\|^2 \le \sum_k \left| \langle x, f_k \rangle \right|^2 \le B \, \|x\|^2 \, .$$

Any numbers A, B for which this inequality is valid are called *frame bounds*. They are not unique. The optimal frame bounds are the largest possible value for A and the smallest possible value for B for which the inequality holds. A and B can thus be seen as the bounds within which the *normalized energy of the coefficients*, defined as

$$\frac{\sum_k \left| \langle x(t), f_k(t) \rangle \right|^2}{\|x(t)\|^2} = \frac{\sum_k \left| \langle x, f_k \rangle \right|^2}{\|x\|^2} \, ,$$

must fall. This explains the word "frame": A and B frame the normalized energy of the coefficients.

If we can choose $A = B$, then we have a *tight frame*:

$$A \, \|x\|^2 = B \, \|x\|^2 = \sum_k \left| \langle x, f_k \rangle \right|^2 \, ,$$

or

$$A \, \|x(t)\|^2 = B \, \|x(t)\|^2 = \sum_k \left| \langle x(t), f_k(t) \rangle \right|^2 \, ,$$

a relation that expresses a *generalized form of Parseval's theorem* valid *for tight frames*. This is the most general case in which a partitioning of the energy of the signal among the coefficients can be written. Finally, if a tight frame has $A = B = 1$, then it is an orthogonal basis.

For a tight frame we can write the generalized synthesis relation

$$x(t) = \frac{1}{A} \sum_k c_k f_k(t),$$

where the factor $1/A$ measures the *redundancy* of the signal's representation. The larger A, the more redundant the representation: $1/A$ is small, which means that the value of $\sum_k c_k f_k(t)$ must be strongly reduced to make it equal to the value of $x(t)$. The case $A = 1$, which identifies an orthogonal basis, thus corresponds to no redundancy: *an orthogonal basis is a non-redundant tight frame.*

If the adopted frame is not tight, nothing similar to Parseval's theorem can be written, and the energy of the signal in the coefficient domain cannot be partitioned exactly. However, the closer A and B, the easier an approximate energy partitioning, and the more accurate the way in which the signal can be approximated, i.e., reconstructed, combining coefficients and basis functions:

$$x(t) \approx \frac{2}{A + B} \sum_k c_k f_k(t).$$

For this approximate *reconstruction*, a reconstruction signal-to-noise ratio SNR_{rec} is defined, which is bounded by

$$\text{SNR}_{\text{rec}} > \frac{B/A + 1}{B/A - 1}.$$

We can thus see that for a tight frame, when $A = B$, the reconstruction is perfect. No noise is added to the signal in the reconstruction process. A tight frame behaves exactly as an orthogonal basis, even if the basis functions $f_k(\theta)$ may even lack linear independence.

If we use a frame which is not tight, we can nevertheless make the reconstruction exact, i.e., the analysis and the synthesis can take place without reconstruction noise *as if* we were using a tight frame. The process of finding the coefficients needed to combine the frame vectors f_k, in such a way that an exact reconstruction $x = \sum_k c_k f_k$ is possible, is tackled by considering it as a frame transformation, an operator that maps the vector x to a coefficient vector c. Following this approach, it is found that we must allow for a *dual set* $\tilde{f}_k$ *of frame vectors*, or *dual frame*, to appear. A vector $x \in H$ can be equivalently represented by either of the two dual frames, f_k or $\tilde{f}_k$:

$$x = \sum_k \left\langle x, \tilde{f}_k \right\rangle f_k = \sum_k \left\langle x, f_k \right\rangle \tilde{f}_k.$$

Thus, a dual set allows for exact signal reconstruction in the case of a frame. The dual set is not unique.

In the special case in which the vectors in a frame f_k actually are linearly independent, the corresponding set of functions is called a *Riesz basis*. A Riesz basis $\{f_k\}$ and its dual $\{\tilde{f}_k\}$ form a pair of biorthogonal bases—often collectively indicated as a *biorthogonal basis*[19]—satisfying

$$\langle f_k, \tilde{f}_l \rangle = \delta[k - l], \qquad k, l \in \mathbb{Z}.$$

The signal representation by a set of linearly independent and orthogonal basis vectors is now generalized, so that the signal is represented by a set of frame vectors, which are in general neither linearly independent nor orthogonal. The biorthogonal case is special because the representation behaves exactly as an expansion in an orthogonal basis (uniqueness, no redundancy), except for the presence of a dual set of basis vectors.

3.8.7 Summary and Complements

We summarize below the essential points discussed in this appendix.

- A signal can be considered as a vector x in a Hilbert space, the specific type of which depends on the nature of the signal. For example, an infinite-duration continuous-time, finite-energy signal $x(t)$ is a vector $x \in L^2(\mathbb{R})$, and its discrete samples form a vector in $\ell^2(\mathbb{Z})$. Let us call S the considered space.
- A signal vector x given in the default form, either as a time function or a sequence of time samples, can be considered as a linear combination of a set of weighted and shifted time impulses:

$$x(t) = \int x(\tau)\delta(\tau - t)\mathrm{d}\tau$$

or

$$x[n] = \sum_m x[m]\delta[m - n].$$

Here, $\delta(\tau - t)$ or $\delta[m - n]$ respectively represent the standard basis that spans the corresponding signal space. In other words, the default form of a signal $x(t)$ or $x[n]$ is actually a weighting function or a countable set of coefficients of the standard basis, which is always implicitly used in the default representation of a signal in the time domain.

---

[19]"Biorthogonal" means that this type of "orthogonality" requires two sets of vectors.

- The signal vector x can alternatively be represented by any of the bases that also span the same vector space, such as an orthogonal basis obtained by performing a transformation in the form of a rotation applied to the vectors of the standard basis. This is, for example, the case of signal representations in the frequency domain. Such a basis is composed of a set of either countable vectors u_k or uncountably infinite vectors $u(f)$. If the basis is orthogonal, we get

$$x = \sum_k c_k u_k = \sum_k \langle x, u_k \rangle u_k, \quad \text{with} \quad c_k = \langle x, u_k \rangle,$$

or

$$x = \int c(f)u(f)\mathrm{d}f = \int \langle x, u(f) \rangle u(f)\mathrm{d}f, \quad \text{with} \quad c(f) = \langle x, u(f) \rangle.$$

The equation for the coefficients $\{c[k]\}$ or weighting function $c(f)$ represents the analysis of the signal by which the signal is decomposed into a set of components $c[k]u_k$ or $c(f)u(t)$. The inverse relation, involving a summation or integration, is the synthesis of the signal, by which the signal is reconstructed from its components.

- The representations of the signal under different orthogonal bases are equivalent, in the sense that the total amount of energy or information contained in the signal, represented by its norm, is conserved by the rotation relating the two orthogonal bases before and after the transformation, due to Parseval's theorem.

- In addition to the orthogonal transforms based on a set of orthogonal basis vectors, each of which carries some independent information on the signal, we can also consider transforms based on a set of frame vectors that may be non-orthogonal and even non-independent. These frame vectors may be correlated and there may exist certain redundancy in terms of the signal information that each of them carries.

- The special case in which the frame vectors are actually independent on one another leads to a signal representation in a biorthogonal basis.

We must also remember that:

- a set $\{u_k(t)\}$ is a *basis* for the vector space S if the coefficients $\{c_k\}$ are *unique* for any particular $x(t) \in S$;
- a basis in S is *orthogonal* if, moreover, we have $\langle u_k(t), u_l(t) \rangle = 0$ for $k \neq l$;
- an orthogonal basis in S is also *orthonormal* (ON) if $\|u_k(t)\| = 1$ for any k;
- if the basis is orthogonal, we can express any element $x(t)$ in the vector space S by

$$x(t) = \sum_k \langle x(t), u_k(t) \rangle u_k(t),$$

since by taking the inner product of $u_k(t)$ with both sides of the previous equation we get

$$c_k = \langle x(t), u_k(t) \rangle.$$

The inner product of the function $x(t)$ with the basis vector $u_k(t)$ thus "picks out" the corresponding coefficient c_k;

- more generally, the basis may be *biorthogonal*: in this case, the expansion can still be written but only if we use, together with a set $\{f_k(t)\}$, a *dual basis set* $\left\{\tilde{f}_k(t)\right\}$. The elements of each sets will not be orthogonal to each other, but to the corresponding element of the other set:

$$\langle f_k(t), f_l(t)\rangle \neq 0 \quad \text{and} \quad \left\langle \tilde{f}_k(t), \tilde{f}_l(t)\right\rangle \neq 0, \quad \text{but} \quad \left\langle f_k(t), \tilde{f}_l(t)\right\rangle = \delta(k - l).$$

This allows for writing a synthesis relation in the form

$$x(t) = \sum_k \left\langle x(t), \tilde{f}_k(t)\right\rangle f_k(t).$$

The inner products $\left\langle x(t), \tilde{f}_k(t)\right\rangle$ are the coefficients in the analysis relation for $x(t)$. A biorthogonal system is more complicated than an orthogonal basis, in that it requires not only handling the expansion set, but finding, calculating, and storing a dual set of functions. On the other hand, it is very general and allows a larger class of expansions.

The space S that we are considering may be a subspace of $L^2(\mathbb{R})$, or even coincide with $L^2(\mathbb{R})$, since our ultimate goal is to find expansions for square-integrable functions. Therefore the norm of a function $x(t) \in S$ is finite.

This *expansion* or *representation* of a function, or signal, or vector $x(t)$ belonging to $L^2(\mathbb{R})$ is extremely valuable. The inner product operates on $x(t)$ to produce a set of coefficients that, when used to linearly combine the basis vectors, give back the original signal $x(t)$. It is the foundation of Parseval's theorem, which states that the energy of $x(t)$ can be partitioned among the expansion coefficients c_k. For a Fourier series, the (orthogonal) basis functions are sinusoidal, with frequencies integer multiples of a fundamental frequency ω_0. For a Taylor's series, the (non-orthogonal) basis functions are simple monomials t^k, etc.[20]

If we allow for more general synthesis relations, we can use a spanning set that is not a basis but a generalization of the concept of basis, i.e., a *frame*. A frame still allows representing a signal in a similar way as we did above; however, the coefficients are not necessarily unique. This has several advantages: the lack of

---

[20] In order to better understand what a basis is, we may mention that in finite dimensions, analysis and synthesis operations can be represented as matrix-vector multiplications. If the expansion functions (vectors) are a basis, the synthesis matrix has these basis vectors as columns; it is square and non-singular. If the basis is orthogonal, then the synthesis matrix is orthogonal, since its rows and columns are orthogonal to one another; its inverse is equal to its transpose, and the matrix multiplied by its transpose gives the identity matrix. If the synthesis matrix is not orthogonal, then the identity matrix is the synthesis matrix multiplied by its inverse, and the dual basis consists of the rows of the inverse. If the synthesis matrix is singular, then its columns (the basis vectors) are not independent, and therefore do not form a basis: the uniqueness of the coefficients is lost. This leads to the concept of frame.

uniqueness opens up the possibility of choosing the frame leading to coefficients that fit a certain application best, and it also makes the representation of a signal less sensitive to noise. Since the orthogonal-basis condition is strong, it might be difficult to find a basis satisfying the extra conditions that a certain application may require; the frame condition is weaker, and one can often find a frame enjoying special properties that are impossible for a basis. For frames we can write a relation using the square of the norm of $x(t)$ in $L^2(\mathbb{R})$:

$$A \, \|x(t)\|^2 \leq \sum_k |\langle x(t), f_k(t)\rangle|^2 \leq B \, \|x(t)\|^2 \, .$$

Note that

- a frame is an *overcomplete* version of a basis set, where the word "overcomplete" refers to the representation redundancy; it allows for exact signal reconstruction if we resort to a dual frame; a frame can be a biorthogonal basis if the frame vectors are independent of each other;
- a tight frame is an *overcomplete* version of an orthogonal basis set.

To complete the picture, we explain what is meant by an *unconditional basis*. The concept of an unconditional basis is of particular interest for wavelet expansions. It was used by Donoho 1993 and other authors to give an explanation of which wavelet basis systems are best for a particular class of signals and why wavelet basis systems can be very good for a wide variety of signal classes and processing tasks.

Broadly speaking, the definition of an unconditional basis given by Donoho 1993 is the following. Let us consider a function class $\mathscr{F}$ with a norm defined in some way, and a basis set $f_k(t)$ such that every function $x(t) \in \mathscr{F}$ has a unique representation of the form $x(t) = \sum_k \alpha_k f_k(t)$. Let us then consider the infinite expansion $x(t) = \sum_k m_k \alpha_k f_k(t)$. If for all $x(t) \in \mathscr{F}$, the infinite sum converges for all $|m_k| \leq 1$, the basis is called an unconditional basis. If the convergence depends on the condition $m_k = 1$ for some $x(t)$, then the basis is called a conditional basis.

For an unconditional basis, it may be shown that convergence of the infinite expansion depends neither on the order of the terms in the summation, nor on the sign of the coefficients. This means a very robust basis where the coefficients drop off rapidly for all members of the function class $\mathscr{F}$. That is indeed the case for wavelets, which are unconditional bases for a very wide set of function classes (Daubechies 1992).

The fundamental idea of bases or frames is representing a continuous function by a sequence of expansion coefficients. For orthogonal bases and tight frames, the respective forms of the Parseval theorem relate the L^2-norm of the expanded function to the ℓ^2-norm of coefficients. Different function spaces are characterized by different norms of the continuous function. If we have an unconditional basis for the considered function space, not only can the norm of the function in the space be related to some norm of the coefficients in the expansion, but also the *absolute values of the coefficients* contain sufficient information to establish the relation. So there is no condition on the sign information of the expansion coefficients (the phase if they are complex), if we only care about the norm of the function: the basis is thus named "unconditional".

References

Bochner, S., Chandrasekharan, K.: Fourier Transforms. Princeton University Press, Princeton (1949)
Bracewell, R.N.: The Fourier Transform and Its Applications. McGraw-Hill, Boston (2000)
Churchill, R.V.: Complex Variables and Applications. McGraw-Hill, New York (1975)
Churchill, R.V., Brown, J.W.: Introduction to Complex Variables and Applications. McGraw-Hill, New York (1984)
Cooley, J.W., Tukey, J.W.: An Algorithm for the Machine Calculation of Complex Fourier Series. Math. Comput. **19**, 297–301 (1965)
Daubechies, I.: Ten lectures on wavelets. In: CBMS-NSF Regional Conference Series in Applied Mathematics, Society for Industrial and Applied Mathematics (SIAM), Philadelphia, PA, USA (1992)
Diniz, P.S.R., da Silva, E.A.B., Netto, S.L.: Digital Signal Processing—System Analysis and Design. Cambridge University Press, Cambridge (2002)
Donoho, D.L.: Unconditional bases are optimal bases for data compression and for statistical estimation. Appl. Comput. Harmonic Anal. **1**(1), 100–115 (1993)
Duhamel, P., Piron, B., Etcheto, J.M.: On Computing the Inverse DFT. IEEE Trans. Acoust. Speech Signal Process. **36**(2), 285–286 (1988)
Kanasewich, E.R.: Time Sequence Analysis in Geophysics. Prentice-Hall, Englewood Cliffs (1981)
Lighthill, M.J.: Introduction to Fourier Analysis and Generalized Functions. Cambridge University Press, Cambridge (1958)
Oppenheim, A.V., Schafer, R.W.: Discrete-time Signal Processing. Prentice Hall, Englewood Cliffs (2009)
Parseval des Chênes, M.-A.: Mémoire sur les séries et sur l'intégration complète d'une équation aux différences partielles linéaire du second ordre, à coefficients constants. Mémoires présentées à lInstitut des Sciences, Lettres et Arts, par divers savants, et lus dans ses assemblées. Sciences, mathématiques et physiques (Savants étrangers) **1**, 638–648 (1806)
Plancherel, M.: Contribution a l'etude de la representation d'une fonction arbitraire par les integrales définies. Rendiconti del Circolo Matematico di Palermo **30**, 298–335 (1910)
Powell, M.J.D.: Approximation Theory and Methods. Cambridge University Press, Cambridge (1981)
Ragazzini, J.R., Zadeh, L.A.: The Analysis of Sampled-Data Systems. Trans. Am. Inst. Electron. Eng. **71**(II), 225–234 (1952)
Rayleigh, J.W.S.: On the character of the complete radiation at a given temperature. Philos. Mag. **27**, 460–469 (1889)
Wang, R.: Introduction to Orthogonal Transforms. Cambridge University Press, Cambridge (2013)

Chapter 4
Sampling of Continuous-Time Signals

4.1 Chapter Summary

This chapter deals with the proper choice of the sampling interval T_s that leads to a correct representation of an analog signal by an infinite set of samples extracted periodically from it. Here "correct representation" means "representation allowing for the reconstruction of the continuous-time signal from its discrete-time version". A theorem known as the sampling theorem prescribes a lower limit for T_s, depending on the upper bandlimit of the spectrum of the analog signal. The theorem also highlights the fact that a representative sampling of a continuous-time signal is possible if, and only if, the signal does not contain frequencies higher than the Nyquist frequency $1/(2T_s)$: in other words, no finite-rate sampling can capture the variations of a continuous-time signal which is not bandlimited, i.e., that contains periodic components of all frequencies, up to infinity.

Other issues related to analog signals, such as the signal's concentrations in the time and frequency domains and their mutual inverse dependence (uncertainty principle), as well as the definition of bounded and compact support in both domains, are also discussed. These topics will be especially useful when dealing with the wavelet transform (Chaps. 13 and 14). The Appendix to this chapter finally provides a summary of the relations among the variables used to express the concept of frequency in the continuous-time and discrete-time cases.

4.2 Sampling Theorem

In this chapter we deal with discrete-time signals resulting from periodic sampling of analog signals. When we decide to sample a continuous-time signal $x(t)$ at regular time intervals T_s (*periodic sampling*), so as to get a discrete-time form $x(nT_s) = x[n]$ of the signal, we must select T_s in a way that the sampled signal represents correctly the original one: in fact,

© Springer International Publishing Switzerland 2016
S.M. Alessio, *Digital Signal Processing and Spectral Analysis for Scientists*, Signals and Communication Technology,
DOI 10.1007/978-3-319-25468-5_4

Fig. 4.1 The sampling of an analog signal viewed as the amplitude modulation of a periodic impulse train. **a** Analog signal; **b** periodic impulse train; **c** sampled signal as a train of modulated impulses

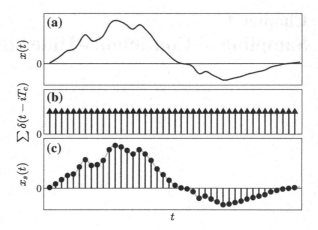

- if T_s is too large, loss of information about the signal details occurs;
- if T_s is too small, a large number of data is stored and processed beyond necessity.

T_s must be selected so as to allow for the reconstruction of $x(t)$ from $x[n]$ with the desired accuracy, but it must not be redundant. The optimal sampling interval can be determined studying the relation between the Fourier transform of the analog signal, that is, the spectrum of $x(t)$, and the DTFT of the sampled signal.

The sampling process may be viewed as an amplitude modulation, in which a *periodic train of impulses*

$$\sum_{i=-\infty}^{+\infty} \delta(t - iT_s)$$

is transformed into a train of modulated impulses, as shown in Fig. 4.1.

We must note here that the sampled signal has been indicated in Fig. 4.1 as $x_s(t)$, rather than as $x[n]$: the following discussion becomes simpler if we use the language and the notation of continuous-time signals for the result of the sampling process. In this notation, each impulse is a Dirac δ, and the individual shifted impulse $\delta(t - iT_s)$ has a unit subtended area concentrated around $t = iT_s$. Of course, $x_s(t) \equiv x(nT_s)$. The modulated impulse train can be written as

$$x_s(t) = x(t) \sum_{i=-\infty}^{+\infty} \delta(t - iT_s).$$

We now turn from the time domain to the frequency domain, using continuous-time Fourier transforms. The time-domain product of $x(t)$ with the impulse train becomes a continuous convolution, i.e., a convolution integral, in the frequency domain. Let us denote by $\Delta(f)$ the *spectrum of the periodic impulse-train,*

Fig. 4.2 The spectrum of
the time-domain impulse
train is a frequency-domain
impulse train

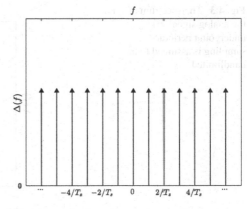

$$\Delta(f) = \frac{1}{T_s} \sum_{i=-\infty}^{+\infty} \delta\left(f - \frac{i}{T_s}\right).$$

The spectrum of the time-domain impulse train is a frequency-domain impulse train
(Fig. 4.2), in which the individual impulses are separated by $1/T_s$. Then we have

$$X_s(f) = X(f) * \Delta(f) = \int_{-\infty}^{+\infty} X(f - f')\Delta(f')\mathrm{d}f' =$$

$$= \int_{-\infty}^{+\infty} X(f - f')\frac{1}{T_s} \sum_{i=-\infty}^{+\infty} \delta\left(f' - \frac{i}{T_s}\right)\mathrm{d}f' = \frac{1}{T_s} \sum_{i=-\infty}^{+\infty} X\left(f - \frac{i}{T_s}\right).$$

The last step is justified by the fact that the sum of infinite terms defining $\Delta(f)$
converges in such a way that integration and summation can be interchanged. In
other words, this sum can be integrated term by term.

The sampling process transforms the aperiodic spectrum of the analog signal
$x(t)$ into a periodic spectrum: the spectrum $X(f)$ of $x(t)$, which we assume to be
bandlimited, i.e., such that $X(f) = 0$ for $|f|$ greater than a certain frequency B_f
(Fig. 4.3), is repeated an infinite number of times on the frequency axis, at regular
intervals $1/T_s$, thus becoming periodic in frequency (Fig. 4.4).

Even if formally $X_s(f)$ is the Fourier transform of the continuous-time signal
$x_s(t)$, it coincides with the DTFT of $x[n]$, i.e., $X(\mathrm{e}^{j\omega})$, that is, the DTFT of the
sampled signal viewed as a sequence.[1] The only difference between $X_s(f)$ and
$X(\mathrm{e}^{j\omega})$ lies in the fact that $X(\mathrm{e}^{j\omega})$ is a function of ω, while $X_s(f)$ is a function of f.

---

[1] This statement can be understood as follows:

$$X_s(f) = \int_{-\infty}^{+\infty} x_s(t)\mathrm{e}^{-j2\pi ft}\mathrm{d}t = \int_{-\infty}^{+\infty} \left[x(t) \sum_{n=-\infty}^{+\infty} \delta(t - nT_s)\right]\mathrm{e}^{-j2\pi ft}\mathrm{d}t =$$

$$= \sum_{n=-\infty}^{+\infty} \int_{-\infty}^{+\infty} x(t)\delta(t - nT_s)\mathrm{e}^{-j2\pi ft}\mathrm{d}t = \sum_{n=-\infty}^{+\infty} x(nT_s)\mathrm{e}^{-j2\pi f(nT_s)} = \sum_{n=-\infty}^{+\infty} x[n]\mathrm{e}^{-j\omega n} \equiv X(\mathrm{e}^{j\omega}).$$

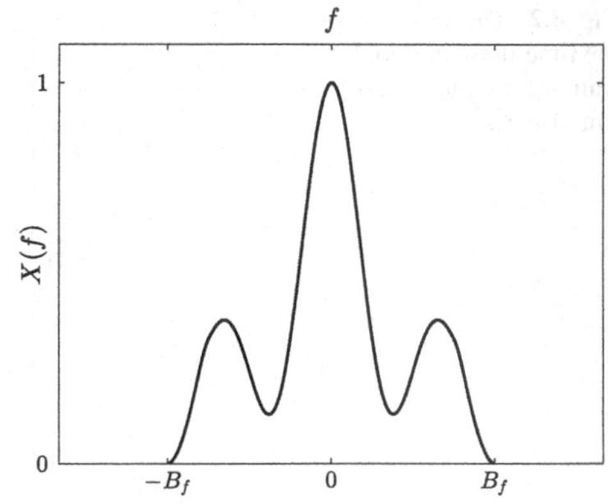

Fig. 4.3 The spectrum of
the analog signal $x(t)$
undergoing periodic
sampling is assumed to be
bandlimited

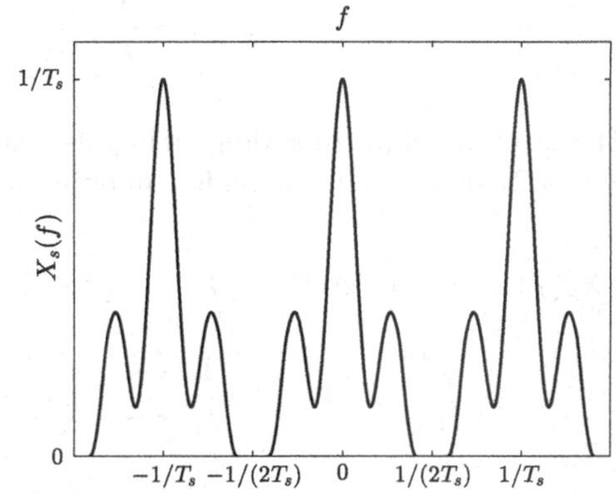

Fig. 4.4 The spectrum of the
sampled signal is periodic

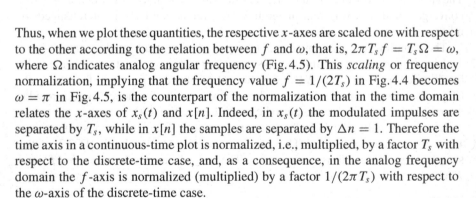

Thus, when we plot these quantities, the respective x-axes are scaled one with respect to the other according to the relation between f and ω, that is, $2\pi T_s f = T_s \Omega = \omega$, where Ω indicates analog angular frequency (Fig. 4.5). This *scaling* or frequency normalization, implying that the frequency value $f = 1/(2T_s)$ in Fig. 4.4 becomes $\omega = \pi$ in Fig. 4.5, is the counterpart of the normalization that in the time domain relates the x-axes of $x_s(t)$ and $x[n]$. Indeed, in $x_s(t)$ the modulated impulses are separated by T_s, while in $x[n]$ the samples are separated by $\Delta n = 1$. Therefore the time axis in a continuous-time plot is normalized, i.e., multiplied, by a factor T_s with respect to the discrete-time case, and, as a consequence, in the analog frequency domain the f-axis is normalized (multiplied) by a factor $1/(2\pi T_s)$ with respect to the ω-axis of the discrete-time case.

For the sampled signal to correctly represent the original signal, the shifted copies of $X(f)$ in Fig. 4.4 must not overlap: the situation depicted in Fig. 4.6 must be avoided, because in that case, in the overlapping interval, the spectrum of the sampled signal is the sum of the overlapping tails (see the *dashed curves* in Fig. 4.6) and is therefore corrupted with respect to the shape of the "true" spectrum. A situation like

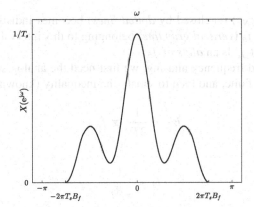

Fig. 4.5 The spectrum of the sampled signal represented as the DTFT of $x[n]$

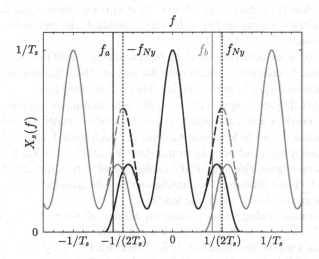

Fig. 4.6 A case in which in $X_s(f)$ the shifted copies of $X(f)$ (*solid curves*) partially overlap: the overlapping produces the spectrum drawn as a *dashed curve*. On the abscissa, $1/(2T_s)$ represents the Nyquist frequency f_{Ny} (see text). *Vertical dotted lines* indicate the interval $[-f_{Ny}, +f_{Ny}]$; *vertical black and gray lines* respectively indicate a frequency f_a falling *outside* the range $[-f_{Ny}, +f_{Ny}]$, and a frequency f_b falling *inside* the same range. The frequency f_a is an alias of f_b

the one illustrated in Fig. 4.6 is referred to as *aliasing in the frequency domain*, or simply *frequency aliasing*.

If we define the *Nyquist frequency* (Grenader 1959) as

$$f_{Ny} = \frac{1}{2T_s},$$

then, observing Fig. 4.6, we can state that, in case of frequency aliasing, a given frequency f_a (*vertical black line*) of the "true" spectrum that is located outside the interval $[-f_{Ny}, +f_{Ny})$ (enclosed by *dotted lines*) becomes indistinguishable from another frequency f_b (*vertical gray line*) belonging to this interval. We can express this fact saying that f_a is an *alias* of f_b.

In order to avoid frequency aliasing, we first need the analog signal's bandlimit B_f (Fig. 4.3) to be finite, and then to satisfy the inequality (known as the *Nyquist's criterion*)

$$B_f < \frac{1}{2T_s} \equiv f_{Ny},$$

that is,

$$T_s < \frac{1}{2B_f}.$$

This is the conceptual content of the *sampling theorem* (Nyquist 1928; Shannon 1948, 1949)[2] that is enunciated as follows: "A continuous-time signal that does not contain frequency components $|f| > B_f$ can, in principle, be reconstructed from its sampled version, if $T_s < 1/(2B_f)$."

Note that so far we have discussed the representativeness of the sampling, which is a prerequisite to reconstruction, but not the way in which the original signal can actually be reconstructed from its samples. This will be done later.

The Nyquist frequency appears to be half of the minimum *sampling frequency* $f_s = 1/T_s$ needed to avoid frequency aliasing. In other words, in order to avoid aliasing we must record at least two samples per each period $T_{min} = 1/B_f$ of the highest-frequency sinusoid present in the data: $T_{min}/T_s > 2$. Clearly, frequency-domain aliasing is avoidable only if the analog signal is bandlimited. This is analogous to the fact that time-domain aliasing can only be avoided if the sequence of which we sample the DTFT has finite length.

If a bandlimited analog signal is sampled without aliasing, the shape of $X_s(f)$ in the range $-f_{Ny} \leq f < f_{Ny}$ will be identical to the shape of $X(f)$, except that the ordinate axis will be multiplied by $1/T_s$. Therefore, $[-f_{Ny}, f_{Ny})$ is the *principal interval* expressed in terms of analog frequency, i.e., the frequency range that contains all information in the sense that it remains "visible" when we sample $x(t)$ at multiples of T_s.

Before proceeding, let us examine explicitly the case of a *border-line* sinusoid, that is, a signal given by

$$x(t) = \cos(2\pi f_0 t - \phi_0)$$

that is sampled with a time step $T_s = 1/(2f_0)$. Is aliasing present?

- If $\phi_0 = 0$, then $x[n] = \cos(\pi n) = (-1)^n$: aliasing seems to be absent, but
- if $\phi_0 = \pi/2$, then $x[n] = \sin(\pi n) = 0$ and therefore there is aliasing.

---

[2]The sampling theorem actually was originally demonstrated neither by Nyquist, nor by Shannon. The original proof of the sampling theorem is due to Cauchy (1841), even if Cauchy's paper does not explicitly contain the statement of the sampling theorem. Shannon himself recognized it (Shannon 1949): "If a function $f(t)$ contains no frequencies higher than W cycles per second, it is completely determined by giving its ordinates at a series of points spaced $1/2W$ seconds apart. This is a fact which is common knowledge in the communication art."

We can thus state that sampling at a rate $1/T_s = 2f_0$ is forbidden. Indeed, since the spectrum of $x(t)$ contains a Dirac δ centered at $f = \pm f_0$, sampling with $T_s = 1/2f_0$ violates the sampling theorem.

4.3 Reconstruction of a Continuous-Time Signal from Its Samples

Let us assume that we sampled our analog signal using the maximum sampling step T_s that meets Nyquist's criterion. To reconstruct the continuous-time signal $x(t)$ from its sampled version $x_s(t)$, an *analog reconstruction filter* must be applied to $x_s(t)$, able to transmit equally, without any attenuation, all the components of $X_s(f)$ in $[-1/\{2T_s\}, 1/\{2T_s\})$, and to completely discard the other components. We can imagine a frequency-selective analog system as the one shown in Fig. 4.7a, with an ideal frequency response implying no delay between output and input. The required value in the passband is T_s, to compensate for the factor $1/T_s$ appearing in the expression of $X_s(f)$.

The impulse response of such a filter is

$$h(t) = \int_{-\infty}^{+\infty} H(f)e^{j2\pi ft}df = \int_{-\frac{1}{2T_s}}^{+\frac{1}{2T_s}} T_s e^{j2\pi ft}df =$$

$$= T_s \frac{1}{2\pi jt}\left(e^{j2\pi \frac{1}{2T_s}t} - e^{-j2\pi \frac{1}{2T_s}}\right) = T_s \frac{1}{2\pi jt}2j\sin\left(\pi\frac{t}{T_s}\right),$$

i.e.,

$$h(t) = \text{Sinc}\left(\frac{t}{T_s}\right) = \frac{\sin(\pi\frac{t}{T_s})}{\pi\frac{t}{T_s}}.$$

The *cardinal sine function* $\text{Sinc}\,(t/T_s)$ crosses the time axis zero every T_s time units. The impulse response $h(t)$ is shown in Fig. 4.8.

Fig. 4.7 a The real frequency response of the ideal analog reconstruction filter; **b** amplitude of the complex frequency response of a realizable analog reconstruction filter

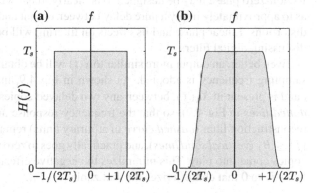

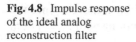

Fig. 4.8 Impulse response
of the ideal analog
reconstruction filter

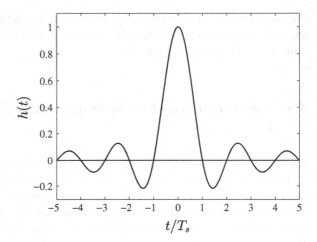

The reconstructed signal is given by a convolution integral:

$$x(t) = x_s(t) * h(t) = \int_{-\infty}^{+\infty} x_s(t')h(t-t')\mathrm{d}t' = \int_{-\infty}^{+\infty} x_s(t')\frac{\sin\left(\pi\frac{t-t'}{T_s}\right)}{\pi\frac{t-t'}{T_s}}\mathrm{d}t' =$$

$$= \int_{-\infty}^{+\infty} x_s(t')\mathrm{Sinc}\left(\frac{t-t'}{T_s}\right)\mathrm{d}t'.$$

From this equation we see that each non-zero value of the sampled signal produces in output to the reconstruction filter a weighted version of $h(t)$, i.e., of the Sinc($\cdot$) function. This means that the reconstructed $x(t)$ value at $t = nT_s$ is not influenced by adjacent values, since $h(t)$ crosses zero at any multiple of T_s. In other words, since we must have $x(nT_s) = x_s(nT_s)$, the reconstructed waveform must go through such points, and it would not do so if the contributions of adjacent samples were not zero.

The ideal analog filter we described above is not computationally realizable, but it can be approximated by an analog filter having the frequency response qualitatively shown in Fig. 4.7b. This frequency response will be discontinuity-free and complex. Its non-zero phase may be designed to be nearly linear with respect to frequency, so as to approximately imply a pure delay between output and input, without significant distortions. Linear phase and its effects on filtering will be treated in Chap. 6, when discussing digital filters.

Even better, an output more similar to $x(t)$ will be obtained if a slightly redundant sampling frequency is adopted. As shown in Fig. 4.9, in this case an empty safe-band is present in $X_s(f)$, between any two delayed copies of $X(f)$ (see the *vertical dotted lines* in Fig. 4.9), so that the frequency-response amplitude of the realizable reconstruction filter (*dashed curve* in arbitrary units) remains close to T_s in the band $|f| \le B_f$ (*vertical solid lines*), and practically goes to zero where the delayed spectral copies come into play. This minimizes the negative effects of the gradual transition from T_s to 0 that characterizes the realizable $|H(f)|$.

Fig. 4.9 Safe reconstruction of the analog signal by a realizable filter (see text). The amplitude of the frequency response is drawn as a *dashed curve* in arbitrary units. The *vertical dotted lines* identify two empty symmetric safe-bands bordering the spectral band $|f| \leq B_f$ (see text), which is indicated by *vertical solid lines*

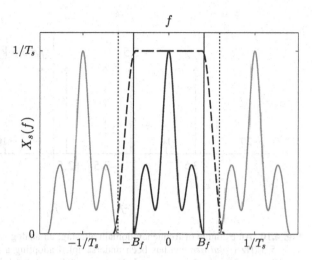

4.4 Aliasing in the Frequency Domain and Anti-Aliasing Filter

If $x(t)$ is not bandlimited, that is, if its spectrum does not go to zero beyond some finite frequency value B_f, frequency aliasing cannot be avoided. Any time a band-unlimited signal is sampled and then reconstructed, frequency aliasing occurs, simply because the frequency content of a band-unlimited signal cannot be captured by a sampling performed at a finite rate. Considering that analog finite-duration signals are known to be band-unlimited (see Sect. 4.6), at first sight the issue appears insurmountable. In practice, however, the spectrum must decrease rapidly enough for $|f| \to \infty$, so that most of the signal's energy falls within a certain finite B_f.

   Thus we can avoid aliasing only if:

- we know the natural bandlimit of the analog signal, at least in the wider sense explained above, and we correctly choose T_s according to Nyquist's criterion;
- we *impose* a band limit, by lowpass filtering $x(t)$ before sampling—provided that this is experimentally possible—and then choosing T_s accordingly. The frequency-truncation error is usually smaller than the error caused by aliasing. The *analog anti-aliasing filter* is equal to the reconstruction filter described above. In practical applications, a security factor will be applied to T_s: the sampling interval will be chosen at least $\approx 10\%$ smaller than that required by Nyquist's criterion.[3]

To understand if a sampled signal has been recorded with the proper sampling interval, we can look at its spectrum, extending over $[-f_{Ny}, +f_{Ny}]$: does the spectrum decrease and tend to zero for $|f|$ approaching f_{Ny}? If not, probably some components with frequency $|f| > f_{Ny}$ have been "reflected" onto the principal interval, as in Fig. 4.6. This is very common: noise often flattens the spectrum near f_{Ny}.

---

[3]Of course the precise amount of advisable oversampling is dictated by the shape of the designed frequency response of the realizable analog filter.

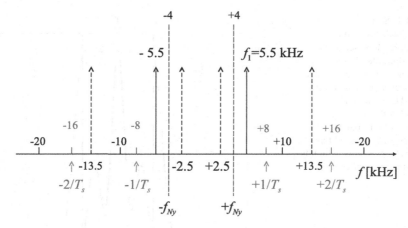

Fig. 4.10 An example of frequency-domain aliasing: an analog sinusoidal signal with frequency $f_1 = 5.5$ kHz (*solid arrows*) has been undersampled adopting a sampling rate of 8 kHz, corresponding to a Nyquist frequency of 4 kHz (*vertical dashed lines*). Frequency aliasing produces an observable spectrum with two spurious spectral lines falling inside the principal interval (*dashed arrows*)

We will now give an example of an undersampled signal. The effects of undersampling are clear if we refer to a sinusoidal signal, i.e., a single frequency component that produces a pair of spectral lines, symmetrically located at the two sides of the frequency origin. In Fig. 4.10, the two *solid arrows* describe the "true" spectrum of an analog sinusoid $x(t) = \sin(2\pi f_1 t)$ with $f_1 = 5.5$ kHz. If this signal is undersampled, adopting a sampling interval such that $f_s = 1/T_s = 8$ kHz, the Nyquist frequency assumes the value $f_{Ny} = 4$ kHz (*vertical dashed lines*). In this way, $f_1 > f_{Ny}$: f_1 is external to the principal interval and cannot be detected.

Due to frequency aliasing, the signal appears instead as a pair of "fake" spectral lines at frequencies $\pm f_0$ located inside the principal interval, as shown by the *dashed arrows* in Fig. 4.10. These arrows are generated by the first right-hand side spectral copy on the positive frequency half-axis, and by the first left-hand side spectral copy on the negative frequency half-axis. So, when a sinusoid is undersampled, the two "true" spectral lines that would represent it are eliminated from the visible spectrum, and two spurious lines located inside the principal interval are generated by spectral periodization.

The frequency f_0 is equal to $(1/T_s) - f_1 = 8 - 5.5 = 2.5$ kHz. In other words, $f_1 = (1/T_s) - f_0$ is an alias of the frequency f_0 belonging to the principal interval. The first spectral copy on the positive half-axis does not only give a line at $f_0 = 2.5$ kHz, but also a second line at $(1/T_s) + f_1 = 8 + 5.5$ kHz $= 13.5$ kHz $= (2/T_s) - f_0$. The first copy on the negative half-axis does not only give a line at $-f_0 = -2.5$ kHz, but also a second line at -13.5 kHz (see the *dashed arrows* in Fig. 4.10 that fall outside the principal interval). Thus also $(2/T_s) - f_0$ is an alias of f_0.

Actually, all the infinite spectral copies on the two half-axes—and not only the first two—generate aliases of f_0. Focusing on the positive frequency half-axis, the frequencies that are alias of a frequency f_0 belonging to the principal interval can be expressed as

$$f_m = \frac{m}{T_s} \pm f_0 = m f_s \pm f_0 = 2m f_{Ny} \pm f_0,$$

where m is an integer in the range $m = [1, \infty)$. Therefore an infinite number of continuous-time sinusoids correspond to the same discrete-time signal.

Let us check the formula for f_m with a numerical experiment. Suppose we set $f_s = 8\,\text{kHz}$ and generate five sinusoidal sequences, one with frequency $f_0 = 2.5\,\text{kHz}$ contained in $[0, f_{Ny}]$, and the remaining four *external* to the principal interval: $f_1 = \frac{1}{T_s} - f_0 = 5.5\,\text{kHz}$, $f_2 = \frac{1}{T_s} + f_0 = 10.5\,\text{kHz}$, $f_3 = \frac{2}{T_s} - f_0 = 13.5\,\text{kHz}$, and $f_4 = \frac{2}{T_s} + f_0 = 18.5\,\text{kHz}$. Figure 4.11a shows that if we compute and plot the amplitude spectrum, all sinusoids appear superimposed to one another, so that we see a single peak at 2.5 kHz. Now suppose we keep the sinusoids' frequencies unaltered but increase the sampling rate to 80 kHz, and plot the spectra again. No sinusoid is undersampled any longer, and the spectra exhibit peaks that are distinct and correctly positioned (Fig. 4.11b). To prove that the effects of aliasing are evident also in the

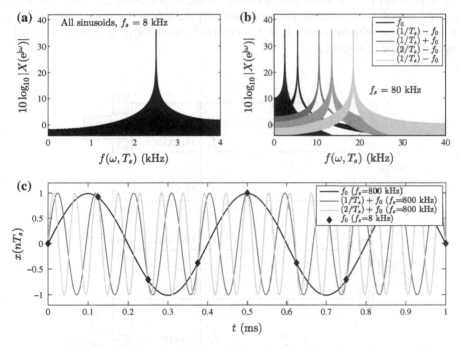

Fig. 4.11 Frequency aliasing visualized in both frequency and time domains. **a** Amplitude spectra of all the sinusoids listed in the text, as they appear when the sampling frequency is 8 kHz, i.e., superimposed to one another; **b** amplitude spectra of the same sinusoids, as they appear when the sampling frequency is 80 kHz, i.e., correctly separated; **c** the sinusoids viewed in the time domain, as they appear when the sampling frequency is 800 kHz

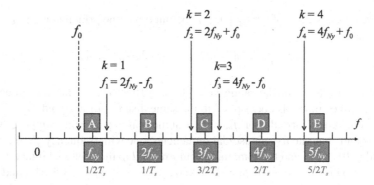

Fig. 4.12 The positive frequency half-axis with the positions of the frequencies f_m, with $m = [1, \infty)$, that are alias of a given frequency f_0 belonging to the principal interval. The points A, B, C, D mark integer multiples of the Nyquist frequency

time domain, let us finally increase the sampling rate to 800 kHz, so as to be able to plot the sinusoids in a smooth way. If we plot the five signals and superimpose to them the samples of the 2.5 kHz sinusoid generated with $f_s = 8$ kHz (Fig. 4.11c) we understand that the latter samples (*black diamonds*) actually fall on all the five curves, though they evidently represent correctly the behavior of only one of them, namely the one with frequency $f_0 = 2.5$ kHz. The formula for f_m is further illustrated by the scheme of Fig. 4.12.

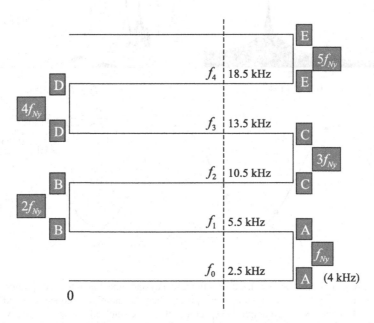

Fig. 4.13 The positive frequency half-axis folded up at any multiple of f_{Ny}: all the frequencies f_m, with $m = [1, \infty)$, that are aliases of a given f_0 line up vertically above f_0. The numerical frequency values refer to the example of Fig. 4.10

Aliasing is also referred to as *spectral folding*, because it corresponds to a folding-up of the f-axis, as sketched in Fig. 4.13: if the axis of Fig. 4.12 is folded at the points A, B, C, D, etc., all the aliases of f_0 line up above f_0.

4.5 The Uncertainty Principle for the Analog Fourier Transform

In our discussion on sampling we mentioned the fact that all analog signals with finite duration are known to be band-unlimited. This statement deserves some more attention. A brief discussion on analog signals that illustrates in a more detailed way this and other related concepts is inserted here. We will also introduce a few definitions that will be useful when dealing with the wavelet transform (Chap. 13): in fact, wavelets are treated using the continuous-time formalism.

We start from the generic case of an analog real signal $x(t)$ with infinite duration but with finite energy, having a Fourier transform (CTFT) $X(f)$. Finite energy implies that although the duration is infinite, the signal energy is concentrated in time to a lesser or greater extent, i.e., its variations around the mean value cease to be significant sooner or later, as time approaches infinity in either direction. Let us further assume that the signal is centered around the origin in the frequency domain, a condition that can always be satisfied by proper modulation. In order to quantify the signal's effective extensions in the dual domains of time and frequency and to describe the amount of *concentration* that characterizes the signal in each domain, the following parameters are introduced:

1. *center*:

$$t^* = \frac{\int_{-\infty}^{+\infty} t x^2(t) \mathrm{d}t}{\int_{-\infty}^{+\infty} x^2(t) \mathrm{d}t},$$

   where the denominator integral represents the energy of $x(t)$. If the signal were complex, in the formula we would have a squared modulus;
2. *root mean square (RMS) duration* or *RMS length* or *radius*:

$$\Delta_t = \sqrt{\frac{\int_{-\infty}^{+\infty} (t - t^*)^2 x^2(t) \mathrm{d}t}{\int_{-\infty}^{+\infty} x^2(t) \mathrm{d}t}};$$

3. *RMS bandwidth*:

$$\Delta_f = \sqrt{\frac{\int_{-\infty}^{+\infty} f^2 |X(f)|^2 \mathrm{d}f}{\int_{-\infty}^{+\infty} |X(f)|^2 \mathrm{d}f}},$$

   where the denominator integral represents the energy of the signal in the frequency domain, which for Parseval's theorem for the CTFT equals the energy in the time domain.

Fig. 4.14 **a** Two Gaussian
signals with different widths
at half height, **c** a sinusoidal
signal, **e** an impulse, and the
corresponding spectra (**b**, **d**,
and **f**, respectively)

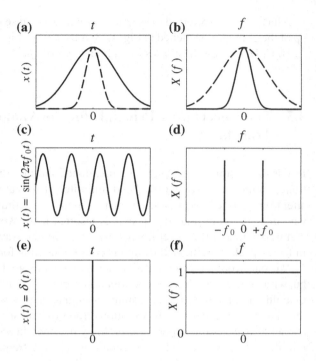

The RMS length and bandwidth are not independent of one another. It can
be demonstrated that for finite-energy analog signals $x(t)$ allowing for a Fourier
representation $X(f)$, a form of *uncertainty principle* holds[4]: $\Delta_t \Delta_f$ is always greater
than, or equal to, a fixed numerical constant. The precise value of this constant is not
important here; what is important is the meaning of this statement, i.e., that fact that
a finite-energy analog signal cannot be localized both in time and frequency. The
equal sign holds only if $x(t)$ is a Gaussian function of time, i.e., if it is proportional
to $e^{-t^2/(2T^2)}$, where the parameter T determines the width of the Gaussian bell at
half maximum, and therefore the related Δ_t. The Fourier transform of a Gaussian
signal is still a Gaussian signal, with a parameter F related to the analogous time-
domain parameter T in such a way that the product FT is a constant; therefore Δ_f
is inversely proportional to Δ_t, and their product is a constant for a Gaussian signal
(see, e.g., Vetterli and Kovačević 1995).[5]

The uncertainty principle implies that an analog signal with infinite length but
finite energy

---

[4]We should call this a theorem and not a principle, since it can be demonstrated. However, this is
the name by which this result is usually referred to.

[5]The proof of the uncertainty principle for Fourier transforms actually assumes that the signal
vanishes faster that $1/\sqrt{t}$ as $t \to \pm\infty$, so that $\lim_{t\to\pm\infty} t x^2(t)$ is zero. In mathematical literature,
this constraint is often expressed by saying that the uncertainty principle holds for Schwartz functions
on the real line—a class of functions that can be thought of as smooth functions that decay rapidly
towards infinity.

- has a large spectral bandwidth if it is very concentrated in time;
- is very dispersed in time if its spectral bandwidth is narrow.

In the limit of infinite energy, a pure sinusoid would have an infinite radius and an impulse would have an infinite bandwidth and a vanishing radius. To illustrate these concepts, Fig 4.14 shows two Gaussian signals with different widths at half maximum, a sinusoid and an impulse, together with the corresponding spectra.

4.6 Support of a Continuous-Time Signal in the Time and Frequency Domains

In mathematics, the *support* of a function is the set of points where the function is not zero-valued. For simplicity we will focus on a function $f(t)$ of a real variable t, for which the support is an interval on the real line. For our purposes we need to know that the support, i.e., the interval of t outside which $f(t)$ is identically zero, can be

- *bounded*, e.g., (I, J), where I and J are finite real numbers;
- *unbounded*, e.g., $(-\infty, J]$;
- *open*, e.g., (I, J);
- *closed*, e.g., $[I, J]$;
- *compact, i.e., bounded and closed.*

Thus, in the previous examples,

- (I, J) is not compact because it is not closed,
- $(-\infty, J]$ is not compact because it is unbounded,
- $[I, J]$ is compact because it is closed and bounded.

We will now discuss the case of an analog signal with *bounded time support*. The spectrum of such a signal can be shown to have unlimited bandwidth. On the other hand, any analog signal that has bounded support in frequency, i.e., a *limited bandwidth*, cannot have bounded time support.

In order to see this, we start from an analog signal $x_\infty(t)$ with unbounded time support, and we observe it through an *analog symmetric rectangular window* $w_T(t)$ of width T that frames a segment of the signal, spanning the interval from $-T/2$ to $+T/2$ (Fig. 4.15a). The signal we can "see" through the window is $x(t) = x_\infty(t)w_T(t)$, and has the same bounded support as the window itself.

The continuous-time Fourier transform (CTFT) of $x(t)$ is

$$X(f) = X_\infty(f) * W_T(f) = X_\infty(f) * \frac{\sin(\pi f T)}{\pi f} \equiv X_\infty(f) * T\operatorname{Sinc}(fT),$$

where $W_T(f)$, shown in Fig. 4.15b, is the spectrum (CTFT[6]) of the window $w_T(t)$, and the symbol $*$ indicates a convolution in the analog frequency domain, that is, a

---

[6]Since the analog window $w_T(t)$ is non-causal and its support is symmetrical around the origin of the time axis, this transform is real.

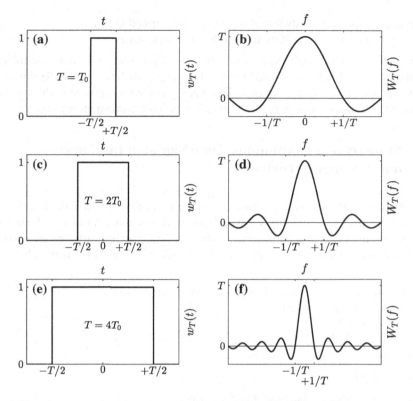

Fig. 4.15 **a** A non-causal analog rectangular window with width T_0, and **b** its spectrum. **c–d** The same for a window width of $2T_0$; **e–f** the same for a window width of $4T_0$

convolution integral. Given that $\text{Sinc}(\cdot)$ has unbounded support, $X(f)$ has unbounded support too.

From the expression of $X(f)$ we understand that the spectrum of $x(t)$ depends not only—as it is obvious—on $X_\infty(f)$, but also on the spectrum of the window. In turn, $W_T(f)$ varies according to the window width, as shown in Fig. 4.15.

The spectrum of the rectangular window has a central main lobe and lateral lobes that become lower and lower as the distance from the main lobe increases. As the window width increases, passing from T_0 to $2T_0$ and then to $4T_0$ in Fig. 4.15a, c, e, the support of the window increases, so that the window function becomes less concentrated in the time domain. At the same time, its spectrum becomes more concentrated in the frequency domain (Fig. 4.15b, d, f), even if its frequency support remains unbounded. For example, if the support of w_T doubles, the width of the main lobe of W_T halves, according to the scaling property of the CTFT:

$$x(kt) \iff \frac{1}{|k|} X(f/k).$$

At this point, the difference between the concept of bounded support and the concept of concentration should be clear. The two notions are very distinct, and the uncertainty principle deals exclusively with concentration:

- "bounded support" means that a signal is exactly zero outside a certain time or frequency interval;
- "concentration" means that most of the signal's energy, which we assumed to be finite, is concentrated in a certain time interval that can be quantified through the RMS length, or, if we are speaking about spectral concentration, in a certain frequency interval that can be quantified through the RMS bandwidth.

The rectangular window has bounded time support, and for this obvious reason it is also concentrated in time. Therefore

1. the support of its spectrum is unbounded;
2. the concentration of its spectrum becomes more and more pronounced as the concentration in time diminishes, i.e., as the window's support increases.

A bounded support implies concentration, but concentration *does not* imply bounded support. If a bounded support is also represented by a closed interval, then we can speak about *compact support*, and classify the corresponding function as *compactly supported*.

4.7 Appendix: Analog and Digital Frequency Variables

Here we here review the relations among the variables used to express the concept of frequency, for analog and discrete-time signals. We refer to an analog signal that undergoes sampling, so that $x[n] = x(nT_s)$.

1. A frequency can be expressed as:

    - ω = (discrete) angular frequency. It represents an angle in the z-plane and is therefore adimensional. It is measured in rad/sample;
    - Ω = analog angular frequency. Its physical dimension is (time)$^{-1}$. It is measured in rad/s. These two variables are related by $\omega = \Omega T_s$;
    - ν = (discrete) adimensional or normalized frequency. It is related to ω by $\nu = \omega/(2\pi)$ It is measured in cycles/ample;
    - f = analog frequency. Its physical dimension is (time)$^{-1}$. It is measured in Hz = s^{-1}. These two variables are related by $\nu = fT_s$.

    For example, we can write a sinusoidal signal as

    $$\sin(\Omega t) = \sin(2\pi f t) = \sin(2\pi f n T_s) = \sin(\omega n) = \sin(2\pi \nu n).$$

To obtain the DFT of a discrete-time signal with finite length, the corresponding DTFT is sampled at frequencies ω_k that are equally spaced by $\Delta\omega = (2\pi/N)$, so that

Fig. 4.16 Harmonic number $k \in [0, N-1]$ and corresponding angular frequency ω for each DFT sample

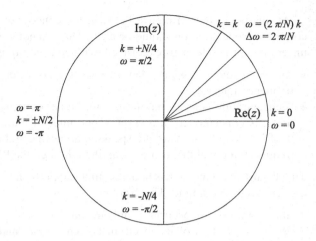

$$\omega_k = k\Delta\omega = \frac{2\pi}{N}k.$$

Figure 4.16 shows the relation between harmonic number (frequency index) k and discrete angular frequency ω.

Angular frequencies ω_k correspond to normalized frequencies ν_k equally spaced by $\Delta\nu = 1/N$ and given by

$$\nu_k = k\Delta\nu = \frac{k}{N}.$$

Given a sampling interval T_s, analog frequencies f_k corresponding to angular frequencies ω_k are equally spaced by $\Delta f = 1/(NT_s)$, where NT_s is the record duration, and are given by

$$f_k = k\Delta f = \frac{k}{NT_s};$$

the corresponding analog angular frequencies are separated by $\Delta\Omega = 1/(2\pi NT_s)$ and their expression is

$$\Omega_k = k\Delta\Omega = k\frac{\Delta\omega}{T_s} = \frac{k}{2\pi NT_s}.$$

2. The principal interval is:

   - in terms of k, $[-N/2, N/2 - 1]$ or $[0, N-1]$;
   - in terms of ω, $[-\pi, \pi)$ or $[0, 2\pi)$;
   - in terms of ν, $[-0.5, 0.5)$ or $[0, 1)$;
   - in terms of Ω, $[-\pi/T_s, \pi/T_s)$ or $[0, 2\pi/T_s)$;
   - in terms of f, $[-1/\{2T_s\}, 1/\{2T_s\})$ or $[0, 1/T_s)$.

3. The Nyquist frequency can be expressed as follows:

   - $k_{Ny} = N/2$;
   - $\omega_{Ny} = \pi$ rad/sample;
   - $\nu_{Ny} = 0.5$ cycles/sample;
   - $\Omega_{Ny} = \pi/T_s$ rad/s, for a sampling interval expressed in seconds;
   - $f_{Ny} = 1/(2T_s)$ Hz, for a sampling interval expressed in seconds.

4. We finally relate harmonic number with analog frequency. When we compute
   the DFT of a sequence $x[n]$ obtained performing periodic measurements of an
   analog signal $x(t)$, we may want to plot the amplitude spectrum $|X(e^{j\omega})|$ as a
   function of analog frequency, i.e., to return to physical dimensional frequencies
   and to the real-world phenomenon that generated the measured values. Therefore
   in applications it is useful to know exactly what frequency f corresponds, given
   a sampling interval T_s, to each DFT point. Recall that we can let f vary in the
   interval $[0, 2f_{Ny})$, or, equivalently, in $[-f_{Ny}, f_{Ny})$. Both cases are considered in
   Fig. 4.17, where values of k and f that correspond to one another are aligned
   vertically. Observe that the above-mentioned equivalent intervals actually are

   - $[0, 2f_{Ny} - \Delta f]$, corresponding to N values of k in $[0, N - 1]$, and
   - $[-f_{Ny}, f_{Ny} - \Delta f]$, corresponding to N values of k in $[-(N/2), N/2 - 1]$,

   where $\Delta f = (NT_s)$ is the distance between the analog frequencies at which the
   DTFT is sampled to give the DFT.

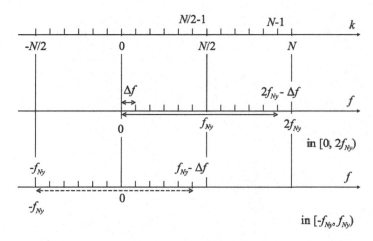

Fig. 4.17 Harmonic number k and corresponding analog frequency for each DFT sample

References

Cauchy, A.-L.: Mémoire sur diverses formules d'analyse. C. R. Acad. Sci. Paris **12**, 283–298 (1841)

Grenander, U.: Probability and Statistics: The Harald Cramér Volume. Almqvist & Wiksell, Stockholm (1959)

Nyquist, H.: Certain Topics in Telegraph Transmission Theory. Trans. AIEE, **47**, 617–644 (1928). Reprint as classic paper in: Proc. IEEE, **90**(2) (2002)

Shannon, C.E.: A Mathematical Theory of Communication. AT&T Tech. J., **27** 379–423, and 623–656 (1948)

Shannon, C.E.: Communication in the Presence of Noise. Proc. Inst. Radio Eng. **37**(1), 10–21 (1949). Reprint as classic paper in: Proc. IEEE, **86**(2) (1998)

Vetterli, M., Kovačević, H.: Wavelets and Subband Coding. Prentice Hall, Englewood Cliffs (1995)

Chapter 5
Spectral Analysis of Deterministic Discrete-Time Signals

5.1 Chapter Summary

In this chapter we will discuss how a discrete-time deterministic signal can be described in terms of elementary sinusoidal oscillations. First we present the practical issues of windowing and spectral sampling. Even if the signal intrinsically has infinite length, we only can record and use a segment of it, i.e., we can observe it through a window of finite width. Furthermore, we cannot compute its DTFT, provided it exists, but only the DFT of the segment framed by the window, i.e., we cannot avoid spectral sampling. The consequences of these limitations are investigated through examples using sinusoidal signals, and the concepts of spectral leakage and loss of spectral resolution are discussed. A description of the main windows used in spectral analysis is also given. We then move to more conceptual issues. Deterministic bounded signals can be energy signals, or power signals. For energy signals, that always allow for a DTFT representation, we can use the squared magnitude of the DTFT to define the energy spectrum, which tells us how the energy of the signal distributes over frequency. The energy spectrum can equivalently be defined introducing the autocorrelation/autocovariance (AC) sequence of the energy signal and using a theorem called the Wiener-Khinchin theorem for energy signals, which states that the energy spectrum is the DTFT of the AC sequence. Autocovariance and autocorrelation both quantify the signal's self-similarity and coincide if the signal has zero mean. The AC sequence essentially describes if, how and to what extent the signal's variations repeat themselves after a certain number of time steps. If it is so, then some variability patterns must exist in the signal that on average are repeated after a number of discrete-time units. This is obviously related to the presence in the signal of periodic components, and therefore to its spectral content. Signals that are not square-summable, i.e., have infinite energy, can have finite average power: these are power signals. For example, periodic signals like complex exponentials are power signals. For power signals, another spectral quantity can be introduced in place of the energy spectrum that does not exist in this case. This is the power spectrum, which describes how the power in the signal distributes over frequency. For zero-mean signals, power is equivalent to variance. General power signals may or may not have a DTFT representation, so their spectral representation necessarily passes through the definition of the power spectrum as the DTFT of the AC sequence (Wiener-Khinchin theorem for power signals). However, AC is defined differently for power signals with respect to energy signals. Broadly speaking, the dimensions are different: in the power-signal case, both the AC and the integral of the spectrum have the dimension of a power (energy/time) and not of an energy. Of course, the power spectrum can be defined only if the DTFT of the AC sequence exists: the AC sequence cannot have infinite energy.

5.2 Issues in Practical Spectral Analysis

The discrete-time deterministic signals we have been studying so far in theory have infinite length. Let us denote by $x_\infty[n]$ an infinitely-long discrete-time deterministic signal. In practical applications, two problems arise when we decide to analyze it:

- even if the signal $x_\infty[n]$ has infinite length, we can record and analyze only a segment $x[n]$, with $n = [0, N-1]$. The operation that from a theoretically infinitely persistent signal $x_\infty[n]$ leads to the extraction of a finite-length segment is referred to as *windowing* or *tapering*. In digital signal processing, a *window* or *taper* is a sequence that is identically zero outside some discrete-time interval, i.e., a signal whose time support is bounded. For example, a sequence whose values are 1 inside this interval and 0 outside is a *rectangular window*. The word *taper* is instead used to indicate window shapes that decrease more gradually toward zero. Mathematically, windowing is described as the multiplication of the signal by the discrete-time window $w[n]$. The product $w[n]x_\infty[n]$ is zero outside the window width. Only the segment of $x_\infty[n]$ that is framed by the window remains: this is *windowing*. The concept has a number of applications: it is useful in spectral analysis, in digital filter design, etc. In Sect. 5.3, a brief review of the most popular windows is presented;
- even if spectral analysis of the signal would require calculating its DTFT, which here is assumed to exist, we only can compute the DFT of the segment $x[n]$: we cannot avoid *spectral sampling*, meaning that we can only compute a finite number of spectral samples.

So, what are the consequences of these limitations? What happens when we study the DTFT of an infinite-length signal through the DFT of a signal's segment? These issues are usually discussed using sinusoidal signals. The conclusions drawn in this way have general validity and provide guidelines that are useful any time we have an idea of what elementary oscillatory modes contribute to the variability of a signal. Every time the basic idea of Fourier analysis comes into play, sinusoidal waveforms are seen as the building blocks of any signal, and are used to test algorithms and methods and to understand the way they work in the frequency domain. This occurs in spite of the fact that a sinusoidal signal has a DTFT representation only in the wide sense discussed in Sect. 3.3.2.

Let us take a sinusoidal signal with frequency ω_0 and initial phase ϑ, for example $x_\infty[n] = A \cos(\omega_0 n + \vartheta)$, with $n \in (-\infty, +\infty)$. We know that the DTFT of this sequence, $X_\infty(e^{j\omega})$, is a couple of impulses located at $\pm\omega_0$ and repeated periodically with period 2π. In this context,

1. windowing smooths and broadens these impulsive peaks, so that peak frequencies are less exactly defined. When two or more sinusoidal contributions are present in the signal, windowing also reduces the possibility of resolving—i.e., distinguishing, separating—individual sinusoidal contributions at frequencies that are close to each other;
2. spectral sampling potentially produces an inaccurate—or even misleading—picture of the spectrum of the sinusoidal signal.

These issues can be understood observing that

- according to the convolution property of the DTFT, windowing, which in the time domain is a multiplication, in the frequency domain is the continuous periodic convolution of the spectrum of the infinitely-long signal with the DTFT of the window sequence:

$$X(e^{j\omega}) = \frac{1}{2\pi} \int_{-\pi}^{+\pi} X_\infty(e^{j\omega}) W(e^{j[\omega-\theta]}) d\theta.$$

In turn, the spectrum of the window typically is a function of ω that in the principal interval exhibits a mainlobe centered at $\omega = 0$, and a certain number of sidelobes that can be more or less pronounced (see next subsection);

- the DFT of the windowed signal $x[n]$, i.e.,

$$X[k] = \sum_{n=0}^{N-1} x[n] e^{-j\frac{2\pi}{N}kn}, \qquad k = [0, N-1],$$

is obtained taking N samples of $X(e^{j\omega})$ equally spaced in angle:

$$X[k] = X(e^{j\omega})|_{\omega=\frac{2\pi}{N}k}.$$

We will now discuss separately the effects of windowing and spectral sampling through examples.

5.2.1   The Effect of Windowing

Let us consider the signal

$$x_\infty[n] = A_0 \cos(\omega_0 n + \vartheta_0) + A_1 \cos(\omega_1 n + \vartheta_1),$$

with $-\infty < n < +\infty$, made up of two real sinusoids with different amplitudes and frequencies, and let us compute the DTFT of the segment $x[n]$ given by

$$x[n] = A_0 w[n] \cos(\omega_0 n + \vartheta_0) + A_1 w[n] \cos(\omega_1 n + \vartheta_1),$$

where $w[n]$ is a generic window with length N.

Note that we are examining a sequence containing two oscillatory contributions arbitrarily separated in frequency, in order to be able to discuss also the resolution issue. For this purpose, and to facilitate our calculations, we write $x[n]$ as a combination of complex exponentials:

$$x[n] = \frac{A_0}{2} w[n] e^{+j\vartheta_0} e^{+j\omega_0 n} + \frac{A_0}{2} w[n] e^{-j\vartheta_0} e^{-j\omega_0 n} +$$
$$+ \frac{A_1}{2} w[n] e^{+j\vartheta_1} e^{+j\omega_1 n} + \frac{A_1}{2} w[n] e^{-j\vartheta_1} e^{-j\omega_1 n}.$$

We can thus immediately write, using the modulation property of the DTFT,

$$X(e^{j\omega}) = \frac{A_0}{2} e^{+j\vartheta_0} W[e^{+j(\omega-\omega_0)}] + \frac{A_0}{2} e^{-j\vartheta_0} W[e^{+j(\omega+\omega_0)}] +$$
$$+ \frac{A_1}{2} e^{+j\vartheta_1} W[e^{+j(\omega-\omega_1)}] + \frac{A_1}{2} e^{-j\vartheta_1} W[e^{+j(\omega+\omega_1)}],$$

where $W[e^{+j(\omega-\omega_0)}]$ has a maximum at $\omega = +\omega_0$, $W[e^{+j(\omega+\omega_0)}]$ has a maximum at $\omega = -\omega_0$, etc. The value of each maximum is given by $\sum_{n=0}^{N-1} w[n]$. In fact,

$$W(e^{j\omega}) = \sum_{n=-\infty}^{+\infty} w[n] e^{-j\omega n} \quad \rightsquigarrow \quad W(e^{j0}) = \sum_{n=-\infty}^{+\infty} w[n] = \sum_{n=0}^{N-1} w[n].$$

If the window has a rectangular shape,

$$w[n] = \begin{cases} 1 & \text{for } 0 \leq n \leq N-1, \\ 0 & \text{elsewhere,} \end{cases}$$

then we have

$$\sum_{n=0}^{N-1} w[n] = N.$$

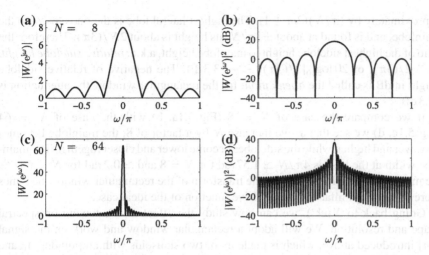

Fig. 5.1 **a–b** Magnitude of the DTFT of the rectangular window with length $N = 8$ in linear units and squared magnitude in dB units, respectively; **c–d** the same for the rectangular window with length $N = 64$

We thus see that $X(e^{j\omega})$ is formed by four copies of $W(e^{j\omega})$ centered at $\pm\omega_0$ and $\pm\omega_1$. Each replica of $W(e^{j\omega})$ is scaled by the complex amplitude of the corresponding exponential contribution present in the signal. For example, the copy of $W(e^{j\omega})$ located at $+\omega_0$ is scaled by $(A_0/2)e^{+j\theta_0}$. We may note that $X(e^{j\omega})$ contains the initial phases of the two sinusoids. However, when we take the absolute value of $X(e^{j\omega})$, i.e., when we pass to the amplitude spectrum, or to the squared amplitude, i.e., the energy spectrum, the initial phases disappear.

Before proceeding further, let us focus on $W(e^{j\omega})$ of the simplest window available, i.e., the rectangular one:

$$W(e^{j\omega}) = \sum_{n=0}^{N-1} e^{-j\omega n} = \frac{1 - e^{j\omega N}}{1 - e^{j\omega}} = e^{j\omega \frac{N-1}{2}} \frac{\sin(N\frac{\omega}{2})}{\sin(\frac{\omega}{2})} = e^{j\omega \frac{N-1}{2}} N D_N(\omega).$$

Phase is linear in ω. Apart from the phase factor, the window transform is proportional to a Dirichlet function; its magnitude is shown in Fig. 5.1a for $N = 8$, and Fig. 5.1c for $N = 64$. Figures 5.1b, d show the corresponding square magnitudes in dB units.[1]

The spectrum of the digital rectangular window exhibits a mainlobe and a certain number of smaller sidelobes that decrease in height with increasing distance from the origin. A large percentage of the spectral content is thus concentrated around $\omega = 0$. The zeros of this function are found at $\omega = (2\pi/N)m$, with $m = \pm1, \pm2, \pm3\ldots$, hence the mainlobe extends from $-2\pi/N$ to $+2\pi/N$. Its width at the base is thus $4\pi/N$. The height of the mainlobe is N. The number of lateral lobes on each side of the main one is given by $(N/2) - 1$. Sidelobes have their maxima at frequencies given

---

[1]$20 \log_{10} a = 10 \log_{10} a^2$ is the square of a expressed in dB.

approximately by $(\pi/N)(2m + 1)$. The highest lateral lobe is the one closest to the mainlobe, and is found at about $\pm 3\pi/N$; its height is about $2N/(3\pi)$. Therefore the ratio of the highest sidelobe height to mainlobe height, a.k.a. *relative sidelobe height*, is $\cong 2/(3\pi)$, or $20\log_{10}[2/(3\pi)] \cong -13.3\,\text{dB}$. The negative of relative sidelobe height in dB is called the *attenuation*: for the rectangular window the attenuation is $13.3\,\text{dB}$.

If we compare the case of $N = 8$ (Fig. 5.1a, b) with the case of $N = 64$ (Fig. 5.1c, d) we see that as we increase N by a factor of 8, the mainlobe becomes narrower and higher, while the sidelobes become lower and closer together. The main-lobe width at the base is $4\pi/N \simeq 1.57\,\text{rad}$ for $N = 8$ and $\simeq 0.2\,\text{rad}$ for $N = 64$. As the number of samples increases, the transform of the rectangular window becomes more and more similar to the impulsive function of the ideal case.

Going back to $X(e^{j\omega})$, we can now study the effect of windowing on spectral shape and resolution. We will adopt a rectangular window and work on the signal $x[n]$ introduced above, which is made up of two sinusoids with amplitudes A_0 and A_0, and frequencies ω_0 and ω_1 separated by $\Delta\omega = \omega_1 - \omega_0$. Keeping the window length fixed at $N = 64$, and starting from a value of $\Delta\omega$ that allows for observing well-separated spectral peaks, we will progressively reduce the value of $\Delta\omega$, and observe the amplitude spectrum of the windowed signal. It is reasonable to expect that well-separated spectral peaks are observed when $\Delta\omega$ is much larger than the width of the mainlobe of the transform of the 64-point rectangular window, i.e., $0.2\,\text{rad}$. This will therefore be our starting point.

1. Figure 5.2a:

$$\omega_0 = \frac{\pi}{3}, \qquad \omega_1 = \frac{2\pi}{3} \qquad \rightarrow \qquad \Delta\omega = \frac{\pi}{3} \simeq 1.05\,\text{rad} \gg 0.2\,\text{rad},$$

$$A_0 = 1, \qquad A_1 = 0.75 \qquad \rightarrow \qquad \frac{A_1}{A_0} = \frac{3}{4}.$$

As expected, the two sinusoids are very well resolved. The heights of the peaks are 32 and 24, respectively. The first value derives from $(A_0/2)N$ with $A_0 = 1$ and $N = 64$; the second derives from $(A_1/2)N$ with $A_1 = 0.75$ and $N = 64$. Thus these heights correctly reflect the amplitude ratio, $A_1/A_0 = 3/4$.

2. Figure 5.2b:

$$\omega_0 = \frac{\pi}{7}, \qquad \omega_1 = \frac{4\pi}{15} \qquad \rightarrow \qquad \Delta\omega \simeq 0.389\,\text{rad} > 0.2\,\text{rad},$$

$$A_0 = 1, \qquad A_1 = 0.75 \qquad \rightarrow \qquad \frac{A_1}{A_0} = \frac{3}{4}.$$

Also in this case the two sinusoids are fairly well resolved. The peak heights are approximately 32 and 24, and therefore reproduce with acceptable accuracy the ratio $A_1/A_0 = 3/4$.

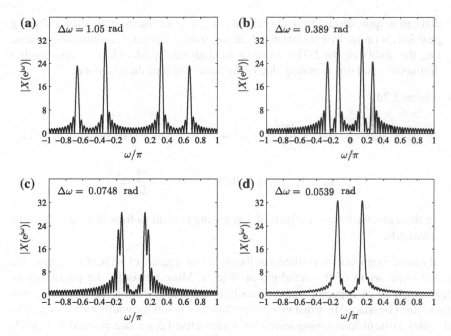

Fig. 5.2 Amplitude spectra of a signal with $N = 64$ samples, containing two sinusoids with frequencies separated by a $\Delta\omega$ that is **a** much larger than the width of the mainlobe of the transform of the 64-point rectangular window, **a** larger than the width of the mainlobe, **c** smaller than the width of the mainlobe, and **d** much smaller than the width of the mainlobe

3. Figure 5.2c:

$$\omega_0 = \frac{\pi}{7}, \qquad \omega_1 = \frac{\pi}{6} \qquad \rightarrow \qquad \Delta\omega \simeq 0.0748 \text{ rad} < 0.2 \text{ rad},$$

$$A_0 = 1, \qquad A_1 = 0.75 \qquad \rightarrow \qquad \frac{A_1}{A_0} = \frac{3}{4}.$$

In this case the two sinusoids are no longer well resolved. Moreover, the sidelobes, adding out-of-phase, influence the peak heights, so that their ratio no longer reflects the amplitude ratio.

This behavior, due to windowing, is described in terms of

- loss of resolution,
- leakage.

Leakage is the word used to express the fact that the sinusoidal component at frequency ω_0 "leaks" in the neighborhood of ω_1 and vice-versa. Leakage is mainly due to the presence and prominence of sidelobes. Sidelobes spread the spectral content that would compete to a given sinusoid, so that it contributes to frequencies that can be quite distant from the central one. They are also responsible for altering peak height ratios. In association to leakage, *reduced spectral resolution* is present, i.e., the ability of the DTFT to separate sinusoidal peaks close to one another diminishes, and this is mainly due to the finite width of the mainlobe.

4. Figure 5.2d:

$$\omega_0 = \frac{\pi}{7}, \qquad \omega_1 = \frac{4\pi}{25} \qquad \rightarrow \qquad \Delta\omega \simeq 0.0539 \,\text{rad} \ll 0.2 \,\text{rad},$$

$$A_0 = 1, \qquad A_1 = 0.75 \qquad \rightarrow \qquad \frac{A_1}{A_0} = \frac{3}{4}.$$

In this case, the negative effects of windowing prevent us from distinguishing the sinusoids.

By these examples we verified that the ability to separate close peaks is connected to the finite width of the mainlobe of $W(e^{j\omega})$. More precisely, the possibility of distinguishing two spectral peaks actually relies mainly on the fact that these two peaks must be spaced by a distance greater than the mainlobe width, but other factors also play a role in determining resolution. Given a fixed $\Delta\omega$ value, resolution depends on:

- the amplitude ratio. If one sinusoid's amplitude is much smaller than the other, the corresponding peak—for example, let us imagine it is the peak at frequency ω_1—can be submerged by the sidelobes of the other peak at ω_0;
- the rapidity with which sidelobes decrease with increasing distance from the mainlobe. In other words, a critical factor for resolution is whether or not the copy of $W(e^{j\omega})$ centered on ω_0 becomes small enough in the neighborhood of ω_1, and vice-versa.

In summary, resolution is sufficient and leakage is small if

- the mainlobe of $|W(e^{j\omega})|$ is narrow,
- the sidelobes are low and decrease rapidly with increasing distance from the mainlobe.

Our previous examples were relative to the use of a rectangular window. Using more gradual windows with the same N, a lowering of the sidelobes can be obtained, but this improvement is accompanied by a widening of the mainlobe. Thus a reduction of leakage, which is mainly due to sidelobes, is inevitably accompanied by a further loss of resolution, which is related to mainlobe width. The condition on the frequency spacing of the sinusoids that can be resolved becomes more restrictive. Of course, this drawback can be overcome if N can be arbitrarily increased, since increasing the number of samples makes the transform of any window more similar to an ideal frequency-domain impulse. We saw this for the rectangular window in Fig. 5.1, and will soon see it for different, more gradual windows.

5.2.2 The Effect of Spectral Sampling

This effect is, once again, well-illustrated by an example. We consider the same type of signal $x[n]$ with $N = 64$ samples used earlier, and set, as in Fig. 5.2b,

$$\omega_0 = \frac{\pi}{7}, \qquad \omega_1 = \frac{4\pi}{15} \qquad \rightarrow \qquad \Delta\omega \simeq 0.389 \,\text{rad},$$

$$A_0 = 1, \qquad A_1 = 0.75 \qquad \rightarrow \qquad \frac{A_1}{A_0} = \frac{3}{4}.$$

We compare its DTFT (Fig. 5.3, continuous *gray curve*) with the DFT calculated over $N = 64$ points (Fig. 5.3, *black stem plot*).

Resolution is a priori sufficient, as witnessed by the fact that in the DTFT (continuous *gray curve*) the two peaks are well resolved. However, since the DTFT peaks fall between one sample of the DFT and the next one, when we can only observe the DFT (*black stem plot*) we are not in a position to correctly evaluate either the frequency, or the amplitude of the sinusoids contained in the signal.

In general, the "true" peaks will not coincide exactly with any of the frequencies ω_k, and we will get a potentially distorted picture of the underlying DTFT. If, while keeping the other parameters constant (N, A_0, A_1), we force this coincidence to exist by slightly moving ω_0 and ω_1, the DFT of the signal over 64 samples appears quite different (Fig. 5.4a).

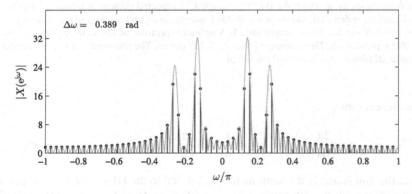

Fig. 5.3 Amplitude spectrum of a signal with $N = 64$ samples, containing two sinusoids separated by a $\Delta\omega$ larger than the width of the mainlobe of the transform of the 64-point rectangular window. Resolution appears satisfactory if we observe the shape of the DTFT (continuous *gray curve*). However, when we compute the DFT over $N = 64$ samples we can actually only observe the *black stem plot*. The fact that the frequencies of the sinusoids composing the signal, in general, do not coincide with any member of the ω_k set over which the DTFT is sampled alters the picture. The accuracy with which frequency and amplitude of the sinusoidal components can be estimated is poor

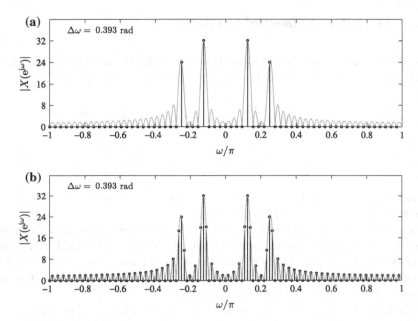

Fig. 5.4 **a** Amplitude spectrum of a signal with $N = 64$ samples, containing two sinusoids separated by a $\Delta\omega$ larger than the width of the mainlobe of the transform of the 64-point rectangular window, so that resolution is satisfactory (see the DTFT, continuous *gray curve*). The black stem plot represents the DFT computed over $N = 64$ points: a rather poor frequency sampling. In spite of this, the DFT deceptively seems to reproduce the shape of the underlying DTFT very well, because the frequencies of the sinusoids composing the signal were chosen in such a way that they exactly coincide with two of the frequencies ω_k at which the DTFT is sampled. No spectral content at frequencies different from ω_0 and ω_1 is detected, and the non-zero DFT samples at ω_0 and ω_1 assume the value expected on the basis of the sinusoids' amplitudes. **b** Amplitude spectrum of the same signal. The black stem plot represents the DFT computed over $N = 128$ points. The presence of a non-zero spectral content at all frequencies is correctly detected

In this example,

$$\omega_0 = \frac{\pi}{8} = 4\frac{2\pi}{64}, \qquad \omega_1 = \frac{\pi}{4} = 8\frac{2\pi}{64} \qquad \rightarrow \qquad \Delta\omega \simeq 0.393\,\text{rad},$$

so that the $\Delta\omega$ is nearly the same as in Fig. 5.4, but in the DFT over $N = 64$ points, ω_0 coincides with the 5th frequency (ω_k with $k = 4$) and ω_1 coincides with the 9th frequency (ω_k with $k = 8$). The DFT appears very "clean" and free from any spectral content at frequencies other than ω_0 and ω_1, but this is an illusion due to the poor frequency sampling. Actually, a non-zero spectral content is present at all frequencies, as indicated by the underlying DTFT (continuous *gray curve*). This can be verified by making the spectral sampling denser through zero padding up to a length of 128 points (Fig. 5.4b).

Let us underline once again that zero-padding does not improve resolution: the latter depends only on the shape and the width of the window used to observe the considered segment of the original infinite-length signal. By zero-padding we simply performed an interpolation that makes more details of the underlying DTFT visible in the DFT plot, within the limits of the available resolution. Zero-padding and the consequent spectral oversampling are most often applied in order to obtain more readable and easily interpretable spectral plots. Frequency interpolation provides smooth spectral curves on which the position and the height of any peak can be better determined.

5.3 Classical Windows

We briefly mentioned above the spectral effects related to the use of more gradual windows in comparison to the rectangular one. In this section we will describe more precisely the characteristics of the most popular windows.

The windows that are most often applied, besides the rectangular one, are the Bartlett window, the Von Hann or *Hanning* window,[2] the Hamming window, and the Blackman window (Harris 1978; Nuttall 1981). These "historical" or "classical" windows were introduced finding empirical ways to reduce relative sidelobe height with respect to the rectangular window case. Each window is characterized by

- mainlobe width, measured, for example, at the lobe base, or at half height. This parameter is expressed as relative width with respect to the rectangular window case: measuring the width at the lobe base we will write $(4\pi/N)\eta$, where the value of η depends on the window, and is 1 for the rectangular window;
- attenuation of the highest sidelobe.

These characteristics are more easily discussed observing the shape of the window squared-transform plotted on a logarithmic scale, i.e., expressed in dB: this choice allows for clearly visualizing the behavior of characteristic parameters as the time-domain window shape varies (Fig. 5.5). We will describe separately each classical window in both domains, starting from the rectangular one.

- *Rectangular window*
  The squared magnitude of the transform of the rectangular window is shown on a dB scale in Fig. 5.5a, b. It may be seen that for a given N, the mainlobe is the narrowest among all considered windows (compare Fig. 5.5a with Fig. 5.5c–f; these panels are relative to different window types, all having the same length). However, the highest sidelobe, adjacent to the main one, is only ≈ 13 dB smaller than the mainlobe, and the decrease of sidelobes with increasing distance from the mainlobe is slow. Doubling the window length (compare Fig. 5.5a, in which $N = 51$, with Fig. 5.5b, in which $N = 101$), the mainlobe narrows and the number of sidelobes increases, but the attenuation of about 13 dB and the slow sidelobe decrease remain unchanged.

---

[2]Often, the Von Hann window is called *hanning*, with the initial letter in lower case.

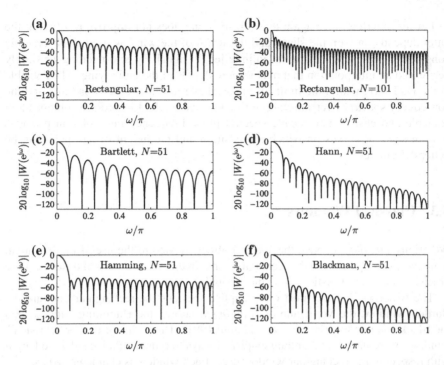

Fig. 5.5 a–f Squared amplitude of the DTFT in logarithmic units (dB) for various classical windows. In all panels, the quantity $10 \log_{10} |W(e^{j\omega})|^2 = 20 \log_{10} |W(e^{j\omega})|$ has been normalized in such a way to be zero at zero frequency. This normalization makes the comparison among different windows easier, and means that the magnitude $|W(e^{j\omega})|$ of the transform is normalized to 1 at $\omega = 0$. In turn, this implies that the time-domain samples of a given window are normalized so as to give 1 when summed, instead than being normalized so as to have 1 as the maximum value. The last choice is the one normally adopted. All the windows represented here have a length of $N = 51$ samples, except the rectangular window shown in panel **b**, which is 101-samples long

- *Bartlett window*

  The transform of the Bartlett window is the square of the rectangular window transform. In fact, the Bartlett window can be seen as the result of the convolution between two rectangular windows (see the example of linear convolution in Sect. 3.5.4). It is, therefore, always non-negative. An increase of the sidelobe attenuation up to about 26 dB (the double of 13 dB) is obtained, as can be seen in Fig. 5.5c. However, the mainlobe width measured at the base increases to about $8\pi/N$, so that $\eta = 2$. The decrease of sidelobes is quite slow. Note that this window, in different contexts, can include a zero sample at each edge, or not. It is common use to call the version that includes zeros a "Bartlett window" and the version that does not include zeros at the edges a "triangular window". Here we will neglect this distinction and use the two names as synonyms.

- *Hanning window*

  The transform of the Hanning window is the weighted sum of three Dirichlet functions variously shifted in frequency, so as to obtain a partial cancellation of sidelobes. More precisely, two symmetrically-positioned Dirichlet functions with weight 1/4 each are added to a central Dirichlet function with weight 1/2. The attenuation becomes 32 dB, while the mainlobe width stays approximately at $8\pi/N$, corresponding to $\eta = 2$. The sidelobe decrease is rapid.

- *Hamming window*

  The transform of the Hamming window is again the weighted sum of three Dirichlet functions variously shifted in frequency, so as to obtain a partial cancellation of sidelobes. The weights were determined empirically, looking for values minimizing the height of the highest sidelobe. The attenuation attains about 44 dB, while the mainlobe width stays approximately at $8\pi/N$, so that we still have $\eta = 2$. Note that in this case the highest sidelobe is not the most central one. The attenuation is remarkable but the decrease of sidelobes is slow.

- *Blackman window*

  The transform of the Blackman window is the weighted sum of five Dirichlet functions variously shifted in frequency, so as to further reduce the relative sidelobe height with respect to the Hamming window; however, no optimization procedure was performed to determine the weights. The attenuation is as high as 58 dB, but the mainlobe width increases to about $12\pi/N$, giving $\eta = 3$. Attenuation is very high and the decrease of sidelobes quite fast, but the mainlobe is the widest one among all classical windows with the same length.

Table 5.1 shows the functional form of the windows mentioned in this discussion. Since, in the various contexts in which these windows are applied, a causal form of the window is sometimes required, while at some other times the non-causal form (symmetrical about the origin) is needed, both forms are shown in Table 5.1. In all cases, the sole free parameter is the window length. For a window in non-causal form, the length N is always an odd number. For a window in causal form, in principle, the length N may be either odd, or even. However, if it is even, of course the window maximum falls at a half-integer value of the time index. The causal forms given in

Table 5.1 Functional form of the most popular classical windows, in their causal and non-causal forms

Window	Non-causal form	Causal form
	$-M \leq n \leq M$	$0 \leq n \leq N-1$
Bartlett	$1 - |n|/M$	$2n/(N-1)$ for $0 \leq n \leq (N-1)/2$ $2 - 2n/(N-1)$ for $(N-1)/2 \leq n \leq N-1$
Hanning	$0.5[1 + \cos(\pi n/M)]$	$0.5\{1 - \cos[2\pi n/(N-1)]\}$
Hamming	$0.54 + 0.46\cos(\pi n/M)$	$0.54 - 0.46\cos[2\pi n/(N-1)]$
Blackman	$0.42 + 0.5\cos(\pi n/M) +$ $+0.08\cos(2\pi n/M)$	$0.42 - 0.5\cos[2\pi n/(N-1)] +$ $+0.08\cos[(4\pi n/(N-1))]$

Fig. 5.6 Time behavior of
the most popular classical
windows in their causal form

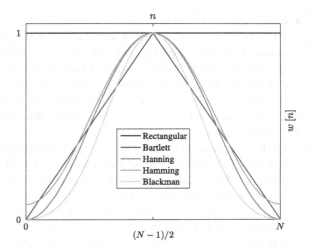

Table 5.1 refer to odd values of N. Here all windows are normalized so as to have 1 as their maximum value.

Figure 5.6 shows the shapes of these windows, in their causal version. For visual clarity, each window is plotted as if it were a function of continuous time, but of course it is a sequence defined only for integer values of n. It can be observed that all windows have zeros at the edges, except the Hamming window.

Classical tapers are often used, in place of the rectangular window, for spectral analysis of deterministic and random signals. From what we have learned about the behavior of classical windows in this subsection, and about the effects of windowing in the previous one, we can understand that no "best" window for spectral analysis exists. The choice of the window always represents a compromise: if resolution is important and a minimal mainlobe width is desired for a fixed value of N, then the rectangular window is the best choice. If minimizing leakage is important and therefore large attenuation and rapid decrease of the sidelobes is desired, then Hanning, Hamming and Blackman windows are preferable.

So far, we have quantified the mainlobe width using the width at the base, i.e., the frequency interval between the zero crossings of the window spectrum which are located at each side of the origin. However, in spectral analysis resolution is often expressed in terms of mainlobe width at half height. There are two ways of quantifying this parameter: in some cases the transform amplitude is required to be half of its maximum value; in some other cases, the transform squared magnitude is required to be half of its maximum value (see Chap. 10). In a plot showing the squared magnitude in dB, this means considering the level $20 \log_{10}(0.5) = -6\,\mathrm{dB}$, or the level $20 \log_{10}(0.707) = 10 \log_{10}(0.5) = -3\,\mathrm{dB}$, respectively. These widths are referred to as the "3-dB width" and "6-dB width", meaning, "3 or 6 dB below maximum level". Table 5.2 shows the 3-dB and 6-dB widths of the main classical windows, expressed in units of $4\pi/N$, together with the relative height of the highest sidelobe expressed in dB.

Table 5.2 3-dB and 6-dB width in units of $4\pi/N$, and relative height of the highest sidelobe in dB, for the main classical windows used in spectral analysis

Window	3-dB width	6-dB width	Relative height of sidelobes
Rectangular	0.4430	0.6034	−13.3
Bartlett	0.6511	0.9042	−26.4
Hanning	0.7347	1.0200	−31.5
Hamming	0.6600	0.9203	−43.5
Blackman	0.8383	1.1724	−58.1

These widths (though halved) and the relative sidelobe heights can also be deduced from Fig. 5.7a–e, where the squared magnitude of the transforms of the above-mentioned windows is shown in logarithmic units (dB), for $N = 51$. In these plots,

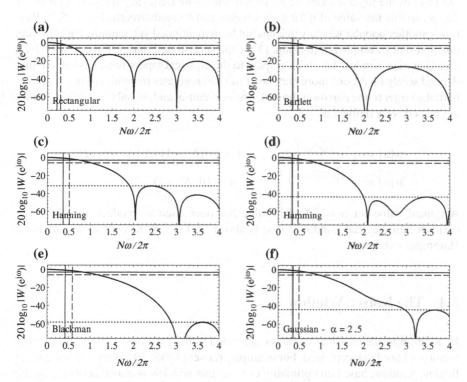

Fig. 5.7 Squared amplitude in dB of the transform of different classical windows with $N = 51$: **a** rectangular window, **b** Bartlett window, **c** Hanning window, **d** Hamming window, **e** Blackman window, **f** Gaussian window with parameter $\alpha = 2.5$. In all panels, the quantity $10\log_{10}|W(e^{j\omega})|^2 = 20\log_{10}|W(e^{j\omega})|$ has been normalized in such a way to be zero at zero frequency. The 3-dB and 6-dB levels are marked by *horizontal solid lines* and *dashed lines*, respectively; the *horizontal dotted lines* indicate the relative height in dB of the highest sidelobe, i.e., the ratio between the height of the highest sidelobe and the height of the mainlobe, expressed in dB

Fig. 5.8 Time behavior of
the causal Gaussian window
with $\alpha = 2.5$, in comparison
with the Hamming window

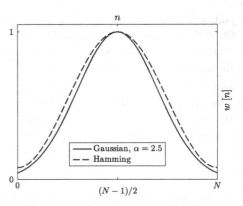

the 3-dB level is shown by *horizontal solid lines*, the 6-dB level by *dashed lines*, and
the level of the highest sidelobe by *dotted lines*. The frequency axes are in units of
$2\pi/N$, so that the value of η for each window can be read directly. Figure 5.7f illustrates another popular window that has not been mentioned yet, namely, the Gaussian
one that will be employed in Chap. 13 for the short-time Fourier transform (STFT).
The Gaussian window has, with respect to the classical tapers presented above that
depend solely on N, one more parameter (α) that regulates the width of the Gaussian
bell: the larger α, the narrower the bell. The non-causal and causal Gaussian windows
are respectively defined as

$$w[n] = e^{-\frac{1}{2}\left[\alpha\frac{n}{(N-1)/2}\right]^2}, \qquad n = [-(N-1)/2, (N-1)/2],$$

$$w[n] = e^{-\frac{1}{2}\left[\alpha\frac{n-(N-1)/2}{(N-1)/2}\right]^2}, \qquad n = [0, N-1].$$

A typical value for α is 2.5; this value has been used to produce Fig. 5.7f. The
corresponding causal window shape is shown in Fig. 5.8, in comparison with the
Hamming window.

5.4 The Kaiser Window

In addition to the classical windows presented in the previous section, many other
windows have been proposed. For example, Kaiser (1966, 1974) introduced a more
flexible window, based on optimality criteria: this window is aimed at obtaining the
narrowest mainlobe, while respecting certain constraints. More precisely, the Kaiser
window minimizes the mainlobe width under the condition of keeping the window
length fixed and constraining the energy of sidelobes to be less than, or equal to,
a fixed percentage of the spectrum total energy. The energy of sidelobes is defined
calculating the integral of the transform squared magnitude over $[0, \pi]$, excluding
the frequency interval pertaining to the mainlobe. The samples $w[n]$ of the Kaiser

window obtained in this way depend, for a given length N, on the energy percentage granted to sidelobes.

The solution of Kaiser's optimization problem can be expressed using the modified Bessel function with zero order, which is defined by the series

$$I_0(x) = \sum_{k=0}^{\infty} \left(\frac{x^k}{2^k k!} \right).$$

The causal Kaiser window with length N is in fact given by

$$w[n] = \frac{I_0\left[\alpha \sqrt{1 - \left(\frac{|2n-N+1|}{N-1} \right)^2} \right]}{I_0(\alpha)},$$

where α—the additional parameter that this window presents with respect to classical windows—serves to adjust the mainlobe width and the relative height of sidelobes. As α increases while N remains constant, the mainlobe widens and the height of sidelobes decreases. On the other hand, the value $\alpha = 0$ produces a rectangular window. If we instead let N increase while keeping α constant, the mainlobe narrows while the maximum height of sidelobes remains unchanged.

The Kaiser window is particularly useful (Kaiser 1974) for designing FIR filters by the method of impulse response truncation using windows (see, for example, Oppenheim and Schafer 2009)—a simple design technique widely used, especially in past decades, and briefly presented in Sect. 7.5. The Kaiser window has been also used in spectral analysis (Kaiser and Schafer 1980).

Figure 5.9 shows the shape of the causal Kaiser window for different values of the parameter α. As α increases, the window becomes more tapered, that is, the windows descends more rapidly from the central maximum towards the minima at the edges.

Fig. 5.9 Shape of the Kaiser window of length N, for various values of the paramenter α

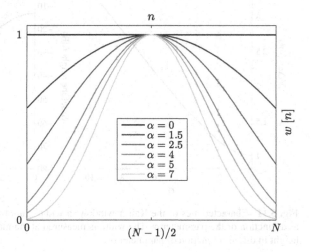

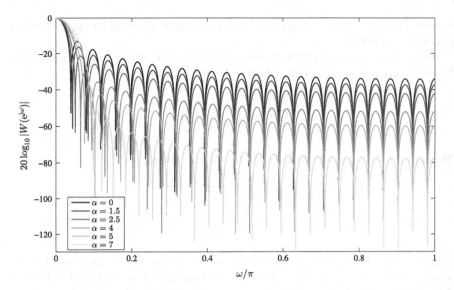

Fig. 5.10 Squared magnitude in dB of the transform of the Kaiser window, for various values of the parameter α and for $N = 51$. The plotted quantity, i.e., $10 \log_{10} |W(e^{j\omega})|^2 = 20 \log_{10} |W(e^{j\omega})|$, has been normalized in such a way to be zero at zero frequency

Figure 5.10 shows the squared magnitude of the Kaiser window transform in dB, for various values of α and for $N = 51$. The mainlobe width, with respect to the rectangular window case, increases almost linearly with α, as shown in Fig. 5.11a. The mainlobe width plotted in this figure is measured at the lobe base. More precisely,

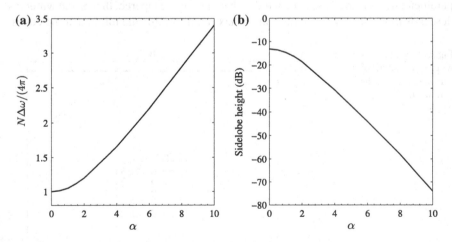

Fig. 5.11 Characteristics of the Kaiser window: **a** width $\Delta\omega$ of the mainlobe in units of $4\pi/N$, as a function of the parameter α. The width is measured at the mainlobe base; **b** relative sidelobe height in dB, as a function of parameter α

the ordinates show the values of $N\Delta\omega/4\pi$, i.e., the ratio between the mainlobe width $\Delta\omega$ of the Kaiser window with parameter α and the mainlobe width $4\pi/N$ of the rectangular window. Note that $N\Delta\omega/4\pi = 1$ for $\alpha = 0$, as expected since then the Kaiser window coincides with a rectangular window. Also the relative level in dB of sidelobes varies with α in a nearly-linear way, as shown in Fig. 5.11b. The relative level is $-13.3\,\mathrm{dB}$ for $\alpha = 0$, as expected for a rectangular window.

5.5 Energy and Power Signals and Their Spectral Representations

A deterministic bounded signal $x[n]$ of infinite length can be an energy signal, or a power signal. The value of $|x[n]|^2$ at a given n can be seen as the *instantaneous power* of the signal $x[n]$ at time n: the energy in a unit time interval. Summing over all unit time intervals we get total energy,

$$\mathcal{E} = \sum_{n=-\infty}^{+\infty} |x[n]|^2.$$

If $\mathcal{E}$ is finite, i.e., if the sequence is square-summable, then we have an energy signal. If $\mathcal{E}$ is infinite, but the average power

$$\mathcal{P} = \lim_{M\to\infty} \frac{1}{2M+1} \sum_{n=-M}^{+M} |x[n]|^2$$

is finite, then we have a power signal. Note that $\mathcal{P} = 0$ for energy signals. Note also that the absolute value in the previous formulas has no effect if the signal is, as in all cases of interest to us, real-valued, but makes the definitions work even for complex-valued signals.

It may be useful to recall that the DTFT of an infinite-length deterministic signals exists

1. in the sense of absolute convergence for the class of absolutely summable signals,
2. in the sense of mean-square convergence for the larger class of square-summable signals, and
3. in the sense of generalized functions or distributions (Dirac δ; discrete spectrum) for signals that are a linear combination of sinusoids/complex exponentials with discrete frequencies.

If a signal is absolutely summable, then it is also square summable, so in cases 1 and 2 we have an energy signal. The periodic signals of case 3 are instead a special case of general power signals. Finally, if a signal has finite length, then it certainly allows for a DTFT representation. This case includes the real-world case of windowed signals. The DTFT can then be represented by the DFT.

Unless stated otherwise, in the following discussion we will assume that the signal has zero mean.

1. **Deterministic Energy Signals**
   For energy signals, the conservation of energy is expressed by Parseval's theorem for energy signals,

$$\mathcal{E} = \sum_{n=-\infty}^{+\infty} |x[n]|^2 = \frac{1}{2\pi} \int_{-\infty}^{+\infty} |X(e^{j\omega})|^2 d\omega,$$

which states that the energy in $x[n]$ equals the energy in $X(e^{j\omega})$, provided that $\mathcal{E} < \infty$. Note that the units of $|X(e^{j\omega})|^2$ are "energy per units of ω" so that the integral has units of energy. We can see the quantity $|X(e^{j\omega})|^2$ as the *energy spectral density* (ESD) of $x[n]$ at frequency ω. Its graph is known as the *energy spectrum* of $x[n]$, and shows how the energy of $x[n]$ distributes over frequencies. The information provided by the real function $|X(e^{j\omega})|^2$ in this sense is equivalent to the information provided by the amplitude spectrum $|X(e^{j\omega})|$, *but* the physical meaning is more precise. Therefore, spectral analysis of energy signals is usefully performed in terms of the energy spectrum. Note that very often, the ESD function itself is referred to as the energy spectrum of $x[n]$, even if, strictly speaking, the energy spectrum is the *plot* of ESD versus frequency. Obviously, $|X(e^{j\omega})|^2$ is always real and non-negative; if $x[n]$ is real, then $|X(e^{j\omega})|^2$ is an even function of frequency.

The energy spectrum that we have just defined as the squared magnitude of the DTFT can also be defined by introducing the notions of *autocorrelation* and *auto-covariance* of an energy signal. Autocorrelation and autocovariance are sequences that quantify the signal's self-similarity: they describe if, how and to what extent the signal's variations repeat themselves after a certain delay. If the signal has zero mean, autocorrelation and autocovariance coincide and often will collectively be indicated by the acronym AC in the rest of the book. If AC is present in a signal, then in the signal some variability patterns exist that on average repeat themselves after a number of discrete-time units, and this can be intuitively related to the presence in the signal of periodic components—the same components that can be detected by spectral analysis. Indeed, a theorem called *Wiener-Khinchin theorem for energy signals*[3] states that the energy spectrum is equal to the DTFT of the AC sequence. This approach will be discussed in Sect. 5.7.1.

---

[3]The name of this theorem refers to Norbert Wiener (1894–1964) and Aleksandr Khinchin (1894–1959). Norbert Wiener proved this theorem for the case of a deterministic function in 1930 (Wiener 1930); Aleksandr Khinchin later formulated an analogous result for stationary stochastic processes and published it in 1934 (Khinchin 1934). Albert Einstein explained, without proofs, the idea in a brief two-page memo in 1914 (see Jerison et al. 1997). Note that the name of the second author of the theorem, a Russian mathematician, is sometimes transliterated from the Cyrillic alphabet as "Khintchine".

2. Deterministic Power Signals

If $x[n]$ is a power signal, then its energy is infinite. Periodic signals like sinusoids and complex exponentials, for example, are power signals, and for them, as for any power signal, the notion of energy spectrum is meaningless. The average power, however, will be finite. Let us define

$$x_M[n] = \begin{cases} x[n] & \text{for} \quad |n| \le M, \\ 0 & \text{for} \quad |n| > M. \end{cases}$$

The DTFT of $x_M[n]$ certainly exists, due to the fact that M is finite, and is given by

$$X_M(e^{j\omega}) = \sum_{n=-\infty}^{+\infty} x_M[n]e^{-j\omega n} = \sum_{n=-M}^{+M} x[n]e^{-j\omega n}.$$

The *power spectral density* (PSD) of $x[n]$ at frequency ω is then defined as the real and non-negative quantity

$$P_{xx}(e^{j\omega}) = \lim_{M \to \infty} \frac{1}{2M+1} |X_M(e^{j\omega})|^2.$$

The graph of PSD is known as the *power spectrum* of $x[n]$, and shows how the power of $x[n]$ is distributed over frequency. By extension, very often the PSD function itself is referred to as the power spectrum of $x[n]$. If $x[n]$ is real, then $P_{xx}(e^{j\omega})$ is an even function of frequency.[4]

Parseval's theorem for power signals is, with obvious notation,

$$\mathscr{P} = \lim_{M \to \infty} \frac{1}{2M+1} \sum_{n=-M}^{+M} |x[n]|^2 = \frac{1}{2\pi} \int_{-\infty}^{+\infty} P_{xx}(e^{j\omega})d\omega =$$

$$= \lim_{M \to \infty} \frac{1}{2\pi(2M+1)} \int_{-\infty}^{+\infty} |X_M(e^{j\omega})|^2 d\omega = \lim_{M \to \infty} \frac{1}{2M+1} \mathscr{E}_M,$$

where $\mathscr{E}_M$ represents the energy of $x_M[n]$. The units of $P_{xx}(e^{j\omega})$ are "power per units of ω", i.e., "variance per units of ω", since for zero-mean signals power is variance.

This definition of power spectrum containing a limit is highly unpractical, and we must look for another definition. As we underlined above, infinitely-long power signals may or may not have a DTFT. Thus, their spectral analysis necessarily passes through the definition of the power spectrum via the DTFT of the AC

---

[4]Note that the reality of $P_{xx}(e^{j\omega})$ suggests the possibility of denoting PSD simply by $P_{xx}(\omega)$, rather than by $P_{xx}(e^{j\omega})$.

sequence for power signals (Sect. 5.6.2). This derives from the application of
the Wiener-Khinchin theorem for power signals (Sect. 5.7.2). The theorem states
that the signal's power spectrum is the DTFT of the AC sequence, *but* AC is
defined differently for power signals, with respect to energy signals. Broadly
speaking, the dimensions are different: in the power-signal case, both the AC and
the integral of the spectrum have the dimension of a power (energy/time) and not
of an energy. This may be better understood reasoning, for a moment, in analog
terms, but it applies to discrete-time signals too, even if in this case time is an
adimensional index. Of course, the DTFT of the AC sequence must exist in order
to be able to define the power spectrum: the AC sequence cannot have infinite
energy. We will introduce briefly the spectrum of deterministic general power
signals in Sect. 5.7.2.

Before turning to the Wiener-Khinchin theorem, we must introduce autocorrelation
and autocovariance sequences.

5.6   Correlation of Deterministic Discrete-Time Signals

Correlation, also called *cross-correlation*, is an operation mathematically similar
to a convolution that is performed on a couple of signals to measure their degree
of similarity. *Autocorrelation* is correlation of a signal coupled and compared with
itself. Cross- and autocorrelation have important applications, especially in the field
of random signal analysis, where these quantities allow for a spectral representation
of the signal that would otherwise be impossible. However, here we introduce these
concepts for deterministic signals.

5.6.1   Correlation of Energy Signals

Let us consider two deterministic real sequences $x[n]$ and $y[n]$, both having finite
energy. The signals are assumed to have zero mean. We define the *cross-correlation*
of $x[n]$ with $y[n]$ as the sequence

$$r_{xy}[l] = \sum_{n=-\infty}^{+\infty} x[n]y[n-l] = x[l] * y[-l],$$

where l represents a discrete-time delay or *lag* of any sign. Here we used the definition of linear convolution, i.e., $\sum_{n=-\infty}^{+\infty} x[n]y[l-n]$, to express $r_{xy}[l]$ as the linear
convolution of the first sequence with the folded version of the second sequence.[5] In

---

[5]For complex sequences, $r_{xy}[l]$ is the linear convolution of the first sequence with the folded- and
complex-conjugated version of the second sequence. We consider real signals in order to simplify

the same way, cross-correlation of $y[n]$ with $x[n]$ is

$$r_{yx}[l] = \sum_{n=-\infty}^{+\infty} y[n]x[n-l] = y[l] * x[-l].$$

What we have called a "correlation" in the present discussion could also be called a "covariance". In fact, the word "correlation" refers to multiplying $x[n]$ by $y[n-l]$, while the word "covariance" refers to multiplying the respective deviations from the mean; since we assumed zero mean, there is no difference between these two concepts in the present case. If the sequences were not centered, we should distinguish between correlation and covariance.

These quantities can have any sign or be zero, and quantify how much the variations of one variable are related to those of the other variable. For example, $r_{xy}[l]$ will be positive and large if the variations of $y[n-l]$ are similar to those of $x[n]$. A value $r_{xy}[l] < 0$ is often referred to as an anticorrelation, meaning that while one variable increases, the other one decreases, and vice-versa. Since $y[n-l]$ with $l > 0$ is a delayed version of the signal $y[n]$, in $r_{xy}[l]$ we couple the sequence $x[n]$ with the sequence $y[n]$ delayed by l time steps; in $r_{yx}[l]$ we couple the sequence $y[n]$ with the sequence $x[n]$ delayed by l time steps. It follows that

$$r_{xy}[l] = r_{yx}[-l],$$

and therefore the sequences $r_{xy}[l]$ and $r_{xy}[l]$ provide the same information concerning the similarity between the two signals.

If $y[n] \equiv x[n]$ we have *autocorrelation*:

$$r_{xx}[l] = \sum_{n=-\infty}^{+\infty} x[n]x[n-l] = x[l] * x[-l].$$

Autocorrelation at lag l describes how the variations of $x[n]$ repeat themselves after l time steps. Any signal will be perfectly autocorrelated with itself at zero lag.

Correlation has the following properties:

1. autocorrelation at zero lag gives the signal's energy,

$$\mathcal{E} = r_{xx}[0] = \sum_{n=-\infty}^{+\infty} |x[n]|^2 \, ;$$

2. the value of cross-correlation can never exceed the square root of the product of individual energies,

---

the notation; if we considered complex signals, all the formulas for correlation should contain a conjugation sign, e.g., $r_{xy}[l] = \sum_{n=-\infty}^{+\infty} x[n]y^*[n-l]$. This is necessary, for instance, if we are dealing with a complex exponential signal.

$$r_{xy}[l] \leq \sqrt{r_{xx}[0]r_{yy}[0]};$$

3. autocorrelation has a maximum at zero lag because then the signal is coupled with itself,

$$r_{xx}[l] \leq r_{xx}[0];$$

4. autocorrelation is an even sequence, and therefore we only need to compute $r_{xx}[l]$ at lag $l \geq 0$ to know the entire autocorrelation sequence:

$$r_{xy}[l] = r_{yx}[-l] \qquad \rightsquigarrow \qquad r_{xx}[l] = r_{xx}[-l].$$

Sometimes, instead of $r_{xy}[l]$ and $r_{xx}[l]$, standardized quantities are preferred, called the *cross-correlation coefficient* and the *autocorrelation coefficient* at lag l, respectively:

$$\rho_{xy}[l] = \frac{r_{xy}[l]}{\sqrt{r_{xx}[0]r_{yy}[0]}}, \qquad |\rho_{xy}[l]| \leq 1,$$

$$\rho_{xx}[l] = \frac{r_{xx}[l]}{r_{xx}[0]}, \qquad |\rho_{xx}[l]| \leq 1.$$

As an illustration of the notion of auto- and cross-correlation, let us take three sampled signals, which for convenience we will however indicate as analog signals $x(t)$, $y(t)$, and $z(t)$, with $t = nT_s$ expressed in seconds. Each signal has $N = 2^9$ samples and the sampling frequency is $f_s = 400\,\text{Hz}$, corresponding to a sampling time $T_s = 0.0025\,\text{s}$. Each signal contains a single triangular pulse. In signal $x(t)$, which will be our reference signal, the pulse occurs at a time $t = 0.6\,\text{s}$ (240th sample). In signal $y(t)$, the pulse occurs at a time $t = 0.1\,\text{s}$ (40th sample). In signal $z(t)$, the pulse occurs at a time $t = 1.1\,\text{s}$ (440th sample). These three signals are shown in Fig. 5.12a as *solid, dashed* and *dotted lines*, respectively. We can state that

- $y(t) = x(t + 0.5)$ leads $x(t)$;
- $z(t) = x(t - 0.5)$ lags $x(t)$.

This means that

- the *delay* of $y(t)$ with respect to $x(t)$ is $-0.5\,\text{s}$;
- the *delay* of $z(t)$ with respect to $x(t)$ is $+0.5\,\text{s}$.

Now, *the delay between two signals is given by the negative of the lag for which the normalized cross-correlation has the largest absolute value.* The normalized autocorrelation of $x(t)$, denoted by $\rho_{xx}(t)$, is obviously maximum at zero lag (Fig. 5.12b). The normalized cross-correlation of $x(t)$ with $y(t)$, denoted by $\rho_{xy}(t)$ (Fig. 5.12c), is maximum at lag $+0.5\,\text{s}$ (delay $-0.5\,\text{s}$), correctly indicating that $y(t)$ leads $x(t)$. The cross-correlation of $x(t)$ with $z(t)$, denoted by $\rho_{xz}(t)$ (Fig. 5.12d), is maximum at lag $-0.5\,\text{s}$ (delay $+0.5\,\text{s}$), correctly indicating that $z(t)$ lags $x(t)$. This can be understood thinking that when computing the cross-correlation of $x(t)$ with $y(t)$, we must *delay* $y(t)$ by a lag of $+0.5\,\text{s}$ to make it equal to $x(t)$.

Fig. 5.12 Illustration of the notions of auto-and cross-correlation. **a** Three signals, each containing a single triangular pulse. The signal $y(t) = x(t + 0.5)$ leads $x(t)$, taken as the reference signal; the signal $z(t) = x(t - 0.5)$ lags $x(t)$. **b** Autocorrelation coefficient of $x(t)$, denoted by $\rho_{xx}(t)$; **c** cross-correlation coefficient of $x(t)$ with $y(t)$, denoted by $\rho_{xy}(t)$; **d** cross-correlation coefficient of $x(t)$ with $z(t)$, denoted by $\rho_{xz}(t)$

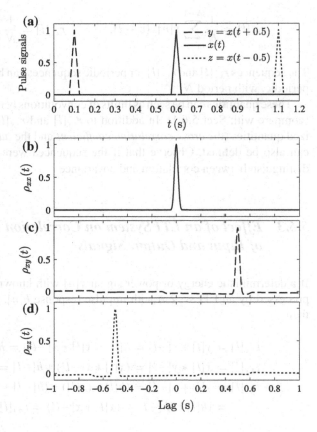

5.6.2   *Correlation of Power Signals*

If we consider two centered sequences having infinite energy, provided they have finite power we can still define a cross-correlation between them by writing

$$r_{xy}[l] = \lim_{M \to \infty} \frac{1}{2M + 1} \sum_{n=-M}^{+M} x[n]y[n - l].$$

Autocorrelation consequently can be written as

$$r_{xx}[l] = \lim_{M \to \infty} \frac{1}{2M + 1} \sum_{n=-M}^{+M} x[n]x[n - l].$$

This is the case, for example, of periodic sequences with period N. For these sequences, the limits in the definitions given above coincide with mean values of *cross-products* over a single period:

$$r_{xy}[l] = \frac{1}{N} \sum_{n=0}^{N-1} x[n]y[n-l], \qquad r_{xx}[l] = \frac{1}{N} \sum_{n=0}^{N-1} x[n]x[n-l].$$

The sequences $r_{xy}[l]$ and $r_{yx}[l]$ for periodic sequences can be shown to be themselves periodic, with period N.

These quantities can be seen as limits of convolutions between finite-length signals (compare with Sect. 5.6.1). In addition to $r_{xy}[l]$ and $r_{xx}[l]$, corresponding normalized quantities (the *cross-correlation coefficient* and the *autocorrelation coefficient*) can also be defined. Observe that if the sequences were not centered, we should distinguish between correlation and covariance.

5.6.3   Effect of an LTI System on Correlation Properties of Input and Output Signals

If a deterministic energy or power signal $x[n]$ with known autocorrelation $r_{xx}[l]$ is processed by an LTI system with impulse response $h[n]$, and if $y[n]$ is the output, then

$$r_{yx}[l] = y[l] * x[-l] = h[l] * (x[l] * x[-l]) = h[l] * r_{xx}[l],$$
$$r_{xy}[l] = x[l] * y[-l] = (x[l] * x[-l]) * h[-l] = r_{xx}[l] * h[-l],$$
$$r_{yy}[l] = y[l] * y[-l] = (h[l] * (x[l])) * (h[-l] * x[-l]) =$$
$$= (h[l] * (h[-l])) * (x[l] * x[-l]) = r_{hh}[l] * r_{xx}[l],$$

where $r_{hh}[l] = h[l] * h[-l]$ exists if the system is stable, that is, if $h[n]$ is absolutely summable. From the last equation we deduce that the output energy is equal to

$$r_{yy}[0] = \sum_{k=-\infty}^{+\infty} r_{hh}[k]r_{xx}[k],$$

where we took into account the even symmetry of the autocorrelation sequence.

5.7   Wiener-Khinchin Theorem

We can now introduce the Wiener-Khinchin theorem. We will treat energy and power signals separately.

5.7.1 Energy Signals and Energy Spectrum

Given two energy signals, both allowing for a DTFT representation:

$$x[n] \Longleftrightarrow X(e^{j\omega}), \qquad y[n] \Longleftrightarrow Y(e^{j\omega}),$$

it can be shown, using the convolution theorem for the DTFT and the folding property of the DTFT, that

$$r_{xy}[l] \Longleftrightarrow X(e^{j\omega})Y(e^{-j\omega}) \equiv cross - energy\ spectral\ density\ \text{(CESD)},$$

which is often simply referred to as the *cross-energy spectrum*. In the case of $y[n] \equiv x[n]$ we thus have

$$r_{xx}[l] \Longleftrightarrow X(e^{j\omega})X(e^{-j\omega}) = X(e^{j\omega})X^*(e^{j\omega}) = \left|X(e^{j\omega})\right|^2,$$

representing the *energy spectrum*. In fact, since

$$r_{xy}[l] = \sum_{n=-\infty}^{+\infty} x[n]y[n-l] = \sum_{n=-\infty}^{+\infty} x[n+l]y[n],$$

passing to the frequency domain and observing that both $r_{xy}[l]$ and $r_{xx}[l]$ certainly have finite energy for two finite-energy signals $x[n]$ and $y[n]$ we obtain the DTFTs,

$$R_{xy}(e^{j\omega}) = \sum_{l=-\infty}^{+\infty} r_{xy}[l]e^{-j\omega l} = \sum_{l=-\infty}^{+\infty} (x[l] * y[-l])\,e^{-j\omega l} = X(e^{j\omega})Y(e^{-j\omega}),$$

and

$$R_{xx}(e^{j\omega}) = \sum_{l=-\infty}^{+\infty} r_{xx}[l]e^{-j\omega l} = \sum_{l=-\infty}^{+\infty} (x[l] * x[-l])\,e^{-j\omega l} = \left|X(e^{j\omega})\right|^2.$$

For Parseval's theorem for the DTFT we can write

$$\mathscr{E} = \sum_{n=-\infty}^{+\infty} |x[n]|^2 = \frac{1}{2\pi} \int_{-\pi}^{+\pi} \left|X(e^{j\omega})\right|^2 d\omega = \frac{1}{2\pi} \int_{-\pi}^{+\pi} R_{xx}(e^{j\omega})d\omega,$$

i.e., the quantity $R_{xx}(e^{j\omega}) = \left|X(e^{j\omega})\right|^2$ describes the frequency distribution of the signal's finite energy. Therefore we can state that the DTFT of the autocorrelation sequence gives the frequency distribution of the signal's energy, representing the energy spectrum: this is the content of the Wiener-Khinchin theorem for energy signals (Wiener 1930; Khinchin 1934). Note that since $r_{xx}[l]$ is an even sequence, its DTFT is purely real, as required for it to represent an energy spectrum.

5.7.2 Power Signals and Power Spectrum

We defined

$$P_{xx}(e^{j\omega}) = \lim_{M\to\infty} \frac{1}{2M+1}|X_M(e^{j\omega})|^2 = \lim_{M\to\infty} \frac{1}{2M+1}\left|\sum_{n=-M}^{+M} x[n]e^{-j\omega n}\right|^2.$$

We can thus write

$$P_{xx}(e^{j\omega}) = \lim_{M\to\infty} \frac{1}{2M+1} \sum_{m=-M}^{+M} \sum_{n=-M}^{+M} x[n]x[m]e^{-j\omega(m-n)}.$$

It can be shown that the double sum appearing above can be written in terms of the AC sequence: setting $k = m - n$ we have

$$\sum_{m=-M}^{+M} \sum_{n=-M}^{+M} x[n]x[m]e^{-j\omega(m-n)} = \sum_{k=-2M}^{+2M} (2M+1-|k|)\, r_{xx}[k]e^{-j\omega k}.$$

It follows that

$$P_{xx}(e^{j\omega}) = \lim_{M\to\infty} \frac{1}{2M+1} \sum_{k=-2M}^{+2M} (2M+1-|k|)r_{xx}[k]e^{-j\omega k} =$$

$$= \lim_{M\to\infty} \sum_{k=-2M}^{+2M} \left(1 - \frac{|k|}{2M+1}\right) r_{xx}[k]e^{-j\omega k} \simeq \sum_{k=-\infty}^{+\infty} r_{xx}[k]e^{-j\omega k},$$

provided that the (finite-energy) sequence $r_{xx}[k]$ decays sufficiently rapidly with lag. This is an alternative definition of power spectrum as the DTFT of the AC sequence, and is the conceptual content of the Wiener-Khinchin theorem for power signals. Note again the reality of the AC's DTFT, related to the even symmetry of the AC sequence.

For some signals, the AC sequence does not decay quickly enough for the theorem to hold. Examples are sequences with non-zero mean (a case which we already excluded) and sequences which are periodic, such as linear combinations of complex exponentials. The definition of power spectrum can then be accommodated through the use of the Dirac δ function (discrete spectrum).

For a signal containing periodic and non-periodic variability components, we can write the power spectrum in the general form

$$P_{xx}(e^{j\omega}) = P_{xx}^c(e^{j\omega}) + \sum_i 2\pi P_i \delta(\omega - \omega_i),$$

where $P_{xx}^c(e^{j\omega})$ represents the *continuous* part of the spectrum, while the weighted sum of impulses represents the discrete part, i.e., the spectral lines. Note that any signal whose mean value is non-zero has an impulse in the spectrum at zero frequency, corresponding to a constant term that appears in the AC sequence. Thus, in order for a signal with no periodic components to have a purely continuous spectrum, its mean must be zero. This explains why it is advisable to always center a signal before analyzing it.

The concept of power spectrum will be resumed and deepened later, when discussing the case of random signals. Typically, random signals are power signals, and their spectral description uses the DTFT of the AC sequence. In that case, as we will see in Chap. 9, the conceptual substitution of energy with power involves an additional averaging step over all possible values of the random signal, which in the presence of a property of the random signal, known as ergodicity, actually becomes a time average over the record length.

We can summarize our discussion on the spectral representation of infinitely-long deterministic energy- and power- signals as follows:

1. an energy signal

- always has a DTFT;
- always has an energy spectral density (ESD) which is the squared magnitude of the signal's DTFT;
- always has an AC sequence;
- always has an energy spectral density expressed as the DTFT of the AC sequence;

2. a power signal

- *may* have a DTFT in the sense of distributions (periodic power signals);
- *may* have a power spectral density related to the squared magnitude of the DTFT of the signal;
- always has an AC sequence, but this sequence is defined differently than in the energy-signal case;
- has a power spectral density defined as the DTFT of the AC sequence, provided that this DTFT converges, which happens if the AC sequence has finite energy, i.e., decays with increasing lag in a sufficiently rapid manner.

References

Harris, F.J.: On the use of windows for harmonic analysis with the discrete fourier transform. Proc. IEEE **66**(1), 51–83 (1978)

Jerison, D., Singer, I.M., Stroock, D.W.: The Legacy of Norbert Wiener: a Centennial Symposium. In: Proceedings of Symposia in Pure Mathematics. American Mathematical Society (SIAM) (1997)

Kaiser, J.F.: Digital filters, Chap. 7. In: Kuo, F.F., Kaiser, J.K. (eds.) System Analysis by Digital Computer. Wiley, New York (1966)

Kaiser, J.F.: Nonrecursive digital filter design using the I_0-sinh Window Function. In: Proceedings 1974 IEEE International Symposium on Circuits and Systems, San Francisco, pp. 20–23 (1974)

Kaiser, J.F., Schafer, R.W.: On the use of the I_0-sinh window for spectrum analysis. IEEE Trans. Acoust. Speech Signal Process. **ASSP-28**(1), 105–107 (1980)

Khinchin, A.Y.: Korrelationstheorie der stationären stocastischen Prozesse. Math. Annalen **109**, 604–615 (1934)

Nuttall, A.H.: Some windows with very good sidelobe behavior. IEEE Trans. Acoust. Speech Signal Process. **29**(1), 84–91 (1981)

Oppenheim, A.V., Schafer, R.W.: Discrete-time Signal Processing. Prentice Hall, Englewood Cliffs (2009)

Riley, K.F., Hobson, M.P., Bence, S.J.: Mathematical Methods for Physics and Engineering. Cambridge University Press, Cambridge (2006)

Wiener, N.: Generalized harmonic analysis. Acta Math. **55**, 117–258 (1930)

Part II
Digital Filters

Part II
Digital Filters

Chapter 6
Digital Filter Properties and Filtering Implementation

6.1 Chapter Summary

In this chapter we will focus on frequency-selective filters that are LTI stable systems with a real and causal impulse response and a rational transfer function. We will examine their properties in detail. The frequency response of these filters is classified according to four prototypes: the lowpass, highpass, bandpass, and bandstop ideal filters. However, ideal filters, which are real—i.e., their frequency response has zero phase—and present jump discontinuities at band edges, are not computationally realizable, and must be approximated by continuous complex functions. The conditions for realizability are discussed in terms of magnitude and phase of the frequency response. We will see that a realizable filter still presents jump discontinuities in the phase that can be eliminated passing to another representation of the frequency response, the so-called continuous-phase representation. Linear phase (LP) and generalized linear phase (GLP) filters are then studied, which allow to filter a sequence without phase distortion between input and output waveforms. The absence of phase distortion is a feature that is appreciated in many applications. We will show that only FIR filters of four types can have exactly LP/GLP, and that their impulse response must satisfy symmetry conditions. The results obtained in this discussion will be used in the following chapters on filter design. The last part of the chapter deals with implementing digital filtering, a goal that is achieved by arranging the difference equation representing the input-output relation of the filter into the most convenient implementation structure. Finally, the possibilities offered by increasing the signal's sampling interval before filtering will be explored, as well as the precautions to be adopted before downsampling a signal, in order to avoid frequency aliasing issues.

© Springer International Publishing Switzerland 2016
S.M. Alessio, *Digital Signal Processing and Spectral Analysis for Scientists*, Signals and Communication Technology, DOI 10.1007/978-3-319-25468-5_6

6.2 Frequency-Selective Filters

Filtering a signal means, in general, changing its characteristics in the frequency domain. In its most general form, filtering is a process that will transform the spectrum of a signal according to some rule of correspondence: any time a processing operation performed on a signal entails modifying, reshaping, or transforming the spectrum of a signal in some way, then the processing involved will be referred to as filtering.

Filters can be classified on the basis of their operating signals as analog or digital. In analog filters, the input, output, and internal signals are in the form of continuous-time signals, whereas in digital filters they are in the form of discrete-time signals. In this book, we will deal with digital filters; analog filters will be involved in our discussion only marginally, in connection to the fact that, classically, digital IIR filters are derived from their analog counterparts. Digital IIR filters evolved as a natural extension of analog filters and are often designed through the use of analog-filter methodologies.

There are many ways in which filters can be classified according to the task for which they are designed. In this book, we focus *frequency-selective filters*.

Frequency selection means that if the input signal contains contributions in some frequency intervals, we want to modify them, attenuating or suppressing, for example, the components pertaining to a certain frequency band, and preserving or enhancing those pertaining to another frequency band, etc. In other words, this type of filtering is used to select one or more desirable bands of frequency components and simultaneously reject one or more undesirable bands. For example, we could use lowpass filtering to select a band of preferred low frequencies and reject a band of undesirable high frequencies; use highpass filtering to select a band of preferred high frequencies and reject a band of undesirable low frequencies; use bandpass filtering to select a band of intermediate frequencies and reject low and high frequencies; or use band-stop filtering to reject a band of intermediate frequencies and select low and high frequencies.

Frequency-selective filters are classified referring to four ideal real and zero-phase prototypes, which are not realizable in practice but represent a useful theoretical reference. The typical frequency responses of the four prototypes are visible in Fig. 6.1. Due to their shape, sometimes these filters are colloquially referred to as brick-wall filters, or top-hat filters. A frequency interval in which the frequency response of a filter is different from zero—and is equal to 1 in the ideal case of Fig. 6.1—is called a *passband*; a frequency interval in which the amplitude response is ideally equal to zero is called a *stopband*. Note that the bandpass filter also summarizes in itself, as particular cases, the lowpass and the highpass filters. Therefore when we speak of a frequency-selective filter without providing any further specification, we usually mean a bandpass system. More general cases may include *multiband filters*, with several stopbands and passbands. In each band, $|H(e^{j\omega})|$ is normally assumed to be constant, but not necessarily equal to 1.

Each of the frequency responses shown in Fig. 6.1 may be seen as the *desired* amplitude response of the filter to be designed, but in reality, no filter can have such

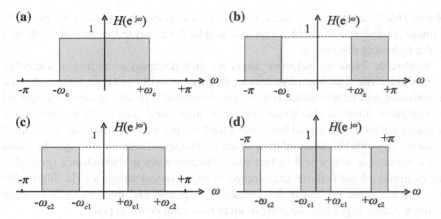

Fig. 6.1 The four prototypes of frequency-selective filters: **a** lowpass, **b** highpass, **c** bandpass, and **d** bandstop

a boxy shape. Indeed, a theorem known as the *Paley-Wiener theorem*[1] implies that for a causal filter to be computationally realizable, $|H(e^{j\omega})|$ can possibly be zero at a number of isolated ω values, but cannot be zero over a continuous ω band of finite width. Therefore, no ideal filter can be causal. A computationally realizable causal filter will have a $|H(e^{j\omega})|$ characterized by

- non-constant behavior in the passband and in the stopband: typically, monotonic or oscillatory behavior.[2] If the behavior is oscillatory, this is also referred to as the frequency response presenting *ripple* in either band;
- the presence of a *transition band* with non-zero width, interposed between the stopband and the passband, in which monotonic behavior is usually requested.

Concerning the phase response, evidently an ideal filter should imply no phase difference between the input and the output, but a causal system cannot have zero

---

[1]This theorem was enunciated by Raymond Paley (1907–1933) and Norbert Wiener (1894–1964) and yielded a method for determining whether or not a causal impulse response exists for a given magnitude frequency response (Paley and Wiener 1933, 1934).

The Paley-Wiener Theorem (see, e.g., Paarmann 2001) states that given a magnitude frequency response that is square-integrable, then a necessary and sufficient condition for it to be the magnitude frequency response of a causal filter is that the following inequality be satisfied:

$$\int_{-\infty}^{+\infty} \frac{\left|\log|H(e^{j\omega})|\right|}{1 + \omega^2} d\omega \le \infty.$$

As a consequence of the Paley-Wiener Theorem, it can be shown that causal filters, having impulse response constrained to be identically zero over the whole negative half n-axis, cannot have continuous bands where the frequency response has zero amplitude.

[2]As we shall see in Chap. 8, some filters can be nearly flat in the passband.

phase: $H(e^{j\omega})$ must be complex. A non-zero phase is thus accepted; when possible, a linear dependence of the phase on ω is sought, for a reason that will be explained in the following discussion.

Realizable frequency-selective filters are then designed according to a number of methods. The rejection of undesired frequencies will not be perfect; undesired frequencies will be attenuated rather than eliminated. For this reason the stopband of realizable filters is also called the *attenuation band*. The realizable frequency response generally has a fixed form with adjustable parameters, which are determined in such a way that the designed filter meets certain specifications. The specifications are somewhat arbitrary, and certain filter characteristics are sometimes ignored— for example, phase features are ignored in the classical design of IIR filters. No discontinuities are present in the frequency response, and the transition between different bands is gradual, i.e., one or more transition bands appear.

6.3   Real-Causal-Stable-Rational (RCSR) Filters

We now restrict our attention to realizable frequency-selective filters that are *LTI stable systems* with a *real* and *causal* impulse response and with a *rational* transfer function. These systems are referred to as RCSR filters, where RCSR is the acronym for "Real-Causal-Stable-Rational". We must recall that for a real filter, i.e., a filter with a real impulse response, the magnitude of $H(e^{j\omega})$ is an even function of frequency and the phase is an odd function. Due to this symmetry, the discussion of RCSR filters can be restricted to the frequency interval $0 \leq \omega \leq \pi$ (recall that $-\pi$ and π are indistinguishable frequencies).

A digital RCSR filter is specified giving its rational transfer function, together with its ROC. This is equivalent to giving its LCCDE, together with initial rest conditions. Since we assumed the system to be stable, $H(z)$ absolutely converges on the unit circle and there represents the filter frequency response $H(e^{j\omega})$. This implies dealing with functions of frequency without any discontinuity, as is true for all realizable filters. The *ideal* frequency-selective filters, that present jump discontinuities at band edges, are limit cases in which the frequency response converges only in a wider sense, and the transfer function does not exist (see Chap. 3).

The frequency response $H(e^{j\omega})$ can be separated into:

- the *amplitude response* or *magnitude response* $|H(e^{j\omega})|$, and
- the *phase response* $\arg[H(e^{j\omega})]$.

We must always remember that even in applications concerning digital signals deriving from analog signals via periodic sampling, the sampling interval T_s does not play any role in digital filter design. Digital filters are always specified as functions of ω, or ν. We will normally use ω.

6.4   Amplitude Response

The amplitude response is a real function of frequency, intrinsically non-negative, which in ideal cases can have jump discontinuities, while in realizable cases must be a continuous function. It can be expressed and plotted using linear or logarithmic units; in the case of logarithmic units, the square amplitude $|H(e^{j\omega})|^2$ in dB is used,

$$10\log_{10}|H(e^{j\omega})|^2 = 20\log_{10}|H(e^{j\omega})|.$$

$|H(e^{j\omega})| = 1$ corresponds to zero dB, $|H(e^{j\omega})| = 10^m$ corresponds to $20m$ dB, and $|H(e^{j\omega})| = 2^m$ corresponds to about $6m$ dB, because

$$20\log_{10} 2^m = 20\log_{10} 2 \cdot \log_2 2^m \simeq 20 \times 0.3 \times m = 6m.$$

At frequencies where $|H(e^{j\omega})| < 1$, i.e., in the attenuation band, the corresponding amplitude in dB is negative. In this case, often the same quantity reversed in sign is used, i.e. $-20\log_{10}|H(e^{j\omega})|$. This is called *attenuation* in dB and is positive if $|H(e^{j\omega})| < 1$. For example, $|H(e^{j\omega})| = 0.0001$ corresponds to an attenuation of $-20 \times (-4) = +80$ dB.

The input-output relation of an LTI system in the frequency domain, $Y(e^{j\omega}) = H(e^{j\omega})X(e^{j\omega})$, when expressed using square amplitudes in dB becomes a sum: the squared amplitude response in dB added to the squared amplitude in dB of the DTFT of the input signal gives the squared amplitude in dB of the DTFT of the output signal. For example, if a given component of the input signal amounts to 60 dB, a filter with an attenuation of 40 dB will reduce it to $60 - 40 = 20$ dB in output.

6.5   Phase Response

The phase response of an RCSR filter can exhibit discontinuities that are elimi- nated by passing to another representation of the frequency response, the so-called continuous-phase representation.

6.5.1   Phase Discontinuities and Zero-Phase Response

Let us consider the frequency response of an LTI-RCSR filter. The phase of $H(e^{j\omega})$ at a given ω is the *phase angle* of the complex number $H(e^{j\omega})$ that, when calculated by a suitable software algorithm, is defined as the *principal value* of the angle itself, i.e., the angle $\in (-\pi, \pi]$ such that

$$\psi(\omega) \equiv \arg[H(e^{j\omega})] = \begin{cases} \arctan 2\left\{\operatorname{Im}[H(e^{j\omega})], \operatorname{Re}[H(e^{j\omega})]\right\} & \text{if } |H(e^{j\omega})| \neq 0, \\ \text{undefined} & \text{if } |H(e^{j\omega})| = 0. \end{cases}$$

Here, $\arctan 2(y, x) \equiv \theta$ is by definition the unique angle $\theta \in (-\pi, \pi]$ for which

$$\cos\theta = \frac{x}{\sqrt{x^2 + y^2}}, \qquad \sin\theta = \frac{y}{\sqrt{x^2 + y^2}}.$$

Any other angle related to $\arg[H(e^{j\omega})]$ by

$$\angle H(e^{j\omega}) = \arg[H(e^{j\omega})] + 2\pi r(\omega),$$

with $r(\omega)$ being a positive or negative integer that can vary with ω, gives, together with $|H(e^{j\omega})|$, the correct complex value of $H(e^{j\omega})$.

The angle $\psi(\omega)$ can be a discontinuous function. More precisely, since the filter is RCSR, $H(e^{j\omega})$ is a continuous function and this implies that $\psi(\omega)$ also is a continuous function, except for two cases:

1. at frequency values ω_0 for which $\operatorname{Re}[H(e^{j\omega})] < 0$ and $\operatorname{Im}[H(e^{j\omega})] = 0$, the value of $\psi(\omega_0)$ is set to $+\pi$, due to the definition of arctan 2. Therefore, if $\operatorname{Im}[H(e^{j\omega})]$ at ω_0^- or at ω_0^+ (or at both frequencies) is slightly negative, then a 2π phase jump occurs, because $\psi(\omega_0^+)$ and/or $\psi(\omega_0^-)$ are set by arctan 2 to values close to $-\pi$ (see Figs. 6.2 and 6.3);

2. at frequency values ω_0 for which $|H(e^{j\omega})| = 0$, $\arg[H(e^{j\omega})]$ is undefined. This fact may imply a π jump as the function $\arg[H(e^{j\omega})]$ passes through ω_0. These jumps are related to the fact that $|H(e^{j\omega})|$ is forcedly non-negative; when the function passes through a zero with multiplicity m, then for odd m-values the phase undergoes a π jump, while for even m-values the phase remains continuous. In Fig. 6.4, zeros of the frequency response corresponding to π jumps in the phase are marked by *arrows*.

Fig. 6.2 The neighborhood of $\pm\pi$ in which the phase-angle principal value, $\arg[H(e^{j\omega})]$, may undergo 2π jumps (see text)

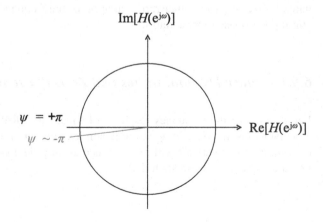

Fig. 6.3 Typical behavior of
the phase-angle principal
value

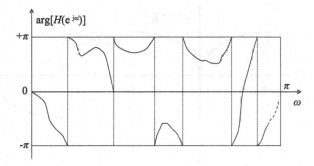

Fig. 6.4 Zeros of the
frequency response that
correspond to π phase jumps

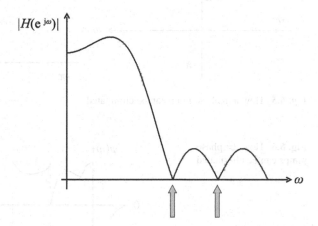

It can be shown that for rational frequency responses, the number of discontinuity
points in the principal interval of ω is necessarily finite.

In order to eliminate these discontinuities, a *continuous-phase representation* is
introduced for the frequency response: the expression

$$H(e^{j\omega}) = H(\omega)e^{j\phi(\omega)},$$

where $H(\omega)$ is real but not necessarily non-negative and $\phi(\omega)$ is a continuous func-
tion, replaces the representation $H(e^{j\omega}) = |H(e^{j\omega})|e^{j\psi(\omega)}$. The function $H(\omega)$ is
referred to as the *amplitude function* or *zero-phase (frequency) response* of the filter.

The functions $H(\omega)$ and $\phi(\omega)$ are constructed in such a way as to preserve the
continuity of $\phi(\omega)$ at any discontinuity point of $\psi(\omega)$. Figure 6.5 shows how a π
jump can be eliminated: the very fact of allowing $H(\omega)$ to be negative makes these
jumps disappear. Figure 6.6 shows how a 2π jump can be eliminated, by adding
integer multiples of 2π to $\psi(\omega)$. The value $\phi(\omega)$ is made unique by imposing the
condition $0 \leq \phi(0) < \pi$; of course, $\phi(\omega)$ can assume values outside the interval
$(-\pi, \pi]$.

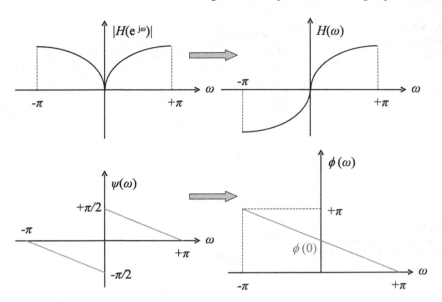

Fig. 6.5 How a π phase jump can be eliminated

Fig. 6.6 How 2π phase jumps can be eliminated

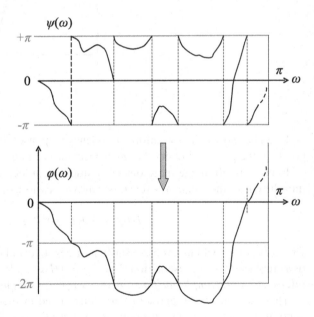

6.5.2   Linear Phase (LP)

A filter for which

$$\phi(\omega) = -\omega\tau,$$

with $\tau = $ constant, i.e., a filter with

$$H(e^{j\omega}) = H(\omega)e^{-j\omega\tau},$$

is referred to as a filter having *linear phase* (LP) in ω. The constant $\tau = -\phi(\omega)/\omega$ is a real number, called the (constant) *phase delay* and measured in number of samples.

A filter having $H(e^{j\omega}) = e^{-j\omega\tau}$ is an *ideal delayer*. This name derives from the fact that a broadband bandpass filter that in the passband has this frequency response, merely applies a pure delay to the input signal. In fact, if we choose a frequency response with an ideal piecewise-constant shape in $[0, \pi]$, with ω_{c1} close to zero and ω_{c2} close to π,

$$H(e^{j\omega}) = \begin{cases} e^{-j\omega\tau} & \text{for } \omega_{c1} \leq \omega \leq \omega_{c2}, \\ 0 & \text{elsewhere,} \end{cases}$$

an input signal made up of elementary sinusoidal contributions with frequencies contained in the filter passband produces the output

$$y[n] = \frac{1}{2\pi} \int_{-\infty}^{+\infty} H(e^{j\omega})X(e^{j\omega})e^{j\omega n}d\omega =$$

$$= \frac{1}{2\pi} \int_{\omega_{c1}}^{\omega_{c2}} X(e^{j\omega})e^{j\omega(n-\tau)}d\omega \approx x[n - \tau].$$

Therefore the output is just a copy of the input, delayed by τ samples. If τ is a fraction, then the system interpolates the output to fractional values of discrete time, while if τ is an integer, a simple delay by τ time steps occurs. In any case, the output is a delayed, but undistorted, copy of the input signal: no *phase distortion* occurs.

The absence of phase distortion is often desired, for example when the behaviors of two different signals in a certain frequency band must be compared. Imagine we filter two related signals using a passband filter and then want to understand which signal leads the variations in that band and which one lags them. We ultimately are investigating cause-effect relations among these two signals. In this case we do not want the relative phase of the corresponding sinusoidal components in the two signals to be altered by filtering. For this reason, considering that zero-phase filters are not computationally realizable, linear-phase filters are well accepted, especially with integer or half-integer delay τ.

6.5.3   Generalized Linear Phase (GLP)

It is sometimes useful to consider phase responses that are not strictly proportional to frequency, but are still linearly dependent on ω:

$$\phi(\omega) = \phi_0 - \omega\tau_g,$$

where ϕ_0 and τ_g are constants. The corresponding ideal delayer has the following frequency response in $[0, \pi]$:

$$H(e^{j\omega}) = \begin{cases} e^{j(\phi_0 - \omega\tau_g)} & \text{for} \quad \omega_{c1} \leq \omega \leq \omega_{c2}, \\ 0 & \text{elsewhere}, \end{cases}$$

that is, once again, a broadband frequency response with an ideal piecewise-constant shape. In this case, in place of a constant phase delay τ we have a constant *group delay* τ_g, measured in number of samples.

The meaning of the term "group delay" emerges if we consider the effect that this ideal delayer has on the phase of a narrow-band signal, i.e., a *modulated signal* of the kind

$$x_1[n] = x[n]\cos(\omega_0 n),$$

in which $x[n]$ is the *modulation signal*—some "lowpass" signal with spectrum concentrated at low frequency—and $\omega_0 \neq 0$ is the frequency of the *carrier wave*. The spectrum of the modulated signal is concentrated in the neighborhood of $\omega = \omega_0$, as it may be understood thinking of the modulation/frequency shift property of the DTFT.[3] Suppose we apply an ideal delayer to the modulated signal. It can be shown that the output $y[n]$ is still a modulated signal, such that its modulation signal is a delayed but undistorted copy of $x[n]$. The phase constant ϕ_0 affects the carrier signal but not the modulation signal: in analog and digital terms,

$$y[t] = x\left(t - \tau_g T_s\right)\cos\left[f_0\left(t - \tau_g T_s\right) + \phi_0\right],$$
$$y[n] = x\left[n - \tau_g\right]\cos\left[\omega_0\left(n - \tau_g\right) + \phi_0\right].$$

This behavior is illustrated in Fig. 6.7. Using the modulation signal shown in Fig. 6.7a ($f = 50\,\text{Hz}$), with sampling time $T_s = 5 \times 10^{-5}$ s, and the carrier wave shown in Fig. 6.7b ($f_0 = \omega_0/(2\pi T_s) = 500\,\text{Hz}$), the modulated signal shown in panel Fig. 6.7c has been created. The modulated signal can be expressed in analog and digital forms as

---

[3] Modulated signals are important in telecommunication engineering. Suppose we want to transmit a narrow-band lowpass signal through a network. This network only lets signals with frequency close to a certain ω_0 pass through (with ω_0 not close to zero) and filters out the rest. If we properly modulate the lowpass signal, we can shift its spectral content to the vicinity of ω_0 and allow signal transmission. Later we will be able to perform the inverse operation and recover the original signal.

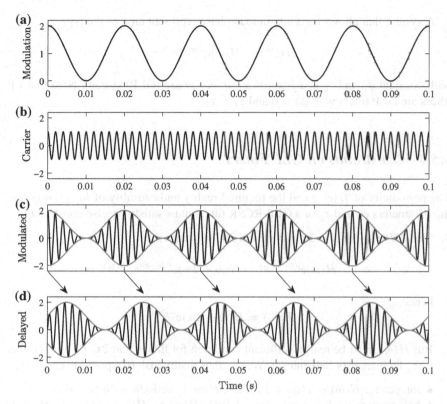

Fig. 6.7 Illustration of the concept of group delay. **a** Low-frequency modulation signal; **b** carrier wave; **c** modulated signal; **d** the modulated signal after passing through an ideal delayer with $\phi_0 = \pi/2$ and $\tau_g = 100$ samples, corresponding in this case to 0.005 seconds (see text for details)

$$x_1(t) = [1 + \cos(2\pi f t)] \cos(2\pi f_0 t),$$
$$x_1[n] = [1 + \cos(2\pi \omega n)] \cos(2\pi \omega_0 n),$$

where t is time in seconds, with $x_1[n] = x_1(nT_s)$. Note the envelope in Fig. 6.7c (*gray curves*). Now suppose we pass this modulated signal through an ideal delayer with $\phi_0 = \pi/2$ and $\tau_g = 100$ samples, corresponding to 0.005 seconds. Figure 6.7d shows the output signal: the delay of the envelope is evident and actually amounts to 0.005 seconds.

The group delay τ_g is given by the reversed-in-sign slope of the linear function $\phi(\omega)$:

$$\tau_g = -\frac{d\phi(\omega)}{d\omega}.$$

If τ_g is an integer, a simple delay of the modulation signal by τ_g time steps occurs; if τ_g is a fraction, then the system interpolates the output signal to fractional values of discrete time.

A system that allows for a continuous-phase representation of the type

$$H(e^{j\omega}) = H(\omega)e^{j(\phi_0 - \omega\tau_g)}$$

with constant ϕ_0 and τ_g is a *generalized linear phase* (GLP) system. Note that LP filters are GLP filters with $\phi_0 = 0$ and $\tau_g = \tau$.

6.5.4 Constraints on GLP Filters

The periodicity of $H(e^{j\omega})$ and the required reality and causality of $h[n]$ imply that the parameters ϕ_0 and τ_g of a GLP-RCSR filters must satisfy precise constraints.

1. Since $H(e^{j\omega})$ is *periodic* with period 2π, we can write

$$H(\omega)e^{j(\phi_0 - \omega\tau_g)} = H(\omega + 2\pi)e^{j(\phi_0 - \omega\tau_g - 2\pi\tau_g)},$$

hence

$$H(\omega) = H(\omega + 2\pi)e^{-j2\pi\tau_g}.$$

But $H(\omega)$ must be real; a sufficient condition for this is that $2\tau_g$ be integer, that is, τ_g be integer or half-integer. We can thus distinguish two possible cases:

- integer τ_g: $H(\omega) = H(\omega + 2\pi)$, i.e., $H(\omega)$ is periodic with period 2π;
- half-integer τ_g: $H(\omega) = -H(\omega + 2\pi)$, $H(\omega) = H(\omega + 4\pi)$, i.e., $H(\omega)$ is periodic with period 4π.

2. Since $h[n]$ is required to be *real*, the corresponding symmetry condition on its transform must hold: $H(e^{-j\omega}) = H^*(e^{j\omega})$. We can thus write

$$H(-\omega)e^{j(\phi_0 + \omega\tau_g)} = H(\omega)e^{-j(\phi_0 - \omega\tau_g)},$$

from which we get

$$e^{j2\phi_0} = \frac{H(\omega)}{H(-\omega)}.$$

We can therefore distinguish two possible cases:

- $\phi_0 = 0$: $H(\omega) = H(-\omega)$, i.e., $H(\omega)$ is an even function of frequency, and $\tau = \tau_g = $ constant;
- $\phi_0 = \pi/2$: $H(\omega) = -H(-\omega)$, i.e., $H(\omega)$ is an odd function of frequency, and $\tau_g = $ constant.

In conclusion, GLP-RCSR digital filters come in four types:

- Type I, with integer $\tau = \tau_g$ and $\phi_0 = 0$;
- Type II, with half-integer $\tau = \tau_g$ and $\phi_0 = 0$;

- Type III, with integer τ_g and $\phi_0 = \pi/2$;
- Type IV, with half integer τ_g and $\phi_0 = \pi/2$.

Types I and II have constant phase delay, while Types III and IV have constant group delay.
3. Requiring *causality* imposes further restrictions on GLP-RCSR filters. This time the restrictions concern the impulse response.

- If $\phi_0 = 0$ (Types I and II), then $H(e^{j\omega}) = H(\omega)e^{-j\omega\tau}$, and therefore

$$h[2\tau - n] = \frac{1}{2\pi} \int_{-\pi}^{\pi} H(e^{j\omega})e^{j\omega(2\tau-n)}d\omega = \frac{1}{2\pi} \int_{-\pi}^{\pi} H(\omega)e^{j\omega(\tau-n)}d\omega.$$

By conjugating this expression and requiring reality for $h[n]$ we get

$$h[2\tau - n] = \frac{1}{2\pi} \int_{-\pi}^{\pi} H(\omega)e^{j\omega(n-\tau)}d\omega = \frac{1}{2\pi} \int_{-\pi}^{\pi} H(e^{j\omega})e^{j\omega n}d\omega = h[n].$$

Now, since the filter must be causal, we must impose the condition $h[n] = 0$ for $n < 0$. But then, the condition $h[n] = 0$ for $n > 2\tau$ must also hold. In conclusion,
  - the filter must be an FIR filter;
  - its length must be $N = 2\tau + 1$ and its order must be $N - 1 = 2\tau$;
  - its impulse response must satisfy the following *symmetry condition*:

$$h[n] = h[N - 1 - n].$$

An example of $h[n]$ satisfying these constraints is given in Fig. 6.8a.
- If $\phi_0 = \pi/2$ (Types III and IV), then $e^{j\phi_0} = e^{j\pi/2} = j$, $H(e^{j\omega}) = jH(\omega)$ $e^{-j\omega\tau_g}$. We can thus write

$$h[2\tau_g - n] = \frac{1}{2\pi} \int_{-\pi}^{\pi} H(e^{j\omega})e^{j\omega(2\tau_g-n)}d\omega = \frac{j}{2\pi} \int_{-\pi}^{\pi} H(\omega)e^{j\omega(\tau_g-n)}d\omega.$$

By conjugating this expression and imposing reality of the impulse response we get

$$h[2\tau_g - n] = -\frac{j}{2\pi} \int_{-\pi}^{\pi} H(\omega)e^{j\omega(n-\tau_g)}d\omega = -\frac{1}{2\pi} \int_{-\pi}^{\pi} H(e^{j\omega})e^{j\omega n}d\omega = -h[n].$$

The filter must be causal, so we must impose the condition $h[n] = 0$ for $n < 0$. But then $h[n] = 0$ for $n > 2\tau_g$ must also hold. In conclusion,
  - the filter must be an FIR filter;
  - its length must be $N = 2\tau_g + 1$ and its order must be $N - 1 = 2\tau_g$;
  - its impulse response must satisfy the following *antisymmetry condition*:

Fig. 6.8 **a** An example of
$h[n]$ satisfying the
constraints on LP-RCSR
filters with $\phi_0 = 0$ (Types I
and II). This Type I filter has
$\tau_g \equiv \tau$; its order is given by
$N - 1 = 2\tau$. **b** An example
of $h[n]$ satisfying the
constraints with $\phi_0 = \pi/2$
(Types III and IV). The order
of this Type III filter is given
by $N - 1 = 2\tau_g$

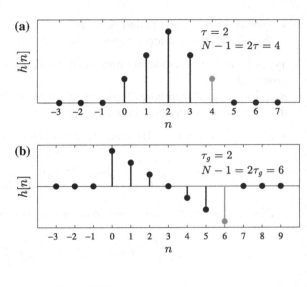

$$h[n] = -h[N - 1 - n].$$

An example of $h[n]$ satisfying these constraints is given in Fig. 6.8b.

In summary, only FIR filters can have exactly linear or generalized linear phase.

6.6 Digital Filtering Implementation

In the next two chapters, we will discuss design methods for FIR and IIR filters. Once
the desired filter has been designed, how will we actually filter our data sequence?
We will do it by implementing the LCCDE.

The design process provides the transfer function,

$$H(z) = \frac{\sum_{k=0}^{M} b_k z^{-k}}{1 + \sum_{k=1}^{N} a_k z^{-k}} = \frac{b(z)}{a(z)}$$

that characterizes the input/output relation, since its coefficients determine the
LCCDE, which here can conveniently be written in the form

$$y[n] = -\sum_{k=1}^{N} a_k y[n - k] + \sum_{k=0}^{M} b_k x[n - k],$$

with finite N and M. For FIR filters, the coefficients of the LCCDE coincide with
the samples of the impulse response; the LCCDE is not recursive, and no initial
conditions are required. For IIR filters, the LCCDE is "naturally" recursive, and

therefore a set of initial conditions is required: the input samples prior to $n = 0$ are assumed to be zero, and for the output samples prior to $n = 0$, initial conditions are provided. Usually, initial rest conditions are assumed. Filtering a signal of length N by any implementation of a given LCCDE will lead to a filtered sequence with the same length N, because, typically, the output is computed starting from the first input sample, and obviously stopping at the last sample. At the beginning of the output sequence, a segment with length equal to the filter order will be affected by the assumption of initial rest conditions. This segment can be seen as an initial transitory filtering phase.

The LCCDE can be implemented by converting it into a software algorithm with a structure in which the three basic operations of addition, multiplication by a constant and delay are properly interconnected. An unlimited variety of algorithms/structures can be devised that correspond to the same input-output relation, i.e., to the same system. The best structure in each particular case is dictated by considerations concerning both efficiency and sensitivity to effects of finite-precision arithmetic. One consideration in choosing among different structures that are a priori equivalent is, in fact, computational complexity: structures with the fewest constant multipliers (high computational speed) and the fewest delay branches (reduction of memory use) are often most desirable. But another major consideration is the potentially negative effect of finite-precision arithmetic. This effect depends on the way in which the computations are organized, i.e., on the structure of the interconnection scheme. Sometimes it is better to use a structure that does not have the minimum number of multipliers and delayers, if that structure is less sensitive to this effect.

More precisely, although two structures may be a priori equivalent with regards to their input/output relation for infinite precision representations of coefficients and variables, i.e., in theory, they may have dramatically different behavior when the numerical precision is limited. Indeed, different structures correspond to different algorithms and therefore imply different effects due to the representation of variables with finite precision and to the round-off of the results of intermediate computations. If all computations were exact, it would not make any difference which of the equivalent structures was used. However, coefficients are stored to finite precision and so are not exact: the filter is actually quantized, and ultimately incorrect. Also, arithmetic calculations are not exact.[4] Arithmetic errors introduce noise that is then filtered by the transfer function between the point of noise creation and the output. These effects are very important for IIR filter implementations, while FIR systems are less affected by them. An in-depth discussion of LTI system realizations and quantization error analysis can be found, for example, in Jones (2005), Schlichthärle (2011), Oppenheim and Schafer (2009).

But why does an unlimited variety of algorithms/structures correspond to the same input-output relation, i.e., to the same system? Once a structure has been devised, we can manipulate it to get an equivalent structure without changing the overall system

---

[4]The worst case for arithmetic errors occurs when calculating the difference between two similar values.

transfer function: why? We already know the answer. All manipulations are based on the linearity of the system and the algebraic properties of the transfer function. Indeed, since the LCCDE set is linear, equivalent sets of difference equations can be obtained simply by linear transformations of the variables. Thus, an unlimited number of equivalent realizations of any given system exists. Each realization represents a different computational way for implementing the system in software.

6.6.1 Direct Forms

Let us consider an LTI system satisfying an LCCDE and therefore having a rational $H(z)$. To make it unique, let us also suppose it is causal. As mentioned above, the input-output relation of such a system can be expressed as an algorithm combining three fundamental operations, namely addition, delay by k time steps (usually $k = 1$) and multiplication by a numerical factor. Such an algorithm defines the system's internal structure, and is represented as a *block diagram* (BD) interconnecting operational blocks of the three types. The pictorial symbols used for the basic blocks are shown Fig. 6.9.

Note that delayers and multipliers are elementary LTI systems. Their transfer function is inscribed in the geometrical symbol that represents them: a rectangle for the delayers and a triangle for the multipliers. In fact, the transfer function of the elementary LTI system that multiplies the input by a factor of a is $H(z) = a$: if $y[n] = ax[n]$, choosing $x[n] = z^n$ we get $y[n] = z^n H(z) = ax[n] = az^n$, from which we deduce that $H(z) = a$. The transfer function of the elementary LTI system that delays the input by k time steps is $H(z) = z^{-k}$. Indeed, if $y[n] = x[n - k]$, choosing $x[n] = z^n$ we get $y[n] = z^n H(z) = x[n - k] = z^{n-k} = z^n z^{-k}$, hence $H(z) = z^{-k}$. The individual delay operations included in a BD are normally unit delays that can be implemented by providing a storage register for each unit delay that is required.

In order to directly represent the most general LCCDE by a BD, the auxiliary sequence $v[n] = \sum_{k=0}^{M} b_k x[n - k]$ is usefully introduced. The LCCDE thus becomes

$$y[n] = - \sum_{k=1}^{N} a_k y[n - k] + v[n],$$

corresponding to the BD shown in Fig. 6.10. Note that the coefficient $a_0 = 1$ is implicit. This implementation is referred to as the *direct form I* implementation: direct forms use coefficients a_k and b_k directly. From the point of view of the transfer

Fig. 6.9 Pictorial symbols for **a** adders, **b** multipliers, and **c** delayers by k time steps

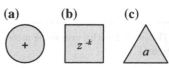

(a)   (b)   (c)

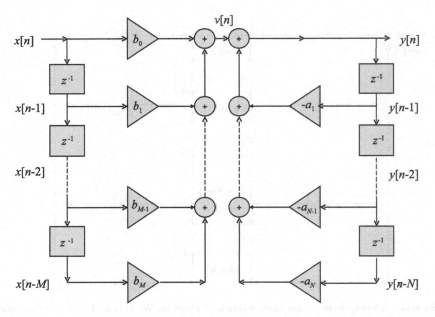

Fig. 6.10 Block diagram for the direct form I implementation of the most general Nth order LCCDE

function, this diagram first implements $b(z)$, i.e., the zeros of $H(z)$, and then $1/a(z)$, i.e., the poles.

This BD can be rearranged and modified in a variety of ways. Let us stress once again that each rearrangement represents a *different* computational algorithm for implementing the *same* system. For example, the BD of Fig. 6.10 can be seen as a cascade of two systems, the first representing the computation of $v[n]$ from $x[n]$, and the second representing the computation of $y[n]$ from $v[n]$. Since both systems are LTI systems (assuming initial rest-conditions for the 1-step-delay elementary systems), the order in which the two systems are cascaded can be reversed, as shown in Fig. 6.11, where for convenience it has been assumed that the two delay chains have a common length N: this implies no loss of generality, since if $M \neq N$, some of the coefficients a_k or b_k in Fig. 6.11 will simply be zero, and the symbol N in Fig. 6.11 will actually indicate max (M, N).

We may further note that exactly the same signal $w[n]$ is stored in the two chains of delay elements in Fig. 6.11. Consequently, the two chains can be collapsed into one, as in Fig. 6.12. This implementation has the minimum possible number of delay elements, namely, max(N, M). An implementation with the minimum number of delays is called a *canonic form* implementation; therefore this structure is commonly referred to as the *direct form II* or *canonic direct form* implementation. This time, from the point of view of the transfer function, the diagram first implements $1/a(z)$, i.e., the poles of $H(z)$, and then $b(z)$, i.e., the zeros.

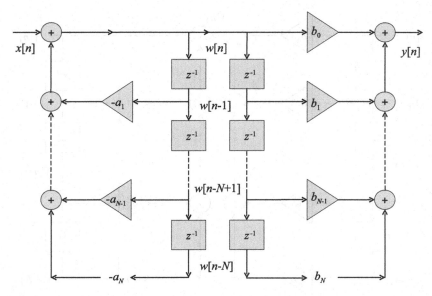

Fig. 6.11 Rearrangement of the block diagram of Fig. 6.10. We assume for convenience that $M = N$. If in reality $M \neq N$, some of the coefficients will be zero

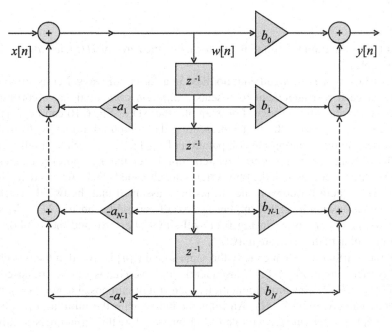

Fig. 6.12 Block diagram for the direct form II or canonic direct form implementation of the Nth order LCCDE

6.6.2   Transposed-Direct Forms

Other important and useful equivalent structures for the general system $H(z) = b(z)/a(z)$ can be derived introducing another way of representing LCCDE implementation. A *linear signal flow graph*, for which we adopt the acronym SFG and that is also known as a *Mason graph*, can be used to represent a difference equation essentially in the same way as a BD, except for a few notational differences. Formally, an SFG is a network of directed branches that connect at nodes. Associated with each node is a variable or node value. An SFG represents multiplications and additions: multiplications are represented by the weights of the branches, also called transmittances, while additions are represented by multiple branches going into one node. An SFG has a one-to-one relation with a system of linear equations (Chen 1967).

As an example of how an SFG appears compared to the corresponding BD, in Fig. 6.13 the two representations of the direct form-II realization of a first-order digital filter are shown. This example can be found in Oppenheim et al. (1999). The SFG in Fig. 6.13b corresponds to the equations

$$w_0[n] = x[n],$$
$$w_1[n] = x[n] - a_1 w_4[n],$$
$$w_2[n] = w_1[n]$$

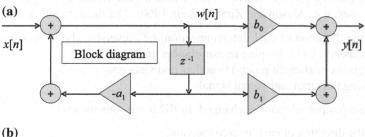

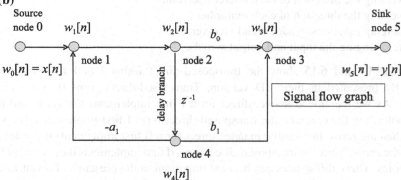

Fig. 6.13 **a** Block diagram for the direct form II of a first-order digital filter. **b** Signal flow graph corresponding to the same block diagram

$$w_3[n] = b_0 w_2[n] + b_1 w_4[n],$$
$$w_4[n] = w_2[n-1],$$
$$w_5[n] = y[n] = w_3[n].$$

There is direct correspondence between branches in the BD and branches in the SFG. In fact, the important difference between the two representations is that nodes in the flow graph represent both branching points and adders, whereas in the block diagram a special symbol is used for adders. A branching point in the BD is represented in the SFG by a node that has only one incoming branch and one or more outgoing branches. An adder in the BD is represented in the SFG by a node that has two or more incoming branches.

SFGs are therefore totally equivalent to BDs. Like BDs, they can be manipulated to gain insight into the properties of a given system. The advantage of using SFGs lies in the fact that a large body of SFG theory exists that can be directly applied to discrete-time systems when they are represented in this way (Chow and Cassignol 1962; Mason and Zimmermann 1960). Alternative structures for LTI causal systems can thus be easily derived: SFG theory provides a variety of procedures for transforming a given SFG into different forms, while leaving the overall input-output relation unchanged.

One of these procedures, called *flow graph transposition* or *flow graph reversal*, leads to transposed system structures that provide useful alternatives to the ones previously presented. Transposed direct forms result from the application of the *transposition theorem*, which derives from Mason's gain formula of signal flow graph theory (see, e.g., Mason and Zimmermann 1960). The theorem states that

- reversing the direction of each interconnection, i.e., reversing the arrows on all network branches, while keeping transmittances unchanged,
- changing junctions (branch points) to adders and vice-versa,
- interchanging the input and output signals,

leaves the input/output relations unchanged. In BD terms, this means

- reversing the direction of each interconnection,
- reversing the direction of each multiplier,
- changing junctions to adders and vice-versa,
- interchanging the input and output signals.

Figures 6.14 and 6.15 show the transposed-direct forms I and II, respectively, directly presented in their BD version. Transposed-direct form II is a canonic form. As we mentioned above, direct form I first implements the zeros and then the poles. On the contrary, the transposed-direct form I first implements the poles and then the zeros. In a similar manner, direct form II first implements the poles and then the zeros, while the transposed-direct form II first implements the zeros and then the poles. These differences can become important in the presence of quantization: a very small change in the coefficients can heavily affect the positions of poles and zeros.

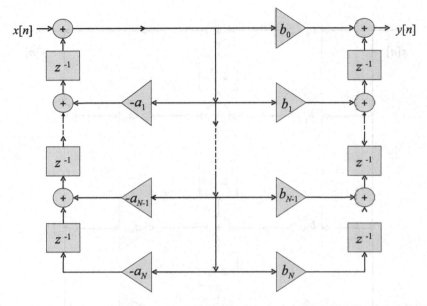

Fig. 6.14 Block diagram for the transposed-direct form I implementation of the Nth order LCCDE. This form results from applying the transposition theorem to the signal flow graph corresponding to the block diagram in Fig. 6.10

6.6.3 FIR Direct and Transposed-Direct Forms

In the case of FIR systems, we have

$$y[n] = \sum_{k=0}^{M} b_k x[n-k] = \sum_{k=0}^{M} h[k]x[n-k],$$

and the two direct-form I and II structures of Figs. 6.10 and 6.12 reduce to the direct form FIR structure shown in Fig. 6.16a. This BD is referred to as a *transverse* or *tapped* filter structure. The corresponding transposed-direct form appears in Fig. 6.16b.

It is important to note that the structure in Fig. 6.16a simply implements $y[n] = h[n] * x[n]$, i.e., the linear convolution of the signal with the filter impulse response. We may recall the fact that FIR filtering via linear convolution can also be implemented in the frequency domain (see Sect. 3.5.4). This may seem an involute approach to filtering, but it is convenient and is often adopted. Efficient implementations of this approach exist, as the *overlap-add method* (Oppenheim et al. 1999), a technique that combines successive frequency domain filtered blocks of an input sequence. More material on filtering implementation via LCCDE (for both IIR and FIR filters) and via linear convolution or overlap-add method (for FIR filters) is provided in Sect. 16.4.1.1, where the issue of the transients present at the edge(s) of the filtered sequence is illustrated in detail.

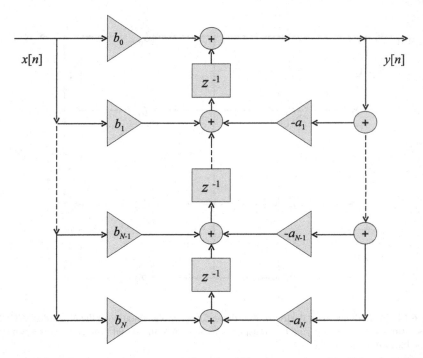

Fig. 6.15 Block diagram for the transposed-direct form II implementation of the Nth order LCCDE. This form results from applying the transposition theorem to the signal flow graph corresponding to the block diagram in Fig. 6.12

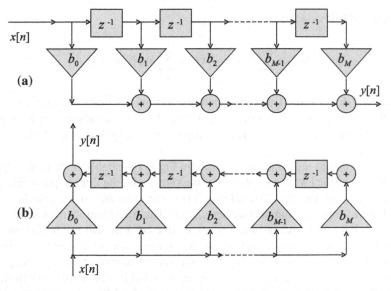

Fig. 6.16 Block diagrams for **a** direct form realization and **b** transposed-direct form realization of an FIR filter

6.6.4 Direct and Transposed-Direct Forms for LP FIR Filters

If the considered FIR system has *linear phase*, the sequence $h[n]$ has particular symmetry properties: for LP filters of the Types I and II, $h[n] = h[N - 1 - n]$. This essentially allows us to halve the number of multipliers in the structure. Figures 6.17 and 6.18 show the corresponding direct and transposed-direct forms, respectively. In these examples, $M = 6$.

6.6.5 Cascade and Parallel Forms

The direct-form structures were obtained from the transfer function written as a ratio of polynomials in z^{-1}. But, as we saw in Sect. 3.2.8, $H(z)$ can be factorized in a way that evidences the positions of zeros and poles. Here we write that factorized form as

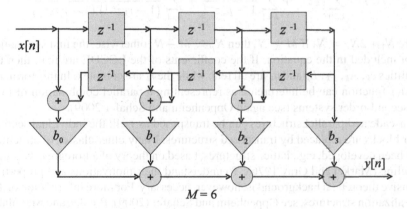

Fig. 6.17 Block diagram for the direct form implementation of a linear phase FIR filter

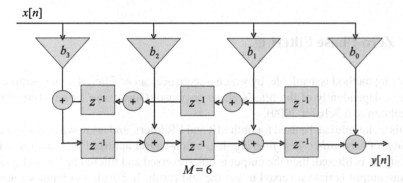

Fig. 6.18 Block diagram for the transposed-direct form implementation of a linear phase FIR filter

$$H(z) = A \frac{\prod_{k=1}^{M_1} \left(1 - w_k z^{-1}\right) \prod_{k=1}^{M_2} \left(1 - u_k z^{-1}\right) \left(1 - u_k^* z^{-1}\right)}{\prod_{k=1}^{N_1} \left(1 - p_k z^{-1}\right) \prod_{k=1}^{N_2} \left(1 - r_k z^{-1}\right) \left(1 - r_k^* z^{-1}\right)},$$

where $M_1 + 2M_2 = M$ and $N_1 + 2N_2 = N$, and where the first-order factors represent real zeros at w_k and real poles at p_k, while the second-order factors represent complex conjugate pairs of zeros at u_k, u_k^*, and complex conjugate pairs of poles at r_k, r_k^*. This is the most general distribution of poles and zeros when all the coefficients a_k and b_k are real.

This factorization suggests a class of structures consisting of a cascade of first- and second-order systems (see, e.g., Oppenheim and Schafer 2009). A variety of theoretically equivalent systems can be obtained by simply pairing the poles and the zeros in different ways.

Alternatively, a partial fraction expansion of the rational $H(z)$ can be exploited. This time we write the partial fraction expansion as

$$H(z) = \sum_{k=0}^{N_p} c_k z^{-k} + \sum_{k=1}^{N_1} \frac{A_k}{1 - p_k z^{-1}} + \sum_{k=1}^{N_2} \frac{B_k \left(1 - q_k z^{-1}\right)}{\left(1 - r_k z^{-1}\right) \left(1 - r_k^* z^{-1}\right)},$$

where $N_1 + 2N_2 = N$. If $M \geq N$, then $N_p = M - N$; otherwise, the first summation is not included in the equation. If the coefficients of the LCCDE are real, then the quantities c_k, A_k, p_k, B_k and q_k are all real, while the r_k are complex. In this form, the transfer function can be interpreted as representing a parallel combination of first- and second-order systems (see again Oppenheim and Schafer 2009).

Cascade and parallel structures can be transposed as well: the individual second-order blocks are replaced by transposed structures. Many other classes of structures have been developed, e.g., lattice structures, based on theory of autoregressive signal modeling (Markel and Gray 1976). To understand their motivations and properties, extensive theoretical background is, however, necessary. For more information on filter realization structures, see Oppenheim and Schafer (2009), Proakis and Manolakis (2007), So (2010).

6.7 Zero-Phase Filtering

A filtering method is available, by which a sequence can be filtered with overall zero phase, independently of the filter's phase response (Gustafsson 1996; Mitra 2001; Oppenheim and Schafer 2009).

This technique can be used for both FIR and IIR filters, and is known as *zero-phase filtering*, or *forward-reverse filtering*: it is a non-causal approach in which first the input signal is filtered, then the output is time-reversed and filtered again, and finally the new output is time-reversed to get the end result. If $x[n]$ is the input sequence and $h[n]$ is the filter's impulse response, we can write

$$x[n] * h[n] = g[n];$$
$$g[-n] * h[n] = r[n];$$
$$y[n] = r[-n].$$

We can best look at this procedure in the frequency domain. The result of the first filter pass is $X(e^{j\omega})H(e^{j\omega})$. Time reversal corresponds to replacing ω by $-\omega$, so after time-reversal we get $X(e^{-j\omega})H(e^{-j\omega})$. The second filter pass corresponds to another multiplication by $H(e^{j\omega})$, i.e., $X(e^{-j\omega})H(e^{-j\omega})H(e^{j\omega})$, which after another time-reversal finally gives $Y(e^{j\omega}) = X(e^{j\omega})H(e^{j\omega})H(e^{-j\omega}) = X(e^{j\omega})|H(e^{j\omega})|^2$ for the spectrum of the output signal. In fact, for real-valued filter coefficients we have $H(e^{-j\omega}) = H^*(e^{j\omega})$. Hence the output spectrum is obtained by filtering the input sequence with a filter with frequency response $|H(e^{j\omega})|^2$, which is purely real-valued, i.e., has zero phase and consequently causes no phase change between input and output. The process doubles the overall filter order, as well as the passband ripple and the stopband attenuation.

This method is particularly interesting for IIR filters: they cannot have exactly linear phase, and forward-reverse filtering can compensate for the related phase distortion. Moreover, in this technique the start-up and ending transients can be minimized, so as to get a filtered sequence with the same usable length as the input sequence. This is achieved by solving a system of linear equations to determine the first filter's initial conditions, and then extrapolating the beginning and the end of $x[n]$ by "reflected values", in such a way that the slopes of original and extrapolated sequences match at the end points. Together with the initial conditions, this reduces the edge effects, which is interesting both for FIR and IIR filters. After the expanded signal is filtered in forward and backward directions, the added parts are cut off.

6.8 An Incorrect Approach to Filtering

We insisted several times on the fact that a brick-wall frequency response, such as that of an ideal lowpass filter, is not computationally realizable. At first sight, however, it might appear very easy to apply such a filter to a sequence: couldn't we simply set the samples of the DFT of the data that correspond to undesired frequencies to zero and then invert the transform? No, we can't. This filtering procedure, though appealing at first sight, is conceptually incorrect for the following reasons.

The ideal filter has an impulse response that is an infinite-length, non-causal sequence $h[n]$. Its DTFT, i.e., the frequency response $H(e^{j\omega})$, thus cannot be represented by a finite number of samples. When we set to zero the samples of the DFT $X[k]$ of an N-samples-long signal, we implicitly assume, on the contrary, that the filter frequency response can be represented by a DFT, $H[k]$, with length N: in practice we multiply two N-points DFTs term by term, i.e., we compute $X[k]H[k]$, but $H[k]$ makes no sense. The inverse transform of $H[k]$ is a sequence containing N samples: hence we actually truncated the infinite impulse response $h[n]$ to N samples. The

Fig. 6.19 A lowpass
example of the effective
frequency response
$H_{\text{eff}}\left(e^{j\omega}\right)$ obtained by
setting the DFT samples of a
data sequence corresponding
to undesired frequencies to
zero and then inverting the
transform. The *black line*
represents the desired filter;
the *gray curve* represents the
effective filter. **a** The
neighborhood of the cutoff
frequency only; **b** The
complete range of
frequencies $[0, \pi]$

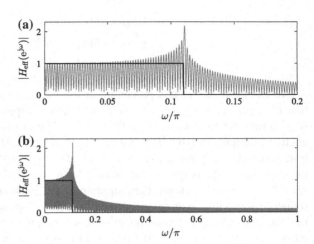

DTFT of this truncated sequence no longer has a boxy shape; it passes through the
desired values at any ω_k, with $k = [0, N - 1]$, but oscillates elsewhere. In other
words, in the frequency domain we implicitly applied to our data an effective frequency response altered by the Gibbs phenomenon. Indeed, truncating (windowing)
the impulse response $h[n]$ corresponds, in the frequency domain, to a continuous
periodic convolution of the ideal boxy frequency response with the DTFT of the
rectangular window.

Figure 6.19 shows a lowpass example of the effective frequency response $H_{\text{eff}}\left(e^{j\omega}\right)$
obtained in this way in a case with $N = 1024$. The cutoff frequency is 0.11π. The
gray curve in Fig. 6.19a shows, over a reduced frequency range around the cutoff
frequency, the effective response, while the *black line* depicts the frequency response
of the desired filter. Figure 6.19b shows the whole frequency range $[0, \pi]$. We can
see that the actual filter does not match any reasonable specifications.

6.9 Filtering After Downsampling

Let us consider a discrete-time signal obtained by sampling an analog signal at regular
time steps T_s. Now imagine that we want to apply a frequency-selective digital filter
to this discrete-time signal—e.g., a bandpass filter. Most often, we will think of
the desired bandpass filters in analog terms, and then we will translate the analog
bandlimits into the corresponding ω values. In some applications, it may be useful
to modify the sampling interval of the input sequence—for instance, to increase it
by downsampling the signal—so as to make the analog-frequency interval $[f_{p1}, f_{p2}]$
we are interested in, which corresponds to an interval $[\omega_{p1}, \omega_{p2}]$ when the sampling
step is T_s, correspond to a different discrete-frequency interval, in order to have a
well-conditioned filter design problem.

A typical example of the usefulness of this approach may be the case in which a very selective bandpass filtering in analog terms is desired, which would lead to a narrow discrete-frequency passband very close to the origin of the ω axis. In this instance, it is often preferred to preliminarily downsample the input, in order to obtain a wider and better-positioned passband in terms of ω. This allows managing filter design more easily, especially for what concerns the transition bandwidth, and therefore the required filter order. Figure 6.20 schematically illustrates how, for a bandpass filter, the passband edges $[f_{p1}, f_{p2}]$ we may have in mind in analog terms translate into a digital passband ω_{p1} and ω_{p2} when the signal $x[n]$ has a sampling interval T_s, but can lead to a different interval $[\omega'_{p1}, \omega'_{p2}]$ when the sampling step is modified to T'_s.

From a completely opposite point of view, we may want to change the sampling interval of the signal from T_s to T'_s in order to shift a discrete-frequency band on which a given digital filter operates (e.g., the passband of an existent frequency-selective digital filter we want to use) from a given position on the analog-frequency axis to another position along the same axis.

In general, a variation of the sampling interval of a signal may imply

- a decrease of T_s, that is, an *interpolation* or *upsampling* of the data sequence, or
- an increase of T_s, that is, a *decimation* or *downsampling* of the data sequence.

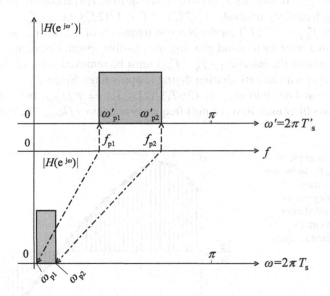

Fig. 6.20 A bandpass filtering operation to be performed on a real-world sampled signal will probably be thought of in terms of analog frequencies of interest, $[f_{p1}, f_{p2}]$. Given the signal's sampling interval T_s, these cutoff analog frequencies lead to digital specifications $[\omega_{p1}, \omega_{p2}]$ that may be inconvenient for some reason (see text). If we modify the sampling interval to $T'_s \neq T_s$, we can design a digital filter with a different passband $[\omega'_{p1}, \omega'_{p2}]$ that we may be able to handle in a better way

A common reason for upsampling is rate matching: for instance, we may want to mix two signals with different sampling rates. A common reason for downsampling is simply to reduce the amount of data to be processed, when the sampling rate is excessive for the application at hand, so that the whole amount of data is not actually needed, and can even be a hindrance, as in the previous example.

Here we will mainly focus on downsampling. Are we free to downsample a signal without precautions? As usual, the answer can be found in the study of the effects of downsampling in the frequency domain.

6.9.1 Theory of Downsampling

Let us take a signal $x[n]$, with $n = [0, N - 1]$, obtained by sampling an analog signal $x(t)$ at regular steps T_s. Suppose we want to downsample it by an integer downsampling (decimation) factor $k_{dec} > 1$, in order to get a signal with a new sampling interval $T_s' = k_{dec} T_s$. We may recall that this operation is linear but not time-invariant. Can we simply take $x_d[n] = x[k_{dec} n]$, as in the example shown in Fig. 6.21? No, we can't. Recall that for $x[n]$ to be aliasing-free, the analog signal $x(t)$ must be band-limited in the analog frequency interval $-1/(2T_s) \leq f < 1/(2T_s)$, that is, $-f_{Ny} \leq f < f_{Ny}$. In turn, for $x_d[n]$ to be aliasing-free, $x[n]$ should be band-limited in the analog frequency interval $-1/(2T_s') \leq f < 1/(2T_s')$, i.e., $-f_{Ny}' \leq f < f_{Ny}'$, where we set $f_{Ny}' = 1/(2T_s')$ for the Nyquist frequency of the downsampled signal. This shows that in order to avoid aliasing, the possible spectral content of the input signal $x[n]$ outside the interval $[-f_{Ny}', f_{Ny}')$ must be removed *prior to decimation*, by filtering $x[n]$ with an anti-aliasing digital lowpass filter. Since $f_{Ny}' = 1/(2T_s') = 1/(2k_{dec} T_s)$ translates into $\omega_{Ny}' = (2\pi T_s)/(2k_{dec} T_s) = \pi/k_{dec}$, the anti-aliasing digital lowpass filter must have a cutoff frequency equal to π/k_{dec}. The output of the

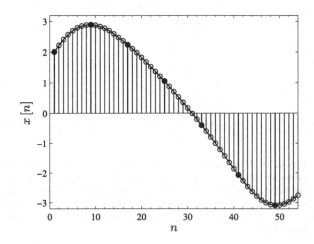

Fig. 6.21 An example of downsampling by an integer factor $k_{dec} > 1$ (here $k_{dec} = 8$). Empty circles represent discarded input samples; filled circles represent selected samples

anti-aliasing filter, $x_f[n]$, will then be safely downsampled, by taking one sample every k_{dec} samples: $x_d[n] = x_f[k_{dec}n]$.

We now need the expression of the spectrum of the downsampled signal to study the effects of decimation. It can be demonstrated that[5]

$$X_d(z) = \frac{1}{k_{dec}} \sum_{i=0}^{k_{dec}-1} X_f\left(e^{-j2\pi i/k_{dec}} z^{1/k_{dec}}\right),$$

_____

[5]Let us define an impulse-train signal as

$$y_{kdec}[n] = \sum_{i=-\infty}^{+\infty} \delta[n - k_{dec}i] = \begin{cases} 1 & \text{for } n \in k_{dec}\mathbb{Z}, \\ 0 & \text{otherwise}, \end{cases}$$

where $\mathbb{Z}$ indicates the set of integer numbers. This sequence is 1 at one out of every k_{dec} samples, and zero everywhere else. Equivalently, the impulse train can be written as

$$y_{kdec}[n] = \frac{1}{k_{dec}} \sum_{i=0}^{k_{dec}-1} e^{j2\pi in/k_{dec}} = \begin{cases} \frac{1}{k_{dec}} \sum_{i=0}^{k_{dec}-1} 1 = 1 & \text{for } n \in k_{dec}\mathbb{Z}, \\ \frac{1}{k_{dec}} \frac{1-e^{j2\pi in}}{1-e^{j2\pi in/k_{dec}}} = 0 & \text{otherwise}, \end{cases}$$

where for the second case we used the expression for the finite sum of a geometric series and the fact that $e^{j2\pi in} = 1$. Now, let us calculate the z-transform of $x_d[n] = x_f[k_{dec}n]$, i.e.,

$$X_d(z) = \sum_{n=-\infty}^{+\infty} x_f[k_{dec}n]z^{-n}.$$

We apply the substitution $m = nk_{dec}$, keeping in mind that this makes the summation run only over indexes m that are integer multiples of k_{dec}:

$$X_d(z) = \sum_{m \in k_{dec}\mathbb{Z}} x_f[m]z^{-m/k_{dec}}.$$

We can now use the above impulse train sequence $y_{kdec}[n]$ to rewrite this as a summation over all integers:

$$X_d(z) = \sum_{n=-\infty}^{+\infty} y_{kdec}[n]x_f[n]z^{-n/k_{dec}} = \sum_{n=-\infty}^{+\infty} \left(\frac{1}{k_{dec}} \sum_{i=0}^{k_{dec}-1} e^{j2\pi in/k_{dec}}\right) x_f[n]z^{-n/k_{dec}} =$$

$$= \frac{1}{k_{dec}} \sum_{i=0}^{k_{dec}-1} \sum_{n=-\infty}^{+\infty} e^{j2\pi in/k_{dec}} x_f[n]z^{-n/k_{dec}} =$$

$$= \frac{1}{k_{dec}} \sum_{i=0}^{k_{dec}-1} \sum_{n=-\infty}^{+\infty} x_f[n]\left(e^{-j2\pi i/k_{dec}} z^{1/k_{dec}}\right)^{-n} = \frac{1}{k_{dec}} \sum_{i=0}^{k_{dec}-1} X_f\left(e^{-j2\pi i/k_{dec}} z^{1/k_{dec}}\right).$$

This is the formula for the z-transform of the downsampler.

hence

$$X_d(e^{j\omega}) = \frac{1}{k_{dec}} \sum_{i=0}^{k_{dec}-1} X_f\left[e^{j(\omega-2\pi i)/k_{dec}}\right].$$

The spectrum of the decimated signal results from the superposition of k_{dec} *shifted and stretched* images of $X_f(e^{j\omega})$. Stretching is connected to the fact that the argument of the exponential term is divided by k_{dec} in the right-hand side of this equation: this means an expansion of the frequency axis by a factor k_{dec}. Shifting is represented by the term $\omega - 2\pi i$ in the exponent, which implies that two adjacent shifted and stretched copies of $X_f(e^{j\omega})$ are separated by 2π on the ω axis. The expression of $X_d(e^{j\omega})$ tells us that in order to avoid aliasing, $X_f(e^{j\omega})$ must vanish at any frequency from π/k_{dec} to π: thus we should filter $x[n]$ by an ideal anti-aliasing lowpass filter with cutoff at $\omega_c = \pi/k_{dec}$, that is, at $\nu_c = 1/(2k_{dec})$, i.e., at $f_s = 1/(2k_{dec}T_s) = f'_{Ny}$ in analog terms.

The ideal anti-aliasing filter is not computationally realizable, so a lowpass approximation with a non-zero transition bandwidth is employed instead. A safety factor on the cutoff frequency will help in achieving better results. For example, we may want to set $\omega_s = \pi/k_{dec}$, where ω_s is the upper bound of the lowpass-filter's transition band, rather than $\omega_p = \pi/k_{dec}$, where ω_p is the upper passband limit.

In Figs. 6.22 and 6.23, the decimation process is illustrated by an example. Both aliasing-free and aliasing-affected cases are considered. In Fig. 6.22, an anti-aliasing filter has been applied to the signal prior to downsampling, so that in the interval $[-\pi, \pi)$ the spectrum $|X_f(e^{j\omega})|$ of the signal that undergoes decimation is not identically zero only in $|\omega| < \pi/k_{dec}$ (Fig. 6.22a); the example is for $k_{dec} = 3$. This ensures that the downsampled signal is aliasing-free. Note that since the formula for $X_d(e^{j\omega})$ contains a summation over $i = [0, k_{dec} - 1]$, for $k_{dec} = 3$ we need to consider $i = [0, 2]$. The frequency-scaled shape of $|X_f(e^{j\omega})|$ for $i = 0$ (stretching only, no shift), shown in Fig. 6.22b, exhibits equally spaced copies of the original spectral shape as it appears in the principal interval in Fig. 6.22a, but these copies are stretched with respect to this original shape, so that in the principal interval the spectrum now occupies all the range $[-\pi, \pi)$. Figure 6.22c, d show stretching associated with shift ($i = 1$ and 2, respectively). Finally, the shape of $|X_d(e^{j\omega})|$ is visible in Fig. 6.22e: this spectrum is given by the sum of the spectra appearing in panels b, c, and d, multiplied by per $1/k_{dec}$. We can observe that aliasing is actually absent: the shape of $|X_f(e^{j\omega})|$ is recognizable in $|X_d(e^{j\omega})|$.

Figure 6.23 shows a case in which no anti-aliasing filter has been used: decimation has been performed directly on the original signal $x[n]$. The spectrum of the original signal occupies the whole principal interval $[-\pi, \pi)$ (Fig. 6.23a). This is a condition in which aliasing occurs. The frequency-scaled shape of $|X(e^{j\omega})|$ for $i = 0$, shown in Fig. 6.23b, actually exhibits stretched copies of the original spectral shape that now extend outside the principal interval and invade the whole frequency axis. For example, the copy centered at $\omega = 0$ occupies the range $[-3\pi, 3\pi)$. Finally, Fig. 6.23c, d illustrate the effects of both stretching and shifting ($i = 1$ and 2, respectively). Figure 6.23e shows the shape of $|X_d(e^{j\omega})|$, determined by the sum of the

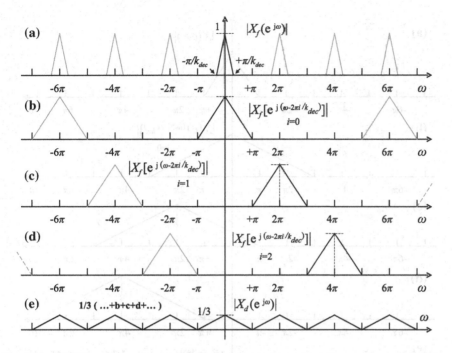

Fig. 6.22 Spectral representation of a bandlimited discrete-time signal that undergoes decimation by a factor $k_{dec} = 3$. Anti-aliasing filtering of the original signal prior to decimation confines the spectrum $|X_f(e^{j\omega})|$ to a part of the principal interval, namely to $|\omega| < \pi/k_{dec}$. **a** Spectrum of the filtered signal; **b** stretching of the spectrum; **c–d** stretching and shifting; **e** spectrum of the downsampled signal. No aliasing occurs

spectra shown in panels b, c, and d, multiplied by per $1/k_{dec}$. We can see that aliasing is actually present in this case: the shape of $|X(e^{j\omega})|$ can no longer be deduced observing $|X_d(e^{j\omega})|$.

It may be useful to observe that the formula we gave for $X_d(e^{j\omega})$ expresses a time-scaling property of the DTFT, which is similar to the following property of the Fourier transform of analog signals: given a signal $x(t)$ with Fourier transform $\mathcal{F}\{x(t)\} \equiv X(f)$, and given a real non-zero constant a,

$$\mathcal{F}\{x(at)\} = \frac{1}{|a|}X\left(\frac{f}{a}\right).$$

Indeed, $x_f[k_{dec}n]$ is a time-domain scaling of $x_f[n]$ by a factor $a = k_{dec}$, and results in the described operation in the frequency domain.

In the Fourier theory of analog signals, a frequency-scaling property also exists: if we indicate by $\mathcal{F}^{-1}\{\cdot\}$ the inverse continuous-time Fourier transform, then

$$\frac{1}{|a|}x\left(\frac{t}{a}\right) = \mathcal{F}^{-1}\{X(af)\}, \quad \text{i.e.,} \quad \mathcal{F}\left\{\frac{1}{|a|}x\left(\frac{t}{a}\right)\right\} = X(af).$$

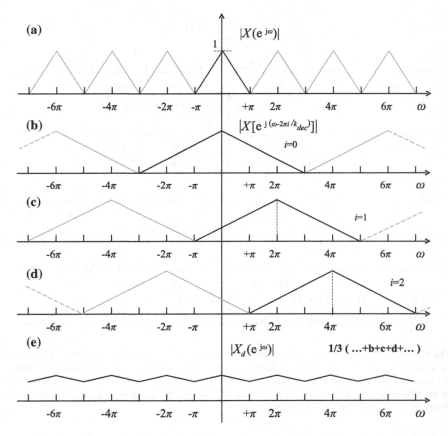

Fig. 6.23 Spectral representation of a bandlimited discrete-time signal that undergoes decimation by a factor $k_{dec} = 3$. No anti-aliasing filter has been applied to the original signal prior to decimation, and **a** the spectrum occupies the whole principal interval. **b** After stretching due to decimation, the spectrum exits the principal interval; **c–d** stretching is combined with shifting to give **e** the spectrum of the downsampled signal, contaminated by aliasing

Now, looking at the last two formulas we see that

- for a equal to the inverse of an integer, i.e., for $a = 1/k_{dec}$, evidently we are brought back to the previous case of downsampling;
- for an integer $a > 1$ this property can be brought back to the case of upsampling by an integer factor k_{int}, i.e., to the case in which an interpolated signal with reduced sampling interval is defined as

$$x_i[n] = \begin{cases} x[\frac{n}{k_{int}}] & \text{for} \quad n = 0, \pm k_{int}, \pm 2k_{int} \dots, \\ 0 & \text{otherwise.} \end{cases}$$

Thus, interpolation is obtained by inserting the proper number of zeros between adjacent samples. Later, the discontinuities in the result can be smoothed out using a lowpass filter (Oppenheim and Schafer 2009). It is not difficult to show that $X_u(z) = X(z^{k_{int}})$.

Downsampling and upsampling are central issues in the frame of the implementation of the discrete wavelet transform (DWT) through the applicaton to a signal of a two-channel filter bank (Chap. 14). In DWT implementation, the downsampling and upsampling are by a factor of 2. In Chap. 14 we will therefore particularize the results obtained here for the special case of $k_{dec} = 2$ and $k_{int} = 2$.

6.9.2 An Example of Filtering After Downsampling

Imagine we recorded a magnetoencephalogram (MEG)[6] of a resting–state normal healthy subject. Typically these data sets include the recordings of over 100 channels, corresponding to different sensors distributed on the scalp. We would like to extract from one of more channels typical brain waves: δ, θ, α, β, and γ waves, which are characterized by frequencies in the ranges 0.5–4, 4–8, 8–12, 12–30 and 30–100 Hz, respectively.[7]

---

[6]Magnetoencephalography is technique aimed at detecting magnetic fields produced by the brain.

At the cellular level, individual neurons in the brain have electrochemical properties that result in the flow of electrically charged ions. Electromagnetic fields are generated by the net effect of this slow ionic current flow. While the magnitude of fields associated with an individual neuron is negligible, the effect of multiple neurons (for example, $5 \times 10^4 - 1 \times 10^5$ neurons) excited together in a specific area generates a measurable magnetic field outside the head. These neuromagnetic signals generated by the brain are extremely weak. Therefore, MEG scanners require superconducting sensors (SQUID, superconducting quantum interference device). The SQUID sensors are bathed in a large liquid helium cooling unit at approximately $-269\,^{\circ}$C. Due to low impedance at this temperature, the SQUID device can detect and amplify magnetic fields generated by neurons a few centimeters away from the sensors located on the patient's head. MEG measurements span field magnitudes from about 10 femtoTesla (fT) for spinal cord signals to about several picoTesla (pT) for brain rhythms. Recall that the symbols micro (μ), nano, (n), pico (p) and femto (f) in front of a measurement unit mean 10^{-6}, 10^{-9}, 10^{-12}, and 10^{-15}, respectively. To appreciate how small the MEG signals are, it should be recalled that Earth's magnetic-field magnitude is about $0.5\,\mu$T and the urban magnetic noise about $1\,$nT–$1\,\mu$T, or about a factor of 1 million to 1 billion larger than MEG signals. For this reason, the subject being studied and the MEG instrument must be in a magnetically shielded room, to mitigate external interference. MEG measurements are performed using magnetometers and/or gradiometers. The gradiometers values are expressed in T/m (Tesla/meter), while magnetometers measure the magnetic field in Tesla. As a rule of thumb, gradiometer measurements can be roughly converted into magnetometers units by multiplying them by the size of the sensor (typically, about 4 cm).

[7]Here we are referring to the classical categorization of brain waves (see Buzsáki 2011). This classification was introduced by International Federation of Societies for Electroencephalography and Clinical (1974). See also Steriade et al. (1990).

Let us focus on a single sensor. Suppose that we know from preliminary analysis that the corresponding signal ($N \approx 4 \times 10^5$; duration of about 7 minutes; sampling frequency 1025 Hz) has shown variability in all the above-mentioned bands. Now we want to extract brain waves by band pass filtering. We decide to design proper LP FIR filters to extract the variability in each band. A segment of the signal (with values expressed in fT) can be see in the upper panel of Fig. 6.24a. The signal is the result of measurements affected by errors, and must be considered as a random power signal whose power spectrum must be estimated from the available set of samples (Chaps. 9–11). In the present case, a simple estimation method called the periodogram (Sect. 10.3) has been used to calculate the power spectrum of the single-channel MEG signal. This spectrum is shown in the upper panel of Fig. 6.24b.

The Nyquist frequency is $1025/2$ Hz $= 512.5$ Hz and corresponds to π in terms of ω. The sampling interval is $1/1025 = 9.7561 \times 10^{-4}$ s. To extract the variability in each band, we thus would need passband filters with the following adimensional passbands:

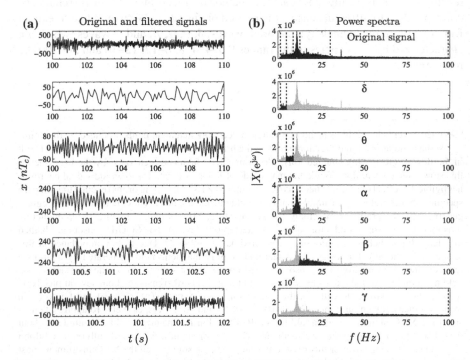

Fig. 6.24 **a** *Upper panel* a segment of a single-channel MEG signal, expressed in fT; *lower panels* extracted waveforms (values in fT). **b** *Upper panel* power spectrum of the original signal (see text); *lower panels* power spectra of filtered MEG signals in the δ to γ frequency bands (*black curves*), superimposed to the spectrum of the original signal (*gray curves*). To facilitate the comparison, the factor $1/k_{dec}$ reducing the spectral power of the decimated signals has been removed. In each panel, *vertical dotted lines* enclose the passband of the filter employed

- $0.0010\pi–0.0078\pi$ for the δ range;
- $0.0078\pi–0.0156\pi$ for the θ range;
- $0.0156\pi–0.0234\pi$ for the α range;
- $0.0234\pi–0.0585\pi$ for the β range;
- $0.0585\pi–0.1951\pi$ for the γ range.

This makes no sense: the sampling interval of the data is exceedingly small for our purposes—the Nyquist frequency is exceedingly large—and these filters would be poorly conditioned, with a very narrow passband positioned in the neighborhood of the zero frequency. We have no choice but decimating our data, in order to pose our filter design problem in the correct terms. We want to be able to use filters with a passband width representing a reasonable fraction of π, and located more or less centrally with respect to the interval $[0, \pi]$. A simple reasoning based on these requirements leads to estimating proper decimation factors k_{dec} to be 120, 40, 25, 12 and 4 for the δ, θ, α, β and γ bands, respectively. These downsamplings lead to the following desired adimensional passbands:

- $0.1171\pi–0.9366\pi$ for the δ range;
- $0.3122\pi–0.6244\pi$ for the θ range;
- $0.3902\pi–0.5854\pi$ for the α range;
- $0.2810\pi–0.7024\pi$ for the β range;
- $0.2341\pi–0.7805\pi$ for the γ range.

They appear very reasonable. We now set the so-called filter *tolerances*, limiting the ripple that can be present in the passband (δ_p) and stopband (r_s). See Chap. 7 for details about filter tolerances. In this example we set: $\delta_p = 0.095$, $r_s = 60$ dB, and width of the transition bands 0.02π. Then we design the filters using the minimax approach that will be described in the Chap. 7 and that produces LP FIR filters. The frequency responses of these filters are shown in Fig. 6.25, both in dB units (Fig. 6.25a) and linear units (Fig. 6.25b). The order is 184: a high value, but harmless since we have such a large amount of data.

We are not done yet: we still have to solve one issue, namely the fact that the first three decimation factors are very high (120, 40 and 25). We should design and apply lowpass anti-aliasing filters that would again be ill-conditioned: indeed, their theoretical cutoffs $\omega_c = \pi/k_{dec}$ would be 0.0262π, 0.0785π, and 0.1257π, respectively. The simplest way to solve this issue is to decimate twice by smaller factors, applying each time the proper anti-aliasing filter, which this time will have a cutoff frequency easier to handle. So we decide to downsample as follows:

- $k_{dec,1} = 12$ the first time and $k_{dec,2} = 10$ the second time in the δ range;
- $k_{dec,1} = 8$ and $k_{dec,2} = 5$ in the θ range;
- $k_{dec,1} = 5$ and $k_{dec,2} = 5$ in the α range.

Now we design anti-aliasing LP FIR filters with order 30 (figure not shown) and apply them to the signal. Then we downsample, and finally apply the passband filters of Fig. 6.25.

Figure 6.24 shows the result of this procedure for all frequency bands. The lower panels of Fig. 6.24 show: in column a, the extracted wave (*black curve*; values in fT);

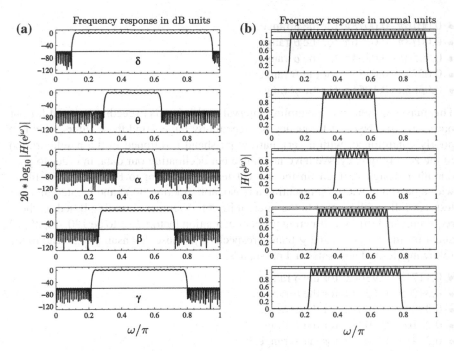

Fig. 6.25 Frequency responses of the bandpass filters used to extract δ, θ, α, β, and γ waves from a single-channel MEG recording: **a** dB units, **b** normal units

in column b, the power spectrum estimated via periodogram of the filtered signal (*black curve*), superimposed to that of the original signal (*gray curve*). For a better visualization, the factor $1/k_{dec}$ reducing the spectral power of the downsampled signals has been removed. Annotations specify which frequency band each panel is relative to; *vertical dotted lines* enclose the passband in each plot.

This example clearly explains how the downsampling approach can be exploited to get a well-posed filtering problem even when the sampling rate is too high for our purposes.

References

Buzsáki, G.: Rhythms of the Brain. Oxford University Press, Oxford (2011)

Chen, W.K.: On flow graph solutions of linear algebraic equations. Society for industrial and applied mathematics (SIAM). J. Appl. Math. **15**(1), 136–142 (1967)

Chow, Y., Cassignol, E.: Linear Signal Flow Graphs and Applications. Wiley, New York (1962)

Electroencephalogr. Clin. Neurophysiol. Neurophysiology: report of the committee on cessation of cerebral function. **37**(5), 521 (1974)

Gustafsson, F.: Determining the Initial States in Forward-Backward Filtering. IEEE Trans. Sig. Process. **44**(4), 988–992 (1996)

Jones, D.: Digital Filter Structures and Quantization Error Analysis. Available via OpenStax-CNX Web site. http://cnx.org/content/col10259/1.1/ (2005)

Mason, S.J., Zimmermann, H.J.: Electronic Circuits, Signals, and Systems. Wiley, New York (1960)

Markel, J.D., Gray, A.H. Jr: Linear Prediction of Speech. Springer, New York (1976)

Mitra, S.K.: Digital SignalProcessing, 2nd edn. McGraw-Hill, New York (2001)

Oppenheim, A.V., Schafer, R.W.: Discrete-Time Signal Processing. Prentice Hall, Englewood Cliffs (2009)

Oppenheim, A.V., Schafer, R.W., Buck, J.R.: Discrete-Time Signal Processing, 2nd edn. Prentice Hall, Upper Saddle River (1999)

Paarmann, L.D.: Design and Analysis of Analog Filters. A signal processing perspective. Kluwer, Dordrecht (2001)

Paley, R.E.A.C., Wiener, N.: Notes on the theory and application of fourier transforms. I-II. Trans. Am. Math. Soc. **35**, 348–355 (1933)

Paley, R.E.A.C., Wiener, N.: Fourier Transforms in the Complex Domain, vol. 19. American Mathematical Society (SIAM) Colloquium Publications, New York (1934)

Proakis, J.G., Manolakis, D.G.: Digital Signal Processing: Principles, Algorithms and Applications. Prentice-Hall, Englewood Cliffs (2007)

Schlichthärle, D.: Digital Filters: Basics and Design. Springer, New York (2011)

So, H.C.: Digital Signal Processing: Foundations, Transforms and Filters, with Hands-on MATLAB Illustrations. McGraw-Hill, New York (2010)

Steriade, M., Gloor, P., Llinás, R., da Silva, F.H.L.: Basic mechanisms of cerebral rhythmic activity. Electroencephalogr. Clin. Neurophysiol. **76**(6), 481–508 (1990)

Chapter 7
FIR Filter Design

7.1 Chapter Summary

This chapter begins with general considerations about the design process of a digital filter. We will describe the form in which the filter specifications must be expressed by the designer, and will examine the reasons why FIR filters might be preferred in applications, in comparison with the arguments in favor of IIR filters. Then the discussion will focus on FIR filter design, leaving the topic of IIR design to the next chapter. The most popular design methods for FIR filters will be presented, and the quantitative approximation criteria that may be established to judge the resemblance of the approximated filter, of which four types exist, with the desired ideal one will be discussed. The properties of LP/GLP FIR filters will then be examined in detail, and a factorization of the zero-phase response, useful to unify the symmetry condition for the coefficients of all four filter types, will be performed. In this factorization, the zero-phase response is split into a fixed factor, depending on the filter type but not on specifications, and an adjustable factor, with coefficients to be determined according to specifications. This mathematical work paves the way to the description of the optimum and most flexible design method for LP/GLP FIR filters, namely the minimax method, which ensures that filters will meet specifications with the minimum possible order and produces equiripple filters, i.e., filters with uniform oscillations in the passband(s) and stopband(s). The properties of these optimum FIR filters will finally be studied.

7.2 Design Process

The design of digital filters is a process that involves the three fundamental stages listed below.

© Springer International Publishing Switzerland 2016
S.M. Alessio, *Digital Signal Processing and Spectral
Analysis for Scientists*, Signals and Communication Technology,
DOI 10.1007/978-3-319-25468-5_7

1. Specification of the desired properties of the system. For example, we may want an LTI-RCSR system operating a certain kind of frequency selection. On the frequency response, amplitude constraints and, possibly, phase constraints are imposed.
2. Approximation of these specifications using a realizable discrete-time system. A design method must be selected, allowing us to approximate the desired ideal filter by a realizable filter with the least possible complexity, i.e., the minimum possible order, under the condition of satisfying the specifications according to a certain quantitative criterion. The criterion adopted characterizes the design method. This step implies:

   - selecting a class of filters suited for approximating the desired frequency response. For example, we may want a linear phase (LP) FIR filter;
   - individuating a design method suited for finding, within the selected class, the member that best satisfies the specifications, according to the criterion that characterizes the method, and keeping complexity fixed;
   - synthesizing this best member, i.e., realizing the system, by calculating its impulse response and its frequency response;
   - analyzing the performances of the designed filter, and rectifying either the order, or the specifications, if the provided specifications actually are not met at the chosen order.

3. Realization the system, i.e., actuation of the desired digital filtering by a software implementation, according to the lines previously described in Sect. 6.6.

7.3 Specifications of Digital Filters

Digital filters are specified giving constraints on their frequency response. The constraints concern both their magnitude and phase responses (see Oppenheim and Schafer 2009).

7.3.1 Constraints on the Magnitude Response

The constraints (i.e., specifications) on $|H(e^{j\omega})|$ include a set of band limits, a set of desired values in each band, and a set of *tolerance limits*, that is, a *tolerance scheme*.

Figure 7.1 illustrates the example of the specifications for an FIR lowpass filter. Here the specifications are expressed giving the passband and stopband limits,[1] ω_p and ω_s, the desired values of $|H(e^{j\omega})|$ in the passband and in the stopband (usually these values are 1 and 0 respectively), and the tolerances δ_p and δ_s. The *black curve* represents an example of amplitude response that would not exceed these *absolute*

---

[1]In this discussion we use the symbol ω_p for the upper passband limit, in place of the symbol ω_c previously used to indicate the cutoff frequency of the ideal filter.

Fig. 7.1 Specifications for the magnitude response of an FIR filter: the example of a lowpass filter. The *black curve* represents an example of amplitude response that would meet the prescribed specifications. The *dashed line* depicts the desired ideal response, while *horizontal solid lines* represent absolute tolerances

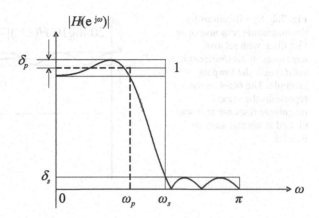

tolerances, represented by *horizontal solid lines*. Observe the transition band interposed between the passband and the stopband: its width can never vanish if the filter must be computationally realizable. In the case of IIR filters, for which the classical design approach starts from an analog filter, the specifications are expressed in a slightly different way, as we will see in Chap. 8.

The quantities δ_p and δ_s in Fig. 7.1 represent the maximum tolerable deviation of $|H(e^{j\omega})|$, in the passband and stopband respectively, with respect to the *desired amplitude response* $|H_d(e^{j\omega})|$ (*dashed line* in Fig. 7.1). In the lowpass case, the desired magnitude response is

$$|H_d(e^{j\omega})| = \begin{cases} 1 & \text{for} \quad \omega \in [0, \omega_p], \\ 0 & \text{for} \quad \omega \in [\omega_s, \pi], \end{cases}$$

and the specifications can thus be expressed as

$$\begin{cases} 1 - \delta_p \leq |H(e^{j\omega})| \leq 1 + \delta_p & \text{for} \quad \omega \in [0, \omega_p], \\ |H(e^{j\omega})| \leq \delta_s & \text{for} \quad \omega \in [\omega_s, \pi], \end{cases}$$

under the constraint that $\omega_s - \omega_p > 0$. In many cases, $|H(e^{j\omega})|$ exhibits oscillatory behavior in the passband and/or in the stopband. Then δ_p is also referred to as the passband *ripple*; δ_s is the stopband ripple.

Sometimes, the tolerances are expressed in logarithmic units, i.e., in dB:

$$\begin{cases} r_p = 20 \log_{10} \left(\frac{1+\delta_p}{1-\delta_p} \right), \\ r_s = -20 \log_{10}(\delta_s). \end{cases}$$

These values are called passband ripple in dB and stopband attenuation in dB, respectively. These values represent *relative tolerances*, in the sense that they contain the

Fig. 7.2 Specifications for
the magnitude response of an
FIR filter, with relative
tolerances in dB (*horizontal
solid lines*): the lowpass
example. The *black curve*
represents the same
magnitude response that was
plotted in normal units in
Fig. 7.1

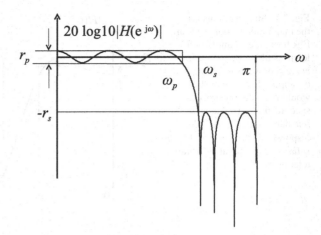

explicit values of $|H_d(e^{j\omega})|$, that is, 0 and 1. Figure 7.2 illustrates the meaning of
these relative tolerances.

Since

$$\log(1 + \delta_p) \simeq \delta_p, \qquad \log(1 - \delta_p) \simeq -\delta_p,$$

we can also approximately write

$$r_p = 20 \log_{10} \left(\frac{1 + \delta_p}{1 - \delta_p} \right) = 20 \left[\log_{10}(1 + \delta_p) - \log_{10}(1 - \delta_p) \right] =$$

$$= 20 \log_{10} e \left[\log(1 + \delta_p) - \log(1 - \delta_p) \right] \simeq 20 \log_{10} e \times 2\delta_p, =$$

$$= (40 \times 0.434)\delta_p = 17.37\delta_p,$$

where $e \approx 2.71828$ is the base of the natural logarithm. The inverse relations are

$$\begin{cases} \delta_p = \frac{10^{r_p/20} - 1}{10^{r_p/20} + 1}, \\ \\ \delta_s = 10^{-r_s/20}. \end{cases}$$

Let us make a numerical example:

$$\delta_p = 0.095 \qquad \rightsquigarrow \qquad r_p \simeq 17.37 \times 0.095 \simeq 1.65 \, \text{dB},$$
$$r_s = 30 \, \text{dB} \qquad \rightsquigarrow \qquad \delta_s = 10^{-30/20} = 0.0316.$$

So far we have considered the lowpass case, in which only one passband and
only one stopband are present. However, even the simplest passband filter has two
stopbands: in the general multiband case, one or more passbands and one or more
stopbands will be present. If

- all passbands are assigned the same tolerance δ_p and if the desired passband response is $H_{dp}(\omega)$, usually piecewise constant,
- if all stopbands are assigned the same tolerance δ_s and the same desired response $H_{ds}(\omega) = 0$,

then we can write, indicating by B_p the union of all passbands and by B_s the union of all stopbands,

$$H_d(e^{j\omega}) = \begin{cases} H_{dp}(\omega) & \text{for } \omega \in B_p, \\ 0 & \text{for } \omega \in B_s, \end{cases}$$

and can express the specifications as

$$\begin{cases} -\delta_p \leq \left[|H(e^{j\omega})| - |H_{dp}(\omega)| \right] \leq \delta_p & \text{for } \omega \in B_p, \\ |H(e^{j\omega})| \leq \delta_s & \text{for } \omega \in B_s, \end{cases}$$

that is,

$$\begin{cases} |H_{dp}(\omega)| - \delta_p \leq |H(e^{j\omega})| \leq |H_{dp}(\omega)| + \delta_p & \text{for } \omega \in B_p, \\ |H(e^{j\omega})| \leq \delta_s & \text{for } \omega \in B_s. \end{cases}$$

For the purpose of filter design, it is more convenient to express the constraints globally, by only one inequality, rather than separately for B_p and B_s. In order to do this, a *global weighting functions* is introduced, which is defined as

$$W(\omega) = \begin{cases} 1 & \text{for } \omega \in B_p, \\ \delta_p/\delta_s & \text{for } \omega \in B_s. \end{cases}$$

Defining a *weighted error function* as

$$E(\omega) = W(\omega) \left[|H(e^{j\omega})| - |H_d(e^{j\omega})| \right],$$

we can write specifications in a global form, valid for the union of B_p and B_s:

$$|E(\omega)| \leq \varepsilon \quad \text{for } \omega \in B = B_p \cup B_s, \quad \text{with } \varepsilon = \delta_p.$$

The constraint on the weighted error function can be expressed as follows: if the maximum absolute value of the weighted error function does not exceed ε in the union of all passbands and stopbands, then the magnitude response $|H(e^{j\omega})|$ meets the specifications.

In Fig. 7.3 we can observe the simple case of a passband filter with only one passband and two stopbands, for which B_p corresponds to

$$\omega_{p1} \leq \omega \leq \omega_{p2},$$

Fig. 7.3 Specifications for the magnitude response of an FIR filter: the example of a bandpass filter. The *black curve* represents an amplitude response that would meet the prescribed specifications

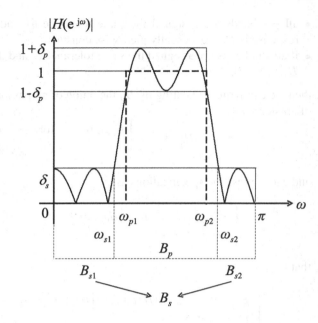

and B_s corresponds to

$$\begin{cases} 0 \le \omega \le \omega_{s1}, \\ \omega_{s2} \le \omega \le \pi. \end{cases}$$

We may mention that in the most general case, each band of a multiband filter will have its tolerance; moreover, the maximum tolerable deviation of $|H(e^{j\omega})|$ from $|H_d(e^{j\omega})|$ in each band will depend on ω. This frequency-dependent deviation is called *approximation error*, and indicated by $e_p(\omega)$ or $e_s(\omega)$. Two different *weighting functions* for the errors in B_p and B_s are then introduced. Maximum tolerances independent of frequency, δ_p and δ_s, are set, and the weighting functions are defined as

$$\begin{cases} W_p(\omega) = \delta_p/e_p(\omega), \\ W_s(\omega) = \delta_s/e_s(\omega). \end{cases}$$

The specifications become

$$\begin{cases} -\delta_p \le W_p(\omega)\left[|H(e^{j\omega})| - |H_{dp}(\omega)|\right] \le \delta_p & \text{for } \omega \in B_p, \\ W_s(\omega)|H(e^{j\omega})| \le \delta_s & \text{for } \omega \in B_s. \end{cases}$$

with an arbitrary $H_{dp}(\omega)$. The *global weighting function* is then defined as

$$W(\omega) = \begin{cases} W_p(\omega) & \text{for } \omega \in B_p, \\ (\delta_p/\delta_s)\, W_s(\omega) & \text{for } \omega \in B_s. \end{cases}$$

Introducing the *weighted error function*

$$E(\omega) = W(\omega)\left[|H(e^{j\omega})| - |H_d(e^{j\omega})|\right],$$

we can write specifications in the union of B_p and B_s as

$$|E(\omega)| \le \varepsilon \quad \text{for} \quad \omega \in B = B_p \cup B_s, \quad \text{with} \quad \varepsilon = \delta_p.$$

Finally, we can also express the weighted error function in terms of the approximated and desired zero-phase responses (Sect. 6.5) $H(\omega)$ and $H_d(\omega)$ (see, for example, Mitra and Kaiser 1993):

$$E(\omega) = W(\omega)\left[H(\omega) - H_d(\omega)\right].$$

This formula will be the starting point for the minimax design, explained in Sect. 7.7. Of course, $E(\omega)$ in this formula can assume both signs not only in the passband of a standard lowpass filter, but also in the stopand. However, this is unessential in our discussion, since the constraint on $E(\omega)$ only involves $|E(\omega)|$.

7.3.2 Constraints on the Phase Response

The phase response of a digital filter is first of all constrained by the requirements of causality and stability: all the poles of $H(z)$ must be contained inside the unit circle. In many applications concerning FIR filters, since zero-phase is impossible for computationally realizable filters, phase linearity is sought too. The shape of the input signal, as determined by the dominant frequency components contained in B_p, is thus preserved in the output signal (see Sect. 6.5.2). Of course, only the phase in the passband(s) is important. In the stopband(s), the elementary sinusoidal contributions to the input signal should ideally be completely eliminated, and therefore the phase dependence on frequency is unimportant there.

In some specialized applications, generalized linear phase is sought. For example, this is the case of Hilbert filters: wideband lowpass or bandpass filters designed to rotate by $\pi/2$ (a quarter of a cycle) the phase of any sinusoidal component of the input signal contained in the passband. In this way, an output signal is obtained, which is in quadrature with the input signal.

Hilbert filters find applications—to mention just a few examples—in modulators, in differentiators, in voice signal processing, in medical image signal processing, etc., as well as in multivariate analysis methods like complex principal component analysis. Complex principal component analysis or CPCA (Horel 1984; Jolliffe 2002; Saff and Snider 2002) represents an extension to the complex domain of the more classical principal component analysis (PCA), a statistical method, proposed in 1901 by Karl Pearson and developed by Harold Hotelling in 1933, which belongs to the vast domain of factorial analysis. The main goal of PCA is the reduction of a high number of mutually dependent variables, representing as many correlated features of the considered phenomenon, to a few independent variables, expressing uncorrelated features. The extension to the complex domain, known as complex principal component analysis, has been used, for example, in geophysics, to detect propagating disturbances in scalar spatial-temporal fields, such as temperature and precipitation fields.

Essentially, Hilbert filters are useful any time that it is advantageous to process a complex signal instead of the original real signal. The complex signal is derived from the real one in a univocal way, without altering the original spectral features of the signal. Such a complex signal is referred to as the analytic signal associated with the original real signal. The Hilbert filter provides the quadrature signal that is necessary to construct the analytic signal: the real part of the analytic signal is the original real signal, while the imaginary part is obtained by filtering the latter using a wideband Hilbert filter. The spectrum of an analytic signal is identically zero at negative frequencies; the spectrum over non-negative frequencies is twice the spectrum of the original real signal over the same frequency interval. Since the latter is an even function of frequency, we can see the spectrum of the analytic signal as the result of reflecting the spectral content of $x[n]$ pertaining to $\omega < 0$ over $\omega \geq 0$. These applications are beyond the scope of our present discussion. In what follows, we will mainly focus on filters having strictly linear phase and therefore our attention will be principally reserved to Type I- and Type II- FIR filters.

Whenever linear phase is unnecessary—or unattainable, as in the case of IIR filters that can have quasi-linear phase but can never have exactly linear phase—tolerance limits on phase quasi-linearity can be established.

It must also be noted that accepting a linear phase in place of the desired zero phase does not make the top-hat frequency response of an ideal filter realizable: indeed, substituting

$$H_d(\mathrm{e}^{\mathrm{j}\omega}) = \begin{cases} 1 & \text{for } |\omega| \leq \omega_p, \\ 0 & \text{for } \omega_p < |\omega| \leq \pi \end{cases}$$

that corresponds to an impulse response which has infinite length and is non-causal,

$$h_d[n] = \frac{\sin(\omega_p n)}{\pi n},$$

with

$$H_d(\mathrm{e}^{\mathrm{j}\omega}) = \begin{cases} \mathrm{e}^{-\mathrm{j}\omega\tau} & \text{for } |\omega| \leq \omega_p, \\ 0 & \text{for } \omega_p < |\omega| \leq \pi \end{cases}$$

that corresponds to an impulse response given by

$$h_d[n] = \frac{\sin[\omega_p(n - \tau)]}{\pi(n - \tau)}$$

does not solve the problem, because this impulse response is still infinite in length and non-causal.

7.4 Selection of Filter Type: IIR or FIR?

Once the specifications have been expressed, a decision must be made about approximating $H_d(e^{j\omega})$ by a rational function, typical of an IIR system, or by a polynomial function, typical of an FIR system. FIR filters

- can have exactly linear phase;
- are always stable;
- are flexible because excellent design methods exist for them that can reproduce almost arbitrary desired frequency responses;
- are less sensitive than IIR filters to the effects of finite-precision arithmetic in software implementations.

However, they also present some drawbacks. FIR filters

- do not allow for any closed-form design equations: no formulas exist in which we can directly insert the given specifications to immediately get the approximated frequency response and consequently the impulse response. The design of FIR filters is performed, as we shall see, through iterative procedures requiring powerful computational facilities; on the contrary, closed-form design equations exist for IIR filters;
- approximate any given specifications with a higher order with respect to IIR filters;
- exhibit oscillations in the passband, while an IIR filter can be practically flat, as explained in Chap. 8.

In summary, the greatest disadvantages of FIR filters with respect to IIR filters lie in the fact that for equal specifications FIR filters have higher order than IIR filters, and that FIR filters cannot have a flat passband. On the other hand, IIR filters cannot have exactly linear phase. However, this is not dramatic if we consider only applications in which data can be processed off-line: if the whole data sequence is recorded before filtering, a zero-phase non-causal approach can be followed that eliminates the distortions due to nonlinear phase (Sect. 6.7). In conclusion, the strongest argument in favor of FIR filters is probably the flexibility of the minimax design method, presented in Sect. 7.7.

In the present discussion, we will concentrate on the design of LP/GLP frequency-selective FIR filters (and ultimately on LP filters). Therefore we will now assume that the decision has been made to approximate $H_d(e^{j\omega})$ by a polynomial function. If just LP is sought, are we then free to choose between Type-I and Type-II filters?

Fig. 7.4 Zero-phase
response for FIR filters of
a Type I, **b** Type II, **c** Type
III, and **d** Type IV

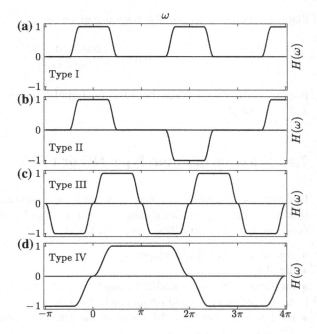

To answer this question, we must recall the restrictions imposed by causality on the
zero-phase response of LP/GLP filters:

- when $\phi_0 = 0$ (Types I and II), then $H(\omega)$ is an even function;
- when $\phi_0 = \pi/2$ (Types III and IV), then $H(\omega)$ is an odd function;

moreover,

- when τ_g is an integer (Types I and III), then $H(\omega)$ is periodic with period 2π;
- when τ_g is a half-integer (Types II and IV), then $H(\omega)$ is periodic with period 4π.

Thus, the picture exemplified in the schemes of Fig. 7.4 emerges:

- Type-I filters: $H(\omega)$ is periodic with period 2π and is an even function (Fig. 7.4a);
- Type-II filters: $H(\omega)$ is periodic with period 4π and is an even function (Fig. 7.4b);
- Type-III filters: $H(\omega)$ is periodic with period 2π and is an odd function (Fig. 7.4c);
- Type-IV filters: $H(\omega)$ is periodic with period 4π and is an odd function (Fig. 7.4d).

Observing Fig. 7.4 we note that

- for Type-I filters, $H(\omega)$ presents even symmetry both around $\omega = 0$ and $\omega = \pi$;
  therefore it can have any shape, i.e., it can represent a lowpass filter, as well as a
  highpass, or a bandpass, or a bandstop filter;
- for Type-II filters, $H(\omega)$ is even around $\omega = 0$ and $\omega = 2\pi$, but is odd around
  $\omega = \pi$ and $\omega = 3\pi$. Therefore it must be zero at $\omega = \pi$ and cannot assume the
  shape of a highpass or a bandstop filter;
- for Type-III filters, $H(\omega)$ is odd around $\omega = 0$ and $\omega = \pi$, so it must be zero at
  these frequencies. Therefore it can only assume the shape of a passband filter;

- for Type-IV filters, $H(\omega)$ is odd around $\omega = 0$ and $\omega = 2\pi$, while it is even around $\omega = \pi$ and $\omega = 3\pi$. At $\omega = 0$ it must vanish, and therefore cannot represent a lowpass or a bandstop filter.

We can conclude that only Type-I filters are suited to approximate all kinds of ideal frequency-selective filter. Type-II filters are suited to approximate lowpass and bandpass filters, but they are associated with half-integer time delays of the output with respect to the input signal, which can be inconvenient. If no particular indications are provided in favor of a Type-II filter, we will then stick to Type-I filters, characterized by an even order. Type-III and IV filters are used for special filters, like Hilbert filters and differentiators, about which we will not go into details.

7.5 FIR-Filter Design Methods and Approximation Criteria

FIR filter design methods can be divided into two categories:

1. methods that require only relatively simple calculations. The most popular is the design by *impulse response truncation using windows*, also known as *windowing method*; another example is provided by the *least-squares (LS) design method*. These methods (see, e.g., Parks and Burrus 1987; Oppenheim and Schafer 2009) provide good results but are not optimized in terms of filter order;
2. methods based on numerical optimization procedures that require iterative calculations and sophisticated design software. The most interesting is the *minimax design method* that produces LP filters meeting specifications at the lowest possible order. For this reason, FIR filters designed by this method are said to be *optimum filters*.

A crucial point for the ease of use of a design method is the availability of some formula allowing an approximate a priori estimate of the filter order required to meet given specifications. In general, it is intuitive that more stringent specifications lead to higher orders. For example, let us consider an FIR passband filter: if we require a narrow transition band and a small ripple in the passband and/or stopband, we will have to design a high-order filter, which implies a long impulse response, many calculations when we apply the filter to a sequence, and an extended transitory filtering phase. If we choose both the specifications and the order in an arbitrary way, very likely the designed filter will not have the desired behavior. We will thus be forced to modify some parameter—normally we will increase the filter order—and then repeat the design procedure until specifications are met. If, for a certain design method, an a priori estimate of the order required to meet specifications is possible, even if this estimate is only approximated, a lot of time and computational resources can be saved. This possibility exists in the minimax design method, and is based on empirical formulas derived from numerical simulations, while it is limited to a very particular case in the impulse-response-truncation method: a further argument in favor of the minimax approach.

Many FIR design methods are based on the definition or an *error measure* through which the similarity between the frequency response of the designed filter and the desired one can be quantified. Three design methods are normally employed in FIR filter design. The first two are named after their characteristic error measure.

1. *Minimax design method*: the samples of $h[n]$, which also are the coefficients of the FIR transfer function, are optimized in such a way as to minimize the absolute maximum error of the realized frequency response with respect to the desired one, in the passband and/or stopband. The approximation thus obtained is called minimax approximation or Chebyshev approximation. Let us focus our attention on the passband, and suppose the designed filter has an oscillatory behavior (Fig. 7.1). The error attains its maximum value where the approximated magnitude response touches the tolerance limits, i.e., at the relative maxima and minima (i.e., peaks) of the magnitude response. The absolute maximum passband error is thus an *absolute peak error*. The same is true in the stopband.

   Giving a tolerance scheme thus means specifying the maximum value that the absolute peak error can assume in each band. The minimum order required to respect these limits can then be estimated. If a Type-I filter is desired and this minimum order is an odd number, it will be increased by one unit. Then the coefficients of the filter, having this order and the specified band limits, are optimized numerically, so that the absolute peak error in the passband and stopband is minimized. The final result can have, in each band, an absolute peak error equal to the corresponding pre-fixed maximum value, or even slightly lower if the chosen order was actually redundant. However, the empirical formulas available for minimum order estimation normally underestimate it, so it is likely that at the first attempt the minimized error exceeds the tolerance limits. The order is then increased by two units (so as to remain an even number) and the procedure repeated until the minimized error turns out to be contained inside the tolerance belt in each band.

   The filters obtained by minimax-error design are found to be *equiripple*, i.e., characterized by uniform oscillations in each band, as in Fig. 7.1.

2. *Least-squares (LS) design method*: the square error between the effective frequency response and the desired one, integrated over the passband(s) and the stopband(s), is minimized. The minimization can possibly be weighted differently on the various bands. A variation of this method, called *constrained least-squares (LS) design*, allows for performing the minimization of the integrated square error while imposing upper and lower limits for the error in the various bands.

Simpler design methods also exist that do not explicitly use any of the error measures mentioned above. The most popular is the *design by impulse response truncation using windows*.

This method starts from a generic ideal desired frequency response $H_d(e^{j\omega})$ that, being a periodic function, can be represented by a Fourier series expansion. The coefficients of the expansion are the samples of the impulse response $h_d[n]$, which is non-causal and infinite in length, because $H_d(e^{j\omega})$ has jumps discontinuities. The most straightforward approach to obtaining a causal FIR approximation to such a system is to truncate the ideal impulse response, and this is done multiplying it by a rectangular or tapered window. The window shape and length must be chosen in such a way as to obtain an approximation respecting some given specifications and tolerances concerning the frequency response of the designed filter. In the case of a rectangular window, the windowed design implicitly uses the least-squares error measure. Note that the impulse response $h_d[n]$ is a bilateral sequence and as such, it must be framed by a window centered at $n = 0$, i.e., non-causal; later, the truncated impulse response is delayed to get causality. Alternatively, a phase factor can be incorporated into $H_d(e^{j\omega})$, so as to be able to use directly a causal truncation window.

Typically, $H_d(e^{j\omega})$ is piecewise-constant, with discontinuities at the boundaries between bands. The Fourier series expressing $H_d(e^{j\omega})$ thus converges only in the mean-square sense and the truncation of the summation to a finite number of terms involves the Gibbs phenomenon; the approximated frequency response exhibits oscillations.

This design method is simple but not optimal: the designed filter is not guaranteed to have the minimum order compatible with the specifications. Since the window length N must be chosen before designing the filter and later the designed filter must be examined to check if it actually does not exceed the pre-fixed tolerances, the availability of an empirical formula for estimating N a priori on the basis of these specifications is a critical factor in terms of ease of use. No empirical formula of this type is available for classical windows: N must be found by trial and error, and this turns out to be very time-consuming. An exception is the Kaiser window (Sect. 5.4): Kaiser (1974) found, by numerical simulations performed on a usefully wide range of conditions, empirical formulas to estimate a priori with reasonable accuracy the values of α and N required for the windowing-designed filter to satisfy given specifications. Of course, using a Kaiser window in filter design does not eliminate the intrinsic limitations of the windowing method: most importantly, its lack of optimality.

Another disadvantage of the windowing method is that the oscillations

- in each band are not equiripple: they are more pronounced in the vicinity of the discontinuities. In this way, the specifications are met at two precise frequencies but turn out to be too restrictive at all other frequencies;
- are equal in the passband and in the stopband, the distance from a discontinuity being equal. If, as is normally the case, the specifications are more stringent in the stopband ($\delta_s < \delta_p$), an unnecessary accuracy in the passband is involved that pushes the filter order up.

Figure 7.5 shows three examples of lowpass Type-I FIR filter with cutoff at $\omega_p = 0.5\pi$, obtained via the windowing method using a rectangular window and different orders: $N - 1 = 14, 28$ and 42. It can be seen that the filter is not equiripple and that the maximum oscillations, located at each side of the discontinuity of the ideal frequency response, have the same amplitude in both bands. This is clearly visible in the insets of Fig. 7.5, representing expanded views of the filters' behaviors in the passband/stopband frequency ranges in which the maximum oscillations take place.

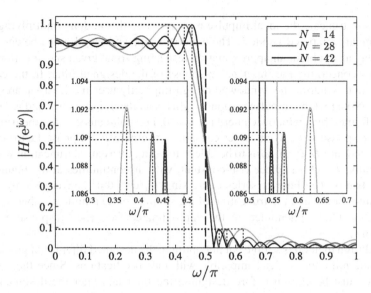

Fig. 7.5 Lowpass Type-I FIR filters with cutoff at $\omega_p = 0.5\pi$, obtained via the windowing method, using a rectangular window and different orders: $N-1 = 14, 28$ and 42. The insets present expanded views of the filters' behaviors in the frequency ranges in which the maximum oscillations occur

It is intuitive that if the approximation error is uniformly distributed over frequency and if the ripples in the passband and in the stopband are adjusted separately, any given specifications can be met at a lower order. This qualitative observation is rigorously confirmed by a theorem—the *alternation theorem*—on which the minimax design method is based. Before going deeper into the minimax method, we must however review the properties of GLP/LP FIR filters and define the quantities that will be used in the minimax design algorithm.

Even if for frequency selection we will normally use Type-I filters, we need to maintain generality for the moment, since the minimax method is formulated for the design of any type of GLP/LP filter.

7.6 Properties of GLP FIR Filters

Let us first recall a few fundamental points. The transfer function of an FIR filter is a polynomial function in z^{-1}, with degree equal to the filter order $N-1$,

$$H(z) = \sum_{n=0}^{N-1} h[n]z^{-n},$$

which has

- $N-1$ poles in $z = 0$,
- $N-1$ zeros anywhere in the z-plane at finite distance from the origin.

$H(z)$ is the z-transform of the impulse response $h[n]$, having finite length N. There-fore $H(z)$ converges on the unit circle, and the frequency response

$$H(e^{j\omega}) = \sum_{n=0}^{N-1} h[n]e^{-j\omega n}$$

certainly exists. Since a sequence with length N is completely determined by N samples of its DTFT, the design of an FIR filter can equivalently be aimed at finding

- the N samples of its impulse response $h[n]$, or
- N samples of its frequency response $H(e^{j\omega})$.

Moreover, we know that

- a LP-RCSR-FIR filter is expressed in the continuous-phase representation by its real zero-phase response $H(\omega)$, such that

$$H(e^{j\omega}) = H(\omega)e^{j\phi(\omega)} = H(\omega)e^{-j\omega\tau},$$

where $\tau = $ constant. A GLP-RCSR-FIR filter is expressed in a similar way, except that the group delay $\tau_g = $ constant comes into play (Sect. 6.5.3).
- $H(\omega)$ can assume positive, zero, and negative values. Consequently, the filter phase response is linearized: $\phi(\omega)$ does not contain those π jumps that are found in correspondence to odd-multiplicity zeros of the magnitude response $|H(e^{j\omega})|$ (Fig. 7.6; see also Fig. 6.4).

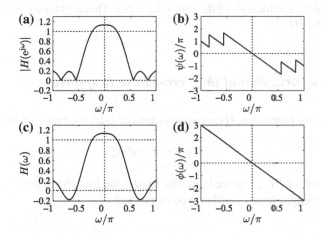

Fig. 7.6 Passing from the magnitude response (**a**) to the zero-phase response (**c**) linearizes the phase response (**d**), that otherwise exhibits π jumps (**b**)

Fig. 7.7 Symmetry of the impulse response for LP/GLP-RCSR-FIR filters of **a** Type I, **b** Type II, **c** Type III, and **d** Type IV. The center of symmetry of each sequence, indicated by the *vertical dashed line* in the corresponding panel, is $k = (N-1)/2$, and can be an integer number (panels a and c) or a half-integer number (panels b and d)

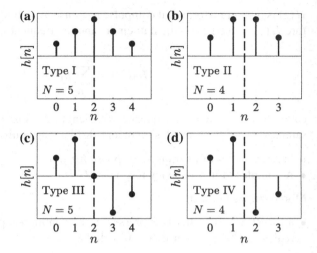

- GLP-RCSR filters come in four types, and for each type the coefficients $h[n]$ must satisfy certain symmetry conditions (see Sect. 6.5.4). The symmetry center, $k = (N-1)/2$, is

  - an integer number for Type-I and Type-III filters,
  - a half-integer number for Type-II and Type-IV filters, as summarized in Fig. 7.7.

As for the magnitude of the impulse response samples, it is worth noting that it is always possible to design a FIR filter using coefficients with magnitude of less than 1.0. This is not true for IIR filters.

We will now work on $H(z)$ and $H(\omega)$, with the aim of unifying the symmetry relations of the coefficients of the four filter types. This is obtained by a factorization of $H(z)$ and $H(\omega)$.

7.6.1   Factorization of the Zero-Phase Response

For each of the four types, $H(z)$ can be factorized (see, for example, Mitra and Kaiser 1993) as

$$H(z) = F(z)G(z),$$

where the factor $F(z)$, referred to as the *fixed term*, has a fixed functional form that, however, changes from one filter type to another:

$$F(z) = \begin{cases} 1 & \text{for Type-I filters,} \\ \frac{1+z^{-1}}{2} & \text{for Type-II filters,} \\ \frac{1-z^{-2}}{2} & \text{for Type-III filters,} \\ \frac{1-z^{-1}}{2} & \text{for Type-IV filters,} \end{cases}$$

while the second factor $G(z)$, referred to as the *adjustable term*, has a polynomial form,

$$G(z) = \sum_{n=0}^{2M} g[n]z^{-n},$$

with

$$M = \begin{cases} \frac{N-1}{2} & \text{for Type-I filters,} \\ \frac{N}{2} - 1 & \text{for Type-II filters,} \\ \frac{N-3}{2} & \text{for Type-III filters,} \\ \frac{N}{2} - 1 & \text{for Type-IV filters.} \end{cases}$$

The adjustable term contains $2M + 1$ coefficients $g[n]$ that are to be determined on the basis of filters specifications.

It must be noted that $G(z)$

- is a Type-I transfer function: for a Type-I filter, $F(z) = 1$, hence we have $H(z) \equiv G(z)$;
- is a polynomial function in z^{-1} with even degree $2M$.

Of course, the coefficients $g[n]$ are related to the coefficients $h[n]$, and consequently are symmetric, but the factorization presented above unifies the symmetry conditions that the new coefficients $g[n]$ must satisfy for Type I–IV filters. Indeed, for all filter types,

$$g[2M - n] = g[n].$$

In this way, for the description of the design method we can focus on $G(z)$, i.e., on Type-I filters; the design of Type-II, -III and -IV filters is brought down to Type-I design. Table 7.1 shows how the coefficients $h[n]$ and the coefficients $g[n]$ are related to one another. The relation is different for each filter type. Type-I filters have $h[n] \equiv g[n]$.

Figure 7.8 illustrates the relation among N and M for each filter type. Note that only $M + 1$ samples of $h[n]$—and therefore of $g[n]$—are independent of one another. They are indicated by *black dots*, while *gray dots* symbolize samples that can be deduced by symmetry from the other ones. For the antisymmetric Type-III impulse response of Fig. 7.8c, the $(M + 2)$th coefficient is always zero, since the center of symmetry is an integer, $M + 2 = (N + 1)/2$.

Table 7.1 Relation between the coefficients $h[n]$ and the coefficients $g[n]$ for each filter type

Coefficient	Type I	Type II	Type III	Type IV
$h[0]$	$g[0]$	$\frac{g[0]}{2}$	$\frac{g[0]}{2}$	$\frac{g[0]}{2}$
$h[1]$	$g[1]$	$\frac{g[1]+g[0]}{2}$	$\frac{g[1]}{2}$	$\frac{g[1]-g[0]}{2}$
$h[n]$	$g[n]$	$\frac{g[n]+g[n-1]}{2}$	$\frac{g[n]-g[n-2]}{2}$	$\frac{g[n]-g[n-1]}{2}$
$h[N-2]$	$g[N-2]$	$\frac{g[N-2]+g[N-3]}{2}$	$-\frac{g[N-4]}{2}$	$\frac{g[N-2]-g[N-3]}{2}$
$h[N-1]$	$g[N-1]$	$\frac{g[N-2]}{2}$	$-\frac{g[N-3]}{2}$	$-\frac{g[N-2]}{2}$
	$M=\frac{N-1}{2}$	$M=\frac{N-2}{2}$	$M=\frac{N-3}{2}$	$M=\frac{N-2}{2}$

Fig. 7.8 Examples of the relation between M and N for GLP filters of **a** Type I, **b** Type II, **c** Type III, and **d** Type IV. The samples represented by *black dots* are independent of each other, while the ones shown as *gray dots* can be deduced by symmetry from the ones in black

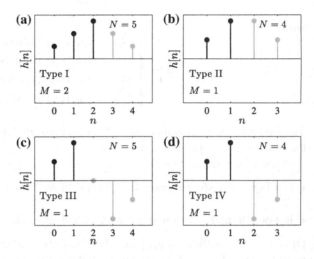

Exploiting the symmetry of the $g[n]$ we can write the adjustable term as

$$G(z) = z^{-M}\{g[M] + g[M-1](z+z^{-1}) + \cdots + g[0](z^M + z^{-M})\},$$

that on the unit circle ($z = e^{j\omega}$) becomes

$$G(e^{j\omega}) = e^{-jM\omega}\{g[M] + g[M-1]2\cos\omega + \cdots + g[0]2\cos(M\omega)\}.$$

This last expression derives from the fact that on the unit circle we have

$$z + z^{-1} = e^{j\omega} + e^{-j\omega} = \cos\omega + j\sin\omega + \cos\omega - j\sin\omega = 2\cos\omega,$$

and that similar relations hold for higher powers of z. Neglecting for the moment the phase factor, we thus have

$$G(\omega) = g[M] + g[M-1]2\cos\omega + g[M-2]2\cos(2\omega) + \cdots + g[0]2\cos(M\omega),$$

that can be rewritten as

$$G(\omega) = \sum_{n=0}^{M} a[n] \cos(\omega n),$$

having defined

$$a[n] = \begin{cases} 2g[M - n] & \text{for } n \neq 0, \\ g[M] & \text{for } n = 0. \end{cases}$$

All $M + 1$ coefficients $a[n]$ are independent of one another.

In a similar way, after some algebra we get

$$F(e^{j\omega}) = \begin{cases} 1 & \text{for Type-I filters,} \\ e^{-j\frac{\omega}{2}} \cos(\frac{\omega}{2}) & \text{for Type-II filters,} \\ e^{-j(\omega - \frac{\pi}{2})} \sin \omega & \text{for Type-III filters,} \\ e^{-j(\frac{\omega}{2} - \frac{\pi}{2})} \sin(\frac{\omega}{2}) & \text{for Type-IV filters,} \end{cases}$$

and neglecting the phase terms we can write

$$F(\omega) = \begin{cases} 1 & \text{for Type-I filters,} \\ \cos(\frac{\omega}{2}) & \text{for Type-II filters,} \\ \sin \omega & \text{for Type-III filters,} \\ \sin(\frac{\omega}{2}) & \text{for Type-IV filters.} \end{cases}$$

Combining the expressions of $G(\omega)$ and $F(\omega)$ we finally get the factorized zero-phase response, $H(\omega) = F(\omega)G(\omega)$. In the expression of the adjustable term, the $M + 1$ independent coefficients $a[n]$ must be determined on the basis of filter specifications. This will be the aim of the design procedure.

Resuming and combining the phase terms momentarily discarded above, we can also derive the form of the phase factor:

$$\phi(\omega) = \begin{cases} -\frac{N-1}{2}\omega & \text{for Type-I and -II filters,} \\ -\frac{N-1}{2}\omega + \frac{\pi}{2} & \text{for Type-III and -IV filters.} \end{cases}$$

We thus have a simple time delay by $(N - 1)/2$ samples for Type-I filters (integer delay) and Type-II filters (half-integer delay).

Before proceeding, we may note that the form of the factor $F(\omega)$ in the four cases is consistent with what we said earlier concerning the fact that only Type-I filters are suited for all kinds of frequency selectivity. Indeed, the behavior of $H(\omega)$ at $\omega = 0$ and at $\omega = \pi$ is constrained by the factor $F(\omega)$, and

- for Type-I filters, $F(\omega) = 1$, so that $F(\omega)$ never vanishes; $H(\omega)$ is free from constraints deriving from $F(\omega)$, and can assume any shape;
- for Type-II filters, $F(\omega) = \cos(\omega/2)$, so that at $\omega = \pi$ we have $F(\omega) = 0$ and $H(\omega) = 0$: the filter cannot be a highpass or a bandstop filter;
- for Type-III filters, $F(\omega) = \sin \omega$, so that at $\omega = 0$ and at $\omega = \pi$ we have $F(\omega) = 0$ and $H(\omega) = 0$: the filter cannot be a lowpass, a highpass or a bandstop filter;
- for Type-IV filters, $F(\omega) = \sin(\omega/2)$, so that at $\omega = 0$ we have $F(\omega) = 0$ and $H(\omega) = 0$: the filter cannot be a lowpass or a bandstop filter.

7.6.2   Zeros of the Transfer Function

An FIR transfer functions has only zeros and no poles, except a multiple pole in the origin of the z-plane. The zeros of the transfer function $H(z)$ of LP/GLP FIR filters are determined by both factors $F(z)$ and $G(z)$. As for the fixed term,

$$F(z) = \begin{cases} 1 & \text{for Type-I filters provides no zeros,} \\ \frac{1+z^{-1}}{2} & \text{for Type-II filters provides a zero at } z = -1, \\ \frac{1-z^{-2}}{2} & \text{for Type-III filters provides two zeros at } z = \pm 1, \\ \frac{1-z^{-1}}{2} & \text{for Type-IV filters provides a zero at } z = +1. \end{cases}$$

Concerning the zeros of $G(z)$, from

$$G(z) = z^{-M}\{g[M] + g[M-1](z + z^{-1}) + \cdots + g[0](z^M + z^{-M})\}$$

we can deduce that

$$G(z^{-1}) = z^M\{g[M] + g[M-1](z^{-1} + z) + \cdots + g[0](z^{-M} + z^M)\}$$

and therefore

$$G(z^{-1}) = z^{2M}G(z).$$

So, $G(z)$ and $G(z^{-1})$ share the same zeros. This property implies that for a $G(z)$ with real coefficients, the zeros occur in

- quadruplets off the unit circle: $r_i e^{\pm j\theta_i}$ and $1/(r_i e^{\pm j\theta_i})$; if we set $a = r_i e^{j\theta_i}$, this means a, a^*, $1/a$ and $1/a^*$;
- conjugate pairs on the unit circle: $e^{\pm j\theta_i}$;
- reciprocal pairs on the real axis: r_i' and $1/r_i'$.

This is illustrated in Fig. 7.9. Note that the zeros of $G(z)$ are also the zeros of a Type-I transfer function $H(z)$, while the other filter types have additional zeros due to $F(z)$.

Fig. 7.9 An example of the positions of the zeros of $G(z)$

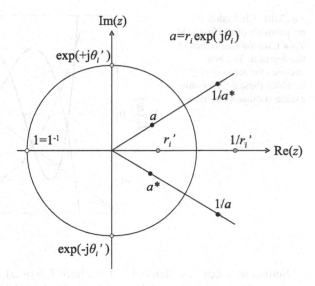

7.6.3 Another Form of the Adjustable Term

The form of $G(\omega)$ presented above is a possible representation of the zero-phase response of Type-I filters. Another representation of $G(\omega)$ that is more useful in the frame of minimax design can be obtained using the Chebyshev polynomials of the first kind, with degree n in the real variable w:

$$T_n(w) = \begin{cases} \cos\left[n \arccos(w)\right] & \text{for} \quad |w| \le 1, \\ \cosh\left[n \operatorname{arccosh}(w)\right] & \text{for} \quad w \ge 1, \\ (-1)^n \cosh\left[n \operatorname{arccosh}(-w)\right] & \text{for} \quad w \le 1. \end{cases}$$

They can be generated recursively by the formula

$$\begin{cases} T_0(w) = 1, \\ T_1(w) = w, \\ \cdots \\ T_n(w) = 2w T_{n-1}(w) - T_{n-2}(w). \end{cases}$$

The behavior of these polynomials is shown in Fig. 7.10 for $n = 1, 2, 3$, and 4. For $|w| \le 1$, $T_n(w)$ cannot exceed 1 in absolute value and oscillates between -1 and 1 a number of times proportional to n; for $|w| > 1$, $T_n(w)$ increases monotonically with $|w|$.

Chebyshev polynomials are even or odd functions of w for even or odd n, respectively. When n is even, $T_n(0) = \pm 1$; when n is odd, $T_n(0) = 0$.

Fig. 7.10 Chebyshev
polynomials of the first kind,
$T_n(w)$, for various values of
the degree n. The box
encloses the region $|w| \leq 1$
in which these polynomials
exhibit oscillatory behavior

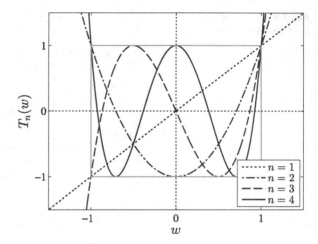

Setting $w = \cos\omega$, so that $|w| \leq 1$, we have $T_n(\cos\omega) = \cos[n\arccos(\cos\omega)] = \cos(n\omega)$. Thus, using these polynomials, $\cos(n\omega)$ can be expressed as a polynomial with degree n in $\cos\omega$. Consequently, $G(\omega)$ can be written as a polynomial of degree M in $\cos\omega$:

$$G(\omega) = \sum_{n=0}^{M} a[n]\cos(\omega n) = \sum_{n=0}^{M} \alpha[n]\cos^n\omega.$$

The coefficients $\alpha[n]$ in this expression can be derived from the coefficients $a[n]$ that are, in turn, related to the coefficients $g[n]$, and therefore to the samples of $h[n]$. Only $M+1$ of them are independent of each other.

7.7   Equiripple FIR Filter Approximations: Minimax Design

The *equiripple* or *minimax* method serves to design GLP-FIR filters of all four types. It is based on the factorization $H(\omega) = F(\omega)G(\omega)$, where only $G(\omega)$ contains adjustable coefficients to be determined according to the specifications of the desired filter. Working on the zero-phase response, rather than on the magnitude response, is convenient and non-restrictive, since we already know the phase response of this kind of filters (Sect. 7.6.1). From previous reasoning it is clear that we can focus on Type-I filters, for which $H(\omega) = G(\omega)$. The zero-phase response $H(\omega)$ for the other three filter types can, if necessary, be obtained successively, using the known form of $F(\omega)$.

We will discuss the method referring to a Type-I lowpass filter. The case of a highpass filter is perfectly equivalent. Some differences arise when a passband or a stopband filter are considered, because these filters have a greater number of bands.

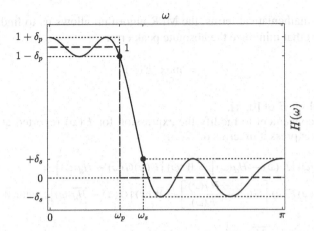

Fig. 7.11 Specifications for the design of a Type-I lowpass filter with $N = 15$ and $M = 7$. The *dashed lines* represent the desired zero-phase response, while the *solid curve* shows the zero-phase response of a filter that would meet specifications. Tolerance limits are represented by *horizontal dotted lines*. $H(\omega)$ presents a number of local extrema (minima and maxima) in the interval $0 \le \omega \le \pi$

Let us consider a Type-I lowpass filter, with even order $N - 1$ and odd impulse-response length N. The filter is required to meet the specifications shown in Fig. 7.11. Note that the figure illustrates specifications for the zero-phase response. In the example of Fig. 7.11, $N = 15$ and $M = (N-1)/2 = 7$. Observe that constraints are set in the passband and in the stopband only; in the transition band, no constraints are imposed on $H(\omega)$. However, the transition band can never vanish for realizable filters. The constrained band are collectively classified as *approximation bands*.

When setting the filter specifications, we must provide the values of five parameters: the order $N - 1$ that for each type automatically determines M; ω_p and ω_s; δ_p and δ_s. However, we cannot set all these values arbitrarily, since they are not independent of each other. Several minimax design algorithms have thus been proposed, in which some of these parameters are fixed and the remaining ones are optimized by an iterative procedure aimed at minimizing the maximum absolute approximation error over the intervals $[0, \omega_p]$ and $[\omega_s, \pi]$. Here we will discuss the approach in which M, ω_p, ω_s, and the ratio $\kappa = \delta_p/\delta_s$ are fixed, while the effective value of δ_p, and consequently the effective value of δ_s, are minimized.

Among the many variants of the minimax algorithm that can be found in literature, the most efficient is the *Parks-McClellan algorithm* (Parks and McClellan 1972), also known as *multiple exchange algorithm* for reasons that will soon become clear. The Parks-McClellan algorithm derives from the *Remez exchange algorithm* (Remez 1934a, b, c), adapted for the design of FIR filters. It has become a standard method for FIR filter design and has been implemented in software in different ways; the implementation by McClellan, Parks and Rabiner (McClellan et al. 1973; Rabiner et al. 1975), referred to as the *MPR algorithm*, is widely used, and we shall focus on it.

In general mathematical terms, the MPR algorithm allows us to find the coefficients of $G(\omega)$ that minimize the absolute peak error

$$\varepsilon = \max_{\omega \in B} |E(\omega)|$$

in a closed subset B of $[0, \pi]$.

Here it is convenient to modify the expression for $E(\omega)$ reported at the end of Sect. 7.3.1, to express it in terms of $G(\omega)$:

$$E(\omega) = W(\omega) \left[H(\omega) - H_d(\omega) \right] = W(\omega) \left[F(\omega)G(\omega) - H_d(\omega) \right] =$$
$$= W(\omega)F(\omega) \left[G(\omega) - \frac{H_d(\omega)}{F(\omega)} \right] = \overline{W}(\omega) \left[G(\omega) - \overline{H_d}(\omega) \right], \text{ with } \overline{W}(\omega) > 0.$$

The desired response $\overline{H_d}(\omega)$ is assumed to be (piecewise) continuous in B. In our lowpass case, the subset B will include both the passband and the stopband:

$$B = 0 \le \omega \le \omega_p \cup \omega_s \le \omega \le \pi.$$

The desired zero-phase response will be

$$\overline{H_d}(\omega) = \frac{H_d(\omega)}{F(\omega)}, \quad \text{with} \quad H_d(\omega) = \begin{cases} 1 & \text{for } 0 \le \omega \le \omega_p, \\ 0 & \text{for } \omega_s \le \omega \le \pi, \end{cases}$$

and the global weighting function will be expressed as

$$\overline{W}(\omega) = W(\omega)F(\omega), \quad \text{with} \quad W(\omega) = \begin{cases} 1 & \text{for } 0 \le \omega \le \omega_p, \\ \kappa & \text{for } \omega_s \le \omega \le \pi, \end{cases}$$

where $\kappa = \delta_p/\delta_s$ is the ratio between the maximum absolute deviations admitted in the passband and in the stopband, respectively. Moreover, we will have $\varepsilon = \delta_p$.

Considering only Type-I filters for which $F(\omega) = 1$, we simply have

$$G(\omega) = H(\omega), \quad \overline{W}(\omega) = W(\omega) \quad \text{and} \quad \overline{H_d}(\omega) = H_d(\omega).$$

Thus, in the Type-I lowpass case, the minimization of $\max |E(\omega)|$ in the frequency intervals $0 \le \omega \le \omega_p$ and $\omega_s \le \omega \le \pi$ is equivalent to minimizing δ_p, under the condition that the final value of δ_p, i.e., the actual value of δ_p in the designed filter, cannot exceed the corresponding pre-fixed value, while keeping M, ω_p, ω_s, and $\kappa = \delta_p/\delta_s$ constant. Consequently, also δ_s is minimized.[2]

An important point in minimax design is the *number of local extrema* that $H(\omega)$ presents in $0 \le \omega \le \pi$. Figure 7.12 shows, for our lowpass filter with $M = 7$, the

---

[2]In principle, we should use different symbols for the pre-fixed values of δ_p and δ_s and the corresponding optimized values, but this would weigh down the notation too much.

Fig. 7.12 Frequencies at which $H(\omega)$ attains tolerance limits and therefore the absolute error is maximum. In this Type-I lowpass example with $M = 7$ (the same as in Fig. 7.11) there are $M + 2 = 9$ such frequencies, numbered from 0 to $M + 1 = 8$, with $\omega_0 = 0$, $\omega_3 = \omega_p$, $\omega_4 = \omega_s$, and $\omega_8 = \pi$. They are highlighted by *dashed vertical lines*

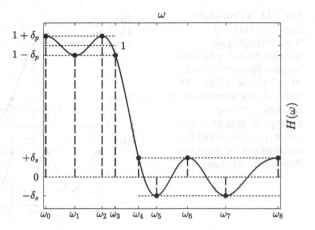

frequencies at which local minima and maxima are found. In this example, these are also frequencies at which $H(\omega)$ touches the tolerance limits. Moreover, $H(\omega)$ attains these limits also at ω_p and ω_s: we thus see $7 + 2 = 9 = M + 2$ frequencies at which the absolute error is maximum (see the *vertical dashed lines* in Fig. 7.12).

The number of local extrema of $H(\omega)$ in the interval $0 \le \omega \le \pi$ actually depends on M, because a trigonometric polynomial with degree M, such as

$$H(\omega) = G(\omega) = \sum_{n=0}^{M} \alpha[n] \cos^n \omega,$$

can have up to $M - 1$ local extrema in the open interval $0 < \omega < \pi$. Moreover,

$$\frac{dH(\omega)}{d\omega} = -\sin \omega \sum_{n=0}^{M} n\alpha[n] \cos^{n-1} \omega = 0 \quad \text{for} \quad \omega = \begin{cases} 0, \\ \pi, \end{cases}$$

so that a maximum or a minimum is always present at $\omega = 0$ and at $\omega = \pi$. In conclusion, in the closed interval $0 \le \omega \le \pi$ there can be up to $M + 1$ local extrema of $H(\omega)$, but maxima and minima are not necessarily all equal. More precisely, one of the two extrema of $H(\omega)$ occurring at $\omega = 0$ or π can be smaller than the other extrema. When this happens, at that point $H(\omega)$ is extremal but does not attain the tolerance limit: the absolute error is not a maximum.

We now must understand what conditions a filter must satisfy in order to be an optimum filter, i.e., to meet specifications at the lowest possible order. The conditions for optimality of a GLP filter designed according to the minimax criterion are given by the *alternation theorem*: given any closed subset B of the closed interval $0 \le \omega \le \pi$, a polynomial function $H(\omega)$ is the unique optimum approximation to $H_d(\omega)$ in B if, and only if, the error function $E(\omega)$ has in B at least $M + 2$ alternations, i.e., if, and only if, at least $M + 2$ frequencies ($\omega_0 \le \omega_1 \le \omega_2 \le \cdots \omega_{M+1}$) exist in B, at which $E(\omega)$ attains its maximum absolute value with alternating signs:

Fig. 7.13 a Zero-phase
response $H(\omega)$ and
b corresponding error
function $E(\omega)$, for the same
lowpass Type-I filter with
$M = 7$ shown in Figs. 7.11
and 7.12. The alternations of
the error function,
corresponding to *black dots*,
are $M + 2 = 9$, so that the
conditions of the alternation
theorem are satisfied

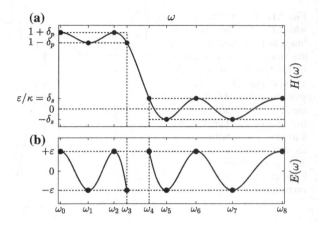

$$E(\omega_i) = -E(\omega_{i-1}) = \pm \max |E(\omega)| = \pm\varepsilon,$$

with $i = [1, M + 1]$ and with $\varepsilon > 0$.

This theorem is attributed to Chebyshev (see, e.g., Ollin 1979). Figure 7.13 shows these alternations for our lowpass example: Fig. 7.13a depicts $H(\omega)$, while Fig. 7.13a shows the error function $E(\omega)$. The filter is optimum according to the alternation theorem, since there are $M + 2 = 9$ alternations: see the *black dots* in Fig. 7.13b. If we now also observe Fig. 7.13a and the *black dots* on the $H(\omega)$ curve, we understand that

- the frequencies ω_i correspond to $E(\omega)$ extrema and to $H(\omega)$ values which are equal to the tolerance limits; their number depends on M;
- ω_p and ω_s correspond to $E(\omega)$ extrema, irrespective of M; they are part of the ω_i set but do not correspond to $H(\omega)$ extrema.

In summary,

- $H(\omega)$ can have up to $M - 1$ local extrema in $0 < \omega < \pi$; the same turns out to be true in the combined open intervals $0 < \omega < \omega_p$, $\omega_s < \omega < \pi$. Moreover, $H(\omega)$ always has a local minimum or maximum at $\omega = 0$ and $\omega = \pi$. Finally, in ω_p and ω_s the tolerance limits are attained, i.e., $H(\omega_p) = 1 - \delta_p$, $H(\omega_s) = \delta_s$. In our lowpass case we have only one ω_p and only one ω_s, so we can have up to $M - 1 + 2 + 2 = M + 3$ frequencies at which $|E(\omega)|$ is maximum;
- the alternation theorem requires at least $M + 2$ extrema of $E(\omega)$ with alternating signs.

We can conclude that the optimum approximation to $H_d(\omega)$ can produce an error function $E(\omega)$ with $M + 2$ or with $M + 3$ alternations.

In ω_p and ω_s, $|E(\omega)|$ is always maximum, so M or $M + 1$ local extrema of $H(\omega)$ must contribute to alternations. When both frequencies $\omega = 0$ and $\omega = \pi$ contribute one alternation, we have a total of $M + 3$ or $M + 2$ alternations. When only one of

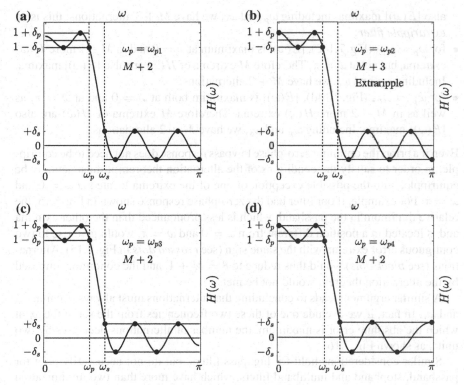

Fig. 7.14 $H(\omega)$ changes its shape when we gradually increase ω_p, while keeping M, κ and the width $\Delta\omega$ of the transition band constant. Four different frequencies, $\omega_{p1} < \omega_{p2} < \omega_{p3} < \omega_{p4}$, characterize the filters shown in panels **a**, **b**, **c**, and **d**, respectively. Each panel shows the number of alternations. In only one case there are $M + 3$ alternations: the corresponding filter is an extraripple filter (panel **b**)

them contributes one alternation, we can just attain a total of $M + 2$ alternations. But when do $M + 3$ alternations occur?

In order to understand this point, let us look at Fig. 7.14 that shows all possible Type-I lowpass cases that can occur when M, κ, and the width $\Delta\omega$ of the transition band are kept constant, while ω_p varies. Obviously, ω_s varies with ω_p. As ω_p varies, the shape of $H(\omega)$ changes. We consider four values of ω_p: $\omega_{p1} < \omega_{p2} < \omega_{p3} < \omega_{p4}$ (evidently, when we set a fixed value for ω_p too, the solution to the minimax approximation problem becomes unique). The corresponding filters appear in Fig. 7.14a, b, c, and d, respectively. We can observe that:

- for $\omega_p = \omega_{p1}$ (Fig. 7.14a), $|E(\omega)|$ is maximum at $\omega = \pi$ and in $M - 1$ more $H(\omega)$ extrema, but not at $\omega = 0$. Therefore M extrema of $H(\omega)$ are also $|E(\omega)|$ maxima. Including ω_p and ω_s, we have $M + 2$ alternations;
- for $\omega_p = \omega_{p2}$ (Fig. 7.14b), $|E(\omega)|$ is maximum both at $\omega = 0$ and at $\omega = \pi$, as well as in $M - 1$ more $H(\omega)$ extrema. Therefore $M + 1$ extrema of $H(\omega)$ are

also $|E(\omega)|$ maxima. Including ω_p and ω_s, we have $M + 3$ alternations: this is an *extraripple filter*;

- for $\omega_p = \omega_{p3}$ (Fig. 7.14c), $|E(\omega)|$ is maximum at $\omega = 0$ and in $M - 1$ more $H(\omega)$ extrema, but not at $\omega = \pi$. Therefore M extrema of $H(\omega)$ are also $|E(\omega)|$ maxima. Including ω_p and ω_s, we have $M + 2$ alternations;
- for $\omega_p = \omega_{p4}$ (Fig. 7.14d), $|E(\omega)|$ is maximum both at $\omega = 0$ and at $\omega = \pi$, as well as in $M - 2$ more $H(\omega)$ extrema. Therefore M extrema of $H(\omega)$ are also $|E(\omega)|$ maxima. Including ω_p and ω_s, we have $M + 2$ alternations.

Even if a priori the optimum zero-phase lowpass response does not need to be equiripple, in order to satisfy the conditions of the alternation theorem it must actually be equiripple, with the possible exception of one of the extrema located at $\omega = 0$ and $\omega = \pi$. For example, if our filter had the zero-phase response shown in Fig. 7.15, the relative minimum in the passband, which is less pronounced than the other extrema and is located in a position different from $\omega = 0$ and $\omega = \pi$, would give rise to two contiguous error extrema with the same sign (see *crossed dots* in Fig. 7.15). Alternations (see *black dots*) would thus reduce to $8 = M + 1$, and the conditions imposed by the alternation theorem would not be met.

A similar argument leads to concluding that alternations must always occur at ω_p and ω_s. In fact, if we exclude one of these two frequencies from the set of the ω_i at which the absolute error is maximum, the number of alternations decreases by two units, as shown Fig. 7.16.

Similar considerations hold for highpass filters, but do not necessarily hold for passband, stopband and multiband filters, which have more than two minimization bands and therefore have a higher number of band edges. This issue will be discussed in Sect. 7.11.

Fig. 7.15 The optimum zero-phase lowpass response $H(\omega)$ must be equiripple, except possibly at $\omega = 0$ and $\omega = \pi$. If, for example, one of the minima in the passband is smaller than the other local extrema, contiguous error extrema with the same sign occur (*crossed dots*). The number of alternations ($8 = M + 1$; *black dots*) is not sufficient for optimality

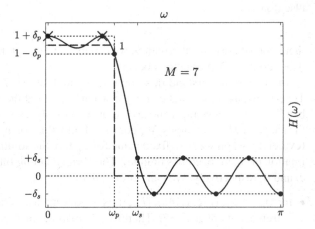

Fig. 7.16 A hypothetical filter in which at ω_p (*empty circle*) the absolute error is not maximum. Two contiguous error extrema with the same sign appear (*crossed dots*). The number of alternations (*black dots*) is $8 = M + 1$ and is not enough for optimality

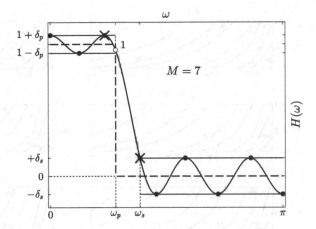

7.8   Predicting the Minimum Filter Order

The filter order, which determines M, is an input parameter to the design algorithm: the iterative process leading to the optimum approximation starts by setting the initial values of the $M + 2$ frequencies ω_i at which $|E(\omega)|$ *is supposed* to be a maximum. Therefore an a priori estimate of the minimum filter order is needed.

For the lowpass case, Herrmann et al. (1973) performed an extensive set of simulations to study the relations among the filter order, the tolerances δ_p and δ_s and the frequencies ω_p and ω_s. They proposed the following empirical formula for the filter length:

$$N = \lceil \frac{2\pi D}{\Delta\omega} - \frac{f(\kappa)\Delta\omega}{2\pi} \rceil + 1,$$

where the symbols $\lceil \; \rceil$ indicate the *ceiling function* which maps a real number to the next higher integer; $\Delta\omega = |\omega_p - \omega_s|$, $\kappa = \delta_p/\delta_s$;

$$D = \left(A_2 \log_{10} \delta_p^2 + A_1 \log_{10} \delta_p + A_0 \right) \log_{10} \delta_s - \left(B_2 \log_{10} \delta_p^2 + B_1 \log_{10} \delta_p + B_0 \right),$$

and

$$f(\kappa) = C_1 \log_{10} \kappa + C_0.$$

The constants appearing in the above equations have the following values:

- $A_0 = -4.761 \times 10^{-1}$; $A_1 = 7.114 \times 10^{-2}$; $A_2 = 5.309 \times 10^{-3}$;
- $B_0 = -2.660 \times 10^{-3}$; $B_1 = -5.941 \times 10^{-1}$; $B_2 = -4.278 \times 10^{-1}$;
- $C_0 = 11.01217$; $C_1 = 0.51244$.

This empirical formula holds for δ_p and δ_s that do not exceed 0.1 and for $\omega_p > 0.08\pi$, $\omega_s < 0.92\pi$. It also works for highpass filters (i.e., $\omega_p > \omega_s$). Hermann's formula is illustrated in Fig. 7.17, in which each panel refers to a different value of $\kappa = \delta_p/\delta_s$ in the range 1–500, and N is represented by *labeled contour lines* in the plane δ_p,

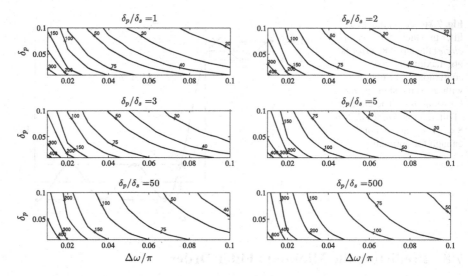

Fig. 7.17 Minimum length N of a lowpass equiripple filter according to the empirical formula by Herrmann et al. (1973). N is represented by *labeled contour lines* in the plane δ_p, $\Delta\omega/\pi$. Each panel refers to a different value of $\kappa = \delta_p/\delta_s$

$\Delta\omega/\pi$. Values of δ_p in the range 0.01–0.1 are considered; $\Delta\omega$ varies from 0.005 to 0.05 π. Note that in all panels, $\kappa > 1$, since normally in applications the stopband tolerance is smaller than the passband tolerance.

Kaiser (1974) subsequently proposed the following simplified formula as a fit to Hermann et al.'s data:

$$N \simeq \lceil \frac{-10\log_{10}(\delta_p\delta_s) - 13}{2.324\,\Delta\omega} \rceil + 1,$$

that clearly shows how N increases with decreasing $\delta_p\delta_s$ and $\Delta\omega$. As a further illustration of Hermann's and Kaiser's formulas, Fig. 7.18 shows, for $\kappa = \delta_p/\delta_s = 3$ and values of δ_p from 0.01 to 0.09, the values of N obtained from Hermann's equation versus $\Delta\omega/\pi$ (*solid lines*), and those obtained from Kaiser's approximation (*dashed lines*).

Once N and the order $N - 1$ have been obtained through these formulas, M is deduced according to the desired filter type (I, II, III, or IV). Note that if we desire a Type-I lowpass filter, we will round the predicted order up to the nearest higher even value. In a passband or stopband case, these formulas still provide useful guidelines, provided different lowpass filters are tried, which go through the bandpass specifications, and conclusions about N are drawn on the basis of the most demanding lowpass filter.

Fig. 7.18 Minimum length N of a lowpass equiripple filter with $\kappa = \delta_p/\delta_s = 3$, computed according to the empirical formula by Herrmann et al. (1973) as a function of $\Delta\omega/\pi$, for different values of δ_p (*solid lines*). The *dashed lines* represent the values of N obtained from Kaiser's approximation (see text)

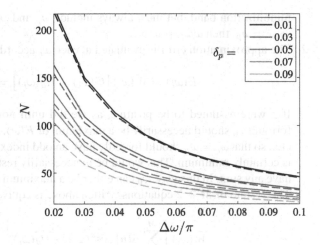

At this point, we must cast light on an apparent contradiction. We stated that we must pre-select the order, which determines M, the values of ω_p, ω_s, and $\kappa = \delta_p/\delta_s$, and that the effective δ_p and δ_s values are then found in the design process. But for predicting the order we do not simply need $\kappa = \delta_p/\delta_s$: we need the values of the individual pre-fixed tolerances δ_p and δ_s. In practice, we will actually start by giving ω_p, ω_s, δ_p and δ_s, and evaluating the order $N - 1$. We will also compute M and $\kappa = \delta_p/\delta_s$. In the following optimization process, which receives $N - 1$, ω_p, ω_s and κ as inputs, the effective value of $\varepsilon = \delta_p$ is minimized, and, as a consequence, the value of δ_s is minimized as well, while their ratio κ remains constant. In the optimum solution to the minimax problem, the final error must not exceed the pre-set values δ_p and δ_s, and this is possible only if a sufficiently high order was chosen. If this is the case, the design process is complete and the effective minimized δ_p value is equal to, or slightly smaller than, its pre-set limit. If, on the contrary, the order was too low, the optimum solution will exceed the pre-set tolerance limits. The order will then be increased (by two units in order to still have a Type-I filter) and the design procedure repeated until specifications are met: each order thus requires a complete optimization process.

7.9   MPR Algorithm

Examining the classical MPR algorithm implementation will lead us to appreciating its efficiency. As in the previous section, we will refer to the Type-I lowpass case.

1. The process is initialized by assigning initial-guess values to the $M + 2$ frequencies ω_i, with $i = [0, M + 1]$, at which $|E(\omega)|$ *is supposed* to be maximum. Note that $M + 2$ is the minimum number of alternations prescribed by the alternation theorem. The frequencies ω_i can, for example, be equally distant on each

minimization band, but must always include ω_p and ω_s in adjacent positions: if $\omega_{i=l} = \omega_p$, then $\omega_{i=l+1} = \omega_s$.

2. The approximation error is evaluated at each ω_i according to the formula

$$E(\omega_i) = \overline{W}(\omega_i)\left[G(\omega_i) - \overline{H_d}(\omega_i)\right] = (-1)^i \varepsilon.$$

If ε were assumed to be positive, as we did until now, then according to this formula ω_0 should necessarily be a maximum of $E(\omega)$, ω_1 should be a minimum, etc., so that $\omega_l = \omega_p$ should forcedly have an odd index l, because at ω_p the error is certainly minimum. This would be unnecessarily restrictive, so ε is allowed to have any sign. In this way, ω_0 can either be a maximum or a minimum, and so on. The system of $M + 2$ equations written above is equivalent to the system

$$\overline{W}(\omega_i)\left[\sum_{n=0}^{M}\alpha[n]\cos^n(\omega_i) - \overline{H_d}(\omega_i)\right] = (-1)^i \varepsilon,$$

containing $M + 2$ unknown variables that are the coefficients $\alpha[n]$, with $n = [0, M]$, and ε. In the lowpass case, we thus simply have

$$1 \cdot \left[\sum_{n=0}^{M}\alpha[n]\cos^n(\omega_i) - 1\right] = (-1)^i \varepsilon \quad \text{in the passband,}$$

$$\kappa \cdot \left[\sum_{n=0}^{M}\alpha[n]\cos^n(\omega_i) - 0\right] = (-1)^i \varepsilon \quad \text{in the stopband.}$$

The frequencies ω_i will be adjusted during the optimization process: at the end of the process they must be determined in such a way that the solution, which is certainly optimum at the start on the initial-guess ω_i set, is actually optimum over the whole continuous B. In other words, the absolute value of the optimized ε must represent the maximum absolute error at any frequency in B. When this goal is attained, the frequencies ω_i—excluding ω_p and ω_s—actually identify the local extrema of the optimum polynomial function $G(\omega)$, i.e., the polynomial having the correct $\alpha[n]$ coefficients.

3. Using the initial guesses for the ω_i, the system must now be solved with respect to ε and $\alpha[n]$, thus determining the trigonometric polynomial $G(\omega) = H(\omega)$ that assumes the correct values at the pre-set ω_i frequencies. These frequencies, however, in general will not be exactly the positions of the local $H(\omega)$ extrema, since we chose them arbitrarily. The extrema of $H(\omega)$ will be somewhere else (let us call $\overline{\omega}_i$ their actual positions). This is shown in see Fig. 7.19a, presenting an example with order $N - 1 = 20$, $\kappa = 1$, $\omega_p = 0.4\pi$ and $\Delta\omega = 0.1\pi$. The order

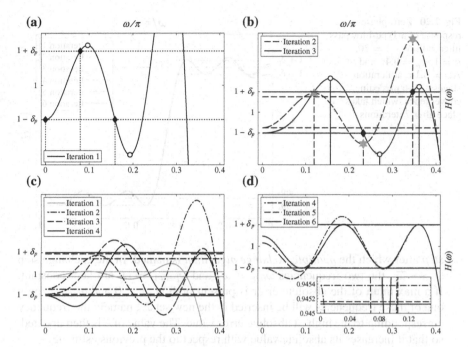

Fig. 7.19 **a** First iteration of the MPR algorithm for the design of a Type-I lowpass filter with order $N - 1 = 20$, $\kappa = 1$, $\omega_p = 0.4\pi$ and $\Delta\omega = 0.1\pi$. The polynomial function (*solid line*) has the correct values at the arbitrarily fixed ω_i (*vertical dotted lines, black diamonds*), but its local minima and maxima are found at some other frequencies $\overline{\omega}_i$ (*empty circles*). **b** The filter passband in the iteration stages n. 2 (*dashed curve*) and n. 3 (*solid curve*). *Horizontal lines* in the same style indicate deviations from 1. The frequencies of local minima and maxima of stage n. 2 (*gray stars*) are assumed as new guessed frequencies for stage n. 3 (*black diamonds*). The local extrema of the polynomial of iteration n. 3 are represented by *empty circles*. **c** The filter passband in the iteration stages n. 1, 2, 3, and 4. The deviations from 1 calculated at each iteration stage are indicated by *horizontal lines*. **d** The filter passband in the iteration stages n. 4, 5, and 6 (see text). The *horizontal lines* representing deviations from 1 at each stage are so close to be indistinguishable. The inset shows a very expanded view of a part of the passband

was evaluated by setting preliminary $\delta_p = \delta_s = 0.06$, which gave $N - 1 = 19$, and then passing to order $N - 1 = 20$ to have a Type-I filter.[3]

Since the initial-guess frequencies do not represent the exact positions of the local $H(\omega)$ extrema, the approximation error evaluated from the equation system given at point 2. turns out to be too small in absolute value: see in Fig. 7.19a how the *solid curve* representing $H(\omega)$ exceeds the limits marked by the *horizontal dotted lines* located at ordinates $1 - \delta_p$ and $1 + \delta_p$.

4. The positions $\overline{\omega}_i$ of the extrema are determined with high accuracy, by evaluating the polynomial on a dense set of frequencies. These $\overline{\omega}_i$ are then chosen as the updated frequency set on which the calculations are repeated. This is the *exchange*

---

[3]The value $\kappa = 1$ was chosen for graphical convenience, i.e., for getting ripples in the two approximation band which are equally large and therefore equally visible.

Fig. 7.20 Zero-phase
response of a Type-I lowpass
filter with $N - 1 = 20$,
$\kappa = 1$, $\omega_p = 0.4\pi$ and
$\Delta\omega = 0.1\pi$, at iteration
stages n. 1–6 preceding
convergence, which takes
place after 7 iterations

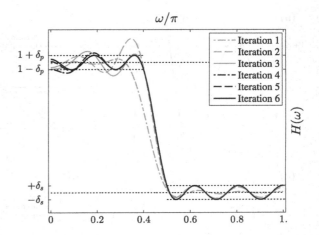

step after which the *multiple exchange algorithm* is named. In the exchange step
$\omega_i \rightarrow \overline{\omega}_i$, the two frequencies $\omega_{i=l} = \omega_p$ and $\omega_{i=l+1} = \omega_s$ are kept constant.
If a maximum of the absolute error is present both at $\omega = 0$ and $\omega = \pi$, only
one of these frequencies will be inserted in the new set $\overline{\omega}_i$, namely the frequency
corresponding to the highest absolute error value. The value of ε is then updated,
so that it increases its absolute value with respect to the previous estimate.

5. The process is iterated until, for all indexes i, $\omega_i - \overline{\omega}_i$ descends below a small
   pre-set value, and ε stabilizes.[4] At the end of the optimization, the frequency
   response and the impulse response of the optimum filter are obtained.

These steps are illustrated in Figs. 7.19b–d and 7.20. They all concern the Type-I
lowpass filter of Fig. 7.19a.

Figure 7.20 shows the shape of the zero-phase response at the iterative stages n.
1–6 preceding convergence of the MPR algorithm, which occurs at stage n.7. We
can see that at each iteration, the positions of local extrema change.

Figure 7.19b shows a close-up of the filter passband, in the iteration stages n. 2
(*dashed curve*) and n. 3 (*solid curve*). *Horizontal lines* in the same style represent the
deviations from 1 at each stage. The abscissas of the local minima and maxima of
stage n. 2 (*gray stars*) are assumed as new guessed abscissas of minima and maxima
for iteration n. 3 (*black diamonds*). The solid curve of iteration n. 3, however, still
exceeds the tolerances: its local extrema are elsewhere (*empty circles*). Therefore, a
new exchange step will be performed in the next iteration stage.

Figure 7.19c shows the filter passband and the deviations from 1 (indicated by
horizontal lines) in the iteration stages n. 1, 2, 3, and 4. This figure is meant to show
how the errors calculated at each iteration stage gradually increase in magnitude.
However, the increase from one iteration to the next becomes slower and slower
as stages progress. At the end, the absolute error will not increase any longer and
convergence will be achieved.

---

[4]The process is actually iterated a maximum number of allowed times, within which *hopefully*
convergence is achieved.

Finally, Fig. 7.19d shows the passband in the iteration stages n. 4, 5, and 6. The deviations from 1 at each stage are so close to one another that the corresponding *horizontal lines* are indistinguishable; the inset shows an expanded view of the low-frequency range. The absolute error is increasing more and more slowly towards convergence. The final passband deviation from 1 at stage 7 (convergence) turns out to be 0.0548753; at stage n.6 the passband deviation from 1 is already as high as 0.0548744.

Figure 7.21 shows the final result of the design process, i.e., the optimum Type-I lowpass filter, viewed in the in the frequency domain (Fig. 7.21a) and in the time domain (Fig. 7.21b).

This is the MPR algorithm in principle; however, the calculations can be simplified in practice. Indeed, it can be shown that at each stage, ε in the system of $M + 2$ equations given at point 2. can be calculated explicitly as a function of the known values of $\overline{H_d}(\omega_i)$ and $\overline{W}(\omega_i)$ and of the quantity

$$\eta_i = \prod_{k=0}^{M+1} (\cos\omega_i - \cos\omega_k)^{-1} \quad \text{with} \quad k \neq i,$$

which can be easily computed. This quantity contains the ω_i that in the initial and intermediate iteration steps will be incorrect. As a result, the calculated error will be too small in magnitude, and more iterations will be needed.

Once the value of ε has been obtained at some iteration stage, in order to evaluate $G(\omega)$ on a dense set of frequency and thus find the new frequencies for the exchange step there is no need to actually solve the system, i.e., to find the coefficients $\alpha[n]$. A Lagrange interpolation formula can be exploited instead, which uses only the values $G(\omega_i)$. These values are known because

$$G(\omega_i) = \overline{H_d}(\omega_i) + \frac{(-1)^i \varepsilon}{\overline{W}(\omega_i)}.$$

Fig. 7.21 Optimum Type-I lowpass filter with order $N - 1 = 20$ represented **a** in the frequency domain and **b** in the time domain. Specifications are: $\kappa = 1$, $\omega_p = 0.4\pi$, and $\Delta\omega = 0.1\pi$. The convergence of the MPR algorithm was attained after 7 iterations and produced an optimized (final) deviation $\delta_p = 0.05488$; the pre-fixed value specified for finding the minimum order was $\delta_p = 0.06$

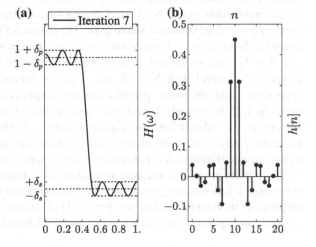

In this way, only the final values of $\alpha[n]$ must be computed, so as to deduce the filter impulse response $h[n]$. In conclusion, we will just

- start with an arbitrary ω_i set;
- compute η_i, $\overline{H_d}(\omega_i)$ and $\overline{W}(\omega_i)$ at each i and evaluate the corresponding value of ε;
- compute the values of $G(\omega_i)$, interpolate these values to get $G(\omega)$ on a dense set of frequencies and locate the extrema of $G(\omega)$;
- perform the exchange step by setting the new ω_i and iterate from point n. 2 until convergence is achieved;
- solve the system to get the $\alpha[n]$ coefficients. But even this is unnecessary: we can simply perform an inverse DFT of the $G(\omega)$ obtained in the final stage, to get the impulse response of the filter that has the minimum possible δ_p for the given values of N, ω_p, $\Delta\omega$ and κ.

The process requires 4–8 iterations for a lowpass filter, and 2–3 times as much for multiband filters. If the final solution exceeds the tolerances, the order must be increased and the design process repeated.

7.10   Properties of Equiripple FIR Filters

It is interesting to examine the dependence of the optimized absolute maximum error from the cutoff frequency ω_p, the other design parameters κ and $\Delta\omega$ being equal, for different orders. In this section, by ε we mean the *optimized maximum absolute error*.

Figure 7.22 shows this dependence for lowpass filters with $\kappa = 1$ and $\Delta\omega = 0.2\pi$, having orders from 8 to 11 ($N = 9 - 12$). Note that filters with $N = 9$ and 11 are Type-I filters, while filters with $N = 10$ and 12 are Type-II filters. For each order, as ω_p increases, ε attains several local minima. The curves related to different orders intersect each other, so that for certain values of ω_p, a shorter lowpass filter ($N = 9$ for example) turns out to be "better" than a longer filter ($N = 10$), since it has a lower optimized maximum absolute error. The reason for this apparent contradiction lies in the fact that $N = 9$ corresponds to a Type-I filter, while $N = 10$ corresponds to a Type-II filter: we cannot directly compare different types of filters. Of course, a Type-I filter with length $N = N_0$ and given specifications cannot be better than a Type-I filter with the same specifications and length $N = N_0 + 2$. This is evident from Fig. 7.23, where only Type-I filters are considered, with $\Delta\omega = 0.1\pi$ and $\kappa = 1$. The curves for ε shown here correspond to orders 14, 16, and 18 ($N = 15$, 17, and 19, respectively). There are no intersections between curves. However, the curves representing different orders touch each other in correspondence with local minima.

These plots are useful because they allow us to understand how in some cases extraripple solutions emerge, while in other cases normal solutions are found. To this purpose, let us examine a close-up (Fig. 7.24) of the plot area contained in the gray rectangle in Fig. 7.23. The black squares and the labels a–f in Fig. 7.24 identify six

Fig. 7.22 Dependence of the optimized absolute maximum error ε on the cutoff frequency ω_p for Type-I and -II lowpass filters, with $\kappa = 1$ and $\Delta\omega = 0.2\pi$. Each curve is relative to a different filter length N, in the range from 9 to 12

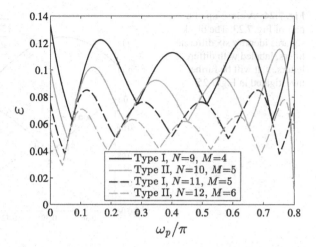

Fig. 7.23 Dependence of the optimized absolute maximum error ε on the cutoff frequency ω_p for Type-I lowpass filters, with $\kappa = 1$ and $\Delta\omega = 0.1\pi$. Each curve is relative to a different filter length N, in the range from 15 to 19. The region enclosed by the gray rectangle is shown in greater detail in Fig. 7.24

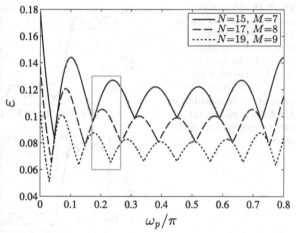

filters that differ from one another only for the precise value of their cutoff frequency ω_p. They all lie on the *dashed curve*, which is relative to the order $N - 1 = 16$ ($M = 8$). However, in c the *dashed curve* with order 16 touches the *dotted curve* with order 18; in e the *dashed curve* with order 16 touches the *solid curve* with order 14; in f the *dashed curve* with order 16 touches again the *dotted curve* with order 18. What is the meaning of points such as c, e, and f? This can be understood observing Fig. 7.25, where each panel shows the zero-phase response of one of the six filters a–f. A *black dot* highlights the value of ω_p.

- Let us start from filter c: this is an *extraripple solution* for the order 16, with a zero-phase response presenting $M + 1$ equal extrema. The number of alternations is $M + 3$. Actually, as we shall see, all the local minima of each curve in Fig. 7.23 correspond to extraripple filters for the corresponding order.

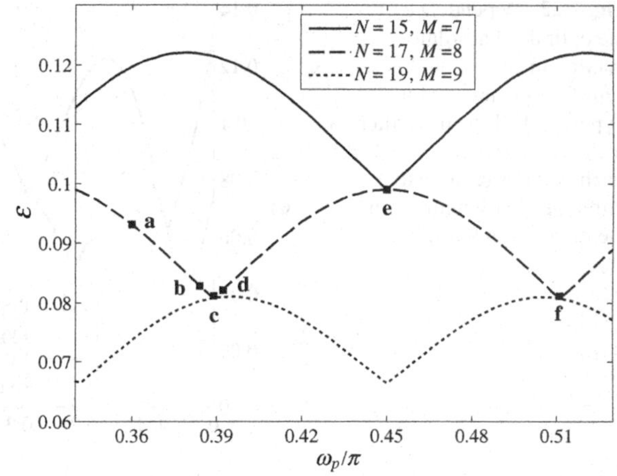

Fig. 7.24 A close-up of a part of Fig. 7.23. The black squares identify six different filters, labeled with different letters, that will be further investigated in Fig. 7.25

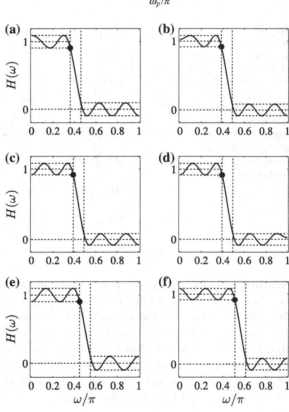

Fig. 7.25 Zero-phase response of each of the six Type-I lowpass filters identified by the letters a–f in Fig. 7.24. The order is $N - 1 = 16 \ (M = 8)$; $\kappa = 1$, $\Delta\omega = 0.1\pi$. The value of ω_p, indicated by a *black dot* in each panel, varies from one filter to the other

- If we slightly decrease ω_p, we get filter b: the number of extrema of $H(\omega)$ still is $M + 1$, but the minimum at $\omega = 0$ is smaller than the other extrema. The number of alternations is $M + 2$: this is a normal solution.
- If we further decrease ω_p, we can reach filter a: the minimum at $\omega = 0$ that in filter b was less pronounced than the other extrema has now disappeared, leaving M equal extrema and $M + 2$ alternations. This is again a normal solution.
- Starting again from c, let us increase ω_p slightly to get filter d. The number of extrema is $M + 1$ as in filter c, but the minimum at $\omega = \pi$ is smaller than the other extrema. The number of alternations is $M + 2$: once again, we get a normal solution.
- Now we increase ω_p until we reach filter e. The minimum at $\omega = 0$ that in filter d was less pronounced than the other extrema has now disappeared, leaving M equal extrema and $M + 2$ alternations. So, this is another normal solution.
- Finally, a further consistent increase of ω_p can lead us to filter f. A maximum equal to the other ones has appeared at $\omega = 0$. The number of extrema is $M + 1$ and the number of alternations is $M + 3$: we attained again an *extraripple solution* for the order 16, as expected since point f is, like c, a local minimum of the curve with order 16 in Figs. 7.23 and 7.24.

We may note that points c and f in Fig. 7.24 also belong to the order-18 curve, and that point e also belongs to the order-14 curve. How can a filter have two different orders at the same time? The answer is, filters c and f are extraripple solutions for order 16, but are also normal solutions for order 18. Similarly, filter e is a normal solution for order 16, but is also an extraripple solution for order 14. For example, the minimax design method provides the same filter c if—all other parameters being equal—we set order 18, or order 16. The only difference is that the impulse response of the optimum filter has two zero samples at $n = 0$ and at $n = N - 1$ when the order is set to 18, and does not have them when the order is set to 16.

Type-II filters have similar properties, except for the fact that the zero-phase response always vanishes at $\omega = \pi$ (Fig. 7.4).

7.11 The Minimax Method for Bandpass Filters

We discussed lowpass filters that, as highpass filters, have only two approximation bands. Passband and stopband filters have three such bands, and the alternation theorem has different implications in these cases.

The alternation theorem does not set any limit on the number of approximation bands, so that the minimum number of alternations required for optimality always is $M + 2$. However, while a lowpass filter has four band limits $(0, \omega_p, \omega_s, \pi)$, multiband filters have more than band limits. For instance, a bandpass filter has six limits $(0, \omega_{s1}, \omega_{p1}, \omega_{p2}, \omega_{s2}, \pi)$, and therefore these filters can exhibit more than $M + 3$ alternations. This means that some statements we made for lowpass filters are no longer valid. For example, it is not necessary that all local minima and maxima of $H(\omega)$ fall inside the approximation bands: there may be local extrema in the transition bands. Also, the approximation is not necessarily equiripple in the approximation bands. These filters are optimum in the sense of the alternation theorem, but are normally discarded in practical applications.

In general, when the design of a multiband filter is performed, there is no guarantee that the transitions bands will be monotonic, because the MPR algorithm leaves them completely free from constraints. However, when a certain choice of the design parameters leads to a solution that presents non-monotonic transition bands, or that is not equiripple in the approximation bands, the filter is designed again after adjusting the order, or one or more band limits, or even the weighting error function, until a regular solution is obtained.

References

Herrmann, O., Rabiner, L.R., Chan, D.S.K.: Practical design rules for optimum finite impulse response low-pass digital filters. Bell Syst. Tech. J. **52**(6), 769–799 (1973)

Horel, J.D.: Complex principal component analysis: theory and examples. J. Clim. Appl. Meteorol. **23**, 1660–1673 (1984)

Jolliffe, I.T.: Principal Component Analysis. Springer, New York (2002)

Kaiser, J.F.: Nonrecursive digital filter design using the I_0-sinh window function. In: Proceedings of the IEEE International Symposium on Circuits and Systems, San Francisco, CA, USA pp. 20–23 (1974)

McClellan, J.H., Parks, T.W., Rabiner, L.R.: A computer program for designing optimum FIR linear phase digital filters. IEEE Trans. Audio Electroacoust. **AU–21**(6), 506–526 (1973)

Mitra, S.K., Kaiser, J.F. (eds.): Handbook for Digital Signal Processing. Wiley, New York (1993)

Ollin, H.Z.: Best polynomial approximation to certain rational functions. J. Approx. Theory **26**, 382–392 (1979)

Oppenheim, A.V., Schafer, R.W.: Discrete-time Signal Processing. Prentice Hall, Englewood Cliffs (2009)

Parks, T., McClellan, J.: Chebyshev approximation for nonrecursive digital filters with linear phase. IEEE Trans. Circuit Theory **19**(2), 189–194 (1972)

Parks, T.W., Burrus, C.S.: Digital Filter Design. Wiley, New York (1987)

Rabiner, L.R., McClellan, J.H., Parks, T.W.: FIR digital filter design techniques using weighted chebyshev approximations. Proc. IEEE **63**(4), 595–610 (1975)

Remez, E.Y.: Sur la détermination des polynômes d'approximation de degré donné. Commun. Soc. Math. Kharkov **10**, 41–63 (1934a)

Remez, E.Y.: Sur un procédé convergent d'approximations successives pour déterminer les polynômes d'approximation. C. R. Acad. Sci. **198**, 2063–2065 (1934b)

Remez, E.Y.: Sur le calcul effectiv des polynômes d'approximation de Tschebyscheff. C. R. Acad. Sci. **199**, 337–340 (1934c)

Saff, E.B., Snider, A.D.: Fundamentals of Complex Analysis with Applications to Engineering and Science. Prentice Hall, Upper Saddle River (2002)

Chapter 8
IIR Filter Design

8.1 Chapter Summary

Digital IIR filters have an infinitely long impulse response and therefore can be
associated with analog filters that have the same characteristics. For this reason, the
classical method for digital IIR filter design is based on the design of an analog filter,
which is later transformed into equivalent digital filter through a mapping in the
complex plane. The advantage of such a technique lies in the fact that analog filter
design and mapping methods for analog-to-digital (A/D) transformation are well
known and have sound theoretical foundations. The process is based on a lowpass
filter, and frequency transformation methods are later applied if a different type of frequency selectivity is desired. These frequency transformations are again well-known
mapping procedures in the complex plane. In this chapter, we will first discuss the
design of the main types of lowpass analog filters, namely Butterworth, Chebyshev
and elliptic lowpass filters, all of which represent different ways of approximating the
desired ideal frequency response. Then we will learn how to transform the analog
lowpass filter into an equivalent IIR digital lowpass filter via bilinear transformation, and finally how to transform the IIR lowpass filter into a highpass, bandpass
or bandstop filter, if needed. The appendix to this chapter provides deeper insight
into the mathematical facets of elliptic design, discussing elliptic integrals of the first
kind, Jacobi elliptic functions, and the elliptic rational function on which the transfer function of the analog elliptic filter is based. It must be explicitly noted that all
classically-designed lowpass IIR filters are inadequate for specifications with cutoff
frequency close to zero and narrow transition band. In such cases, it may be advisable
to preliminarily downsample the signal and then design a filter with a less extreme
cutoff frequency, allowing for a wider transition band.

© Springer International Publishing Switzerland 2016   263
S.M. Alessio, *Digital Signal Processing and Spectral
Analysis for Scientists*, Signals and Communication Technology,
DOI 10.1007/978-3-319-25468-5_8

8.2   Design Process

Two approaches to the design of digital IIR filters are found in literature, and are illustrated in Fig. 8.1. The first approach consists of designing an analog lowpass filter, transforming it into an analog highpass, or bandpass, or bandstop filter if required, and finally converting it into a digital filter. The frequency transformation thus takes place in the analog domain. This approach is widely employed in software implementations, for example in Matlab. However, we will focus on the second approach that consists in transforming first a lowpass analog filter into a lowpass digital filter, and then, if necessary, transforming the lowpass digital filter into a highpass, or bandpass, or bandstop digital filter. The frequency transformation here takes place in the digital domain. The reason for this choice is that from a methodological point of view this procedure is conceptually simple, and also because in this book the focus is on the techniques concerning digital filters, rather than analog filters.

Suppose we want to design a digital IIR filter with one of the four prototype-frequency-selectivity responses, and satisfying some given specifications. These specifications will be translated into proper specifications for an analog lowpass filter, which will then be designed. Later, the analog lowpass filter will be transformed into a digital lowpass IIR filter. Finally, this filter can be converted, if desired, into a digital IIR highpass, or bandpass, or bandstop filter, satisfying the original specifications. Since the frequency response of an analog filter is defined over $-\infty < \Omega < +\infty$, while the frequency response of a digital filter is restricted to the range $-\pi \leq \omega < \pi$ and repeats itself periodically outside this interval, the two responses cannot be identical. The similarity of the responses is imposed over a limited interval of analog frequencies, starting from $\Omega = 0$.

Approach I

| Design of analog lowpass filter | Frequency transformation in the analog domain: $s \rightarrow s$ | A/D conversion: $s \rightarrow z$ | IIR digital filter |

Approach II

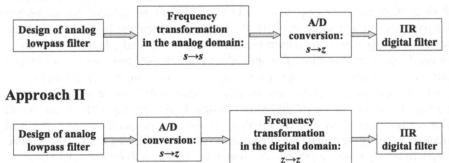

| Design of analog lowpass filter | A/D conversion: $s \rightarrow z$ | Frequency transformation in the digital domain: $z \rightarrow z$ | IIR digital filter |

Fig. 8.1 Two viable approaches for the transformation of an analog lowpass filter into a digital lowpass, highpass, bandpass or bandstop IIR filter. Here z represents the complex independent variable for the transfer function of the digital filter, while s represents the complex independent variable used to express the transfer function of the analog filter

The procedure described above requires studying:

- the design method for analog lowpass filters;
- the transformation method for converting an analog lowpass filter into a digital IIR lowpass filter. Among the many methods for analog to digital (A/D) conversion available in literature, the most popular are the method of impulse-response invariance that however has limited applicability, and the *bilinear transformation*, which is widely used and will be presented in this book;
- the transformation method for converting a digital lowpass IIR filter into a highpass, bandpass or bandstop filter of the same kind.

In all these approaches, the phase features of the designed filters are not considered. These methods are therefore *magnitude-only design* methods. Actually, more sophisticated techniques exist that can simultaneously approximate desired amplitude and phase responses, but they require advanced optimization methods that are beyond the scope of our discussion.

First of all, we must examine the design of analog filters. For this task, techniques exist that have been thoroughly studied and apply to the three main analog filter types: Butterworth, Chebyshev (of which two subtypes exist, I and II) and elliptic filters, which are also referred to as Cauer or Zolotarev filters. From now on we will simply indicate Chebyshev filters of types I and II as Chebyshev-I and Chebyshev-II filters, respectively.

These analog filter types are characterized by different functional forms of the transfer function, and represent different approximations of the ideal piecewise-constant response; their classical design deals with the lowpass case. Starting from Butterworth filters, passing to Chebyshev and finally to elliptic ones, the design complexity increases while, all specifications being equal, progressively lower orders can be adopted to satisfy them. The order being equal, more stringent specifications can be met. For given specifications, elliptic filters have the minimum possible order, but also exhibit a phase that is the most distant from linearity.

As we mentioned above, IIR filter design focuses on amplitude response and considers phase response as a minor issue. In any case, these filters always exhibit nonlinear phase response[1]; indeed, their phase response is difficult to control. This is acceptable only in filtering applications in which the phase distortion undergone by the signal does not represent a negative implication, or in *off-line* applications in which the zero-phase filtering technique can be employed (Sect. 6.7).

In any application we may be interested in, we may prefer to adopt an IIR filter rather than an FIR filter because, all specifications being equal, we desire

- a lower order; for example, a narrow transition band even at a low order;
- as flat a frequency response as possible in the passband and/or in the stopband, a feature that cannot be achieved with an FIR filter.

By the classical IIR design method, we can obtain filters that approximate the four ideal prototypes, with desired unitary amplitude response in the passband and desired

---

[1] Some particular IIR filters, such as polyphase filters, can have quasi-linear phase response, but they will not be described in this book.

zero amplitude response in the stopband, but we cannot design filters with arbitrary characteristics, e.g., multiband filters. Multiband IIR filters can be obtained only by those direct design methods implying advanced optimization methods that we mentioned above. On the other hand, by the classical IIR design method we can design a filter with given specifications just inserting the values of certain parameters in closed-form design equations: a pocket calculator could be enough. Of course, in our computer era this is not a decisive element.

Let us start with some fundamental notions concerning analog filters.

8.3 Lowpass Analog Filters

Lowpass analog filters are specified in a similar manner to that adopted for digital filters. The main differences are the following:

- frequencies are given as Ω in rad/s; for example, a lowpass filter will be specified giving the limits of the passband and stopband, Ω_p and Ω_s respectively, as well as the absolute tolerances δ_p and δ_s in the two bands. These are the basic design parameters;
- the *transfer function* is expressed as $H(s)$, representing the *Laplace transform* of the analog impulse response $h(t)$;
- the amplitude response in the passband is normally constrained to $[1 - \delta_p, 1]$, the desired passband magnitude being 1; this is mainly due to formal convenience and consolidated use, since the value of the passband magnitude response could be easily adjusted in such a way to make the (possibly existing) ripple symmetric about 1, as for FIR filters.

8.3.1 Laplace Transform

Let us briefly recall that the Laplace transform is a generalization of the *continuous-time Fourier transform* (CTFT). The CTFT is defined (see, e.g., Oppenheim et al. 1996; Bochner and Chandrasekharan 1949; Bracewell 2000) through the analysis and synthesis relations:

$$X(j\Omega) = \int_{-\infty}^{+\infty} x(t)e^{-j\Omega t}\,dt,$$

$$x(t) = \frac{1}{2\pi} \int_{-\infty}^{+\infty} X(j\Omega)e^{j\Omega t}\,d\Omega.$$

Here we used $X(j\Omega)$ to indicate the CTFT, in place of the more usual symbol $X(\Omega)$, because this is useful in the frame of the connection between the CTFT and Laplace transforms. Note that $\Omega = 2\pi f$, where f is analog frequency in Hz. Analog angular frequency is measured in rad/s.

The Laplace transform is introduced because it is useful for many purposes: first of all, the CTFT does not converge for many continuous-time signals for which the Laplace transform exists. Instead of using, as a basis for signal expansion, complex exponential functions with the form $e^{j\Omega t}$, with a purely imaginary exponent, the Laplace transform uses the more general functions e^{st}, with

$$s = \sigma + j\Omega,$$

where σ and Ω are real variables. In the so-called s-plane, where a Cartesian reference frame is established having $\mathrm{Re}(s)$ as the abscissa and $\mathrm{Im}(s)$ as the ordinate, the function e^{st} represents a general set of complex exponential functions.

The (bilateral)[2] *Laplace transform* of $x(t)$ is defined as (see, e.g., Bracewell 2000)

$$X(s) = \int_{-\infty}^{+\infty} x(t)e^{-st}dt.$$

The area in the s-plane composed by the values of s such that the Laplace transform converges is referred to as the *Region of Convergence* (ROC). Inside the ROC, $X(s)$ is an analytic function.

The ROC of $X(s)$ has the following properties:

- it is made of strips that are parallel to the imaginary axis;
- if $x(t)$ has finite duration and is absolutely integrable, then the ROC extends over the whole s-plane;
- if $x(t)$ is a causal (right-hand sided) function and if the straight line $\mathrm{Re}(s) = \alpha$ is contained inside the ROC, then all values of s for which $\mathrm{Re}(s) > \alpha$ are contained inside the ROC as well;
- if $x(t)$ is an anticausal (left-hand sided) function and if the straight line $\mathrm{Re}(s) = \alpha$ is contained inside the ROC, then all values of s for which $\mathrm{Re}(s) < \alpha$ are contained inside the ROC as well;
- if $x(t)$ is a two-sided function and if the straight line $\mathrm{Re}(s) = \alpha$ is contained inside the ROC, then the ROC is a strip of the s-plane including the straight line;
- if $X(s)$ is a rational function, with zeros and poles, obviously the ROC does not contain any pole; the ROC is delimited by poles, or extends to infinity;
- if $X(s)$ is a rational function and if $x(t)$ is a causal (right-hand sided) function, the ROC is the region of the s-plane extending to the right of the pole which is located to the right of all other poles;
- if $X(s)$ is a rational function and if $x(t)$ is an anticausal (left-hand sided) function, the ROC is the region of the s-plane extending to the left of the pole which is located to the left of all other poles.

---

[2]The monolateral Laplace transform is defined over $t \in [0, \infty)$.

The *inverse Laplace transform* is defined by the limit of an integral in the s-plane, in which the integration contour is a straight line parallel to the imaginary axis, given by the equation $\mathrm{Re}(s) = \beta$:

$$x(t) = \frac{1}{2\pi j} \lim_{R \to \infty} \int_{\beta-jR}^{\beta+jR} X(s)e^{st}\mathrm{d}s.$$

Here β is a real number such that the integration contour is contained inside the ROC of the Laplace transform, and R is a real parameter.

Finally, for an analog real signal $x(t)$ the following property holds:

$$X(-s) = X^*(s) \quad \rightsquigarrow \quad |X(s)|^2 = X(s)X(-s).$$

8.3.2   Transfer Function and Design Parameters

From now on, we focus on LTI *stable and causal analog systems* represented by *proper rational transfer functions*,

$$H(s) = \frac{b_0 s^q + b_1 s^{q-1} + \cdots + b_q}{s^p + a_1 s^{p-1} + \cdots + a_p}$$

with $q \leq p$; this condition on q and p is related to requiring that

$$\lim_{s \to \infty} H(s) = 0.$$

Note that we set $a_0 = 1$. These systems have zeros and poles, the latter being the zeros of the denominator of $H(s)$. This form of the transfer function corresponds to input-output relations in the form of linear differential equations with constant coefficients.

The ROC of a stable $H(s)$ is always determined by a condition of the form

$$\mathrm{Re}(s) > -\alpha,$$

with α a real positive number; therefore the ROC is a part of the s-plane delimited by a straight line parallel to the imaginary axis, located in the left-hand half-plane ($\sigma = -\alpha < 0$). Therefore, the ROC of an analog transfer function always contains the imaginary axis of the s-plane, i.e., the straight line identified by $\sigma = 0$, $s = j\Omega$. On the imaginary axis, $H(s)$ represents the filter's frequency response, i.e., the CTFT of the filter's impulse response. The ROC extends to the right of the straight line, as in Fig. 8.2, and since the ROC cannot contain any poles, $-\alpha$ coincides with the abscissa of the pole(s) that are located to the right of all other poles in the s-plane. This means that for a stable system, all poles lie in the left-hand half-plane ($\sigma < 0$),

Fig. 8.2 Complex s-plane
and region of convergence
(ROC; *gray-shaded area*) of
the transfer function of an
analog filter that is causal,
stable and rational. The *dots*
indicate poles

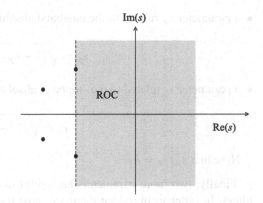

as in the example shown in Fig. 8.2. In this case the CTFT of the impulse response
$h(t)$, i.e., $H(j\Omega) \equiv H(s = j\Omega)$, exists and represents the *frequency response* of the
system.

The design of an analog filter requires the specification of *absolute tolerances* δ_p
and δ_s; the corresponding *relative tolerances* expressed in dB are

$$\begin{cases} r_p = -20\log_{10}\left(1 - \delta_p\right), \\ r_s = -20\log_{10}(\delta_s). \end{cases}$$

Moreover, the design of an analog filter requires introducing four more parameters
that are functions of the basic design parameters:

- the *selectivity factor k* that must satisfy $0 < k < 1$ (it would be 1 for an ideal filter
  with no transition band),

$$k = \frac{\Omega_p}{\Omega_s};$$

- the *discrimination factor* k_1, which must be non-negative,

$$k_1 = \left[\frac{\left(1 - \delta_p\right)^{-2} - 1}{\delta_s^{-2} - 1}\right]^{1/2} = \left(\frac{10^{r_p/10} - 1}{10^{r_s/10} - 1}\right)^{1/2}.$$

For $\delta_p \to 0$, the numerator of k_1 goes to zero, hence k_1 vanishes. For $\delta_s \to 0 = 0$,
the denominator of k_1 goes to infinity, hence k_1 also vanishes in this case. Therefore
for an ideal filter $k_1 = 0$, while in practical cases any $\delta_p > 0$ implies $k_1 > 0$. A
narrow transition band implies $k \lesssim 1$ and that a strong attenuation in the stopband
and/or a small ripple in the passband imply $k_1 \ll 1$. In most cases we thus have

$$k_1 \ll k \lesssim 1;$$

- a parameter ε_p related to the passband absolute tolerance,

$$\varepsilon_p = \sqrt{(1 - \delta_p)^{-2} - 1};$$

- a parameter ε_s related to the stopband absolute tolerance,

$$\varepsilon_s = \sqrt{\delta_s^{-2} - 1}.$$

Note that $\varepsilon_p / \varepsilon_s = k_1$.

Finally, two more parameters are useful in describing the properties of analog filters. In order to introduce them we must pass from the transfer function to the frequency response, which is deduced from $H(s)$ by setting $\sigma = 0$, $s = j\Omega$:

$$H(j\Omega) = \frac{b_0(j\Omega)^q + b_1(j\Omega)^{q-1} + \cdots + b_q}{(j\Omega)^p + a_1(j\Omega)^{p-1} + \cdots + a_p}.$$

The additional parameters are

- the $-3\,dB$ *cutoff frequency*, defined as the frequency Ω_{3dB} at which the magnitude response of the filter is $0.707 = 1/\sqrt{2}$ times its ideal unitary passband value, since $20\log_{10}(0.707) = -3$. Therefore the squared amplitude of the filter's frequency response at Ω_{3dB} is 0.5;
- the *asymptotic attenuation at high frequency*, a parameter determined by the difference between the denominator and the numerator degrees of the proper rational function $H(s)$. From the formula for $H(j\Omega)$ we can see that for $\Omega \to \infty$ the behavior of the frequency response is given by

$$H(j\Omega) \approx b_0(j\Omega)^{q-p};$$

in practice, $\Omega \to \infty$ must be understood as Ω greater than about 5 times the maximum absolute value of the frequencies of all poles and zeros. In logarithmic units we can write, with $q \leq p$,

$$20\log_{10}|H(j\Omega)| \approx 20\log_{10}|b_0| - 20(p - q)\log_{10}\Omega;$$

this is expressed by saying that the asymptotic attenuation is about $20(p - q)$ dB/decade.

Classical analog lowpass filters are designed starting from the functional form of the squared-magnitude frequency response, which can generally be expressed as

$$|H(j\Omega)|^2 = \frac{1}{1 + \varepsilon_p^2 \Lambda_N^2(w)} \qquad \text{with} \qquad w = \Omega/\Omega_p.$$

In this expression, w is a *normalized frequency*. The function $\Lambda_N(w)$ is the *attenuation function*, constructed in such a way to obtain in the passband and in the stopband the desired filter behavior. In particular, where $\Lambda_N(w)$ is monotonic, the same is true also for $|H(j\Omega)|^2$; where $\Lambda_N(w)$ exhibits an oscillatory behavior—and this can happen either in the passband, or in the stopband, or in both bands—$|H(j\Omega)|^2$ exhibits ripples in the same band(s). If $\Lambda_N(w)$ is constrained to be a polynomial or a rational function, then $H(s)$ is constrained to be rational; the order of the polynomial or rational function $\Lambda_N(w)$ determines the order of $H(s)$. The attenuation function depends on the independent variable w, but may also depend on parameters, such as k and k_1, which serve to impose to $|H(j\Omega)|^2$ the desired behavior.

The various types of classical analog filters that will be discussed in this chapter are built in this way, and differ from one another only for the choice of the functional form of $\Lambda_N(w)$, which determines

- a flat passband and a flat stopband for Butterworth filters, i.e., no ripples,
- an oscillatory passband and a flat stopband for Chebyshev-I filters,
- a flat passband and an oscillatory stopband for Chebyshev-II filters,
- an oscillatory passband and an oscillatory stopband for elliptic filters.

Note that the word "flat" in the above statements must be meant as "nearly flat". Now, let us call

- $T_N(w)$ the *Chebyshev polynomial* of the first kind, with degree N in the normalized frequency w;
- $R_N(w)$ the *elliptic rational function* with degree N in the variable w;
- cd $(\cdot)$ the *Jacobi elliptic function* known as the cd function;
- K the *complete elliptic integral of the first kind* with parameter k;
- K_1 the complete elliptic integral with parameter k_1;

the meaning of the quantities listed here will be explained in the following sections.[3] The function $\Lambda_N(w)$ is then defined as follows:

$$\Lambda_N(w) = \begin{cases} w^N & \text{for Butterworth filters,} \\ T_N(w) & \text{for Chebyshev-I filters,} \\ [k_1 T_N(k^{-1}w^{-1})]^{-1} & \text{for Chebyshev-II filters,} \\ R_N(w) = \text{cd}(NuK_1, k_1), \text{ with } w = \text{cd}(uK, k), & \text{for elliptic filters.} \end{cases}$$

In the last formula, u is a complex variable which is constrained to assume values leading to real values of w. Further mathematical details can be found in the appendix to this chapter, in which the functions mentioned above, in particular Jacobi elliptic functions, elliptic integrals and the elliptic rational function, are described.

With the notation introduced above, a unified design method for analog lowpass filters of the four types can be formalized. We will not pursue this approach com-

---

[3]The Chebyshev polynomial of the first kind has already been defined in Chap. 7.

pletely: for our purposes, clarity is more important than conciseness. Nevertheless, we will underline the analogies among Butterworth, Chebyshev and elliptic design procedures.

The use of the normalized frequency w makes calculations easier and simplifies the notation. The value $w = 1$ corresponds to the upper limit Ω_p of the passband, while the value $w = \Omega_s/\Omega_p = k^{-1}$ corresponds to the lower limit of the stopband.

If we require filter specifications in the passband and in the stopband to be met at the edges, $\Omega_p \leadsto w = 1$ and $\Omega_s \leadsto w = k^{-1}$ respectively, recalling the definitions of ε_p and ε_s we get the following constraints for the function $\Lambda_N(w)$:

$$|H(j\Omega_p)|^2 = \frac{1}{1 + \varepsilon_p^2 \Lambda_N^2(1)} = (1 - \delta_p)^2 = \frac{1}{1 + \varepsilon_p^2} \Rightarrow \Lambda_N(1) = 1,$$

$$|H(j\Omega_s)|^2 = \frac{1}{1 + \varepsilon_p^2 \Lambda_N^2(k^{-1})} = \delta_s^2 = \frac{1}{1 + \varepsilon_s^2} \Rightarrow \Lambda_N(k^{-1}) = \frac{\varepsilon_s}{\varepsilon_p} = k_1^{-1}.$$

We can deduce that any type of $\Lambda_N(w)$

- must be normalized to 1 at $w = 1$;
- must be normalized to k_1^{-1} at $w = k^{-1}$.

The second condition expresses the general form of the so-called *degree equation* (Orfanidis 2006). For Butterworth, Chebyshev-I and elliptic filters, this equation establishes a relation among the parameters N, k and k_1. As for Chebyshev-II filters, with respect to the case of Chebyshev-I filters the constraints at Ω_p and Ω_s exchange their roles: for Chebyshev-II filters, the relation among N, k and k_1 is provided by the condition $\Lambda_N(1) = 1$. In fact,

- by imposing $\Lambda_N(1) = \left[k_1 T_N(k^{-1})\right]^{-1} = 1$ we get $k_1 T_N(k^{-1}) = 1$, hence

$$T_N(k^{-1}) = k_1^{-1}$$

(degree equation);
- by imposing $\Lambda_N(k^{-1}) = \varepsilon_s/\varepsilon_p = k_1^{-1}$ we get

$$\left[k_1 T_N(k^{-1}k\right]^{-1} = [k_1 T_N(1)]^{-1} = k_1^{-1},$$

which simply means $T_N(1) = 1$, a condition that is satisfied by Chebyshev polynomials of the first kind (Fig. 7.10).

The existence of the degree equation implies that the values of the parameters N, k and k_1 cannot all be prescribed arbitrarily: given two values, the remaining one is automatically determined. The equation that relates N, k and k_1 for a given filter type represents the specific form of the degree equation for that filter type.

The name "degree equation" derives from the fact that this equation determines the filter order required to meet some given specifications, i.e., the order associated with given values of k and k_1. The existence of a relation among N, k and k_1 has other remarkable implications, especially in the design of elliptic filters (see the appendix).

We started from $|H(j\Omega)|^2$, the squared amplitude of the frequency response. We can go back to the squared amplitude of the transfer function by substituting

$$j\Omega = s, \qquad \Omega = s/j = -js$$

into $|H(j\Omega)|^2$. We thus get

$$|H(s)|^2 = H(s)H(-s).$$

Therefore, $|H(j\Omega)|^2$ determines the product $H(s)H(-s)$ but does not determine the rational transfer function $H(s)$ in a unique way. The set of poles of $H(s)H(-s)$ obviously includes both the poles of $H(s)$ and the poles of $H(-s)$. The position of the poles of $H(-s)$ can be deduced from the position of the poles of $H(s)$ by changing the sign of both real and imaginary parts.

The transfer function can be determined univocally by calling upon stability: a stable $H(s)$ will have, as its poles, all the poles of $H(s)H(-s)$ that lie in the left-hand half-plane. A stable $H(s)$ cannot have poles elsewhere. When not only poles but also zeros are present, as in the Chebyshev-II and elliptic cases, these zeros are found—as we will see later—on the imaginary axis, in symmetric positions with respect to the real axis. Hence, when s undergoes a sign reversal, nothing changes: $H(s)$ and $H(-s)$ share the same set of zeros.

The rational transfer function of an analog filter can be written in factorized form, using the positions z_m of its zeros and the positions p_n of its poles, and a gain G:

$$H(s) = G \frac{\prod_m (s - z_m)}{\prod_n (s - p_n)}.$$

When no zeros are present, as in the case of Butterworth and Chebyshev-I filters, the product in the numerator is substituted by 1:

$$H(s) = G \frac{1}{\prod_n (s - p_n)}.$$

An important value of $H(s)$ is the value assumed at $s = 0$, which we will indicate by H_0:

$$H_0 = G \frac{\prod_m (-z_m)}{\prod_n (-p_n)}$$

if both zeros and poles exist, and

$$H_0 = G \frac{1}{\prod_n (-p_n)}$$

in the case of poles only. Therefore we can write

$$G = H_0 \frac{\prod_n (-p_n)}{\prod_m (-z_m)}, \qquad \text{or} \qquad G = H_0 \prod_n (-p_n),$$

respectively. Note that H_0 also is the value assumed by the frequency response $H(\mathrm{j}\Omega)$ at $\Omega = 0$.

The filter's *cutoff frequency*, indicated by Ω_c, is identified on the basis of considerations related to the constraints imposed in the passband and in the stopband:

- for Butterworth filters, $\Omega_c > \Omega_p$: these two frequencies are related to one another but are not coincident. More precisely,

$$\Omega_c = \Omega_p \left[\left(1 - \delta_p \right)^{-2} - 1 \right]^{-1/2N} = \Omega_p \varepsilon_p^{-1/N};$$

- for Chebyshev-I and elliptic filters, $\Omega_c = \Omega_p$;
- for Chebyshev-II filters, $\Omega_c = \Omega_s$.

The rationale behind these choices will be explained later. Finally, an *analog prototype filter* for which no cutoff frequency is explicitly specified is meant as a *normalized filter* having $\Omega_c = 1$ rad/s. Figure 8.3 illustrates the main parameters involved in lowpass analog filter design; Table 8.1 gives the definition of each parameter and its relation with the other ones.

Fig. 8.3 Specifications for an analog filter, including absolute tolerances δ_p and δ_s, as they appear in a linear plot of the frequency-response magnitude

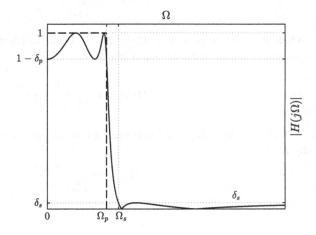

Table 8.1 The parameters involved in analog filter design

Parameter	Expression	Expression	Expression
k	$= \frac{\Omega_p}{\Omega_s}$		
k_1	$= \sqrt{\frac{(1-\delta_p)^{-2}-1}{\delta_s^{-2}-1}}$	$= \sqrt{\frac{10^{r_p/10}-1}{10^{r_s/10}-1}}$	$= \frac{\varepsilon_p}{\varepsilon_s}$
ε_p	$= \sqrt{(1-\delta_p)^{-2}-1}$	$= \sqrt{10^{r_p/10}-1}$	$= \sqrt{G_p^{-2}-1}$
ε_s	$= \sqrt{\delta_s^{-2}-1}$	$= \sqrt{10^{r_s/10}-1}$	$= \sqrt{G_s^{-2}-1}$
δ_p	$= 1 - \frac{1}{\sqrt{\varepsilon_p^2+1}}$	$= 1 - 10^{-r_p/20}$	
r_p	$= -20\log_{10}(1-\delta_p)$	$= 10\log_{10}(1+\varepsilon_p^2)$	$= -20\log_{10} G_p$
δ_s	$= \frac{1}{\sqrt{\varepsilon_s^2+1}}$	$= 10^{-r_s/20}$	
r_s	$= -20\log_{10}\delta_s$	$= 10\log_{10}(1+\varepsilon_s^2)$	$= -20\log_{10} G_s$

8.4 Butterworth Filters

A lowpass Butterworth filter is defined as

$$|H(j\Omega)|^2 = \frac{1}{1+\varepsilon_p^2 w^{2N}},$$

where $w = \Omega/\Omega_p$, and the integer N represents the filter order. Note that with $\Omega_c = \Omega_p \varepsilon_p^{-1/N}$, i.e., $\Omega_p = \Omega_c \varepsilon_p^{1/N}$, the Butterworth filter can also be written as

$$|H(j\Omega)|^2 = \frac{1}{1+\varepsilon_p^2 w^{2N}} = \frac{1}{1+\varepsilon_p^2 \left(\frac{\Omega}{\Omega_p}\right)^{2N}} = \frac{1}{1+\left(\frac{\Omega}{\Omega_c}\right)^{2N}},$$

a formula that does not contain any parameter ε_p. This is reasonable, considering the actual absence of any ripple in Butterworth filters. Moreover, in this expression the cutoff frequency appears in place of the passband limit, and Ω replaces w. Similar formulas can be written for the other filter types: it is always possible to use Ω_c in place of Ω_p (Chebyshev-I and elliptic filters) or Ω_s (Chebyshev-II filters) to write $|H(j\Omega)|^2$ in terms of Ω, rather than in terms of w (e.g., Porat 1996).

Figure 8.4 shows the shape of the Butterworth frequency response for a few N values. This figure shows the square magnitude of the frequency response; the cutoff frequency Ω_c as a function of Ω_p, δ_p and N has been defined in the previous subsection.

The main properties of a lowpass Butterworth filter are the following:

- $|H(j\Omega)|^2$ is a monotonically decreasing function of frequency that goes to zero for $\Omega \to \infty$;
- its maximum value is 1 and occurs at $\Omega = 0$;

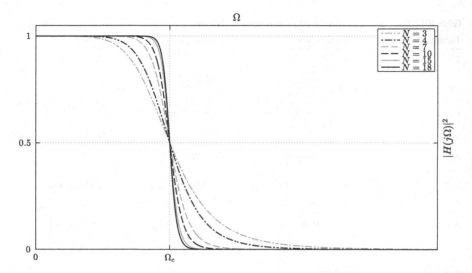

Fig. 8.4 Squared magnitude of the frequency response of an analog Butterworth filter with cutoff frequency $\Omega_c = 10$ rad/s and with $N = 3, 4, 7, 10, 15$, and 18

- with the choice of Ω_c given in the previous subsection, $|H(j\Omega_c)|^2 = 0.5$, so that $\Omega_c = \Omega_{3\text{dB}}$: the cutoff frequency Ω_c of a Butterworth filter is the -3 dB cutoff frequency. In fact, with $\Omega_c = \Omega_p \varepsilon_p^{-1/N}$, i.e., $\Omega_c / \Omega_p = \varepsilon_p^{-1/N}$, we have

$$|H(j\Omega_c)|^2 = \frac{1}{1 + \varepsilon_p^2 \varepsilon_p^{-2}} = \frac{1}{2},$$

from which we actually see that $\Omega_c = \Omega_{3\text{dB}}$.

This is not true for the other analog filters types, for which the cutoff frequency is assumed to be the abscissa of the point in which, as frequency increases starting from zero, the frequency response attains the passband maximum tolerance for the last time (Chebyshev-I and elliptic filters), or the stopband maximum tolerance for the first time (Chebyshev-II filters);

- the asymptotic attenuation at high frequency is $20N$ dB/decade, because $q = 0$ and $p = N$;
- it can be shown that the derivatives of any order $1 \leq l \leq 2N - 1$ vanish at Ω_c:

$$\frac{\partial^l |H(j\Omega)|^2}{\partial \Omega^l} \Big|_{\Omega = \Omega_c} = 0, \qquad 1 \leq l \leq 2N - 1;$$

in other words, Butterworth filters are *maximally flat* in the passband.

Substituting $\Omega = s/j = -js$ in the formula for $|H(j\Omega)|^2$ gives the squared magnitude of the Butterworth transfer function: it is convenient to use the expression containing only Ω and Ω_c, from which we get

$$H(s)H(-s) = \frac{1}{1 + \left(\frac{s}{j\Omega_c}\right)^{2N}} = \frac{1}{1 + (-1)^N \left(\frac{s}{\Omega_c}\right)^{2N}}.$$

This function has no zeros. Its poles are the $2N$ complex solutions p_n of the equation $1 + (-1)^N (s/\Omega_c)^{2N} = 0$; their expression is

$$p_n = \Omega_c \exp\left[j\frac{\pi(N+1+2n)}{2N}\right], \qquad 0 \le n \le 2N - 1.$$

The poles p_n of $H(s)H(-s)$ that correspond to $0 \le n \le N - 1$ are located in the left-hand half s-plane, and for stability are assigned to $H(s)$. The poles of $H(s)$ for the Butterworth filter are thus

$$p_n = \Omega_c \exp\left[j\frac{\pi(N+1+2n)}{2N}\right], \qquad 0 \le n \le N - 1,$$

while the remaining poles belong to $H(-s)$. Figure 8.5 shows the position of the poles of $H(s)H(-s)$ for an even and an odd value of N. There are N poles of $H(s)$ (*black dots*) equispaced on a half-circle of radius Ω_c, centered at the origin; the angle between two adjacent poles is π/N.

The poles of $H(s)$ satisfy the condition $\prod_{n=0}^{N-1}(-p_n) = \Omega_c^N$. Once the positions of the poles are known, the transfer function can be written in factorized form, as

$$H(s) = G \frac{1}{\prod_{n=0}^{N-1}(s - p_n)},$$

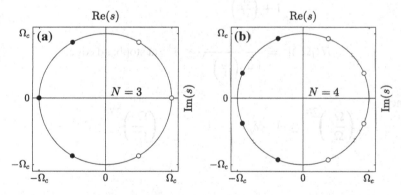

Fig. 8.5 Poles of $H(s)H(-s)$ for a Butterworth analog filter with a $N = 3$ and b $N = 4$. These poles all lie on a circle of radius Ω_c centered at the origin. The *black dots* represent the poles of $H(s)$, while the *empty circles* represent the poles of $H(-s)$

with

$$G = H_0 \prod_{n=0}^{N-1} (-p_n) = H_0 \Omega_c^N = \Omega_c^N,$$

since for Butterworth filters we have $H_0 = 1$. In summary, the factorized transfer function of a Butterworth filter is

$$H(s) = \Omega_c^N \prod_{n=0}^{N-1} \frac{1}{s - p_n} = \prod_{n=0}^{N-1} \frac{(-p_n)}{s - p_n}.$$

This all-pole transfer function can also be expressed as

$$H(s) = \frac{\Omega_c^N}{s^N + a_1 s^{N-1} + \cdots + a_{N-1} s + a_N},$$

where the coefficients a_i are those of the monic polynomial[4] having the p_n as poles and having $a_N = \Omega_c^N$, i.e., a_N equal to the product of reversed-in-sign poles.[5] Note that at $s = 0$, $H(s)$ assumes the correct unitary value.

The design process for a Butterworth filter required to meet given specifications proceeds as follows.

- Once the values of Ω_p, Ω_s, δ_p and δ_s have been chosen, the selectivity factor k and the discrimination factor k_1 are computed.
- These two factors determine the minimum required order. In fact, imposing that tolerances are not exceeded at the edges of the approximation bands we have

$$|H(j\Omega_p)|^2 = \frac{1}{1 + \left(\frac{\Omega_p}{\Omega_c}\right)^{2N}} \geq (1 - \delta_p)^2 \quad \text{at passband edge,}$$

$$|H(j\Omega_s)|^2 = \frac{1}{1 + \left(\frac{\Omega_s}{\Omega_c}\right)^{2N}} \leq \delta_s^2 \quad \text{at stopband edge,}$$

hence

$$\left(\frac{\Omega_p}{\Omega_c}\right)^{2N} \leq (1 - \delta_p)^{-2} - 1, \qquad \left(\frac{\Omega_s}{\Omega_c}\right)^{2N} \geq \delta_s^{-2} - 1.$$

---

[4] A monic polynomial is such that the coefficient a_0 of the highest-degree term is unitary.

[5] The poles are N, but a relation exists among them, so that there are only $N - 1$ degrees of freedom (DOF); the monic polynomial with degree N has N coefficients and therefore one of the coefficients must actually be fixed.

The ratio of these two inequalities gives

$$\left(\frac{\Omega_s}{\Omega_p}\right)^{2N} \geq \frac{\delta_s^{-2} - 1}{(1 - \delta_p)^{-2} - 1},$$

and substituting the definitions of $k = \Omega_p / \Omega_s$ and $k_1 = \varepsilon_p / \varepsilon_s$ (see Table 8.1) we get $k^{-2N} \geq k_1^{-2}$, i.e., $k^{-N} \geq k_1^{-1}$, representing the degree equation for the Butterworth filter. Passing to natural logarithms we obtain the condition

$$N \geq \frac{\log\left(k_1^{-1}\right)}{\log\left(k^{-1}\right)} = \frac{\log(\varepsilon_s / \varepsilon_p)}{\log(\Omega_s / \Omega_p)}.$$

The minimum order found from this relation is the solution of the degree equation for the Butterworth filter.

In general, however, the right-hand term of the last inequality is not an integer number, and the nearest larger integer is chosen as the minimum order N.

In practice, we will normally specify Ω_p, Ω_s, δ_p and δ_s, which determine k and k_1. Then we will compute N and round it up to the nearest larger integer. Since in this way we slightly increased N with respect to the (non-integer) value corresponding to k and k_1, we must then compute k_1 again from the old k and the rounded N; we could also compute k again from the old k_1 and the rounded N. Since k is an increasing function of N (a higher order leads to a narrower transition band), while k_1 is a decreasing function of N (a higher order leads to a greater attenuation in the stopband and/or a smaller deviation from the desired unitary value in the passband), the designed filter will have slightly better characteristics than those originally requested. Usually, k_1 is the quantity that is re-evaluated; δ_p is left unchanged and δ_s is re-evaluated accordingly, so as to exactly reflect the filter behavior in the stopband. In the case of Butterworth filters, there is no ripple and the important parameter is k, which regulates the width of the transition band; the re-evaluated k_1 simply represents that value which, when inserted in the formula expressing the solution of the degree equation, gives an integer value. The re-evaluated δ_s that derives from it gives the value of the designed frequency response at $\Omega = \Omega_s$.

For re-evaluating k_1 after rounding N up, we will use the degree equation, $k_1^{-1} = k^{-N}$; the formula for re-evaluating δ_s is then

$$\delta_s = \left[1 + \frac{(1 - \delta_p)^{-2} - 1}{k_1^2}\right]^{-1/2}.$$

If, on the contrary, we decided to re-evaluate k, then Ω_s would change, making the transition band slightly narrower, and at this new abscissa the frequency response would equal the original δ_s, which would remain unaltered, like k_1.

- In principle, the cutoff frequency Ω_c of the Butterworth filter can be chosen arbitrarily in the range

$$\Omega_p \left[\left(1 - \delta_p\right)^{-2} - 1 \right]^{-1/2N} \leq \Omega_c \leq \Omega_s \left[\delta_s^{-2} - 1 \right]^{-1/2N}.$$

This inequality derives from the conditions imposed separately at the passband and stopband edges (see the previous point). Actually, Ω_c is set equal to the lower limit of this range.

• The positions of the poles p_n are calculated.
• The transfer function $H(s)$ is calculated from its factorized form. The frequency response $H(j\Omega)$ follows, by setting $s = j\Omega$.

An example of the result of this procedure is shown in Figs. 8.6 (dB scale) and 8.7 (linear scale). This plot was obtained setting $\Omega_p = 1$ rad/s, $\Omega_s = 2$ rad/s, $\delta_p = 0.001$. The filter order is 15. The re-evaluated k_1 and δ_s after rounding up N to the integer value of 15 are 3.05×10^{-5} and 6.82×10^{-4}, respectively. In Fig. 8.7, *horizontal dotted lines* represent $(1 - \delta_p)^2$, the original value of δ_s^2, and the re-evaluated squared stopband tolerance that we indicate by $\delta_{s,\text{fin}}^2$. Had we chosen to re-evaluate k keeping k_1 fixed, the new Ω_s would fall at the intersection of the of δ_s^2 line with the frequency response curve.

It can be seen in Fig. 8.7 that the frequency response of the Butterworth filter is flat (i.e., has no ripples) in the passband. In the stopband it rolls off towards zero. When viewed on a dB plot (Fig. 8.6), the response slopes off linearly towards negative infinity at 20 N dB per decade. Butterworth filters thus have a magnitude frequency response that changes monotonically with ω, unlike other filter types, which exhibit ripples and therefore are non-monotonic in the passband and/or stopband.

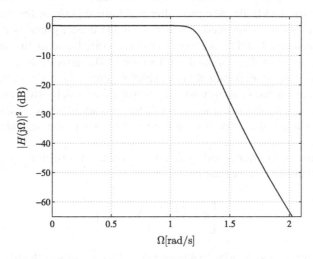

Fig. 8.6 Squared magnitude of the frequency response in dB of an analog Butterworth filter satisfying the following specifications: $\delta_p = \delta_s = 0.001$, $\Omega_p = 1$ rad/s, $\Omega_s = 2$ rad/s; these specifications lead to $k = 0.5$ and $k_1 = 4.4755 \times 10^{-5}$. The minimum order is $N = 15$. Ω_c is chosen as $\Omega_p[(1 - \delta_p)^{-2} - 1]^{-1/2N} = 1.23$ rad/s

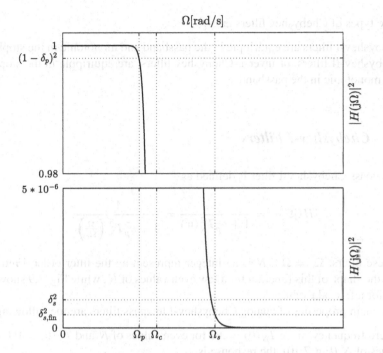

Fig. 8.7 An expanded view of the same squared-magnitude frequency response shown in Fig. 8.6, in normal units and with tolerance limits: the *horizontal dotted lines* indicate $(1 - \delta_p)^2$, the original value of δ_s^2, and the re-evaluated squared stopband tolerance $\delta_{s,\text{fin}}^2$ (see text)

8.5 Chebyshev Filters

Butterworth filters may seem "too good": due to their monotonic frequency response and their maximally flat behavior in the passband, they appear "glued" to 1 over the whole passband and very close to 0 over the whole stopband, thus attaining the maximum approximation error only at the edge of the two bands. Even if at times this behavior can be useful, in most cases it is enough if the deviation of the effective response from the desired one does not exceed some chosen δ_p and δ_s. Intuitively, we can understand that spreading the approximation error uniformly over all the passband (or over the stopband), the filter specifications can be met at a lower order. This idea is realized in Chebyshev filters.

Chebyshev filters are built using the polynomial with the same name and the same order of the filter: $T_N(w)$. These filters exhibit ripple in the passband or in the stopband and are equiripple. Accepting the existence of some ripple allows a more rapid transition from the passband to the stopband, with respect to requiring a monotonic behavior. As a consequence, the order of a Chebyshev filter meeting some given specifications is normally lower than the order of a Butterworth filter.

Two types of Chebyshev filters exist:

- Chebyshev-I filters are equiripple in the passband and monotonic in the stopband;
- Chebyshev-II filters, or inverse-Chebyshev filters, are equiripple in the stopband and monotonic in the passband.

8.5.1   Chebyshev-I Filters

The lowpass Chebyshev-I filter is defined as

$$|H(j\Omega)|^2 = \frac{1}{1 + \varepsilon_p^2 T_N^2(w)} = \frac{1}{1 + \varepsilon_p^2 T_N^2\left(\frac{\Omega}{\Omega_c}\right)}.$$

For these filters, $\Omega_c = \Omega_p$; N is an integer representing the filter order. Figure 8.8 shows the shape of this function for a few even values of N, while Fig. 8.9 shows the shape for a few odd values N.

The main properties of analog Chebyshev-I lowpass filters are the following:

- at zero frequency, since $T_N(0) = \pm 1$ for even values of N and $T_N(0) = 0$ for odd values of N (Fig. 7.10), the response is

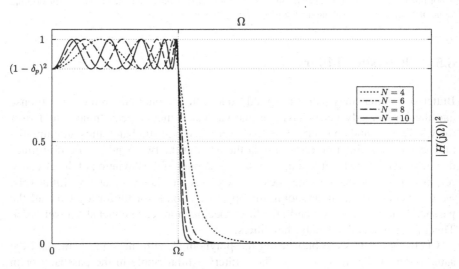

Fig. 8.8 Squared modulus of the frequency response of an analog Chebyshev-I filter with cutoff frequency $\Omega_c = \Omega_p = 10\,\text{rad/s}$ and passband tolerance $\delta_p = 0.075$, for $N = 4, 6, 8$, and 10

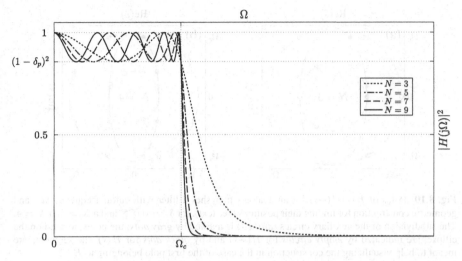

Fig. 8.9 Squared modulus of the frequency response of an analog Chebyshev-I filter with cutoff frequency $\Omega_c = \Omega_p = 10\,\text{rad/s}$ and passband tolerance $\delta_p = 0.075$, for $N = 3, 5, 7$, and 9

$$H_0 = \begin{cases} \dfrac{1}{\sqrt{1+\varepsilon_p^2}} & \text{for } N \text{ even,} \\[3mm] 1 & \text{for } N \text{ odd;} \end{cases}$$

- for $0 \le \Omega \le \left(\Omega_c = \Omega_p\right)$, i.e., for $w = \Omega/\Omega_p \le 1$, Chebyshev polynomials $T_N(w)$ (Fig. 7.10) never exceed 1 in absolute value, hence

$$\frac{1}{1+\varepsilon_p^2} \le |H(\mathrm{j}\Omega)|^2 \le 1, \quad \text{with} \quad \frac{1}{1+\varepsilon_p^2} = (1-\delta_p)^2;$$

- for $\Omega > \Omega_c$, i.e., for $w = \Omega/\Omega_p \ge 1$, Chebyshev polynomials $T_N(w)$ increase monotonically, hence the frequency response decreases monotonically;
- asymptotically, for $\Omega \to \infty$, $|H(\mathrm{j}\Omega)|^2 \to 0$, because $T_N^2(\Omega/\Omega_c) \to \infty$ for $\Omega \to \infty$;
- the asymptotic attenuation is the same as the asymptotic attenuation of a Butterworth filter with the same order: $20N$ dB/decade.

These filters have no zeros. The positions of the poles are obtained by setting the denominator of $|H(\mathrm{j}\Omega)|^2$ to zero; the result is more easily explained using a geometric construction, rather than formulas. We will explain this construction with reference to Fig. 8.10, which illustrates a case with odd N and a case with even N.

The poles of $H(s)H(-s)$ lie on an ellipse in the s-plane, which is defined through two auxiliary circles with radii equal to the major an minor ellipse axes. The minor axis has a length $a\Omega_c$, with

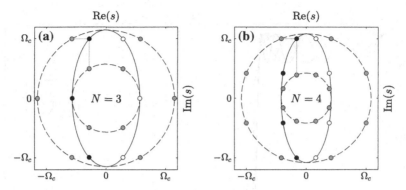

Fig. 8.10 Poles of $H(s)H(-s)$ for an analog Chebyshev-I filter with cutoff frequency Ω_c, and geometric construction for finding their positions (see text), for $\delta_p = 0.075$, and **a** $N = 3$, **b** $N = 4$. The subdivision of the auxiliary circles (*dashed*) is marked by *gray dots*; the poles, located on the ellipse, are indicated by *empty circles* for $H(-s)$ and by *black dots* for $H(s)$; the segments are meant to help visualizing the construction in the case of the first pole belonging to $H(s)$

$$a = \frac{1}{2}\left(\alpha^{1/N} - \alpha^{-1/N}\right), \qquad \alpha = \varepsilon_p^{-1} + \sqrt{1 + \varepsilon_p^{-2}};$$

the major axis has a length $b\Omega_c$, with

$$b = \frac{1}{2}\left(\alpha^{1/N} + \alpha^{-1/N}\right).$$

In order to locate the poles of $H(s)H(-s)$, which obviously depend on ε_p and hence on the passband ripple (N and analog cutoff frequency being fixed), the two circles are divided by points equally spaced in angle by π/N, in such a way that the points are symmetrically distributed with respect to the imaginary axis and one point falls on the real axis if N is odd, but does not if N is even. This subdivision of the major and minor auxiliary circles corresponds exactly to the way in which the circle with radius Ω_c, related to the poles of a Butterworth filter, is subdivided in Fig. 8.5. Finally, the poles are located on the ellipse: the ordinates of these points are those of the points we identified on the major auxiliary circle, and the abscissas are those of the points we identified on the minor auxiliary circle. The poles p_n located in the left-hand half s-plane are those having $0 \le n \le N - 1$; for stability they are assigned to $H(s)$; their expression is (see, e.g., Shenoi 2005)

$$p_n = -\Omega_c\left\{\sinh(\phi)\sin\left[\frac{\pi(2n+1)}{2N}\right] + \mathrm{j}\cosh(\phi)\cos\left[\frac{\pi(2n+1)}{2N}\right]\right\},$$

with $0 \le n \le N - 1$, and with

$$\phi = \frac{1}{N}\sinh^{-1}\left(\varepsilon_p^{-1}\right).$$

The number of poles of $H(s)$ is therefore N; these poles satisfy

$$\prod_{n=0}^{N-1}(-p_n) = \begin{cases} \dfrac{\Omega_c^N \sqrt{1+\varepsilon_p^2}}{2^{N-1}\varepsilon_p} & \text{for } N \text{ even,} \\[3ex] \dfrac{\Omega_c^N}{2^{N-1}\varepsilon_p} & \text{for } N \text{ odd.} \end{cases}$$

Recalling that in the absence of zeros we can write

$$H(s) = G\frac{1}{\prod_n (s - p_n)} \qquad \text{with} \quad G = H_0 \prod_n (-p_n),$$

from the expressions given above for H_0 and $\prod_{n=0}^{N-1}(-p_n)$ we can deduce that

$$G = \frac{\Omega_c^N}{2^{N-1}\varepsilon_p},$$

hence the factorized form of the transfer function is

$$H(s) = \frac{\Omega_c^N}{2^{N-1}\varepsilon_p} \prod_{n=0}^{N-1} \frac{1}{s - p_n} = H_0 \prod_{n=0}^{N-1} \frac{(-p_n)}{s - p_n}.$$

This all-pole transfer function can also be written as

$$H(s) = \frac{\Omega_c^N}{2^{N-1}\varepsilon_p \left(s^N + a_1 s^{N-1} + \cdots + a_{N-1}s + a_N\right)},$$

with a_N equal to the product of the reversed-in-sign poles.

The design process for a Chebyshev-I filter required to meet given specifications proceeds as follows.

- Once Ω_p, Ω_s, δ_p and δ_s have been chosen, the selectivity factor k and the discrimination factor k_1 are calculated.
- The cutoff frequency Ω_c, the parameter ε_p and the minimum order N are calculated. For this purpose, we impose the conditions at the edges of the approximation bands:

$$\left|H(j\Omega_p)\right|^2 = \frac{1}{1 + \varepsilon_p^2 T_N^2\left(\frac{\Omega_p}{\Omega_c}\right)} \geq (1 - \delta_p)^2 \qquad \text{at passband edge,}$$

$$\left|H(j\Omega_s)\right|^2 = \frac{1}{1 + \varepsilon_p^2 T_N^2\left(\frac{\Omega_s}{\Omega_c}\right)} \leq \delta_s^2 \qquad \text{at stopband edge.}$$

The passband-edge condition can be satisfied with the equal sign, choosing $\Omega_c = \Omega_p$, since $T_N(1) = 1$ and $1 + \varepsilon_p^2 = (1 - \delta_p)^{-2}$. The stopband-edge condition is satisfied

- observing that $\Omega_s/\Omega_c = \Omega_s/\Omega_p = k^{-1} > 1$, which implies $T_N(k^{-1}) > 1$;
- observing that normally $k_1 \ll 1$, i.e., $k_1^{-1} \gg 1$;
- imposing

$$T_N(k^{-1}) = \cosh\left[N \operatorname{arc\,cosh}(k^{-1})\right] \geq k_1^{-1} = \left[\frac{\delta_s^{-2} - 1}{\left(1 - \delta_p\right)^{-2} - 1}\right]^{1/2} = \frac{\varepsilon_s}{\varepsilon_p},$$

which provides

$$|H(\mathrm{j}\Omega_s)|^2 \leq \frac{1}{1 + \varepsilon_s^2} = \delta_s^2.$$

From the inequality concerning $T_N(k^{-1})$ we get

$$N \geq \frac{\operatorname{arc\,cosh}(k_1^{-1})}{\operatorname{arc\,cosh}(k^{-1})} = \frac{\operatorname{arc\,cosh}(\varepsilon_s/\varepsilon_p)}{\operatorname{arc\,cosh}(\Omega_s/\Omega_p)}.$$

Since both k^{-1} and k_1^{-1} are never less than 1, we can write, equivalently,

$$N \geq \frac{\log\left(k_1^{-1} + \sqrt{k_1^{-2} - 1}\right)}{\log\left(k^{-1} + \sqrt{k^{-2} - 1}\right)},$$

where the equality $\operatorname{arc\,cosh}(x) = \log\left(x + \sqrt{x^2 - 1}\right)$, valid for $x \geq 1$, was used. The minimum order found in this way is the solution of the degree equation for Chebyshev-I filters, which is $T_N(k^{-1}) = k_1^{-1}$. Comparing the Chebyshev-I filter's formula for N with the Butterworth filter's one, it can be shown that the values of k and k_1 being equal, the Chebyshev-I filter's order is always smaller than the Butterworth filter's order.

Once again, the right-hand inequality term is not an integer number in general, and to have the minimum order we must round it up the nearest largest integer; as a consequence, the designed filter will have slightly better characteristics than required by the original specifications. Ripple is present in the passband only, and the maximum deviation specified for the stopband is not so important. If we re-evaluate k_1 using $T_N(k^{-1}) = k_1^{-1}$, while leaving δ_p unaltered, we can use the same formula for re-evaluating δ_s already given for the Butterworth filter:

$$\delta_s = \left[1 + \frac{(1 - \delta_p)^{-2} - 1}{k_1^2}\right]^{-1/2}.$$

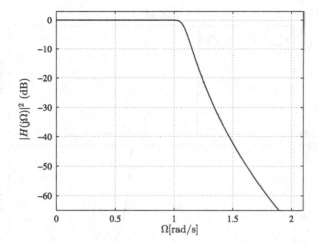

Fig. 8.11 Squared magnitude of the frequency response in dB of an analog Chebyshev-I filter satisfying the following specifications: $\delta_p = \delta_s = 0.001$, $\Omega_p = 1$ rad/s, $\Omega_s = 2$ rad/s; such specifications imply $k = 0.5$ and $k_1 = 4.4755 \times 10^{-5}$, from which the minimum order $N = 9$ is deduced; Ω_c is chosen equal to $\Omega_p = 1$ rad/s

- The positions p_n of the poles are subsequently computed.
- The transfer function $H(s)$ is finally computed from its factorized expression, keeping into account that H_0 varies according to N being odd or even. The frequency response follows directly, by setting $s = j\Omega$.

The result of this procedure is illustrated by Figs. 8.11 (dB units) and 8.12 (normal units). In Fig. 8.12, *horizontal dotted lines* represent $(1 - \delta_p)^2$, the original value of δ_s^2, and the re-evaluated squared stopband tolerance $\delta_{s,\text{fin}}^2$. These plots were obtained setting $\Omega_p = 1$ rad/s, $\Omega_s = 2$ rad/s, and $\delta_p = 0.001$. The filter order is 9. The re-evaluated k_1 and δ_s after rounding up N to the integer value of 9 are 1.42×10^{-5} and 3.18×10^{-4}, respectively. Had we chosen to re-evaluate k keeping k_1 fixed, the new Ω_s would fall at the intersection of the δ_s^2 line with the frequency response curve.

In Fig. 8.11, which has ordinates expressed in dB, the passband ripple is so small that it can hardly be seen; in the stopband the response goes off towards negative infinity but the decrease is not linear.

8.5.2 Chebyshev-II Filters

With respect to the Chebyshev-I filter, the Chebyshev-II filter represents an inverse way of approximating the specifications, a way that assigns the ripple to the stopband. To get this result,

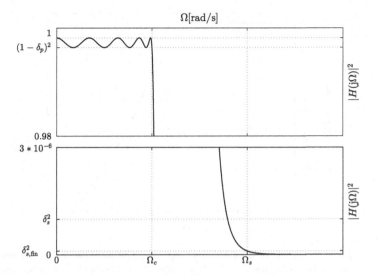

Fig. 8.12 An expanded view of the same squared-magnitude frequency response shown in Fig. 8.11, in normal units and with tolerance limits: the *horizontal dotted lines* indicate $(1 - \delta_p)^2$, the original value of δ_s^2, and the re-evaluated squared stopband tolerance $\delta_{s,\text{fin}}^2$ (see text)

- we take the squared amplitude of the Chebyshev-I frequency response having, as its ε_p, the inverse of that value of ε_s which derives from the tolerances we expressed for the Chebyshev-II filter we want to build;
- we reverse the argument of $T_N(\cdot)$, i.e., we pass from Ω/Ω_c to Ω_c/Ω. This operation exchanges the filter behavior around the origin with the filter's asymptotic behavior, while leaving the behavior at $\Omega = \Omega_c$ unchanged; so it transforms a lowpass filter into a highpass filter;
- we subtract from 1 the squared modulus of the highpass frequency response thus obtained, and get back a lowpass filter.

For a Chebyshev-II filter, the cutoff frequency is chosen as $\Omega_c = \Omega_s$, and since we set $w = \Omega/\Omega_p$, $k^{-1} = \Omega_s/\Omega_p$, we have

$$w = \frac{\Omega}{\Omega_p} = \frac{\Omega}{\Omega_s}\frac{\Omega_s}{\Omega_p} = \frac{\Omega}{\Omega_c}\frac{\Omega_s}{\Omega_p} = \frac{\Omega}{\Omega_c}k^{-1},$$

hence

$$\frac{\Omega_c}{\Omega} = k^{-1}w^{-1}.$$

We thus obtain the equation that defines the lowpass Chebyshev-II filter:

$$|H(j\Omega)|^2 = 1 - \frac{1}{1 + \varepsilon_s^{-2}T_N^2(k^{-1}w^{-1})} = \frac{\varepsilon_s^{-2}T_N^2(k^{-1}w^{-1})}{1 + \varepsilon_s^{-2}T_N^2(k^{-1}w^{-1})} =$$

$$= \frac{1}{1 + 1/\left[\varepsilon_s^{-2}T_N^2(k^{-1}w^{-1})\right]} = \frac{1}{1 + \varepsilon_s^2/T_N^2(k^{-1}w^{-1})}.$$

Using $k_1 = \varepsilon_p/\varepsilon_s$, i.e. $\varepsilon_s^2 = \varepsilon_p^2/k_1^2$, we can write

$$|H(j\Omega)|^2 = \frac{1}{1 + \varepsilon_p^2 / \left[k_1^2 T_N^2(k^{-1}w^{-1})\right]} = \frac{1}{1 + \varepsilon_p^2 \left[k_1 T_N\left(k^{-1}w^{-1}\right)\right]^{-2}}.$$

Therefore the substitution

$$T_N(w) \Rightarrow \left[k_1 T_N(k^{-1}w^{-1})\right]^{-1}$$

transforms the Chebyshev-I filter into a Chebyshev-II filter, and the equiripple behavior in the passband is transformed into an equiripple behavior in the stopband. Figure 8.13 shows the shape of $|H(j\Omega)|^2$ for a few even values of N; Fig. 8.14 shows the shape for a few odd values of N.

The main properties of Chebyshev-II filters are the following:

- at $\Omega = 0, k^{-1}w^{-1} \to \infty$, hence $|T_N(k^{-1}w^{-1})| \to \infty, \left[T_N(k^{-1}w^{-1})\right]^{-2} \to 0$; therefore

$$H_0 = 1$$

for any values of the order and of the parameters;
- for $0 \leq \Omega < (\Omega_c = \Omega_s)$, corresponding to $0 \leq w \leq \Omega_s/\Omega_p = k^{-1}$ with $k_1^{-1} > 1$ (and therefore to arguments of $T_N(\cdot)$ decreasing from ∞ to 1), the frequency response decreases monotonically, because the absolute value of Chebyshev polynomials decreases monotonically when the argument goes from ∞ to 1;

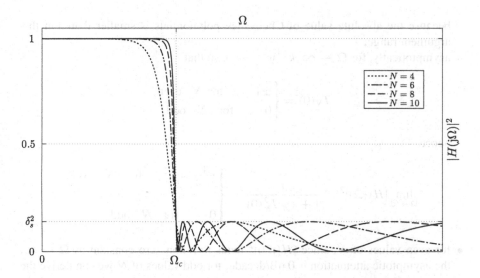

Fig. 8.13 Squared modulus of the frequency response of an analog Chebyshev-II filter, with cutoff frequency $\Omega_c = \Omega_s = 10\,\text{rad/s}$ and $\delta_s = 0.375$, for $N = 4, 6, 8,$ and 10

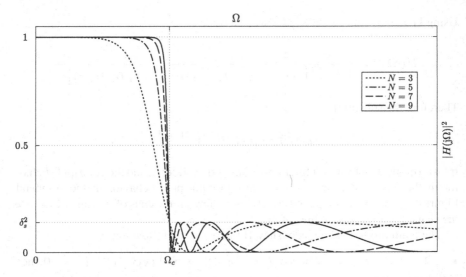

Fig. 8.14 Squared modulus of the frequency response of an analog Chebyshev-II filter, with cutoff frequency $\Omega_c = \Omega_s = 10\,\text{rad/s}$ and $\delta_s = 0.375$, for $N = 3, 5, 7$, and 9

- for $\Omega \geq (\Omega_c = \Omega_s)$, corresponding to $w \geq \Omega_s/\Omega_p = k^{-1}$ and therefore to arguments of $T_N(\cdot)$ smaller than $k^{-1}k = 1$, the frequency response satisfies

$$0 \leq |H(j\Omega)|^2 \leq \frac{1}{1 + \varepsilon_s^2} = \delta_s^2,$$

because the absolute value of Chebyshev polynomials is smaller than 1 in this argument range;

- asymptotically, for $\Omega \to \infty$, $k^{-1}w^{-1} \to 0$, so that

$$T_N(0) = \begin{cases} \pm 1 & \text{for } N \text{ even,} \\ 0 & \text{for } N \text{ odd;} \end{cases}$$

hence

$$\lim_{\Omega \to \infty} |H(j\Omega)|^2 = \frac{1}{1 + \varepsilon_s^2/T_N^2(0)} = \begin{cases} \frac{1}{1+\varepsilon_s^2} = \delta_s^2 & \text{for } N \text{ even,} \\ 0 & \text{for } N \text{ odd;} \end{cases}$$

- for even values of N, since $|H(j\Omega)|^2$ tends to a non-zero constant as $\Omega \to \infty$, the asymptotic attenuation is 0 dB/decade; for odd values of N we can derive the asymptotic attenuation as follows.

A Chebyshev polynomial $T_N(w)$ with odd N is given approximately, in the vicinity of $w = 0$, by $T_N(w) \cong K_N w$, where K_N is a constant. Hence at high frequencies, at which $\Omega_c/\Omega \to 0$, we can write

$$|H(j\Omega)|^2 \cong \frac{\varepsilon_s^{-2} K_N^2 \Omega_c^2/\Omega^2}{1 + \varepsilon_s^{-2} K_N^2 \Omega_c^2/\Omega^2} \approx \varepsilon_s^{-2} K_N^2 \frac{\Omega_c^2}{\Omega^2}$$

and

$$20 \log_{10} |H(j\Omega)| \approx 20 \log_{10} |K_N \Omega_c/\varepsilon_s| - 20 \log_{10} \Omega.$$

Therefore for odd values of N the asymptotic attenuation is 20 dB/decade.

Let us now turn to the positions of the poles. They can be found using the following property: the poles of the transfer function $H(s)$ of a Chebyshev-II low-pass normalized filter ($\Omega_c = 1$ rad/s) with a given parameter ε_s are the inverses of the poles of a Chebyshev-I lowpass normalized filter, having the same order and parameter $\varepsilon_p = 1/\varepsilon_s$. Then, let v_n ($0 \le n \le N - 1$) be the Chebyshev-II poles and p_n ($0 \le n \le N - 1$) the Chebyshev-I poles of non-normalized filters. Passing from normalized prototypes to general filters with $\Omega_c \ne 1$ rad/s, the positions of poles and zeros are multiplied by Ω_c. Therefore, with obvious notation, if

$$v_n^{\text{norm}} = 1/p_n^{\text{norm}},$$

then

$$v_n = \Omega_c v_n^{\text{norm}} = \frac{\Omega_c}{p_n^{\text{norm}}} = \frac{\Omega_c^2}{p_n}, \qquad 0 \le n \le N - 1.$$

Figure 8.15a, b shows the poles of a Chebyshev-II $H(s)$ filter (*black dots*) for a case with odd N and a case with even N, respectively. The poles of $H(-s)$ are not shown to avoid cluttering. These poles do not fall on an ellipse and are found both internally and externally to the circle of radius Ω_c. Their position, for a given N and a given cutoff analog frequency, depends on the parameter ε_s, i.e., from the stopband ripple.

Recalling that for a Chebyshev-I filter with poles p_n we have

$$\prod_{n=0}^{N-1}(-p_n) = \begin{cases} \frac{\Omega_c^N \sqrt{1+\varepsilon_p^2}}{2^{N-1}\varepsilon_p} & \text{for even } N, \\[2ex] \frac{\Omega_c^N}{2^{N-1}\varepsilon_p} & \text{for odd } N, \end{cases}$$

and that for the Chebyshev-I filter with respect Chebyshev-II filter the equality $\varepsilon_p = \varepsilon_s^{-1}$ holds, it can be shown that the poles v_n of a Chebyshev-II filter satisfy

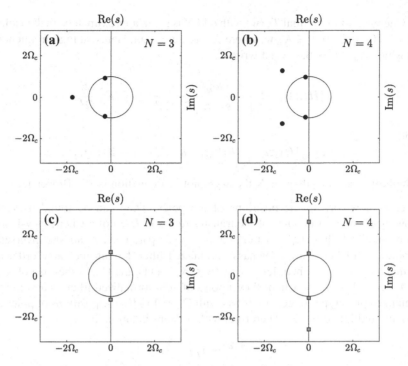

Fig. 8.15 **a, b** Poles (*black dots*) and **c, d** zeros (*gray squares*) of the transfer function $H(s)$ of an analog Chebyshev-II filter, for $\delta_s = 0.375$ and for $N = 3$ and 4, respectively

$$\prod_{n=0}^{N-1}(-\nu_n) = \begin{cases} \dfrac{\Omega_c^N 2^{N-1}}{\varepsilon_s \sqrt{1+\varepsilon_s^{-2}}} & \text{for even } N, \\[4mm] \dfrac{\Omega_c^N 2^{N-1}}{\varepsilon_s} & \text{for odd } N. \end{cases}$$

Chebyshev-II filters have zeros: they are shown as *gray squares* in Fig. 8.15c, d, and fall at those frequencies at which $T_N(k^{-1}w^{-1}) = T_N(\Omega_c/\Omega) = 0$. The zeros of $T_N(w)$ are found at $w = \cos\left[(2m+1)\,\pi/2N\right]$, with $0 \le m \le N-1$, as it can be deduced from the definition of these polynomials. This leads to deducing that the zeros of $H(s)$ are located on the imaginary axis, at values of s given by

$$u_m = \frac{j\Omega_c}{\cos\left[\frac{(2m+1)\pi}{2N}\right]}, \qquad 0 \le m \le N-1.$$

When N is even, there are N zeros at finite distance from the origin; when N is odd, there are only $N-1$ zeros at finite distance, because $m = (N-1)/2$ corresponds to $\cos(\pi/2)$ in the denominator of U_m, hence one of the zeros goes to infinity. It must be noted that the position of the zeros of $H(s)$ is independent of the parameters ε_s, ε_p, and of the ripple; it depends only on Ω_c and N. The zeros are related by

$$\prod_{m=0}^{N-1} (-u_m) = \begin{cases} \Omega_c^N 2^{N-1} & \text{for } N \text{ even,} \\ \Omega_c^{N-1} 2^{N-1}/N & \text{for } N \text{ odd.} \end{cases}$$

We can now write the factorized form of the transfer function: recalling that $H_0 = 1$ in this case,

- for even N, there are N zeros and N poles at finite distance from the origin, hence

$$H(s) = G \prod_{n=0}^{N-1} \frac{s - u_n}{s - v_n} \quad \text{with} \quad G = H_0 \prod_{n=0}^{N-1} \frac{(-v_n)}{(-u_n)} = H_0 \prod_{n=0}^{N-1} \frac{v_n}{u_n};$$

- for odd N, there are $N - 1$ zeros and N poles at finite distance, hence

$$H(s) = G \frac{\prod_{m=0}^{N-2}(s - u_m)}{\prod_{n=0}^{N-1}(s - v_n)} \quad \text{with} \quad G = H_0 \frac{\prod_{n=0}^{N-1}(-v_n)}{\prod_{m=0}^{N-2}(-u_m)} = -H_0 \frac{\prod_{n=0}^{N-1} v_n}{\prod_{m=0}^{N-2} u_m}.$$

Introducing $r = N \bmod 2$ (remainder of the division of N by 2) we can write a unified formula holding for any N:

$$H(s) = G \frac{\prod_{m=0}^{N-r-1}(s - u_m)}{\prod_{n=0}^{N-1}(s - v_n)} \quad \text{with} \quad G = H_0 \frac{\prod_{n=0}^{N-1}(-v_n)}{\prod_{m=0}^{N-r-1}(-u_m)}.$$

This transfer function with poles and zeros can also be expressed as

$$H(s) = \begin{cases} G \dfrac{s^N + b_1 s^{N-1} + \cdots + b_{N-1}s + b_N}{s^N + a_1 s^{N-1} + \cdots + a_{N-1}s + a_N} & \text{for } N \text{ even,} \\[2mm] G \dfrac{s^{N-1} + b_1 s^{N-2} + \cdots + b_{N-2}s + b_{N-1}}{s^N + a_1 s^{N-1} + \cdots + a_{N-1}s + a_N} & \text{for } N \text{ odd,} \end{cases}$$

with b_N or b_{N-1} equal to the product of the zeros reversed in sign, for N even or odd respectively, and with a_N equal to the product of the poles reversed in sign.

The design process for a Chebyshev-II filter that must satisfy some given specifications is the following.

- Once Ω_p, Ω_s, δ_p and δ_s have been chosen, the selectivity factor k and the discrimination factor k_1 are computed accordingly.
- The cutoff frequency Ω_c, the parameter ε_s and the minimum order N are calculated by imposing at the edges of the approximation bands

$$\left| H(j\Omega_p) \right|^2 = \frac{1}{1 + \varepsilon_s^2 / T_N^2\left(\frac{\Omega_c}{\Omega_p}\right)} \geq (1 - \delta_p)^2 \quad \text{at passband edge,}$$

$$|H(j\Omega_s)|^2 = \cfrac{1}{1 + \varepsilon_s^2 / T_N^2\left(\frac{\Omega_c}{\Omega_s}\right)} \leq \delta_s^2 \quad \text{at stopband edge.}$$

The stopband condition can be satisfied with the equal sign, choosing $\Omega_c = \Omega_s$; then, since $T_N(1) = 1$, we have

$$|H(j\Omega_s)|^2 = \frac{1}{1 + \varepsilon_s^2} = \delta_s^2.$$

The passband condition can be considered satisfied if

– we observe that $\Omega_c / \Omega_p = \Omega_s / \Omega_p = k^{-1} > 1$, which implies $T_N(k^{-1}) > 1$;
– we recall that normally $k_1 \ll 1$, hence $k_1^{-1} \gg 1$;
– we impose

$$T_N(k^{-1}) = \cosh\left[N \operatorname{arc cosh}(k^{-1})\right] \geq k_1^{-1},$$

which implies, as required,

$$\left|H(j\Omega_p)\right|^2 = \cfrac{1}{1 + \varepsilon_s^2 / T_N^2\left(\frac{\Omega_c}{\Omega_p}\right)} \geq \frac{1}{1 + \varepsilon_s^2 k_1^2} = \frac{1}{1 + \varepsilon_p^2} = (1 - \delta_p)^2.$$

From the constraint we set, the condition on the order is derived:

$$N \geq \frac{\operatorname{arc cosh}(k_1^{-1})}{\operatorname{arc cosh}(k^{-1})}.$$

The minimum order is the solution of the degree equation, for Chebyshev-II filters, which is $T_N(k^{-1}) = k_1^{-1}$, i.e., the same degree equation that holds for Chebyshev-I filters.

In general, the right-hand member of this inequality does not lead to an integer value of N and the calculated value needs to be rounded up to the nearest larger integer. The expression of the minimum order is identical to the one we found for Chebyshev-I filters: the values of k and k_1 being equal, the order of the Chebyshev-II filters is always equal to the order of the Chebyshev-I filter. Therefore the choice between types I and II is dictated only by whether we prefer to allow for the existence of some ripple in the passband, or in the stopband.

The need for re-evaluating k or k_1 after rounding N up obviously also exists in this case. Let us suppose we re-evaluate k_1 from the degree equation. Since the ripple lies in the stopband, once the value of k_1 that makes N integer has been found, we can leave δ_s unaltered and re-evaluate δ_p, which is relatively unessential in this case. We will use the formula

$$\delta_p = 1 - \left[1 + k_1^2\left(\delta_s^{-2} - 1\right)\right]^{-\frac{1}{2}} \quad \text{with} \quad k_1 = \left[T_N(k^{-1})\right]^{-1}.$$

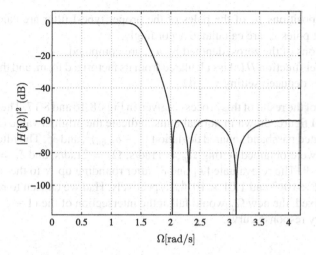

Fig. 8.16 Squared modulus of the frequency response in dB of a Chebyshev-II lowpass filter satisfying the following specifications: $\delta_p = \delta_s = 0.001$, $\Omega_p = 1\,\mathrm{rad/s}$, $\Omega_s = 2\,\mathrm{rad/s}$; this leads to $k = 0.5$ and $k_1 = 4.4755 \times 10^{-5}$, from which a minimum order $N = 9$ is found; Ω_c is set equal to $\Omega_s = 2\,\mathrm{rad/s}$

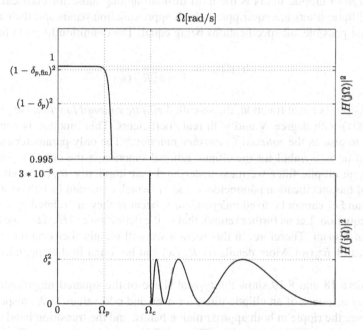

Fig. 8.17 A close-up of the squared modulus of the frequency response shown in Fig. 8.16, in normal units and with tolerance limits: the *horizontal dotted lines* indicate the original value of $(1 - \delta_p)^2$, the re-evaluated squared stopband tolerance $\delta_{s,\mathrm{fin}}^2$ (see text) $(1 - \delta_{p,\mathrm{fin}})^2$, and δ_s^2

- Next, the positions p_n of the poles of the proper type-I filter are calculated, and the type-II poles v_n are calculated accordingly.
- The positions of the zeros, denoted by u_m, are computed.
- The transfer function $H(s)$ is calculated from its factorized form, and the frequency response is deduced setting $s = j\Omega$.

An example of the result of this process is given in Figs. 8.16 and 8.17. The *dotted lines* in the second figure, drawn in normal units, indicate the original value of $(1 - \delta_p)^2$, the re-evaluated passband squared deviation $(1 - \delta_{p,\text{fin}})^2$, and δ_s^2. The filter order is 9. These plots were obtained setting $\Omega_p = 1\,\text{rad/s}$, $\Omega_s = 2\,\text{rad/s}$, and $\delta_s = 0.001$. The filter order is 9. The re-evaluated k_1 and δ_p after rounding up N to the integer value of 9 are 1.42×10^{-5} and 1.01×10^{-4}, respectively. Had we chosen to re-evaluate k keeping k_1 fixed, the new Ω_p would fall at the intersection of the $(1 - \delta_p)^2$ line with the frequency response curve.

8.6 Elliptic Filters

The design of elliptic filters is the most difficult among those for classical analog filters. Elliptic filters are equiripple in both approximation bands and their order is the lowest possible, all specifications being equal. The definition formula for order N is

$$|H(j\Omega)|^2 = \frac{1}{1 + \varepsilon_p^2 R_N^2 (w)}$$

where $R_N(w)$ is a real function, the so-called *elliptic rational function* (see Lutovac et al. 2001) with degree N and with real coefficients. This function is sometimes referred to also as the *rational Chebyshev function*.[6] The only parameter explicitly indicated in the symbol for the elliptic rational function is the order N, but when defining an elliptic filter we must understand that implicitly $R_N(w)$ contains the values of the specification parameters k and k_1, which constrain its behavior.

N, k and k_1 cannot be fixed independently, because they are related by a specific degree equation. Let us further remark that in the definition of $|H(j\Omega)|^2$ we find the square of $R_N(w)$. Therefore, in this section we will mainly focus on the essential properties of $R_N^2(w)$. More details on $R_N(w)$ can be found in the appendix to this chapter.

Figures 8.18 and 8.19 show the typical shape of the squared magnitude of the frequency response of an elliptic filter, for even and odd values of N, respectively. We can see the ripple in both approximation bands, and the transition band that can be rather narrow even at low orders. The cutoff frequency is chosen as $\Omega_c = \Omega_p$.

---

[6]The last name is ambiguous, because other functions exist that are called in the same way.

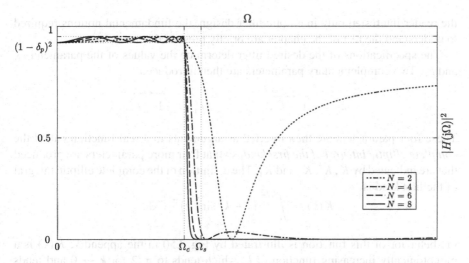

Fig. 8.18 Squared modulus of the frequency response of an analog elliptic filter with cutoff frequency $\Omega_c = \Omega_p = 10\,\text{rad/s}$ and with $\Omega_s = 11.3333\,\text{rad/s}$, $\delta_p = 0.015$, for $N = 2, 4, 6$ and 8

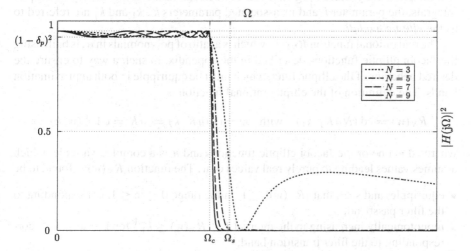

Fig. 8.19 Squared modulus of the frequency response of an analog elliptic filter with cutoff frequency $\Omega_c = \Omega_p = 10\,\text{rad/s}$ and with $\Omega_s = 11.3333\,\text{rad/s}$, $\delta_p = 0.015$, for $N = 3, 5, 7$ and 9

The study the function $R_N(w)$ requires some fundamental notions about elliptic integrals[7] and Jacobi elliptic functions. The reader who is not familiar with these mathematical concepts and is interested in getting a general idea of these topics may prefer to go through the appendix to this chapter at this point in the discussion. For a more detailed treatment, see, for example, Antoniou (1993) and Orfanidis (2006). For

---

[7]This name derives from the fact that these integrals were originally studied in the frame of the problem of calculating the length of an arc of ellipse.

the reader interested only in elliptic filter design, the fundamental notions required
to understand the rest of the chapter are briefly presented here.

The specifications of the desired filter determine the values of the parameters k
and k_1. Two complementary parameters are then introduced,

$$k' = \sqrt{1 - k^2}, \qquad\qquad k_1' = \sqrt{1 - k_1^2}.$$

These four parameters are then inserted as arguments in a real function called the
complete elliptic integral of the first kind, so that four more parameters are produced
that are indicated by K, K', K_1 and K_1'. The definition of the complete elliptic integral
of the first kind is

$$K(k) = \int_0^{\pi/2} \left(1 - k^2 \sin^2 \theta\right)^{-1/2} d\theta;$$

the behavior of this function is illustrated by Fig. 8.50 in the appendix. $K(k)$ is a
monotonically increasing function of k, which tends to $\pi/2$ for $k \to 0$ and tends
to infinity for $k \to 1$. In the same figure, the behavior of $K'(k)$ is also shown: this
is a monotonically decreasing function of k. In the frame of the theory of elliptic
integrals, the parameter k and its associated parameters k', k_1 and k_1' are referred to
as the *elliptic moduli*.

The real rational function $R_N(w)$, which is a ratio of polynomials in w, is built using
the Jacobi elliptic functions described in the appendix, in such a way to ensure the
desired behavior of the elliptic filter, which must be equiripple in both approximation
bands. The definition of the elliptic rational function is

$$R_N(w) = \text{cd}\,(NuK_1, k_1) \quad \text{with} \quad w = \text{cd}(uK, k) \Rightarrow uK = \text{cd}^{-1}(w, k),$$

where cd is one of the Jacobi elliptic functions and u is a complex variable, which
assumes values leading to purely real values of w. The function $R_N(w)$ is found to be

- equiripple, and such that $|R_N(w)| \leq 1$, in the range $0 \leq w \leq 1$, corresponding to
  the filter passband;
- monotonically increasing in the interval $1 \leq R_N(w) \leq k_1^{-1}$ for $1 \leq w \leq k^{-1}$, corresponding to the filter transition band;
- characterized by an equiripple inverse $1/R_N(w)$, with $|1/R_N(w)| \leq k_1$, in the range
  $k^{-1} \leq w < \infty$, corresponding to the filter stopband.
  This feature, which ensures an equiripple behavior of the filter in the stopband, is
  a consequence of imposing the so-called *inversion relation*

$$R_N(w) = \frac{1}{k_1 R_N(w^{-1}k^{-1})},$$

which in turn implies that the elliptic filter satisfies the degree equation $R_N(k^{-1}) =
k_1^{-1}$. This fact can be seen substituting $w = k^{-1}$ into the inversion relation and
keeping into account that $R_N(w)$ is normalized to 1 at $w = 1$. Since $R_N(w)$ satisfies
the inversion relation, we can also state that

Fig. 8.20 Overall view of the squared elliptic rational function, $R_N^2(w)$, for $k = 0.96$ and a $N = 3$, **b** $N = 4$. The corresponding values of k_1 are shown in each panel. *Horizontal solid lines* mark ordinates equal to k_1^{-2}, *vertical solid lines* mark abscissas equal to k^{-1}; *dotted lines* represent 1 on both axes

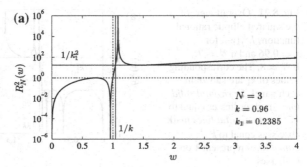

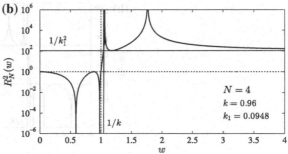

$$|H(\mathrm{j}\Omega)|^2 = \frac{1}{1 + \varepsilon_p^2 R_N^2(w)} = \frac{1}{1 + \varepsilon_s^2 / R_N^2(w^{-1}k^{-1})}$$

with $w^{-1}k^{-1} = (\Omega_p/\Omega)(\Omega_s/\Omega_p) = \Omega_s/\Omega$. This equation highlights the connection between the filter ripples in the passband and in the stopband and the fact that the stopband behavior of the filter is better understood in terms of $1/R_N(w)$.

The squared magnitude of the filter's frequency response is determined by $R_N^2(w)$. The shape of the squared elliptic function[8] is illustrated in Figs. 8.20 (for $k = 0.96$, $N = 3$ and 4, and $k_1 = 0.2385$ and 0.0948, respectively) and 8.21 (for $k = 0.96$, $N = 7$ and 8, and $k_1 = 0.00575$ and 0.00226, respectively). Inspection of these figures reveals that poles and zeros are present, and that their number increases with order. It can also be noted that the squared function is actually equiripple in the passband, between $w = 0$ and $w = 1$, and that it is monotonically increasing between 1 and k_1^{-2} in the transition band, between $w = 1$ and $w = k^{-1}$. For a better visualization of the stopband behavior, Fig. 8.22 gives $1/R_N^2(w)$ for the same parameter values used in Figs. 8.20 and 8.21. The equiripple behavior is evident. The function $1/R_N^2(w)$ oscillates between 0 and k_1^2. The previous figures represent only non-negative values of w, but $R_N^2(w)$, and the filter's frequency response, are symmetric around $w = 0$.

---

[8]Plotting the square instead of the plain function avoids many sources of confusion. Of course, we could use $|R_N(w)|$ as well. Plots of $R_N(w)$ can be found in the appendix to this chapter.

Fig. 8.21 Overall view of
the squared elliptic rational
function, $R_N^2(w)$, for
$k = 0.96$ and **a** $N = 7$,
b $N = 8$. The corresponding
values of k_1 are shown in
each panel. *Horizontal solid
lines* mark ordinates equal to
k_1^{-2}, *vertical solid lines* mark
abscissas equal to k^{-1};
dotted lines represent 1 on
both axes

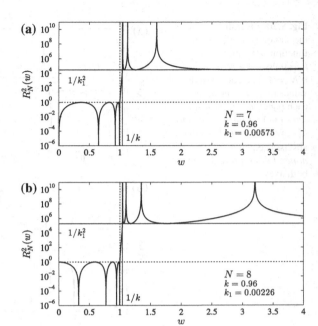

It is also important to mention that the inversion relation leads to the constraint

$$\frac{N K'}{K} = \frac{K_1'}{K_1},$$

which is a form of degree equation that can be used directly to obtain the minimum
filter order on the basis of the specifications. Finally, *modular equations* can be
derived from the degree equation. They allow calculating k_1 in terms of N and k, or
k in terms of N and k_1. These equations are useful, for example, during the elliptic
design process, to re-evaluate k_1 from N and k after the calculated minimum order
has been rounded up to the nearest larger integer.

Once the desired filter specifications have been expressed, the parameters k and
k_1 can be calculated. In turn, these parameters determine the values of the complementary parameters $k' = \sqrt{1 - k^2}$ and $k_1' = \sqrt{1 - k_1^2}$. The four values of the elliptic
integral, K, K', K_1 and K_1', can thus be found. The minimum order N then follows
from the degree equation. Subsequently, after N has been rounded up to the nearest
larger integer, k_1 is re-evaluated by means of the corresponding modular equation,
so as to exactly satisfy the degree equation in association with the new integer N,
and produce the correct δ_s value. Alternatively, k, or both parameters, can be slightly
altered. The cutoff frequency is chosen as $\Omega_c = \Omega_p$.

Thanks to the behavior of $R_N(w)$, the filter exactly meets specifications at $\Omega = \Omega_c = \Omega_p$, i.e. at $w = 1$, and at $\Omega = \Omega_s$, i.e. at $w = k^{-1}$, which are the edges of the
approximation bands:

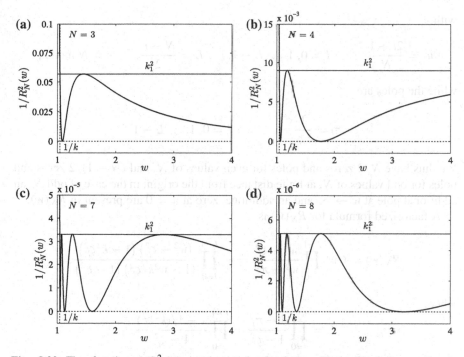

Fig. 8.22 The function $1/R_N^2(w)$ in the stopband of the elliptic filter, for **a** $N = 3$, **b** $N = 4$, **c** $N = 7$, and **d** $N = 8$. The values of k and k_1 associated to each N are the same reported in the panels of Figs. 8.20 and 8.21. *Horizontal solid lines* mark ordinates equal to k_1^2

$$\left|H(j\Omega_p)\right|^2 = \frac{1}{1 + \varepsilon_p^2 R_N^2(1)} = \frac{1}{1 + \varepsilon_p^2} = (1 - \delta_p)^2 \quad \text{since} \quad R_N(1) = 1;$$

$$\left|H(j\Omega_s)\right|^2 = \frac{1}{1 + \varepsilon_p^2 R_N^2(k^{-1})} = \frac{1}{1 + \varepsilon_s^2} = \delta_s^2 \quad \text{since} \quad R_N(k^{-1}) = \frac{1}{k_1} = \frac{\varepsilon_s}{\varepsilon_p}.$$

The transfer function of the elliptic filter is calculated on the basis of a factorized form, given its poles and zeros. More precisely, from the squared magnitude of the frequency response, the product $H(s)H(-s)$ is defined; then $H(s)$ is determined selecting, among the zeros and poles of $H(s)H(-s)$, those that lie in the left-hand half s-plane, so as to ensure a stable $H(s)$.

Observing the definition $|H(j\Omega)|^2 = 1/\left[1 + \varepsilon_p^2 R_N^2(w)\right]$, we can see that the filter's zeros, which will be denoted by z_m, derive from the poles of $R_N(w)$, while the poles derive from the zeros of the denominator. The zeros (ζ_l) and poles (π_l) of $R_N(w)$ are related; in fact, the zeros are

$$\zeta_l = \text{cd}(u_l K, k),$$

with

$$u_l = \frac{2l+1}{N}, \qquad l = 0, 1, \ldots L - 1, \qquad L = \frac{N-r}{2}, \qquad r = N \bmod 2,$$

while the poles are

$$\pi_l = \frac{1}{k\zeta_l} \qquad l = 0, 1, \ldots L - 1.$$

We thus have $N/2$ zeros and poles for even values of N, and $(N-1)/2$ zeros and poles for odd values of N, at finite distance from the origin; in the case of odd N, an additional pole at $w \to \infty$ and an additional zero at $w = 0$ are present in $R_N(w)$.

A factorized formula for $R_N(w)$ is

$$R_N(w) = Cw^r \prod_{l=0}^{L-1} \frac{w^2 - \zeta_l^2}{w^2 - \pi_l^2} = w^r \prod_{l=0}^{L-1} \frac{\left(w^2 - \zeta_l^2\right)\left(1 - k^2 \zeta_l^2\right)}{\left(1 - w^2 k^2 \zeta_l^2\right)\left(1 - \zeta_l^2\right)},$$

where

$$C = \prod_{l=0}^{L-1} \frac{1 - \pi_l^2}{1 - \zeta_l^2} = \prod_{l=0}^{L-1} \frac{1 - 1/(k^2 \zeta_l^2)}{1 - \zeta_l^2}$$

is a normalization constant chosen in such a way to have $R_N(1) = 1$.

From the poles and zeros of $R_N(w)$, the poles and zeros of the elliptic transfer function are derived. Recalling that we can pass from $|H(j\Omega)|^2$ to $H(s)H(-s)$ by setting $s = j\Omega$, we first consider a normalized filter ($\Omega_c = 1$ rad/s, $w = \Omega$) so as to be able to substitute $w = \Omega = -js$ into the factorized formula for $R_N(w)$. We thus get

$$R_N(-js) = C(-js)^r \prod_{l=0}^{L-1} \frac{s^2 + \zeta_l^2}{s^2 + \pi_l^2}.$$

Inserting this formula into $H(s)H(-s) = 1/[1 + \varepsilon_p^2 R_N^2(-js)]$ we find, for the normalized filter,

$$H(s)H(-s) = \frac{\prod_{l=0}^{L-1}\left(s^2 + \pi_l^2\right)^2}{\prod_{l=0}^{L-1}\left(s^2 + \pi_l^2\right)^2 + \left(C\varepsilon_p\right)^2 (-s^2)^r \prod_{l=0}^{L-1}\left(s^2 + \zeta_l^2\right)^2}.$$

From this expression, the zeros of $H(s)$ and $H(-s)$ of the normalized filter can be found; they are located at $\pm j\pi_l$. The zeros of the non-normalized elliptic filter are finally derived:

$$z_m = \pm j\Omega_c \pi_m = \pm \Omega_c \frac{j}{k\zeta_m}, \qquad m = 0, 1, \ldots L - 1.$$

As for the poles, we must require the denominator of $H(s)H(-s)$ of the normalized filter to vanish; this occurs when $R_N(w) = \pm j/\varepsilon_p$. This equation has complex solutions w_n, to which the poles of the non-normalized $H(s)H(-s)$ are connected by $p_n = j\Omega_c w_n$. The poles of the non-normalized elliptic filter that lie in the left-hand half s-plane and belong to $H(s)$ are found to be

$$p_n = \pm j\Omega_c \,\text{cd}[(u_n - jv_0)K, k] \quad \text{with} \quad u_n = \frac{2n+1}{N}, \quad n = 0, 1, \ldots L - 1.$$

Here the quantity v_0 is the real solution of the equation

$$\text{sn}(jv_0 N K_1, k_1) = \frac{j}{\varepsilon_p},$$

sn being another Jacobi elliptic function.[9] The expression of v_0 is

$$v_0 = -\frac{j}{NK_1}\text{sn}^{-1}\left(\frac{j}{\varepsilon_p}, k_1\right).$$

If N is even, $r = N \bmod 2 = 0$, and $L = N/2$. If N is odd, $r = 1$ and $L = (N - 1)/2$; in this case an additional *real* pole exists in the left-hand half s-plane, which can be obtained from the formula for p_n setting $u_n = 1$; this pole receives the index $n = L$:

$$p_L = j\Omega_c \text{cd}\left[(1 - jv_0)K, k\right] = j\Omega_c \text{sn}(jv_0 K, k), \quad with \, L = \frac{N-1}{2}.$$

In conclusion, in the left-hand half s-plane the transfer function of the elliptic filter has

- for N even, $2N/2 = N$ zeros and poles,
- for N odd, $2(N - 1)/2 = N - 1$ zeros and $2[(N - 1)/2] + 1 = N$ poles, that is, $N - r$ zeros and N poles.

Figure 8.23 shows the positions of the poles (*black dots*) and zeros (*gray squares*) of an elliptic $H(s)$ in cases with $N = 3$ and 4. The corresponding frequency response-magnitudes are those presented in Figs. 8.19 (*dotted line*) and 8.18 (*dash-dotted line*), respectively.

The factorized form of the transfer function is thus

$$H(s) = G\frac{\prod_{m=0}^{N-r-1}(s - z_m)}{\prod_{n=0}^{N-1}(s - p_n)},$$

where

---

[9]For a description of the Jacobi elliptic functions cd and sn and of their inverse functions the reader is referred to the appendix.

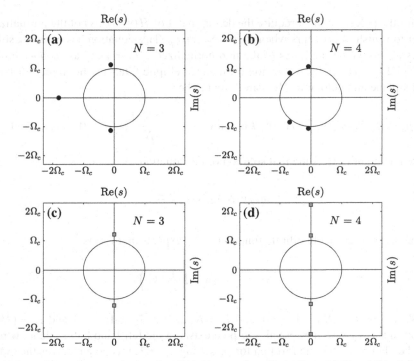

Fig. 8.23 a Poles of the transfer function $H(s)$ for an analog elliptic filter with $N = 3$ (*black dots*);
b the same for $N = 4$; **c, d** corresponding zeros (*gray squares*)

$$G = H_0 \frac{\prod_{n=0}^{N-1}(-p_n)}{\prod_{m=0}^{N-r-1}(-z_m)}$$

and

$$H_0 = (1 - \delta_p)^{1-r} = \left(\frac{1}{\sqrt{1 + \varepsilon_p^2}}\right)^{1-r} = \begin{cases} \frac{1}{\sqrt{1+\varepsilon_p^2}} & \text{for even } N, \\ 1 & \text{for odd } N. \end{cases}$$

To get the factorized form of the frequency response it is now sufficient to substitute $s = \mathrm{j}\Omega$ into the formula for $H(s)$. We could also write $H(s)$ as the gain multiplied by a ratio of polynomials in s, but we omit this formula.

In summary, the design process for an elliptic filter satisfying some given specification is the following.

- We choose Ω_p, Ω_s, δ_p and δ_s and calculate the elliptic moduli, the corresponding complete elliptic integrals and the parameter ε_p.
- Using the degree equation in the form $N = (K K_1')/(K_1 K')$ we compute N and round it up to the nearest larger integer.
- We set $\Omega_c = \Omega_p$.
- We re-evaluate k_1 from the corresponding modular equation and deduce the effective values of $\varepsilon_s = \varepsilon_p/k_1$ and $\delta_s = 1/\sqrt{\varepsilon_s^2 + 1}$.

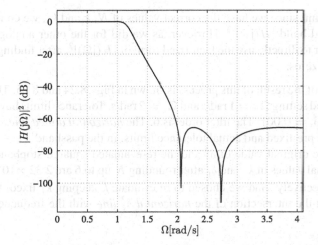

Fig. 8.24 Squared magnitude of the frequency response in dB of an analog elliptic filter meeting the following specifications: $\delta_p = \delta_s = 0.001$, $\Omega_p = 1\,\text{rad/s}$, $\Omega_s = 2\,\text{rad/s}$; consequently, $k_1 = 2.32 \times 10^{-5}$ and $k = 0.5$, giving $N = 6$; $\Omega_c = \Omega_p = 1\,\text{rad/s}$. The final value of the stopband ripple is 5.19×10^{-4}

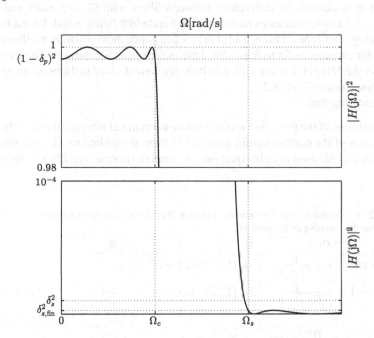

Fig. 8.25 Squared magnitude of the frequency response shown in Fig. 8.24, in normal units: expanded views of the passband and the stopband. The *horizontal dotted lines* visualize the pre-fixed and actual tolerance limits: in the passband, $(1 - \delta_p)^2$; in the stopband, the original value of δ_s^2 and the final value $\delta_{s,\text{fin}}^2$

- At this point, since we have the correct values of N, k and k_1, we could compute $R_N(w)$ and build $|H(j\Omega)|^2$. However, as we did for the other analog filters, we may prefer to directly use the factorized form of $|H(j\Omega)|^2$, after finding the filter's poles and zeros.

An example of the result of this process is shown in Figs. 8.24 and 8.25. The filter has been designed setting $\Omega_p = 1$ rad/s and $\Omega_s = 2$ rad/s. Tolerance limits have been chosen as $\delta_p = \delta_s = 0.001$. The filter order is 6. The *horizontal dotted lines* in Fig. 8.25 visualize the pre-fixed and actual tolerance limits: in the passband, $(1 - \delta_p)^2$; in the stopband, the original value of δ_s^2 and the re-evaluated squared stopband tolerance $\delta_{s,\mathrm{fin}}^2$. The final values of k_1 and δ_s after rounding N up to 6 are 2.32×10^{-5} and 5.19×10^{-4}, respectively. Had we chosen to re-evaluate k keeping k_1 fixed, the new Ω_s would fall at the intersection of the *horizontal δ_s^2 line* with the frequency response curve.

8.7   Normalized and Non-normalized Filters

We will now discuss the differences between filters with $\Omega_c = 1$ rad/s and filters with $\Omega_c \neq 1$ rad/s, tolerances and transition bandwidth being equal, for each of the four analog filter types. This is useful in practice, since most software toolboxes, like Matlab for example, refer to normalized filters for some calculations. The factorized forms for the $H(s)$ of non-normalized filters that were derived in the previous sections are summarized in Table 8.2.

   Considering that

- the positions of the poles and zeros of a non-normalized filter are those of the poles and zeros of the corresponding normalized filter, multiplied by Ω_c, and that
- the value of H_0 must remain equal passing from the normalized filter to the generic one,

Table 8.2 Factorized forms of the transfer function $H(s)$ for analog filters with generic cutoff Ω_c; see text for the meaning of the symbols

Filter	$H(s)$	G		H_0	
Butterworth	$G\dfrac{1}{\prod_{n=0}^{N-1}(s-p_n)}$	$H_0 \prod_{n=0}^{N-1}(-p_n) = \Omega_c^N$		1	
Chebyshev-I	$G\dfrac{1}{\prod_{n=0}^{N-1}(s-p_n)}$	$H_0 \prod_{n=0}^{N-1}(-p_n) = \dfrac{\Omega_c^N}{2^{N-1}\varepsilon_p}$	$\dfrac{1}{\sqrt{1+\varepsilon_p^2}}$	For N even
				1	For N odd
Chebyshev-II	$G\dfrac{\prod_{m=0}^{M-1}(s-u_m)}{\prod_{n=0}^{N-1}(s-v_n)}$	$H_0 \dfrac{\prod_{n=0}^{N-1}(-v_n)}{\prod_{m=0}^{M-1}(-u_m)}$		1	
Elliptic	$G\dfrac{\prod_{m=0}^{M-1}(s-z_m)}{\prod_{n=0}^{N-1}(s-p_n)}$	$H_0 \dfrac{\prod_{n=0}^{N-1}(-p_n)}{\prod_{m=0}^{M-1}(-z_m)}$	$\dfrac{1}{\sqrt{1+\varepsilon_p^2}}$	For N even
				1	For N odd

$M = N - r$, with $r = N \bmod 2$

if we call $H_{norm}(s)$ the transfer function of the normalized filter and write

$$H(s) = \mu H_{norm}(s),$$

where μ is a factor by which we must multiply the transfer function when Ω_c passes from 1 rad/s to a generic value, observing Table 8.2 we can draw the conclusions that follow.

- Butterworth and Chebyshev-I filters:
  there are N poles and no zeros; $\mu = \Omega_c^N$, to compensate for the variation of the factor $\prod_n(-p_n)$.
- Chebyshev-II and elliptic filters:
  there are N poles and $N - r$ zeros; if N is even, $\mu = 1$; if N is odd, then $\mu = \Omega_c$, because of the existence of one unpaired pole.

Note that the zeros that seem to be "missing" with respect to the number of poles actually exist, but are found at $\Omega \to \infty$. Therefore, in all cases, the number of poles equals the number of zeros, but some zeros can be at infinite distance from the origin in the s-plane.

Still observing Table 8.2, we can write a unified factorized formula valid for all filter types:

$$H(s) = G \frac{\prod_{m=0}^{M-1}(s - z_m)}{\prod_{n=0}^{N-1}(s - p_n)} \quad \text{with} \quad G = \frac{H_0 \prod_{n=0}^{N-1}(-p_n)}{\prod_{m=0}^{M-1}(-z_m)},$$

where H_0 and M vary according to the filter type and the value of the order, which may be even or odd; more precisely,

- for Butterworth and Chebyshev-I filters we have $M = 0$;
- for Chebyshev-II and elliptic filters we have $M = N - r$, with $r = N \bmod 2$.

In the unified factorized formula for $H(s)$, the symbols z_m and p_n have been used to respectively indicate zeros and poles of any analog filter type.

8.8 Comparison Among the Four Analog Filter Types

As a conclusion of our work on lowpass analog filters, we will compare Butterworth, Chebyshev-I, Chebyshev-II and elliptic filters, all designed on the basis of the same specifications (those adopted in the previous discussion): this comparison appears in Fig. 8.26. It can be seen that the Butterworth filter has the widest transition band and the highest order, while the elliptic filter has the narrowest transition band and the lowest order.

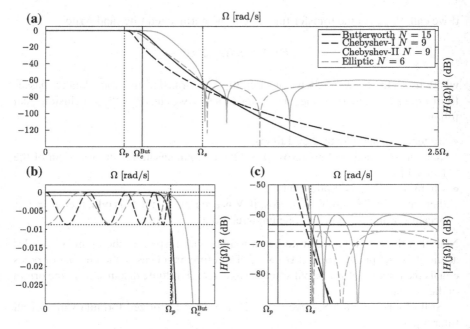

Fig. 8.26 Squared amplitude of the frequency response in dB of Butterworth, Chebyshev-I, Chebyshev-II and elliptic filters, all designed according to the same specifications, i.e., those adopted for Figs. 8.6, 8.11, 8.16 and 8.24. **a** The four responses over the range $0 \le \Omega \le 2.5\,\Omega_s$. A *vertical solid line* marks the cutoff frequency of the Butterworth filter, which is different from both Ω_p and Ω_s; the latter are represented by *vertical dotted lines*. **b** Expanded view of the filters' passbands. *Horizontal dotted lines* lines visualize tolerance limits: 0 dB (i.e. 1) and -0.0087 dB, corresponding to the original absolute tolerance $\delta_p = 0.001$. The ripple corresponds to the originally specified δ_p for all filters but the Chebyshev-II filter, for which δ_p is re-evaluated after rounding up the order to the nearest larger integer (see the *horizontal solid gray line*). **c** Expanded view of the filters' stopbands. *Horizontal lines* visualize δ_s^2 for each filter, in the same line style adopted for the frequency response curve. The stopband ripple generally does not correspond to the originally specified stopband tolerance ($\delta_s = 0.001$, i.e., 60 dB attenuation), because δ_s is re-evaluated after rounding N up to the nearest higher integer; the Chebyshev-II filter is an exception to this rule (see the *horizontal solid gray line*), because in this case δ_p is re-evaluated instead

8.9 From the Analog Lowpass Filter to the Digital One

We discussed how analog lowpass filters of four types can be designed according to some desired specifications. We are now ready to transform them into digital lowpass filters. These transformations are forms of *mapping in the complex domain*, aimed at preserving different features of the filter. For example, a method exists, aimed at preserving the shape of the impulse response, and is referred to as *impulse invariance transformation*. The analog impulse response is sampled periodically with a sampling step T_s, so as to obtain the impulse response of the digital filter; the sampling step is chosen so as to capture correctly the shape of the analog impulse response, but since no analog filter with finite order can be exactly bandlimited, frequency response distortion may occur due to aliasing. This method is useful only when the analog lowpass filter to be transformed is free from ripple in the stopband, a case in which aliasing turns out to be tolerable. Another method is aimed at converting a representation of the analog filter in the form of a differential equation into a corresponding representation in the form of an LCCDE, and is referred to as the *finite difference approximation technique*. However, the method most often adopted in practice is the *bilinear transformation* that preserves the characteristics of the transfer function when $H(s)$ is transformed into the corresponding $H(z)$.

8.9.1 Bilinear Transformation

In the following discussion we will denote the frequency response of the analog filter that is converted into a digital filter by $H_a(j\Omega)$. The corresponding gain will be indicated by G_a. Digital IIR filters are named after their analog parents.

The bilinear transformation uses the variable change[10]

$$s = \frac{2}{T_s} \frac{1 - z^{-1}}{1 + z^{-1}} = \frac{2}{T_s} \frac{z - 1}{z + 1} \quad \Rightarrow \quad z = \frac{1 + sT_s/2}{1 - sT_s/2},$$

where T_s is a parameter that represents the sampling interval with which continuous time is supposed to be converted into discrete time. The value of T_s is irrelevant to our purposes, and we can as well set $T_s = 1$ s. This transformation is also referred to as the *linear fractional transformation*, for obvious reasons. If we eliminate fractions from its expression we get

$$\left(\frac{T_s}{2}\right) sz + \left(\frac{T_s}{2}\right) s - z + 1 = 0,$$

which is linear in each complex variable when the other one is fixed. The transformation is thus bi-linear in s and z. The mapping in the complex domain operated by the bilinear transformation is sketched in Fig. 8.27. Inserting $s = \sigma + j\Omega$ into the transformation we get

$$z = \frac{1 + \sigma T_s/2 + j\Omega T_s/2}{1 - \sigma T_s/2 - j\Omega T_s/2};$$

---

[10]This transformation, and other analogous transformations discussed later, are often expressed in literature in terms of z^{-1} (unit delay), a form that is useful when the $H(z)$ to be obtained is more conveniently written as the ratio of polynomials in z^{-1}, or even in the corresponding factorized form. We will also, however, write the form in terms of z, considering that in this book, in most cases ratios of polynomials in z or corresponding factorized forms are used.

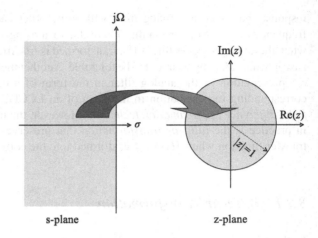

Fig. 8.27 *Mapping* from the s-plane to the z-plane in the bilinear transformation

we thus see that

$$\sigma < 0 \quad \Rightarrow \quad |z| = \left| \frac{1 + \sigma T_s/2 + j\Omega T_s/2}{1 - \sigma T_s/2 - j\Omega T_s/2} \right| < 1,$$

$$\sigma = 0 \quad \Rightarrow \quad |z| = \left| \frac{1 + j\Omega T_s/2}{1 - j\Omega T_s/2} \right| = 1,$$

$$\sigma > 0 \quad \Rightarrow \quad |z| = \left| \frac{1 + \sigma T_s/2 + j\Omega T_s/2}{1 - \sigma T_s/2 - j\Omega T_s/2} \right| > 1.$$

Therefore

1. the whole left-hand half s-plane is mapped into the interior of the unit circle of the z-plane. As a consequence, the bilinear transformation is a stable transformation, in the sense that a stable analog filter leads to a stable digital filter;
2. the imaginary axis of the s-plane is mapped onto the unit circle of the z-plane. As Ω varies from $-\infty$ to $+\infty$, a single rotation around the origin of the z-plane is performed along the unit circle, so that ω varies from $-\pi$ to π, and the relation between Ω and ω is univocal: there is no aliasing in the frequency domain. Substituting $\sigma = 0$ into the transformation and observing that on the unit circle we can write

$$z = e^{j\omega} = \frac{1 + j\Omega T_s/2}{1 - j\Omega T_s/2},$$

we get

$$\omega = 2 \arctan \left(\frac{\Omega T_s}{2} \right), \qquad \Omega = \frac{2}{T_s} \tan \left(\frac{\omega}{2} \right).$$

The relation between Ω and ω is nonlinear, and Ω is *warped into* ω, as illustrated, with $T_s = 1$ s, in Fig. 8.28.

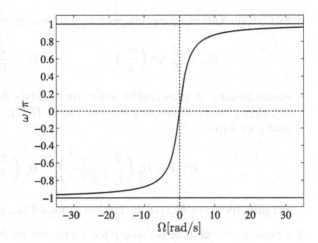

Fig. 8.28 Warping of the Ω-axis into the ω-axis in the bilinear transformation

As a first example, let us consider the transfer function

$$H_a(s) = (s + 1)/s^2 + 5s + 6$$

and transform it into $H(z)$ using the bilinear transformation with $T_s = 1$ s:

$$H(z) = H_a\left(\frac{2}{T_s}\frac{1 - z^{-1}}{1 + z^{-1}}\right) = H_a\left(2\frac{1 - z^{-1}}{1 + z^{-1}}\right) =$$

$$= \frac{2\left(1 - z^{-1}\right)/\left(1 + z^{-1}\right) + 1}{\left[2\left(1 - z^{-1}\right)/\left(1 + z^{-1}\right)\right]^2} + 5\left[2\left(1 - z^{-1}\right)/\left(1 + z^{-1}\right)\right] + 6.$$

Simplifying we obtain

$$H(z) = \frac{3 + 2z^{-1} - z^{-2}}{20 + 4z^{-1}} = \frac{0.15 + 0.1z^{-1} - 0.05z^{-2}}{1 + 0.2z^{-1}}.$$

As an alternative, we could have started from $s = (2/T_s)[(z - 1)/(z + 1)]$ to get a ratio of polynomials in z.

8.9.2 Design Procedure

Suppose we want to design an IIR digital lowpass filter of a certain type (e.g., a Butterworth filter) using the Butterworth analog design procedure and then applying the bilinear transformation. Then we must proceed as follows:

- we specify ω_p, ω_s, δ_p, and δ_s of the desired digital lowpass filter;
- we choose a value for T_s. We can arbitrarily set $T_s = 1$ s;

- we perform a *pre-warping*, i.e., we transform ω_p and ω_s into Ω_p and Ω_s:

$$\Omega_p = \frac{2}{T_s} \tan\left(\frac{\omega_p}{2}\right), \qquad \Omega_s = \frac{2}{T_s} \tan\left(\frac{\omega_s}{2}\right);$$

- we design the analog lowpass filter of the desired type that meets the specifications Ω_p, Ω_s, δ_p and δ_s; in this way we obtain some $H_a(s)$;
- finally we write

$$H(z) = H_a\left(\frac{2}{T_s}\frac{1 - z^{-1}}{1 + z^{-1}}\right) = H_a\left(\frac{2}{T_s}\frac{z - 1}{z + 1}\right)$$

and simplify to get $H(z)$ in the form of a rational function in z^{-1} or in z.

It is however useful to derive an explicit expression for $H(z)$, in the zero-pole-gain form, as a function of the analog gain G_a and of the zeros z_m and poles p_n of the analog filter. For this purpose we go back to the general factorized form

$$H_a(s) = G_a \frac{\prod_{m=0}^{M-1} (s - z_m)}{\prod_{n=0}^{N-1} (s - p_n)}$$

with

$$G_a = H_0 \frac{\prod_{n=0}^{N-1} (-p_n)}{\prod_{m=0}^{M-1} (-z_m)},$$

where H_0 and M depend on the filter type and on the order being even or odd, and substitute, assuming $T_s = 1$ s,

$$s = 2\frac{1 - z^{-1}}{1 + z^{-1}} = 2\frac{z - 1}{z + 1}.$$

We get

$$H(z) = G_a \frac{\prod_{m=0}^{M-1} [2(z - 1)/(z + 1) - z_m]}{\prod_{n=0}^{N-1} [2(z - 1)/(z + 1) - p_n]} =$$

$$= G_a \frac{(z + 1)^{-M} \prod_{m=0}^{M-1} [2(z - 1) - z_m(z + 1)]}{(z + 1)^{-N} \prod_{n=0}^{N-1} [2(z - 1) - p_n(z + 1)]} =$$

$$= G_a (z + 1)^{N-M} \frac{\prod_{m=0}^{M-1} [z(2 - z_m) - (2 + z_m)]}{\prod_{n=0}^{N-1} [z(2 - p_n) - (2 + p_n)]},$$

that is,

$$H(z) = G_a \frac{\prod_{m=0}^{M-1} (2 - z_m)}{\prod_{n=0}^{N-1} (2 - p_n)} (z + 1)^{N-M} \frac{\prod_{m=0}^{M-1} [z - (2 + z_m)/(2 - z_m)]}{\prod_{n=0}^{N-1} [z - (2 + p_n)/(2 - p_n)]}.$$

In conclusion we can write

$$H(z) = G\,(z+1)^{N-M}\,\frac{\prod_{m=0}^{M-1}\left(z - z_m^d\right)}{\prod_{n=0}^{N-1}\left(z - p_n^d\right)},$$

having set

$$G = G_a\,\frac{\prod_{m=0}^{M-1}\left(2 - z_m\right)}{\prod_{n=0}^{N-1}\left(2 - p_n\right)}$$

for the gain of the digital filter, and z_m^d, p_m^d for the zeros and the poles of the digital filter, respectively. The zeros and poles of the digital filter are related to the zeros and poles of the analog filter by

$$z_m^d = \frac{2 + z_m}{2 - z_m}, \qquad\qquad p_n^d = \frac{2 + p_n}{2 - p_n}.$$

Note that in $H(z)$ the factor $(z+1)^{N-M}$ introduces $N - M$ zeros at $z = -1$ that correspond to $N - M$ zeros of the analog filter at $s = \infty$, so that we have as many zeros as poles: N zeros and N poles. This factorized form also highlights the fact that the filter order does not change passing from the analog to the digital domain.

The frequency responses of all the examples described in the next subsection can be obtained from this general factorized formula, by setting $z = e^{j\omega}$.

8.9.3   Examples

By the bilinear transformation we can transform the analog Butterworth, Chebyshev and elliptic lowpass filters designed in the previous sections into digital IIR filters. All the analog-filter-design examples given in Sects. 8.4, 8.5 and 8.6 started from the following digital specifications: $\delta_p = \delta_s = 0.001$, equivalent to $r_p = 0.0087$, and $r_s = 60$; $\Omega_p = 1$ rad/s and $\Omega_s = 2$ rad/s. These analog frequencies correspond, with $T_s = 1$ s, to choosing for the target digital filters the characteristic frequencies $\omega_p = 2\arctan(1/2) = 0.2952\pi$ and $\omega_s = 2\arctan(1) = 0.5\pi$.

1. Applying the bilinear transformation to the analog Butterworth lowpass filter shown in Figs. 8.6 and 8.7 we obtain a digital IIR Butterworth filter with order $N = 15$. Details of the squared-magnitude frequency response of the designed digital filter are shown in Fig. 8.29a, b as *solid lines*. The phase response, the phase delay and the group delay are shown as *solid lines* in Fig. 8.29c–e, respectively.
2. Applying the bilinear transformation to the analog Chebyshev-I lowpass filter shown in Figs. 8.11 and 8.12 we obtain a digital IIR Chebyshev-I filter with order $N = 9$. Details of the squared-magnitude frequency response of the designed digital filter are shown in Fig. 8.29a, b as *dashed lines*. The phase response, the phase delay and the group delay are shown as *dashed lines* in Fig. 8.29c–e, respectively.

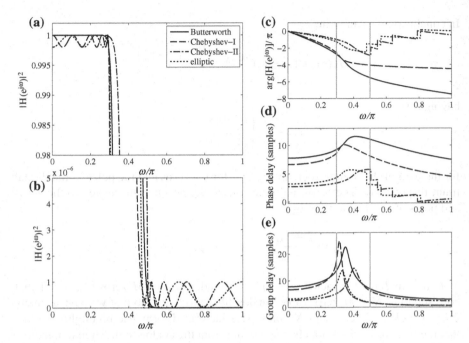

Fig. 8.29 Bilinear transformation. **a, b** Details of the squared-magnitude frequency responses of the IIR digital filters obtained transforming the corresponding analog filters designed in Sects. 8.4–8.6 (see text). Butterworth filter: order 15 (*solid line*); Chebyshev-I filter: order 9 (*dashed line*); Chebyshev-II filter: order 9 (*dot-dashed line*); elliptic filter: order 6 (*dotted line*). Corresponding **c** phase response, **d** phase delay and **e** group delay. In panels c, d and e, the *vertical gray lines* mark ω_p and ω_s

3. Applying the bilinear transformation to the analog Chebyshev-II lowpass filter shown in Figs. 8.16 and 8.17 we get a digital IIR Chebyshev-II filter with order $N = 9$. Details of the squared-magnitude frequency response of the designed digital filter are shown in Fig. 8.29a, b as *dot-dashed lines*. The phase response, the phase delay and the group delay are shown as *dot-dashed lines* in Fig. 8.29c–e, respectively.

4. Applying the bilinear transformation to the analog elliptic lowpass filter shown in Figs. 8.24 and 8.25 we obtain a digital IIR elliptic filter with order $N = 6$. Details of the squared-magnitude frequency response of the designed digital filter are shown in Fig. 8.29a, b as *dotted lines*. The phase response, the phase delay and the group delay are shown as *dotted lines* in Fig. 8.29c–e, respectively.

Inspection of Fig. 8.29a, b reveals a perfect correspondence between the digital squared-magnitude responses and the analog ones presented in the previous sections. The phase responses (Fig. 8.29c) appear markedly non-linear. The π-jumps in the Chebyshev-II and elliptic cases are due to zeros of the magnitude response in the stopband, which for these two types of filters is not monotonically decreasing. Phase and group delay (Fig. 8.29d and e, respectively) is, in all cases, far from being constant in the filter's passband.

8.10 Frequency Transformations

We learned how to design digital IIR lowpass filters from the corresponding analog lowpass filters. Now we must see how filters with different frequency selectivity, i.e., highpass, bandpass and bandstop filters, can be designed. They are obtained by transforming the frequency axis of an IIR lowpass filter, through a proper transformation applied to the independent variable z. These algebraic transformations, which are very similar to the bilinear one, were introduced by Constantinides (1970).

The procedure for the design of such filters includes the following steps:

- in general, from the specifications given for the desired filter, the specification of the intermediate lowpass digital IIR filter are deduced. From these specifications, those of the analog lowpass filter from which the process must start are calculated. Except in particular cases that will be discussed separately, we are free to choose the lowpass analog cutoff frequency Ω_c arbitrarily, and normally we will set $\Omega_c = 1$ rad/s;
- the analog lowpass filter is designed according to the transformed specification, using one of the approximation approaches discussed in Sect. 8.3;
- through bilinear transformation, the analog lowpass filter is converted into an intermediate digital IIR lowpass filter;
- the algebraic transformations mentioned above are applied to get the final digital IIR highpass, or bandpass, or bandstop filter.

Obviously, a highpass Chebyshev-I filter is obtained starting from a lowpass Chebyshev-I filter, etc.

We will now investigate what properties these frequency transformations must possess. Let us consider the transfer function of a lowpass digital IIR filter: in order to avoid ambiguities, we will write it as $H_{\text{LP}}(Z)$, with Z indicating the corresponding independent variable. We will then call $H(z)$ the transfer function of the final IIR filter (a highpass filter, for example). We seek a transformation of the kind[11]

$$Z = R(z),$$

such that

$$H(z) = H_{\text{LP}}(Z) \quad \text{for} \quad Z = R(z).$$

The transformation from the lowpass to the final filter consists of substituting $Z = R(z)$ in $H_{\text{LP}}(Z)$. Assuming that $H_{\text{LP}}(Z)$ is a rational function with real coefficients, representing a stable and causal filter, we want $H(z)$ to possess the same features. This implies that

1. $R(z)$ must be a *rational function* of z with real coefficients;
2. the unit circle in the Z-plane must be mapped onto the unit circle in the z-plane;

---

[11]Usually, in literature, these transformation are given in terms of z^{-1} rather than in terms of z, because the form of $H(z)$ as a rational function in z^{-1} is considered. Here we prefer to use the variable z directly.

3. the interior of the unit circle in the Z-plane must be mapped onto the interior of the unit circle in the z-plane;
4. the exterior of the unit circle in the Z-plane must be mapped onto the exterior of the unit circle in the z-plane.

The functional form of the $R(\cdot)$ that satisfies these requirements turns out to be

$$Z = R(z) = \pm \prod_{i=1}^{n} \frac{z - c_i^*}{1 - c_i z},$$

with n zeros in c_i^* and n poles in $1/c_i$. The coefficients c_i are not necessarily all complex and must satisfy $|c_i| < 1$ for stability and for ensuring the correspondence between the interiors of the two unit circles; choosing n and the coefficients c_i in proper ways, a wide range of transformations can be achieved.

In order to justify these statements, let us observe that $R(z)$ can be seen as the transfer function of a filter: an *allpass filter*, i.e., a filter with unitary amplitude response at any frequency, but with a phase response able to change the position along the frequency axis of the lowpass filter characteristics and turn them into those required for the final filter. For example, the bandpass of the lowpass filter may be turned into a stopband, etc. In other words, the substitution of Z with $R(z)$ in $H_{\text{LP}}(Z)$, which modifies the lowpass digital filter, can be interpreted as a filtering operation.

Let us assume that the poles $p_i = 1/c_i$ of $R(z)$ are strictly *external* to the unit circle in the z-plane: $|c_i| < 1$. Then we write $Z = re^{j\theta}$, $z = \rho e^{j\omega}$ and $c_i = C_i e^{j\phi_i}$, thus obtaining

$$re^{j\theta} = \pm \prod_{i=1}^{n} \frac{\rho e^{j\omega} - C_i e^{-j\phi_i}}{1 - C_i \rho e^{j(\omega+\phi_i)}}.$$

The squared modulus of this expression is

$$r^2 = \prod_{i=1}^{n} \frac{\rho^2 + C_i^2 - 2\rho C_i \cos(\omega + \phi_i)}{1 + \rho^2 C_i^2 - 2\rho C_i \cos(\omega + \phi_i)},$$

to which each pole of $R(z)$ contributes by one factor. Then we get the following picture:

$$r > 1 \Rightarrow \rho^2 + C_i^2 > 1 + \rho^2 C_i^2 \rightarrow (1 - C_i^2)\rho^2 > (1 - C_i^2) \Rightarrow \rho > 1,$$
$$r = 1 \Rightarrow \rho^2 + C_i^2 = 1 + \rho^2 C_i^2 \rightarrow (1 - C_i^2)\rho^2 = (1 - C_i^2) \Rightarrow \rho = 1,$$
$$r < 1 \Rightarrow \rho^2 + C_i^2 < 1 + \rho^2 C_i^2 \rightarrow (1 - C_i^2)\rho^2 < (1 - C_i^2) \Rightarrow \rho < 1,$$

that describes exactly the behavior we desire.

If we think of the digital lowpass transfer function in the zero-pole-gain form, we understand that a frequency transformation of this kind replaces each pole and each zero of the lowpass filter with a number of poles and zeros that is equal to the order of the *mapping filter* with transfer function $R(z)$. Thus poles and zeros of the final filter will be found at different positions with respect to the lowpass intermediate filter,[12] but they will always be internal to the unit circle in the z-plane, thanks to the constraints we imposed on the transformation. So the transformation actually turns an IIR stable filter into another IIR stable filter.

The transformations that are most widely used correspond to the following parameters:

- $n = 1$ for lowpass $\rightarrow$ lowpass[13] and lowpass $\rightarrow$ highpass transformations. In this case, only one coefficient is present, which we will denote by α and that must be real:

$$R(z) = \pm \frac{z - \alpha}{1 - \alpha z};$$

the plus sign gives a lowpass $\rightarrow$ lowpass transformation, while the minus sign produces a lowpass $\rightarrow$ highpass transformation;

- $n = 2$ for lowpass $\rightarrow$ bandpass and lowpass $\rightarrow$ bandstop transformations. In this case, there are two coefficients, c_1 and c_2, and

$$R(z) = \pm \frac{(z - c_1^*)(z - c_2^*)}{(1 - c_1 z)(1 - c_2 z)};$$

the plus sign gives a lowpass $\rightarrow$ bandstop transformation, while the minus sign produces a lowpass $\rightarrow$ bandpass transformation. Indeed, the sign in front of $R(z)$ determines if the frequency $\theta = 0$ of the lowpass filter is moved elsewhere, or if the Nyquist frequency $\theta = \pi$ is moved instead: more precisely,

- the minus sign in front of $R(z)$ produces the condition known as *zero-frequency mobility*, or *DC mobility*[14];
- the plus sign in front of $R(z)$ produces the condition known as *Nyquist mobility*.

$R(z)$ has two zeros in $c_{1,2}^*$ and two poles in $1/c_{1,2}$. If we set the denominator of R_z to zero, we find the poles $1/c_{1,2}$. We expand the product in the denominator and set $\alpha_1 = -(c_1 + c_2)$, $\alpha_2 = c_1 c_2$. We require α_1 and α_2 to be real and write:

$$(1 - c_1 z)(1 - c_2 z) = c_1 c_2 z^2 - (c_1 + c_2)z + 1 = \alpha_2 z^2 + \alpha_1 z + 1 = 0.$$

---

[12]The new positions of poles and zeros can be found considering a generic factor $(Z - q_i)$ contributing to the lowpass transfer function in the zero-pole gain-form. Here q_i represents either a zero, or a pole, and of course the factor can be in the numerator or in the denominator, respectively. We can then substitute Z with $R(z)$ and set the factor equal to zero, to find q_i.

[13]The final lowpass is obviously required to have a different cutoff frequency.

[14]DC stands for *direct current*: here it means "zero frequency".

The roots of this equations are the poles of $R(z)$:

$$\frac{1}{c_{1,2}} = \frac{-\alpha_1 \pm \sqrt{\Delta}}{2\alpha_2} \quad \text{with} \quad \Delta = \alpha_1^2 - 4\alpha_2.$$

But since we imposed $\alpha_2 = c_1 c_2$ we can write

$$c_{1,2} = \frac{-\alpha_1 \mp \sqrt{\Delta}}{2}.$$

We can thus see that

- if $\Delta > 0$, then the coefficients $c_{1,2}$ are real and distinct;
- if $\Delta = 0$, then the coefficients $c_{1,2}$ are real and coincident;
- if $\Delta < 0$, then the coefficients $c_{1,2}$ are complex, but form a complex-conjugate pair:

$$c_1 = c_2^*, \qquad c_2 = c_1^*.$$

Hence in general we can write

$$\alpha_1 = -(c_1 + c_2) = -(c_1^* + c_1^*) = -(c_2^* + c_2^*) = -2\mathrm{Re}(c_1) = -2\mathrm{Re}(c_2),$$
$$\alpha_2 = c_1 c_2 = c_1 c_1^* = |c_1|^2 = c_2 c_2^* = |c_2|^2.$$

The numerator of $R(z)$ can be written as

$$\pm (z - c_1^*)(z - c_2^*) = \pm \left[z^2 - (c_1^* + c_2^*)z + c_1^* c_2^*\right] = \pm (z^2 + \alpha_1 z + \alpha_2),$$

and in conclusion we can write for $n = 2$,

$$R(z) = \pm \frac{z^2 + \alpha_1 z + \alpha_2}{\alpha_2 z^2 + \alpha_1 z + 1}.$$

Note that when $|c_{1,2}| < 1$ we always have $|\alpha_1| < 2$ and $|\alpha_2| < 1$.

In summary, the transformations of a lowpass filter into another lowpass filter with different cutoff frequency (a case that we mention for completeness but that will not be discussed) and of a lowpass filter into a highpass filter are first-order transformations, while those of a lowpass filter into a bandpass or bandstop filter are second-order transformations. In terms of frequency, in the four cases the functional form of $R(z)$ implies a different relation between θ and ω, i.e., a different mapping of the frequency axis. Moreover, the order of the transformation determines the order of the final filter, which does not vary with respect to the intermediate lowpass case when the transformation has order 1, while it doubles when the transformation has order 2.

Table 8.3 Frequency transformations for frequency-selective digital IIR filters: θ_c is the cutoff frequency of the intermediate digital lowpass filter; ω_c concerns final lowpass or highpass filters; the pair ω_{c1}, ω_{c2} concerns final bandpass or bandstop filters; the transformation from a lowpass filter to another lowpass filter with different cutoff frequency is reported for completeness but is not discussed in the text

Filter type	Transformation	Associated parameters
Lowpass	$Z = \frac{z-\alpha}{1-\alpha z}$	$\alpha = \frac{\sin[(\theta_c-\omega_c)/2]}{\sin[(\theta_c+\omega_c)/2]}$
Highpass	$Z = -\frac{z-\alpha}{1-\alpha z}$	$\alpha = \frac{\cos[(\theta_c+\omega_c)/2]}{\cos[(\theta_c-\omega_c)/2]}$
Bandpass	$Z = -\frac{z^2+\alpha_1 z+\alpha_2}{\alpha_2 z^2+\alpha_1 z+1}$	$\alpha_1 = -2\alpha\beta/(\beta+1)$
		$\alpha_2 = (\beta-1)/(\beta+1)$
		$\alpha = \frac{\cos[(\omega_{c2}+\omega_{c1})/2]}{\cos[(\omega_{c2}-\omega_{c1})/2]}$
		$\beta = \cot\left(\frac{\omega_{c2}-\omega_{c1}}{2}\right)\tan\frac{\theta_c}{2}$
Bandstop	$Z = \frac{z^2+\alpha_1 z+\alpha_2}{\alpha_2 z^2+\alpha_1 z+1}$	$\alpha_1 = -2\alpha/(\beta+1)$
		$\alpha_2 = -(\beta-1)/(\beta+1)$
		$\alpha = \frac{\cos[(\omega_{c2}+\omega_{c1})/2]}{\cos[(\omega_{c2}-\omega_{c1})/2]}$
		$\beta = \tan\left(\frac{\omega_{c2}-\omega_{c1}}{2}\right)\tan\frac{\theta_c}{2}$

The four frequency transformations are summarized in Table 8.3. In the case of a second-order $R(z)$, α_1 and α_2 are usually expressed as functions of two other parameters, which are indicated with α and β. The last column in Table 8.3 gives the expressions of these two parameters that obviously depend on the design characteristic frequencies: the cutoff frequency θ_c of the intermediate lowpass filter and

- the cutoff frequency ω_c of the final lowpass/highpass filter in the first two transformations,
- the cutoff frequencies ω_{c1} and ω_{c2} of the final bandpass/bandstop filter in the last two transformations.

The formulas listed in the third column of Table 8.3 will be derived in the following subsections.

8.10.1 From a Lowpass to a Highpass Filter

This transformation implies the mapping sketched in Fig. 8.30; thus on the two frequency axes, $\theta = -\pi$ leads to $\omega = 0$, $\theta = -\theta_c$ leads to $\omega = \omega_c$, and $\theta = 0$ leads to $\omega = \pi$. Since this is a first-order transformation, a value of θ corresponds to a single value of ω; the negative half θ-axis covers the positive half ω-axis, and vice-versa.

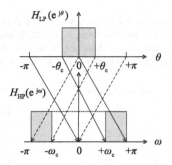

Fig. 8.30 Mapping of the frequency axis in the transformation from a lowpass to a highpass IIR filter, illustrated using the shapes of ideal filters: the values of θ and ω that correspond to one another are connected by *arrows*. This scheme derives from the functional form of $R(z)$ and holds for any α with $|\alpha| < 1$

Let us now see how the dependence of α on θ_c and ω_c (Table 8.3) can be derived. We start writing the transformation

$$Z = -\frac{z - \alpha}{1 - \alpha z}$$

on the unit circles, obtaining

$$e^{j\theta} = -\frac{e^{j\omega} - \alpha}{1 - \alpha e^{j\omega}}.$$

We then form

$$e^{j\theta} \pm 1 = -\frac{e^{j\omega} - \alpha}{1 - \alpha e^{j\omega}} \pm 1 = \frac{-e^{j\omega} + \alpha \pm \left(1 - \alpha e^{j\omega}\right)}{1 - \alpha e^{j\omega}} =$$

$$= \frac{-\left(e^{j\omega} \mp 1\right) \mp \alpha \left(e^{j\omega} \mp 1\right)}{1 - \alpha e^{j\omega}} = -(1 \pm \alpha)\frac{e^{j\omega} \mp 1}{1 - \alpha e^{j\omega}}$$

from which we get

$$\frac{e^{j\theta} - 1}{e^{j\theta} + 1} = \frac{(1 - \alpha)\left(e^{j\omega} + 1\right)}{(1 + \alpha)\left(e^{j\omega} - 1\right)}.$$

Now, if we take the imaginary part of the left-hand member we have

$$\text{Im}\left(\frac{e^{j\theta} - 1}{e^{j\theta} + 1}\right) = \frac{\sin \theta}{1 + \cos \theta} = \tan \frac{\theta}{2}.$$

In a similar way, the imaginary part of the factor $\frac{e^{j\omega} + 1}{e^{j\omega} - 1}$ in the right-hand member gives a term $-\frac{\sin \omega}{1 - \cos \omega} = -\cot \frac{\omega}{2}$. Therefore we can write

$$\tan \frac{\theta}{2} = \frac{\alpha - 1}{\alpha + 1} \cot \frac{\omega}{2}.$$

This equation must be solved with respect to α:

$$(\alpha + 1) \frac{\sin \theta/2}{\cos \theta/2} = (\alpha - 1) \frac{\cos \omega/2}{\sin \omega/2},$$

$$\frac{\cos \omega/2}{\sin \omega/2} + \frac{\sin \theta/2}{\cos \theta/2} = \alpha \left(\frac{\cos \omega/2}{\sin \omega/2} - \frac{\sin \theta/2}{\cos \theta/2} \right),$$

hence

$$\alpha = \frac{\cos \theta/2 \cos \omega/2 + \sin \theta/2 \sin \omega/2}{\cos \theta/2 \cos \omega/2 - \sin \theta/2 \sin \omega/2} =$$

$$= \frac{\cos \left(\frac{\theta - \omega}{2} \right)}{\cos \left(\frac{\theta + \omega}{2} \right)}.$$

The parameter α relates any pair (θ, ω) of corresponding values. The value of α is then determined establishing a pair of values, known a priori, that must correspond to one another: the design frequencies $-\theta_c$ and ω_c (see the *solid arrows* in Fig. 8.30). In other words, the relation

$$e^{-j\theta_c} = -\frac{e^{j\omega_c} - \alpha}{1 - \alpha e^{j\omega_c}},$$

must hold. The formula shown in Table 8.3 can thus be immediately deduced:

$$\alpha = \frac{\cos \left(\frac{-\theta_c - \omega_c}{2} \right)}{\cos \left(\frac{-\theta_c + \omega_c}{2} \right)} = \frac{\cos \left(\frac{\theta_c + \omega_c}{2} \right)}{\cos \left(\frac{\theta_c - \omega_c}{2} \right)}.$$

We may note that as expected, the transformation as appears on the unit circles,

$$e^{j\theta} = -\frac{e^{j\omega} - \alpha}{1 - \alpha e^{j\omega}},$$

also implies that

- $\omega = 0$ corresponds to $e^{j\theta} = -1, \theta = \pm\pi$;
- $\omega = \pm\pi$ corresponds to $e^{j\theta} = 1, \theta = 0$.

We may incidentally mention that the transformation of a lowpass filter into another lowpass filter with different cutoff frequency proceeds along exactly the same lines. The only difference is that the correspondence is between the interval $[0, \pi]$ of θ and the interval $[0, \pi]$ of ω; θ_c leads to ω_c, the two zero-frequencies correspond to each other and the two Nyquist frequencies correspond to each other.

Fig. 8.31 Curves of $\theta(\omega)$ in the transformation from a lowpass to a highpass IIR filter, for various values of α; to avoid cluttering, only the range $[0, \pi]$ of ω is plotted

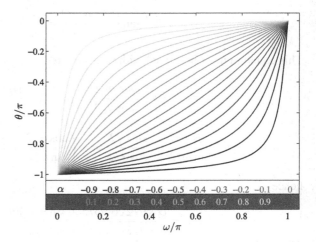

The relation between ω of the highpass filter and θ of the lowpass filter is non-linear. This can be seen graphically, considering a particular value of α, letting ω vary from 0 to π and deriving the corresponding θ values from the equations $Z = -(z - \alpha)/(1 - \alpha z)$, $\theta = \arg(Z)$. Repeating this procedure for various values of α, the curves shown in Fig. 8.31 can be found that represent another way of illustrating the present mapping. The case $\alpha = 0$ is special, because it produces a linear relation between ω and θ: indeed, $\alpha = 0$ implies $\cos[(\theta - \omega)/2] = 0$. Since for $0 \le \omega \le \pi$ we have $-\pi \le \theta \le 0$, the cosine in the numerator of α will vanish for $\theta - \omega = -\pi$, $\theta = \omega - \pi$. The cutoff frequencies $-\theta_c$ and ω_c are then related by $\theta_c + \omega_c = \pi$, and the transformation is reduced to $Z = -z$.

The value of α for a particular design depends on θ_c and ω_c. However, if in the analog domain we start assigning $\Omega_c = 1$ rad/s, we get $\theta_c = 2 \tan(\Omega_c/2) = 0.2952\,\pi$, and α turns out to depend on ω_c only. Figure 8.32 shows α as a function of ω_c in this case; according to the prescribed value of ω_c, α can be positive, zero or negative, but its absolute value is always less than 1. The transition from positive to negative values of α takes place at $\omega_c = \pi - \theta_c = 0.7048\,\pi$ (see the *dotted lines* in Fig. 8.32), as can be understood remembering that $\alpha = 0$ gives $\theta_c + \omega_c = \pi$.

Now let us make an example of this frequency transformation. Imagine that we want a highpass Chebyshev-I filter satisfying the following specifications: $\omega_c = \omega_p = 0.6\,\pi$; transition bandwidth $\Delta\omega = \omega_c - \omega_s = 0.1\,\pi$, hence $\omega_s = 0.5\,\pi$; $\delta_p = \delta_s = 0.001$, hence $r_p = 0.0087$, $r_s = 60$. We start from a lowpass analog Chebyshev-I filter with $\Omega_c = \Omega_p = 1$ rad/s. The bilinear transformation with $T_s = 1$ s leads us to a digital lowpass IIR filter having $\theta_c = \theta_p = 2 \arctan(\Omega_c/2) = 0.2952\,\pi$. We can now substitute into α the values $\theta_c = 0.2952\,\pi$ and $\omega_c = 0.6\,\pi$, thus getting $\alpha = 0.1847$. At this point we need θ_s, from which to deduce Ω_s for the analog design; the frequency θ_s of the intermediate digital lowpass filter must be such that after frequency transformation we have $\omega_s = \omega_c - \Delta\omega = 0.5\,\pi$, as desired. We can write, in agreement with the mapping of Fig. 8.30,

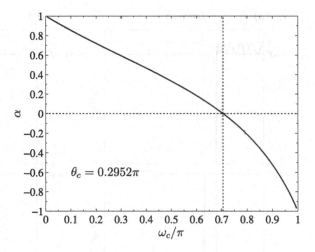

Fig. 8.32 The parameter α for the transformation from a lowpass to a highpass IIR filter, as a function of ω_c: θ_c is constant, because $\Omega_c = 1\,\text{rad/s}$ is assumed for the analog lowpass filter. The *dotted lines* mark the coordinates $\alpha = 0$, $\omega_c = 0.7048\,\pi$

$$e^{-j\theta_s} = -\frac{e^{j\omega_s} - \alpha}{1 - \alpha e^{j\omega_s}}.$$

This gives $\theta_s = -\arg\left(e^{-j\theta_s}\right) = 0.3837\,\pi$, from which we get $\Omega_s = 2\tan\left(\theta_s/2\right) = 1.3764\,\text{rad/s}$.

We can also adopt a different approach to get θ_s and then Ω_s. A given value of α relates univocally any value of ω to the corresponding value of θ. Hence the pair of values $(-\theta_s, \omega_s)$, when substituted into the expression of α, must give $\alpha = 0.1847$. We can thus solve the transcendental equation

$$\alpha - \frac{\cos\left(\frac{\theta_s + \omega_s}{2}\right)}{\cos\left(\frac{\theta_s - \omega_s}{2}\right)} = 0$$

to find the unknown θ_s. This can be done numerically by iterative methods, for example by the Newton-Raphson method (Pao 1999).[15]

Similar considerations hold for filters that are not Chebyshev-I. However, in the case of a Butterworth filter we must face another difficulty: $\Omega_c \neq \Omega_p$, and precisely, $\Omega_c > \Omega_p$. On the other hand, the analog Butterworth lowpass filter design starts from Ω_p, Ω_s and from the tolerances. But in order to find Ω_p from some given Ω_c (in the present case, 1 rad/s) through the formula given in Sect. 8.4, the filter order should already be known. A simple way of finding Ω_p is to proceed by trial-and-error: for instance, we can consider a dense set of values of Ω_p between 0 and $\Omega_c = 1\,\text{rad/s}$. From the given values of δ_p and δ_s, and for each value of Ω_p, we can find the integer N and the corresponding Ω_c guessed value. We will proceed until this guessed value becomes equal to 1; the corresponding Ω_p and the order N found in this way are the correct values for the analog design process.

---

[15]This method also requires writing the derivative of the transcendental equation with respect to θ_s.

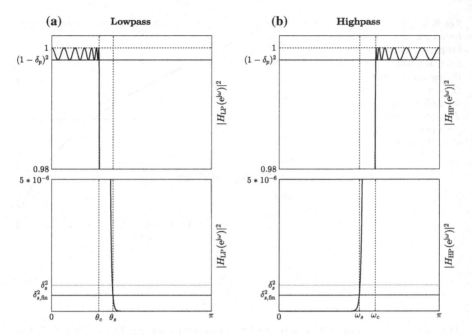

Fig. 8.33 Details of **a** the frequency response of a lowpass Chebyshev-I IIR filter and **b** the frequency response of the corresponding highpass Chebyshev-I IIR filter satisfying the following specifications: $\omega_c = \omega_p = 0.6\,\pi$; transition bandwidth $\Delta\omega = \omega_c - \omega_s = 0.1\,\pi$, hence $\omega_s = 0,5\,\pi$; $\delta_p = \delta_s = 0.001$, hence $r_p = 0.0087$, $r_s = 60$; order $N = 13$

Now we know what is needed to design the analog lowpass filter from which to start and obtain $|H_a(\mathrm{j}\Omega)|^2$, convert it into a digital IIR lowpass filter by bilinear transformation, thus getting $\left|H_{\mathrm{LP}}(\mathrm{e}^{\mathrm{j}\theta})\right|^2$, and finally, using the proper frequency transformation, obtain the final highpass IIR filter. For this purpose, given the frequencies ω_p and ω_s of the highpass filter that we desire, the corresponding θ-frequencies of the intermediate lowpass IIR filter will be deduced, and then the frequencies Ω of the analog filter representing the starting point will be calculated (with $\Omega_c = 1$ rad/s). Then, after computing α, we will substitute Z with $R(z) = -(z - \alpha)/(1 - \alpha z)$ in the factorized form of the lowpass transfer function obtained by bilinear transformation. For all design steps, the deviations δ_p and δ_s remain the same.

The result of such a process is illustrated in Fig. 8.33; the resulting highpass-filter order is $N = 13$ and is the same as the lowpass-filter order. Figure 8.33a shows the details of the frequency response of the lowpass IIR filter, while Fig. 8.33b shows the same for the highpass IIR filter. The correspondence between the values of θ and ω of interest, shown in Fig. 8.34, is exactly the one described in the foregoing discussion.

Another possible approach consists in fixing a priori a convenient value of α, instead of fixing $\Omega_c = 1$ rad/s. E.g., for the lowpass $\rightarrow$ highpass case, the choice $\alpha = 0$ implies, as we already mentioned, a very simple transformation,

Fig. 8.34 Frequency mapping in the design of the highpass IIR filter shown in Fig. 8.33

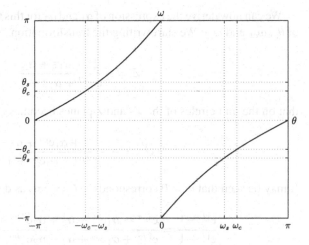

Fig. 8.35 Mapping of the frequency axis in the transformation from a lowpass to a bandpass IIR filter, illustrated using the shapes of ideal filters: values of θ and ω that correspond to each other are connected by *arrows*; this scheme follows from the functional shape of $R(z)$ and holds for any α_1, α_2

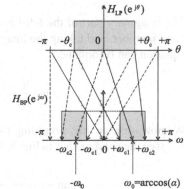

$$Z = -z \qquad \text{and} \qquad \theta = \omega - \pi.$$

8.10.2 From a Lowpass to a Bandpass Filter

In the bandpass and bandstop cases, the frequency transformation problem is more involved, because there are more free parameters.

The transformation from a lowpass to a bandpass filters implies the mapping sketched in Fig. 8.35. In this second-order transformation, a single value of θ in $[-\pi, \pi]$ corresponds to a pair of negative and positive values of ω. We can visualize this mapping imagining a first rotation of θ on its unit circle from $-\pi$ to π that covers the values of ω in $[-\pi, 0]$, and a second rotation of θ that covers the values of ω in $[0, \pi]$. In Fig. 8.35, a further characteristic frequency is introduced with respect to the highpass case. This is $\omega_0 = \arccos(\alpha)$, with α defined as in Table 8.3.

We can now derive the expression of α_1 and α_2 for this transformation, as functions of θ_c, ω_{c1} and ω_{c2}. We start writing the transformation,

$$Z = -\frac{z^2 + \alpha_1 z + \alpha_2}{\alpha_2 z^2 + \alpha_1 z + 1},$$

that on the unit circles of the Z- and z-planes becomes

$$e^{j\theta} = -\frac{e^{2j\omega} + \alpha_1 e^{j\omega} + \alpha_2}{\alpha_2 e^{2j\omega} + \alpha_1 e^{j\omega} + 1}.$$

It may be seen that $\omega = 0$ corresponds to $\theta = \pm\pi$, as desired. We then form

$$\frac{e^{j\theta} - 1}{e^{j\theta} + 1} = \frac{e^{2j\omega} + \alpha_1 e^{j\omega} + \alpha_2 + \alpha_2 e^{2j\omega} + \alpha_1 e^{j\omega} + 1}{e^{2j\omega} + \alpha_1 e^{j\omega} + \alpha_2 - \alpha_2 e^{2j\omega} - \alpha_1 e^{j\omega} - 1}.$$

The imaginary part of the left-hand member equals $\tan \theta/2$. Working on the right-hand member, and taking the imaginary part of it, after some algebra the following equation can be obtained:

$$\tan\frac{\theta}{2} = -\frac{\alpha_1 + (1 + \alpha_2)\cos\omega}{(1 - \alpha_2)\sin\omega}.$$

We now observe that, according to the mapping, $+\theta_c$ leads to ω_{c2} and $-\theta_c$ leads to ω_{c1} (see the *solid arrows* in Fig. 8.35, relative to the range $\omega \geq 0$); we thus can write

$$\tan\frac{\theta_c}{2} = -\frac{\alpha_1 + (1 + \alpha_2)\cos\omega_{c2}}{(1 - \alpha_2)\sin\omega_{c2}},$$

$$\tan\left(\frac{-\theta_c}{2}\right) = -\tan\frac{\theta_c}{2} = -\frac{\alpha_1 + (1 + \alpha_2)\cos\omega_{c1}}{(1 - \alpha_2)\sin\omega_{c1}},$$

that is an equation system in the two unknowns α_1 and α_2. Solving the first equation for α_1,

$$\alpha_1 = -\tan\frac{\theta_c}{2}(1 - \alpha_2)\sin\omega_{c2} - (1 + \alpha_2)\cos\omega_{c2},$$

and substituting into the second equation, we get

$$\alpha_2 = \frac{\tan\frac{\theta_c}{2}(\sin\omega_{c1} + \sin\omega_{c2}) + (\cos\omega_{c2} - \cos\omega_{c1})}{\tan\frac{\theta_c}{2}(\sin\omega_{c1} + \sin\omega_{c2}) - (\cos\omega_{c2} - \cos\omega_{c1})},$$

that with some algebra provides

$$\alpha_2 = \frac{\tan\frac{\theta_c}{2}\cot\left(\frac{\omega_{c2} - \omega_{c1}}{2}\right) - 1}{\tan\frac{\theta_c}{2}\cot\left(\frac{\omega_{c2} - \omega_{c1}}{2}\right) + 1}.$$

If we then set

$$\beta = \tan \frac{\theta_c}{2} \cot \left(\frac{\omega_{c2} - \omega_{c1}}{2} \right),$$

we finally obtain the expected equation,

$$\alpha_2 = \frac{\beta - 1}{\beta + 1}.$$

Then we substitute again α_2 into α_1, thus getting

$$\alpha_1 = -\frac{2}{\beta + 1} \tan \frac{\theta_c}{2} \left[\sin \omega_{c2} + \frac{\cos \left(\frac{\omega_{c2} - \omega_{c1}}{2} \right)}{\sin \left(\frac{\omega_{c2} - \omega_{c1}}{2} \right)} \cos \omega_{c2} \right].$$

But since

$$\sin \omega_{c2} + \frac{\cos \left(\frac{\omega_{c2} - \omega_{c1}}{2} \right)}{\sin \left(\frac{\omega_{c2} - \omega_{c1}}{2} \right)} \cos \omega_{c2} = \frac{\cos \left(\frac{\omega_{c2} + \omega_{c1}}{2} \right)}{\sin \left(\frac{\omega_{c2} - \omega_{c1}}{2} \right)},$$

we derive, after setting

$$\alpha = \frac{\cos \left(\frac{\omega_{c2} + \omega_{c1}}{2} \right)}{\cos \left(\frac{\omega_{c2} - \omega_{c1}}{2} \right)},$$

$$\begin{aligned}
\alpha_1 &= -\frac{2}{\beta + 1} \tan \frac{\theta_c}{2} \left[\frac{\cos \left(\frac{\omega_{c2} + \omega_{c1}}{2} \right)}{\sin \left(\frac{\omega_{c2} - \omega_{c1}}{2} \right)} \right] = \\
&= -\frac{2}{\beta + 1} \tan \frac{\theta_c}{2} \left[\frac{\cos \left(\frac{\omega_{c2} + \omega_{c1}}{2} \right)}{\cos \left(\frac{\omega_{c2} - \omega_{c1}}{2} \right)} \frac{\cos \left(\frac{\omega_{c2} + \omega_{c1}}{2} \right)}{\sin \left(\frac{\omega_{c2} - \omega_{c1}}{2} \right)} \right] = \\
&= -\frac{2}{\beta + 1} \frac{\cos \left(\frac{\omega_{c2} + \omega_{c1}}{2} \right)}{\cos \left(\frac{\omega_{c2} - \omega_{c1}}{2} \right)} \tan \frac{\theta_c}{2} \cot \left(\frac{\omega_{c2} - \omega_{c1}}{2} \right) = \\
&= -\frac{2\alpha\beta}{\beta + 1},
\end{aligned}$$

as reported in Table 8.3.

The relation between the frequencies ω of the bandpass IIR filter and θ of the lowpass IIR filter is nonlinear. We can see it graphically starting, as in the case of the highpass filter, by setting $\Omega_c = 1$ rad/s, hence $\theta_c = 2 \tan(\Omega_c/2) = 0.2952\pi$. Then considering a particular pair of values of ω_{c1} and ω_{c2}, we calculate α and β and hence the coefficients α_1 and α_2. By taking values of ω from 0 to π, we calculate the corresponding values of θ from the equations $Z = -(z^2 + \alpha_1 z + \alpha_2)/(\alpha_2 z^2 + \alpha_1 z + 1)$, $\theta = \arg(Z)$. Repeating this process for several pairs of $\{\omega_{c1}, \omega_{c2}\}$ values, we can find the curves shown in Fig. 8.36.

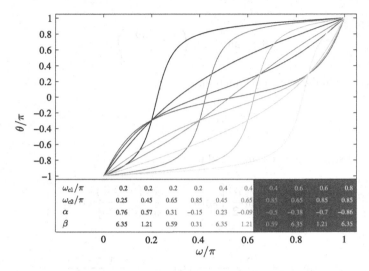

ω_{c1}/π	0.2	0.2	0.2	0.2	0.4	0.4	0.4	0.6	0.6	0.8
ω_{c2}/π	0.25	0.45	0.65	0.85	0.45	0.65	0.85	0.65	0.85	0.85
α	0.76	0.57	0.31	−0.15	0.23	−0.09	−0.5	−0.38	−0.7	−0.86
β	6.35	1.21	0.59	0.31	6.35	1.21	0.59	6.35	1.21	6.35

Fig. 8.36 Curves of $\theta(\omega)$ in the transformation from a lowpass to a bandpass IIR filter, for values of α and β that derive from considering various pairs of cutoff frequencies ω_{c1}, ω_{c2}: θ_c is constant, because $\Omega_c = 1$ rad/s is assumed for the analog lowpass filter; to avoid cluttering, only the interval $[0, \pi]$ of ω is plotted

Fig. 8.37 The parameter α in the transformation from a lowpass to a bandpass IIR filter, as a function of ω_{c1} and ω_{c2}: θ_c is constant, because we assume $\Omega_c = 1$ rad/s for the analog lowpass filter

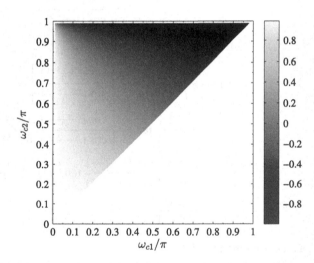

The value of α for a particular filter design depends on ω_{c1} and ω_{c2}. The map of Fig. 8.37 shows α as a function of the passband cutoff frequencies.[16] The parameter β depends only on the passband width, $\omega_{c2} - \omega_{c1}$. In Fig. 8.38, the *solid curve* shows β as a function of $\omega_{c2} - \omega_{c1}$: β increases steeply as passband width decreases.

---

[16]Obviously we assume that $\omega_{c1} < \omega_{c2}$: α can then be positive, zero or negative, but its absolute value never exceeds 1.

Fig. 8.38 The parameter β in the transformation from a lowpass to a bandpass IIR filter (*solid curve*) and in the transformation from a lowpass to a bandstop IIR filter (*dashed curve*), as a function of $\omega_{c2} - \omega_{c1}$: θ_c is constant, because $\Omega_c = 1$ rad/s is assumed for the analog lowpass filter. Note that a logarithmic y-axis has been used for this plot

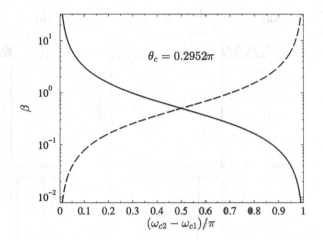

We now make a transformation example. Imagine that we want a passband filter meeting some specifications ω_{c1}, ω_{c2}, transition bandwidth $\Delta\omega$, hence $\omega_{s1} = \omega_{c1} - \Delta\omega$, $\omega_{s2} = \omega_{c2} + \Delta\omega$, tolerances δ_p and δ_s. From ω_{c1} and ω_{c2} we can compute α; setting $\Omega_c = 1$ rad/s we immediately get θ_c; on this basis we can calculate β and then α_1 and α_2. We still must determine θ_s, and consequently Ω_s.

Normally, when specifying the desired bandpass IIR filter we would give two symmetric transition bands of equal width. However, in this design method this result cannot be achieved: the transformation produces a final filter with transition bands of different width. The widest one turns out to be $\omega_{s2} - \omega_{c2}$. Therefore we will specify ω_{s2}, so as to constrain the less favorable case. According to the frequency mapping, ω_{s2} corresponds to $+\theta_s$. We thus write

$$e^{j\theta_s} = -\frac{e^{2j\omega_{s2}} + \alpha_1 e^{j\omega_{s2}} + \alpha_2}{\alpha_2 e^{2j\omega_{s2}} + \alpha_1 e^{j\omega_{s2}} + 1},$$

from which we find

$$\theta_s = \arg\left(e^{j\theta_s}\right).$$

The frequency ω_{s1} corresponds instead to $-\theta_s$ and in the final filter its effective value ω_{s1}^{eff} will be greater than the originally foreseen value $\omega_{c1} - \Delta\omega$, thus producing a narrower left-hand transition band.

Once θ_s has been calculated in this way, we can deduce Ω_s. All parameters needed for the analog lowpass design are now available. Later we apply the bilinear transformation to convert the analog lowpass filter into a digital lowpass filter and finally apply our frequency transformation to get the final bandpass filter. An example of the result of a Chebyshev-I design of this kind is shown in Fig. 8.39. The filter meets the following specifications: $\omega_{p1} = \omega_{c1} = 0.2\,\pi$; $\omega_{p2} = \omega_{c2} = 0.4\,\pi$; $\omega_{s2} = 0.44\,\pi$; $\delta_p = \delta_s = 0.001$, hence $r_p = 0.0087$, $r_s = 60$. The resulting values of the design parameters are: $\alpha = 0.6180$, hence $\omega_0 = 0.2879\,\pi$; $\beta = 1.5388$, $\alpha_1 = -0.7492$,

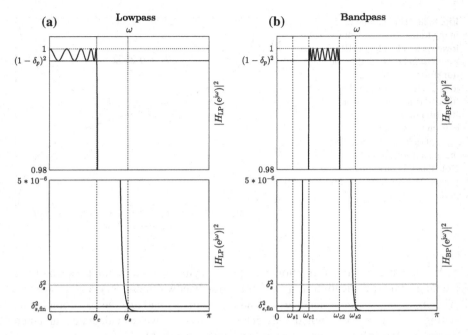

Fig. 8.39 Details of **a** the frequency response of the Chebyshev-I lowpass IIR filter and **b** the frequency response of the corresponding bandpass Chebyshev-I IIR filter satisfying the following specifications: $\omega_{p1} = \omega_{c1} = 0.2\,\pi$; $\omega_{p2} = \omega_{c2} = 0.4\,\pi$; $\omega_{s2} = 0.44\,\pi$; $\delta_p = \delta_s = 0.001$, hence $r_p = 0.0087$, $r_s = 60$; lowpass order $N = 14$; bandpass order $N = 28$

$\alpha_2 = 0.2122$; $\theta_s = 0.3778\,\pi$, hence $\Omega_s = 1.3493\,$rad/s. The order is found to be $N = 14$ for the lowpass IIR filter and $N = 28$ for the final passband IIR filter.

The correspondence between the values of θ and ω of interest is shown in Fig. 8.40. Inspection of this plot confirms what was described earlier. Note that the *gray lines* refer to the first θ rotation along the unit circle of the Z-plane from $-\pi$ to π, which covers the negative range of ω values, while the *black lines* refer to the second θ rotation, covering the positive range of ω values.

In order to reduce the number of parameters and make the design easier, we can fix a value for β: the usual choice is then $\beta = 1$. This means (see Table 8.3) assuming

$$\theta_c = \omega_{c2} - \omega_{c1},$$

i.e., fixing θ_c on the basis of the passband width of the final filter. Then $\Omega_c = 2\tan(\theta_c/2) \neq 1$ rad/s, and

$$\alpha = \frac{\cos\left(\frac{\omega_{c2}+\omega_{c1}}{2}\right)}{\cos\frac{\theta_c}{2}}.$$

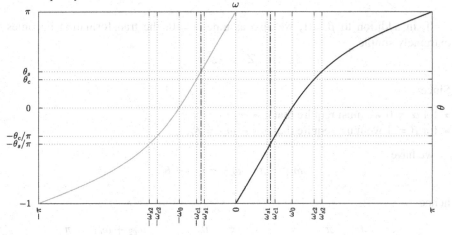

Fig. 8.40 Frequency mapping in the design of the bandpass filter shown in Fig. 8.39: *gray lines* describe the first θ rotation along the unit circle of the Z-plane from $-\pi$ to π that covers the negative range of ω values, while *black lines* describe the second θ rotation, covering the positive range of ω values

Fig. 8.41 Curves of $\theta(\omega)$ in the transformation from a lowpass to a bandpass IIR filter, for values of α that derive from considering various pairs of cutoff frequencies ω_{c1}, ω_{c2}, and for $\beta = 1$: as α increases, the curves gradually pass from being concave up to being concave down; to avoid cluttering, only the interval $[0, \pi]$ of ω is plotted

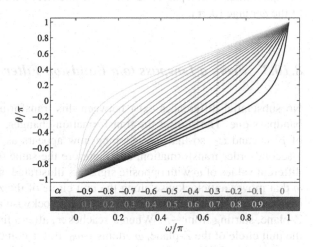

Since with $\beta = 1$ we have $\alpha_1 = -\alpha$ and $\alpha_2 = 0$, the transformation is reduced to

$$Z = -\frac{z^2 - \alpha z}{1 - \alpha z};$$

for computing θ_s we will use

$$\theta_s = -\arg\left(\frac{e^{2j\omega_{s2}} - \alpha e^{j\omega_{s2}}}{1 - \alpha e^{j\omega_{s2}}}\right).$$

In Fig. 8.41, the curve $\omega(\theta)$ for various values of α and for $\beta = 1$ is shown.

If, in addition to $\beta = 1$, we also assume $\alpha = 0$, the transformation becomes extremely simple:

$$Z = -z^2.$$

Since

- for $\alpha = 0$ we must require that $\omega_{c2} + \omega_{c1} = \pi$,
- for $\beta = 1$ we must require that $\omega_{c2} - \omega_{c1} = \theta_c$,

we have

$$\omega_{c1} = \pi - \omega_{c2} = \omega_{c2} - \theta_c,$$

hence

$$\omega_{c2} = \frac{\theta_c}{2} + \frac{\pi}{2}, \qquad \omega_{c1} = \frac{\pi}{2} - \frac{\theta_c}{2}, \qquad \frac{\omega_{c2} + \omega_{c1}}{2} = \frac{\pi}{2},$$

as is needed for α to vanish. This is a very particular case because it corresponds to a bandpass filter with a passband that is symmetrical with respect to the middle point of the ω range $[0, \pi]$.

8.10.3   From a Lowpass to a Bandstop Filter

No substantial differences exist between this transformation and the lowpass $\rightarrow$ bandpass one. The sign of the transformation changes, as well as the expressions of β, α_1 and α_2, so that the mapping now appears as in Fig. 8.42. This is again a second-order transformation, and therefore the same θ in $[-\pi, \pi]$ produces two different values of ω with opposite signs. As illustrated in Fig. 8.43, we can imagine ω that moves anticlockwise along the unit circle of the z-plane, starting from $-\omega_0$ (marked by the *black cross*), as θ rotates anticlockwise along the unit circle of the Z-plane, starting from $-\pi$. When θ reaches π, after a first complete rotation along the unit circle of the Z-plane, ω attains $+\omega_0$. But π coincides with $-\pi$, and θ starts a new anticlockwise rotation along the unit circle of the Z-plane; ω, continuing its anticlockwise motion in the z-plane, reaches again $-\omega_0$ when this second θ rotation is completed.

The expressions for α_1 and α_2 for this transformation (see Table 8.3) are derived from the given values of θ_c, ω_{c1} and ω_{c2}, in a way that is similar to what we did in the passband case. We start writing the transformation,

$$Z = \frac{z^2 + \alpha_1 z + \alpha_2}{\alpha_2 z^2 + \alpha_1 z + 1},$$

that on the unit circles of the Z- and z-planes becomes

$$e^{j\theta} = \frac{e^{2j\omega} + \alpha_1 e^{j\omega} + \alpha_2}{\alpha_2 e^{2j\omega} + \alpha_1 e^{j\omega} + 1}.$$

We may note that $\omega = 0$ correctly corresponds to $\theta = 0$.

The right-hand member of the equation

$$\frac{e^{j\theta} - 1}{e^{j\theta} + 1} = \frac{e^{2j\omega} + \alpha_1 e^{j\omega} + \alpha_2 - \alpha_2 e^{2j\omega} - \alpha_1 e^{j\omega} - 1}{e^{2j\omega} + \alpha_1 e^{j\omega} + \alpha_2 + \alpha_2 e^{2j\omega} + \alpha_1 e^{j\omega} + 1}$$

is worked out algebraically and its imaginary part is isolated. This provides the relation

$$\tan\frac{\theta}{2} = \frac{(1 - \alpha_2)\sin\omega}{(1 + \alpha_2)\cos\omega + \alpha_1}.$$

Fig. 8.42 Mapping of the frequency axis in the transformation from a lowpass to a bandstop IIR filter, illustrated using the shapes of ideal filters: values of θ and ω that correspond to each other are connected by *arrows*; this scheme follows from the functional shape of $R(z)$, independently of the particular values assumed by α_1 and α_2

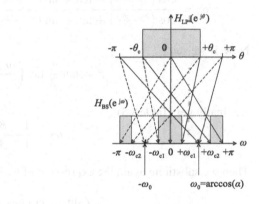

Fig. 8.43 The path in the z-plane followed by ω when θ performs two subsequent complete rotations on the unit circle of the Z-plane: the *black cross* indicates the beginning and the end of the path; the first θ rotation is represented by the *light gray arrow* and the corresponding θ values are highlighted by a frame; the second θ rotation is represented by *darker gray arrows* (see text)

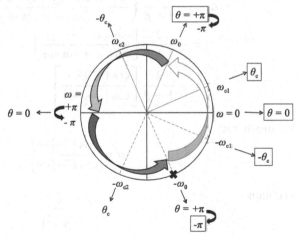

The mapping prescribes correspondence between $+\theta_c$ and ω_{c1}, as well as correspondence between $-\theta_c$ and ω_{c2}. Thus, the following equation system must hold:

$$\tan\frac{\theta_c}{2} = \frac{(1-\alpha_2)\sin\omega_{c1}}{(1+\alpha_2)\cos\omega_{c1}+\alpha_1},$$

$$\tan\left(\frac{-\theta_c}{2}\right) = -\tan\left(\frac{\theta_c}{2}\right) = \frac{(1-\alpha_2)\sin\omega_{c2}}{(1+\alpha_2)\cos\omega_{c1}+\alpha_1}.$$

Solving the first equation for α_1,

$$\alpha_1 = \frac{(1-\alpha_2)\sin\omega_{c1} - \tan\frac{\theta_c}{2}(1+\alpha_2)\cos\omega_{c1}}{\tan\frac{\theta_c}{2}},$$

and substituting into the second equation gives

$$\alpha_2 = \frac{\cos\left(\frac{\omega_{c2}-\omega_{c1}}{2}\right) - \tan\frac{\theta_c}{2}\sin\left(\frac{\omega_{c2}-\omega_{c1}}{2}\right)}{\cos\left(\frac{\omega_{c2}-\omega_{c1}}{2}\right) + \tan\frac{\theta_c}{2}\sin\left(\frac{\omega_{c2}-\omega_{c1}}{2}\right)} = \frac{1 - \tan\frac{\theta_c}{2}\tan\left(\frac{\omega_{c2}-\omega_{c1}}{2}\right)}{1 + \tan\frac{\theta_c}{2}\tan\left(\frac{\omega_{c2}-\omega_{c1}}{2}\right)}.$$

Now, if we set

$$\beta = \tan\frac{\theta_c}{2}\tan\left(\frac{\omega_{c2}-\omega_{c1}}{2}\right)$$

we finally get (Table 8.3)

$$\alpha_2 = \frac{1-\beta}{1+\beta}.$$

Then we substitute again the expression of α_2 in the expression of α_1 and obtain

$$\alpha_1 = -\frac{2}{1+\beta}\left(\frac{\sin\omega_{c1}\cos\omega_{c2} + \cos\omega_{c1}\sin\omega_{c2}}{\sin\omega_{c1}+\sin\omega_{c2}}\right) =$$

$$= -\frac{2}{1+\beta}\left[\frac{\sin(\omega_{c1}+\omega_{c2})}{\sin\omega_{c1}+\sin\omega_{c2}}\right] =$$

$$= -\frac{2}{1+\beta}\left[\frac{\cos\left(\frac{\omega_{c2}+\omega_{c1}}{2}\right)}{\cos\left(\frac{\omega_{c2}-\omega_{c1}}{2}\right)}\right].$$

We finally set

$$\alpha = \frac{\cos\left(\frac{\omega_{c2}+\omega_{c1}}{2}\right)}{\cos\left(\frac{\omega_{c2}-\omega_{c1}}{2}\right)},$$

to obtain

$$\alpha_1 = -\frac{2\alpha}{\beta+1},$$

Fig. 8.44 Curves of $\theta(\omega)$ in the transformation from a lowpass to a bandstop IIR filter, for values of α and β that derive from considering various pairs of cutoff frequencies ω_{c1} and ω_{c2}; to avoid cluttering, only the interval $[0, \pi]$ of ω is plotted

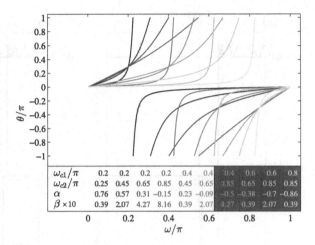

ω_{c1}/π	0.2	0.2	0.2	0.2	0.4	0.4	0.4	0.6	0.6	0.8
ω_{c2}/π	0.25	0.45	0.65	0.85	0.45	0.65	0.85	0.65	0.85	0.85
α	0.76	0.57	0.31	−0.15	0.23	−0.09	−0.5	−0.38	−0.7	−0.86
$\beta \times 10$	0.39	2.07	4.27	8.16	0.39	2.07	4.27	0.39	2.07	0.39

as expected (Table 8.3).

The nonlinear relation between ω and θ for the stopband case, deduced from $\theta = \arg(Z) = \arg\left(z^2 + \alpha_1 z + \alpha_2\right) / \left(\alpha_2 z^2 + \alpha_1 z + 1\right)$, appears in Fig. 8.44.

In this transformation, the expression of α is the same found in the lowpass $\rightarrow$ bandpass case, and therefore as ω_{c1} and ω_{c2} vary, α assumes the same values shown in Fig. 8.37; the behavior of β as a function of $\omega_{c2} - \omega_{c1}$ appears as a *dashed curve* in Fig. 8.38. The parameter β increases rapidly with increasing passband width.

Let us now turn to a design example. We start establishing the specifications of the final filter, including only one stopband limit, namely ω_{s2}. We set $\Omega_c = 1$ rad/s and immediately get θ_c; we compute α from ω_{c1} and ω_{c2}, as well as β from ω_{c1}, ω_{c2} and θ_c; then we evaluate α_1 and α_2. At this point we only need to fix θ_s. Also in this case, the transformation produces, in the final filter, two transition bands of different widths. The widest one turns out to be the right-hand one, $\omega_{c2} - \omega_{s2}$, and this is the reason why we only specified ω_{s2}. According to the mapping, ω_{s2} corresponds to $-\theta_s$, so we can write

$$e^{-j\theta_s} = \frac{e^{2j\omega_{s2}} + \alpha_1 e^{j\omega_{s2}} + \alpha_2}{\alpha_2 e^{2j\omega_{s2}} + \alpha_1 e^{j\omega_{s2}} + 1},$$

from which we get

$$\theta_s = -\arg\left(e^{-j\theta_s}\right).$$

The frequency ω_{s1} corresponds instead to $+\theta_s$, and in the final filter its effective value ω_{s1}^{eff} will be smaller than the value originally foreseen, i.e., $\omega_{c1} + \Delta\omega$, thus producing a narrower left-hand transition band. Once θ_s has been found in this way, we can compute Ω_s and design the analog lowpass filter of the desired type, which is then converted into a lowpass digital filter via bilinear transformation; at last, the frequency transformation will be applied to produce the final filter. An example of Chebyshev-I filter designed according to this procedure is shown in Fig. 8.45. Specifications are: $\omega_{p1} = \omega_{c1} = 0.2\,\pi$; $\omega_{p2} = \omega_{c2} = 0.4\,\pi$; $\omega_{s2} = 0.44\,\pi$; $\delta_p = \delta_s = 0.001$, hence

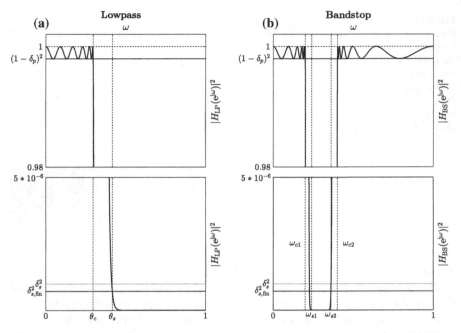

Fig. 8.45 Details of **a** the frequency response of the lowpass Chebyshev-I IIR filter and **b** the frequency response of the corresponding bandstop Chebyshev-I IIR filter satisfying the following specifications: $\omega_{c1} = 0.2\,\pi$; $\omega_{c2} = 0.4\,\pi$; $\omega_{s2} = 0.36\,\pi$; $\delta_p = \delta_s = 0.001$, hence $r_p = 0.0087$, $r_s = 60$; lowpass order $N = 11$; bandstop order $N = 22$

$r_p = 0.0087$, $r_s = 60$. The resulting parameter values are: $\alpha = 0.6180$, hence $\omega_0 = 0.2879\,\pi$; $\beta = 0.1625$, $\alpha_1 = -1.0633$, $\alpha_2 = 0.7205$; $\theta_s = 0.4156\,\pi$, hence $\Omega_s = 1.5292$ rad/s. The order is found to be $N = 11$ for the lowpass filter and $N = 22$ for the final bandstop filter.

Figure 8.46 shows the correspondence between the values of θ and ω of interest, which is in agreement with what was described above. In Fig. 8.46, the *gray lines* are relative to the first θ rotation along the unit circle of the Z-plane from $-\pi$ to π that covers the negative range of ω values, while the *black lines* are relative to the second θ rotation, covering the positive range of ω values.

To reduce the number of parameters we can directly set $\beta = 1$. Since $\beta = \tan\left[(\omega_{c2} - \omega_{c1})/2\right]\tan(\theta_c/2)$, assuming $\beta = 1$ means choosing

$$\theta_c = \pi - (\omega_{c2} - \omega_{c1}).$$

In fact, if we call γ and δ any two angles, then $\tan\gamma\,\tan\delta = 1$ if $\tan\gamma = 1/\tan\delta = \cot\delta$, which is true if $\delta = \pi/2 - \gamma$. As a consequence, α assumes the form

$$\alpha = \frac{\cos\left(\frac{\omega_{c2}+\omega_{c1}}{2}\right)}{\cos\left(\frac{\pi-\theta_c}{2}\right)}.$$

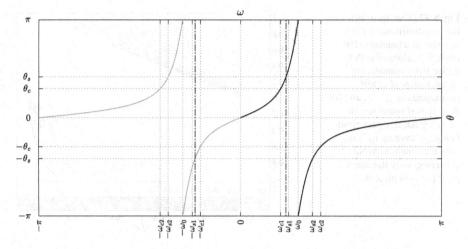

Fig. 8.46 Frequency mapping in the design of the bandstop IIR filter shown in Fig. 8.45: the *gray lines* refer to the first θ rotation along the unit circle of the Z-plane from $-\pi$ to π that covers the negative range of ω values, while the *black lines* refer to the second θ rotation, covering the positive range of ω values

But $\beta = 1$ also implies $\alpha_1 = -\alpha$, $\alpha_2 = 0$, and the transformation simplifies to

$$Z = \frac{z^2 - \alpha z}{1 - \alpha z}.$$

From this equation, we can understand that in order to compute θ_s we can use

$$\theta_s = \arg\left(\frac{e^{2j\omega_{s2}} - \alpha e^{j\omega_{s2}}}{1 - \alpha e^{j\omega_{s2}}}\right).$$

In Fig. 8.47, the curve $\omega(\theta)$ is given for several values of α and for $\beta = 1$. The very special choice $\beta = 1$, $\alpha = 0$ is considered too:

- for $\alpha = 0$ we must require that $\omega_{c2} + \omega_{c1} = \pi$;
- for $\beta = 1$ we must require that $\theta_c = \pi - (\omega_{c2} - \omega_{c1})$.

Then

$$\omega_{c1} = \pi - \omega_{c2} = \omega_{c2} - \pi + \theta_c$$

and therefore

$$\omega_{c2} = \pi - \frac{\theta_c}{2}, \qquad \omega_{c1} = \pi - \omega_{c2} = \frac{\theta_c}{2}.$$

The transformation reduces to

$$Z = z^2.$$

Fig. 8.47 Curves of $\theta(\omega)$ in the transformation from a lowpass to a bandstop IIR filter, for values of α that derive from considering various pairs of cutoff frequencies ω_{c1}, ω_{c2}, and for $\beta = 1$: as α increases, the curves gradually pass from being concave up to being concave down; to avoid cluttering, only the interval $[0, \pi]$ of ω is plotted

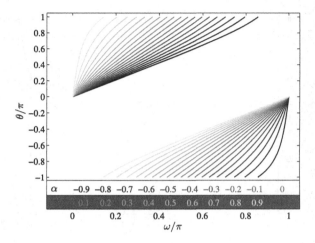

8.11   Direct Design of IIR Filters

The design of non-standard IIR filters, e.g., multiband filters, cannot be dealt with using the methods described in the previous sections.[17] The filter must be designed directly in the discrete-time domain, requiring it approximates the desired piecewise-constant desired response. The phase issue is completely neglected. These methods are referred to as "direct methods" because no analog filter comes into play and the design takes place in the discrete-time domain only. The optimality of the designed filter is not discussed.

Among the many methods available in literature, including the Prony method, the Dezky method, the Steiglitz-McBride method etc., the most widely applied is probably the Yule-Walker method (Porat and Friedlander 1984). Its name reflects the way by which the coefficients of the filter's frequency response are calculated: they are found solving a set of equations referred to as *modified Yule-Walker equations*. We will come across this name, and the *ARMA modeling* that forms the basis of the method, when we discuss parametric methods for the estimate of the power spectrum of a random signal. Direct methods of IIR filter design are beyond the scope of this book.

8.12   Appendix

This appendix discusses elliptic integrals of the first kind, Jacobi elliptic functions, and the elliptic rational function, on the basis of which the transfer function of analog elliptic filters is constructed.

---

[17] Actually it is possible to use frequency transformations to pass from a digital lowpass filter, derived from an analog lowpass filter, to a digital multiband filter, but these are high-order transformations that push the order of the final filter up considerably. Moreover, direct design techniques avoid constraints and compromises that are unnecessary.

8.12.1 Trigonometric Functions with Complex Argument

Ordinary (circular) trigonometric functions and hyperbolic trigonometric functions can be defined not only for real arguments, but also for complex arguments (Saff and Snider 2002).

Consider a complex variable $z = x + iy$, where x and y are real. The complex exponential function e^z is periodic, with imaginary period $j2\pi$, and this same imaginary period is shared by the hyperbolic trigonometric functions $\sinh z$ and $\cosh z$ that are linear combinations of e^z and e^{-z}:

$$\sinh z = \frac{e^z - e^{-z}}{2}, \qquad \cosh z = \frac{e^z + e^{-z}}{2}.$$

Using $\sinh z$ and $\cosh z$, other hyperbolic trigonometric functions can be defined in the complex domain, such as the hyperbolic tangent. Using the Euler formula

$$e^{jy} = \cos y + j \sin y,$$

which holds for any real y, the exponential function e^z can be written as

$$e^z = e^x e^{iy} = e^x (\cos y + j \sin y).$$

From Euler's formula it also follows that

$$e^{-jy} = \cos y - j \sin y,$$

and that the inverse relations hold:

$$\sin y = \frac{e^{jy} - e^{-jy}}{2j}, \qquad \cos y = \frac{e^{jy} + e^{-jy}}{2}.$$

These formulas for real y suggest extending trigonometric functions to *complex angles z*, by writing

$$\sin z = \frac{e^{jz} - e^{-jz}}{2j}, \qquad \cos z = \frac{e^{jz} + e^{-jz}}{2}.$$

These functions of the complex argument $z = x + iy$ are thus obtained directly from the exponential function, by substituting z with jz; like e^z, they are periodic functions. Their period is real and equal to 2π. Through $\sin z$ and $\cos z$, the other trigonometric functions, for example the tangent, can be extended to the complex domain. Unlike their real counterparts, the functions $\sin z$ and $\cos z$ are not limited to 1 in absolute value, and can assume any real or complex value. However, they still satisfy

$$\sin^2 z + \cos^2 z = 1.$$

Fig. 8.48 Period-strips for **a**
$\sin z$, $\cos z$ and **b** $\sinh z$,
$\cosh z$

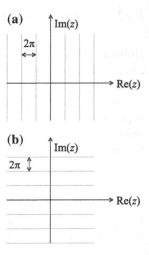

Clearly, $\sin z$ and $\cos z$ are real if $y = 0$, i.e., if z is real. For complex z, $\sin z$ is real if $\cos y = 0$, and $\cos z$ is real if $\sin y = 0$.

These functions of complex variable are said to be *singly periodic*, because they possess a single period in the complex plane that we shall call T. For them, the complex z-plane can be divided into an infinite number of strips of width T and infinite length, called period-strips. Given any point z inside one of these period-strips, a corresponding point $z + mT$ can be found inside any other period-strip, such that the values of the considered function at the two points (congruent points) are identical. The period-strips for $\sin z$ and $\cos z$ have width 2π and are juxtaposed to one another, being all oriented parallel to the imaginary axis (Fig. 8.48a), while those for $\sinh z$ and $\cosh z$, having the same width 2π, are oriented parallel to the real axis (Fig. 8.48b). The functions $\tan z$ and $\tanh z$ have different periods: π and $j\pi$, respectively.

If $w = f(z)$, where $f(z)$ is a sine or a cosine, circular or hyperbolic, it can be shown that an infinite-length strip of width π, lying inside a period-strip and centered on a zero of the considered function—i.e., a half period-strip—will be mapped onto the whole complex w-plane with biunivocal correspondence. If $f(z)$ represents a circular or hyperbolic tangent, then a whole period-strip (instead of a half period-strip) will be mapped onto the whole complex w-plane.

It can be shown that the following properties hold:

$$\sin(jz) = j \sinh z,$$
$$\sinh(jz) = j \sin z,$$
$$\cos(jz) = \cosh z,$$
$$\cosh(jz) = \cos z.$$

Fig. 8.49 Complex sine:
a amplitude, **b** real part, and
c imaginary part

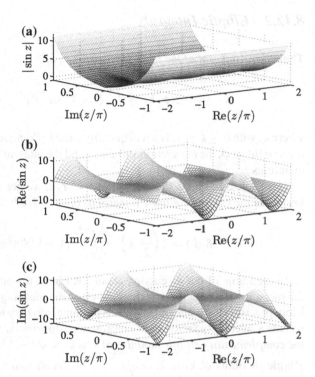

This implies, in particular, that

$$\sin z = -j \sinh(jz),$$
$$\cos z = \cosh(jz).$$

Observing that

$$\sin z = \sin(x + jy) = \sin x \cos(jy) + \cos x \sin(jy) = \sin x \cosh y + j \cos x \sinh y,$$
$$\cos z = \cos(x + jy) = \cos x \cos(jy) - \sin x \sin(jy) = \cos x \cosh y - j \sin x \sinh y,$$

we can understand that since $\cosh y$ never vanishes, and since $\sin x$ and $\cos x$ never vanish simultaneously, the zeros of $\sin z$ and $\cos z$ are located on the real axis. For example, Fig. 8.49 shows the behavior of $\sin z$. Observe the single periodicity with respect to the real axis, with period 2π.

8.12.2   Elliptic Integrals

The *elliptic integral of the first kind* is defined as

$$z(\phi, k) = \int_0^\phi \left(1 - k^2 \sin^2 \theta\right)^{-1/2} d\theta,$$

where k, with $0 \le k \le 1$, is a real quantity called *elliptic modulus*. In this expression, the *amplitude* ϕ, the integration variable θ and the integral itself, $z(\phi, k)$, are complex quantities.

If we substitute for ϕ the *real* value $\phi = \pi/2$, we get a *real* value for z, which is known as the *complete elliptic integral of the first kind*:

$$K(k) = z\left(\frac{\pi}{2}, k\right) = \int_0^{\pi/2} \left(1 - k^2 \sin^2 \theta\right)^{-1/2} d\theta.$$

In the frame of our discussion about IIR filters, the elliptic modulus can assume the meaning of a selectivity factor, as well as the meaning of a discrimination factor. More precisely, four integrals of this type come into play in IIR filter design: if k represents a selectivity factor and k_1 represents a discrimination factor, and if we define the *complementary elliptic modulus* of k as $k' = \sqrt{1 - k^2}$, while the complementary elliptic modulus of k_1 is $k_1' = \sqrt{1 - k_1^2}$, we need four complete elliptic integrals: $K(k)$, $K(k')$, $K(k_1)$ and $K(k_1')$, which we will simply indicate as K, K', K_1 and K_1', respectively. Figure 8.50a shows the behavior of $K(k)$ (*solid curve*). This integral is a monotonically increasing function of k, which goes to $\pi/2$ for $k \to 0$ and goes to infinity for $k \to 1$. Figure 8.50a also shows K' as a function of k: this is a monotonically decreasing function of k (*dashed curve*). The two curves intersect each other at $k = 1/\sqrt{2} = 0.707$. If K and K' are plotted versus k^2 instead of k, these curves become symmetric, as visible in Fig. 8.50b. Note that $K(k = 0) = K'(k = 1) = \pi/2$ and $K(k = 1) = K'(k = 0) \to \infty$.

8.12.3   Jacobi Elliptic Functions

Jacobi elliptic functions are a set of twelve functions; the basic ones are the *Jacobi elliptic sine* $\mathrm{sn}(z, k)$, the *Jacobi elliptic cosine* $\mathrm{cn}(z, k)$ and the function $\mathrm{dn}(z, k)$. They are defined implicitly through the elliptic integral of the first kind: they are obtained from the elliptic integral of the first kind by inverting, first of all, the function $z(\phi, k)$, so that the amplitude ϕ is written as a function of z, with k as a parameter, i.e., $\phi(z, k)$. Then the basic Jacobi elliptic functions are defined as

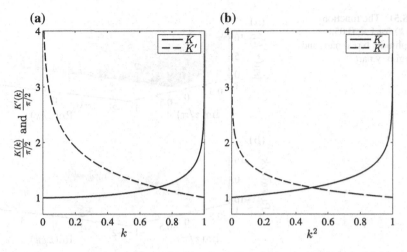

Fig. 8.50 **a** Complete elliptic integrals of the first kind, K (*solid line*) and K' (*dashed line*), plotted versus k; **b** the same plotted versus k^2

$$\mathrm{sn}(z, k) = \sin[\phi(z, k)],$$
$$\mathrm{cn}(z, k) = \cos[\phi(z, k)],$$
$$\mathrm{dn}(z, k) = \sqrt{1 - k^2 \sin^2[\phi(z, k)]}.$$

The remaining nine functions are constructed by combining the three basic functions. For our discussion on IIR filters, we need the function

$$\mathrm{cd}(z, k) = \frac{\mathrm{cn}(z, k)}{\mathrm{dn}(z, k)}.$$

This is the most important Jacobi elliptic function in the frame of elliptic filter design theory. The only other Jacobi elliptic function used in the same theory is $\mathrm{sn}(z, k)$. Note that for $\phi = \pi/2$, $z = K(k)$, and we have $\mathrm{sn}(K, k) = \sin(\pi/2) = 1$, $\mathrm{cd}(K, k) = \cos(\pi/2) = 0$.

For $k \to 0$ and $k = 1$, the Jacobian elliptic functions have limiting forms consisting of circular and hyperbolic trigonometric functions, respectively. For example, the four functions sn, cn, dn and cd reduce to circular trigonometric and hyperbolic trigonometric functions, respectively:

$$\mathrm{sn}(z, 0) = \sin z, \qquad\qquad \mathrm{sn}(z, 1) = \tanh z;$$
$$\mathrm{cn}(z, 0) = \cos z, \qquad\qquad \mathrm{cn}(z, 1) = \mathrm{sech}\, z;$$
$$\mathrm{dn}(z, 0) = 1, \qquad\qquad\quad \mathrm{dn}(z, 1) = \mathrm{sech}\, z;$$
$$\mathrm{cd}(z, 0) = \cos z, \qquad\qquad \mathrm{cd}(z, 1) = 1.$$

Fig. 8.51 The function
$\mathrm{sn}(z, k)$ for $k = 0.05$:
a amplitude, **b** real part, and
c imaginary part

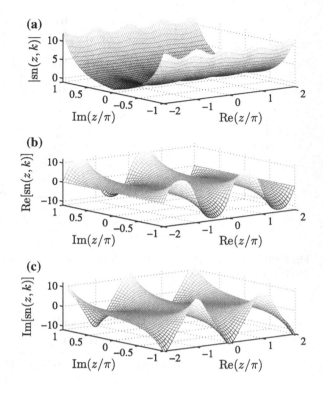

At low values of k, the two functions sn and cd that we are most interested in are very similar to $\sin z$ and $\cos z$, respectively for instance, compare Fig. 8.49 with Fig. 8.51, which shows the function $\mathrm{sn}(z, k)$ for $k = 0.05$.

Jacobi elliptic functions possess properties that are similar to the properties of trigonometric and hyperbolic functions. For example, the three basic functions sn, cn and dn satisfy

$$\mathrm{sn}^2(z, k) + \mathrm{cn}^2(z, k) = 1,$$
$$k^2 \mathrm{sn}^2(z, k) + \mathrm{dn}^2(z, k) = 1.$$

As a function of z, with fixed k, each of the Jacobi elliptic functions is a *doubly periodic* function in the z-plane (Abramowitz and Stegun 1972; Olver et al. 2010). Indeed, the existence of singly-periodic functions in the complex plane naturally leads to exploring if functions exist that have two, or even more, distinct periods. It can be shown that

- no functions exist with three, or more, periods;
- if a function possesses two periods, their ratio cannot be real; in other words, the two periods must be relative to different directions in the complex plane. They cannot lie along the same straight line.

Within these restrictions, a single-valued function being doubly periodic with periods equal to any two given complex numbers a and b can be constructed, so as to have $f(z + ma + nb) = f(z)$ for any z and for any pair of integer numbers m, n. A function of this kind is called an *elliptic function*.

The double periodicity associates with each z-value an infinite two-dimensional grid of points $z + ma + nb$ in the z-plane, at which the function assumes identical values. Around these grid points we can visualize an infinitely extended tiling of the z-plane, made up of period-parallelograms with sides a and b such that, for any z, the grid identifies a set of congruent points inside the parallelograms. Since the grid points $z + ma + nb$ can also be written as $z + (m - n)a + n(a + b)$, or as $z + m(a - b) + (n + m)b$, it is clear that the periods of a function of this kind are not unique and can be set, for example, not only equal to a, b, but also to $a, a + b$, or $a + b, b$, etc. Changing the definition of the function periods in this way changes the shape, but not the area, of the period-parallelograms. Therefore a proper choice of the periods allows for restricting the attention to rectangular tilings, i.e., to *period-rectangles*. In this way, it is possible to work with pairs of periods including *one purely real period and one purely imaginary period*.

Inside each period-rectangle, the function $f(z)$ must possess at least one singularity, because otherwise it would have no singularities at all and therefore, for Liouville's theorem of complex analysis (see, e.g., Kreyszig 2011), it would be a constant. Moreover, the line integral of the function along the rectangle's contour must vanish, because the contributions to the integral given by opposite sides of the rectangle are equal in absolute value and opposite in sign, due to the double periodicity of the function. Cauchy's residue theorem then implies that the sum of the residues in the singularities contained inside the rectangle must vanish too. Since a single pole with zero residue is a removable singularity, the simplest doubly-periodic elliptic function $f(z)$ will either have a double pole with zero residue, or two simple poles with residues equal in magnitude and opposite in sign, inside each period-rectangle. The complexity, a.k.a the *order*, of the function is measured by the sum of the orders of the poles contained inside a period-rectangle. Therefore, no first-order elliptic functions exist, and two kinds of second-order elliptic functions exist. Moreover, if $f(z)$ is an elliptic function, then $1/f(z)$ is also an elliptic function, with the same order and the same periods. This tells us that a second-order elliptic function also has either a double zero, or two simple zeros inside each period-rectangle.

A second-order elliptic function, with one double pole with zero residue in each period-rectangle, was originally described by Weierstrass, and is named after this author. It is a very simple function that can be used to systematically build second-order elliptic functions having two simple poles per period-rectangle. These derived functions, properly normalized, are the Jacobi elliptic functions that possess two simple zeros and two simple poles per period-rectangle.

We know it is convenient to define six different circular trigonometric functions: sine, cosine, tangent and their respective reciprocal functions. In a similar manner, twelve different Jacobi elliptic functions naturally emerge from the theory. The functions sine/cosine and tangent have periods 2π and π respectively, i.e., have periods that differ by a factor of 2. Similarly, Jacobi elliptic functions can be properly grouped according to their periods, and these periods differ from one group to the other by a factor of 2. It is standard practice to describe these periods in terms of a pair of parameters, one real and one imaginary: K and jK', respectively. These parameters play the role that in the trigonometric and hyperbolic cases belongs to $\pi/2$.

These quarter periods define a rectangle located in the first quadrant of the complex z-plane, having the points 0, K, jK' and $K + jK'$ as vertexes which, is referred to as the *fundamental rectangle*. Each vertex is indicated by a letter (see Fig. 8.52): S (*starting point*), C (*corner on the real axis*), D (*diagonally opposite point*), and N (*normal to the real axis*).

The fundamental rectangle has one pole and one zero on its contour. The positions of the pole and of the zero on the contour univocally identify one of the twelve functions. Indeed, if we denote by p one of the symbols s, c, d, n, and by q one of the remaining symbols, we can indicate any Jacobi elliptic function by pq(z, k); the number of combinations of the symbols p and q is $4 \times 3 = 12$. Then we will call P the z-plane point corresponding—in the sense that will immediately be explained— to the symbol p, S the z-plane point corresponding to the symbol s, and so on. Each Jacobi elliptic function pq(z, k) is built in such a way to have a simple zero in the vertex P and a simple pole in the vertex Q of the fundamental rectangle: for example, sn(z, k) has a simple zero in S ($z = 0$) and a simple pole in N ($z = jK'$); cd(z, k) has a simple zero in C ($z = K$) and a simple pole in D ($z = K + jK'$), etc.

We must explicitly note here an important fact: only one of the two parameters K, K' can be chosen arbitrarily. For example, if we set a value for K, then K' is automatically determined. This is evident from the definition we gave of the Jacobi elliptic functions in terms of inversion of the elliptic integral of the first kind, and also from what we said about K and K' in Sect. 8.12.2. This remark clarifies that the fundamental rectangle has dimensions that vary with K, while its shape is constrained by the ratio K'/K. In other words, the elliptic modulus k alone determines area and shape of the fundamental rectangle. This implies that the six Jacobi elliptic functions that have a zero or a pole in the origin of the complex plane, i.e., sc, sd, sn, cs, ds, and ns, are odd functions, while the remaining six functions are even functions. Another consequence of this relation between K and K'—that we may see as a normalization of Jacobi elliptic functions—is that the following properties hold:

Fig. 8.52 The fundamental rectangle in the complex z-plane for Jacobi elliptic functions. The points S, C, D and N appear in an anticlockwise sequence at the corners of the fundamental rectangle

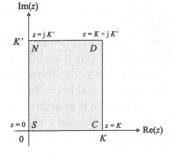

Table 8.4 Periods, zeros and poles of the twelve Jacobi elliptic functions

Periods	z-position of a pole				z-position of a zero			
	jK'	$K+jK'$	K	0	0	K	$K+jK'$	jK'
$4K, 2jK'$	sn	cd	dc	ns	sn	cd	dc	ns
$4K, 4jK'$	cn	sd	nc	ds	sd	cn	ds	nc
$2K, 4jK'$	dn	nd	sc	cs	sc	cs	dn	nd

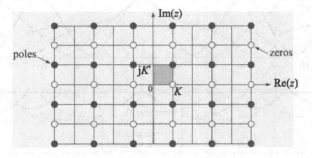

Fig. 8.53 Poles and zeros of the Jacobi elliptic function $cd(z, k)$. The *gray-shaded rectangle* is the fundamental one

$$pq(z, k) = \frac{1}{qp(z,k)} = \frac{pr(z,k)}{qr(z,k)} = \frac{rq(z,k)}{rp(z,k)},$$

where r is one of the symbols s, c, d, n, provided it is different from both p and q. The three functions sn, cn, and dn are the original elliptic functions that Jacobi obtained in 1827 inverting the elliptic integral of the first kind, while the other nine functions were introduced in 1882 by Glasher as reciprocal and quotients of the first three functions.

Table 8.4 lists, for each of the twelve functions, the real period and imaginary periods in the z-plane, the position of the zero on the edge of the fundamental rectangle, and the position of the pole. The other poles are found at congruent points with respect to the indicated point, i.e., in the set of points that can be obtained performing translations by $2mK + 2njK'$ in the z-plane, with integer values of m and n. Similarly, the other zeros can be found performing translations by $2mK + 2njK'$. E.g., for $cd(z, k)$, the poles and zeros are given by

$$\text{zeros} : z = K + 2mK + 2njK' = (2m + 1)K + 2njK',$$
$$\text{poles} : z = K + jK' + 2mK + 2njK' = (2m + 1)K + (2n + 1)jK',$$

where n and m are arbitrary integer numbers (positive, negative or zero). The resulting grid of poles and zeros is shown in Fig. 8.53.

The periods listed in Table 8.4 can be explained as follows.

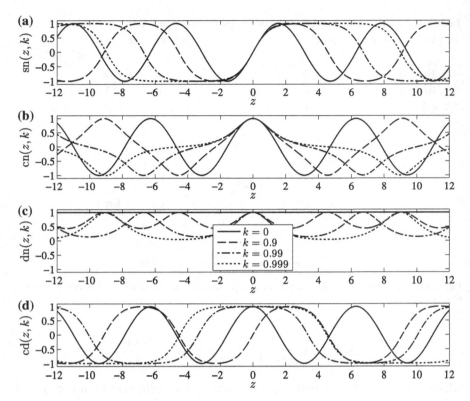

Fig. 8.54 Jacobi elliptic functions for real values of z and for several values of k: **a** $\mathrm{sn}(z, k)$, **b** $\mathrm{cn}(z, k)$, **c** $\mathrm{dn}(z, k)$, and **d** $\mathrm{cd}(z, k)$

The difference between the z-position of a zero and that of the nearest pole is a *half-period* of $\mathrm{pq}(z, k)$. This half-period is plus or minus a member of the triplet K, $\mathrm{j}K'$, $K + \mathrm{j}K'$. The other two members of the triplet are *quarter-periods* of $\mathrm{pq}(z, k)$. For instance, for sn the above-mentioned difference is $\mathrm{j}K$, so that actually we can list three periods: not only $2\mathrm{j}K'$ and $4K$, but also $4K + 4\mathrm{j}K'$. The same is true for cd. To make another example, let us consider cn: the half-period is $K + \mathrm{j}K'$, so that we have the periods $2K + 2\mathrm{j}K'$, $4K$ and $4\mathrm{j}K'$, and so on. In Table 8.4, only real and imaginary periods are reported; the complex ones are not included.

From Table 8.4 we see that for sn and cd we need $2 \times 4 = 8$ fundamental rectangles to cover one period-rectangle. Yet another concept is that of *fundamental region*, i.e., the minimum region of the complex z-plane that through the mapping $z \rightarrow w$ covers the whole complex w-plane. We will soon see that for $w = \mathrm{cd}(z, k)$, the fundamental region is made up of 4 fundamental rectangles: hence it is a half period-rectangle.

We can start our study of Jacobi elliptic functions by observing their periodicity in the case of real argument z: in this case the functions are real. Figure. 8.54 shows, in separate panels, the behavior of the four functions $\mathrm{sn}(z, k)$, $\mathrm{cn}(z, k)$, $\mathrm{dn}(z, k)$ and

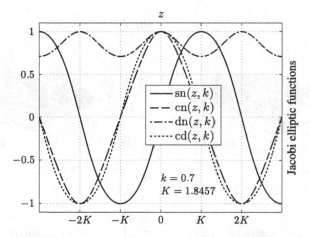

Fig. 8.55 Comparison among the four Jacobi elliptic functions $sn(z, k)$, $cn(z, k)$, $dn(z, k)$, and $cd(z, k)$, for real values of z and for $k = 0.7$

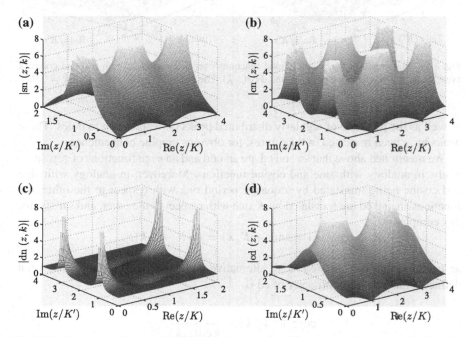

Fig. 8.56 Magnitude of the Jacobi elliptic functions: **a** $sn(z, k)$, **b** $cn(z, k)$, **c** $dn(z, k)$, and **d** $cd(z, k)$. The figure illustrates the case of $k = 0.2$

$cd(z, k)$ for real values of z and for several values of k, while Fig. 8.55 compares these four functions among them, for a single value of k and for real z values.

When the argument z is complex, the behavior of the magnitude of these four functions appears as shown in Figs. 8.56 and 8.57, in which $k = 0.2$, and $k = 0.99$, respectively. A double periodicity with respect to $Re(z)$ and $Im(z)$ is clearly visible,

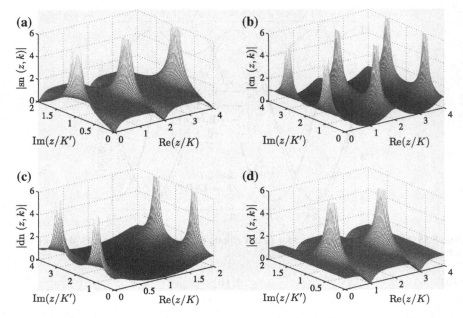

Fig. 8.57 Magnitude of the Jacobi elliptic functions over one period on both real and complex axes: **a** sn(z, k), **b** cn(z, k), **c** dn(z, k), and **d** cd(z, k). The figure illustrates the case of $k = 0.99$

as well as the presence of regularly distributed peaks corresponding to poles. These peaks have been truncated in the figures, for obvious graphic convenience.

We mentioned above that sn and cd are an odd and an even functions of z, respectively, in analogy with sine and cosine functions. Moreover, in analogy with sine and cosine being translated by a quarter period one with respect to the other, the functions sn and cd are translated by K one with respect to the other, and satisfy, for any complex z,

$$\text{cd}(z, k) = \text{sn}(z + K, k) = \text{sn}(K - z, k),$$

an equation that can be used as an alternative definition of cd, once sn has been defined. Another useful property of cd is

$$\text{cd}(z + jK', k) = \frac{1}{k\,\text{cd}(z, k)}.$$

Let us now examine, for $w = \text{cd}(z, k)$, the mapping between the z-plane and the w-plane. We defined the fundamental region as the smallest region of the z-plane that, when mapped through the function $w = \text{cd}(z, k)$, covers the whole w-plane. This region is centered on the simple zero in C, and the point C is surrounded by four fundamental rectangles that form the fundamental region. Each of these four rectangles is mapped onto a different quadrant of the w-plane, according to the scheme shown in Fig. 8.58.

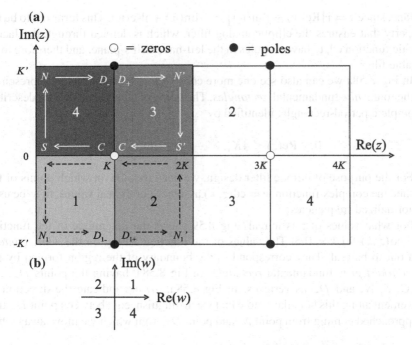

Fig. 8.58 **a** The function $w = \text{cd}(z, k)$ in the z-plane: fundamental region (union of the four *gray-shaded* fundamental *rectangles*), period rectangle (union of the eight *gray-shaded* and *white* fundamental *rectangles*), and mapping of z onto the four quadrants of the complex w-plane. *Arrows* indicate the direction of movement along the boundary of the region formed by the two *darkest-gray* fundamental *rectangles*, in connection with the discussion reported in the text about purely real w values. **b** Numbering of the complex w-plane quadrants. Note that the poles are in common with adjacent period-rectangles, so that a period-rectangle actually possesses two poles and two zeros

The fundamental region of $w = \text{cd}(z, k)$ is

$$0 \leq \text{Re}(z) \leq 2K, \qquad -K' \leq \text{Im}(z) \leq K'.$$

In Fig. 8.58a, the numbers 1–4 indicate the quadrants of the complex w-plane, drawn in Fig. 8.58b, onto which the individual *gray-shaded* fundamental *rectangles* in the z-plane are mapped: for instance, the two lower fundamental rectangles are mapped onto the first and second quadrant of the w-plane.

Let us now think of the complex s-plane with respect to which the transfer functions of analog filters are defined, and of the equation $s = j\Omega$ that relates the variable s with analog angular frequency Ω. Now, let us assume that some real values of the variable w represent frequency: $s = jw$. We can simply think of a normalized elliptic filter with $\Omega_c = \Omega_p = 1$ rad/s, $w = \Omega$ and $s = j\Omega = jw$: this is not restrictive, since the conversion of a normalized filter into a filter with any cutoff frequency is straightforward (Sect. 8.7). Now, let us extend our attention to the whole complex w-plane: the first and second quadrant of the w-plane correspond to the left-hand half s-plane, since $s = j\left[\mathrm{Re}(w) + j\mathrm{Im}(w)\right] = -\mathrm{Im}(w) + j\mathrm{Re}(w)$. This turns out to be the property that ensures the elliptic analog filter, which is defined through the Jacobi elliptic function cd, to have its poles in the left-hand half s-plane, and therefore to be a stable filter.

In Fig. 8.58a we can also see one more copy of fundamental region, represented by the four *white* fundamental *rectangles*. The union of all eight rectangles describes a complete period-rectangle, identified by

$$0 \leq \mathrm{Re}(z) \leq 4K, \qquad -K' \leq \mathrm{Im}(z) \leq K'.$$

For the purpose of elliptic filter design, we need to know in which points of the z-plane the complex function $w = \mathrm{cd}(z, k)$ assumes purely real values, fit to be used as normalized frequencies.

For what values of z is w real? Fig. 8.59 shows the magnitude of the function $w = \mathrm{cd}(z, k)$ for $k = 0.5$. The values of the function falling on the *white contour* turn out to be real. They correspond to the boundary of the region formed by the two *darkest-gray* fundamental *rectangles* in Fig. 8.58a, having the points D_+, N', S', C, S, N, and D_- as vertexes. In Fig. 8.58a, *arrows* indicate the direction of movement along this boundary; note that we distinguish, e.g., between point D_- that is approached coming from point N, and point D_+ from which we move away when

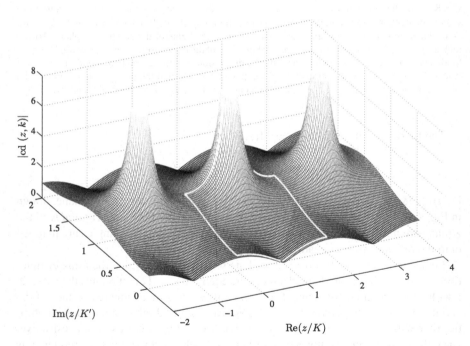

Fig. 8.59 The magnitude of $\mathrm{cd}(z, k)$ for $k = 0.5$. This complex function assumes real values along the *white contour* (see text)

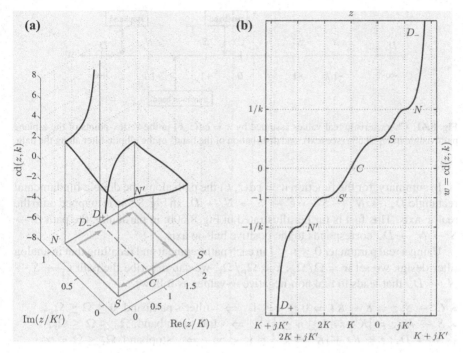

Fig. 8.60 Real values assumed by $w = \mathrm{cd}(z, k)$ while going through the path $D_+ \to N' \to S' \to C \to S \to N \to D_-$ in Fig. 8.58a. **a** Values plotted in 3-D above the complex z-plane; **b** values plotted in 2-D versus the value assumed by z along the above-mentioned path, indicated by *arrows* in panel a. The z-values shown on the horizontal axis correspond to the z-plane coordinates of the points D_+, N', S', C, S, N, and D_-. These are complex values, so the horizontal axis in this panel is not an ordinary real Cartesian axis. In this example, $k = 0.5$

approaching N'. Halfway through the segment $N \to N'$ there is a pole. The path we are considering begins immediately at the right of the pole and ends immediately at the left of it, so that the pole is not included in the contour: the symbols D_-, D_+ used in Fig. 8.58a actually are meant to suggest this fact.

Now observe Fig. 8.60a, which shows a three-dimensional plot of the real values assumed by $w = \mathrm{cd}(z, k)$ along the above-mentioned contour. Note that here we see $\mathrm{cd}(z, k)$ and not just its absolute value. Starting from point D_+ at which $\mathrm{cd}(z, k)$ is close to $-\infty$, while we go through the prescribed path in the direction indicated by the arrows in Figs. 8.58a and 8.60a, the values of the function w increase, and tend to $+\infty$ as we get close to D_-. The function reverses in sign at point C: it is negative before C and positive after C. This appears logical if we observe the numbers representing quadrants of the w-plane in Fig. 8.58a and b: the negative $\mathrm{Re}(w)$ half axis borders the third quadrant of the w-plane, while the positive $\mathrm{Re}(w)$ half axis borders the fourth quadrant.

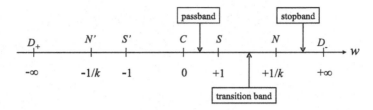

Fig. 8.61 Characteristic real values assumed by $w = \mathrm{cd}(z, k)$ in the vertex-points of the z-plane path visible in Fig. 8.58a (see text), and distribution of the bands of the elliptic filter along the path

In summary, for the function $w = \mathrm{cd}(z, k)$ the path along the double fundamental rectangle $D_+ \to N' \to S' \to C \to S \to N \to D_-$ in Fig. 8.58a is mapped onto the real w-axis. This fact is further illustrated in Fig. 8.60b: in particular, the path $C \to S \to N \to D_-$ corresponds to the positive half-w-axis.

Using a real parameter $0 \leq t \leq 1$ in each path segment and recalling that in analog filter design we set $w = \Omega/\Omega_p$, $k = \Omega_p/\Omega_s$, we can describe the path $C \to S \to N \to D_-$ that leads to real non-negative w-values as follows.

- $C \to S$: $z = K - Kt \Rightarrow 0 \leq w \leq 1 \quad \Rightarrow \quad$ filter's passband, $0 \leq \Omega \leq \Omega_p$;
- $S \to N$: $z = jK't \Rightarrow 1 \leq w \leq k^{-1} \quad \Rightarrow \quad$ transition band, $\Omega_p \leq \Omega \leq \Omega_s$;
- $N \to D_-$: $z = Kt + jK' \Rightarrow k^{-1} \leq w < \infty \quad \Rightarrow \quad$ stopband, $\Omega_s \leq \Omega < \infty$.

Similar expressions for the negative half-axis can be written along the path $D_+ \to N' \to S' \to C$; these values can be derived using the properties of $\mathrm{cd}(z, k)$ (Orfanidis 2006). The distribution of the bands of the elliptic filter along the path is further illustrated in Fig. 8.61.

If instead of the path described above, which is located in the upper quadrants of the z-plane, we select a similar path in the lower quadrants, we obtain exactly the same real w-values: indeed, the negative part of the real w-axis does not belong only to the third quadrant of the w-plane, but also to the second quadrant; the positive part of the real w-axis does not belong only to the fourth quadrant of the w-plane, but also to the first quadrant.

We thus understood in which points of the z-plane the function cd assumes those real values that are fit to represent the normalized frequencies w in the design of analog elliptic filters.

In the calculations concerning analog elliptic filters, the inversion of Jacobi elliptic functions $w = \mathrm{cd}(z, k)$ and $w = \mathrm{sn}(z, k)$ is also required (Orfanidis 2005, 2006): for example, we need to compute $z = \mathrm{cd}^{-1}(w, k)$, so as to find, for a given k, the value of z that corresponds to a certain value of w. In general, such a z value is not univocally defined, but it becomes univocally defined if we restrict ourselves to the fundamental region of the z-plane corresponding to the *gray-shaded rectangles* in Fig. 8.58a, i.e., $0 \leq \mathrm{Re}(z) \leq 2K$, $-K' \leq \mathrm{Im}(z) \leq K'$. In fact, in a period-rectangle *two* values of z exist that correspond to a fixed value of $w = \mathrm{cd}(z, k)$. One of these values is found in the fundamental region, and the other one in the adjacent region (the union of the *white rectangles* in Fig. 8.58a). For example, if $z = \mathrm{cd}^{-1}(w, k)$ lies in the fundamental region, then $z_1 = 4K - z$ lies in the adjacent region and we have $w = cd(z, k) = cd(z_1, k)$. Similar considerations hold for $sn(z, k)$: in this case, the fundamental region is

$$-K \le \text{Re}(z) \le K, \qquad -K' \le \text{Im}(z) \le K'.$$

However, if we remember that $sn(z, k) = cd(K - z, k)$, we understand that actually we only need to devise a method to invert cd; then we will use $z = sn^{-1}(w, k) = K - cd^{-1}(w, k)$ to invert sn, if needed. This inversion is performed by the *Landen-Gauss transformation*, presented in the next subsection.

We can conclude this discussion by noting that often z values are expressed in units of real quarter period K, by writing

$$uK = z, \qquad u = \frac{z}{K}.$$

The variable u is thus a complex variable that can be represented in yet another plane: the u-plane. In terms of u, the fundamental region of $cd(z, k)$ becomes

$$0 \le \text{Re}(u) \le 2, \qquad -K'/K \le \text{Im}(u) \le K'/K,$$

and the positions of the zero and of the pole on the edge of the fundamental rectangle are determined by K'/K only: the pole in D corresponds to $u = 1 + j(K'/K)$, while the zero in C corresponds to $u = 1$. The function $w = cd(z, k)$ is then written as $w = cd(uK, k)$, and similar expressions are written for the other Jacobi elliptic functions. Moving in the u-plane along a path corresponding to the z-plane path used so far, $w = cd(uK, k)$ will obviously be real. This is actually the notation that we saw in Sect. 8.3.2.

8.12.4 Landen-Gauss Transformation

We must now explain how the Jacobi elliptic functions are computed numerically, with focus on cd and sn. When $k \to 0$, the ratio K'/K tends to infinity, and Jacobi elliptic functions degenerate into trigonometric circular functions; their period-rectangles degenerate into period-strips. When $k \to 1$, then $K'/K \to 0$, and Jacobi elliptic functions degenerate into trigonometric hyperbolic functions; their period-rectangles degenerate into the corresponding period-strips. A symmetry exists between the two limit cases:

- in the trigonometric limit ($k = 0$) we have $k' = 1$, $K = \pi/2$ and $K' \to \infty$;
- in the hyperbolic limit ($k = 1$) we have $k' = 0$, $K \to \infty$ and $K' = \pi/2$.

Therefore the Jacobi elliptic functions lie, in a sense, along a continuous path that extends from trigonometric functions to hyperbolic functions; the parameter k and its complementary value k' measure the "position" of the considered function along this path, i.e., they give the "distance" of the function from the path ends. At the middle of this path, the fundamental rectangle is a square, with $K = K' \approx 1.854$ and $k = k' = 1/\sqrt{2}$.

The *Landen-Gauss transformation* (Orchard and Willson 1997) is a method for moving along this path by discrete steps, in one direction or in the opposite one, by modifying k or k' in such a way that at each step the ratio K'/K is doubled or halved. The values of the considered function before and after this transformation are algebraically related. In this way, we can move from any value of k to a value for which the considered Jacobi elliptic function is numerically indistinguishable from the corresponding limit-function that represents the beginning or the end of the path. Then we can evaluate the proper limit function and successively we can thread the same path in the opposite direction, calculating the intermediate elliptic function at each step from the previous one through the algebraic relations that connect them. We thus can go back until the desired elliptic function is found that corresponds to the value originally given for k. Usually, since it is simpler to compute trigonometric functions rather than hyperbolic ones, the procedure is applied towards and then backwards from the trigonometric path end.

A first transformation, referred to as the *Landen transformation*, allows mapping the parameters and the values of the elliptic integral. Starting from a given value of k, a sequence of values k_n rapidly decreasing toward zero is generated by the recursion formula (Orchard and Willson 1997; Orfanidis 2006)

$$ k_n = \left(\frac{k_{n-1}}{1 + k'_{n-1}} \right)^2 , \qquad n = 1, 2, \ldots M, \qquad \text{with} \qquad k'_{n-1} = \left(1 - k_{n-1}^2 \right)^{1/2} , $$

which is initialized by setting $k_0 = k$. The recursion is stopped at $n = M$, when k_M, which is zero in the trigonometric limit, has become smaller than a specified tolerance level.[18] For all values of k encountered in the practice of elliptic filter design, which are normally included in the range $0 \leq k \leq 0.999$, the recursion can stop at $M = 5$, since the following k_n values turn out to be less than 10^{-15}; for $k \leq 0.99$ they are less than 10^{-20}. The inverse recursion formula is

$$ k_{n-1} = \frac{2\sqrt{k_n}}{1 + k_n}, \qquad n = M, M - 1, \ldots 1. $$

The recursion formula for k corresponds to a recursion formula for the complete elliptic integral K: if we set $K_n = K(k_n)$, then we have

$$ K_{n-1} = (1 + k_n) K_n, $$

---

[18]This tolerance level can be taken equal to the *machine epsilon* ε, which is defined as the absolute value of the difference between a positive floating point number and the closest higher floating point number. The machine epsilon depends on the particular software used: for example, in Matlab $\varepsilon = 2^{-52} = 2.2204 \times 10^{-16}$.

and therefore, starting from the initial value $K_0 = K(k = k_0)$, we have $K_0 = (1 + k_1)K_1 = (1 + k_1)(1 + k_2)K_2$ etc.:

$$K = (1 + k_1)(1 + k_2) \cdots (1 + k_M)K_M, \qquad K_M = \frac{\pi}{2}.$$

In fact, k_M is close to zero, hence K_M is very close to $\pi/2$. By applying the Landen recursion to k', we can compute K' in a similar way.

The Landen recursion also allows for evaluating efficiently the function $\mathrm{cd}(uK, k)$, as well as the function $\mathrm{sn}(uK, k)$. To this purpose, the Landen transformation must be associated with the retrograde recursion referred to as the *Gauss transformation* (Byrd and Friedman 1971):

$$\frac{1}{\mathrm{cd}(uK_{n-1}, k_{n-1})} = \frac{1}{1 + k_n} \left[\frac{1}{\mathrm{cd}(uK_n, k_n)} + k_n \mathrm{cd}(uK_n, k_n) \right], \qquad n = M, M - 1, \ldots 1.$$

The recursion is initialized with $n = M$, where k_M is so small that the function cd is indistinguishable from a cosine: $\mathrm{cd}(uK_M, k_M) \approx \cos(u\pi/2)$. In such a way, the computation of $w = \mathrm{cd}(uK, k)$ for any complex value of u proceeds evaluating $w_n = \mathrm{cd}(uK_n, k_n)$, with start at $w_M = \mathrm{cd}(uK_M, k_M)$ and end at $w_0 = \mathrm{cd}(uK_0, k_0) = \mathrm{cd}(uK, k) = w$:

$$w_{n-1}^{-1} = \frac{1}{1 + k_n} \left(w_n^{-1} + k_n w_n \right), \qquad n = M, M - 1, \ldots 1.$$

The recursive procedure described above can also be used to calculate the corresponding inverse functions: it is sufficient to proceed in the opposite direction, i.e., onwards from $n = 1$ to $n = M$:

$$w_n = \frac{2w_{n-1}}{(1 + k_n) \left(1 + \sqrt{1 - k_{n-1}^2 w_{n-1}^2} \right)}, \qquad n = 1, 2, \ldots M.$$

Starting from a given complex value $w = \mathrm{cd}(uK, k)$, and setting $w_0 = w$, we will carry on the recursion till $w_M = \cos(u\pi/2)$ that can be easily inverted to yield $u = (2/\pi) \arccos(w_M)$, and therefore $z = uK$. However, in this case care must be taken about the fact that u is not univocally defined, and that in order to make it unique, we must select that value that falls inside the fundamental region, $0 \leq \mathrm{Re}(u) \leq 2$ and $-K'/K \leq \mathrm{Im}(u) \leq K'/K$. In a similar way, we can use this transformation to invert $w = \mathrm{sn}(uK, k)$, except that we will write $u = (2/\pi) \arcsin(w_M)$ and that u will be constrained to lie inside the range $- \leq \mathrm{Re}(u) \leq 1$, $-K'/K \leq \mathrm{Im}(u) \leq K'/K$.

8.12.5 Elliptic Rational Function

The *elliptic rational function* $R_N(w)$ is a real rational function in the real variable w, with order N, which is even for even N and odd for odd N. It is defined as follows:

$$R_N(w) = \mathrm{cd}\,(NuK_1, k_1) \quad \text{with} \quad w = \mathrm{cd}(uK, k) \Rightarrow uK = \mathrm{cd}^{-1}\,(w, k).$$

Let us underline that in this definition, it is meant that the variable w is real, and therefore composed by values of $\mathrm{cd}(uK, k)$ computed along the path shown in Fig. 8.60a.

Let us set

$$u_1 = Nu, \qquad R_N(w) = \mathrm{cd}(u_1 K_1, k_1).$$

Now we must understand on what path in the complex u- and u_1-planes the function $\mathrm{cd}\,(NuK_1, k_1) = \mathrm{cd}\,(u_1 K_1, k_1)$ is computed, and what values it assumes.

Figure 8.62a shows, for $N = 3$, the path in the complex u-plane that is followed to compute w-values and corresponding $R_N(w)$-values. The real and imaginary parts of u are reported on the inner axes; the outer axes show the corresponding values of $\mathrm{Re}(u_1)$, $\mathrm{Im}(u_1)$. Also characteristic values of R_N are shown. The path borders a large rectangle which is the union of three smaller rectangles. The large rectangle is a fundamental one for $w = \mathrm{cd}(uK, k)$, while the smaller rectangles are a division of the large rectangle into $N = 3$ equal parts. Each of them is a fundamental rectangle for $R_N(w) = \mathrm{cd}(u_1 K_1, k_1)$. This figure is useful to visualize what happens to the variable u_1, and therefore to $R_N(w) = \mathrm{cd}(u_1 K_1, k_1)$, as u goes around the edge of the fundamental rectangle belonging to $w = \mathrm{cd}(uK, k)$. The path is exactly the one indicated as $C \rightarrow S \rightarrow N \rightarrow D_-$ in Figs. 8.58a and 8.60; the only difference is that we are now using $u = z/K$ instead of z. In Fig. 8.62b, the case $N = 4$ is illustrated in a similar way.

From Fig. 8.61 we can see that the values assumed by w along the path depicted in Fig. 8.62a for $N = 3$ and in Fig. 8.62b for $N = 4$ are: 0 at the beginning of the path, 1 in the origin of the u-plane, k^{-1} on the upper-left corner of the large rectangle in Fig. 8.62a or b, and infinity at the path end. This is shown in Fig. 8.62c. This figure is meant to show which band of the elliptic filter corresponds to each different stretch of the path in the u-plane (see the *solid arrow* for the passband, the *dashed arrow* for the transition band, and the *dash-dotted arrow* for the stopband).

As shown in Fig. 8.62, the parameter k_1 is constrained by the relation $NK'/K = K'_1/K_1$ that can be understood as follows. An elliptic filter must be equiripple both in the passband and in the stopband. This is achieved by defining the function $R_N(w)$ in such a way that it is equiripple in the filter's passband, and then imposing on $R_N(w)$ the condition known as *inversion relation*,

$$R_N(w) = \frac{1}{k_1 R_N(w^{-1}k^{-1})},$$

which ensures an equiripple filter behavior in the stopband too (Sect. 8.6). The inversion relation incorporates in itself the elliptic degree equation, $R_N(k^{-1}) = k_1^{-1}$: indeed, by substituting $w = k^{-1}$ into the inversion relation we get

$$R_N(k^{-1}) = k_1^{-1} R_N^{-1}(k^{-1}k) = k_1^{-1} R_N^{-1}(1) = k_1^{-1},$$

since $R_N(1)$ is normalized to 1.

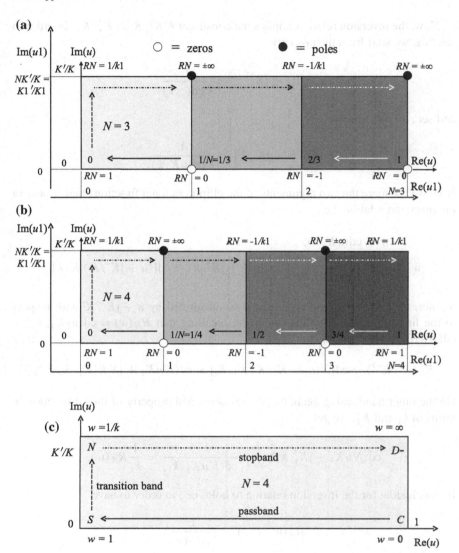

Fig. 8.62 **a** Mapping of $R_N(w) = \text{cd}(u_1 K_1, k_1)$ in the u-plane, with $u_1 = Nu$, under the condition $NK'/K = K_1'/K_1$, for $N = 3$. The inner axes represent $\text{Re}(u)$ and $\text{Im}(u)$; the outer axes add the information about $\text{Re}(u_1)$ and $\text{Im}(u_1)$. The *arrows* depict the path that is followed; this path in the u-plane corresponds to the z-plane path $C - S - N - D_-$ visible in Fig. 8.58a. See text for details. **b** The same for $N = 4$. **c** Real values assumed by w along the same path in the u-plane and corresponding bands of the elliptic filter whose frequency response is built using $R_N(w)$. The *arrows* in different styles highlight this band distribution—they are *solid* in the passband, *dashed* in the transition band, and *dash-dotted* in the stopband

Now, the inversion relation implies the constraint $NK'/K = K_1'/K_1$. In order to see this, we start from the property

$$\text{cd}(z + \text{j}K', k) = \frac{1}{k\,\text{cd}(z, k)}$$

and set $z = uK$ to write

$$\text{cd}(uK + \text{j}K', k) = \frac{1}{k\,\text{cd}(uK, k)}.$$

Now we compare the two arguments of the elliptic rational functions that appear in the inversion relation, i.e.,

$$w = \text{cd}(uK, k),$$
$$w^{-1}k^{-1} = \frac{1}{k\,\text{cd}(uK, k)} = \text{cd}(uK + \text{j}K', k) = \text{cd}[(u + \text{j}K'/K)\,K, k],$$

we note that in the second argument u is substituted by $u + \text{j}K'/K$ with respect to the first argument. We thus can write, recalling that $R_N(w) = \text{cd}(u_1 K_1, k_1) = \text{cd}(NuK_1, k_1)$ and setting $z = NuK_1$,

$$R_N(w^{-1}k^{-1}) = \text{cd}[N(u + \text{j}K'/K)\,K_1, k_1] = \text{cd}(NuK_1 + \text{j}NK'K_1/K, k_1).$$

On the other hand, using again the above-mentioned property of the cd function in terms of k_1 and K_1', we get

$$\text{cd}(NuK_1 + \text{j}K_1', k_1) = \frac{1}{k_1\,\text{cd}(NuK_1, k_1)} = \frac{1}{k_1}R_N(w).$$

In conclusion, for the inversion relation to hold, i.e., in order to have

$$R_N(w^{-1}k^{-1}) = \frac{1}{k_1 R_N(w)},$$

the following equation must hold for any u:

$$\text{cd}(NuK_1 + \text{j}NK'K_1/K, k_1) = \text{cd}(NuK_1 + \text{j}K_1', k_1).$$

But then we must impose the condition $NK'K_1/K = K_1'$, i.e.,

$$\frac{NK'}{K} = \frac{K_1'}{K_1}, \qquad \text{q.e.d.}$$

This constraint gives the fundamental rectangle of $R_N(w) = \text{cd}\,(u_1 K_1, k_1)$ the shape and the dimensions in the u-plane that are needed for it to be contained horizontally

inside the fundamental rectangle of $w = \text{cd}(uK, k)$ exactly N times. With reference
to Fig. 8.62b ($N = 4$), as u starts its path and moves along its own real axis from 1 to
$1 - 1/N = 3/4$, thus covering a distance $1/N = 1/4$, u_1 moves along its real axis
from 4 to 3, thus covering a distance 1 and completing one side of a first fundamental
rectangle of $\text{cd}(u_1 K_1, k_1)$. In the following stretch of path, u goes from 3/4 to 1/2,
while u_1 goes from 3 to 2, i.e., it goes by one more unit distance, and completes
one side of a second fundamental rectangle of $\text{cd}(u_1 K_1, k_1)$. Then, u goes from
1/2 to 1/4, while u_1 goes from 2 to 1, i.e., it goes by one more unit distance, and
completes one side of a third fundamental rectangle of $\text{cd}(u_1 K_1, k_1)$. Finally, u goes
from 1/4 to 0 while u_1 goes from 1 to 0, i.e., it goes by one more unit distance,
and completes one side of a fourth and last fundamental rectangle of $\text{cd}(u_1 K_1, k_1)$.
Now the path becomes vertical: u moves from 0 to jK'/K, while u_1 goes from 0 to
$jNK'/K = jK_1'/K_1$ (note that the imaginary axes in Fig. 8.62b are scaled in such a
way to show this correspondences clearly). A similar return journey follows, with u_1
that travels a unit distance for four times, while u travels a single unit distance. Along
the path, real values of $\text{cd}(u_1 K_1, k_1)$ are found, in a one-to-one correspondence with
real w values. This produces the real attenuation function $R_N(w)$ used to construct
the squared frequency response of the elliptic filter. The choice of k_1 dictated by the
degree equation and the simple geometric relation between the u- and u_1-fundamental
rectangles that is a consequence from that choice are the ingredients for getting an
$R_N(w)$ which is a rational function of $w = \text{cd}(uK, k)$.

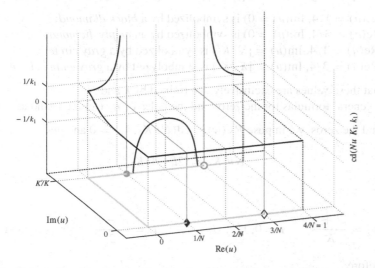

Fig. 8.63 The elliptic rational function for $N = 4$ and $k = 0.5$, drawn in 3-D above the complex u-
plane. The *gray segments* on the plot floor describe the path followed in the u-plane, and correspond
to the passband, the transition band and the stopband of the elliptic filter. The two *circles* mark the
points at which the functions goes to infinity, while the two *diamonds* mark the points at which the
function vanishes

Figure 8.62b also illustrates, for $N = 4$, how the poles and zeros and the characteristic values of $R_N(w) = \mathrm{cd}(u_1 K_1, k_1)$ are distributed with respect to the fundamental rectangle of $\mathrm{cd}(uK, k)$. A three-dimensional view of $R_N(w)$ in the u-plane for $N = 4$ is shown in Fig. 8.63, for $k = 0.5$. The *gray segments* on the plot floor describe the path followed in the u-plane, and correspond to the passband, the transition band and the stopband of the elliptic filter. The two *circles* mark the points at which the functions goes to infinity, while the two *diamonds* mark the points at which the function vanishes.

The w-values corresponding to the zeros and poles of $R_N(w)$ for $N = 4$, visible in Figs. 8.62b and 8.63, correspond to the following values of w:

$$w \equiv \zeta_0 = \mathrm{cd}(K/N, k) = \mathrm{sn}\left(\frac{N-1}{N}K, k\right) \quad \text{(a zero)},$$

$$w \equiv \zeta_1 = \mathrm{cd}\left(\frac{N-1}{N}K, k\right) = \mathrm{sn}(K/N, k) \quad \text{(another zero)},$$

$$w \equiv \pi_0 = \mathrm{cd}(K/N + jK', k) = \frac{1}{k\,\mathrm{cd}(K/N, k)} \quad \text{(a pole)},$$

$$w \equiv \pi_1 = \mathrm{cd}\left(\frac{N-1}{N}K + jK', k\right) = \frac{1}{k\,\mathrm{cd}\left(\frac{N-1}{N}K, k\right)} = \frac{1}{k\,\mathrm{sn}(K/N, k)} \quad \text{(another pole)},$$

where we adopted the symbols ζ_l and π_l, with $i = 0$ or 1, to indicate zeros and poles, respectively. In Fig. 8.63, these points are distinguished by different markers:

- ζ_0 ($\mathrm{Re}(u) = 1/4$, $\mathrm{Im}(u) = 0$) is symbolized by a *black diamond*;
- ζ_1 ($\mathrm{Re}(u) = 3/4$, $\mathrm{Im}(u) = 0$) is symbolized by an *empty diamond*;
- π_0 ($\mathrm{Re}(u) = 1/4$, $\mathrm{Im}(u) = jK/K'$) is symbolized by a *gray circle*;
- π_1 ($\mathrm{Re}(u) = 3/4$, $\mathrm{Im}(u) = jK/K'$) is symbolyzed by a *gray empty circle*.

Note that these values are exclusively determined by k and N.

The general formulas for the zeros and the poles of $R_N(w)$ are found as follows.

- To find the zeros, we impose $R_N(w) = \mathrm{cd}(NuK_1, k_1) = 0$ and get

$$Nu_l K_1 = (2l + 1)K_1,$$

with

$$u_l = \frac{2l + 1}{N}, \qquad l = 0, 1, \ldots L - 1, \qquad L = \frac{N - r}{2}, \qquad r = N \bmod 2.$$

Therefore
$$\zeta_l = \mathrm{cd}(u_l K, k), \qquad l = 0, 1, \ldots L - 1.$$

- To find the poles π_l, we can use the inversion relation, $R_N(w^{-1}k^{-1}) = 1/[k_1 R_N(w)]$, from which we deduce

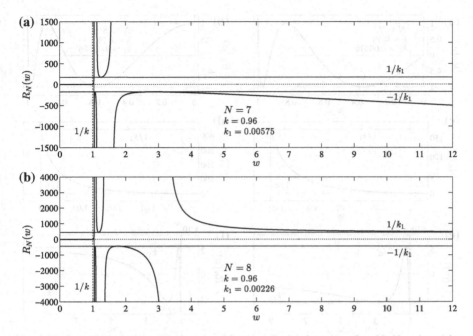

Fig. 8.64 Overall view of the elliptic rational function $R_N(w)$ for **a** $N = 7$ and **b** $N = 4$, and for the indicated values of k and k_1. In both panels, *horizontal solid lines* mark ordinates equal to $\pm k_1^{-1}$, while *vertical dotted* and *solid lines* indicate the abscissas $w = 1$ and $w = k^{-1}$, respectively (see text)

$$\pi_l = \frac{1}{k\zeta_l}.$$

All the zeros of $R_N(w)$ fall in the filter's passband; all the poles fall in the stopband. Note that in the case of $N = 4$, $L = 2$, so that l assumes only the values 0 and 1 and we only have $u_0 = 1/N$ and $u_1 = 3/N = (N-1)/N$. We thus find exactly the u-values of the example of Fig. 8.63.

We therefore have $N/2$ zeros and poles of $R_N(w)$ for even N and $(N-1)/2$ zeros and poles for odd N at finite distance from the origin; for odd N, an additional pole is present at $w \to \infty$, and an additional zero is present at $w = 0$, as can be seen observing Fig. 8.62a, which is relative to $N = 3$.

The rational function $R_N(w)$ is illustrated in Figs. 8.64 and 8.65, for $N = 7$ and 8 and $k = 0.96$. Note that when N changes, k being equal, in agreement with the degree equation, k_1 also changes. These moduli are involved in the calculation of the values of $R_N(w)$, even if they do not appear explicitly in the function's name.

Figure 8.64a, b shows the function for $N = 7$ and 8, respectively. The adopted values of k and k_1 are displayed. Only positive values of w are included in these plots; for negative w the function repeats itself symmetrically or antisymmetrically, according to N being even or odd. *Horizontal solid lines* mark ordinates equal to $\pm k_1^{-1}$, which identify a belt in which no values of $R_N(w)$ are found for $w > k^{-1}$

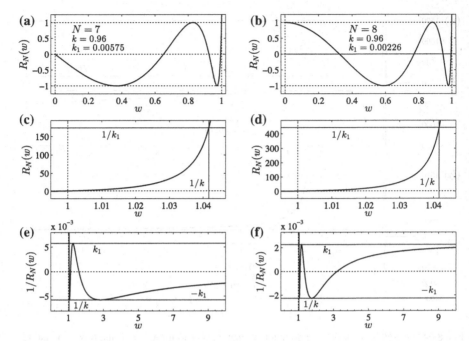

Fig. 8.65 Detailed behavior of the elliptic rational function $R_N(w)$ for $N = 7$ and 8 and for the indicated values of k and k_1: close-ups of **a**, **b** $R_N(w)$ in the passband of the corresponding elliptic filter, **c**, **d** $R_N(w)$ in the transition band, and **e**, **f** $1/R_N(w)$ in the stopband. *Horizontal solid lines* in panels **c**, **d** mark the ordinate k_1^{-1}. In panels **c–f**, *vertical solid lines* correspond to $w = k^{-1}$; *horizontal solid lines* mark the ordinates k_1^{-1} (panels **c**, **d**) and $\pm k_1$ (panels **e**, **f**; see text)

(stopband of the elliptic filter); the function assumes the value 1 at $w = 1$ (*vertical dotted line*) and the value k_1^{-1} at $w = k^{-1}$ (*vertical solid line*). Expanded views of the function are given by Fig. 8.65a, b, for $N = 7$ and 8, respectively:

- Figure 8.65a shows $R_N(w)$ for $N = 7$ and w comprised between 0 and 1 (*vertical dotted lines*), i.e., in the passband of the corresponding elliptic filter;
- Figure 8.65b shows the same for $N = 8$;
- Figure 8.65c shows $R_N(w)$ for $N = 7$ and w comprised between 1 (*vertical dotted line*) and k^{-1} (*vertical solid line*), i.e., in the transition band of the elliptic filter; the *horizontal solid line* marks the ordinate k_1^{-1}, which is the value assumed by $R_N(w)$ for $w = k^{-1}$; moreover the function assumes the value 1 at $w = 1$;
- Figure 8.65d shows the same for $N = 8$;
- Figure 8.65e shows $1/R_N(w)$ for $N = 7$ and $w > k^{-1}$ (*vertical solid line*), i.e., in the filter's stopband; *horizontal solid lines* mark the ordinates $\pm k_1$, identifying a belt inside which $1/R_N(w)$ is contained when $w > k^{-1}$;
- Figure 8.65f shows the same for $N = 9$.

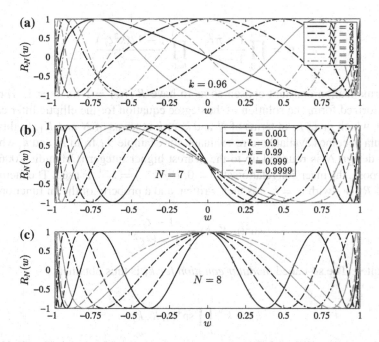

Fig. 8.66 The elliptic rational function in the range $w = [-1, 1]$, for **a** different values of N from 3 to 8, and $k = 0.96$; **b** different values of k, and $N = 7$; **c** different values of k, and $N = 8$

Figure 8.64 illustrates how, as we learned in the foregoing discussion, the function exhibits an increasing number of poles and zeros, as N increases; Fig. 8.65 shows that the function is, as anticipated,

- equiripple and such that $-1 \leq R_N(w) \leq 1$ in the range $0 \leq w \leq 1$ (filter passband),
- monotonically increasing in the interval $1 \leq R_N(w) \leq k_1^{-1}$ for $1 \leq w \leq k^{-1}$ (transition band),
- characterized by an equiripple inverse $1/R_N(w)$, with $|1/R_N(w)| \leq k_1$, in the range $k^{-1} \leq w < \infty$.

The symmetry/antisymmetry of the function around $w = 0$, according to N being even or odd, is illustrated by Fig. 8.66a that shows, for $N = 3$ to 8 and in the range $w = [-1, 1]$, the dependence of $R_N(w)$ on N at fixed $k = 0.96$. Figure 8.66b, c illustrate instead the dependence of $R_N(w)$ on k at fixed N over the same w range; considered values of N are 7 (Fig. 8.66b) and 8 (Fig. 8.66c).

Having thus found the expressions of the zeros and poles of the rational elliptic function, we can finally build $R_N(w)$ as a rational function from its zeros and poles, obtaining the factorized form already encountered in Sect. 8.6:

$$R_N(w) = C w^r \prod_{l=0}^{L-1} \frac{w^2 - \zeta_l^2}{w^2 - \pi_l^2} = w^r \prod_{l=0}^{L-1} \frac{\left(w^2 - \zeta_l^2\right)\left(1 - k^2\zeta_l^2\right)}{\left(1 - w^2 k^2 \zeta_l^2\right)\left(1 - \zeta_l^2\right)},$$

where

$$C = \prod_{l=0}^{L-1} \frac{1 - \pi_l^2}{1 - \zeta_l^2} = \prod_{l=0}^{L-1} \frac{1 - 1/(k^2 \zeta_l^2)}{1 - \zeta_l^2}$$

is a normalization constant chosen so as to have $R_N(w) = 1$ for $w = 1$. Through this factorized form, the solution of the degree equation for the elliptic filter can be derived, which gives k_1 in terms of N and k, or k in terms of N and K_1. In this way a formula becomes available to re-evaluate, for example, k_1 from N and k, when in elliptic design N is rounded up to the nearest higher integer (Orfanidis 2006). To this purpose, the inversion relation at $w = 0$, $R_N(k^{-1}) = k_1^{-1}$, is used. The factorized form of $R_N(w)$ with $w = k^{-1}$ is then written, and a property of the sn function,

$$\text{sn}^2(u_l K, k) = \frac{1 - \zeta_l^2}{1 - k^2 \zeta_l^2},$$

is exploited. The so-called *modular equation for k_1* is thus obtained,

$$k_1 = k^N \prod_{l=0}^{L-1} \text{sn}^4(u_l K, k).$$

Moreover, the fact that the degree equation is invariant with respect to the substitution $k \to k_1'$, $k_1 \to k'$ can be exploited. This invariance is shown by the fact that with such a substitution, the formula $NK'/K = K_1'/K_1$ becomes $NK_1/K_1' = K/K'$, i.e., $NK'/K = K_1'/K_1$ as before. Due to the degree-equation invariance, the modular equation found for k_1 also serves to write k in terms of k_1 and N, through the complementary moduli:

$$1 - k = k' = \left(k_1'\right)^N \prod_{l=0}^{L-1} \text{sn}^4(u_l K_1', k_1).$$

This equation is known as the *modular equation for k*.

References

Abramowitz, M., Stegun, I.A. (eds.): Handbook of Mathematical Functions with Formulas, Graphs, and Mathematical Tables. Dover, New York (1972)
Antoniou, A.: Digital Filters: Analysis, Design, and Applications. McGraw-Hill, Blacklick (1993)
Bochner, S., Chandrasekharan, K.: Fourier Transforms. Princeton University Press, Princeton (1949)
Bracewell, R.N.: The Fourier Transform and Its Applications. McGraw-Hill, Boston (2000)
Byrd, P.F., Friedman, M.D.: Handbook of Elliptic Integrals for Engineers and Scientists. Springer, New York (1971)
Constantinides, A.G.: Spectral transformations for digital filters. Proc. IEEE 117, 1585–1590 (1970)

Kreyszig, E.: Advanced Engineering Mathematics. Wiley, Hoboken (2011)

Lutovac, M.D., Tosic, D.V., Evans, B.L.: Filter Design for Signal Processing Using MATLAB and Mathematica. Prentice-Hall, Upper Saddle River (2001)

Olver, F.W.J., Lozier, D.W., Boisvert, R.F., Clark, C.W.: NIST Handbook of Mathematical Functions. Cambridge University Press, New York (2010)

Orchard, H.J., Willson, A.N.: Elliptic functions for filter design. IEEE Trans. Circuits Syst. I **44**, 273–287 (1997)

Orfanidis, S.J.: High-order digital parametric equalizer design. J. Audio Eng. Soc. **53**, 1026–1046 (2005)

Orfanidis, S.J.: Lecture Notes on Elliptic Filter Design. http://eceweb1.rutgers.edu/~orfanidi/ece521/notes.pdf (2006)

Oppenheim, A.V., Willsky, A.S., Hamid, S.: Signals and Systems. Prentice Hall, Englewood Cliffs (1996)

Pao, Y.C.: Engeneering Analysis. CRC Press LLC, Boca Raton (1999)

Porat, B.: A Course in Digital Signal Processing. Wiley, New York (1996)

Porat, B., Friedlander, B.: The modified Yule-Walker method of ARMA spectral estimation. IEEE Trans. Aerosp. Electron. Syst. **AES-20**(2), 158–173 (1984)

Saff, E.B., Snider, A.D.: Fundamentals of Complex Analysis with Applications to Engineering and Science. Prentice Hall, Upper Saddle River (2002)

Shenoi, B.A.: Introduction to Digital Signal Processing and Filter Design. Wiley, New York (2005)

Part III
Spectral Analysis

Chapter 9
Statistical Approach to Signal Analysis

9.1 Chapter Summary

So far we have considered deterministic signals, for which each sample is univocally determined by a mathematical function or by some rule, so that past and present values of the signal are perfectly known and future values are exactly predictable. In reality, however, we often analyze sequences that derive from periodically repeated measurements of some quantity—e.g., temperature, gross domestic product, voltage at an EEG electrode—performed during a finite time interval, and we are interested in the characteristic of the quantity, rather than in those of the particular discrete-time signal we measured. In fact, the ultimate subject of the investigation is the process that generates the measured values: for example, we typically may want to get information about its spectral features. The generating process may, in general, depend on time.

We must keep in mind that the generating process may have—and usually does have—a temporal persistence that goes beyond the finite time limits of our measurements. From a theoretical point of view, the process is an infinitely-long signal, typically an analog signal, of which we can measure only a finite number of samples over a finite time span. We must also consider that even if we were measuring an intrinsically deterministic process, any measurement would be affected by random fluctuations/errors. The resulting discrete-time signal should thus be interpreted as the superposition of a deterministic signal and a random signal. Most times, anyway, the process that generates the measured quantity is so complicated that we do not know it enough to be able to express it in deterministic terms, i.e., through a well-defined rule or formula; and even if we were able to do so, the deterministic description would be too complicated to be of practical use.

On the other hand, processing and analyzing a signal requires a mathematical description of the signal—a mathematical model. In the present case, a statistical-probabilistic model is adopted: a record of measurements is classified as a discrete-time random signal, or process, derived by the sampling of some analog random

© Springer International Publishing Switzerland 2016
S.M. Alessio, *Digital Signal Processing and Spectral Analysis for Scientists*, Signals and Communication Technology, DOI 10.1007/978-3-319-25468-5_9

signal, or process. Note that the words signal and process in this sense are used interchangeably, even if strictly speaking, the process generates the signal. The value of any random signal is not exactly specified at a given past or current instant, and future values are not predictable with certainty on the basis of past behavior. In other words, the sequence we are analyzing would not be exactly reproducible by repeating the set of measurement (Blackman and Tukey 1958; Hannan 1960).

This chapter provides a brief introduction to the basic theory of discrete-time random processes.

A discrete-time random process is represented as an indexed sequence of random variables, with the index representing discrete time, to allow for time dependence. Each random variable at some given time n can be described by theoretical average quantities, like the mean, the variance, and the autocorrelation/autocovariance. These are called ensemble averages, because the averaging is performed over the infinite number of values that in principle could be the outcome of a measurement performed at time n. To compute these theoretical averages, we should specify the probabilistic laws characterizing the random variable. Since these laws are unknown in most cases, the problem is simplified assuming for the random process both stationarity, i.e., no dependence on time, and ergodicity, a property that allows for substituting ensemble averages with other quantities: time averages that can be calculated even if only a single infinite-length sequence derived from the process is—in principle— accessible. These time averages are then estimated in practice from a finite-length data record. A more detailed treatment of these topics may be found in Hayes (1996).

A discrete-time random signal of infinite length has infinite energy and therefore does not allow for a DTFT representation. However, it can be a power signal, with finite-energy autocorrelation/autocovariance. The DTFT of autocovariance then provides a spectral representation of the discrete-time random power signal. This representation is the power spectrum, introduced at the end of the chapter. Also the representation of the common spectral content of two random signals and of the phase relations of the respective frequency components is briefly discussed. This representation is the cross-power spectrum.

9.2 Preliminary Considerations

A sequence $x[n]$ made of measurements of a quantity, performed at different times n for a finite time duration ($n = [0, N - 1]$), stems from an infinitely-long random process (Fig. 9.1). The latter is typically a continuous-time, i.e., analog, process, even if it could be intrinsically discrete-time as well. Unless stated otherwise, hereafter we will assume that the original process is continuous-time.

An *analog random process* generates an infinitely-long *analog random signal*, i.e., some persistent and measurable continuous-time random quantity, the characteristics of which, in general, vary with time. In essence, this signal *is* the random process.

Due to its nature, when measurements are performed, this analog random signal can realize itself into an infinite number of different infinitely-long sequences,

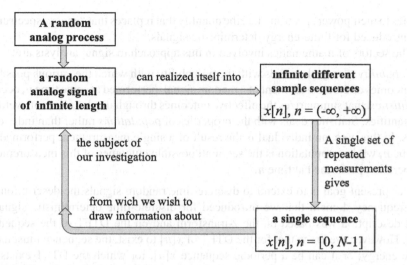

Fig. 9.1 The relation between a random process and the sequence $x[n]$, $n = [0, N - 1]$ collecting the results of its repeated measurements. Based on $x[n]$ we seek information about the generating random process

referred to as *sample sequences* or *realizations of the process*. This means that if we were able to measure the random quantity over an infinite time duration, and were able to perform the measurement process an infinite number of times (imagine an infinite number of experimenters simultaneously measuring the same quantity, each experimenter being able to work for an infinite amount of time), we would obtain an infinite number of different infinitely-long sequences. The *infinite set of realizations*, i.e., the collection of all sample sequences that can be realizations of the given process, is referred to as the *ensemble of sample sequences* (*ensemble of realizations*). This ensemble represents a *discrete-time random signal, or process*. It is characterized by some *probability laws* and is statistically representable in terms of *average properties*; the averages, as well as the probability laws, may depend on time, as a consequence of the dependence on time of the original analog random process. The averages at time n can—unlike the *individual values* measured at time n—be reproducible (meaning that two different experimenters would find the same averages), and therefore meaningful.

In reality, we can perform measurements only over a finite time span, and we will assume we do so once: the single finite-length sequence we obtain is then a segment of one of the infinite, infinitely-long possible realizations of the process. We will call it *a segment of a sample sequence*. Our goal is to extract from this segment reliable information about the discrete-time random signal, which we assume to faithfully represent the original analog signal. In other words, we assume that any sampling process performed in the frame described above gives a frequency-aliasing-free result, since it meets Nyquist's criterion.

Here we must underline the fact that the discussion on frequency aliasing we made about deterministic signals applies to random power signals as well. More precisely, it applies to their power spectrum, i.e., the quantity that replaces that "energy spectrum" we introduced for finite-energy deterministic signals.[1]

The sectors of mathematics involved in this approach to signal analysis are:

- *probability theory*, that deals with the probability with which the various possible outcomes of the measurement of a random signal, performed at some time n, occur;
- *statistics*, that summarizes the effective outcomes through the definition of average quantities. Statistics deals with the properties of *populations* rather than individuals. In this case, an individual is the result of a single measurement performed at time n, while a population is the set of all possible outcomes of this measurement operation performed at time n.

Our present goal is to extend to discrete-time random signals the description in the frequency domain that we introduced for discrete-time deterministic signals. That description was based on the z-transform and on the DTFT of the sequence $x[n]$. However, we know that for the DTFT of $x[n]$ to exist, the sequence must have finite energy; or it can be a periodic sequence $x[n]$, for which the DTFT exists in the sense of distributions. The process realizations that we called sample sequences, from which we extract a finite-length random record, do not, in general, have finite energy, nor are they periodic—typically they are *aperiodic* power signals. How can we give a spectral representation of these signals?

Many properties of a discrete-time random power signal may be expressed through a related signal, namely one of its characteristic average quantities: its *autocorrelation/autocovariance sequence*. Under some conditions that will be discussed later, this sequence has finite energy, and therefore allows for a DTFT representation. This representation has an interesting interpretation in terms of the frequency distribution of the average power in the random signal—energy per unit of discrete time—and leads to the notion of *power spectrum* of the random signal. The power spectrum is the desired spectral representation, and is also useful to describe the transformation operated by a discrete-time LTI system on random power signals.

9.3 Random Variables

The statistical-probabilistic description of discrete-time random signals requires introducing quantitatively the concept of the *random variable*, with which a continuous probabilistic variable is associated, which is called the *probability density function* (see, e.g., Helstrom 1991).

Let us suppose that a given experiment, i.e., a measurement of a random signal performed at the discrete time n, can a priori give any result among those contained

---

[1] In particular, most often the noise superposed to the useful signal will be a wideband process: its spectral bandwidth will be larger than the signal bandwidth. If we choose the sampling interval on the basis of the bandlimit that the useful signal is supposed to have, then we may get aliasing of the noise component, which can corrupt the spectrum at all frequencies—not only in the vicinity of the Nyquist frequency. Of course, this issue becomes irrelevant if we are in a position to apply an anti-aliasing filter before sampling, since then also the noise bandwidth becomes limited.

inside a certain interval of permitted values. The result we effectively obtain from the measurement is the precise value x_n assumed in that particular experiment by the random variable x_n, representing the time-dependent random process as it behaves at time n. The quantity x_n is a *single value extracted at random from the population of the possible values of* x_n. We will assume that, at any instant n, these possible values cover a continuous range spanning the whole real axis; the corresponding variable will be indicated by $X_n \in (-\infty, +\infty)$. This is not particularly restrictive.[2]

The random variable x_n is associated with some probability law: not all measurement outcomes are equally probable. This probability law is described by the *probability density function* $p_{x_n}(x_n, n)$. Loosely speaking, the probability density function describes the relative likelihood for the random variable x_n related to time n to take on a given value x_n. The probability density function is non-negative everywhere, and its integral over the entire range of admitted values equals one.

The probability of the random variable to take on a value falling within a particular range is given by the integral of the probability density function over that range—it is given by the area under curve representing the probability density function, between the lowest and highest range bounds. We can thus introduce the *probability distribution function* $P_{x_n}(x_n, n)$, which expresses the probability that the measurement outcome is $\leq x_n$, i.e., it is contained in $(-\infty, x_n]$:

$$P_{x_n}(x_n, n) = \int_{-\infty}^{x_n} p_{x_n}(X_n, n)dX_n.$$

We can imagine to hypothetically repeat our experiment at an infinite number of time instants n and form with the outcomes x_n the sequence[3] $x[n]$, with $n = (-\infty, +\infty)$. Each sequence element is a particular value taken on, inside the continuous range of possible values, by the random variable x_n involved in the n-th experiment: $x[n] \equiv x_n$. Now, let us consider two measurement experiments performed at different times n and m. The mutual dependence of the corresponding random variables x_n and x_m is described by the *joint probability density function*, denoted as $p_{x_n, x_m}(x_n, n, x_m, m)$.

The time-ordered set $\{x_n\}$ of the random variables for $n = (-\infty, +\infty)$, together with their complete probabilistic description, including all simple and joint probability laws, *quantitatively defines a random process*, also referred to as a *stochastic process*.[4]

---

[2]Note that if we assume for simplicity that the population of possible values is the same at any time n, as we actually did, we could drop the n index and simply write X. However, we will not drop the index n, because later this might generate confusion.

[3]From now on, and until the end of the next chapter, we will indicate the sequences composed by measurement outcomes by $x[n]$, rather than by $x[n]$, to avoid confusing them with random variables, for which we adopted the symbol x_n.

[4]Stochastic comes from the Greek word $\sigma\tau\acute{o}\chi o\varsigma$, which means "aim". It also denotes a target stick; the pattern of arrows around a target stick stuck in a hillside is representative of what "stochastic" means.

Applying the random-process model to practical signal processing implies that the particular sequence we want to analyze, x[n] with $n = [0, N - 1]$, is interpreted as a segment of one of the members of the ensemble of sample sequences of the underlying random process. A complete probabilistic characterization of the process would require specifying all possible simple and joint probability density functions, i.e., all p_{x_n} and p_{x_n, x_m}. However, probability laws are normally unknown and the analysis focuses on the attempt to estimate some process properties on the basis of a finite-length segment of a typical sample sequence (see, eg., Hayes 1996; Proakis and Manolakis 1996). In the following discussion we will define the conditions under which these estimates can reliably be obtained.

This approach to signal processing is often referred to as *statistical signal processing* (Gray and Davisson 2004; Kay 1993; Scharf and Demeure 1991).

9.4   Ensemble Averages

Statistical averages relevant to component variables x_n of a random process are referred to as *ensemble averages*, in the sense that they are obtained averaging over the ensemble of sample sequences. For example, the *average value*, or *mean value*, or simply the *mean* of a random process is defined as

$$m_{x_n} = E[x_n] = \int_{-\infty}^{+\infty} X_n p_{x_n}(X_n, n) dX_n.$$

This is the first *moment* of the process, and in general depends on n. In this formula, E indicates the *linear operator of ensemble average*. The concept of ensemble average is illustrated in Fig. 9.2.

The mean is also referred to as the *expected value* of the random variable, or the *expectation* relative to the random variable. The meaning of expected values becomes clear if we compare once again deterministic and stochastic processes. While a deterministic process implies

- precise relation between causes and effects,
- repeatability (reproducibility),
- predictability,

a random process is such that

- the relation between causes and effects is not given in a mathematical sense,
- there is a "stochastic regularity" among the different observations, but
- this regularity can be observed only if a large number of observations is carried out, such that the expectations of the involved variables can be inferred.

Moments of higher order are also defined, by means of the same ensemble average operator. A moment of given order $k > 0$ may be computed applying E directly to the k-th power of x_n, or applying E to the k-th power of the deviation of x_n from the mean.

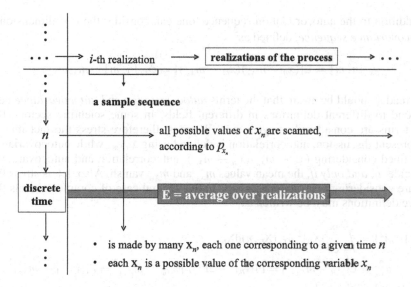

Fig. 9.2 A scheme depicting the framework in which ensemble averages are calculated

In the latter case, the moment is called a *central moment*. The most important central moment is the second, i.e., the *variance*; the corresponding non-central moment is the *square mean* or *mean square value*. In formulas:

$$E[x_n^2] = \int_{-\infty}^{+\infty} X_n^2 p_{x_n}(X_n, n) dX_n,$$

$$\sigma_{x_n}^2 = E[(x_n - m_{x_n})^2] = E[x_n^2] - 2E[x_n]m_{x_n} + m_{x_n}^2 = E[x_n^2] - m_{x_n}^2.$$

Taking the cube of x_n or $x_n - m_{x_n}$, instead than the square, third-order moments can be defined, etc. These higher-order moments are used less often than the mean and the variance. If p_{x_n} is symmetric about the mean value, like a Gaussian function, all odd-order central moments vanish. Intuitively, we may thus understand that they give indications about the asymmetry of p_{x_n}. Moreover, as the order increases, the influence on a given central moment of the values that are far from the mean grows. All these moments, in general, depend on the time index n and can be computed only knowing p_{x_n}. In all cases of practical interest in this book, they are finite quantities.

Another important average can be calculated only if we know p_{x_n, x_m}. This is the two-dimensional *autocorrelation sequence* that for real data is defined as

$$\phi_{xx}[n, m] = E[x_n x_m] = \int_{-\infty}^{+\infty} \int_{-\infty}^{+\infty} X_n X_m p_{x_n, x_m}(X_n, n, X_m, m) dX_n dX_n.$$

In addition to the autocorrelation sequence, one can consider the two-dimensional *autocovariance sequence*, defined as[5]

$$\gamma_{xx}[n, m] = \mathrm{E}[(x_n - m_{x_n})(x_m - m_{x_m})] = \phi_{xx}[n, m] - m_{x_n} m_{x_m}.$$

The reader should be aware that the terms *autocorrelation* and *autocovariance* correspond to different definitions in different fields. In some scientific sectors, the two terms are considered as synonyms. We must therefore stress the fact that in the present discussion, autocorrelation is defined using $x_n x_m$, while autocovariance is defined considering $(x_n - m_{x_n})(x_m - m_{x_m})$: autocorrelation and autocovariance coincide *if, and only if,* the mean values m_{x_n} and m_{x_m} vanish. Also, let us stress the we are considering real random signals. In the general case of complex signals the above definitions must be written as

$$\mathrm{E}[|x_n|^2] = \int_{-\infty}^{+\infty} |X_n|^2 \, p_{x_n}(X_n, n)\mathrm{d}X_n,$$

$$\sigma_{x_n}^2 = \mathrm{E}[|x_n - m_{x_n}|^2] = \mathrm{E}[|x_n|^2 - 2\mathrm{E}[x_n]m_{x_n} + |m_{x_n}|^2 = \mathrm{E}[|x_n|^2] - |m_{x_n}|^2,$$

$$\phi_{xx}[n, m] = \mathrm{E}[x_n x_m^*],$$

etc. For example, we must use these more general definitions for complex exponentials with random features (see Sect. 9.7).

Both autocorrelation and autocovariance measure the mutual dependence among the values of a random process at different times, and in a sense describe the *memory of the process*. If

$$\phi_{xx}[n, m] = \mathrm{E}[x_n x_m] = \mathrm{E}[x_n]\mathrm{E}[x_m] = m_{x_n} m_{x_m},$$

then the variables x_n and x_m are said to be *uncorrelated* or *linearly independent*. Linear independence is a different condition from *statistical independence*, which implies

$$P_{x_n, x_m}(x_n, n, x_m, m) = P_{x_n}(x_n, n)P_{x_m}(x_m, m).$$

Statistical independence means that two random variables convey no information about each other and, as a consequence, receiving information about one of the two does not change our assessment of the probability distribution of the other. It is a stronger constraint that implies linear independence; the contrary is not necessarily true.

---

[5]We must observe, recalling the general discussion on vector spaces reported in the appendix of Chap. 3, that the autocovariance between the real random variables x_n and x_m associated with times m and n can be seen as an inner product: $\langle (x_n - m_{x_n}), (x_m - m_{x_m}) \rangle$. This inner product actually involves the ensemble average operator, as we anticipated in Sect. 3.8.2. The Fourier representation of the random process $\{x_n\}$, which later will be attained using autocovariance, is thus based on a definition of inner product that is not the standard one.

9.5   Stationary Random Processes and Signals

The model described above is quite complex. A simplification of the problem occurs when probability density functions are invariant with respect to a translation of the time origin, i.e., they do not depend on n. This is the case of *stationary random signals*. The moments become constant and are denoted simply by m_x, σ_x^2, $\mathrm{E}[x^2]$, and so on. The joint probability density function now depends only on the time difference, also called *lag*, $l = n - m$, and therefore ϕ_{xx} and γ_{xx} become monodimensional sequences:

$$\phi_{xx}[n, n - l] = \phi_{xx}[l] = \mathrm{E}[x_n x_{n-l}],$$
$$\gamma_{xx}[n, n - l] = \gamma_{xx}[l] = \mathrm{E}[(x_n - m_x)(x_{n-l} - m_x)].$$

This suggests the introduction of a less stringent form of stationarity that is very useful in practice. Very often we are faced with random process that are not stationary in the strict sense defined above—or at least we cannot know if they are truly stationary or not, since we do not know their probability laws—but are *stationary in a wide sense* or *weakly stationary*, meaning that mean and variance, which are assumed to be finite, are constant, and both autocorrelation and autocovariance are functions of lag only. From now on, we will abbreviate the expressions "wide sense stationary" and "wide sense stationarity" by the acronym WSS.

An example of a WSS process is *white noise*, a random process in which all x_n are linearly independent. This means that even if we know a white-noise sequence up to the $(n - 1)$-th sample, based on these past data we have no clue about the n-th sample. For this process, the variance is finite, and a zero mean is normally assumed ($m_x = 0$), so that autocorrelation and autocovariance coincide and are given by

$$\gamma_{xx}[l] = \mathrm{E}[x_n x_{n-l}] = \begin{cases} \mathrm{E}[x_n^2] = \sigma_x^2 & \text{for } l = 0, \\ \mathrm{E}[x_n]\mathrm{E}[x_{n-l}] = m_x^2 = 0 & \text{for } l \neq 0. \end{cases}$$

The second equation derives from the fact that $\mathrm{E}[x_n]$ and $\mathrm{E}[x_{n-l}]$ are both equal to m_x for a stationary process. In a single formula, we can write

$$\gamma_{xx}[l] = \sigma_x^2 \delta[l].$$

Let us now mention some general properties of the autocorrelation and autocovariance sequences of a real stationary random signal.

- $\gamma_{xx}[l] = \phi_{xx}[l] - m_x^2$: for a signal with zero mean, also called a *centered signal*, autocorrelation and autocovariance are coincident;
- $\phi_{xx}[0]$ is equal to the mean square value $\mathrm{E}[x_n^2]$;
- $\gamma_{xx}[0]$ is equal to $\mathrm{E}[(x_n - m_x)^2] = \sigma_x^2$: the variance of a process equals its autocovariance at zero lag;
- $\phi_{xx}[l] = \phi_{xx}[-l]$ and $\gamma_{xx}[l] = \gamma_{xx}[-l]$: autocorrelation/autocovariance is an even function of lag;

- $|\phi_{xx}[l]| \leq \phi_{xx}[0]$ and $|\gamma_{xx}[l]| \leq \gamma_{xx}[0]$: autocorrelation/autocovariance is maximum at zero lag;
- if $y_n = x_{n-n_0}$, then $\phi_{yy}[l] = \phi_{xx}[l]$ and $\phi_{yy}[l] = \phi_{xx}[l]$: a time shift of the signal does not affect its autocorrelation/autocovariance.

Autocovariance is of particular interest for our purposes, because it acts as the intermediary between a WSS random process and its spectral representation, which is referred to as the power spectral density (PSD), or power spectrum. The autocovariance sequence can be standardized dividing it by the variance of the process. In this way it becomes the *autocorrelation coefficient sequence*, a.k.a. *standardized autocovariance*:

$$\rho_{xx}[l] = \frac{E[x_n x_{n-l}]}{E[x_n^2]} = \frac{\gamma_{xx}[l]}{\sigma_x^2}.$$

This quantity varies in the range $[-1, 1]$.

When two different WSS random processes $\{x_n\}$ and $\{y_n\}$ are considered, autocorrelation is replaced by *cross-correlation* and autocovariance is replaced by *cross-covariance*: for real signals[6]

$$\phi_{xy}[n, n - l] = E[x_n y_{n-l}],$$
$$\gamma_{xy}[n, n - l] = E[(x_n - m_x)(y_{n-l} - m_y)].$$

In this frame it is useful to introduce the notion of *jointly wide-sense stationarity*. Two random processes are called jointly WSS if they are individually WSS, and if their cross-correlation and cross-covariance depend only on the lag:

$$\phi_{xy}[n, n - l] = \phi_{xy}[l],$$
$$\gamma_{xy}[n, n - l] = \gamma_{xy}[l].$$

This sequence is often standardized to become the *cross-correlation coefficient sequence*, a.k.a. *standardized cross-covariance*:

$$\rho_{xy}[l] = \frac{E[x_n y_{n-l}]}{\sqrt{E[x_n^2]E[y_n^2]}} = \frac{\gamma_{xy}[l]}{\sigma_x \sigma_y}.$$

As an example of jointly WSS processes, we may mention the input and output of an LTI system, when the input is WSS. Cross-covariance is the basis of a representation of the common spectral content of two jointly WSS random series and of the phase relations of the respective frequency components, which is referred to as the cross-power spectral density (CPSD), or cross-power spectrum, or simply cross-spectrum.

---

[6]For complex signals we would write $\phi_{xy}[n, n - l] = E[x_n y_{n-l}^*]$, etc.

9.6 Ergodicity

Ensemble averages are mathematical entities that in practice must be substituted by other averages we can compute even if we do not know the probability laws governing the process. We may think of calculating *time averages*, like

- the *temporal mean of a random signal*,

$$\overline{x_n} = \lim_{N \to \infty} \frac{1}{2N+1} \sum_{n=-N}^{+N} x_n,$$

and, after reducing the signal to zero mean by subtracting $\langle x_n \rangle$ from each x_n,
- the *temporal autocorrelation/autocovariance sequence*

$$\overline{x_n x_{n-l}} = \lim_{N \to \infty} \frac{1}{2N+1} \sum_{n=-N}^{+N} x_n x_{n-l}.$$

It can be shown that if $\{x_n\}$ is a WSS random process with finite mean, these limits exist. These averages, being functions of an infinite collection of random variables, are random variables as well. However, under a condition known as *ergodicity*, temporal averages are constant and are equal to the corresponding ensemble averages. More precisely, the random variables $\overline{x_n}$ and $\overline{x_n x_{n-l}}$ have mean values equal to m_x and $\phi_{xx}[l]$ respectively, and their variances vanish. In practical applications, normally it is assumed as a working hypothesis that if a sequence belongs to a WSS process, then that process is also ergodic.

The conditions under which a stationary process is ergodic are involute and difficult to apply in practice. The reader interested in a moderately mathematical treatment of ergodicity may consult Hayes (1996). Ergodicity implies that a single part of a sample sequence, provided it is long enough, contains all possible values of the signal, with frequencies of occurrence matching the probability laws that characterize the ensemble of sample sequences and the process itself. Intuitively, for a stationary signal to be ergodic, the statistical dependence between two temporal segments of the process that are separated by some lapse of time must tend to zero rapidly enough, as the time separation increases. We can expect it will be so, provided we remove from the signal the constant component by centering the signal, and we also remove any constant-slope trends and periodic components known a priori, such as daily oscillations in temperature series, seasonal cycles, etc. For a Gaussian process, in which the variables x_n take on values distributed according to a Gaussian probability density function, it can be shown that the only obstacle to ergodicity is the presence of a deterministic component with a line spectrum, as in the examples mentioned above. If the deterministic component is subtracted, ergodicity can be assumed, and a posteriori, the quality of this assumption can supported by the fact that a continuous spectrum is actually observed. Here we are using the word "spectrum" in a wide

sense, since we have not explained yet how a quantitative spectral representation of
a WSS random signal can be obtained.

We must remark that the time averages introduced above still cannot be computed
in reality, since no measurements can be performed for an infinite amount of time. In
practice, we will be able to compute *estimates* of the mean and of the AC, based on
a finite-length data record. We will deal with this issue in Sect. 9.10.3. The shorter
the data segment we possess, the larger the *estimation error* we get. The need for a
large N is often in conflict with the requirement of stationarity: we cannot assume
stationarity if the record duration is larger than the typical time-scale over which the
statistics of the signal varies significantly.

9.7 Wiener-Khinchin Theorem for Random Signals
and Power Spectrum

From now on, we will assume, unless stated otherwise, that we are dealing with a WSS
zero-mean random process—i.e., a centered process—for which autocorrelation and
autocovariance coincide, and will continue our discussion in terms of $\gamma_{xx}[l]$. This
assumption results in no loss of generality, since for any process that has non-zero
mean, a corresponding zero-mean process may always be formed by subtracting the
mean. We must recognize that strictly speaking, in order to center a WSS random
signal we should know the "true" mean m_x, while the best we can do is to estimate the
mean in some way and then subtract it from the data. Nevertheless, we will make this
assumption. Autocorrelation/autocovariance can thus be meaningfully indicated by
the single acronym AC; the words "autocorrelation" and "autocovariance" become
synonyms.

The theoretical AC sequence $\gamma_{xx}[l]$ allows, through its DTFT, to define an average
measure of the spectral properties of an ergodic WSS random signal. Indeed, this
sequence is aperiodic and often has finite energy, so that it allows for a z-transform
and a DTFT representation. This property derives from the fact that for ergodic
processes, the variables x_n become less and less correlated as their time separation
increases, so that

$$\lim_{l \to \infty} \gamma_{xx}[l] = 0,$$

and this guarantees the existence of the z-transform of $\gamma_{xx}[l]$, which we will denote
by $\Gamma_{xx}(z)$. The series that defines $\Gamma_{xx}(z)$ has an ROC characterized by

$$R < |z| < \frac{1}{R}, \qquad |R| < 1,$$

i.e., the upper radius of convergence is the inverse of the lower radius. The ROC
includes the unit circle because $\gamma_{xx}[l] \to 0$ for $l \to \infty$ (see the example shown
in Fig. 9.3). The upper and lower convergence radii are inversely related to one

Fig. 9.3 An example of region of convergence for $\Gamma_{xx}(z)$, the z-transform of the theoretical AC sequence

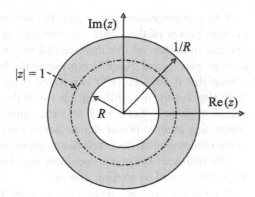

another because $\gamma_{xx}[l]$ is an even real sequence, so that $\Gamma_{xx}(z) = \Gamma_{xx}(1/z)$. Another consequence of this property is that the poles of $\Gamma_{xx}(z)$ come in conjugate pairs.

The DTFT of $\gamma_{xx}[l]$, i.e., $\Gamma_{xx}(z)|_{z=e^{j\omega}} = \Gamma_{xx}(e^{j\omega})$, is the frequency representation sought for the stationary ergodic random process. This is a real and even function of ω: indeed, $\Gamma_{xx}(e^{j\omega}) = \Gamma_{xx}(e^{-j\omega})$, because $\gamma_{xx}[l]$ is an even real function of lag.

From the expression of the inverse z-transform for $n = 0$, it can be shown that since $\gamma_{xx}[0] = \sigma_x^2$, then

$$\sigma_x^2 = \frac{1}{2\pi j} \oint_C \Gamma_{xx}(z) z^{-1} dz,$$

where C is a closed anticlockwise contour contained inside the ROC of $\Gamma_{xx}(z)$. Therefore on the unit circle of Fig. 9.3 (*dash-dotted curve*) we have

$$\sigma_x^2 = \frac{1}{2\pi} \int_{-\pi}^{\pi} \Gamma_{xx}(e^{j\omega}) d\omega = \frac{1}{2\pi} \int_{-\pi}^{\pi} P_{xx}(e^{j\omega}) d\omega,$$

where we indicated by $P_{xx}(e^{j\omega}) = \Gamma_{xx}(e^{j\omega})$ the *power spectral density* (PSD) of the random process. Colloquially, the PSD is called *power spectrum*, even if, strictly speaking, the power spectrum is the plot of PSD versus frequency. This important equation expresses the fact that up to a multiplicative constant [7] $1/(2\pi)$, *the variance of a random process is given by the integral over frequency of its power spectrum.* In other words, the function

$$P_{xx}(e^{j\omega}) \equiv \Gamma_{xx}(e^{j\omega}) = \sum_{l=-\infty}^{+\infty} \gamma_{xx}[l] e^{-j\omega l},$$

which is a real and even function of ω representing the distribution of variance over frequency, is the DTFT of the AC sequence $\gamma_{xx}[l]$. This statement is the conceptual content of the *Wiener-Khinchin theorem for random power signals that are weekly stationary and ergodic.*

---

[7] The factor $1/(2\pi)$ derives from using ω as the frequency variable.

When, as we assumed, the signal has zero mean, $\sigma_x^2 = \mathrm{E}[x_n^2]$ represents the finite *average power* of the signal, i.e., the average of the energy computed over a unit discrete-time interval. We thus see that the area under the curve $P_{xx}(\mathrm{e}^{\mathrm{j}\omega})$ gives the average power, and $P_{xx}(\mathrm{e}^{\mathrm{j}\omega})$ represents the frequency distribution of this average power: this is why $P_{xx}(\mathrm{e}^{\mathrm{j}\omega})$ is referred to as the "power spectrum" of the signal. Note also that the only difference between the Wiener-Khinchin definition of power spectrum given in Sect. 5.7.2 for deterministic power signals and the present one concerning random power signals lies in the definition of the AC sequence, which in the random-signal case includes an expectation operation.

The reality of the power spectrum suggests to simply write $P_{xx}(\omega)$ rather than $P_{xx}(\mathrm{e}^{\mathrm{j}\omega})$; we will do so from now on.

Note that $P_{xx}(\omega)$ is a theoretical quantity that in practice must be substituted by a proper estimate. The estimate must be defined in such a way that we are able to compute it even if we only possess a finite-length data record representing a segment of a typical sample sequence. We will tackle this issue later.

A second definition of power spectrum can be given:

$$P_{xx}(\omega) = \lim_{N \to \infty} \mathrm{E}\left[\frac{1}{N} \left| \sum_{n=0}^{N-1} x_n \mathrm{e}^{-\mathrm{j}\omega n} \right|^2 \right],$$

which is equivalent to the first one, under the mild assumption that the AC theoretical sequence decays sufficiently rapidly, so that

$$\lim_{N \to \infty} \frac{1}{N} \sum_{l=-(N-1)}^{N-1} |l|\, |\gamma_{xx}[l]| = 0.$$

The equivalence of the two definitions can be verified as follows (Stoica and Moses 2005):

$$\lim_{N \to \infty} \mathrm{E}\left[\frac{1}{N} \left| \sum_{n=0}^{N-1} x_n \mathrm{e}^{-\mathrm{j}\omega n} \right|^2 \right] = \lim_{N \to \infty}\left[\frac{1}{N} \sum_{m=0}^{N-1}\sum_{i=0}^{N-1} \mathrm{E}\left\{ x_m x_i^* \right\} \mathrm{e}^{-\mathrm{j}\omega(m-i)} \right] =$$

$$= \lim_{N \to \infty} \frac{1}{N} \sum_{l=-(N-1)}^{N-1} (N - |l|)\, \gamma_{xx}[l] \mathrm{e}^{-\mathrm{j}\omega l} =$$

$$= \sum_{l=-\infty}^{\infty} \gamma_{xx}[l] \mathrm{e}^{-\mathrm{j}\omega l} - \lim_{N \to \infty} \frac{1}{N} \sum_{l=-(N-1)}^{N-1} |l|\, \gamma_{xx}[l] \mathrm{e}^{-\mathrm{j}\omega l} = P_{xx}(\omega),$$

where we used a well-known summation formula:

$$\sum_{m=0}^{N-1}\sum_{i=0}^{N-1} f[m-i] = \sum_{l=-(N-1)}^{N-1} (N - |l|)\, f[l],$$

$f(\cdot)$ being an arbitrary sequence, and for generality assumed a complex set of random variables $\{x_n\}$. The above definition of $P_{xx}(\omega)$ resembles the definition of energy spectral density of the deterministic case. The main differences between the two formulas are the appearance of the expectation operator in the random-signal case, and the normalization factor $1/N$.

To present a first example of a theoretical power spectrum, we can consider a white noise signal. As mentioned above, white noise is a stationary signal or process that from the statistical point of view is made up of variables x_n that are uncorrelated, with zero mean and finite variance. Stationarity requires all x_n to have the same probability distribution. Since white noise is defined only in terms of its AC, $\gamma_{xx}[l] = \sigma_x^2\delta[l]$, an infinite variety of white-noise random processes exist: being stationary and uncorrelated in time does not restrict the values that a signal can assume. Any distribution of values is possible, provided that the mean value is zero. If samples have a normal distribution, the signal is said to be Gaussian white noise. Gaussian white noise is a good approximation of many real-world situations. But even a binary signal, which can only take on the values $+1$ or -1, will be white if the sequence values are statistically uncorrelated. We can find also uniformly distributed white noise, Poisson-distributed white noise, etc.

For white noise we have $\gamma_{xx}[l] = \sigma_x^2\delta[l]$, and since the DTFT of $\delta[n]$ equals 1, $P_{xx}(\omega) = \sigma_x^2$: white noise has a flat spectrum. It contains the same power at all frequencies; no frequency is privileged. Actually, the word "white" refers to white light, in which all colors are equally present.

Another important random signal that is found in many applications is the random-phase sinusoid, a real-valued signal defined as

$$x_n = A\sin(n\omega_0 + \vartheta),$$

where amplitude A and frequency ω_0 are fixed constants and ϑ is a uniformly-distributed random variable between 0 and 2π. The probability density function for ϑ is

$$f_\vartheta(\Theta) = \begin{cases} \frac{1}{2\pi} & \text{for} \quad 0 \le \Theta < 2\pi, \\ 0 & \text{otherwise.} \end{cases}$$

Observe that the presence of a random initial phase makes the sinusoid a random signal. A deterministic sinusoid would have a deterministic phase linearly increasing with n. The signal is stationary and its mean,

$$m_x = \mathrm{E}[x_n] = \mathrm{E}[A\sin(n\omega_0 + \vartheta)],$$

is

$$m_x = \int_{-\infty}^{\infty} A \sin(n\omega_0 + \Theta) f_\vartheta(\Theta) d\Theta = \frac{A}{2\pi} \int_{-\pi}^{\pi} \sin(n\omega_0 + \Theta) d\vartheta = 0,$$

so the process has zero mean. The AC may be determined in a similar fashion: writing

$$\gamma_{xx}[n, m] = \mathrm{E}\{x_n x_m\} = \mathrm{E}\{A \sin(n\omega_0 + \vartheta) A \sin(m\omega_0 + \vartheta)\}$$

and using the trigonometric identity $2 \sin \alpha \sin \beta = \cos(\alpha - \beta) - \cos(\alpha + \beta)$ we get

$$\gamma_{xx}[n, m] = \frac{1}{2}A^2 \mathrm{E}\{\cos[(n - m)\omega_0]\} - \frac{1}{2}A^2 \mathrm{E}\{\cos[(n + m)\omega_0 + 2\vartheta]\}.$$

The first term is the expected value of a deterministic quantity, and the expectation sign does not affect a deterministic quantity; the second term is equal to zero, as we can qualitatively understand by comparing it with the expression of m_x. Therefore, we have

$$\gamma_{xx}[n, m] = \frac{1}{2}A^2 \cos[(n - m)\omega_0] = \frac{1}{2}A^2 \cos[l\omega_0] = \gamma_{xx}[l], \quad \text{with} \quad l = n - m.$$

So the AC is periodic with the same frequency possessed by the sinusoid, and depends only on lag l, thus ensuring the WSS nature of the process. The power spectrum is then

$$P_{xx}(\omega) = \sum_{l=-\infty}^{+\infty} \gamma_{xx}[l] e^{-j\omega l} = \frac{1}{2}A^2 \sum_{l=-\infty}^{+\infty} \cos[l\omega_0] e^{-j\omega l}.$$

Using Euler's relation, we can write

$$P_{xx}(\omega) = \frac{1}{4}A^2 \sum_{l=-\infty}^{+\infty} \left(e^{j\omega_0 l} + e^{-j\omega_0 l}\right) e^{-j\omega l} = \frac{\pi A^2}{2} [\delta(\omega - \omega_0) + \delta(\omega + \omega_0)],$$

where the DTFT relation $e^{j\omega_0 l} \Longleftrightarrow 2\pi \delta(\omega - \omega_0)$ has been used. Thus the process has a line spectrum.

Similarly, we could treat the random-phase complex exponential signal

$$x_n = A e^{j(n\omega_0 + \vartheta)},$$

which has zero mean as well, and has an AC sequence given by

$$\gamma_{xx}[n, m] = \mathrm{E}\{x_n x_m^*\} = \mathrm{E}\left[A e^{j(n\omega_0 + \vartheta)} A e^{-j(m\omega_0 + \vartheta)}\right] = A^2 e^{j(n-m)\omega_0}.$$

This AC depends only on $n - m$: setting $l = n - m$ we get

$$\gamma_{xx}[l] = A^2 e^{jl\omega_0},$$

because $e^{jl\omega_0}$ is a deterministic term and the expectation operation applied to it leaves it unaltered. Thus the process is WSS, and its power spectrum is

$$P_{xx}(\omega) = 2\pi A^2 \delta(\omega - \omega_0).$$

We may note, incidentally, that defining the random-phase sinusoidal signal as

$$x_n = A \cos(n\omega_0 + \vartheta) = \frac{A}{2} \left[e^{j(n\omega_0 + \vartheta)} + e^{-j(n\omega_0 + \vartheta)} \right],$$

which appears perfectly equivalent to using a sine function if we consider the presence of the variable initial-phase term, the result concerning the random-phase complex exponential signal immediately leads to the formula derived above:

$$P_{xx}(\omega) = \frac{\pi A^2}{2} \left[\delta(\omega - \omega_0) + \delta(\omega + \omega_0) \right].$$

Thus on the positive frequency half-axis, the theoretical spectral of a random-phase sine or cosine with amplitude A and of a random-phase complex exponential with amplitude $A/2$ are identical.

It can be shown (Hayes 1996) that if two or more processes are uncorrelated, then the AC of their sum equals the sum of the respective ACs. Due to the linearity of the Fourier transform, the power spectra are also additive. Thus a random-phase sinusoidal or complex exponential signal with white noise $e[n]$ added will have the same power spectrum, augmented by a term equal to the variance σ_e^2 of the additive white noise. The possible presence in the signal of other sinusoids/complex exponentials with different parameters would produce in the power spectrum additive terms similar to the term produced by the single sinusoid/complex exponential considered here. We will exploit this result in Sect. 10.3.3.

9.8 Cross-Power Spectrum of Two Random Signals

In addition to the spectral features of a single WSS random process, we may be interested in the common spectral characteristics of two processes, $\{x_n\}$ and $\{y_n\}$, which we will assume to be real, individually WSS and jointly stationary, ergodic, and with zero mean. For this purpose, we may define the cross-power spectral density of the two processes by transforming their cross-covariance. Actually, it is convenient to start from the sequence of *cross-correlation coefficients*, also referred to as *normalized cross-covariance* or *standardized cross-covariance*, between two real random variables x_n and y_n:

$$\rho_{xy}[l] = \frac{E[x_n y_{n-l}]}{\sqrt{E[x_n^2]E[y_n^2]}} = \frac{\gamma_{xy}[l]}{\sigma_x \sigma_y}.$$

Standardization cannot be avoided here, since in general we are comparing variables with different ranges of numerical values and different measurement units.[8] Evidently, $\rho_{xy}[l] = \rho_{yx}[-l]$; moreover, most random processes we may be interested in are such that

$$\lim_{l \to \infty} \rho_{xy}[l] = 0,$$

so that the z-transform of $\rho_{xy}[l]$ exists and converges on the unit circle, providing the desired frequency representation of the cross-relation between the two random variables x_n and y_n. This representation is the *cross-power spectral density* (CPSD) or *cross-power spectrum*, often simply called *cross-spectrum*, even if the term "cross-spectrum" should be reserved to the plot of CPSD versus frequency:

$$P_{xy}(e^{j\omega}) = \sum_{l=-\infty}^{+\infty} \rho_{xy}[l] e^{-j\omega l}.$$

Since in general $\rho_{xy}[l]$ is not an even sequence, the cross-spectrum is a complex quantity that contains both amplitude and phase information:

- its modulus gives the *magnitude cross-power spectrum*, which informs about the frequency distribution of the cross-covariance of the two signals;
- its phase gives the *phase cross-power spectrum*, which indicates whether a given cyclicity at some frequency in one of the variables is in or out of phase with the corresponding cyclicity in the second one. Obviously this information is meaningful only at those frequencies at which the modulus of the cross-spectrum between the two variables is substantial. Provided it is so, phase information is essential in evaluating possible cause-effect relations between the two variables.

Often the *magnitude squared coherence* (MSC) is also defined, by taking the squared modulus of the cross-power spectrum and dividing it by the PSDs of the individual variables:

$$MSC_{xy}(\omega) = \frac{\left|P_{xy}(e^{j\omega})\right|^2}{P_{xx}(\omega)P_{yy}(\omega)}.$$

Its values are in the range from 0 to 1.

---

[8]If we considered the general case of complex sequences, the above formula would become $\rho_{xy}[l] = \frac{E[x_n y_{n-l}^*]}{\sqrt{E[|x_n|^2]E[|y_n|^2]}}$.

These theoretical quantities must then be substituted with proper estimates. These estimates must be defined in such a way that we are able to compute them even if we only possess two finite-length data records $x[n]$ and $y[n]$, representing segments of typical sample sequences of the two processes. We will do so later, and we will assume that the available records have the same length N.

9.9 Effect of an LTI System on a Random Signal

The concepts of AC, power spectrum, cross covariance and cross-spectrum allow us to describe the transformation that a random power signal undergoes when processed by a discrete-time LTI system.

Here we need to recall some z-transform properties (see Sect. 3.2.5). Given a sequence $x[n]$ that we assume to be complex in general, and that has a transform $X(z)$, we have

$$x^*[n] \Longleftrightarrow X^*(z^*),$$
$$x[-n] \Longleftrightarrow X(1/z),$$
$$x^*[-n] \Longleftrightarrow X^*(1/z^*).$$

It is useful for future discussion to note that if the sequence

• is real, then

$$x[n] = x^*[n] \Longleftrightarrow X(z) = X^*(z^*);$$

• has even symmetry around the time origin, then

$$x[n] = x[-n] \Longleftrightarrow X(z) = X(1/z);$$

• is both real and even, then

$$x[n] = x^*[-n] \Longleftrightarrow X(z) = X^*(z^*) = X(1/z) = X^*(1/z^*).$$

Now, let us consider a stable discrete-time LTI system with impulse response $h[n]$, transfer function $H(z) = \sum_{n=-\infty}^{\infty} h[n]z^{-n}$ and frequency response $H(e^{j\omega}) = \sum_{n=-\infty}^{\infty} h[n]e^{j\omega n}$. Here it is convenient for future use of the results to consider a general complex system and feed it with a complex input sequence $x[n]$ being amplitude-bounded and representing a sample sequence of a discrete-time complex WSS random process $\{x_n\}$ with zero mean. The output sequence,

$$y[n] = \sum_{k=-\infty}^{+\infty} h[n-k]x[k] = \sum_{k=-\infty}^{+\infty} h[k]x[n-k],$$

belongs to a complex stationary process $\{y_n\}$. Indeed, the mean value of the system's output is

$$m_{y_n} = E\left\{\sum_{k=-\infty}^{+\infty} h[k]x[n-k]\right\} = \sum_{k=-\infty}^{+\infty} h[k]E\{x[n-k]\} = m_x \sum_{k=-\infty}^{+\infty} h[k] = m_x H(e^{j0}),$$

i.e., a constant. Moreover, $m_y = 0$ for a centered input with $m_x = 0$.

Now we compute γ_{yx}, γ_{xy} and γ_{yy}, as well as their z-transforms. Note that for complex sequences we define autocovariance as

$$\gamma_{xx}[l] = E\left\{x[n]x^*[n-l]\right\} = \gamma_{xx}^*[-l]$$

and cross-covariance as
$$\gamma_{xy}[l] = E\left\{x[n]y^*[n-l]\right\}.$$

1. The output-input cross-covariance is given by

$$\gamma_{yx}[l] = E\left\{y[n]x^*[n-l]\right\} = E\left\{\left(\sum_{k=-\infty}^{+\infty} h[k]x[n-k]\right)x^*[n-l]\right\}$$

$$= \sum_{k=-\infty}^{+\infty} h[k]E\left\{x[n-k]x^*[n-l]\right\}.$$

Letting $m = n - k$, hence $n - l = m - (l - k)$, we get

$$\gamma_{yx}[l] = \sum_{k=-\infty}^{+\infty} h[k]E\left\{x[m]x^*[m-(l-k)]\right\} = \sum_{k=-\infty}^{+\infty} h[k]\gamma_{xx}[l-k] = h[l] * \gamma_{xx}[l]:$$

the output-input cross-covariance is equal to the convolution of the impulse response with the input AC sequence.
Passing to the z-transform, and neglecting the issue of cross-covariance standardization that in this case is unessential, we get

$$\Gamma_{yx}(z) \equiv P_{yx}(z) = H(z)\Gamma_{xx}(z) = H(z)P_{xx}(z) \quad \rightsquigarrow \quad P_{yx}(e^{j\omega}) = H(e^{j\omega})P_{xx}(\omega),$$

where we introduced the symbols $P_{xx}(z)$ and $P_{yx}(z)$ to represent power spectra in the z-domain. Thus, *the cross-spectrum between output and input is the product of the input signal's power spectrum with the filter's frequency response.*

2. The input-output cross-covariance is

$$\gamma_{xy}[l] = E\left\{x[n]y^*[n-l]\right\} = E\left\{x[m+l]y^*[m]\right\} = \gamma_{yx}^*[-l]$$
$$= h^*[-l] * \gamma_{xx}^*[-l] = h^*[-l] * \gamma_{xx}[l].$$

Thus *the input-output cross-covariance is equal to the convolution of the folded complex conjugate of the impulse response with the input AC sequence.*
Passing to the z-transform we have

$$P_{xy}(z) = H^*(1/z^*)P_{xx}(z) \quad \rightsquigarrow \quad P_{yx}(e^{j\omega}) = H^*(e^{j\omega})P_{xx}(\omega),$$

i.e., *the cross-spectrum between input and output is the product of the input signal's power spectrum with the complex conjugate of the filter's frequency response.*

3. The AC of the (centered) output is

$$\gamma_{yy}[l] = E\{y[n]y^*[n-l]\} =$$

$$= E\left\{y^*[n-l]\sum_{k=-\infty}^{+\infty}h[n-k]x[k]\right\} =$$

$$= \sum_{k=-\infty}^{+\infty}h[n-k]E\{y^*[n-l]x[k]\} =$$

$$= \sum_{k=-\infty}^{+\infty}h[n-k]\gamma_{xy}[k-n+l] = \sum_{k=-\infty}^{+\infty}h[n-k]\gamma_{xy}[l-(n-k)] =$$

$$= \sum_{m=-\infty}^{+\infty}h[m]\gamma_{xy}[l-m] = h[l]*\gamma_{xy}[l] = h[l]*h^*[-l]*\gamma_{xx}[l] = h^*[-l]*\gamma_{yx}[l],$$

where we used the fact that convolution can be written as $\sum_{k=-\infty}^{+\infty}h[n-k]x[k]$ or as $\sum_{k=-\infty}^{+\infty}h[k]x[n-k]$.
Passing to the z-transform we have

$$P_{yy}(z) = H(z)H^*(1/z^*)P_{xx}(z) \rightsquigarrow P_{yy}(\omega) = H(e^{j\omega})H^*(e^{j\omega})P_{xx}(\omega) = |H(e^{j\omega})|^2 P_{xx}(\omega):$$

the power spectrum of the output signal is equal to the spectrum of the input signal multiplied by the squared magnitude of the filter's frequency response.

From the above result concerning $P_{yy}(\omega)$ we can deduce the average output power:

$$\gamma_{yy}[0] = \frac{1}{2\pi}\int_{-\pi}^{\pi}|H(e^{j\omega})|^2 P_{xx}(\omega)d\omega.$$

By means of this equation we can show an important property of the power spectrum of a real, WSS and ergodic random signal $x[n]$. Let us suppose we apply to $x[n]$ an ideal (real) narrowband filter with arbitrary center frequency ω_0 and bandwidth B:

$$H(e^{j\omega}) = \begin{cases} 1 & \text{for } |\omega - \omega_0| \leq B/2, \\ 0 & \text{otherwise.} \end{cases}$$

The expected power at the filter's output is due to the input frequency components close to ω_0 and is measured by $\gamma_{yy}[0]$. Recalling that $P_{xx}(\omega)$ is an even function of ω and setting $\omega_1 = \omega_0 - B/2$, $\omega_2 = \omega_0 + B/2$ we can write the output power as

$$
\begin{aligned}
\gamma_{yy}[0] &= \frac{1}{2\pi} \int_{-\pi}^{\pi} |H(e^{j\omega})|^2 P_{xx}(\omega) d\omega \\
&= \frac{1}{2\pi} \left[\int_{-\omega_2}^{-\omega_1} P_{xx}(\omega) d\omega + \int_{+\omega_1}^{+\omega_2} P_{xx}(\omega) d\omega \right] \\
&= \frac{1}{2\pi} \int_{+\omega_1}^{+\omega_2} \left[P_{xx}(-\omega) + P_{xx}(\omega) \right] d\omega \\
&= \frac{1}{2\pi} \int_{+\omega_1}^{+\omega_2} 2 P_{xx}(\omega) d\omega = \frac{1}{\pi} \int_{+\omega_1}^{+\omega_2} P_{xx}(\omega) d\omega.
\end{aligned}
$$

But power is always non-negative; therefore if we let the filter bandwidth tend to zero, the inequality

$$
\lim_{\omega_2 - \omega_1 \to 0} \gamma_{yy}[0] \geq 0
$$

must hold. Therefore

$$
P_{xx}(\omega_0) \geq 0
$$

and since ω_0 is arbitrary, we can conclude that the power spectrum of $x[n]$ is a real, even and *non-negative* function of frequency.

9.10   Estimation of the Averages of Ergodic Stationary Signals

We will now discuss how the average properties of a WSS ergodic random process can be estimated from a finite-length segment of a typical sample sequence x[n], $n = [0, N - 1]$. Thanks to ergodicity, ensemble averages can be substituted by time averages. After some general notions concerning estimation theory, we will deal with the estimation of the mean, variance and autocovariance of a random WSS ergodic process, as well as with the estimation of the cross-covariance of two ergodic, individually and jointly WSS processes. The estimation of the power spectrum and of the cross-power spectrum is left to the following two chapters.

9.10.1   General Concepts in Estimation Theory

The estimate of a parameter, which we will call α, of a random signal $\{x_n\}$ is a function of the random variables for which a sample is available:

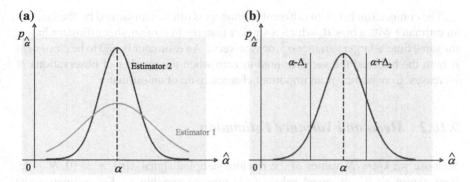

Fig. 9.4 **a** "Goodness" of an estimator: in this example, estimator 2 is better than estimator 1, because for the former the probability of the estimate $\hat{\alpha}$ being close to the true value α is higher. **b** An example of confidence interval for an estimator

$$\hat{\alpha} = F(x_0, x_1, x_2, \ldots x_{N-1}).$$

Therefore the estimate $\hat{\alpha}$ is a random variable, the probability density function of which will be indicated by $p_{\hat{\alpha}}(\hat{\alpha})$. The functional form and the values of $p_{\hat{\alpha}}(\hat{\alpha})$ depend on the choice of the *estimator* $F(\cdot)$ and on the probability densities of the variables x_n. If $p_{\hat{\alpha}}(\hat{\alpha})$ has a symmetric shape, then its center is the "true value" α of the parameter, as in the case of Fig. 9.4a. An estimator is "good" if the probability of $\hat{\alpha}$ being close to α is high: for example, in Fig. 9.4a, estimator 2 is better than estimator 1.

We can quantify how much $p_{\hat{\alpha}}(\hat{\alpha})$ is concentrated around α through the concept of *confidence interval*. Referring to Fig. 9.4b, if we call $1 - \beta$ the probability of $\hat{\alpha}$ being comprised between $\alpha - \Delta_1$ and $\alpha + \Delta_2$, then we can say that $\hat{\alpha}$ falls inside the range $(-\Delta_1, +\Delta_2)$ around its true value "at the confidence level of $1 - \beta$". In order to be able to compute Δ_1 and Δ_2 for a given value of β we should know $p_{\hat{\alpha}}(\hat{\alpha})$; the usual working hypothesis in the absence of such information is that $p_{\hat{\alpha}}(\hat{\alpha})$ is Gaussian, or nearly Gaussian.

Often for a given parameter several estimators are proposed, and the need for a comparison among their performances arises. The properties of estimators that are used to compare them are

- *polarization* or *bias*,

$$B_{\hat{\alpha}} = \alpha - E[\hat{\alpha}],$$

where $E[\hat{\alpha}]$ indicates the expected value of the estimate, and
- *variance*,

$$\sigma_{\hat{\alpha}}^2 = E[(\hat{\alpha} - E[\hat{\alpha}])^2].$$

If $\sigma_{\hat{\alpha}}^2$ is small, then $p_{\hat{\alpha}}(\hat{\alpha})$ is concentrated around the mean value $E[\hat{\alpha}]$, which, if $B_{\hat{\alpha}} = 0$, coincides with the true value α.

The comparison between different estimators is often complicated by the fact that an estimator with a bias $B_{\hat{\alpha}}$ which is smaller than the bias of another estimator has, at the same time, a larger variance $\sigma_{\hat{\alpha}}^2$, or vice-versa. An estimator is said to be *consistent* if both the bias and the variance tend to zero when the number of observations N increases. Consistency is an important characteristic of an estimator.

9.10.2 Mean and Variance Estimation

Suppose we know N values of the random sampled signal x[n], $n = [0, N - 1]$. Each sample is the observed value of the random variable x_n. For estimating the mean value of $\{x_n\}$, the *sample mean* is used, which is the arithmetic mean of the results of our measurements:

$$\hat{m}_x = \frac{1}{N} \sum_{n=0}^{N-1} x[n].$$

Under simplifying hypotheses, namely that the process is Gaussian and that the variables x_n are real and statistically independent, it can be shown that $\hat{m}_x$ is the *maximum likelihood* estimate of m_x.

Maximum likelihood estimators are a widely used class of estimators, based on the joint probability of the observed values as a function of the parameter to be estimated. The maximum likelihood estimate is the value of the parameter that maximizes the probability of getting, when measuring the variables x_n, exactly those $x[n]$ values that have been observed, with $n = [0, N - 1]$.

As a sum of Gaussian variables, $\hat{m}_x$ is a Gaussian variable; therefore it is completely characterized by its mean $E[\hat{m}_x]$ and variance $\sigma_{\hat{m}_x}^2$. But

$$E[\hat{m}_x] = \frac{1}{N} \sum_{i=0}^{N-1} E\{x[i]\} = E\{x_n\} = m_x,$$

and therefore the polarization is zero. It can, moreover, be shown that

$$\sigma_{\hat{m}_x}^2 = \frac{\sigma_x^2}{N},$$

hence $\sigma_{\hat{m}_x}^2 \to 0$ for $N \to \infty$: the sample mean is a consistent estimate.

As for estimating the variance of the process $\{x_n\}$, we must distinguish between two cases:

1. m_x is assumed to be known.
   Then the maximum likelihood estimate of the variance is

$$\hat{\sigma}_x^2 = \frac{1}{N} \sum_{n=0}^{N-1} (x[n] - m_x)^2,$$

that turns out to be consistent;

2. both m_x and σ_x^2 have to be estimated, and this is the most frequent situation. Then the maximum likelihood estimate for m_x is the sample mean, while for σ_x^2 the maximum likelihood estimate is the *sample variance*

$$\hat{\sigma}_x^2 = \frac{1}{N} \sum_{n=0}^{N-1} (x[n] - \hat{m}_x)^2,$$

in which $\hat{m}_x$ replaces m_x that is unknown.
Let us set $\hat{v} = \hat{\sigma}_x^2$. It can be shown that

$$E\left[\hat{\sigma}_x^2\right] = E[\hat{v}] = \frac{N-1}{N}\sigma_x^2,$$

hence $B_{\hat{v}} \neq 0$: the sample variance is biased, but is *asymptotically unbiased*,[9] because as $N \to \infty$, $E[\hat{\sigma}_x^2] = E[\hat{v}] \to \sigma_x^2$.
The variance of the sample variance is found to be

$$\sigma_{\hat{v}}^2 = E\left\{(\hat{v} - E[\hat{v}])^2\right\} = \frac{1}{N}\left[E\left\{x^4[n]\right\} - (E\left\{x^2[n]\right\})^2\right],$$

which goes to zero for $N \to \infty$: the sample variance is a consistent estimate.

9.10.3 Autocovariance Estimation

The sequence to be estimated is $\gamma_{xx}[l] = E[x_n x_{n-l}]$. This suggests to apply the definition of sample mean to $x[n]x[n-l]$. For $l \geq 0$ we get the estimate

$$\hat{\gamma}_{xx}[l] = c'_{xx}[l] = \frac{1}{N-l} \sum_{n=l}^{N-1} x[n]x[n-l].$$

---

[9]Some authors prefer the definition

$$\hat{\sigma}_x^2 = \frac{1}{N-1} \sum_{n=0}^{N-1} (x[n] - \hat{m}_x)^2$$

that makes the sample variance an unbiased estimate of the variance of the random signal.

For $l < 0$ we can write a similar formula, in which we let the summation run from 0 to $N - 1 - |l|$, and divide by $N - |l|$. However, since $c'_{xx}[l]$ is an even sequence, once its values for $l \geq 0$ have been computed, the values for $l < 0$ can be deduced by symmetry. This observation allows writing simply

$$c'_{xx}[l] = \frac{1}{N - |l|} \sum_{n=|l|}^{N-1} x[n]x[n - |l|]$$

for lags of any sign. Under the hypothesis that $x[n]x[n - l]$ belongs to a Gaussian process, this is the maximum likelihood estimate for $\gamma_{xx}[l]$. Note that $c'_{xx}[l]$ exists for $|l| \leq N - 1$, while the theoretical AC exists for any lag in $(-\infty, +\infty)$.

The estimate $c'_{xx}[l]$ is unbiased: indeed,

$$\mathrm{E}\{c'_{xx}[l]\} = \gamma_{xx}[l] \qquad \text{for} \qquad |l| < N.$$

An approximate expression for the variance of $c'_{xx}[l]$ can be derived, which holds for $|l| \ll N$:

$$\sigma^2_{c'_{xx}[l]} \cong \frac{N}{(N - |l|)^2} \sum_{r=-\infty}^{+\infty} \{\gamma_{xx}[r] + \gamma_{xx}[r + l] + \gamma_{xx}[r - l]\}.$$

This expression shows that for moderate lag values, $\sigma^2_{c'_{xx}[l]}$ is approximately proportional to $1/N$. It can thus be deduced that $\sigma^2_{c'_{xx}[l]} \to 0$ for $N \to \infty$: $c'_{xx}[l]$ is a consistent estimate.

Another estimator proposed for $\gamma_{xx}[l]$ is

$$c_{xx}[l] = \frac{1}{N} \sum_{n=|l|}^{N-1} x[n]x[n - |l|] = \frac{N - |l|}{N} c'_{xx}[l].$$

Since the expected value of $c'_{xx}[l]$ is $\gamma_{xx}[l]$,

$$\mathrm{E}\{c_{xx}[l]\} = \frac{N - |l|}{N} \gamma_{xx}[l] \qquad \text{for} \qquad |l| < N.$$

Thus $c_{xx}[l]$ is biased, with polarization

$$B_{c_{xx}[l]} = \gamma_{xx}[l] - \frac{N - |l|}{N} \gamma_{xx}[l] = \frac{|l|}{N} \gamma_{xx}[l],$$

but asymptotically unbiased. For $|l| \ll N$ we can write

$$\sigma^2_{c_{xx}[l]} \cong \frac{(N - |l|)^2}{N^2} \frac{N}{(N - |l|)^2} \sum_{r=-\infty}^{+\infty} \{\gamma_{xx}[r] + \gamma_{xx}[r + l] + \gamma_{xx}[r - l]\}$$

$$= \frac{1}{N} \sum_{r=-\infty}^{+\infty} \{\gamma_{xx}[r] + \gamma_{xx}[r + l] + \gamma_{xx}[r - l]\} .$$

This expression goes to zero for $N \to \infty$, so $c_{xx}[l]$ is a consistent estimate.

The main difference between $c'_{xx}[l]$ and $c_{xx}[l]$ lies in the behavior of the variance of these two estimates when $|l|$ increases, while N is kept constant. As $|l| \to N$, i.e., as $N - |l| \to 0$, $\sigma^2_{c'_{xx}[l]}$ increases remarkably, and this is due to the fact that at large lags only a few samples remain, over which x[n]x[n − l] can be averaged. This conceptual point is reflected formally in the presence of the factor $(N - |l|)^2$ in the denominator of $\sigma^2_{c'_{xx}[l]}$. On the other hand, $\sigma^2_{c_{xx}[l]}$ does not have the same tendency to increase at large lag values. However, as $|l| \to N$, $E[c_{xx}[l]] \to 0$, so that the bias of $c_{xx}[l]$ becomes as large as $\gamma_{xx}[l]$. Thus, for $|l| \to N$ neither of the two estimates is satisfactory.

However, if we examine the behavior of bias and variance of $c_{xx}[l]$ when $|l|$ is kept constant and N increases, as we did above for $c'_{xx}[l]$, we find that both $\sigma^2_{c_{xx}[l]}$ and $B_{c_{xx}[l]}$ decrease as N increases, hence $c_{xx}[l]$ is a consistent estimate; recall that this is also true for $c'_{xx}[l]$, that is unbiased and has a variance $\sigma^2_{c'_{xx}[l]}$ which decreases for increasing N. From this point of view, both estimates thus appear acceptable. In conclusion, using either $c'_{xx}[l]$ or $c_{xx}[l]$ the estimate is expected to improve using a large number N of samples and limiting the estimate to lag values not too close to N. It is useful to note that AC estimates can be computed through the linear convolution of the first sequence with the folded version of the second sequence. We will simply flip the second series from left to right before computing the convolution, and then will apply the proper factor, $1/N$ for $c_{xx}[l]$ and $1/(N - |l|)$ for $c'_{xx}[l]$.

Finally we mention that these AC estimates can be standardized, leading to estimates $\hat{\rho}_{xx}[l]$ of $\rho_{xx}[l]$.

9.10.4 Cross-Covariance Estimation

The cross-covariance $\gamma_{xy}[l]$ of two different individually and jointly WSS random processes can be estimated, assuming $0 \leq l \leq N - 1$, as[10]

$$c_{xy}[l] = \frac{1}{N} \sum_{n=l}^{N-1} x[n]y[n - l] \quad \text{for} \quad l \geq 0, \qquad c_{xy}[l] = c_{yx}[-l] \quad \text{for} \quad l < 0.$$

---

[10] This definition is valid for real signals. If we considered general complex signals, we should write

$$c_{xy}[l] = \frac{1}{N} \sum_{n=l}^{N-1} x[n]y^*[n - l] \quad \text{for} \quad l \geq 0, \qquad c_{xy}[l] = c^*_{yx}[-l] \quad \text{for} \quad l < 0.$$

Of course this applies to $c_{xx}[l]$ and $c'_{xx}[l]$ too.

The corresponding expected value is

$$E\left\{c_{xy}[l]\right\} = \frac{N - |l|}{N}\gamma_{xy}[l] \quad \text{for} \quad |l| < N.$$

This is a biased, but asymptotically unbiased, estimate of $\gamma_{xy}[l]$. The variance $\sigma^2_{c_{xy}}$ is found to be inversely proportional to N, so this is a consistent estimate. Standardization can then be applied to cross-covariance, to obtain an estimate $\hat{\rho}_{xy}[l]$ of $\rho_{xy}[l]$.

9.11 Appendix: A Road Map to the Analysis of a Data Record

Before studying spectral estimation methods in the next two chapters, it may be useful to schematically list what steps are normally taken when a data record is first examined and then analyzed in detail. The steps listed below constitute a road map to what in the past has been referred to as *time-series analysis* (TSA), where "time series" is synonymous with sequence or data record and is a widely used term in statistical signal processing—the processing of random signals. The data in a time series is referred to as monodimensional or bivariate data, meaning that a single dependent variable has been measured versus a single independent variable, which typically is time.[11]

In the following list, the steps for TSA are indicated by $(+)$ if they are considered as essential, and by $(-)$ if they are optional. The list does not have a strict chronological order, and some steps may be closely related or alternative to some other steps. It is not exhaustive and is meant as a mere collection of suggestions for further study:

+ temporal plot and visual inspection;
+ unevenly-spaced data? $\Rightarrow$ interpolation.
  Spectral methods for unequally-spaced data actually exist, like the Lomb-Scargle periodogram (see Scargle 1982), a variation of the DFT in which an unequally-spaced time series is decomposed into a linear combination of sinusoidal functions; however, interpolation is a common practice;
+ detection, counting (as a percentage of total) and substitution of missing data through interpolation techniques (gap filling): e.g., linear interpolation, splines; advanced methods like SSA (singular spectrum analysis; see Chap. 12);
+ elementary statistical analysis (mean, standard deviation; histogram $\Rightarrow$ Gaussian distribution?);
− search for outliers (data very distant from the mean) and noise spikes, and possible exclusion/replacement of related data samples;
− estimation of the AC sequence;

---

[11] In contrast, multivariate data is typically produced by a variable measured; as a function of more than one independent variable, for example, space and time.

- de-noising: filtering, SSA, Discrete Wavelet Transform (DWT; see Chap. 14), etc.;
- trend removal: functional fits, advanced methods like SSA, etc.;
- separation/removal of known periodicities and dominant signals (e.g., seasonality) in order to approach gaussianity/ergodicity; pre-whitening;
+ stationarity?

- Yes ⇒ stationary spectral analysis by classical and advanced methods; significance tests for spectral peaks (ergodicity assumed). See Chaps. 10 and 11;
- No ⇒ evolutionary spectral analysis: Short Time Fourier Transform (STFT), Continuous Wavelet Transform (CWT). See Chap. 13.

Stationarity and *ergodicity* deserve some more words. They are basic assumption for all the spectral estimation methods presented in the next two chapters. Obstacles to stationarity and ergodicity in a series may include trends, changes of series variance, deterministic cycles, etc.

Deterministic cycles of known origin appear in many types of series. For example, meteo-climatic records most often contain the 1 year cycle due to seasons and/or the 1 day cycle related to Earth's rotation, depending on the record duration and on the sampling interval. The subtraction from the data of these deterministic components known a priori often helps spectral analysis, allowing for the detection of spectral features that would otherwise be submerged by these dominant signals. One easy way to remove such cycles in data measured at regular intervals—e.g., to remove the annual cycle of the seasons from data measured once per month—is to calculate the sample mean of every month of the calendar year and then subtract the mean of all Januaries from each January sample, the mean of all Februaries from each February sample, and so on.

This procedure can be considered as a way to get closer to ergodicity before applying spectral estimation methods. It can also be classified as a form of *pre-whitening*: by this term we mean an operation performed on a signal to make it more similar to white noise. and thus more suitable to be analyzed by statistics-based methods. More in general, pre-whitening can be achieved using a *predictor* that models the persistence in the signal. The predictor attempts to forecast the sample at time n on the basis of past samples. Later this persistence is subtracted from the n-th signal sample, so that, ideally, it becomes uncorrelated from previous samples. We will discuss in Chap. 11, in the frame of the parametric approach to spectral estimation, the modeling of persistence in a sequence, performed by *stochastic models*. Their use for pre-whitening before spectral analysis is not common practice in all research fields.

Trends and changes in variance are obstacles to the stationarity assumption. There are research fields in which trends, in particular, may be very pronounced. These trends are usually removed before analysis.

Incidentally, we may mention that the cases in which the goal of time-series processing is spectral analysis must be distinguished from the cases in which the goal is signal modeling. While in the former case, spectral analysis of non-stationary series is simply tackled by evolutionary methods, such as the wavelet transform discussed in Chap. 13, in the latter case the issue of *stationarizing* a non-stationary series is often considered. In several research fields, modeling a signal is a fundamental part of time-series analysis. Modeling a random signal means building a mathematical model for the process that generates the data. This is useful, for example, for predicting the signal's future behavior, and can be done using the above-mentioned stochastic models. Many modeling/forecasting methods are based on the assumption that a non-stationary time series can be rendered approximately stationary (i.e., "stationarized") through the use of mathematical transformations. For example, a stationarized series is relatively easy to predict: you simply predict that its statistical properties will be the same in the future as they have been in the past. The predictions for the stationarized series can then be "untransformed", by reversing whatever mathematical transformations were previously used, to obtain non-obvious predictions for the original series. Thus, finding the sequence of transformations needed to stationarize a time series can provide important clues in the search for an appropriate series model. Another reason for trying to stationarize a time series is to be able to obtain meaningful sample statistics such as means, variances, and correlations with other variables. Such statistics are useful descriptors of future behavior only if the series is stationary. For example, if the series is consistently increasing over time, its sample mean and variance will grow with the size of the sample, and will always underestimate the mean and the variance in future times. Also, if the mean and variance of a series are not well-defined, then neither are its correlations with other variables.

In time series modeling, many different ways to stationarize a series have been devised. A simple but often effective way to stabilize the variance across time is to apply a power transformation (square root, cube root, log, etc.) to the time series. Linear or functional trends, including seasonal cycles, can be identified through series regression on linear or nonlinear functions of time (polynomials, sinusoids etc.) and then subtracted from the data. Another helpful transformation meant at stationarization, widely applied especially to economic time series, is the differencing operation. The d-th differencing operator applied to a time series $x[n]$ creates a new series the value of which at time n is the difference $x[n + d] - x[n]$.

Most business and economic time series are far from stationarity when expressed in their original measurement units. First, they may include the effects of inflation. Inflation adjustment, or "deflation", is accomplished by dividing a monetary time series, e.g., in dollars, by a price index such as the Consumer Price Index (CPI). The deflated series is then said to be measured in "constant dollars", whereas the original series was measured in "nominal dollars" or "current dollars". Inflation is often a significant component of apparent growth in any series measured in dollars, or yen, euros, etc. By adjusting for inflation, the real growth, if any, can be detected.

However, even after deflation and/or possibly seasonal adjustment, most economic time series will typically still exhibit trends, cycles, and other non-stationary behavior. If the series has a stable long-run trend and tends to revert to the trend line following a disturbance, it may be possible to stationarize it by de-trending, perhaps in conjunction with application of a suitable transformation, like the logarithmic one. Such a series is said to be trend-stationary.

Sometimes, even de-trending is not sufficient to make the series stationary. If the mean, variance, and autocorrelations of the series are not constant in time even after the previously-mentioned expedients, perhaps the statistics of the changes $x[n + d] - x[n]$ in the series, with a suitable choice of d, will be constant. Such a series is said to be difference-stationary. In some cases, it can be hard to distinguish between a series that is trend-stationary and one that is difference-stationary, and the so-called "unit root tests" may be used to get a more definitive answer. These are statistical hypothesis-tests of stationarity that are designed for determining whether differencing is required. A number of unit root tests are available, and they are based on different assumptions and may lead to conflicting answers. One of the most popular tests is the Augmented Dickey-Fuller (ADF) test (see Said and Dickey 1984). For more details on these issues see, e.g., Chatfield (2013).

References

Blackman, R.B., Tukey, J.W.: The Measurement of Power Spectra from the Point of View of Communication Engineering. Dover, New York (1958)

Chatfield, C.: The Analysis of Time Series: An Introduction. Chapman and Hall—CRC Press, Boca Raton (2013)

Gray, M., Davisson, L.D.: An Introduction to Statistical Signal Processing. Cambridge University Press, Cambridge (2004)

Hannan, E.J.: Time Series Analysis. Methuen, London (1960)

Hayes, M.H.: Statistical Digital Signal Processing and Modeling. Wiley, New York (1996)

Helstrom, C.W.: Probability and Stochastic Processes for Engineers. MacMillan, New York (1991)

Kay, S.M.: Fundamentals of Statistical Signal Processing. Prentice Hall, Upper Saddle River (1993)

Proakis, J.G., Manolakis, D.G.: Digital Signal Processing: Principles, Algorithms, and Applications. Prentice Hall, Upper Saddle River (1996)

Said, S.E., Dickey, D.A.: Testing for unit roots in autoregressive-moving average models of unknown order. Biometrika **71**(3), 599–607 (1984)

Scharf, L.L., Demeure, L.: Statistical Signal Processing: Detection, Estimation, and Time Series Analysis. Addison-Wesley, Boston (1991)

Scargle, J.D.: Studies in L. II—Statistical aspects of spectral analysis of unevenly spaced data. Astrophys. J., Part 1, **263**, 835–853 (1982)

Stoica, P., Moses, R.: Spectral Analysis of Signals. Prentice Hall, Upper Saddle River (2005)

Chapter 10
Non-Parametric Spectral Methods

10.1 Chapter Summary

The techniques described in this chapter belong to the field of statistical spectral analysis, meaning the analysis of the frequency content of stationary and ergodic random signals. The problem is how to get a good estimate of the true power spectrum on the basis of a finite number of samples of a typical sample sequence. The simplest approach is the periodogram, which can be defined in two equivalent ways: as the Fourier transform of some estimate of the AC sequence—in this case the estimate is the correlogram, or Blackman-Tukey periodogram—or as the square modulus of the Fourier transform of the data sequence, divided by its length—in this case the estimate is called the Schuster's periodogram, or simply the periodogram. Unfortunately, even adopting a consistent estimate for the AC sequence, the periodogram/correlogram does not turn out to be a consistent estimate of the power spectrum. It is also biased, though asymptotically unbiased, but the main issue is its inconsistency and the high variance of the spectral estimates. The search for a stable and consistent estimate of the power spectrum leads to the methods of Bartlett and Welch, based on averaging a number of independent or nearly-independent periodograms obtained dividing the stationary signal into subsequences, and to the Blackman-Tukey method, based on smoothing the correlogram by convolving it with a proper spectral window.

A brief account of the estimation of the cross-spectrum of two random signals, which is useful when we are interested in the common spectral content of two series and on the phase relations of the respective frequency components, is also given.

Next we describe statistical tests that can be used to establish confidence intervals for spectral estimates, as well as significance levels allowing us to judge if a peak detected in a spectrum can be considered as due to a real feature present in the data, or if instead it is compatible with the hypothesis of it just being a fluctuation due to noise.

The multitaper method (MTM) is then presented, which is another Fourier-based approach in which independent periodogram estimates are averaged. However, these

© Springer International Publishing Switzerland 2016   403
S.M. Alessio, *Digital Signal Processing and Spectral*
Analysis for Scientists, Signals and Communication Technology,
DOI 10.1007/978-3-319-25468-5_10

estimates are not obtained segmenting the record, but applying to it a number of tapers orthogonal to one another by construction. The MTM is interesting because of its good performances in terms of resolution and variance of the spectral estimate, even if the statistical test traditionally adopted to establish the significance of peaks detected by the MTM is based on a restrictive assumption: the true spectrum is assumed to be the superposition of a continuous, possibly colored-noise-like component, and one or more spectral lines, corresponding to harmonic processes. This assumption can lead to misleading test results if it is not handled carefully.

We will also discuss the use of the FFT for practical computation of spectral estimates, and the different normalization schemes adopted in literature for the power spectrum. This is an issue that should not be overlooked, because it may cause confusion when one attempts to compare the spectra of a given signal obtained by different authors and methods.

From now on, for simplicity we will return to the usual notation $x[n]$ for sequences, dropping the distinction introduced in Chap. 9 between random variables and sample sequences. We will assume that the signals we analyze have been centered and therefore have zero mean, so that autocorrelation and autocovariance coincide and are simply denoted by the acronym AC. In statistical spectral analysis, the data record is often referred to as a time series. We will hereafter use the terms sequence, (time) series and data record, or simply record, interchangeably.

10.2 Power Spectrum Estimation

The methods for estimating the power spectrum of an ergodic WSS random signal are the subject of the present chapter and of the following one. Here we will try to give the bigger picture of this topic.

Spectral analysis, i.e., the application of the multitudinous methods of power-spectrum estimation that have been proposed in literature, is perhaps the most widely used tool for the study of the characteristics of the random process that is thought to generate the data. Spectral analysis is aimed at investigating the existence of possible cyclicities or pseudo-cyclicities in the signal, or, on the contrary, to ascertain its noise-like nature, with the ultimate goal of relating the process that generates the data with the factors that influence the process and determine its variability. This is of interest in many fields of physics and other disciplines, like social and economic sciences, life sciences, etc.

Here we must first of all clarify what is meant by cyclicity or pseudo-cyclicity, and by noise-like nature.

Cyclicity refers to the discrete component of the theoretical spectrum, i.e., the spectral lines that correspond to phase-coherent harmonic signals (phase-coherent sinusoidal oscillations): the Fourier transform of a clean periodic signal of infinite length yields a spectrum with a Dirac function at the frequency of the signal, i.e., a peak of zero width and infinite magnitude. In spectra computed from windowed signals, these lines will obviously be broadened and smeared out by the effect of windowing, but this is not the point here.

Pseudo-cyclicity or *quasi-periodicity* refers to narrowband spectral features due to anharmonic "quasi-oscillatory" signals in the data, which may exhibit phase and amplitude modulation, and intermittent oscillatory behavior, but can nonetheless be recognized as significant by proper statistical tests, relative to some suitably defined null hypothesis.

Noise is represented by the continuous component of the spectrum. The spectrum of noise does not exhibit prominent peaks and presents an overall shape that can be flat (white noise) or can be slowly variable with frequency (*colored noise*).

In many applications, signals are frequently associated with narrowband, but not strictly harmonic variability, and truly harmonic signals are rarely detected. Actually, the presence of true lines in the theoretical spectrum of a signal is an obstacle to ergodicity, so that these harmonic components should, in principle, be properly removed before spectral analysis. Removing them means, first of all, detecting them and accurately finding their frequency, amplitude, and phase. This is the purpose of *harmonic analysis*. Even neglecting the ergodicity issue mentioned above, classical spectral methods are inadequate for harmonic analysis: they can only give indirect information on the amplitude of the signal component at a given frequency, through the area under the peak centered at that frequency, the width of which is, roughly speaking, inversely proportional to the length N of the available sequence; this area is nearly constant, since the height of the peak is also proportional to N. A method for harmonic analysis attempts, instead, to determine directly the finite amplitude of a pure line in the spectrum of a sequence of finite length. Dedicated *methods of frequency estimation for harmonic processes/signals* exist.

Going back to the estimation of the power spectrum, we must clearly state that application of the statistical techniques of estimation theory nearly always requires more information on the signal nature than is available. As a consequence, the most widely employed spectral methods remarkably rely on empirical considerations. The scenario of these techniques is wide and intricate, and no method is the best method ever: each one has its indications, its advantages and its limitations. In the analysis of a data record it is thus always advisable to apply more than one method. The spectral features recurring independently of the estimation technique can be considered as plausibly true, while those features that are detected by one method only can be suspected to be due to mathematical/numerical artifacts. In all cases, when the analysis detects an outstanding spectral feature, such as a prominent peak at some frequency, it is desirable to have at hand proper statistical tests by which to establish the probability level at which this feature can be considered significant. Such statistical tests are available for several spectral methods, but not for all methods.

There are two main methods of PSD estimation: non-parametric and parametric methods.

Loosely speaking, non-parametric methods are used when little is known about the mechanism that generates the data. Non-parametric methods of spectral estimation are the subject of the present chapter, which also deals briefly with estimating the cross spectrum of two random signals.

Parametric methods are based on data stochastic modeling, and thus assume that some description of the data generation mechanism can be given. They are especially useful when high resolution is needed. They typically have greater computational complexity than non-parametric methods. Parametric methods are the subject of Chap. 11.

The search for almost-periodic phenomena in a data record can also be performed by another non-parametric approach, namely singular spectrum analysis (SSA), which however is not, strictly speaking, a method for estimating the power spectrum: it is aimed at representing the signal as a linear combination of elementary variability modes that are not necessarily harmonic components, but can exhibit amplitude and frequency modulations in time, and are data-adaptive, i.e., modeled on the data. SSA is presented in Chap. 12.

The field of PSD estimation methods should be distinguished from the above-mentioned field of frequency-estimation methods for harmonic processes, which *assume* that a signal is composed of a limited—usually small—number of generating phase-coherent sinusoidal or complex exponential components plus noise, and seek to find the location and intensity of the line spectral features. While in the former no assumption is made about the number of cyclic components in the signal and the aim is to estimate the whole generating spectrum, in the latter, sometimes referred to as *subspace methods* (Marple 1987; Stoica and Moses 2005; Schmidt 1986; Stoica and Moses 2005), the sequence is decomposed into a signal subspace—here "signal" indicates the harmonic content—and a noise subspace. The formalism of subspace methods involves linear algebra. For the basic notions of vector and matrix analysis needed to understand the meaning of the technical terms used in this subsection, the reader is referred to Hayes (1996), Chap. 2, pp. 20–48, where a very good summary is provided.

Exploiting orthogonality between the eigenvectors of the two subspaces allows for a pseudo-spectrum to be formed, where large peaks ascribable to harmonic components can appear. These methods work very well even is the useful signal is submerged in a large amount of noise, but are computationally very expensive. They can be grouped into two categories: noise subspace methods and signal subspace methods. Both categories can be utilized in one of two ways: eigenvalue decomposition of a matrix properly formed using AC samples—the AC matrix—or singular value decomposition of a matrix formed using data samples—the data matrix (see Chap. 12 for the precise meaning of these expressions).

Noise subspace methods attempt to solve for one or more of the noise subspace eigenvectors. Then, the orthogonality between the noise subspace and the signal subspace produces zeros in the denominator of the resulting spectral estimates, resulting in large values or spikes at the true frequencies of the signal components. The number of discrete sinusoids must be determined/estimated or known ahead of time.

Signal subspace methods attempt to discard the noise subspace prior to spectral estimation, improving the signal-to-noise ratio (SNR). A reduced-rank AC matrix is formed with only the eigenvectors that have been recognized to belong to the signal subspace, and then the reduced-rank matrix is used. Subspace methods include, for instance, MUSIC (Multiple Signal Classification) and EIG (Eigenvector Method). A detailed description of these methods is beyond the scope of this book.

Finally, we may mention that even an evolutionary spectral method like the Continuous Wavelet Transform (CWT), described in Chap. 13, can yield an *average estimate* of the power spectrum of a non-stationary signal, namely the global wavelet spectrum (GWS). In the case of effective stationarity, the GWS can simply be considered as a power spectrum estimation method.

Before concluding, let us stress once again some fundamental points. First, the convenience of adopting one particular spectral method or another depends on

- the features of the sequence: whether it is short, long, noisy, noise-like or of the kind "noise plus signal", etc.
- what the analysis is expected to detect: do we expect individual dominant periodic or quasi-periodic features? Are we interested in the overall spectral behavior of a noise-like signal? Etc.
- sometimes, considerations based on a priori information or hypotheses concerning the system from which the signal derives.

Second, it is recommended to use, when possible, several spectral methods to analyze a data record, and then compare the results.

Last, but not least, let us underline it is highly advisable to reduce the signal to zero mean before applying any spectral method, in order to avoid the presence of possible high power at zero frequency in the true spectrum, which in the estimated spectrum would contaminate with its leakage a consistent range of frequencies. In fact, leakage and resolution loss due to windowing are present in power spectrum estimation theory in exactly the same terms as we described them for deterministic signals (Sect. 5.2.1). If the mean is not removed, autocorrelation and autocovariance do not coincide and the power spectrum has an impulse at zero frequency, since zero frequency means the constant component of the signal. If the mean value of the data is relatively high, the low-frequency, low-amplitude spectral components will be obscured by the leakage due to this impulse at zero frequency. Even if the sample mean is only an estimate of the true mean of the process, subtracting it from the data before spectral analysis improves the power spectrum estimate at low frequencies.

10.3 Periodogram

The *periodogram* represents the simplest and most classical approach to spectral estimation, based on the DTFT and DFT. The estimate of the power spectrum must be performed on the basis of a finite-length segment of a sample sequence. The finite length of the available data record affects the quality of the estimate. When stationarity is assumed, as in the present case, the longer the record, the better the estimate that can be extracted from the record itself. However, if the statistical properties of the signal are actually non-stationary at some time scale, we must be careful to select a record length that does not violate the stationarity hypothesis.

The finite record length causes a distortion of the estimated spectrum with respect to the "true" one: this issue, involving leakage and resolution loss due to windowing, appears in the theory of statistical spectral analysis exactly in the same terms as in the case of deterministic signals (Chap. 5). Since the record length is limited by the rapidity of the variations in the signal's statistical properties, the rule will be to choose the shortest record that still allows for resolving spectral features at closely spaced frequencies. At the same time, the use of the DFT in place of the DTFT, and of the FFT as a practical tool, will imply the same issues related to frequency sampling that we discussed in the deterministic case.

The function we must estimate is the theoretical power spectrum; we start from the first definition we gave in the previous chapter for $P_{xx}(\omega)$,

$$P_{xx}(\omega) = \sum_{l=-\infty}^{+\infty} \gamma_{xx}[l] e^{-j\omega l},$$

such that

$$\frac{1}{2\pi} \int_{-\pi}^{\pi} P_{xx}(\omega) d\omega = \sigma_x^2.$$

In the previous chapter we introduced two plausible estimators for the AC sequence $\gamma_{xx}[l]$, but unfortunately the Fourier transforms of such estimates do not turn out to be good estimates of the power spectrum. In fact, they are found to be non-consistent: the variance of the spectral estimate does not tend to zero when the number of available signal samples tends to infinity. However, we will see that a good spectral estimate can be obtained by properly smoothing the Fourier transform of the AC estimate.

Let us take $c_{xx}[l]$ as the AC estimate. We will compute its samples for any possible lag value, from $-(N-1)$ to $N-1$, obtaining

$$c_{xx}[l] = \frac{1}{N} \sum_{n=|l|}^{N-1} x[n]x[n-|l|], \qquad \text{for} \ \ |l| \leq N-1.$$

Under the hypothesis that $x[n]x[n-l]$ belongs to a Gaussian process, this is the maximum likelihood estimate for $\gamma_{xx}[l]$. Applying the DTFT definition equation to $c_{xx}[l]$ we get the *correlogram*, also referred to as the *Blackman-Tukey periodogram* (Blackman and Tukey 1958):

$$I_N(\omega) = \sum_{l=-(N-1)}^{N-1} c_{xx}[l] e^{-j\omega l}.$$

Note that $I_N(\omega)$ certainly exists and is a real quantity, because it is the DTFT of a finite-length non-causal sequence. The correlogram is indicated as $I_N(\omega)$ due to its reality.

Substituting the expression of $c_{xx}[l]$ in $I_N(\omega)$, an equivalent expression for the periodogram is found (Stoica and Moses 2005) that directly descends from the second definition of the theoretical power spectrum given in the previous chapter:

$$I_N(\omega) = \frac{1}{N} \left| X(e^{j\omega}) \right|^2,$$

which is known as the *Schuster's periodogram*, or simply the *periodogram* (Schuster 1898). Again, $X(e^{j\omega})$ certainly exists, being the DTFT of the finite-length causal record $x[n]$, $N = [0, N-1]$. The $X(e^{j\omega})$ that appears here must not be mistaken with the DTFT of the infinitely-long random power signal from which the record derives— a DTFT that does not exist. The above equality is easily proved considering that $c_{xx}[l]$ is $1/N$ times the convolution of $x[n]$ with the folded version of $x[n]$. Taking the Fourier transform and using the convolution theorem, we get $I_N(\omega) = [X(e^{j\omega})X^*(e^{j\omega})]/N = \left| X(e^{j\omega}) \right|^2 /N$, q.e.d.

From Schuster's expression of the periodogram we can see that $I_N(\omega)$

- is a non-negative function of ω,
- can be computed directly from the data $x[n]$,
- is the square modulus of the DTFT of the data sequence, divided by the length of the sequence.

This connection between the periodogram and the DTFT of the data sequence is the central argument in favor of the use of $c_{xx}[l]$, in terms of computational convenience.

If, instead, we take $c'_{xx}[l]$ as the AC estimate, and compute its samples for all possible lag values from $-(N-1)$ a $N-1$,

$$c'_{xx}[l] = \frac{1}{N - |l|} \sum_{n=|l|}^{N-1} x[n]x[n - |l|], \qquad \text{for } |l| \leq N - 1,$$

we obtain a different estimate of the correlogram that we will denote as $P_N(\omega)$:

$$P_N(\omega) = \sum_{l=-(N-1)}^{N-1} c'_{xx}[l]e^{-j\omega l}.$$

In practice, since both AC estimated sequences have with finite length, the transforms we can compute will not be DTFTs, but sequences of DTFT values at discrete frequencies, i.e., DFTs, and we will get *samples* of the spectral estimate, $I_N[k] = I_N(\omega_k)$, or $P_N[k] = P_N(\omega_k)$. However, in the following discussion we will indicate the spectral estimate as a continuous function of ω, i.e., as a DTFT. This is common practice and is convenient.

Another remark we must make is that actually, when we compute the DFT of the AC sequence via FFT, we implicitly treat it as a causal sequence; as a consequence, the FFT output will be complex, and in order to get the spectral samples we will have to take the modulus of the FFT output.

We can thus distinguish two Fourier methods to estimate the power spectrum of a stationary and ergodic random signal from its N samples $x[n]$:

- the *direct method* that implies computing the transform of the data record, and
- the *indirect method* that requires two separate steps: first estimating the AC sequence, and then computing its transform.

Two of the classical spectral methods we will discuss (Bartlett's and Welch's methods) are direct methods, while a third method, named after Blackman and Tukey, is built on AC estimation.

Let us now examine the statistical properties of the periodogram estimator, in its two versions $I_N(\omega)$ and $P_N(\omega)$.

10.3.1 Bias

The expected value of the periodogram $I_N(\omega)$ is

$$E[I_N(\omega)] = \sum_{l=-(N-1)}^{N-1} E[c_{xx}[l]]e^{-j\omega l} = \sum_{l=-(N-1)}^{N-1} \frac{N-|l|}{N}\gamma_{xx}[l]e^{-j\omega l},$$

which is different from $\Gamma_{xx}(e^{j\omega}) = P_{xx}(\omega)$ for two reasons: the presence of the factor $(N-|l|)/N$ and, more importantly, the finite summation limits. We deduce that $I_N(\omega)$ is a biased estimate of $\Gamma_{xx}(e^{j\omega}) = P_{xx}(\omega)$. However, this estimate is asymptotically unbiased because

$$\lim_{N\to\infty} E[I_N(\omega)] = P_{xx}(\omega).$$

The expected value of the periodogram $P_N(\omega)$ is, instead,

$$E[P_N(\omega)] = \sum_{l=-(N-1)}^{N-1} E[c'_{xx}[l]]e^{-j\omega l} = \sum_{l=-(N-1)}^{N-1} \gamma_{xx}[l]e^{-j\omega l}.$$

In terms of bias, $E[P_N(\omega)]$ appears "better" than $E[I_N(\omega)]$ but, due to the finite summation limits, $P_N(\omega)$ is still a biased estimate of $\Gamma_{xx}(e^{j\omega}) = P_{xx}(\omega)$, in spite of the fact that $c'_{xx}[l]$ is an unbiased estimate of $\gamma_{xx}[l]$. Asymptotically, the bias vanishes.

The expected values of $I_N(\omega)$ and $P_N(\omega)$ can be interpreted as Fourier transforms of the theoretical, infinitely-long AC sequence $\gamma_{xx}[l]$, windowed by non-causal windows including $2N-1$ samples (Marple 1987). In other words, the finite-length AC sequence that we can compute when N data samples are available can be viewed as the result of windowing $\gamma_{xx}[l]$ with a non-causal window of finite length $2N-1$, i.e., as the multiplication of $\gamma_{xx}[l]$ by the window:

- for the periodogram $I_N(\omega)$,

$$E[I_N(\omega)] = \sum_{l=-(N-1)}^{N-1} \frac{N - |l|}{N} \gamma_{xx}[l] e^{-j\omega l} = \sum_{l=-\infty}^{\infty} \{w_B[l]\gamma_{xx}[l]\} e^{-j\omega l},$$

where $w_B[l]$ indicates the *triangular* or *Bartlett window*,[1]

$$w_B[l] = \begin{cases} \frac{N-|l|}{N} & |l| \leq N - 1, \\ 0 & \text{elsewhere}, \end{cases}$$

shown in Fig. 10.1a for $2N - 1 = 49$;
- for the periodogram $P_N(\omega)$,

$$E[P_N(\omega)] = \sum_{l=-(N-1)}^{N-1} \gamma_{xx}[l] e^{-j\omega l} = \sum_{l=-(-\infty)}^{\infty} \{w_R[l]\gamma_{xx}[l]\} e^{-j\omega l},$$

where $w_R[l]$ indicates the *rectangular window*,

$$w_R[l] = \begin{cases} 1 & |l| \leq N - 1, \\ 0 & \text{elsewhere}, \end{cases}$$

shown in Fig. 10.1b for $2N - 1 = 49$.

These expectations represent DTFTs of sequences that are non-zero only over a finite number $2N - 1$ of samples, i.e., the product sequences $w_B[l]\gamma_{xx}[l]$ and $w_R[l]\gamma_{xx}[l]$. These products correspond to *continuous periodic convolutions* in the frequency domain:

$$E[I_N(\omega)] = \frac{1}{2\pi} \int_{-\pi}^{+\pi} W_B\left\{e^{j(\omega-\theta)}\right\} \Gamma_{xx}(e^{j\theta}) d\theta =$$

$$= \frac{1}{2\pi} \int_{-\pi}^{+\pi} W_B\left\{e^{j(\omega-\theta)}\right\} P_{xx}(\theta) d\theta,$$

$$E[P_N(\omega)] = \frac{1}{2\pi} \int_{-\pi}^{+\pi} W_R\left\{e^{j(\omega-\theta)}\right\} P_{xx}(\theta) d\theta,$$

---

[1] In this frame we use the two terms indifferently, even if the triangular window and the Bartlett window are actually two different sequences: the former has no zeros at its edges, while the latter includes two zero samples at the beginning and end. In the present case, there must be no zeros at the edges; therefore in principle we should call this a "triangular window".

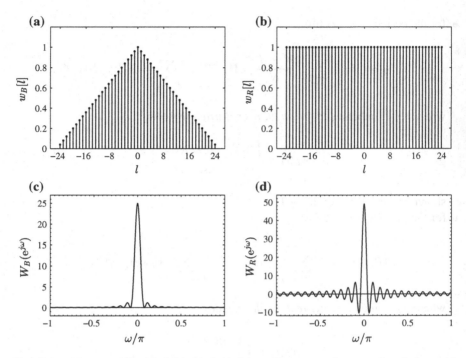

Fig. 10.1 a Non-causal Bartlett (triangular) window and **b** non-causal rectangular window, with lengths of $2N - 1 = 49$ samples; **c–d** corresponding Fourier transforms, known as spectral windows in the frame of power-spectrum estimation theory

where the non-causal-window transforms, also known as *spectral windows*, are real and given by

$$W_B(\mathrm{e}^{\mathrm{j}\omega}) = \sum_{l=-(N-1)}^{N-1} \frac{N - |l|}{N} \mathrm{e}^{-\mathrm{j}\omega l} = \frac{1}{N} \left[\frac{\sin\left(N\frac{\omega}{2}\right)}{\sin\frac{\omega}{2}} \right]^2 = N D_N^2(\omega),$$

$$W_R(\mathrm{e}^{\mathrm{j}\omega}) = \sum_{l=-(N-1)}^{N-1} \mathrm{e}^{-\mathrm{j}\omega l} = \frac{\sin\left[(2N - 1)\frac{\omega}{2}\right]}{\sin\frac{\omega}{2}} = (2N - 1) D_{2N-1}(\omega).$$

$W_B(\mathrm{e}^{\mathrm{j}\omega})$ is also known as the *Fejér kernel*; $W_R(\mathrm{e}^{\mathrm{j}\omega})$ is also known as the *Dirichlet kernel*. The expressions for the Fejér and Dirichlet kernels can be derived as follows:

$$W_B(\mathrm{e}^{\mathrm{j}\omega}) = \sum_{l=-(N-1)}^{N-1} \frac{N - |l|}{N} \mathrm{e}^{-\mathrm{j}\omega l} = \frac{1}{N} \sum_{m=0}^{N-1}\sum_{i=0}^{N-1} \mathrm{e}^{-\mathrm{j}\omega(m-i)} = \frac{1}{N} \left| \sum_{m=0}^{N-1} \mathrm{e}^{\mathrm{j}\omega m} \right|^2 =$$

$$= \frac{1}{N} \left| \frac{\mathrm{e}^{\mathrm{j}\omega N} - 1}{\mathrm{e}^{\mathrm{j}\omega} - 1} \right|^2 = \frac{1}{N} \left| \frac{\mathrm{e}^{\mathrm{j}N\frac{\omega}{2}} - \mathrm{e}^{-\mathrm{j}N\frac{\omega}{2}}}{\mathrm{e}^{\mathrm{j}\omega/2} - \mathrm{e}^{-\mathrm{j}\omega/2}} \right|^2 = \frac{1}{N} \left[\frac{\sin(N\frac{\omega}{2})}{\sin\frac{\omega}{2}} \right]^2 = N D_N^2(\omega);$$

$$W_R(e^{j\omega}) = \sum_{l=-(N-1)}^{N-1} e^{-j\omega l} = 2\text{Re}\left[\frac{e^{j\omega N} - 1}{e^{j\omega} - 1}\right] - 1 =$$

$$= \frac{2\cos\left[\frac{(N-1)\omega}{2}\right]\sin\left[\frac{N\omega}{2}\right]}{\sin\frac{\omega}{2}} - 1 = \frac{\sin(\omega\frac{2N-1}{2})}{\sin\frac{\omega}{2}}(2N-1)D_{2N-1}(\omega).$$

Figure 10.1c, d illustrate the shape of the functions $W_B(e^{j\omega})$ and $W_R(e^{j\omega})$. To emphasize the dependence on the length of the estimated AC sequence, hereafter we will indicate them as $W_{B,2N-1}(e^{j\omega})$ and $W_{R,2N-1}(e^{j\omega})$, respectively.

It is clear that since N is finite, both $I_N(\omega)$ and $P_N(\omega)$ are biased, and provide a distorted version of the true spectrum of the underlying infinite-length random signal. Each spectral estimate at frequency ω also contains the contributions of nearby frequencies, weighted according to the shape of the corresponding spectral window. In particular, in the case of $P_N(\omega)$ the spectral window also assumes negative values (Fig. 10.1d), since the Dirichlet function with odd order does so (Sect. 3.4.2); this may lead to negative values of the spectral estimate.

We now drop $P_N(\omega)$ as a spectral estimator and continue our discussion considering $I_N(\omega)$ only. From now on, by "periodogram" we mean $I_N(\omega)$.

The presence of sidelobes in $W_{B,2N-1}(e^{j\omega})$ can cause diffusion towards a given frequency $\theta = \omega$, of power belonging to frequencies that can be quite far from ω. This power diffusion in the frequency domain is the well-known *leakage*. Moreover, the non-zero width of the main lobe sets an upper limit to *spectral resolution*. For example, if the sequence $x[n]$ contains a sinusoidal component at frequency ω_0, in the true spectrum we would see a spectral line at $\omega = \omega_0$, while observing the periodogram $I_N(\omega)$ we will see a copy of $W_{B,2N-1}(e^{j\omega})$ centered at ω_0. If we need to detect two spectral components with frequencies close to one another, and with equal amplitudes, the widening of the respective peaks—that would ideally be impulsive— can lead to the impossibility of resolving them as their frequency distance decreases. We thus have *resolution loss* due to the finite signal observation time. The problem worsens if the two sinusoids that are close to each other have different amplitudes: in such cases the sidelobes related to the strongest component can completely submerge the peak of the weakest component. Leakage and can be mitigated only reducing the sidelobes, while keeping the main lobe as narrow as possible, in order to preserve resolution. Unfortunately, the use of more gradual windows cannot give both results at the same time: it can reduce the sidelobes but only at the expense of a widening of the main lobe. All other conditions being equal, we can reduce the main lobe width only increasing N, but this cannot be done in all cases. In conclusion, this is a disadvantage inherent in Fourier spectral techniques: the finite duration of $x[n]$ sets a limit to the quality of the spectral estimates we can obtain from the available data samples.

10.3.2   *Variance*

We will now examine the variance of the periodogram and in particular its dependence on the number N of available signal samples. An approximate calculation (Jenkins and Watts 1968) of this variance is based on the hypothesis that the process is Gaussian and white, at least locally. Recall that while a flat spectrum belongs to white noise, spectra with smooth behavior, slowly varying with frequency and with no peaks, represent colored noise. Hereinafter, we will encounter several examples of colored noise. Requiring a signal to be *locally white* means requiring it to be colored noise for which PSD varies with frequency arbitrarily but very slowly, so that we can consider it nearly constant over the bandwidth of each spectral estimate.

This approximate calculation leads to an expression for the covariance between two periodogram values $I_N(\omega_1)$ and $I_N(\omega_2)$ at two frequencies ω_1 and ω_2, which we will indicate by $c_{I_N(\omega)}(\omega_1, \omega_2)$:

$$c_{I_N(\omega)}(\omega_1, \omega_2) \approx P_{xx}(\omega_1)P_{xx}(\omega_2) \left\{ \left[\frac{\sin\left(N\frac{\omega_1-\omega_2}{2}\right)}{N\sin\left(\frac{\omega_1-\omega_2}{2}\right)} \right]^2 + \left[\frac{\sin\left(N\frac{\omega_1+\omega_2}{2}\right)}{N\sin\left(\frac{\omega_1+\omega_2}{2}\right)} \right]^2 \right\}.$$

If we evaluate this expression at discrete ω values equispaced by $2\pi/N$, i.e., at those frequencies $\omega_k = 2\pi k/N$ that form the set of discrete frequencies at which we sample the DTFT of our signal, we find that this covariance vanishes: the values of $I_N(\omega_k)$ are uncorrelated. As N increases, such frequencies get closer and closer to one another, and the fact that the estimated PSD values at closely spaced frequencies are uncorrelated suggests a "wild" variability of the periodogram.

The periodogram variance can then be found setting $\omega_1 = \omega_2 = \omega$ in the covariance expression:

$$\sigma_{I_N(\omega)}^2 \cong P_{xx}^2(\omega) \left\{ 1 + \left[\frac{\sin(N\omega)}{N\sin\omega} \right]^2 \right\}.$$

This variance does not go to zero for $N \to \infty$ and is of the order of the square of the PSD we want to estimate (Marple 1987):

$$\sigma_{I_N(\omega)}^2 \approx P_{xx}^2(\omega).$$

Therefore, the periodogram is not a consistent estimator of the power spectrum, and we can expect it to exhibit remarkable fluctuations around the true spectrum. We can also say that the periodogram is not a stable spectral estimate: *stability* is the degree to which irrelevant spectral details are smoothed out by an estimator, so that the estimates drawn from different segments of a typical sample sequence would be in agreement. Low stability is thus synonymous with large variance of the spectral estimate. These results on periodogram bias, variance and inconsistency are illustrated by the examples given below.

10.3.3 Examples

Periodogram inconsistency: Examples with white noise

Let us consider Gaussian white noise $e[n]$ with variance σ_e^2, for which the true spectrum is constant and equal to variance. From now on, we assume that all white noise processes mentioned in the text of this chapter are Gaussian. We build several white noise sequences with different numbers N of samples and compute the periodogram of each sequence. Figure 10.2 shows three such cases, with $N = 14$, $N = 51$ and $N = 135$. As N increases, the periodograms (*black curves*) exhibit increasing oscillations, and vary more and more rapidly with frequency. The *horizontal lines* indicate the variance, and the fluctuations with respect to this theoretical PSD value do not decrease in amplitude as the amount of data increases. In fact, the variance of the periodogram in this case is constant too: it does not depend on N and is of the order of the square of $P_{xx}(\omega) = \sigma_e^2$.

This behavior can be understood thinking that $I_N(\omega)$ is the transform of $c_{xx}[l]$, and when $|l| \to N$ the AC estimate $c_{xx}[l]$ is inaccurate and presents large variability. This variability at large lags manifests itself in $I_N(\omega)$ as fluctuations at all frequencies, so that for large values of N, $I_N(\omega)$ varies rapidly with ω. To be more clear, we can say that as the amount of available data increases, the number of $I_N(\omega)$ values that are calculated also increases, and therefore we do not use more data to estimate a single PSD value: from a statistical point of view there is no improvement.

In order to better appreciate how the periodogram variance does not decrease with increasing N, we can examine, instead of a single realization of a white noise process,

Fig. 10.2 a–c Periodograms $I_N(\omega)$ (*black curves*) of three sequences of Gaussian white noise with the same variance, including $N = 14$, 51 and 135 samples, respectively. The *horizontal lines* indicate the theoretical PSD value, equal to data variance. A zero padding has been performed with $N_{FFT} = 8192$; the ordinate axis is linear

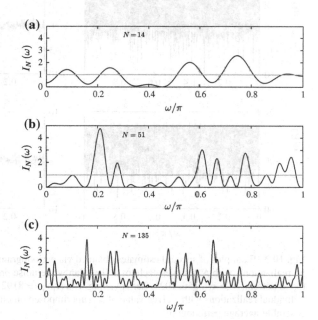

a number of realizations, and then average over realizations. We generate, for example, 50 realizations of white noise with unit variance and N samples. We compute the periodogram of each sequence and then average over the 50 periodograms thus obtained, in order to get a quantity that approximates the expected value of the periodogram. Figure 10.3 shows the result of this simulation: in the panels of Fig. 10.3a we can see, for three different values of N, the 50 periodograms of the individual realizations, superimposed on one another; the panels of Fig. 10.3b show the ensemble average over the realizations. The plots are in dB and a consistent zero padding has been done. Note that hereafter, many figures will be presented having the same structure, and since this is easily recognizable, this description will not be repeated, either in the text or in the captions. From top to bottom, the value of N in Fig. 10.3 varies from 64 to 256 and to 1024. We can observe once again that though the expected

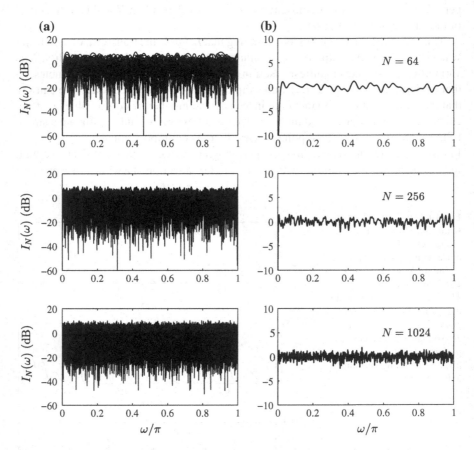

Fig. 10.3 Examples of spectral estimates obtained via periodograms. The analyzed sequences are 50 realizations, with N samples each, of a Gaussian white noise process with unit variance; from top to bottom, $N = 64, 256, 1024$. Zero padding with $N_{FFT} = 8192$. **a** The 50 periodograms of the individual realizations with a given value of N, superimposed on one another; **b** the corresponding ensemble-average periodograms

value, approximated by the averages over 50 realizations, correctly appears close to 1 (0 dB), the spectral variance does not decrease with increasing N.

Periodogram bias: examples with one sinusoid in noise

On the basis of the results of Sect. 9.7 we can state that the theoretical power spectrum of a sample sequence drawn from a random-phase sinusoidal process with frequency ω_0 and amplitude A and with white noise added,

$$x_\infty[n] = A \sin (n\omega_0 + \vartheta) + e[n]$$

where ϑ is a uniformly distributed random variable between 0 and 2π and $e[n]$ is white noise with variance σ_e^2, is

$$P_{xx}(\omega) = \frac{\pi A^2}{2} [\delta(\omega - \omega_0) + \delta(\omega + \omega_0)] + \sigma_e^2,$$

and that the theoretical power spectrum of a sample sequence drawn from a random-phase complex exponential process with frequency ω_0 and amplitude $A/2$ and with white noise added,

$$x_\infty[n] = \frac{A}{2} e^{j(n\omega_0 + \vartheta)} + e[n],$$

is

$$P_{xx}(\omega) = 2\pi \left(\frac{A}{2}\right)^2 \delta(\omega - \omega_0) + \sigma_{e^2} = \frac{\pi A^2}{2} \delta(\omega - \omega_0) + \sigma_{e^2}.$$

Note that on the positive frequency half-axis, the theoretical spectrum of the sinusoid with amplitude A and of the complex exponential with amplitude $A/2$ are identical. This kind of synthetic signal can be considered as a random signal, due to the presence of white noise and of the random initial phase, and as such can be used to test algorithms of statistical spectral estimation.

In reality we only can process a segment of $x_\infty[n]$, i.e., $x[n]$, with $n = [0, N-1]$. The operation that from $x_\infty[n]$ leads to $x[n]$ is a windowing, i.e., a product of $x_\infty[n]$ with a causal rectangular window $w[n]$ of length N. Focusing on the sinusoidal signal with added noise, the windowed signal is

$$x[n] = x_\infty[n]w[n] = A \sin (n\omega_0 + \vartheta)\, w[n] + e[n]w[n]$$

and the expected value of the periodogram is the convolution integral of the "true spectrum" $P_{xx}(\omega)$ with the spectral window (see previous section):

$$E[I_N(\omega)] = \frac{1}{2\pi} \int_{-\pi}^{+\pi} W_{B,2N-1} \left[e^{j(\omega - \theta)}\right] P_{xx}(\theta) d\theta =$$

$$= \sigma_e^2 + \frac{A^2}{4} \left\{ W_{B,2N-1} \left[e^{j(\omega - \omega_0)}\right] + W_{B,2N-1} \left[e^{j(\omega + \omega_0)}\right] \right\},$$

or

$$E[I_N(\omega)] = \frac{N}{2\pi} \int_{-\pi}^{+\pi} D_N^2(\omega - \theta) P_{xx}(\theta) d\theta =$$

$$= \sigma_e^2 + N\frac{A^2}{4} \left\{ D_N^2(\omega - \omega_0) + D_N^2(\omega + \omega_0) \right\}.$$

How come we started multiplying $x_\infty[n]$ by a causal rectangular window and ended up with a convolution integral of $P_{xx}(\omega)$ with the Fejér kernel, i.e., the transform of a Bartlett (triangular) non-causal window of length $2N - 1$? The reason is that when we use the AC sequence to estimate the power spectrum (Blackman-Tukey definition of periodogram), we apply a $2N - 1$-long triangular window to the AC sequence; but the $2N - 1$-long triangular window can be seen, apart from possible normalization of its samples, as the linear convolution of two N-long rectangular windows (see the example of linear convolution between two rectangular sequences in Sect. 3.5.4). The Fourier transform of the window applied to the AC sequence is the square of the transform of the N-long rectangular window. To give a formal proof of the expected value of the periodogram reported above starting from $x[n] = x_\infty[n]w[n]$, we could work on the Schuster's definition of periodogram.

The theoretical spectrum and the expected value of the periodogram for the present example, with $N = 64$ and $\omega_0 = 0.4\pi$, are shown in Fig. 10.4a and b, respectively. Thanks to the symmetry of the power spectrum for real data, it is sufficient to plot the spectrum on the positive ω half-axis. Let us underline that this plot also represents the spectral shape of a random-phase complex exponential signal with amplitude $A/2$.

Fig. 10.4 a Theoretical power spectrum of a sinusoid with amplitude A immersed in white noise with variance σ_e^2, and **b** expected value of the periodogram of $N = 64$ samples of the same signal, as it appears when logarithmic ordinates are used ($10 \log_{10} \{E[I_N(\omega)]\}$). The sinusoid's frequency in this example is $\omega_0 = 0.4\pi$

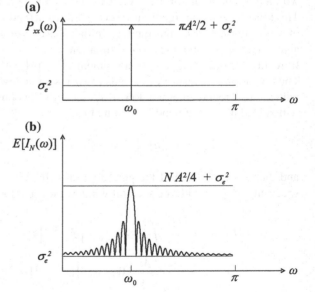

Figure 10.4 shows that where in theory we should see a spectral line, we actually see a copy of the spectral window, scaled by a factor $A^2/4$. In other words, the sinusoid appears as a replica, centered at ω_0 and scaled by a factor $NA^2/4$, of the square of the Dirichlet function $D_N(\omega)$. This is bias. Since $D_N(\omega) = 1$ at $\omega = 0$ (Fig. 3.13), the height of the spectral peak is expected to be $N(A^2/4) + \sigma_e^2$. Also the sidelobes of the spectral window will appear scaled by a factor $A^2/4$ and lifted by σ_e^2. The frequencies of the spectral maxima and zeros will be the same that characterize the function $W_{B,2N-1}(e^{j\omega})$; the width of the main peak at its base will be equal to the mainlobe width of the Bartlett window transform, i.e., $\simeq 8\pi/(2N-1) \simeq 4\pi/N$. In a similar way, we can deal with the case of a signal containing several sinusoidal components, each with different amplitude, frequency and initial phase, thus transposing to the periodogram and to an arbitrary finite number of sinusoids what we said about the amplitude spectrum of a deterministic signal containing two sinusoids (see Sect. 5.2).

In Fig. 10.4 we can observe the two effects we already described in the deterministic case:

- the smoothing produced by $W_{B,2N-1}(e^{j\omega})$ that causes the sinusoid's power to spread over a frequency band whose width is approximately equal to $4\pi/N$;
- the leakage due to sidelobes that generates spurious secondary peaks at frequencies $\omega_k \approx \omega_0 \pm (2\pi/N)k$. If the signal contained more than one sinusoidal component, the leakage due to the sidelobes of a powerful component would be able to mask weak components possibly present nearby the dominant one.

Figure 10.5 shows what we obtain by calculating the periodogram of 50 realizations of this process with $\sigma_e^2 = 1$ and taking the ensemble average. The parameter values are $A = 5$, $\omega_0 = 0.4\pi$ and N variable from 40 to 512, from top to bottom. Though all the periodograms have a peak near 0.4π, a remarkable periodogram variability is observed. The ensemble average approximates the expected periodogram. We can see that as the amount of data increases the sinusoid's power is spread over a narrower and narrower frequency band and the peak rises. The expected peak heights in each panel are, from top to bottom, $10\log_1 0(40 \times 25/4) + 1 = 25$, 27, 33 and 36 for $\sigma_e^2 = 1$ and $N = 40$, 64, 256 and 512, respectively (see the *gray horizontal lines* in the right-hand panels).

Signal-to-noise Ratio

We now examine, using Schuster's definition of periodogram, the influence that the noise has on the peak that represents the sinusoid/complex exponential. In order to approach this issue from a more general point of view, we apply to the signal a real, non-negative and causal window $w[n]$ that is not necessarily rectangular—a generic taper that is not identically zero only in $0 \leq n \leq N - 1$. This is aimed at obtaining a result not strictly limited to the periodogram case, but valid also for the modified periodogram that will be introduced in Sect. 10.5. To make calculations easier, it is convenient to restrict our attention to the positive ω half axis and to a single complex exponential with positive frequency ω_0 and amplitude $A/2$, superimposed to white

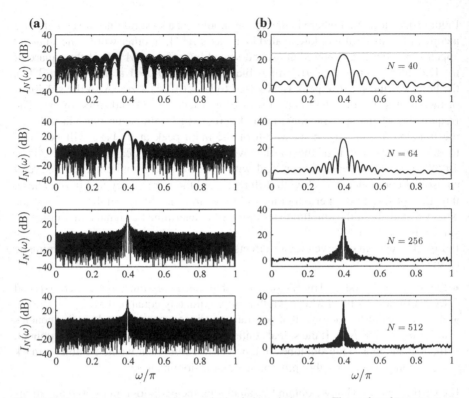

Fig. 10.5 Examples of spectral estimates obtained via periodograms. The analyzed sequences are 50 realizations, with N samples each, of a random-phase sinusoidal signal with amplitude $A = 5$ and frequency $\omega_0 = 0.4\pi$, immersed in Gaussian white noise with unit variance. From top to bottom: $N = 40, 64, 256$ and 512. Zero padding with $N_{FFT} = 8192$. The *gray horizontal lines* in the right-hand panels represent expected peak heights (see text). **a** 50 Realizations. **b** Ensemble average

noise $e[n]$ with variance σ_e^2. We also neglect the random phase for simplicity. The windowed signal is

$$x[n] = x_\infty[n]w[n] = \frac{A}{2}e^{j\omega_0 n}w[n]$$

The spectral peak falls at ω_0; at this precise frequency we can write the transform of the windowed signal as

$$X(e^{j\omega_0}) = \frac{A}{2}\sum_{n=0}^{N-1}e^{j\omega_0 n}e^{-j\omega_0 n}w[n] + \sum_{n=0}^{N-1}e[n]w[n]e^{-j\omega_0 n} =$$

$$= \frac{A}{2}\sum_{n=0}^{N-1}w[n] + \sum_{n=0}^{N-1}e[n]w[n]e^{-j\omega_0 n}.$$

The first right-hand-side term in this expression is real and we denote it as u; the second is complex and we write it as $a + jb$: $X(e^{j\omega_0}) = u + a + jb$. The corresponding periodogram value according to Schuster's definition is

$$I_N(\omega_0) = \frac{1}{N}\left|X(e^{j\omega_0})\right|^2 = \frac{1}{N}[(u+a)^2 + b^2] =$$

$$= \frac{1}{N}[u^2 + a^2 + 2ua + b^2] = \frac{1}{N}[u^2 + 2ua + (a^2 + b^2)] =$$

$$= \frac{1}{N}[u^2 + 2ua + |a + jb|^2] = \frac{1}{N}[u^2 + 2ua + (a + jb)(a - jb)].$$

This gives

$$I_N(\omega_0) = \frac{A^2}{4N}\left(\sum_{n=0}^{N-1} w[n]\right)^2 +$$

$$+ \frac{A}{N}\left(\sum_{n=0}^{N-1} w[n]\right)\mathrm{Re}\left(\sum_{n=0}^{N-1} e[n]w[n]e^{-j\omega_0 n}\right) +$$

$$+ \frac{1}{N}\sum_{n=0}^{N-1}\sum_{m=0}^{N-1} e[n]w[m]e[n]w[m]e^{-j\omega_0(n-m)}.$$

$I_N(\omega_0)$ is a random variable for which we can calculate the expected value:

$$E[I_N(\omega_0)] = E\left[\frac{1}{N}\left|X(e^{j\omega_0})\right|^2\right] =$$

$$= \frac{A^2}{4N}\left(\sum_{n=0}^{N-1} w[n]\right)^2 +$$

$$+ \frac{A}{N}\left(\sum_{n=0}^{N-1} w[n]\right)\mathrm{Re}\left(\sum_{n=0}^{N-1} w[n]E\{e[n]\}e^{-j\omega_0 n}\right) +$$

$$+ \frac{1}{N}\sum_{n=0}^{N-1}\sum_{m=0}^{N-1} w[n]w[m]E\{e[n]e[m]\}e^{-j\omega_0(n-m)}.$$

The expected value of the noise is zero, hence the second addend vanishes; $e[n]$ and $e[m]$ are uncorrelated for $n \neq m$, while for $n = m$ the term $E\{e[n]e[m]\}$ yields σ_e^2, while the exponential term $e^{-j\omega_0(n-m)}$ reduces to 1. Therefore,

$$E[I_N(\omega_0)] = \frac{1}{N}\left[\frac{A^2}{4}\left(\sum_{n=0}^{N-1} w[n]\right)^2 + \sigma_e^2\left(\sum_{n=0}^{N-1} w^2[n]\right)\right],$$

in which the first term is relative to the complex exponential with amplitude $A/2$ and the second represents the noise contribution. The ratio of the two terms is called the *output signal-to-noise ratio* or the output SNR:

$$\text{SNR}_\text{o} = \frac{A^2}{4\sigma_e^2} \frac{\left(\sum_{n=0}^{N-1} w[n]\right)^2}{\sum_{n=0}^{N-1} w^2[n]}.$$

Since the *input signal-to-noise ratio* is defined as

$$\text{SNR}_\text{i} = \frac{A^2}{4\sigma_e^2},$$

we finally have

$$\text{SNR}_\text{o} = \text{SNR}_\text{i} \frac{\left(\sum_{n=0}^{N-1} w[n]\right)^2}{\sum_{n=0}^{N-1} w^2[n]},$$

where the fraction on the right-hand side is a factor that depends on the type and length of the taper applied to $x_\infty[n]$. In the case of the periodogram, the taper is a rectangular window, for which this ratio equals $N^2/N = N$. For non-rectangular windows, which are used in the so-called *modified periodogram*, the dependence of this factor on N is nearly linear. It is thus common practice to introduce a quantity called *processing gain of the window*,

$$\text{PG} = \frac{(\sum_{n=0}^{N-1} w[n])^2}{N \sum_{n=0}^{N-1} w^2[n]},$$

that turns out to be dependent on the taper shape but practically independent of N. For the rectangular window, $\text{PG} = 1$. We can thus finally write[2]

$$\text{SNR}_\text{o} = \text{SNR}_\text{i} \times N \times \text{PG}.$$

The practice shows that a periodic signal can be detected clearly, and its frequency can be estimated accurately, if $\text{SNR}_\text{o} \geq 25$, i.e., if

$$\text{SNR}_\text{o} = \text{SNR}_\text{i} \times N \times \text{PG} \geq 25.$$

---

[2]Note that the terms "input" and "output" here refer to a *power spectrum estimation algorithm*: a signal having a given SNR_i enters the algorithm; in the spectral estimate, the signal-to-noise ratio is modified by a factor depending on the shape and length of the taper and is transformed into SNR_o. Looking at the latter, we can evaluate the capability of the estimator to detect the imprint of the sinusoid/complex exponential buried in noise.

Therefore for the periodogram, for which PG $= 1$, we must have SNR$_i \geq 25/N$ for the sinusoidal signal not to be submerged by the noise. From now on, we will simply write SNR to indicate SNR$_o$, unless otherwise specified.

As an example of the effect of varying SNR, in Fig. 10.6 the periodogram of a signal with $N = 64$ samples is shown. The signal contains a sinusoid with $A = 5$ and $\omega_0 = 0.4\pi$ and white noise with increasing variance: from top to bottom we set $\sigma_e = 1$, 10 and 100. While the parameters of the sinusoid, and its amplitude in particular, remain fixed, as the noise variance increases the spectral background rises, until it completely submerges the peak: the SNR is no longer sufficient to ensure the peak detection. This can be understood observing that SNR$_i$ in the three cases assumes the values 6.25, 0.0625 e 0.000625, and that consequently with PG $= 1$ we have $SNR = 400$, 4.0 and 0.04 respectively, so that only the first value exceeds the threshold of 25.

If in all three cases we get the SNR to exceed the threshold, the sinusoid remains visible. For example, in Fig. 10.7 everything is as in the previous figure, except the

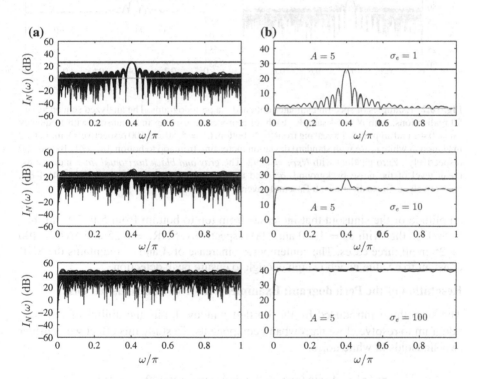

Fig. 10.6 Examples of spectral estimates obtained via periodograms. The analyzed sequences are 50 realizations, with $N = 64$ samples each, of a process consisting in a sinusoid with amplitude $A = 5$ and frequency $\omega_0 = 0.4\pi$, immersed in Gaussian white noise with standard deviation increasing from top to bottom: $\sigma_e = 1$, 10 and 100 respectively. Zero padding with $N_{FFT} = 8192$. The *gray and black horizontal lines* indicate the mean level of the noise background, σ_e^2, and the expected height of the peak, $N(A^2/4) + \sigma_e^2$, respectively. **a** 50 Realizations. **b** Ensemble average

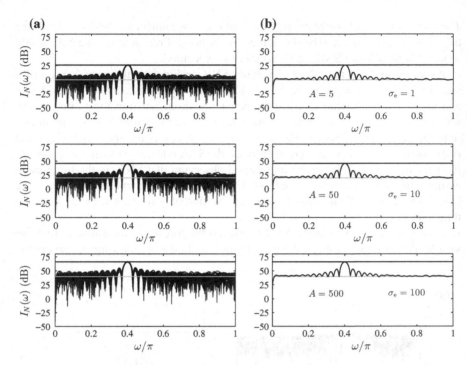

Fig. 10.7 Examples of spectral estimates obtained via periodograms. The analyzed sequences are 50 realizations, with $N = 64$ samples each, of a process consisting in a sinusoid with frequency $\omega_0 = 0.4\pi$ and amplitude increasing from top to bottom ($A = 5, 50$ and 500 respectively), immersed in Gaussian white noise with standard deviation increasing from top to bottom ($\sigma_e = 1, 10$ and 100 respectively). Zero padding with $N_{FFT} = 8192$. The *gray and black horizontal lines* indicate the mean level of the noise background, σ_e^2, and the expected height of the peak, $N(A^2/4) + \sigma_e^2$, respectively. **a** 50 Realizations. **b** Ensemble average

amplitude of the sinusoid that increases from top to bottom from 5 to 50 and then to 500, so that with $\sigma_e = 1, 10$ and 100 respectively, $\text{SNR}_i = 6.25$ and $SNR = 400 \gg 25$ in all three cases. The contemporary increase of A and σ_e maintains the SNR constant to a value which is high enough to ensure peak detection.

Resolution of the Periodogram: Examples with Two Sinusoids in Noise

The smoothing introduced by the Bartlett window limits the ability of the periodogram to resolve close narrowband components. To study this effect we consider two sinusoids in white noise,

$$x_\infty[n] = A_0 \sin(\omega_0 n + \vartheta_0) + A_1 \sin(\omega_1 n + \vartheta_1) + e[n],$$

where ϑ_0 and ϑ_1 are uncorrelated random variables uniformly distributed between 0 and 2π, and $e[n]$ is white noise with variance σ_e^2. We then consider the finite-length signal $x[n]$ extracted from $x_\infty[n]$ by windowing. The theoretical power spectrum of the signal is

$$P_{xx}(\omega) = \sigma_e^2 + \frac{\pi A_0^2}{2}\left[\delta(\omega - \omega_0) + \delta(\omega + \omega_0)\right] + \frac{\pi A_1^2}{2}\left[\delta(\omega - \omega_1) + \delta(\omega + \omega_1)\right],$$

and the expected value of the periodogram is

$$E[I_N(\omega)] = \frac{1}{2\pi}\int_{-\pi}^{+\pi} W_B\left[e^{j(\omega-\theta)}\right]P_{xx}(\theta)\mathrm{d}\theta =$$

$$= \sigma_e^2 + \frac{A_0^2}{4}\left\{W_B\left[e^{j(\omega-\omega_0)}\right] + W_B\left[e^{j(\omega+\omega_0)}\right]\right\} +$$

$$+ \frac{A_1^2}{4}\left\{W_B\left[e^{j(\omega-\omega_1)}\right] + W_B\left[e^{j(\omega+\omega_1)}\right]\right\}.$$

For this signal, the theoretical spectrum and the expected value of the periodogram are shown Fig. 10.8 for $N = 64$, $\omega_0 = 0.4\pi$, $\omega_1 = 0.45\pi$ and $A_0 = A_1 = A$. Since the main lobe width of the spectral window depends only on N, for a given value of N a limit exists for the closeness of the frequencies of two sinusoids—or the central frequencies of two narrowband processes—beyond which the two features are no longer resolved. A way to quantify resolution is to define a $\Delta\omega$ equal to the spectral main lobe width, at its base, i.e., $4\pi/N$, or at half height, i.e., about 6 dB below the level of the maximum, a case in which we find $\Delta\omega = 0.89 \times 2\pi/N$. This value can be assumed as a measure of the periodogram's resolution, even if $\Delta\omega$ actually measures the *reciprocal* of the resolution, meaning that the smaller $\Delta\omega$, the greater the resolution. Even if looking at $\Delta\omega$ represents an approximate and arbitrary rule to quantify resolution, in practice it is difficult to resolve spectral features that

Fig. 10.8 **a** Theoretical power spectrum of two sinusoids with equal amplitude A immersed in white noise with variance σ_e^2, and **b** expected value of the periodogram of $N = 64$ samples of the same signal, as it appears when logarithmic ordinates are used. The value shown for the peaks' heights in $E[I_N(\omega)]$ is approximate because it does not take into account the superposition of the two spectral curves related to the single sinusoids. The frequencies ω_0 and ω_1 in this example are 0.4 e 0.45π respectively

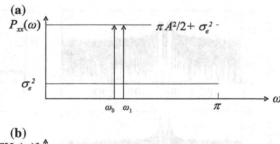

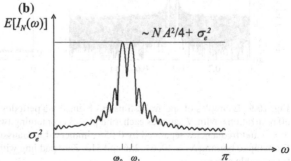

are finer than this. The rule also serves to understand how many data samples are needed to achieve some desired resolution.

Figure 10.9 shows the periodograms of 50 realizations of a process of this kind and their ensemble average, which approximates the expected value. In this example, the following parameters have been adopted: $\Delta\omega = 0.05\pi$, $A_0 = A_1 = A = 5$ and $\sigma_e^2 = 1$. Applying the rule for resolution, with $\Delta\omega = 0.05\pi$ we expect, using the less restrictive half-height main lobe width, to be able to resolve the sinusoids with at least $N \simeq (0.89 \times 2)/0.05 \simeq 36$ samples. With $N = 40$ (upper panels of Fig. 10.9) we actually are on the very margin of detectability for the sinusoids as separate entities. With $N = 64$ (second row of panels) the two sinusoids are already resolved, and are better and better resolved as N further increases in the lower panels.

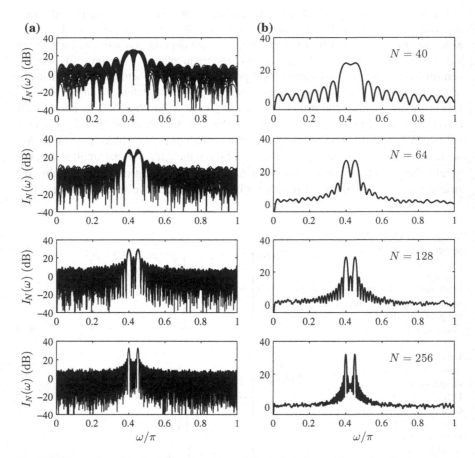

Fig. 10.9 Examples of spectral estimates obtained via periodograms. The analyzed sequences are 50 realizations, with N samples each, of a process containing two sinusoids with equal amplitudes $A = 5$ and frequencies separated by 0.05π, immersed in Gaussian white noise with unit variance. From top to bottom: $N = 40, 64, 256$ and 512. Zero padding with $N_{FFT} = 8192$. **a** 50 Realizations. **b** Ensemble average

Remarks

The problems of leakage and frequency resolution we described above, as well as the great variability of the spectral estimates via periodogram, provide the motivation for the spectral methods described in the next sections. Such methods do not make any assumptions concerning the examined process and are therefore referred to as *non-parametric spectral methods*, in order to distinguish them from other methods that will be introduced in the next chapter and that assume that the signal can be represented approximately by a mathematical model having parameters (parametric methods).

Non-parametric methods aim at obtaining a consistent estimate of the power spectrum through some smoothing of the periodogram. They allow for a reduction of the estimate's variance, but at the cost of some loss of resolution.

10.3.4 Variance Reduction by Band- and Ensemble-Averaging

An immediate and widely used method to reduce the variance of the periodogram consists of performing a running average (moving average) of an odd number L of $I_N[k]$ samples. The choice of an odd number allows for assigning the resulting average to one of the original frequencies ω_k. Usually, L values of 3–5 are employed, but if we want to smooth out the periodogram more drastically we can increase L. This procedure is called *band averaging*, since, as shown in Fig. 10.10 in the case of $L = 3$, it means uniting L frequency bins, thus averaging over a frequency band of width equal to L times the original bin width.

The running average corresponds to applying to the sequence $I_N[k]$ a rectangular FIR filter with length L; the samples of the impulse response are all equal to $1/L$.

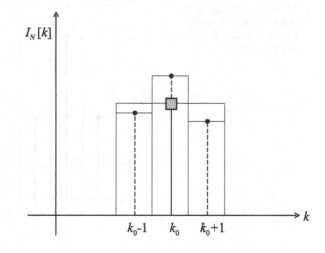

Fig. 10.10 Performing a band averaging of the periodogram by a running average over $L = 3$ spectral samples means uniting L frequency bins, thus averaging over a frequency band of width equal to L times the original bin width

Since the frequency response of such a filter only meets very poor specifications, band averaging must be considered as a naive way to reduce the periodogram variance. As an alternative, which however is only slightly better, some authors have suggested to use a filter with a tapered shape of $h[n]$, like, for instance, a Daniell's filter (Daniell 1946) with length L:

$$h_D[n] \begin{cases} \frac{1}{2(L-1)} & \text{for } n = 0 \text{ and } n = L - 1, \\[2ex] \frac{1}{L-1} & \text{otherwise.} \end{cases}$$

The shape of Daniell's filter for $L = 5$ is shown in Fig. 10.11a, while Fig. 10.11b shows the corresponding rectangular running-average filter.

A more sophisticated approach is referred to as *ensemble averaging* (see Kay 1988). If we had K sequences, each drawn from a different realization of a random process, we could calculate the periodogram of each sequence and then take the ensemble average. The expected PSD value at any given frequency would be the same expected for an individual periodogram but the variance would be reduced by a factor K, since we would average over independent estimates. This approach can be adopted even if we have only one sequence: we can generate a *pseudo-ensemble* by dividing the sequence into a number of equal-length subsequences and then proceed as described above. Of course, in doing so we rely on the hypothesis of stationarity of the process.

This is the rationale behind the methods of Bartlett and Welch. The latter also considers the possibility of applying tapered windows to the signal, i.e., to average a number of *modified periodograms* (Sect. 10.5) rather than ordinary periodograms. We will see that these methods reduce the variance of the spectral estimate but also cause resolution loss.

Fig. 10.11 **a** Daniell filter and **b** running-average filter used for the band averaging of the perodogram, in a case of length $L = 5$

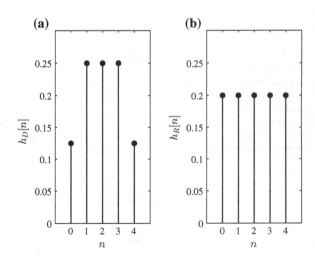

10.4 Bartlett's Method

In Bartlett's method (Bartlett 1948), the data sequence $x[n]$, with $n = [0, N - 1]$, is divided into K adjacent and non-overlapping segments containing M samples each: $N = KM$. We can denote these subsequences by $x^{(i)}[n] = x[n + (i - 1)M]$, where $0 \le n \le M - 1$ and $1 \le i \le K$. The periodograms of the subsequences are then computed:

$$I_M^{(i)}(\omega) = \frac{1}{M} \left| \sum_{n=0}^{M-1} x^{(i)}[n] e^{-j\omega n} \right|^2 ,$$

in which M in the denominator represents the length of each subsequence. Note that when the periodogram is computed at discrete frequencies, the decrease of the data length from N to M causes each periodogram to lose some low-frequency information. In fact, the distance between adjacent frequencies ω_k is now $2\pi/M > 2\pi/N$ and this is also the value of the first non-zero discrete frequency for which we a get a spectral estimate.

If $\gamma_{xx}[l]$ is "small" for $|l| > M$, i.e., if AC decreases rapidly enough with increasing lag, it is reasonable to assume that the periodograms $I_M^{(i)}(\omega)$ are independent on one another. We can then take as the Bartlett spectral estimate the sample mean of these K independent observations of $I_M(\omega)$, i.e., the non-negative function of frequency

$$B_{xx}(\omega) = \frac{1}{K} \sum_{i=1}^{K} I_M^{(i)}(\omega).$$

The expected value of $B_{xx}(\omega)$ is

$$E[B_{xx}(\omega)] = \frac{1}{K} \sum_{i=1}^{K} E[I_M^{(i)}(\omega)],$$

but since the various $E[I_M^{(i)}(\omega)]$ are all equal, we get

$$E[B_{xx}(\omega)] = E[I_M^{(i)}(\omega)] = \frac{1}{2\pi} \int_{-\pi}^{+\pi} W_{B,2M-1} \left\{ e^{j(\omega-\theta)} \right\} P_{xx}(\theta) d\theta$$

where the transform of the Bartlett window of length $2M - 1$ appears, corresponding to periodograms calculated over $M = N/K$ samples:

$$W_{B,2M-1}\left(e^{j\omega}\right) = \frac{1}{M} \left[\frac{\sin(M\frac{\omega}{2})}{\sin\frac{\omega}{2}} \right]^2 .$$

Therefore, the expected value of the Bartlett spectral estimate is given by the convolution integral of the true spectrum with a spectral window $W_{B,2M-1}(e^{j\omega})$, and therefore

$B_{xx}(\omega)$ is a biased estimate. Bartlett's estimate is actually more biased than $I_N(\omega)$, because with $M < N$ the spectral window has a wider main lobe with respect to the spectral window $W_{B,2N-1}(e^{j\omega})$ appearing in $I_N(\omega)$. As a consequence, resolution is smaller. Sidelobes are also responsible for leakage issues.

As the sample mean of K independent observations of $I_N(\omega)$, $B_{xx}(\omega)$ has a variance equal to the variance of the individual $I_M^{(i)}(\omega)$, divided by their number K:

$$\sigma^2_{B_{xx}(\omega)} = \frac{1}{K}\sigma^2_{I_M(\omega)} = \frac{1}{K}P^2_{xx}(\omega)\left\{1 + \left[\frac{\sin(M\omega)}{M\sin\omega}\right]^2\right\} \approx \frac{1}{K}P^2_{xx}(\omega).$$

We thus see that $\sigma^2_{B_{xx}(\omega)} \to 0$ for $K \to \infty$, that is, for $N \to \infty$: by introducing $B_{xx}(\omega)$ we obtained, as desired, a consistent estimate of $P_{xx}(\omega)$.

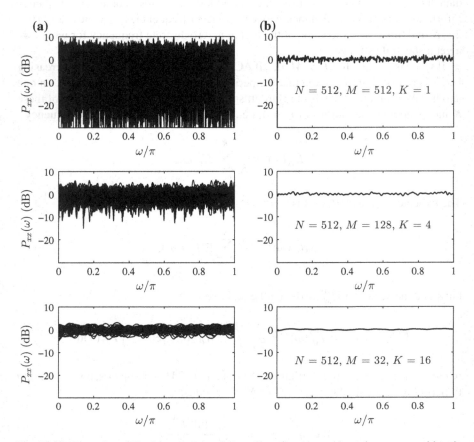

Fig. 10.12 Examples of Bartlett spectral estimates for a Gaussian white noise process with unit variance (50 realizations with $N = 512$ samples each). From top to bottom the following cases are considered: $M = 512, K = 1$ (periodogram); $M = 128, K = 4$; $M = 32, K = 16$. **a** 50 Realizations. **b** Ensemble average

To illustrate the properties of Bartlett's method we can consider 50 realizations of a white noise process with unit variance and take $N = 512$ samples of each realization. We then compute a Bartlett spectral estimate choosing different values for M and K. In Fig. 10.12, from top to bottom, the values of M and K are: $M = 512$, $K = 1$ (periodogram); $M = 128$, $K = 4$; $M = 32$, $K = 16$. We can observe that the variance of the spectral estimate decreases as K increases.

In summary, for a given value of N the variance of $B_{xx}(\omega)$ decreases with increasing K but at the same time the bias increases, so that resolution worsens. In applications we will have to choose a value of K representing the proper compromise between low variance and high resolution. If we expect narrowband features close to one another in the spectrum and resolution is an issue, we will choose a relatively small K and therefore a relatively large M; if we are interested in a smooth spectral estimate and do not need high resolution we will privilege low variance and choose a higher value of K. On the other hand, if we were able to chose an arbitrarily large N—compatibly with the stationarity of the process—after choosing M according to the desired resolution we could deduce from the expression of $\sigma^2_{B_{xx}(\omega)}$ the value of K and hence of N that would lead to an acceptable value of the variance.

10.5   Modified Periodogram

The modified periodogram is a variation of the periodogram which tends to reducing leakage by lowering the sidelobes of the spectral window that appears in $E[I_N(\omega)]$. It is not aimed at reducing the variance of the periodogram.

When computing the periodogram we implicitly apply a Bartlett window to $\gamma_{xx}[l]$, and this corresponds to applying a rectangular window to the random signal. If we apply a more gradual window to the data, the window applied to $\gamma_{xx}[l]$ becomes more tapered than the Bartlett one and has lower sidelobes (see Fig. 5.5).

We may wonder *why* a more tapered window has lower sidelobes. Let us consider a window exhibiting a rapid transition between two amplitude levels: the most rapid transition is that of the rectangular window that passes from 0 to 1 in just one unit of discrete time. This transition actually is an impulse that in the frequency domain has infinite bandwith, meaning that we need all possible frequencies to represent it. If the transition is less rapid, in the frequency domain it can be described by less high-fequency components and this is reflected into a lowering of sidelobes.

The modified periodogram is employed in the Welch method for the reduction of periodogram's variance, in which the data sequence is divided into subsequences of length $M < N$ and the modified periodograms of all of them are averaged to reduce variance. Taking into consideration the use of tapered windows, in Welch's method the subsequences are allowed to partially overlap.

Multiplying the sequence samples by a tapered window affects the signal's average power, because the samples falling at the edges of each subsequence are attenuated. In order to compensate for this effect, in the definition of the modified periodogram a normalization factor is inserted, such that the choice of a particular window does not affect the PSD estimates. Thus, the definition of *modified periodogram* is

$$J_N(\omega) = \frac{1}{NU} \left| \sum_{n=0}^{N-1} w[n]x[n]e^{-j\omega n} \right|^2 ,$$

where the normalization factor U is chosen as $U = (1/N)\sum_{n=0}^{N-1} w[n]^2$, representing the average power of the window.

The expected value of the modified periodogram is

$$E[J_N(\omega)] = \frac{1}{2\pi NU} \int_{-\pi}^{+\pi} \left| W\left[e^{j(\omega-\theta)}\right] \right|^2 P_{xx}(\theta)d\theta,$$

where $1/(NU)\left|W(e^{j\omega})\right|^2$ is the spectral window.

What is the rationale behind the choice of U? We want to choose the factor $1/(NU)$ in such a way that $E[J_N(\omega)] \to P_{xx}(\omega)$ as, for increasing N, the bandwidth of the spectral window becomes narrower and narrower.[3] For this purpose, recalling Parseval's theorem we impose the condition

$$\frac{1}{2\pi NU} \int_{-\pi}^{+\pi} \left| W\left(e^{j\omega}\right) \right|^2 d\omega = \frac{1}{2\pi NU} \int_{-\pi}^{+\pi} \left| \sum_{n=0}^{N-1} w[n]e^{-j\omega n} \right|^2 d\omega = \frac{1}{NU} \sum_{n=0}^{N-1} w[n]^2 = 1$$

which actually requires

$$U = \frac{1}{N} \sum_{n=0}^{N-1} w[n]^2 = \begin{cases} 1 & \text{for a rectangular window,} \\ 0 < U < 1 & \text{for tapered windows normalized so that } \max(w[n]) = 1. \end{cases}$$

Note that if the window is rectangular, $U = 1$ and the modified periodogram becomes an ordinary periodogram.

The fact that U in the denominator of $J_N(\omega)$ is required to be less than 1 in the case of tapered windows is easily understood if we think that when we multiply a sequence by a gradual window, the samples near the edges are attenuated. An alternative to choosing U in the way described above could be to absorb the normalization in the samples of the tapered window in such a way as to get $U = 1$ in all cases. Of course $\max(w[n])$ would no longer be 1. In summary, since only the application of a window with unit average power does not alter the average power of the signal, we can define the window in such a way to get unit average power whatever the window shape, or

---

[3]For $N \to \infty$ the spectral window would become a periodic impulse train, i.e., a train of Dirac δ, but for a finite N the bias can vanish only asymptotically.

use a window with $\max(w[n]) = 1$ and average power possibly different from 1, and then properly correct the spectral estimate.

Since the modified periodogram is simply a periodogram of a windowed sequence, its variance is approximately the same as for the periodogram. Therefore the modified periodogram is not a consistent estimate of the power spectrum and the use of a gradual window does not offer any advantage in terms of variance reduction. The gradual window however ensures the possibility of a trade-off between resolution (mainlobe width of the spectral window) and leakage (height of sidelobes). For instance, if we define resolution through the bandwidth of the spectral window at 3 dB, using a Hamming window in place of the rectangular one for the purpose of reducing sidelobes, we must also accept a decrease of resolution[4] of about 50 %.

Let us see an example of modified periodogram. We take a signal with two sinusoids in a small amount of white noise,

$$x[n] = A_0 \sin(\omega_0 n + \vartheta_0) + A_1 \sin(\omega_1 n + \vartheta_1) + e[n]$$

with $A_0 = 0.1$, $A_1 = 1$, $\omega_0 = 0.225\pi$, $\omega_1 = 0.3\pi$ and random initial phases; $\sigma_e^2 = 2.5 \cdot 10^{-3}$; $N = 128$ samples. We take 50 realizations of this process and investigate if this spectral tool offers advantages for the detection of a sinusoid which is much weaker than the other. In this example, the SNR is good for both components, and their frequency separation would be sufficient to resolve them if they had comparable amplitudes. As we can see in Fig. 10.13, if we choose a rectangular window (upper panels; simple periodograms), the smaller sinusoid is however masked by the dominant one and is almost invisible. If instead we estimate the spectrum by a modified periodogram with Hamming window (lower panels), the weaker sinusoid is clearly visible, because the sidelobes of the spectral window are much lower.

As explained above, by the modified periodogram we reduce leakage but from the point of view of resolution, the widening of the main lobe of the spectral window worsens the situation. Moreover, the same widening/lowering also implies a lesser detectability of a single sinusoid in noise, all other conditions being equal, as whitnessed by the fact that all non-rectangular windows have a processing gain $PG < 1$:

- $PG = 0.75$ for the Bartlett window,
- $PG = 0.67$ for the Hanning window,
- $PG = 0.73$ for the Hamming window,
- $PG = 0.58$ for the Blackman window.

---

[4]Note that this choice for defining resolution is coherent with what we stated about the periodogram: there we talked about the possibility of defining resolution through the bandwidth of the spectral window $W_{B,2N-1}$ at 6 dB, but since $w_{B,2N-1}[n] = (1/N)w_{R,N}[n] * w_{R,N}[n]$, so that $W_{B,2N-1}(e^{j\omega}) \propto |W_R(e^{j\omega})_{B,N}|^2$, this is equivalent to using for the periodogram the bandwidth at 3 dB for $W_R(e^{j\omega})$.

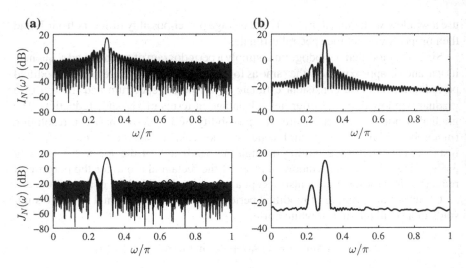

Fig. 10.13 Periodogram (upper panels) and modified periodogram with Hamming window (lower panels) for $N = 128$ samples of a signal containing two sinusoids immersed in Gaussian white noise. The signal's parameters are such that SNR is good for both sinusoids and their frequency distance is large enough to ensure that we could see them resolved if their amplitudes were comparable. In this case, however, one sinusoid has an amplitude 10 times smaller than the other. When the spectrum is estimated via periodogram, the smaller sinusoid is masked by the sidelobes of the spectral window centered at the frequency of the larger-amplitude component. Using a modified periodogram, the picture is definitely more clear. **a** 50 Realizations. **b** Ensemble average

10.6   Welch's Method

Welch's method method (Welch 1967), also known as Welch's overlapped segment averaging (WOSA) method, consists in taking the average of modified periodograms. It associates the Bartlett's idea of averaging periodograms of subsequences in order to reduce the variance of the spectral estimate with the idea of applying to the data a tapered window, in order to reduce leakage. It is the most widely used method among classical ones.

The sequence of N samples is subdivided into K subsequences of M samples each, and a real window $w[n]$ is applied directly to each subsequence $x^{(i)}[n]$ before calculating the periodograms. Welch's method thus performs the direct calculation of the PSD estimate, which is defined as the average of the modified periodograms:

$$B_{xx}^{W}(\omega) = \frac{1}{K} \sum_{i=1}^{K} J_{M}^{(i)}(\omega),$$

with each modified periodogram being defined as

$$J_M^{(i)}(\omega) = \frac{1}{MU} \left| \sum_{n=0}^{M-1} w[n] x^{(i)}[n] e^{-j\omega n} \right|^2,$$

with a normalization factor

$$U = \frac{1}{M} \sum_{n=0}^{M-1} w[n]^2$$

that serves to have asymptotically unbiased estimates.

We may note that each modified periodogram could also be written as a *modified correlogram*:

$$J_M^{(i)}(\omega) = \frac{1}{U} \left| \sum_{l=-(M-1)}^{M-1} c_{vv}^{(i)}[l] e^{-j\omega l} \right|^2,$$

where $c_{vv}^{(i)}[l]$ is the AC of the windowed subsequence $v^{(i)}[n] = w[n] x^{(i)}[n]$.

The expected value of the Welch spectral estimate is given by the convolution integral of the true spectrum with the spectral window

$$W(e^{j\omega}) = \frac{1}{MU} \left| \sum_{n=0}^{M-1} w[n] e^{-j\omega n} \right|^2,$$

that is,

$$E[B_{xx}^W(\omega)] = E[J_M^{(i)}(\omega)] = \frac{1}{2\pi} \int_{-\pi}^{+\pi} W\left[e^{j(\omega-\theta)}\right] P_{xx}(\theta) d\theta =$$

$$= \frac{1}{2\pi MU} \int_{-\pi}^{+\pi} \left| \sum_{n=0}^{M-1} w[n] e^{-j(\omega-\theta)n} \right|^2 P_{xx}(\theta) d\theta.$$

Therefore as in the time domain we apply some causal and non necessarily symmetric window $w[n]$ to the data sequence, in the frequency domain a spectral window $W(e^{j\omega})$ appears, which is proportional to the square modulus of the transform of $w[n]$. This could be demonstrated as we did for the periodogram, i.e., working on the Blackman-Tukey definition of modified correlogram.

These properties lead to the following consequences:

- with any type of time window, the spectral window is always non-negative, and therefore also the Welch estimate $B_{xx}^W(\omega)$ is always non-negative, being $P_{xx}(\omega)$ itself a non-negative quantity. With other estimators, negative spectral estimates can occur at some frequencies. These values are meaningless and in the practice they are set to zero. This cannot happen with the Welch estimator;

- the Welch estimate is biased, but asymptotically unbiased;
- it is smoother than the periodogram, because the true spectrum is convolved in the frequency domain with a spectral window which has a wider main lobe with respect to the spectral window appearing in the periodogram. Indeed, in this case N is substituted by $M < N$.

The variance of the Welch estimator, in the hypothesis that the subsequences do not overlap, can be shown to be

$$\sigma^2_{B^W_{xx}(\omega)} \cong \frac{1}{K} P^2_{xx}(\omega),$$

hence the variance decreases with increasing K and goes to zero as $K \to \infty$, meaning $N \to \infty$: so we have a consistent estimator.

In Welch's method, however, a partial overlap of the subsequences is also possible; indeed, some overlap is nearly always adopted. The subsequences are then written as $x^{(i)}[n] = x[n + iD]$ with $n = [0, M - 1]$ and $i = [0, K - 1]$, where iD is the discrete time at which the ith subsequence starts. The overlap of adjacent subsequences covers $M - D$ samples, so that $N = M + D(K - 1)$. For example, if $D = M$ the segments do not overlap and K is the same as in Bartlett's method, $K = N/M$; if $D = M/2$ there is a 50% overlap and $K = 2(N/M) - 1 \equiv K_{Welch} \cong 2K_{Bartlett}$, M being equal (with obvious notation). In this way the same resolution as Bartlett's method is maintained, in the sense that M is the same, but the number of periodograms over which we average the spectral estimate is doubled, so that variance reduction improves. Note also that with the same 50% overlap we could form $K = (N/M) - 1$ subsequences with length $2M$, so that $K_{Welch} \cong K_{Bartlett}$ but with doubled length segments: resolution would improve due to the doubling of M, while the variance reduction would be the same as in Bartlett's method. The choice of parameters in the Welch method thus offers the possibility of easily handling the resolution-versus-variance trade-off, since we can choose to increase either the number or the length of the data segments.

A general expression for the variance of Welch's estimator in the general case of a varying amount of overlap between segments in association with different tapers is difficult to give, even assuming that this estimator is approximately unbiased to simplify the problem. Indeed, this variance depends on both the window and the amount of overlap, which for a given N determine M and K. For an approximate expression see, e.g., Percival and Walden (1993). In general, we could think that an amount of overlap greater than 50%, with a given N, would result in a more substantial reduction of variance, since we would get a larger K and would average over more periodograms. However, in most cases it would no longer be possible to consider the individual periodograms as independent spectral estimates, not even approximately: so this is not a good argument, and actually 50% is the percentage of overlap normally used in association with moderate tapers, such as the Bartlett or the Hanning window. For stronger tapers, an overlap percentage greater than 50% may be advisable to get the best results in terms of variance reduction (Percival and Walden 1993). As an example, here we report an approximate expression for the variance of the Welch estimator, valid for 50 % overlap and the use of a Bartlett window:

$$\sigma^2_{B^W_{xx}(\omega)} \cong \frac{9}{8K}P^2_{xx}(\omega),$$

which means that K being equal, the variance of the Welch estimator with a Bartlett window is comparable to that of the Bartlett estimator, in fact slightly higher, by a factor 9/8. However, with a fixed number N of data and a fixed value of M, i.e., with a given resolution, a 50 % overlap in the Welch method produces two times as many segments over which we can average, i.e., it yields $K_{Welch} \cong 2K_{Bartlett} = 2N/M$, and therefore halves the variance with respect to Bartlett's method. Expressing the variance in terms of N and M, rather than in terms of K, for the Bartlett window and 50 % overlap we have

$$\sigma^2_{B^W_{xx}(\omega)} \cong \frac{9}{16}\frac{M}{N}P^2_{xx}(\omega),$$

and since $\sigma^2_{B_{xx}(\omega)} \approx (M/N)P^2_{xx}(\omega)$ for a given resolution (fixed M), we get

$$\sigma^2_{B^W_{xx}(\omega)} \cong \frac{9}{16}\sigma^2_{B_{xx}(\omega)}.$$

The behavior of the Welch estimator with respect to those of the periodogram and of the Bartlett estimator is illustrated in the examples of Fig. 10.14. We consider 50 realizations of a signal containing two sinusoids with equal amplitudes and different frequencies, immersed in white noise with unit variance. We set $A_0 = A_1 = 5$, $\omega_0 = 0.4\pi$, $\omega_1 = 0.45\pi$, with $N = 512$ samples. In Fig. 10.14, the two top panels show a Bartlett spectral estimate with $K = 1$, that is, a periodogram. We can observe the great variance of the estimate—the part of the spectrum that does not contain any peak should be a flat line—and the finite width of the peaks that causes resolution to be limited, though more than sufficient in this particular case.

To the same 50 realizations we also apply Bartlett's method with $K > 1$, choosing first $K = N/M = 4$ (that leads, with $N = 512$, to $M = 128$; second row of panels from top), and then $K = 8$ (that leads to $M = 64$; third row). We see that the variance of the estimate decreases with increasing K, while resolution worsens. At last, to the same 50 sequences with $N = 512$ we apply Welch's method adopting a Hanning window, $M = 128$ and a 50 % overlap, so that $K = 7$ (bottom panels). If we compare the bottom panels with those of the third row, we see that K being nearly equal ($K = 8$ in the third row, $K = 7$ in the bottom row), the variance is approximately the same, as expected. Moreover, though the main spectral lobe width of the Hanning window used for the Welch estimate is larger than the main spectral lobe of the rectangular window used for the Bartlett estimate, the resolution we get by Welch's method is similar to the one we get by Bartlett's method, or even better. This is due to the fact that the 50 % overlap in Welch's method allows for having $M = 128$, while in Bartlett's method we had $M = 64$. But basically, what do we gain applying Welch's method instead of Bartlett's method in this case? We obtain a leakage reduction, because by abandoning the rectangular window we gain a reduction in height of the

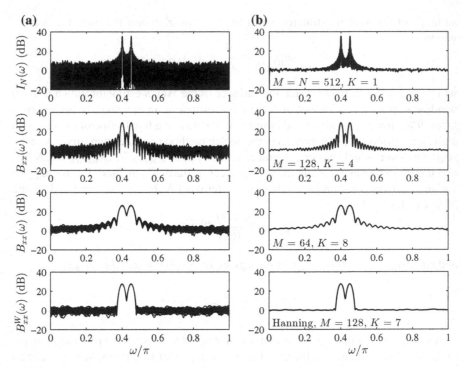

Fig. 10.14 Examples of spectral estimates obtained via periodograms, Bartlett's and Welch's methods. The data are 50 realizations, with $N = 512$ samples each, of a process containing two sinusoids ($A_0 = A_1 = 5$; $\omega_0 = 0.4\pi$, $\omega_1 = 0.45\pi$) in unit-variance Gaussian white noise. From top to bottom: periodogram; Bartlett's method with $M = 128$ and $K = 4$; Bartlett's method with $M = 64$ and $K = 8$; Welch's method with $M = 128$, a Hanning window and 50% overlap, resulting in $K = 7$. **a** 50 Realizations. **b** Ensemble average

spectral window's sidelobes. This is demonstrated by the comparison between the third and fourth panels from top in Fig. 10.14b: observe how flat is the peak-free part of the spectrum in the Welch's estimate with a Hanning window.

10.7   Blackman-Tukey Method

The spectral methods presented above are direct methods based on the DTFT of the sequence. The Blackman-Tukey method (Blackman and Tukey 1958) works on the estimated AC sequence instead. It is particularly useful for short sequences, in that it does not operate any subdivision of the record in smaller segments, but only smooths the periodogram by convolving it with a proper spectral window $W(e^{j\omega})$ which is a real and even function of ω. With respect to the periodogram, this method can provide a significant variance reduction, i.e., an increase of stability, at the cost of a remarkable resolution loss.

The Blackman-Tukey (BT hereafter) spectral estimate is defined as

$$S_{xx}(\omega) = \frac{1}{2\pi} \int_{-\pi}^{+\pi} I_N(\theta) W\left\{ e^{j(\omega-\theta)} \right\} d\theta,$$

where $I_N(\theta)$ is the BT periodogram (correlogram), i.e., the transform of $c_{xx}[l]$, and $W(e^{j\omega})$ is a real spectral window and an even function of ω. $S_{xx}(\omega)$ can therefore be interpreted as the Fourier transform of the product of $c_{xx}[l]$ with the inverse transform of $W(e^{j\omega})$, that is,

$$w[l] = \frac{1}{2\pi} \int_{-\pi}^{+\pi} W(e^{j\omega}) e^{j\omega l} d\omega,$$

representing a real, non-causal tapered window with even symmetry around the time origin. This symmetry is in fact required to ensure the spectral window $W(e^{j\omega})$ to be a real even function of ω; moreover, it would make no sense to apply a non-symmetrical window to a symmetrical AC sequence. Let us assume that the time window $w[l]$ has a length $2M - 1$, with samples going from discrete time $-(M - 1)$ to $M - 1$, and is identically zero elsewhere. We thus can write

$$S_{xx}(\omega) = \sum_{l=-(M-1)}^{M-1} c_{xx}[l] w[l] e^{-j\omega l}.$$

This definition highlights how the BT estimate is nothing but a *modified correlogram*. Therefore the approach also has to do with leakage reduction. The window is applied to $c_{xx}[l]$ and not directly to the data; $c_{xx}[l]$ exists for all lags $|l| < N$, but we cut out a shorter segment, of length $2M + 1$, symmetrical with respect to $l = 0$ in agreement with the symmetry properties of $c_{xx}[l]$, and properly tapered at the edges. This serves to reduce the statistical weight of the estimates $c_{xx}[l]$ at large lags, which are the least accurate and most variable ones and negatively affect the variance of the spectral estimate.

$S_{xx}(\omega)$ is then required to be non-negative, as $P_{xx}(\omega)$. A sufficient condition for this—even if not necessary—is that $W(e^{j\omega}) \geq 0$ over the whole interval $(-\pi, \pi]$, since the periodogram is itself a non-negative estimate of the power spectrum. This is true for the Bartlett window but is not true, for instance, for the Hanning and Hamming windows.[5] Thus these windows, even if able to provide a better behavior of sidelobes, may lead to negative spectral estimates at some frequencies, which are corrected setting them to zero.

The expected value of the modified correlogram is

$$E[S_{xx}(\omega)] = \frac{1}{2\pi} \int_{-\pi}^{+\pi} E[I_N(\omega)] W\left\{ e^{j(\omega-\theta)} \right\} d\theta.$$

---

[5]Recall that here we are speaking of non-causal windows, having a real transform. Note that Fig. 5.5 shows transform squared magnitudes and therefore does not allow to verify the above statement.

We may recall that

$$E[I_N(\omega)] = \frac{1}{2\pi} \int_{-\pi}^{+\pi} P_{xx}(\omega) W_{B,2N-1} \left\{ e^{j(\omega-\theta)} \right\} d\theta,$$

so we have a convolution in the frequency domain between $W(e^{j\omega})$, $W_{B,2N-1}(e^{j\omega})$ and $P_{xx}(\omega)$. In the time domain this corresponds to

$$E[S_{xx}(\omega)] = \sum_{l=-(M-1)}^{M-1} \left\{ \gamma_{xx}[l]w[l]w_B[l] \right\} e^{-j\omega l},$$

where $\gamma_{xx}[l]$ is the theoretical AC sequence and

- $w_B[l] = (N - |l|)/N$ with $|l| < N$ is a causal, $2N - 1$-samples long Bartlett window
- $w[l]$ is a non-causal taper, symmetrical around $l = 0$ and $2M - 1$-samples long.

We assumed $M \leq N$. If N, which is

- the length of the data record, and
- the half-length of $\gamma_{xx}[l]$ and $w_B[l]$,

is large with respect to M, which is the half-length of the window $w[l]$ applied to the AC sequence, then $W_{B,2N-1}(e^{j\omega})$ is narrowband with respect to $W(e^{j\omega})$. Since a convolution with a narrowband spectral window does not greatly alter a given function of frequency,[6] we can write

$$E[S_{xx}(\omega)] \cong \frac{1}{2\pi} \int_{-\pi}^{+\pi} P_{xx}(\omega) W \left\{ e^{j(\omega-\theta)} \right\} d\theta,$$

since with respect to the smoothing operated by $W(e^{j\omega})$, $E[I_N(\omega)] \approx P_{xx}(\omega)$: the expected value of the BT estimate is similar to the convolution of the true spectrum with $W(e^{j\omega})$. Thus choosing a short AC window $w[l]$, by setting $M \ll N$, implies adopting a spectral window $W(e^{j\omega})$ which is definitely wideband with respect to the bandwidth of the intrinsic spectral window $W_{B,2N-1}(e^{j\omega})$, and as a consequence, the BT estimate becomes very smooth and stable, but its resolution becomes very poor with respect to the periodogram.

The variance of $S_{xx}(\omega)$ is found to be (Kay 1988)

$$\sigma^2_{S_{xx}(\omega)} \cong \frac{1}{2\pi N} \int_{-\pi}^{+\pi} P^2_{xx}(\omega) W^2 \left\{ e^{j(\omega-\theta)} \right\} d\theta,$$

and therefore it is essential to study the influence of the shape and length of the window $w[l]$ on the variance of the BT estimate. If M is simultaneously

---

[6]The convolution with a Dirac δ would leave the function unaltered.

- much smaller than N, so that $W(e^{j\omega})$ is wideband with respect to $W_{B,2N-1}(e^{j\omega})$,
- large enough to ensure that at the same time $W(e^{j\omega})$ is narrowband with respect to the typical variations of $P_{xx}(\omega)$, so that a proper frequency resolution is guaranteed,

then we can write, for frequencies not near 0 or $\pm\pi$ (see Kay 1988),

$$\sigma^2_{S_{xx}(\omega)} \simeq \frac{1}{2\pi N} P^2_{xx}(\omega) \int_{-\pi}^{+\pi} W^2(e^{j\omega}) d\omega.$$

We can now study the asymptotic behavior of the BT spectral estimate. We know that $I_N(\omega)$ is asymptotically unbiased. How does $E[S_{xx}(\omega)]$ behave for $N \to \infty$?

If N is large we can choose a large M and make $W(e^{j\omega})$ narrowband with respect to the variations of $P_{xx}(\omega)$, hence

$$E[S_{xx}(\omega)] \sim P_{xx}(\omega) \frac{1}{2\pi} \int_{-\pi}^{+\pi} W(e^{j\omega}) d\omega.$$

To have an asymptotically unbiased estimate we must thus impose the condition

$$w[0] = \frac{1}{2\pi} \int_{-\pi}^{+\pi} W(e^{j\omega}) d\omega = 1,$$

i.e., we must properly normalize the window samples. On the other hand, $\sigma^2_{S_{xx}(\omega)}$ goes to zero for $N \to \infty$, hence the BT spectral estimate is consistent.

The improvements with respect to the periodogram that are achieved using tapered windows in the BT method can be quantified comparing the expressions of $E[S_{xx}(\omega)]$ and $\sigma^2_{S_{xx}(\omega)}$ with the corresponding expressions for the periodogram, which is also a non-modified correlogram. For large values of N the variance of the periodogram can be approximated as

$$\sigma^2_{I_N(\omega))} = \begin{cases} P_{xx}(0) \cdot (1+1) = 2P_{xx}(0) & \text{for} \quad \omega = 0, \\ P_{xx}(\pi) \cdot (1+1) = 2P_{xx}(\pi) & \text{for} \quad \omega = \pi, \\ P^2_{xx}(\omega) \cdot (1+0) = P^2_{xx}(\omega) & \text{elsewhere,} \end{cases}$$

where the second term in parentheses—1 or 0—derives from $\{\sin(N\omega)/[N\sin\omega]\}^2$ in the expression of the periodogram variance (Sect. 10.3.2). Therefore in the open interval $0 < \omega < \pi$ the variance reduction achieved by the BT method with respect to the periodogram is

$$R = \frac{\sigma^2_{S_{xx}(\omega)}}{\sigma^2_{I_N(\omega)}} = \frac{1}{2\pi N} \int_{-\pi}^{+\pi} W(e^{j\omega})^2 d\omega = \frac{1}{N} \sum_{l=-(M-1)}^{M-1} w[l]^2.$$

Table 10.1 Properties of some non-causal windows used to truncate the estimated AC sequence in the Blackman-Tukey method; see text for the meaning of the symbols

Window	Expression		L	R
Rectangular	$w_R[l] = 1$	for $|l| < M$	$\frac{2\pi}{M}$	$2\frac{M}{N}$
	$w_R[l] = 0$	elsewhere		
Bartlett	$w_B[l] = 1 - |l|/M$	for $|l| < M$	$\frac{8\pi}{2M-1} \approx \frac{4\pi}{M}$	$\frac{2M}{3N} \simeq 0.67\frac{M}{N}$
	$w_B[l] = 0$	elsewhere		
Raised cosine	$w_H[l] = \alpha + \beta \cos \frac{\pi l}{M-1}$	for $|l| < M$	$\frac{4\pi}{M}$	$2(\alpha^2 + \frac{\beta^2}{2})\frac{M}{N}$
	$w_H[l] = 0$	elsewhere		
Hanning	$\alpha = \beta = 0.5$			$0.75\frac{M}{N}$
Hamming	$\alpha = 0.54, \beta = 0.46$			$0.80\frac{M}{N}$

To derive this equation, Parseval's theorem was used, according to which

$$\frac{1}{2\pi} \int_{-\pi}^{+\pi} W(e^{j\omega})^2 d\omega = \sum_{l=-(M-1)}^{M-1} w[l]^2.$$

Since we desire a variance reduction in comparison with the periodogram, we must choose M and the window shape in such a way to have $R < 1$. At the same time we will also have to consider the resolution loss connected to the tapered window we intend to adopt. The latter factor can be quantified by the half-width L of the main lobe of $W(e^{j\omega})$, measured at the base of the lobe, or at some other specified height. Table 10.1 illustrates these properties of the windows that are normally used in the BT method—those applied in addition to the implicit window $w_B[l]$. The values of L and R shown in Table 10.1 are approximate and valid for $1 \ll M \ll N$. Obviously, the smallest possible values of L and R are sought. The last column in Table 10.1 clearly reveals that for a fixed N the variance reduction is essentially related to the choice of M, i.e., to the truncation of the AC sequence.

From Table 10.1 we see, for instance, that with $M/N = 1/20$ the Bartlett window provides $R = 1/30$, i.e., the variance is 30 times smaller than that of the periodogram. In practice, Kay (1988) recommends to choose

$$\frac{N}{10} < M < \frac{N}{5},$$

i.e., windows should be no more than one-tenth to one-fifth the total number of data points, in order to obtain the desired estimate-variance reduction, and not too much smaller, in order to retain the ability to resolve peaks at neighboring frequencies and to obtain the desired leakage reduction. More precisely, Chatfield (1975) recommends

- for $100 < N < 200$, $M = N/6$;
- for $1000 < N < 2000$, $M = N/10$;
- for larger values of N, more restrictive choices.

In applications, it is advisable to try more than one value of M and compare the results: for example, if the suggested value is $N/10$, we may want to try $N/5, N/10, N/15$. As for the window shape, the most widely used windows are probably the Bartlett window and the so-called *raised cosine windows* (Table 10.1: Hanning and Hamming).

In summary, under the hypothesis that the length of the window applied to the AC sequence is such that the spectral window $W(e^{j\omega})$ can be considered both narrowband with respect to the typical variations of the spectrum that must be estimated, and wideband with respect to $W_{B,2N-1}(e^{j\omega})$, the following equations hold:

$$E[S_{xx}(\omega)] \sim P_{xx}(\omega) \frac{1}{2\pi} \int_{-\pi}^{+\pi} W(e^{j\omega})d\omega,$$

$$\sigma^2_{S_{xx}(\omega)} \simeq \frac{1}{2\pi N} P^2_{xx}(\omega) \int_{-\pi}^{+\pi} W(e^{j\omega})^2 d\omega \cong \frac{1}{N} P^2_{xx}(\omega) \sum_{l=-(M-1)}^{M-1} w[l]^2.$$

This means that a large M implies a good frequency resolution and a limited smoothing of the true spectrum, but also a large variance of the spectral estimate. We must not forget that a large M is possible only if N is large, so that we can accurately compute the AC estimates for lags $l = [-(M-1), M-1]$. On the contrary, a small M lowers the variance but smooths the spectrum so much that resolution is severely limited. These properties are reflected in the dependence of $\sigma^2_{S_{xx}(\omega)}$ on N and M: the variance decreases with increasing N but for fixed N is found to be directly proportional to M. The search for a proper compromise between small variance, i.e., small spectrum-estimation error, and satisfactory resolution thus turns into the search for the most appropriate M value.

The behavior of the BT estimator is illustrated in Fig. 10.15. All the panels refer to 50 realizations of the same process analyzed in Fig. 10.14. Two different windows are considered, Bartlett and Hamming, and M assumes different values: with a constant number of signal samples ($N = 512$), the values $M = N/6$, $M = N/8$, and $M = N/10$ are considered. For a given window type, variance decreases as M decreases ($R \propto M$; this effect is hardly visible in the dB plots of Fig. 10.15). For a given value of M, the Bartlett window produces a greater variance reduction than the Hamming window. Resolution is nearly the same for both windows and increases with M. In any case: the BT estimate is always a quite smooth and low-resolution estimate. An example of application of the BT spectral method to a real-world signal can be found in Sect. 12.3.

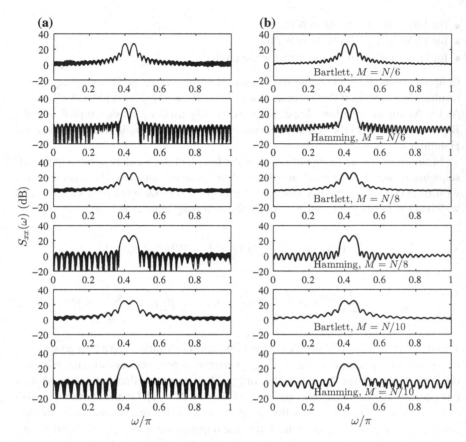

Fig. 10.15 Examples of spectral estimates obtained applying the Blackman-Tukey method. The analyzed series are the same of Fig. 10.14, with $N = 512$. Bartlett and Hamming windows are used, with different values of M. **a** 50 Realizations. **b** Ensemble average

10.8   Statistical Significance of Spectral Peaks

We will now see how confidence intervals can be established for the power spectrum estimate of a data record and how the statistical significance of spectral peaks possibly detected in the spectrum can be evaluated. We will focus on the case of estimates via the periodogram.

Let us take a sequence $x[n]$ and suppose that each random variable x_n, of which $x[n]$ at time n is a sample, has a Gaussian distribution. Then, Re $\{X[k]\}$ and Im $\{X[k]\}$ are also Gaussian. Hence both $|X[k]|^2$ and the periodogram estimate $I_N[k]$ obtained via FFT are expected to approximately follow a χ^2 distribution with $\nu = 2$ *degrees of freedom* (DOF). In fact, the χ^2 distribution is typical of the sum of squares of a number of independent Gaussian variables, and each square provides a DOF.

Fig. 10.16 χ^2 distribution for some values of the number ν of degrees of freedom (DOF)

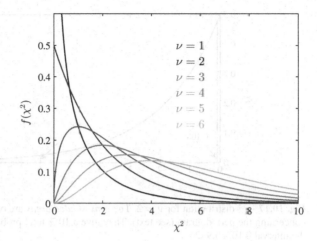

Defining clearly what the expression "degrees of freedom" means in a general statistical sense is definitely beyond the scope of this book. The reader is referred to Stuart (2010) for this and other statistical topics. In the present frame, we can see the DOF as the number of Gaussian variables that contribute to forming the variable $|X[k]|^2$ the distribution of which we are interested in. Here $|X[k]|^2$ is determined by the real and imaginary parts of $X[k]$, so we have $\nu = 2$ DOF.[7] This is why the periodogram has such a large variance: two degrees of freedom are not enough for the periodogram to behave properly from a statistical point of view. The various expedients adopted to reduce the periodogram's variance and hence improve the stability of the estimate are nothing but methods to increase the number of DOF of the related χ^2 distribution.

Figure 10.16 shows the shape of the χ^2 distribution for some values of ν. The distribution is asymmetric and is defined uniquely by the value of ν. Among the properties of the variables distributed as χ^2 we may mention the addition theorem, according to which the sum of a number of independent variables distributed as χ^2 is χ^2-distributed with a number of DOF equal to the sum of the individual ν values. The functional form of this distribution is

$$f\left(\chi^2\right) = \frac{\left(\chi^2\right)^{\frac{\nu}{2}-1} e^{-\frac{\chi^2}{2}}}{2^{\frac{\nu}{2}} \Gamma\left(\frac{\nu}{2}\right)}$$

---

[7]It may be more intuitive to refer to the discrete trigonometric expansion of a signal (Sect. 3.7): the two Gaussian variables in this case are the expansion coefficients a_k and b_k; the former is related to a cosine of $\omega_k n$ and the latter to a sine, and sine and cosine are mutually in phase quadrature. Therefore $a_k^2 + b_k^2 = |X[k]|^2$ has a χ^2 distribution with $\nu = 2$. Note that this conclusion is independent of the sequence length, because the sinusoidal sequences used as the expansion basis are infinitely long.

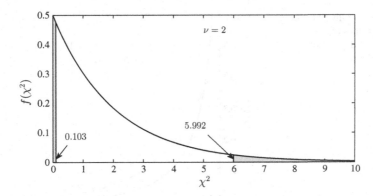

Fig. 10.17 χ^2 distribution for $\nu = 2$. The *gray-shaded areas* are of interest for the statistical test concerning the periodogram (see text). They give a 10 % total probability for χ^2 to be external to the interval 0.103–5.992

where

$$\Gamma(\alpha) = \int_0^\infty x^{\alpha-1} e^{-x} dx.$$

As ν increases, the distribution progressively becomes wider and lower. Figure 10.17 shows the χ^2 distribution for $\nu = 2$ and illustrates the conditions in which statistical tests for power spectrum values are performed.

Let us start with the issue of establishing confidence intervals for PSD values. First, a probability level must be selected for the test. Since in this case we are interested in both tails of the distribution, we indicate the probability level through its half value: we write, for instance, $p/2 = 0.05$. Since $\chi^2_{1-p/2,\nu} = \chi^2_{0.95,2} = 5.99$ is the value of χ^2 that has 95 % of probability of not being exceeded and $\chi^2_{p/2,\nu} = \chi^2_{0.05,2} = 0.103$ is the value of χ^2 that has 95 % of probability of being exceeded, the *gray-shaded areas* in Fig. 10.17 give a total probability of $p = 0.1 = 10\%$ for χ^2 to be external to the interval 0.103–5.99, while $1 - p = 0.9 = 90\%$ is the probability of χ^2 being comprised inside the same interval. We will say that χ^2 falls inside the interval $\chi^2_{p/2,2}$—$\chi^2_{1-p/2,2}$ at the *confidence level* (c.l.) of $100\,(1-p)\,\%$, while $100\,p\,\%$ is referred to as the *significance level* of the test.

In the case of the power spectrum, the variable which is distributed as χ^2 is precisely $\nu I_N[k]/P_{xx}(\omega_k)$, i.e., ν times the ratio between the estimate and the true value. Therefore we can state that with $\nu = 2$

- at the significance level of $100\,p\,\%$
- at the confidence level (c.l.) of $100\,(1-p)\,\%$

the *confidence interval* for the kth true PSD value at frequency ω_k is

$$\frac{2I_N[k]}{\chi^2_{1-p/2,2}} \leq P_{xx}(\omega_k) \leq \frac{2I_N[k]}{\chi^2_{p/2,2}},$$

with $\chi^2_{p/2,2} < \chi^2_{1-p/2,2}$. For example, with $p = 0.90$ as in Fig. 10.17 we have

$$\frac{2}{5.992} I_N[k] \le P_{xx}(\omega_k) \le \frac{2}{0.103} I_N[k],$$

that is,

$$0.33 I_N[k] \le P_{xx}(\omega_k) \le 1.94 I_N[k].$$

The lower bound has 95 % probability of being exceeded and thus the true value of $I_N[k]$ has only 5 % probability of being smaller than this lower bound. The upper bound has 95 % probability of not being exceeded and thus the true value has only 5 % probability of being larger than this upper bound. The bounds are computed for each k, i.e., at each frequency ω_k.

Let us now turn to the evaluation of the statistical significance of a spectral peak detected in the power spectrum. Here the question is: does this PSD value likely represent a real spectral feature related to some periodic or quasi-periodic process, or is it compatible with the hypothesis of a simple noise spike? To answer this question we need

- the estimated PSD sample $I_N[k]$ to be tested
- a continuous background spectrum—a *noise background spectrum*—against which the PSD estimate will be tested. This noise spectrum is the *null hypothesis* for the test.

For instance, let us assume as the null hypothesis that the process is white noise, and therefore has a true flat spectrum equal to σ_x^2. The test consists in checking if at a given probability level, the detected peak is compatible with this assumption. We admit that $I_N[k]$ can fluctuate around its true value σ_x^2, because of its random nature, and we look for the maximum value that $I_N[k]$ can attain while fluctuating. In this case we are interested only in the distribution's upper tail, so we write

$$\frac{2I_N[k]}{\chi^2_{1-p,2}} \le P_{xx}(\omega_k) = \sigma_x^2.$$

We do not know the true variance of the process, but we can approximate it by the variance estimated from the available data record. Then the maximum value I_{signif} that can be attained by the spectral estimate is

$$I_{\text{signif}} = \frac{\sigma_x^2 \, \chi^2_{1-p,2}}{2}.$$

With $p = 0.05$ and $\chi^2_{1-p,2} = 5.992$ we have $I_{\text{signif}} \approx 3\sigma_x^2$. Under the null hypothesis of white noise the significance level I_{signif} for the power spectral estimate is independent of frequency. It is also referred to as the p%-c.l. spectrum (here, the 95 %-c.l. spectrum).

The choice of the background spectrum is not trivial. In many fields—for instance in geophysics (Allen and Smith 1996) and other sectors of physics, but also in

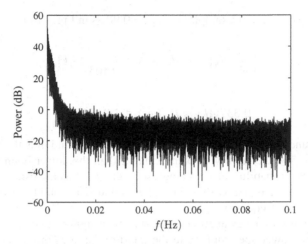

Fig. 10.18 An example of power spectrum, estimated via periodogram, of a sequence of physical measurements that exhibits decreasing power with increasing frequency. Data is atmospheric pressure micro-fluctuations in Pascal (Pa), measured using a microbarometer in an experiment aimed at detecting the influence of atmospheric gravity waves on visibility during fog episodes in the Po Valley, Northern Italy

economics (see, e.g., Hess and Iwata 1997; McConnell and Perez-Quiros 1989; Sella et al. 2013), in medicine and biology (see, e.g., Meltzer et al. 2008), etc.—the spectra of measured variables exhibit a general behavior with decreasing power for increasing frequency, rather than being nearly constant. As an example, Fig. 10.18 shows the periodogram of a record of $N = 25,921$ data of atmospheric pressure micro-fluctuations in Pascal (Pa), measured using a microbarometer in an experiment aimed at detecting the influence of atmospheric gravity waves on visibility during fog episodes in the Po Valley, Northern Italy (Richiardone et al. 1995). In such cases, the most representative background is not white noise but *red noise*, a process that will be discussed in the next chapter and that is characterized by a smooth spectrum with power declining as frequency increases. As we will see in the next chapter, the theoretical power spectrum of a red noise process is given by

$$P_{rr}[k] = \frac{\sigma_x^2(1 - \alpha^2)}{1 + \alpha^2 - 2\alpha \cos(2\pi k/N)}$$

with $\alpha = \rho[1]$, where $\rho[1]$ indicates the AC coefficient (normalized autocovariance) of the process at lag $l = 1$.

As a null hypothesis, we can thus assume that the process is red noise with a parameter α estimated from the data. Alternatively, some authors have suggested to use as the background spectrum a strongly smoothed version of the periodogram itself or of some another spectral estimate, in order to eliminate the peaks and preserve only the general spectral behavior. This could be a better assumption than a pre-imposed functional form (Mann and Lees 1996). Anyway, the background spectrum

Fig. 10.19 Periodogram of a series of physical measurements, with significance levels computed assuming a background of white or red noise (95 % c.l.). The *gray dotted and dashed horizontal lines* represent the white noise background and the related significance level, respectively. The *black dotted and dashed smooth curves* represent the red noise background and the related significance level, respectively

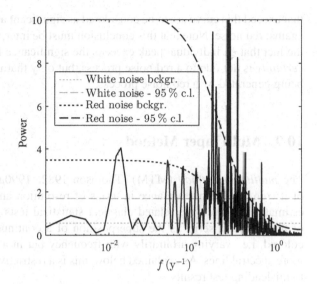

Legend in figure:
...... White noise bckgr.
— — White noise - 95 % c.l.
•••••• Red noise bckgr.
— — Red noise - 95 % c.l.

(Power axis vertical; horizontal axis $f\,(\mathrm{y}^{-1})$)

will now be a function of frequency and will replace σ_x^2 in the formula we gave for the white-noise case: we will write

$$I_{\mathrm{signif}}[k] = \frac{P_{rr}[k]\,\chi^2_{1-p,2}}{2}.$$

Figure 10.19 shows, as an example, the power spectrum of a physical signal[8]—a climatological temperature record with $T_s = 0.25$ y (one season), with the related significance levels evaluated assuming white or red noise as the null hypothesis. The typical decrease of power with increasing frequency is clearly visible in Fig. 10.19 in the region above 0.1 y^{-1}, i.e., for periods smaller than 10 years, a region that in climatology is called "interannual". The *black dotted curve* shows the red-noise background spectrum. It can be seen that this type of background follows the general spectral behavior of this data much better than white noise (*gray dotted line*). It thus allows evaluating peak significance in a more adequate way. The *black dashed curve* represents the significance level against red noise at the 95 % c.l. The spectral peaks

---

[8]The data is seasonal climatic anomalies of sea surface temperature (SST) in °C, spatially averaged over the so-called NINO3 region of the Pacific Ocean. This region is comprised between latitudes of 50° South and 50° North and between longitudes of 90° West and 150° West, and is particularly representative for the study of climatic variability at interannual time scale, and more precisely for the study of the El Niño—La Niña phenomena. The data used here spans the years 1871–2012 and were drawn from monthly NINO3 data available at http://www.esrl.noaa.gov/psd/gcos_wgsp/Timeseries/Data/nino3.long.data.

The term "climatic anomalies" means that the data has been centered by subtracting from each seasonal sample the mean value of the corresponding season, computed over the whole record; for instance: spring of 1900 minus average over all springs. This makes the data free from the annual temperature cycle, the dominant presence of which in the spectrum would obscure weaker cyclicities.

that exceed this curve are to be considered as significant at the 5 % significance level against red noise. Note that this conclusion must be interpreted in a statistical sense: the fact that an individual peak exceeds the significance level does not mean that it *certainly* is not due to a red noise process, but only that it has a small likelihood of being generated by a red noise process.

10.9   MultiTaper Method

The *multitaper method* (MTM) (Thomson 1982, 1990a, b) is interesting because of its remarkable performance in terms of resolution and variance of the spectral estimate. However, the related classical statistical tests assume that the spectrum of the analyzed series is the superposition of a continuous component—possibly colored, i.e., varying arbitrarily with frequency but in a smooth way—and one or more spectral lines. As explained below, this is a restrictive assumption that can lead to misleading test results.

In order to lower the variance of the estimate, independent periodogram estimates are averaged that are not obtained segmenting the record but applying to it a number of tapers that are orthogonal to one another by construction. These tapers are an optimal set of K *eigentapers* $w_k[n]$, $k = [1, K]$, designed to minimize leakage. More precisely, these tapers are the solution of the variational problem of minimization of leakage outgoing from an adimensional frequency band of properly chosen half-width Δv, centered on v. Such a set is collectively referred to as *Slepian tapers* (Slepian 2005) or *prolate spheroidal sequences*. The minimization problem is a suitable Rayleigh-Ritz minimization problem that leads to an eigenvalue problem.[9] The reader who is not acquainted with these terms and with the basic elements of linear algebra may want to simply consult a concise summary like the one appearing in Hayes (1996), or resort to a book presenting linear algebra from first principles, such as MacLane and Birkhoff (1999). Also Trefethen and Bau (1997) is a useful reference.

The Slepian tapers satisfy the following orthogonality constraint:

$$\sum_{n=0}^{N-1} w_j[n]w_k[n] = \delta_{jk},$$

---

[9]The Rayleigh-Ritz method is a classical variational method for finding approximate solutions of differential equations, whose exact solutions are hard to find. The method was first used by Lord Rayleigh in 1870 to solve the vibration problem of organ pipes closed on one end and open at the other. However, the approach did not receive much recognition by the scientific community. Nearly 40 years later, due to the publication of two papers by Ritz, the method came to be called the Ritz method. To recognize the contributions of both authors, the theory was later renamed the Rayleigh-Ritz method.

where N, the length of each Slepian taper, is equal to the number of data samples. The optimal number K of tapers is determined by choosing the half-bandwidth $\Delta\nu$. The tapers are built from the first to the Kth one, imposing the conditions that each new kth taper

- is orthogonal to the previous one(s),
- maximizes the *concentration ratio*

$$\lambda_k = \frac{\int_{-\Delta\nu}^{+\Delta\nu} \left| W_k(e^{j2\pi\nu}) \right|^2 d\nu}{\int_{-\frac{1}{2}}^{+\frac{1}{2}} \left| W_k(e^{j2\pi\nu}) \right|^2 d\nu}$$

of the spectral window $W_k(e^{j2\pi\nu}) \equiv W_k(e^{j\omega})$ in the band of half-width $\Delta\nu$ centered on ν. This means minimizing the outgoing leakage from the same band. The concentration ratios λ_k are the eigenvalues of the variational problem.

$\Delta\nu$ is expressed as a proper multiple of the distance between adjacent frequencies in the DFT of a sequence of length N. This distance, sometimes called the Rayleigh frequency, in terms of ν is $1/N$, and therefore

$$\Delta\nu = \frac{p}{N}.$$

The smaller $\Delta\nu$, the smaller the admitted leakage.

Given a value of N, first the value of p is fixed, so that the value of $\Delta\nu$ is also fixed. Then the members of the set of tapers, $w_k[n]$, are generated iteratively. The first taper is the window with length N that ensures the greatest spectral concentration in a band of half-width $\Delta\nu$ around ν; the second taper is the window with length N that is orthogonal to the first taper and ensures the greatest spectral concentration in a band of half-width $\Delta\nu$ around ν; the third taper is the window with length N that is orthogonal to both the first two tapers and guarantees the greatest spectral concentration, and so on.

The parameter p is called the *time-bandwidth parameter* or *time-frequency product*:

$$p = N\Delta\nu = \frac{N\Delta\omega}{2\pi} = NT_s\Delta f = T_d\Delta f,$$

where $T_d = NT_s$ is the duration of the data record and T_s is its sampling interval. $\Delta\omega$, $\Delta\nu$ and Δf measure, in terms of different frequency variables, the resolution of the method, in the sense that a spectral peak that would ideally be impulsive—a spectral line—will be detected in the MTM spectrum plotted as a function of f as a bump with width $2\Delta f = 2\Delta\nu/T_s = \Delta\omega/(\pi T_s)$.

Only the first $2p - 1$ tapers turn out to be useful for minimizing leakage, and therefore the total number of tapers is constrained by

$$K = 2p - 1.$$

Fig. 10.20 The sets of
Slepian tapers useful for
MTM spectral estimation, for
$p = 2, 2.5, 3, 3.5$ and 4 (from
top to bottom) and $N = 512$

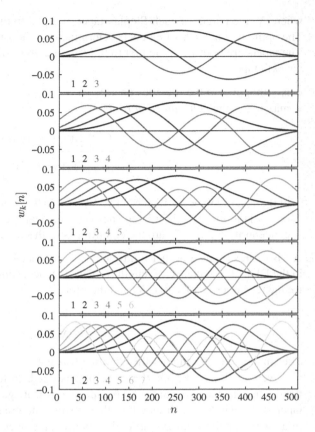

In Fig. 10.20, the sets of Slepian tapers useful for MTM spectral estimation are
shown, for the most common values of p and for $N = 512$. For a given value of p,
the $(k + 1)$th taper shows one more crossing of the zero line with respect to the kth
taper.

The MTM spectral estimate is built considering the individual DTFTs of the K
tapered signals, i.e., of the K sequences obtained multiplying the data by each Slepian
taper:

$$X_k(e^{j\omega}) = \sum_{n=0}^{N-1} w_k[n]x[n]e^{-j\omega n}, \qquad\qquad k = [1, K].$$

The MTM spectral estimator is defined as a weighted average of the corresponding
periodograms, which are referred to as *eigenspectra*.

According to the definition of periodogram, we should take the square modulus
of these DTFTs, divided by N. However, the factor $1/N$ in the MTM disappears,
because a factor $1/\sqrt{N}$ is absorbed in the definition of the tapers $w_k[n]$. The individual
eigenspectra are thus defined as $\left|X_k(e^{j\omega})\right|^2$. Then they are linearly combined with

weights λ_k, to form the *raw MTM spectrum*

$$S^{MT}(\omega) = \frac{\sum_{k=1}^{K} \lambda_k \left| X_k(e^{j\omega}) \right|^2}{\sum_{k=1}^{K} \lambda_k}.$$

In this way the sidelobes, and consequently the outgoing leakage from the spectral band of width $2\Delta\nu$ centered on some frequency ν, are minimized, and, assuming that the various spectral windows are narrowband with respect to the typical variations of the series' power spectrum, a sufficient resolution is ensured.

A further leakage reduction is possible by defining the *adaptive spectrum*, which is a nonlinear superposition of eigenspectra:

$$S_a^{MT}(\omega) = \frac{\sum_{k=1}^{K} b_k^2(\omega)\lambda_k \left| X_k(e^{j\omega}) \right|^2}{\sum_{k=1}^{K} b_k^2(\omega)\lambda_k}.$$

Here $b_k(\omega)$ is a weight function that further protects the spectral estimate from broadband leakage.

The MTM provides a better resolution and a lower estimate variance with respect to those that can be achieved by the methods studied up to now, and even by the parametric methods that will be studied in the next chapter. The desired trade-off between resolution and stability is achieved by properly choosing the number of tapers, i.e., the value of p: as p and K increase, variance decreases because more periodograms are averaged, but leakage and resolution worsen because $\Delta\nu$ increases. For data records with length of the order of hundreds, typical values of p go from 2 to 4—for instance, 2, 5/2, 3, 7/2, 4; for $p > 4$ numerical instabilities can occur. Longer series can "stand" higher values of p, though maintaining an acceptable resolution, i.e., a reasonably narrow $\Delta\nu$. In Fig. 10.21, the adaptive MTM spectrum of 50 realizations of the same process with two sinusoids in noise considered in Fig. 10.14 is shown, for $N = 512$. The values of p range from 2 to 4. It may be clearly seen that resolution and variance decrease with increasing p.

The K periodograms that re combined in the MTM estimate are uncorrelated, so that the raw estimate follows approximately a χ^2 distribution with $2K$ DOF. The adaptive spectrum has an effective number of DOF that usually is not much different from the nominal value $2K$ given for the raw spectrum. For the MTM spectrum, dedicated statistical tests are available. The *F-test* by Fisher-Snedecor (see Stuart 2010) for spectral lines serves to test the significance of isolated spectral peaks detected at some frequency. It determines the probability of the estimated PSD being ascribable to an isolated harmonic component with the same frequency. The test does not depend on the effective PSD value and therefore allows detecting sinusoidal components with low amplitudes, or to reject large-amplitude peaks that fail the test. This test is, however, criticized by the following argument.

The F-test is based on the assumption that the signal is a sum of sinusoids and white noise. In practice, it is sufficient that the spectrum of the noise component in the signal be locally white: we must be in a position to assume that the K eigenspectra

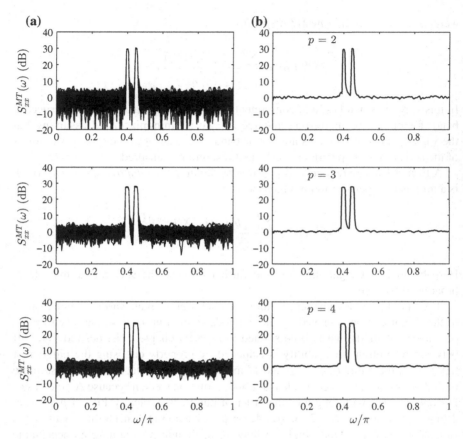

Fig. 10.21 Examples of adaptive MTM spectral estimates for the same series analyzed in Fig. 10.14 ($N = 512$). The value of the time-bandwidth parameter varies: from top to bottom, $p = 2, 3$ and 4. **a** 50 Realizations. **b** Ensemble average

are distributed in each frequency bin as they would be for white noise. Nevertheless, this signal's model is quite restrictive: sinusoids are harmonic signals with coherent phase, while quasi-periodic narrowband components that are not strictly harmonic are widely diffuse in time series. The test is no longer meaningful in such cases, because it actually performs what is known as harmonic analysis—it examines the amplitudes of the sinusoids found by the MTM estimator. If in reality the signal is colored noise, so that its spectrum has a structure as a function of frequency instead of being flat or nearly flat, this colored-noise structure can be artificially resolved into spurious lines, associated with arbitrary frequencies, which can pass the test (Vautard et al. 1992).

Since, presumably, most narrowband variability is not associated with strictly harmonic, purely periodic behaviour, Mann and Lees (1996) proposed to combine conventional harmonic analysis, which moreover works well just in the case of low SNR (MacDonald 1989), with additional criteria allowing for the detection of significant narrowband quasi-oscillatory components as well. Thus, the procedure of Mann and Lees (1996) bears to distinguish among harmonic, anharmonic, and background noise components. This method tests all the detected peaks against a red noise background null-hypothesis. The variance and lag-1 AC of the red noise process are directly estimated from the data and used to produce a red-noise "empirical" spectrum. The spectral background is robustly estimated by minimizing the misfit between the red noise empirical spectrum and a smoothed version of the adaptively-weighted MTM spectrum, which in this way becomes insensitive to outliers. The smoothing is achieved by applying a *median smoothing* filter[10] to the adaptively-weighted MTM spectrum. Then, by a χ^2 test the significance levels against the estimated red-noise background spectrum are determined at some chosen confidence levels; finally, a *reshaped spectrum* is generated, which isolates narrowband, possibly intermittent, amplitude- and-phase-modulated oscillations from harmonic phase-coherent sinusoids.

An example of MTM applied to a real-world signal, with visualization of the reshaped spectrum, is provided in Sect. 12.3.

10.10 Estimation of the Cross-Power Spectrum of Two Random Signals

The theoretical cross-power spectrum, or simply cross-spectrum, of two real WSS random variables x_n and y_n,

$$P_{xy}(e^{j\omega}) = \sum_{l=-\infty}^{+\infty} \rho_{xy}[l]e^{-j\omega l},$$

must be estimated using a finite number of samples of the two signals, i.e., two sequences $x[n]$ and $y[n]$ that we will assume to have a common length N. The cross-covariance $\gamma_{xy}[l]$ is firt estimated as explained in Sect. 9.10.4. Then standardization is applied to cross-covariance, and an estimate $\hat{\rho}_{xy}[l]$ of $\rho_{xy}[l]$ is obtained. The estimate of the cross-spectrum $P_{xy}(e^{j\omega})$ is finally defined in the form of a *cross-correlogram* (also referred to as *cross-periodogram*):

$$C_{xy}(e^{j\omega}) = \sum_{l=-(N-1)}^{N-1} \hat{\rho}_{xy}[l]e^{-j\omega l},$$

---

[10]The median is the numerical value separating the higher half of a data sample, a population, or a probability distribution, from the lower half. The median smoothing filter computes the median of PSD values falling inside a small frequency bin centered on a given frequency and assigns the result to this central frequency.

that is complex and has no particular symmetry properties. Its expected value is

$$
E[C_{xy}(e^{j\omega})] = \sum_{l=-(N-1)}^{N-1} \frac{N-|l|}{N} \rho_{xy}[l] e^{-j\omega l},
$$

hence $C_{xy}(e^{j\omega})$ is a biased, but asymptotically unbiased estimate. However it turns out that the variance of the estimate, $\sigma_{C_{xy}}^2$, does not go to zero for $N \to \infty$. In order to stabilize the estimate, tapers can be applied: the *modified cross-correlogram* (also, *modified cross-periodogram*) is

$$
S_{xy}(e^{j\omega}) = \sum_{l=-(N-1)}^{N-1} \hat{\rho}_{xy}[l] w[l] e^{-j\omega l},
$$

where $w[l]$ indicates a non-causal taper. Under similar assumptions to those made for the BT estimate $S_{xx}(\omega)$, it can be shown that the expected value of $S_{xy}(e^{j\omega})$ is approximately

$$
E[S_{xy}] \cong \frac{1}{2\pi} \int_{-\pi}^{+\pi} P_{xy}(e^{j\theta}) W\left[e^{j(\omega-\theta)}\right] d\theta
$$

and that its variance $\sigma_{S_{xy}}$ decreases with increasing N and with a decreasing length of the window $w[l]$ applied to $\hat{\rho}_{xy}[l]$. Pseudo-ensemble averaging of (modified) cross-periodograms is often applied to get a *Welch cross-spectral estimate* $B_{xy}^W(\omega)$. Magnitude- and phase-spectrum estimates can then be derived, as well as estimates of the magnitude squared coherence (MSC). Recall that phase estimates in cross-spectrum analysis are only useful at those frequencies at which significant coherence exists. MSC is estimated as

$$
\hat{MSC}_{xy}(\omega) = \frac{\left|B_{xy}^W(\omega)\right|^2}{B_{xx}^W(\omega) B_{yy}^W(\omega)},
$$

and assumes values between 0 and 1. The resolution-versus-variance trade-off is similar to that discussed for the estimate of the power spectrum of a single random process.

Let us see an example of how to use the cross-spectrum to obtain the phase difference between sinusoidal components in two series, $x[n]$ and $y[n]$. The individual series consist of two sine waves, with frequencies $\omega_0 = 0.2\pi$ and $\omega_1 = 0.4\pi$, immersed in white noise $e[n]$ with variance $\sigma_e^2 = 0.25$. The sine waves in the x-series both have unit amplitudes. The ω_0 sine wave in the y-series has amplitude of 0.5 and the ω_1 sine wave in the same series has amplitude of 0.35. The individual sine waves in the y-series are out-of-phase with respect to the corresponding ones contained in the x-series, by different amounts at ω_0 and ω_1. Precisely, with obvious notation we set

$$x[n] = A_{x,0} \sin(\omega_0 n) + A_{x,1} \sin(\omega_1 n) + e[n],$$
$$y[n] = A_{y,0} \sin(\omega_0 n - \frac{\pi}{4}) + A_{y,1} \sin(\omega_1 n - \frac{\pi}{2}) + e[n].$$

For each series, $N = 1000$ samples are generated, and Welch's method of cross-spectral estimation is applied, with a Hamming window of length $M = 100$ and overlap of 80%. Figures 10.22 and 10.23 respectively show the magnitude squared coherence and the phase spectrum—here indicated by $\Theta_{xy}(\omega)$—estimated in this way.

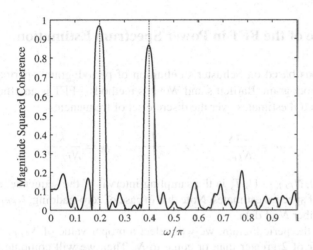

Fig. 10.22 Estimated magnitude squared coherence (MSC) between two 1000-sample-long sequences. Each sequence contains two sinusoids in Gaussian white noise; the sine waves in the second series are phase-lagged with respect to those in the first series (see text for details). The *dotted lines* mark the true frequencies of the sinusoids. The estimate has been obtained using Welch's method with a Hamming window of length $M = 100$ and overlap of 80%

Fig. 10.23 Estimated phase spectrum $\Theta_{xy}(\omega)$ for the two sequences analyzed in Fig. 10.22. The *dotted lines* and the *black dots* mark the true values of phase lag in the data

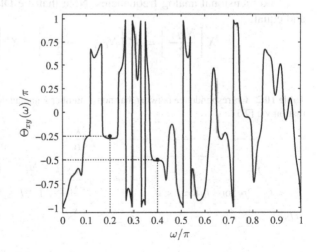

The magnitude squared coherence is greater than 0.8 at $\omega_0 = 0.2\pi$ and $\omega_1 = 0.4\pi$. The phase spectrum provides, at the corresponding frequencies, the phase differences estimated from the cross spectrum, which are close to the true values indicated by *black dots*: indeed, since the ω_0-sinusoid in $y[n]$ lags the ω_1-sinusoid in $x[n]$ by a delay $\pi/4$, the phase-spectrum estimate at ω_0 is close to $-\pi/4$, and so on (compare with what was explained about the correlation of deterministic signals in Sect. 5.6).

Unfortunately, in real-world applications the interpretation of cross-spectral behaviors is often more troublesome than in this simple example.

10.11   Use of the FFT in Power Spectrum Estimation

In the methods based on Schuster's definition of periodogram—periodogram and modified periodogram, Bartlett's and Welch's methods—FFT is an efficient tool to calculate spectral estimates over the discrete set of frequencies

$$\omega_k = \frac{2\pi k}{N_{FFT}}, \qquad \nu_k = \frac{k}{N_{FFT}}, \qquad f_k = \frac{k}{N_{FFT} T_s},$$

where $k = [0, N_{FFT} - 1]$, T_s is the sampling interval of the sequence, and N_{FFT} is the number of spectral estimates. Note that in case of zero-padding, N_{FFT} is different from the number N of data.

To obtain the periodogram, we will select a proper value of N_{FFT}, choosing an integer power of 2 greater than or equal to N. Then we will compute the DFT of the signal via FFT. Table 10.2 gives the precise correspondence between harmonic number k and analog frequency f. This is useful in real-world applications: we perform the analysis in the domain of adimensional frequencies, but when we come to plotting and interpreting our results, we may prefer to go back to the real world, i.e., to dimensional analog frequencies. Note that the DFT values at $\pm f_{Ny}$ are real and equal:

$$X\left[\frac{N_{FFT}}{2}\right] = X^*\left[N_{FFT} - \frac{N_{FFT}}{2}\right] = X^*\left[\frac{N_{FFT}}{2}\right].$$

Table 10.2 Correspondence between harmonic number k and analog frequency f in spectral estimation via FFT

f	k
0	0
$0 < f < f_{Ny}$	$1 \leq k \leq \frac{N_{FFT}}{2} - 1$
$f_{Ny} < f < 2f_{Ny}$ or $-f_{Ny} < f < 0$	$\frac{N_{FFT}}{2} + 1 \leq k \leq N_{FFT} - 1$
$\pm f_{Ny}$	$\frac{N_{FFT}}{2}$

Fig. 10.24 Correspondence
between harmonic number k
and analog frequency f in
spectral estimation via FFT,
in the equivalent intervals
$[0, 2f_{Ny})$ and $[-f_{Ny}, f_{Ny})$: an
example with $N_{FFT} = 8$

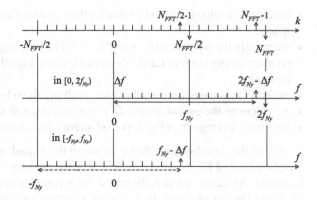

Figure 10.24 shows graphically the same correspondence between harmonic number
and values of f in the equivalent intervals $[0, 2f_{Ny})$ and $[-f_{Ny}, f_{Ny})$. Observe that since
the DFT values at $\pm f_{Ny}$ are equal, we might equivalently write $(-f_{Ny}, f_{Ny}]$.

This figure is very similar to Fig. 4.17, with the only difference that the more general quantity N_{FFT} is used in place of N, to account for possible zero-padding. Once
$X[k]$, $k = [0, N_{FFT} - 1]$, has been calculated, the discrete values of the periodogram
are computed as

$$I_N(\omega)|_{\omega=\omega_k} = \frac{1}{N} |X[k]|^2 = I_N[k],$$

where the length N of the sequence appears. Is it important to keep in mind that in
general $N \neq N_{FFT}$.

For real data, $|X[k]| = |X^*[N_{FFT} - k]|$. For this reason, the spectral content per-
taining to the interval $(f_{Ny}, 2f_{Ny})$ is sometimes reflected into the interval $[0, f_{Ny}]$.
In addition to the previous definition of $I_N(\omega)$, which is referred to as a *two-sided*
spectrum, a *one-sided* spectrum $I'_N[k]$ is then introduced:

$$I'_N[k] = \begin{cases} I_N[k] & \text{for} \quad k = 0 \quad \text{and} \quad k = \frac{N_{FFT}}{2}, \\ 2I_N[k] & \text{for} \quad k = [1, \frac{N_{FFT}}{2} - 1], \end{cases}$$

that is,

$$I'_N[k] = \begin{cases} \frac{1}{N} |X[k]|^2 & \text{for} \quad k = 0 \quad \text{and} \quad k = \frac{N_{FFT}}{2}, \\ \frac{1}{N} \{|X[k]|^2 + |X[N_{FFT} - k]|^2\} = \frac{2}{N_{FFT}} |X[k]|^2 & \text{for} \quad k = [1, \frac{N_{FFT}}{2} - 1]. \end{cases}$$

For constructing a modified periodogram we will specify the desired window and
multiply the sequence by the window before proceeding as described above. For
applying Bartlett's and Welch's methods via FFT we can proceed as follows:

- we choose a value for M and a value for the amount of overlapping. This determines K too;
- we specify the window $w[n], n = [0, M - 1]$. A rectangular window coupled with no overlapping leads to Bartlett's method, while any other choice leads to Welch's method;
- we build the subsequences $x^{(i)}[n]$ and multiply them by $w[n]$;
- we calculate the periodograms $J_M^{(i)}(\omega)$ over a discrete set of frequencies via FFT;
- we finally average over the K periodograms thus obtained.

As for the correlogram, after computing the biased AC estimate $c_{xx}[l]$, we will simply take its FFT. However, in this way we treat the sequence $c_{xx}[l]$ as a causal sequence. As a consequence, the transform is complex and we must take its modulus to obtain the correlogram. In a similar way we can compute BT power spectrum estimates, except that we will apply a (causal) tapered window to $c_{xx}[l]$ before taking the FFT.

10.12 Power Spectrum Normalization

A potential source of confusion in spectral analysis is the fact that in literature, the power spectrum is normalized in several different ways. We will now review the most common normalization schemes.

Let us start by using the harmonic number k or the adimensional frequency ν as the frequency variable. According to Parseval's theorem for the DFT, we can write, for real data and for an FFT over N samples (no zero-padding),

$$\sum_{n=0}^{N-1} |x[n]|^2 = \frac{1}{N} \sum_{k=0}^{N-1} |X[k]|^2.$$

In case of zero-padding, the DFT is computed over $N_{FFT} \geq N$ points and we must write instead

$$\sum_{n=0}^{N-1} |x[n]|^2 = \frac{1}{N_{FFT}} \sum_{k=0}^{N_{FFT}-1} |X[k]|^2.$$

On the other hand,

$$I_N[k] = \frac{|X[k]|^2}{N} \qquad \text{for} \qquad k = [0, N_{FFT} - 1],$$

hence

$$\frac{1}{N_{FFT}} \sum_{k=0}^{N_{FFT}-1} I_N[k] = \frac{1}{N_{FFT}N} \sum_{k=0}^{N_{FFT}-1} |X[k]|^2 = \frac{1}{N} \sum_{n=0}^{N-1} |x[n]|^2,$$

that clarifies how

- for $N_{FFT} = N$ the sum of all spectral estimates is equal to the sum of squares of the data;
- in general, the sum of all spectral estimates, divided by their number N_{FFT}, is equal to the data variance. Thus the mean spectral level is equal to the average power of the sequence, $(1/N) \sum_{n=0}^{N-1} |x[n]|^2$, that for centered data is nothing but the variance.

But since the distance between discrete normalized frequencies is

$$\Delta \nu = \frac{1}{N_{FFT}},$$

what we called the mean spectral level (average power, variance) also appears as the area subtended by the spectral curve, obtained multiplying the sum of the spectral estimates by the constant bin width $1/N_{FFT}$ (or multiplying each spectral estimate by $1/N_{FFT}$ and then summing up).

If we were to use the one-sided spectrum $I_N'[k]$, we would have instead

$$\frac{1}{N_{FFT}} \sum_{k=0}^{N_{FFT}/2} I_N'[k] = \frac{1}{N} \sum_{n=0}^{N-1} |x[n]|^2 =$$

$$= \frac{1}{N_{FFT}} \left(I_N[0] + I_N \left[\frac{N_{FFT}}{2} \right] + 2 \sum_{k=1}^{N_{FFT}/2-1} I_N[k] \right).$$

So far we referred to a power spectrum that we want to plot as a function of k or ν. However, we may want to plot the spectrum as a function of ω; most often, in the case of a real-world discrete-time signal obtained by sampling an analog signal with time step T_s, we may prefer to use analog frequency f. Since $\nu = k/N = fT_s$, spectral normalizations are then used that take into account the fact that

$$\sigma_x^2 = \frac{1}{2\pi} \int_{-\pi}^{+\pi} P_{xx}(\omega)d\omega = \frac{1}{2\pi} \int_{-0.5}^{+0.5} P_{xx}(\nu)2\pi\, d\nu = \int_{-0.5}^{+0.5} P_{xx}(\nu)d\nu =$$

$$= \int_{-f_{Ny}}^{+f_{Ny}} P_{xx}(f)T_s df = \frac{1}{f_s} \int_{-f_{Ny}}^{+f_{Ny}} P_{xx}(f)df.$$

Now, the $I_N[k]$ sequence the we get from FFT could also be indicated as $I_N[\nu_k]$, and is an estimate of $P_{xx}(\nu)$. Therefore, in principle we should adhere to the following normalization rules:

- in spectral plots with ω on the abscissa, on the ordinate we should plot $I_N[k]/(2\pi)$, so that the sum of spectral values multiplied by the bin amplitude $\Delta\omega = 2\pi/N_{FFT}$ gives the data variance;

- in spectral plots with f on the abscissa, on the ordinate we should plot $I_N[k]T_s = I_N[k]/f_s$, so that the sum of spectral values multiplied by the bin amplitude $\Delta f = 1/(N_{FFT}T_s)$ gives the data variance.

Furthermore, in some applications in which spectra of different sequences having different variances must be compared, it may be useful to standardize the spectrum, dividing it by the data variance σ_x^2, so that

- the area subtended to the spectral curve is equal to 1,
- white noise with the same variance of the data would appear—in theory—as a horizontal line at an ordinate of 1.

We are now in a position to conclude specifying the proper measurement units that should appear in a spectral plot. If we call u the measurement unit of the data $x[n]$, the unit for variance and power is u^2. The power spectrum is a distribution of power over frequency, so

- $I_N[k]$ and $I_N(\nu_k)$ plots with k or ν on the abscissa, respectively, should report u^2 as the ordinate unit, since frequency is a bare number in this case; to be precise, in the case of ν, which strictly speaking is expressed in cycles/sample, we should write u^2/(cycles/sample), but this pedantry is normally neglected;
- $I_N(\omega_k)/(2\pi)$ plots with ω on the abscissa should report u^2/(rad/sample) as the ordinate unit;
- $I_N(f_k)T_s = I_N(f_k)/f_s$ plots with f on the abscissa, and with frequencies typically expressed in Hz, should report u^2/Hz as the ordinate unit. Note that the factor $1/f_s = T_s$ serves to bring the spectrum back to its continuous-time-domain definition, which differs from the present discrete-time-domain definition by a factor of T_s (Sect. 3.5).

If a logarithmic scale is used for the ordinate axis in place of the linear scale, then a plot of $10\log_{10}[I_N(\nu_k)]$ should report dB as the ordinate unit.

In the case of $10\log_{10}[I_N(\omega_k)/(2\pi)]$ plotted versus ω, it is common practice to write dB/(radians/sample); in the case of $10\log_{10}[I_N(f_k)/f_s]$ versus f it is usual practice to write dB/Hz, even if this is just a way to loosely refer to the fact that the spectrum $I_N[k]$ was normalized dividing it by the sampling frequency before computing the logarithm.

Finally, a warning: this is what would be conceptually correct. However, for simplicity and internal coherence, *all the spectra appearing in this book are always given in terms of $I_N(\nu) = |X[k]|^2/N$, independently of the frequency variable used in abscissa*. This may be conceptually incorrect, but it is certainly less confusing for the beginner.

References

Allen, M.R., Smith, L.A.: Monte Carlo SSA: detecting irregular oscillations in the presence of colored noise. J. Clim. **9**, 3373–3404 (1996)

Bartlett, M.S.: Smoothing periodograms from time series with continuous spectra. Nature **161**, 686–687 (1948)

Blackman, R.B., Tukey, J.W.: The Measurement of Power Spectra from the Point of View of Communication Engineering. Dover, New York (1958)

Chatfield, C.: The Analysis of Time Series: An Introduction. Chapman and Hall, New York (1975)

Daniell, P.J.: Discussion on symposium on autocorrelation in time series. J. R. Stat. Soc. B **8**, 88–90 (1946)

Hayes, M.H.: Statistical Digital Signal Processing and Modeling. Wiley, New York (1996)

Hess, G.D., Iwata, S.: Measuring and comparing business-cycle features. J. Bus. Econ. Stat. **15**(4), 43244 (1997)

Jenkins, G.M., Watts, D.G.: Spectral Analysis and its Applications. Holden-Day, San Francisco (1968)

Kay, S.M.: Modern Spectral Estimation: Theory and Applications. Prentice-Hall, Englewood Cliffs (1988)

McConnell, M.M., Perez-Quiros, G.: Output fluctuations in the United States: what has changed since the early 1980? Am. Econ. Rev. **90**(5), 1464–1476 (2000)

Mac Lane, S., Birkhoff, G.: Algebra. AMS Chelsea Publishing, New York, NY, USA (1999)

Mann, M.E., Lees, J.M.: Robust estimation of background noise and signal detection in climatic time series. Clim. Change **33**(3), 409–445 (1996)

Marple, S.L.: Digital Spectral Analysis: With Applications. Prentice-Hall, Upper Saddle River (1987)

MacDonald, G.J.: Spectral analysis of time series generated by nonlinear processes. Rev. Geophys. **27**, 449–469 (1989)

Meltzer, J.A., Zaveri, H.P., Goncharova, I.I., Distasio, M.M., Papademetris, X., Spencer, S.S., Spencer, D.D., Constable, R.T.: Effects of working memory load on oscillatory power in human intracranial EEG. Cereb. Cortex **18**, 1843–1855 (2008)

Percival, D.B., Walden, A.T.: Spectral Analysis for Physical Applications. Cambridge University Press, Cambridge (1993)

Richiardone, R., Alessio, S., Canavero, F., Einaudi, F., Longhetto, A.: Experimental study of atmospheric gravity waves and visibility oscillations in a fog episode. Nuovo Cimento C **18**(6), 647–662 (1995)

Schmidt, R.O.: Multiple emitter location and signal parameter estimation. IEEE Trans. Antennas Propag. **34**, 276–280 (1986)

Schuster, S.A.: On the investigation of hidden periodicities. Terr. Mag. **3**, 13–41 (1898)

Sella, L., Vivaldo, G., Groth, A., Ghil, M.: Economic Cycles and Their Synchronization: A Survey of Spectral Properties. Working Paper 105.2013, Fondazione ENI Enrico Mattei (FEEM), Milan, Italy (2013)

Slepian, D.: Prolate spheroidal wave functions, fourier analysis, and uncertainty V: the discrete case. AT&T Tech. J. **57**, 1371–1430 (1978)

Stoica, P., Moses, R.L.: Spectral Analysis of Signals. Prentice Hall, Upper Saddle River (2005)

Stuart, A.: Kendall's Advanced Theory of Statistics. Wiley, New York (2010)

Thomson, D.J.: Spectrum estimation and harmonic analysis. Proc. IEEE **70**, 1055–1096 (1982)

Thomson, D.J.: Time series analysis of Holocene climate data. Phil. Trans. R. Soc. Lond. A **330**, 601–616 (1990a)

Thomson, D.J.: Quadratic-inverse spectrum estimates: applications to palaeoclimatology. Phil. Trans. R. Soc. Lond. A **332**, 539–597 (1990b)

Trefethen, L., Bau, D.: Numerical Linear Algebra. SIAM—Society for Industrial and Applied Mathematics, Philadelphia (1997)

Vautard, R., Yiou, P., Ghil, M.: Singular-spectrum analysis: a toolkit for short, noisy, chaotic signals. Phys. D **58**, 95–126 (1992)

Welch, P.D.: The use of fast fourier transform for the estimation of power spectra: a method based on time averaging over short, modified periodograms. IEEE Trans. Audio Electroacoust. **AU-15**, 70–73 (1967)

Chapter 11
Parametric Spectral Methods

11.1 Chapter Summary

Non-parametric spectral analysis, with its various facets presented in Chap. 10, has both advantages and shortcomings: it is easy to apply and its statistical properties have been studied in depth; on the other hand, the power spectra estimated with non-parametric methods have problems of leakage, and often present insufficient resolution and poor stability. The principal difference between non-parametric methods and those presented in this chapter is that in the former, no assumptions are made on the signal, except for it being WSS and ergodic. Parametric spectral methods assume that the signal satisfies a generating model with known functional form, and then estimate the model's parameters. The signal's spectral estimate is derived from the estimated model. In those cases where the assumed model is a close approximation to the reality, it is no wonder that parametric methods provide more accurate spectral estimates than non-parametric techniques. We will, however, see that in practice parametric methods can be applied without worrying about the true nature of the signal.

Since parametric methods are based on mathematical modeling of persistence, i.e., autocorrelation/autocovariance (AC) in a time series, we will start by briefly presenting the related basic concepts. The models discussed hereafter are known as stochastic models and were originally introduced for the prediction of time-series values ahead into the future. Stochastic modeling means looking at the signal as the hypothetical output of some LTI system, the input of which is white noise.

First, an appropriate class of stochastic models must be selected for the task of spectral analysis. We will focus on autoregressive (AR) models, which are preferred with respect to other model types, such as moving-average (MA) models and the more general autoregressive moving-average (ARMA) models, for reasons that will be explained. An AR model can represent both broadband and narrowband processes, according to the number and values of its parameters, and can therefore fit a wide class of random processes.

© Springer International Publishing Switzerland 2016
S.M. Alessio, *Digital Signal Processing and Spectral
Analysis for Scientists*, Signals and Communication Technology,
DOI 10.1007/978-3-319-25468-5_11

AR models can have different orders—i.e., a different number of parameters—and the next step is to choose the order which is best suited for modeling a given signal. Many methods for order estimation are available in literature, but all are subject to limitations, and above all, they were devised for stochastic modeling aimed at forecasting. In spectral analysis, the order choice is more often based on trial and error, under only very general constraints, such as the fact that the order must not exceed a certain fraction of the number of available data samples. In practice, parametric spectral estimation methods are normally applied to a given data sequence without any prior investigations about the best model order, pragmatically trying different orders and comparing the resulting spectra among them and with estimates obtained by other spectral methods. Nevertheless, we will briefly present the most widely used criteria for model selection, which can be used as indicators in spectral analysis.

The determination of the model parameters then follows. We will examine five different approaches to parameter estimation: the Yule-Walker or autocorrelation method, the covariance and modified covariance methods, Burg's method, and the maximum entropy method (MEM), which is equivalent to the Yule-Walker method but stems from general philosophical considerations in the frame of information theory. We do not include detailed mathematical descriptions of these estimators and do not discuss their statistical properties, since these topics are beyond the scope of an introductory book like the present one.

In stochastic modeling for forecasting purposes, after the order has been chosen and the AR model has been fitted to the data, the model should be tested to check if it satisfies a number of reasonable requirements. In parametric spectral analysis, the pragmatic approach to order selection mentioned above leads to neglecting this diagnostic step in most cases; we will, nonetheless, mention what requirements should be met in principle, and how a model can be tested statistically in this sense.

An AR model of given order and with given parameters is characterized by a precise functional form of the power spectrum, namely, a rational form. Hence, once the modeling procedure is completed, the power spectrum of the signal is estimated by simply incorporating the estimated parameters into this functional form.

Let us observe that when we model a measured WSS random signal for spectral estimation purposes, we do not expect the signal to be exactly that AR signal (process) having the order we selected and the parameter values we determined. If this were true, then we would obtain the true spectrum of the analyzed process. Instead, we aim at an approximate description of our signal, able to provide a good estimate of its power spectrum. Several examples of stochastic models and of their properties, as well as examples of parametric spectral estimates for different kinds of signals, will be given. Comparison with estimates obtained via non-parametric methods will help in highlighting the possibilities offered by parametric methods.

Finally, a warning: this chapter makes use of some basic linear algebra: vectors and matrices and their manipulation, sets of linear equations, special matrix forms, diagonalization of matrices (eigenvalue decomposition), etc. This is because in many of the mathematical developments that will be encountered, it is convenient to use vector and matrix notation, which simplifies equations and makes it easier to understand how they are solved. The reader who has not mastered the basic tools of vector and matrix analysis might find it useful to read the very good summary provided by Hayes (1996), Chap. 2, pp. 20–48, or resort to a book presenting linear algebra from first principles, such as Mac Lane and Birkhoff (1999). As for notation, we will use boldface letters such as $\mathbf{M}$ and $\mathbf{v}$ to indicate matrices and vectors. Superscript T, as in $\mathbf{M}^T$, denotes the transpose of a matrix.

11.2 Signals with Rational Spectra

Rational spectra form a dense set in the class of continuous spectra, but can also represent data records containing sinusoids buried in white noise.

A *rational spectrum* is a rational function of $e^{j\omega}$, i.e., the ratio of two polynomials in $e^{j\omega}$ (Stoica and Moses 2005):

$$P_{xx}(e^{j\omega}) \equiv P_{xx}(\omega) = \frac{\sum_{k=-q}^{q} \gamma_1[k]e^{-j\omega k}}{\sum_{k=-p}^{p} \gamma_2[k]e^{-j\omega k}},$$

where $\gamma_1[k]$ and $\gamma_2[k]$ are two sequences having the even symmetry which characterizes the AC sequence of a real signal, and q and p represent some summation limits. The rational spectral form stems from an extremely general signal decomposition theorem (Wold 1938), known as the *Wold's decomposition theorem* or *Wold's representation theorem*, according to which every WSS causal signal, which we will assume to have zero mean, can be written, under some regularity conditions, as the sum of two components, one deterministic and one stochastic. Formally,

$$x[n] = \sum_{i=0}^{\infty} h_i \zeta[n-i] + \eta[n],$$

where h_i represents a possibly infinite-length sequence of weights or coefficients that we will assume to be absolutely summable, $\zeta[n]$ is a series of uncorrelated variables known as *innovations* (like white noise $e[n]$ with finite variance, which is a WSS process), and $\eta[n]$ is a deterministic signal, i.e., a signal which is exactly predictable from its past, such as a discrete-time periodic sinusoid, or even a constant, like the signal's mean value.

Now, let us consider a purely random zero-mean WSS signal, from which the deterministic component, if any, has already been subtracted. Let us take white noise $e[n]$ as the uncorrelated variables. Then we can write Wold's decomposition in the form

$$x[n] = \sum_{i=0}^{\infty} h_i e[n-i],$$

which is the linear convolution between the series of coefficients h_i and the white noise sequence. The coefficients h_i can now be seen as the samples $h[i]$ of the impulse response of a real and causal LTI filter that is required to be stable. In general, it will be an IIR filter. We further require the filter to have a rational transfer function:

$$H(z) = \frac{B(z)}{A(z)},$$

with

$$A(z) = 1 + a_1 z^{-1} + \cdots + a_p z^{-p},$$
$$B(z) = 1 + b_1 z^{-1} + \cdots + b_q z^{-q}.$$

This is not too restrictive, because even if it were not so, the theory of rational approximation of functions (see, e.g., Powell 1981) states that we can always approximate a continuous function as the transfer function as closely as we want, under relatively large convergence constraints, by a rational function with sufficiently high numerator and the denominator degrees. For stability, the poles of the transfer function must be inside the unit circle.

So, Wold's theorem allows us to represent a WSS random causal signal as the output of an RCSR LTI system that receives WSS white noise as input.[1] The samples $e[n]$ are independent white-noise shocks, so the AC of the input $e[n]$ is $\gamma_{ee}[l] = \sigma_e^2 \delta[l]$, and its power spectrum is constant and equal to σ_e^2. Since the filter transfer function is rational, the output signal spectrum will also be so. We thus see that Wold's decomposition, under the assumption of the RCSR nature of the filter, leads to rational spectra. We will later explain that these rational spectra correspond to modeling the signal using *the most general stochastic model*, indicated as ARMA(p, q). This supports the common practical approach, which is to approximate the spectrum of WSS random signals by the spectrum of signals generated by stochastic models.

Before proceeding, let us summarize our considerations:

- the spectra of a very wide class of causal WSS random signals can be approximated by rational functions;
- this corresponds to seeing the signal as the result of filtering white noise with an LTI RCSR filter with proper order and coefficients;
- this also means viewing the generation of the measured series as a stochastic process, which can be modeled by a general ARMA(p, q) model. The spectrum of the signal is then a rational function of frequency containing the model parameters, so that fitting a model to our data naturally leads to parametric spectral estimation.

Since the power spectrum of a signal is always real and non-negative, the general form of a rational spectrum can be expressed in the parametrized form (see, e.g., Stoica and Moses 2005)

$$P_{xx}(\omega) = \sigma_e^2 \frac{\left|B(e^{j\omega})\right|^2}{\left|A(e^{j\omega})\right|^2} = \sigma_e^2 \left|H(e^{j\omega})\right|^2,$$

---

[1] Since the input is stationary and the filter is stable, the output is also certainly stationary.

where

$$A(e^{j\omega}) = 1 + a_1 e^{-j\omega} + \cdots + a_p e^{-j\omega p},$$
$$B(e^{j\omega}) = 1 + b_1 e^{-j\omega} + \cdots + b_q e^{-j\omega q}$$

are the polynomial numerator and denominator of the filter's frequency response,

$$H(e^{j\omega}) = \frac{B(e^{j\omega})}{A(e^{j\omega})}.$$

This result is referred to as the *spectral factorization theorem* (see, e.g., Stoica and Moses 2005; Kay 1988). In order to prove this result, we can pass to the z-transform domain and use the formulas derived in Sect. 9.9. We consider the case of white noise $e[n]$, with variance σ_e^2 and power spectrum $P_{ee}(\omega) = \sigma_e^2$, being filtered by the filter with transfer function $H(z) = \frac{B(z)}{A(z)}$, thus producing an output $x[n]$. We find indicating the spectrum in the z-plane by $P_{xx}(z)$,

$$P_{xx}(z) = P_{ee}(z)H(z)H^*(1/z^*) = P_{ee}(z)\frac{B(z)B^*(1/z^*)}{A(z)A^*(1/z^*)}.$$

Substituting $z = e^{j\omega}$ we get

$$P_{xx}(\omega) = \sigma_e^2 \frac{B\left(e^{j\omega}\right) B^*\left(e^{j\omega}\right)}{A\left(e^{j\omega}\right) A^*\left(e^{j\omega}\right)} = \sigma_e^2 \frac{\left|B(e^{j\omega})\right|^2}{\left|A(e^{j\omega})\right|^2},$$

q.e.d.

The poles of $P_{xx}(z)$ are determined by the poles of the filter's transfer function. The transfer function $H(z)$ has p poles that occur in complex-conjugate pairs because the coefficients are real (Sect. 3.2.2). Now, if z_i is a pole of $H(z)$ and $P_{xx}(z)$, then $1/z_i^*$ is a pole of $H^*(1/z^*)$, i.e., another pole of $P_{xx}(z)$. Therefore $P_{xx}(z)$ has $2p$ poles that appear in symmetrical (reciprocal complex-conjugate) positions in the z-plane. If $z_i = re^{j\pm\theta}$ is a pair of poles of $H(z)$ and $P_{xx}(z)$, then $1/z_i^* = (1/r)e^{j\pm\theta}$ also represents a pair of poles of $P_{xx}(z)$. For example, if $p = 4$ the poles of $H(z)$ are[2] $z_i = \{re^{\pm j\theta}, r'e^{\pm j\theta'}\}$, with $i = [1, 4]$; the poles of $P_{xx}(z)$ are $z_l = \{re^{\pm j\theta}, r'e^{\pm j\theta'}, (1/r)e^{\pm j\theta}, (1/r')e^{\pm j\theta'}\}$, with $l = [1, 8]$. In $P_{xx}(\omega)$ these poles will produce peaks at $\omega = \{\pm\theta, \pm\theta'\}$, and focusing on the positive frequency half-axis we will see two peaks at $\omega = \theta$ and θ'.

---

[2] We do not need to worry about $B(z)$ and its degree, because the numerator of $H(z)$, representing the polynomial transfer function of an FIR filter, contributes only zeros outside the origin of the z-plane; it has poles in the origin. On the other hand, $1/A(z)$ is an IIR rational transfer function that contributes poles outside the origin and zeros in the origin. In summary, outside the origin the poles of $H(z)$ are exclusively due to the zeros of $A(z)$.

By assumption, $P_{xx}(\omega)$ is continuous, hence it is finite for all ω values; as a result, $A(z)$ must have all its zeros strictly inside the unit circle. The corresponding model is then said to be stable, and corresponds to a stable IIR filter. If we further assume, for simplicity, that $P_{xx}(\omega)$ does not exactly vanish at any ω, then we can choose the polynomial $B(z)$ so that all zeros lie inside the unit circle. The corresponding model is then said to be *minimum phase*; the corresponding filter is minimum-phase,[3] and has a stable inverse.

In summary, an arbitrary rational spectrum can be associated with a WSS causal random signal obtained filtering white noise with variance σ_e^2 through an LTI RCSR filter with transfer function $H(z) = B(z)/A(z)$ and frequency response $H(e^{j\omega}) = B(e^{j\omega})/A(e^{j\omega})$. Such a filtering can be written in the time domain as an LCCDE:

$$x[n] = -\sum_{i=1}^{p} a_i x[n-i] + \sum_{i=0}^{q} b_i e[n-i],$$

and it is assumed that $b_0 = 1$. Note that implicitly we also set $a_0 = 1$. Hence the parametrized model of $P_{xx}(\omega)$ turns out to be a model of the signal itself. The spectral estimation problem can be turned into to a problem of signal modeling, where the model involved is a general ARMA(p, q) model (presented in the next section). The signal's stochastic model includes the filter coefficients, along with a description of the input signal, which is a random signal usually taken to be zero-mean white noise, hence completely described by its variance σ_e^2.

The spectral factorization problem associated with a rational spectrum actually has multiple solutions, with the stable and minimum phase ARMA(p, q) model type being only one of them. We will, however, focus on this kind of solution, since when the final goal is the estimation of the power spectrum, focusing on stable and minimum phase ARMA(p, q) models is not restrictive and is convenient.

11.3 Stochastic Models and Processes

A stochastic model with given order and parameters represents a stochastic, i.e., random process, able to generate a random signal characterized by well-defined spectral properties. *Note that though conceptually the stochastic process and the*

---

[3]A system with rational transfer function is minimum-phase if not only all its poles, but also all its zeros are inside the unit circle, so that both the system and its inverse are causal and stable. A minimum-phase system is called this way because it has an additional useful property: the natural logarithm of the magnitude of the frequency response is related to the phase angle of the frequency response by the Hilbert transform. This implies that for all causal and stable systems that have the same magnitude response, the minimum phase system has its impulse-response energy concentrated near the start of $h[n]$, i.e., it minimizes the delay of the energy in the impulse response. As a result, for all causal and stable systems that have the same magnitude response, the minimum phase system has the minimum group delay.

stochastic model are two distinct entities, in the sense that the model represents the process in an exact or approximated way, the two expressions are often used interchangeably.

Stochastic models/processes play a fundamental role in mathematical modeling of phenomena in many fields of science, engineering, and economics. In particular, the ARMA modeling approach is widely used. ARMA modeling was described in the 1951 thesis of Whittle (1951), and was later popularized by Box and Jenkins in 1970 (see Box et al. 2008). The ARMA model was created as a tool for

- detecting, from a finite-length data record, the characteristics of the underlying random process and thus understanding its nature by revealing something about the mechanism that builds persistence into the series;
- forecasting future values of the series on the basis of past values;
- if desired, removing from the signal the imprint of some known process, so as to get a more random residual signal to which statistical methods, such as methods of spectral estimation, can be applied more pertinently (pre-whitening);
- finally, and more interestingly for our purposes, getting a kind of spectral estimate derived directly from the stochastic model that best fits the signal, etc.

These goals are tackled exploiting and modeling the persistence exhibited by the series. Autocorrelated process have *memory*, that can be long- or short-term, depending on the rate at which absolute AC decreases as the lag between pairs of signal samples increases, i.e., depending on the *persistence* of the AC.[4]

Historically, the first stochastic model dates back to the work of Yule (1927). Studying the motion of a pendulum in a viscous medium, with friction proportional to velocity, Yule observed that the amplitude $s[n]$ of the oscillation could be expressed as a homogeneous difference equation:

$$s[n] + a_1 s[n-1] + a_2 s[n-2] = 0, \qquad n = 0, 1, 2 \ldots$$

the solution of which is a damped harmonic motion. However, the measured values of $s[n]$ are affected by errors. Yule then proposed to drop the usual interpretation of these measured values as the superposition of hypothetical "true" values and random errors, and to describe the pendulum's motion by a non-homogeneous difference equation, in which white noise appears as an external *driving function* determining the pendulum's behavior:

$$s[n] + a_1 s[n-1] + a_2 s[n-2] = e[n], \qquad n = 0, 1, 2 \ldots,$$

an equation in which $p = 2$ delayed signal samples appear. The parameter p is the equation order.

---

[4]When the AC absolute values take much longer to decay than the rate associated with the ARMA class of processes discussed in this chapter, the process is often referred to as having long-term memory or long-range persistence. Therefore the memory associated with ARMA processes is usually classified as short-term memory.

In the language of digital signal processing, Yule's model is the LCCDE of an LTI RCSR filter that transforms the white noise signal $e[n]$ into the observed signal $s[n]$. The filter must be causal and stable. The LCCDE of Yule's system has $M = 0$ and $N = 2$ (see Chap. 3); the filter's frequency response is

$$H(e^{j\omega}) = \frac{1}{A(e^{j\omega})},$$

with

$$A(e^{j\omega}) = 1 + a_1 e^{-j\omega} + a_2 e^{-j2\omega} = 1 + \sum_{k=1}^{p} a_k e^{-j\omega k}, \text{ with } p = N = 2.$$

This is an IIR system corresponding to an *all-pole model*, i.e., a model with poles only.[5] The order $p = 2$ implies the existence of two unknown coefficients a_1 and a_2 that can be determined from the measured values of $s[n]$. Since the LCCDE is given in recursive form, and the filter output $s[n]$ "regresses" onto its own past values, Yule's model is called *autoregressive* and is the prototype of a class of models, referred to as AR(p) models. Their general expression is, for any integer p,

$$x[n] = -\sum_{i=1}^{p} a_i x[n-i] + e[n], \qquad n = 0, 1, 2 \ldots$$

Later on, different types of models were introduced: the *moving average* or MA(q) model and the more general *autoregressive moving-average* or ARMA(p, q) model, also known as the *Box-Jenkins model* (Box et al. 2008). We may also mention that a number of variations of these models exist: for example, an MA model can be seen as the derivative with respect to time of an *integrated moving average* model or IMA model. A model having an AR component and an IMA component is an *autoregressive integrated-moving-average* model, or ARIMA model, which is another kind of Box-Jenkins model, etc. However, since we are only interested in stochastic models for spectral estimation purposes, we will neglect these variations.

11.3.1 Autoregressive-Moving Average (ARMA) Model

The *autoregressive moving-average model* ARMA(p, q) or *Box-Jenkins model* (Box et al. 2008) corresponds to filtering white noise by the most general LTI RCSR system with transfer function having zeros and poles, and with a frequency response given by

---

[5]This expression is common in stochastic model literature, but does not mean that the IIR filter has no zeros. Actually, as we learned in Sect. 3.2.8 (Table 3.1), a rational transfer function with numerator degree $M = 0$ and denominator degree $N = p$ has p poles and no zeros at finite values of z outside the origin of the z-plane, and p zeros in the origin.

$$H(e^{j\omega}) = \frac{B(e^{j\omega})}{A(e^{j\omega})},$$

where the numerator polynomial has order q and the denominator polynomial has order p:

$$A(e^{j\omega}) = \sum_{i=0}^{p} a_i e^{-j\omega i} = 1 + \sum_{i=1}^{p} a_i e^{-j\omega i},$$

$$B(e^{j\omega}) = \sum_{i=0}^{q} b_i e^{-j\omega i} = 1 + \sum_{i=1}^{q} b_i e^{-j\omega i},$$

whith both a_0 and b_0 normalized to 1. Let us stress that the transfer function of an ARMA(p, q) model must have all its poles inside the unit circle for the stability of the filter, and all its zeros inside the unit circle for the existence of a stable inverse filter.

The expression of the ARMA(p, q) model for a zero-mean, causal WSS random signal,

$$x[n] = -\sum_{i=1}^{p} a_i x[n-i] + \sum_{i=0}^{q} b_i e[n-i], \qquad n = 0, 1, 2 \ldots$$

corresponds to a precise functional form of the power spectrum of the output WSS signal $x[n]$. In fact, if we take the z-transform of the previous equation we have

$$A(z)X(z) = B(z)E(z) \qquad \rightsquigarrow \qquad A(e^{j\omega})X(e^{j\omega}) = B(e^{j\omega})E(e^{j\omega}),$$

that leads to the z-plane power spectrum that we already saw in Sect. 11.2,

$$P_{xx}(z) = P_{ee}(z)\frac{B(z)B^*(1/z^*)}{A(z)A^*(1/z^*)} \qquad \rightsquigarrow \qquad \sigma_e^2 \frac{|B(e^{j\omega})|^2}{|A(e^{j\omega})|^2}.$$

This is in agreement with Schuster's definition of periodogram: since $|E(e^{j\omega})|^2 / N = \sigma_e^2$, we have

$$P_{xx}(\omega) = \frac{|X(e^{j\omega})|^2}{N} = \frac{|B(e^{j\omega})|^2 |E(e^{j\omega})|^2}{N|A(e^{j\omega})|^2} = \sigma_e^2 \frac{|B(e^{j\omega})|^2}{|A(e^{j\omega})|^2}.$$

We will now derive the theoretical AC sequence of an ARMA(p, q) signal. Since the output and the input are related by the ARMA model's LCCDE, we can find the AC sequence $\gamma_{xx}[l]$ by multiplying both sides of the LCCDE by $x[n-l]$ and taking the ensemble average. From now on, for brevity we will indicate $\gamma_{xx}[l]$ simply by $\gamma[l]$. In the same way, the normalized AC will be indicated simply by $\rho[l]$. Thus we have

$$\gamma[l] + \sum_{i=1}^{p} a_i \gamma[l - i] = \sum_{i=0}^{q} b_i \mathrm{E}\{e[n - i]x[n - l]\}.$$

It can be shown that $x[n]$ and $e[n]$ are jointly WSS, so that, indicating by $\gamma_{ex}[l]$ the cross-covariance between the filter's input and output, we have

$$\mathrm{E}\{e[n - i]x[n - l]\} = \mathrm{E}\{e[n]x[n - (l - i)]\} = \gamma_{ex}[l - i].$$

We can thus write

$$\gamma[l] + \sum_{i=1}^{p} a_i \gamma[l - i] = \sum_{i=0}^{q} b_i \gamma_{ex}[l - i].$$

The cross-covariance $\gamma_{ex}[l]$ can be expressed in terms of $\gamma[l]$ and of the impulse response $h[n]$ of the ARMA IIR filter, which, being the inverse Fourier transform of the frequency response, is in general a nonlinear function of the coefficients a_i and b_i. In fact, in the time domain the filter output can be written as a linear convolution:

$$x[n] = e[n] \star h[n] = \sum_{m=-\infty}^{+\infty} e[m]h[n - m],$$

with $h[n] = \mathrm{IDFT}\{H(e^{j\omega})\}$. Hence we can write

$$\gamma_{ex}[l - i] = \mathrm{E}\{e[n - i]x[n - l]\} = \mathrm{E}\left\{ \sum_{m=-\infty}^{+\infty} e[n - i]e[m]h[n - l - m] \right\} =$$

$$= \sum_{m=-\infty}^{+\infty} \left(\mathrm{E}\{e[n - i]e[m]\}\right)h[n - l - m] = \sigma_e^2 h[i - l].$$

The last equality follows from the fact that $\mathrm{E}\{e[n - i]e[m]\} = \sigma_e^2 \delta[n - i - m]$, because $e[n]$ is white noise; so, in the summation only the term with $n - m = i$ survives.[6] We then have

$$\gamma[l] + \sum_{i=1}^{p} a_i \gamma[l - i] = \sigma_e^2 \sum_{i=0}^{q} b_i h[i - l].$$

---

[6] The expression for $\gamma_{ex}[l]$ can also be directly derived from the formula for the input-output cross-covariance of an LTI system, reported in Sect. 9.9:

$$\gamma_{ex}[l] = \sum_{k=-\infty}^{+\infty} h[k]\gamma_{ee}[l + k] = \sigma_e^2 \sum_{k=-\infty}^{+\infty} h[k]\delta[l + k] = \sigma_e^2 h[-l] \rightsquigarrow \gamma_{ex}[l - i] = \sigma_e^2 h[i - l].$$

The sum on the right-hand side of this equation, which we denote by $c[l]$, can be written as

$$c[l] = \sum_{i=0}^{q} b_i h[i-l] = \begin{cases} \sum_{i=0}^{q-l} b_{i+l} h[i] & \text{for} \quad l \leq q, \\ 0 & \text{for} \quad l > q. \end{cases}$$

Here we changed $i - l$ into i and took into account that the filter is causal, hence $h[i] = 0$ for $i < 0$. In conclusion, for $l \geq 0$ we get the set of equations

$$\gamma[l] + \sum_{i=1}^{p} a_i \gamma[l-i] = \begin{cases} \sigma_e^2 c[l] & \text{for} \quad 0 \leq l \leq q, \\ 0 & \text{for} \quad l > q, \end{cases}$$

which are called the *Yule-Walker equations* (Yule 1927; Walker 1931) for the ARMA(p, q) process (YW equations hereafter).[7] They define a recursion for the AC sequence in terms of the filter coefficients.

If we have a finite set of estimated samples of $\gamma[l]$, as happens when we are dealing with a finite-length section of a sample sequence, and if we know the filter's coefficients, we can use the YW equations to extrapolate the AC sequence. Conversely, YW equations can be used to estimate the filter coefficients from a known $\gamma[l]$. However, due to the product $b_{i+l} h[i]$ appearing in the expression of $c[l]$, the YW equations for the general ARMA model are nonlinear, and solving them is, in general, a difficult task.

11.3.2 Autoregressive (AR) Model

This special type of ARMA(p, q) process results when $q = 0$. In this case $x[n]$ is generated filtering white noise with an all-pole filter with transfer function

$$H(z) = \frac{1}{1 + \sum_{i=1}^{p} a_i z^{-i}} = \frac{1}{A(z)};$$

the frequency response is

$$H(e^{j\omega}) = \frac{1}{1 + \sum_{i=1}^{p} a_i e^{-j\omega i}} = \frac{1}{A(e^{j\omega})}.$$

The power spectrum of the signal is then

$$P_{xx}(\omega) = \sigma_e^2 \frac{1}{\left| A(e^{j\omega}) \right|^2},$$

---

[7]These equations have been named variously in the literature, and are also referred to as *normal equations*, or *Wiener-Hopf equations*.

or, in terms of the z-transfom,

$$P_{xx}(z) = P_{ee}(z) \frac{1}{A(z)A^*(1/z^*)}.$$

The general AR(p) model of a zero-mean, causal WSS random process $x[n]$ can be written as

$$x[n] = -\sum_{i=1}^{p} a_i x[n-i] + e[n] = \sum_{i=1}^{p} \alpha_i x[n-i] + e[n],$$

where the coefficients α_i are obviously related to the a_i by a simple sign reversal,

$$\alpha_i = -a_i.$$

The YW equations for an AR(p) process may be derived from those of the ARMA(p, q) general case by setting $q = 0$: the sequence $c[l]$ reduces to a single sample $c[0] = b_0 h[0]$, i.e., $c[0] = h[0]$ since $b_0 = 1$, and we get

$$\gamma[l] + \sum_{i=1}^{p} a_i \gamma[l-i] = \sigma_e^2 h[0]\delta[l].$$

Due to the initial value theorem for causal impulse responses we have

$$h[0] = \lim_{z\to\infty} H(z) = 1,$$

hence

$$\gamma[l] + \sum_{i=1}^{p} a_i \gamma[l-i] = \sigma_e^2 \delta[l] \quad \text{for} \quad l \geq 0.$$

If we write these equations in matrix form for $l = 1, \ldots p$, after using the symmetry property of the AC sequence, i.e., $\gamma[l] = \gamma[-l]$, we obtain

$$\begin{bmatrix} \gamma[0] & \gamma[1] & \gamma[2] & \cdots & \gamma[p-1] \\ \gamma[1] & \gamma[0] & \gamma[1] & \cdots & \gamma[p-2] \\ \gamma[2] & \gamma[1] & \gamma[0] & \cdots & \gamma[p-3] \\ \vdots & \vdots & \vdots & \vdots & \vdots \\ \gamma[p-1] & \gamma[p-2] & \gamma[p-3] & \cdots & \gamma[0] \end{bmatrix} \begin{bmatrix} a_1 \\ a_2 \\ a_3 \\ \vdots \\ a_p \end{bmatrix} = - \begin{bmatrix} \gamma[1] \\ \gamma[2] \\ \gamma[3] \\ \vdots \\ \gamma[p] \end{bmatrix}$$

where the matrix on the left, which will be indicated by $\boldsymbol{\Gamma}_p$, has p rows and p columns. If moreover we define the two column vectors

$$\mathbf{a}_p = \begin{bmatrix} a_1 & a_2 & a_3 & \ldots & a_p \end{bmatrix}^T,$$

$$\boldsymbol{\gamma}_p = \begin{bmatrix} \gamma[1] & \gamma[2] & \gamma[3] & \ldots & \gamma[p] \end{bmatrix}^T,$$

we can write synthetically the YW equations as

$$\boldsymbol{\Gamma}_p \mathbf{a}_p + \boldsymbol{\gamma}_p = 0.$$

Note that given all the AC samples that are needed, the YW equations determine a unique set of coefficients a_i, as long as the $p \times p$ matrix $\boldsymbol{\Gamma}_p$ is positive-definite, hence non-singular.[8] If the autocorrelation matrix is singular, then the corresponding random variables $x[n]$ are linearly dependent. This means that if we measure a set of N successive samples of $x[n]$, then all the future samples can be computed recursively; the process is fully predictable. It is the exact opposite of a white noise process, which has no correlation between any pair of samples. In the frequency domain, for a white noise process the power spectrum is constant, whereas for a fully predictable process, the power spectrum can only have lines—it cannot have any continuous component. Without entering into these details, from now on we will simply assume non-singularity of the AC matrix. This is not too restrictive, since in case of singularity, we can think of progressively eliminating the linear dependence among random variables until we finally arrive at a smaller set of random variables, which do not have linear dependence. The correlation matrix of these random variables is then non-singular, and we can solve a smaller set of normal equations in a unique manner. Note also that the matrix $\boldsymbol{\Gamma}_p$ is Toeplitz.[9] This is a consequence of the WSS property of the signal; indeed, it can be shown that any positive definite Toeplitz matrix is the autocorrelation matrix of some WSS process. For a more detailed mathematical description of the properties of the AC matrix in the frame of stochastic modeling see, e.g., Vaidyanathan (2008).

Returning to the YW equations, and considering also $l = 0$, we can augment the σ_e^2 information to the YW equations. This is obtained by simply moving the right-hand side term of the YW equations to the left and adding an extra row at the top of the matrix $\boldsymbol{\Gamma}_p$. As a result, we get the *augmented Yule-Walker equations* in matrix form:

$$
\begin{bmatrix}
\gamma[0] & \gamma[1] & \gamma[2] & \cdots & \gamma[p-1] & \gamma[p] \\
\gamma[1] & \gamma[0] & \gamma[1] & \cdots & \gamma[p-2] & \gamma[p-1] \\
\gamma[2] & \gamma[1] & \gamma[0] & \cdots & \gamma[p-3] & \gamma[p-2] \\
\vdots & \vdots & \vdots & \vdots & \vdots & \vdots \\
\gamma[p] & \gamma[p-1] & \gamma[p-2] & \cdots & \gamma[1] & \gamma[0]
\end{bmatrix}
\begin{bmatrix}
1 \\
a_1 \\
a_2 \\
a_3 \\
\vdots \\
a_p
\end{bmatrix}
=
\begin{bmatrix}
\sigma_e^2 \\
0 \\
0 \\
0 \\
\vdots \\
0
\end{bmatrix}
$$

---

[8] In linear algebra, a square ($n \times n$) symmetric real matrix $\mathbf{M}$ is said to be positive definite if the quadratic form $\mathbf{v}^T \mathbf{M} \mathbf{v}$ is positive for every non-zero column vector $\mathbf{v}$ of n real numbers. The symbol $\mathbf{v}^T$ denotes the transpose of $\mathbf{v}$, i.e., the corresponding row vector.

[9] In linear algebra, a Toeplitz matrix (named after Otto Toeplitz), or diagonal-constant matrix, is a matrix in which each descending diagonal from left to right is constant, i.e., all the elements on any line parallel to the main diagonal are identical.

where this time the matrix on the left has $p + 1$ rows and columns. Synthetically, these augmented equations can be written as

$$\boldsymbol{\Gamma}_{p+1} \begin{bmatrix} 1 \\ \mathbf{a}_p \end{bmatrix} = \begin{bmatrix} \sigma_{e,p}^2 \\ \mathbf{0}. \end{bmatrix}$$

where we used the column vectors of zeros

$$\mathbf{0} = [0 \quad 0 \quad 0 \quad \ldots \quad 0]^T$$

having p elements. To stress the dependence on p, here we wrote $\sigma_{e,p}^2$ for the white-noise variance resulting from modeling at the order p.

We might use the YW equations to generate the AC sequence from a given set of model parameters. On the other hand, if AC estimates are known from lag 0 to lag p, i.e., if the first $p + 1$ samples of the theoretical AC sequence have been estimated, we can exploit the YW equations to derive estimates of the model parameters. The YW equations are linear in the coefficients a_i, so it is a simple matter to find the coefficients:

$$\mathbf{a}_p = -\boldsymbol{\Gamma}_p^{-1} \boldsymbol{\gamma}_p.$$

Later we will find $\sigma_{e,p}^2$ from the first row of the augmented equations:

$$\sigma_{e,p}^2 = \gamma[0] + \sum_{i=1}^{p} a_i \gamma[i].$$

This approach to modeling is referred to as the YW method.

Let us stress that in most applications, the theoretical AC sequence $\gamma[l]$ is unknown, and must be estimated from a sample realization of the process. The best way to do so is taking the biased AC estimate, $c_{xx}[l]$. Then we can derive estimates of the model coefficients. Indeed, it can be assumed (Stoica and Moses 2005) that the YW system has a unique solution, not only when the theoretical AC elements are used, but also when we substitute them with their biased estimates. The estimated quantities should be clearly distinguished from theoretical ones using the symbols $\hat{\gamma}[i]$, $\hat{a}_i$, and $\hat{\sigma}_e$. In the following discussion, however, we will let the context clarify whether we refer to true values or to estimates, and will use these symbols as seldom as possible, to avoid weighing down the notation.

The set of YW equations for $l > 0$, which is a system of linear difference equations with order p:

$$\gamma[l] = -\sum_{i=1}^{p} a_i \gamma[l - i],$$

is often written in terms of the AC coefficient (normalized covariance):

$$\rho[l] = -\sum_{i=1}^{p} a_i \rho[l - i]).$$

If the filter is known, the system can be solved for $\rho[l]$ (or $\gamma[l]$) using the roots of the associated characteristic polynomial. In this way we can gain insight into the behavior of $\rho[l]$ versus l in the AR(p) model.

Let us take, for example, an AR(2) model. A general solution for $\rho[l]$ in the YW set of equations can then be obtained using the roots z_1 and z_2 of the associated characteristic polynomial, i.e., in terms of the solutions $z_{1,2}$ of the equation

$$z^2 + a_1 z + a_2 = 0.$$

The filter's transfer function is

$$H(z) = \frac{1}{1 + a_1 z^{-1} + a_2 z^{-2}} = \frac{1}{z^{-2}\left(z^2 + a_1 z + a_2\right)} = \frac{z^2}{z^2 + a_1 z + a_2};$$

the term z^2 in the numerator gives two zeros of $H(z)$ at $z = 0$, while the term $z^2 + a_1 z + a_2$ in the denominator gives two poles of $H(z)$, the positions of which are given by the roots of $z^2 + a_1 z + a_2 = 0$. If such roots, which are

$$z_{1,2} = \frac{-a_1 \pm \sqrt{a_1^2 - 4a_2}}{2},$$

satisfy the condition $|z_{1,2}| < 1$, then the all-pole filter is stable and the AR(2) process is stationary, i.e., it has constant (zero) mean, constant variance and autocovariance depending only on lag.[10]

The general solution for $\rho[l]$ can behave in different ways. Three cases are possible:

---

[10]If the roots lie *on* the unit circle, the AR process will only be stationary in case of noise being identical to zero. In that case a harmonic process will result, consisting of a sum of cosine functions. We might wonder what will happen if the AR process has poles *very close* to the unit circle. As poles on the unit circle represent a harmonic process, an AR process with poles near the unit circle can be expected to demonstrate some kind of pseudo-periodic behavior (Priestley 1994). In this case the AC can be described as a sum of weakly damped sinusoids. Furthermore, the AR process may exhibit a kind of almost-non-stationary behavior, because the transfer function will be close to instability in the filtering sense. The pole locations will also affect the reliability of the various parameter estimation techniques. It was claimed by Priestley (1994) that YW equations may lead to poor parameter estimates, even for moderately large data samples, if the AR operator has a pole near the unit circle.

1. if z_1 and z_2 are real and distinct roots, then

$$\rho[l] = c_1 z_1^l + c_2 z_2^l \qquad \text{for} \qquad l = 0, 1, 2, \ldots$$

where c_1 and c_2 are constants that can, for instance, be obtained from $\rho[0]$ and $\rho[1]$. Moreover, since stationarity requires $|z_1|, |z_2| < 1$, in this case the sequence of AC coefficients is a mixture of two terms decaying exponentially with increasing lag;

2. if z_1 and z_2 are complex-conjugate roots of the form $r \pm js$, then we have

$$\rho[l] = R^l \left[c_1 \cos(\lambda l) + c_2 \sin(\lambda l) \right] \qquad \text{for} \qquad l = 0, 1, 2, \ldots,$$

where $R = |z_i| = \sqrt{r^2 + s^2}$ and λ is determined by $\cos \lambda = r/R$, $\sin \lambda = s/R$. Therefore we have $r \pm js = R (\cos \lambda \pm j \sin \lambda)$. Again, c_1 and c_2 are constants. In this case the AC coefficient sequence has the form of a damped sinusoid with damping factor R and angular frequency λ;

3. if there are two coincident real roots, setting $z_0 = z_1 = z_2$ we have

$$\rho[l] = (c_1 + c_2 l) z_0^l \qquad \text{for} \qquad l = 0, 1, 2, \ldots,$$

where c_1 and c_2 are constants. In this case the AC coefficient sequence has an exponentially decaying form.

For any order, if all roots are real and distinct, then

$$\rho[l] = c_1 z_1^l + c_2 z_2^l + \cdots + c_p z_p^l \qquad \text{for} \qquad l = 1, 2, \ldots,$$

where $c_1, c_2, \ldots c_p$ are constants. We thus understand that the AC of a stationary AR(p) process can be a mixture of exponential-decay terms and damped sinusoidal terms, depending on the values of the roots.

We may observe that the AR(1) process has $a_2 = 0$, and therefore the characteristic equation is $z + a_1 = 0$; therefore there is only one root, $z_1 = -a_1$. The sequence $\rho[l]$ decays as z_1^l for increasing lag.

11.3.3 Moving Average (MA) Model

Another special case of the ARMA(p, q) process arises when $p = 0$. In this case, the signal is generated by filtering white noise with a filter with transfer function

$$H(z) = B(z) = 1 + \sum_{i=1}^{q} b_i z^{-1}$$

and frequency response

$$H(e^{j\omega}) = B(e^{j\omega}) = 1 + \sum_{i=1}^{q} b_i e^{-j\omega i}.$$

This system is an FIR system or an *all-zero model*.[11] Hence an MA(q) model is always stable. The power spectrum is

$$P_{xx}(\omega) = \sigma_e^2 \left| B(e^{j\omega}) \right|^2,$$

or, in terms of the z-transform,

$$P_{xx}(z) = P_{ee}(z) B(z) B^*(1/z^*).$$

The MA(q) model,

$$x[n] = \sum_{i=0}^{q} b_i e[n - i],$$

expresses the signal $x[n]$ as a linear superposition of delayed white-noise shocks $e[n - i]$, each weighted according to the value of the coefficient b_i.

Why is this model called "moving average"? The name "moving average" is somewhat misleading, because the weights in the sum on left-hand side of the previous equation need not total unity, nor be positive. However, each value of $x[n]$ can be thought of as a weighted moving average of past random shocks; the MA nomenclature is in common use, and therefore we employ it.

The YW equations for an MA(q) process may be found from the AC of the ARMA model by setting $p = 0$, $a_0 = 1$, $a_i = 0$ for $i > 0$, and $h[n] = b_n$. For non-negative values of l not exceeding q we therefore have

$$c[l] = \sum_{i=0}^{q-l} b_{i+l} b_i,$$

hence

$$\gamma[l] = \begin{cases} \sigma_e^2 \sum_{i=0}^{q-|l|} b_{i+|l|} b_i & \text{for} \quad |l| \leq q, \\ 0 & \text{for} \quad |l| > q, \end{cases}$$

where this time we have used the absolute value on l so that the expression for $\gamma[l]$ is valid for lags of both signs, provided that $|l| \leq q$. For $|l| > q$ the AC $\gamma[l]$ vanishes.

---

[11] This does not mean that the FIR filter has no poles. Actually (see Table 3.1), a rational transfer function with numerator degree $M = q$ and denominator degree $N = 0$ has p zeros and no poles at finite values of z outside the origin of the z-plane, and p poles in the origin.

The AC sequence of an MA(q) process is thus equal to zero for all values of l that lie outside the interval $[-q, q]$, i.e., AC "cuts off" after lag q. We thus see that variance is a constant:

$$\sigma_x^2 = \gamma[0] = \left(1 + b_1^2 + b_2^2 + \cdots b_q^2\right) \sigma_e^2.$$

This is a zero-mean process which is always stationary, independently of the weights. In fact, the mean is zero as for $e[n]$, the variance is constant and the AC is a function of lag only.

For the MA(q) model, the AC sequence $\gamma[l]$ depends nonlinearly on the MA parameters b_i. How does this affect the opportunity of using MA modeling for parametric spectral estimation?

One method to estimate an MA(q) spectrum consists of two steps:

- estimating the q parameters b_i and σ_e^2, and
- inserting the estimated parameters in the formula for the spectrum of the MA(q) process.

The difficulty with this approach lies in the first step, which is a nonlinear estimation problem. Therefore, unlike the case of an AR(p) process, estimating the coefficients of an MA process and using them to construct a spectral estimate is a nontrivial problem. AR modeling thus appears easier to tackle than the general ARMA case or the MA case.

Another method to estimate an MA spectrum is based on the re-parametrization of the MA(q) spectrum in terms of the AC sequence. We know that in the MA case $\gamma[l]$ vanishes for lags external to the interval $[-q, q]$. Owing to this simple observation, the definition of the spectrum as a function of AC turns into a finite-dimensional spectral model:

$$P_{xx}(\omega) = \sum_{l=-q}^{q} \gamma[l] e^{-j\omega l}$$

Hence a simple estimator of the MA(q) spectrum is obtained by inserting a proper estimate of the AC sequence into this formula. If the biased estimate $c_{xx}[l]$ is used, then *the MA spectral estimator is the correlogram*, computed using a window of length $2q + 1$. This is not unexpected: the correlogram with a rectangular window of length $2q + 1$ implicitly *assumes* that the AC samples at lags outside the window interval are equal to zero; this is precisely the *effective* behavior of the MA(q) model. Thus the spectral estimate of the spectrum of the MA(q) signal will be unbiased. A disadvantage of this approach to MA(q) spectral estimation is that owing to the (implicit) use of a rectangular window in the formula for $P_{xx}(\omega)$, the estimate we obtain is not necessarily positive at all frequencies. Indeed, it is often noted in applications that this approach to the MA(q) spectrum produces negative estimates. In order to cure this deficiency, we might use another window ensuring only positive spectral estimates; however, the corrected estimator would no longer be an unbiased estimator of the spectrum of an MA(q) signal.

11.3.4 How the AR and MA Modeling Approaches Are Theoretically Related

MA and AR processes are close relatives. Indeed, starting from Wold's decomposition theorem (Sect. 11.2), it can be shown that the theorem's expression represents an infinite-order MA process from which both finite-order MA and AR processes can be derived (Montgomery et al. 2008). In other words, any stationary random signal can be expressed as an infinite weighted sum of past and present white-noise shocks. The MA(∞) process, containing an infinite number of weights, is theoretically important for its generality but it is useful practically only in the following special cases:

- finite-order MA(q) processes, in which only a finite number q of weights are different from zero;
- finite-order AR(p) processes that derive from those infinite-order MA processes the weights of which are not independent of each other and can therefore be generated using a finite number p of parameters;
- processes that are a mixture of a finite-order MA process and a finite-order AR process, i.e., ARMA(p, q) processes, which provide a parsimonious description of a WSS stochastic process, due to the combination of AR and MA characteristics. In fact, Wold's decomposition can be shown to summarize in itself, i.e., in the MA(∞) process, the most general ARMA(p, q) process.

For an MA(∞) process, at each instant, among the infinite past noise shocks, only a finite number will contribute significantly to the present value of the process $x[n]$, and the time window that frames these contributing shocks progressively will advance in time, making the older shocks obsolete for the purpose of determining $x[n]$. Some processes can intrinsically have this dynamic with a reasonably small finite number of contributing past shocks. For other processes, the influence of past shocks can persist for so long that the estimate of a very high number of coefficients—a number tending to infinity—may be necessary: too many coefficients will certainly be needed with respect to the amount of the data that is available for their estimation. A solution to this problem is offered by AR(p) processes that can be derived assuming that the infinite coefficients b_i of the MA(∞) process are not independent of one another, but rather depend on i according to some mathematical formula and are therefore representable on the basis of a finite number of parameters. It is reasonably assumed that, though the MA model order is infinite, the contributions of the more remote shocks are smaller than those provided by more recent ones. Since the shocks are independent random variables with equal probability distributions, it can simply be assumed that the infinite coefficients progressively decrease as their index increases, reflecting the gradually smaller influence of remote shocks as we go back into the past. This leads from the MA(∞) process to finite-order AR processes. Conversely, using the partial AC sequence (PAC) that will be defined in Sect. 11.6.1, it can be shown that an MA(q) process satisfying some conditions allows for a representation in terms of an absolutely summable AR(∞) process; being more general, in the frame of the Wiener-Kolmogorov theory of extrapolation and interpolation of random sequences

and processes (Yaglom 2004), a theorem exists (Kolmogorov 1941), which states
that any WSS random process can be modeled as an AR(∞) process. Since we are
only interested in stochastic models for the purpose of spectral estimation, we will
not discuss these topics.

11.3.5 First-Order AR and MA Models: White, Red and Blue Noise

We will now give some more details on first-order AR and MA processes. This offers
the opportunity to introduce types of noise different from the white one, namely red
and blue noises, which are produced by AR(1) and MA(1) processes.

Let us compare the AC and the power spectrum of the AR(1) and MA(1) processes
with those of white noise.

1. **White noise**:
   a stationary random process represented by the model

$$x[n] = e[n]$$

   with

   - $E\{e[n]\} = 0$,
   - $E\{e[n]^2\} = \sigma_e^2$ = constant,
   - $E\{e[n]e[k]\} = 0$ for $k \neq n$.

2. **AR(1) process**:
   a random process represented by the model

$$x[n] = -a_1 x[n-1] + e[n] = \alpha x[n-1] + e[n],$$

   where we wrote $\alpha = \alpha_1 = -a_1$. This process is stationary if $|\alpha| < 1$; for $|\alpha| \geq 1$
   the process has statistical averages that increase or decrease in time, according to
   α being positive or negative, and therefore is not stationary. This non-stationarity
   is due to the instability of the corresponding all-pole filter and is illustrated in
   Fig. 11.1, in which the considered values of α are shown in each panel. We see
   that for $\alpha = -1.5$, for example, the signal values rapidly fall toward $-\infty$; for
   $\alpha = +1$ the mean value is constant but the variance increases with n and tends
   to infinity; the same happens, a fortiori, in the case of $\alpha = 1.5$, with very large
   oscillations.

   Figure 11.2 shows examples of six stationary signals generated by AR(1) processes. Each of them was obtained by filtering white noise with $\sigma_e^2 = 1$ by an
   AR(1) filters with the value of α shown in the corresponding panel. $N = 1024$
   samples of 50 realizations of each of the six processes were generated for later use,

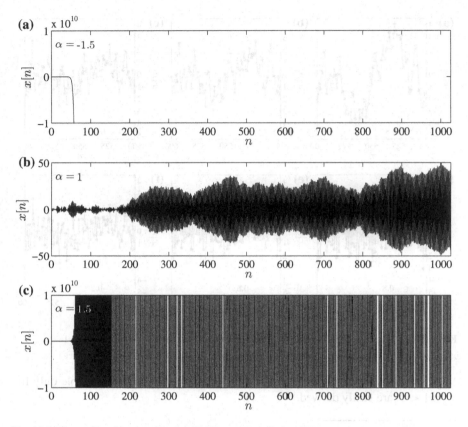

Fig. 11.1 Examples of non-stationary AR(1) processes ($|\alpha| \geq 1$)

but to avoid cluttering, Fig. 11.2 shows only 100 samples of a single realization of each of the six processes. The mean and the variance of these processes visibly remain stable.

For the AR(1) process,

- $E\{x[n]\} = 0$,
- $E\{x[n]^2\} = \sigma_x^2 = \sigma_e^2/\left(1 - \alpha^2\right)$,
- $E\{x[n]x[n-l]\} = \gamma[l] = \sigma_x^2 \alpha^l = \sigma_e^2 \alpha^l/\left(1 - \alpha^2\right)$.

Hence the AC coefficient sequence is

$$\rho[l] = \frac{\gamma[l]}{\sigma_x^2} = \alpha^l = (\rho[1])^l.$$

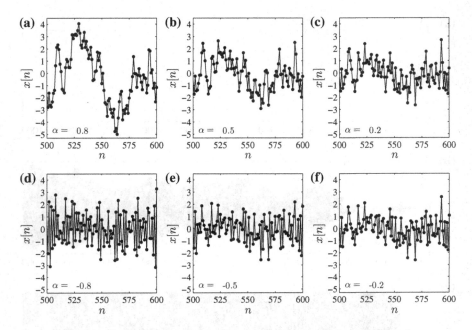

Fig. 11.2 Examples of AR(1) stationary signals with various values of α, with $|\alpha| < 1$

These equations, which highlight that actually the model is reasonable only for $|\alpha| < 1$, are easily derived.[12]

---

[12]The equation $\sigma_x^2 = \sigma_e^2 / \left(1 - \alpha^2\right)$ can be obtained as follows:

$$\mathrm{E}\left\{x[n]^2\right\} = \mathrm{E}\left\{(\alpha x[n-1] + e[n])^2\right\} =$$
$$= \alpha^2 \mathrm{E}\left\{x[n-1]^2\right\} + 2\alpha \mathrm{E}\left\{x[n-1]e[n]\right\} + \mathrm{E}\left\{e[n]^2\right\} = \alpha^2 \mathrm{E}\left\{x[n-1]^2\right\} + \sigma_e^2,$$

because the white-noise input and the output signal are uncorrelated.

If we substitute the model expression into the last equation, i.e., if we use $x[n-1] = \alpha x[n-2] + e[n-1]$, and then iterate, we get

$$\mathrm{E}\left\{x[n]^2\right\} = \sigma_e^2 + \alpha^2 \sigma_e^2 + \alpha^4 \sigma_e^2 + \cdots + \alpha^{2l} \mathrm{E}\left\{x[n-l]^2\right\}$$

that for increasing l becomes, for a centered $x[n]$ for which $\mathrm{E}\left\{x[n]^2\right\} = \sigma_x^2$,

$$\sigma_x^2 = \sigma_e^2(1 + \alpha^2 + \alpha^4 + \cdots) = \frac{\sigma_e^2}{1 - \alpha^2}.$$

In a similar way, the formula for the AC at lag l could be derived. We can also derive these formulas directly from YW equations:

$$\gamma[l] + a_1 \gamma[-1] = \sigma_e^2 \delta[l]$$

The AC sequence of an AR(1) process does not go to zero for $|l| > 0$: the process has memory. However, since $|\rho[1]| \leq 1$, the AC absolute values decrease for increasing l and asymptotically go to zero. This can be seen in Fig. 11.3a–f, where the AC coefficient series, estimated for 50 realizations of each of the processes illustrated in Fig. 11.2a–f, respectively, and then averaged over realizations (*black dots*) is compared with the expected one, calculated as $\rho[l] = \alpha^l = (\rho[1])^l$ (*gray lines*). The AC sequences for the six AR(1) processes show exponentially decreasing absolute values. They are always non-negative for positive α, while they oscillate in sign for negative α. This can be understood observing, for example, the two realizations with $\alpha = 0.8$ and $\alpha = -0.8$ (Fig. 11.3a, d, respectively). For $\alpha = 0.8$, we can observe short intervals in which the signal values tend to move upwards or downwards. Positive values of $\rho[l]$ can therefore be expected. On the contrary, the AR(1) signal having $\alpha = -0.8$ "jumps up and down nervously", and negative values of $\rho[l]$ appear.

AR(1) processes are aperiodic, and have a smooth spectrum given by

$$P_{xx}(\omega) = \frac{1}{N} \left| X(e^{j\omega}) \right|^2 = \frac{1}{N} \frac{\left| E\left(e^{j\omega}\right) \right|^2}{\left| 1 + a_1 e^{-j\omega} \right|^2},$$

where $E\left(e^{j\omega}\right)$ is the DTFT of $e[n]$. As derived by Bartlett (1955), this expression can be rewritten as

---

(Footnote 12 continued)
provides

$$\gamma[0] + a_1 \gamma[l - 1] = \sigma_e^2 = \gamma[0] + a_1 \gamma[1]$$

and since $\gamma[0] = \sigma_x^2$,

$$\sigma_x^2 = \sigma_e^2 - a_1 \gamma[1] = \sigma_e^2 + \alpha \gamma[1].$$

Also,

$$\gamma[1] + a_1 \gamma[0] = 0, \qquad \gamma[1] = -a_1 \sigma_x^2 = \alpha \sigma_x^2,$$

hence

$$a_1 = -\frac{\gamma[1]}{\gamma[0]} = -\rho[1], \qquad \alpha = \rho[1].$$

Proceeding further, we have

$$\gamma[2] + a_1 \gamma[1] = 0, \qquad \gamma[2] = -a_1 \gamma[1] = \alpha \gamma[1] = \alpha^2 \sigma_x^2,$$

and by iterating, we deduce that in general

$$\gamma[l] = \alpha^l \sigma_x^2 = \rho[1]^l \sigma_x^2, \qquad \rho[l] = (\rho[1])^l = \alpha^l.$$

Moreover we can write

$$\sigma_x^2 = \sigma_e^2 + \alpha \gamma[1] = \sigma_e^2 + \alpha^2 \sigma_x^2, \qquad \sigma_x^2 = \frac{\sigma_e^2}{1 - \alpha^2},$$

and finally

$$\gamma[l] = \alpha^l \sigma_x^2 = \frac{\sigma_e^2 \alpha^l}{1 - \alpha^2}.$$

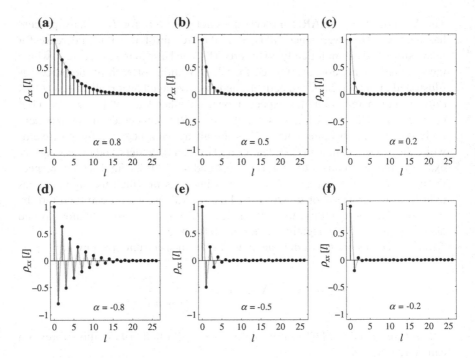

Fig. 11.3 a–f AC coefficient sequences of the AR(1) processes with different values of the parameter α illustrated in Fig. 11.2a–f, respectively. The coefficients (*black dots*) are calculated averaging over 50 realization of each process, each realization being $N = 1024$-samples long. The *gray lines* represent the corresponding theoretical values

$$P_{xx}(\omega) = \frac{\sigma_e^2}{1 + \rho^2[1] - 2\rho[1]\cos\omega}.$$

In many applications it is convenient to replace the numerator σ_e^2, the knowledge of which implies that the AR(1) model that best fits the data has been found, by its expression as a function of σ_x^2 that can be directly estimated from the data record. Therefore we will write $\sigma_e^2 = \sigma_x^2 \left\{ 1 - \rho^2[1] \right\}$, and

$$P_{xx}(\omega) = \frac{\sigma_x^2 \left\{ 1 - \rho^2[1] \right\}}{1 + \rho^2[1] - 2\rho[1]\cos\omega}.$$

Figure 11.4a–f shows the pole-zero plots of the six AR(1) transfer functions involved in this example. As required for filter stability, and therefore for the stationarity of the processes, in all cases we find one pole inside the unit circle. This pole is real and has abscissa equal to $\alpha = -a_1$; in polar form, its magnitude is $r = |\alpha|$ and its frequency is 0 for positive values of α, π for negative values of α.

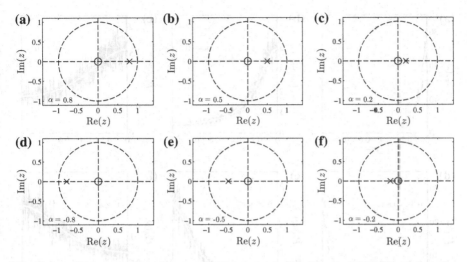

Fig. 11.4 a–f Pole-zero plots of the transfer functions of each of the six AR(1) filters generating the signals that appear in Fig. 11.2a–f, respectively. Zeros are represented by *circles* and poles by *crosses*

Figure 11.5a–f shows the power spectrum estimated averaging over 50 realizations of each of the processes appearing in Fig. 11.2a–f, respectively. Each estimate (*black curve*) is compared with the theoretical expectation (*gray curve*). These plots show that according to the sign of α, low or high frequencies are dominant in the power spectrum, i.e., the spectrum peaks at zero frequency, or at $\omega = \pi$. Thus, the AR(1) filter can act on white noise as a lowpass or a highpass filter. In conclusion,

- if $\alpha > 0$, then the spectrum has dominant low frequency components, and the process is referred to as *red noise*, in analogy with red light;
- if $\alpha < 0$, then the spectrum has dominant high frequency components, and the process is referred to as *blue noise*, in analogy with blue light;
- if $\alpha = 0$ the process is *white noise*, because as in white light, all frequencies are equally represented.[13]

3. **MA(1) process**:
   its model is

   $$x[n] = e[n] + \alpha e[n-1],$$

   were we set $\alpha = b_1$. The process represented by this model is stationary for any value of α, since the corresponding FIR filter is always stable. Figure 11.6a–f

---

[13]Recall that, in general, the noise that is not white is termed "colored noise" and has a smooth spectrum with more power at some frequencies with respect to others, in analogy to colored light. A typical colored noise spectrum does not contain narrowband features associated with periodic or quasi-periodic signal components (Sect. 10.2).

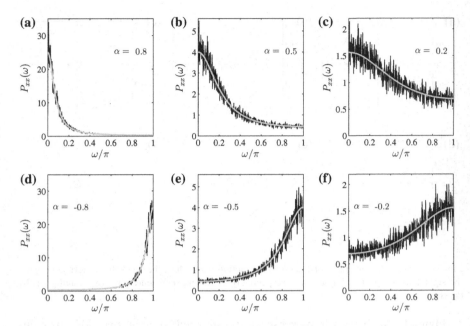

Fig. 11.5 a–f Power spectra of the AR(1) processes with different values of the parameter α illustrated in Fig. 11.2a–f, respectively. These spectra have been calculated via periodograms on 50 realizations of each process (each realization being $N = 1024$-samples long) and then averaged over realizations. The average is represented by the *black curve* in each panel; the *gray curve* shows the theoretical spectrum

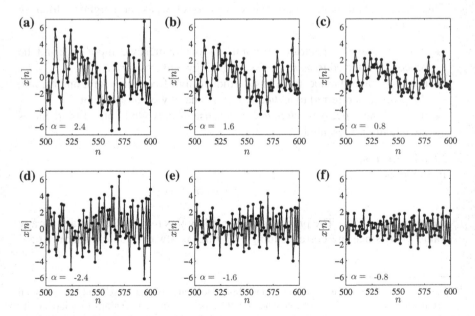

Fig. 11.6 Examples of MA(1) signals with various values of α

shows, for different values of α, examples of signals generated by an MA(1) process. Unit variance was assumed for the white-noise input to each MA(1) filter.

For a centered MA(1) process,

- $E\{x[n]\} = 0,$
- $E\{x[n]^2\} = \sigma_x^2 = \sigma_e^2(1 + \alpha^2),$
- $E\{x[n]x[n - l]\} = \gamma[l] =$
  - for $l = 0$, $\sigma_e^2(1 + \alpha^2)$;
  - for $l = \pm 1$, $\alpha\sigma_e^2$;
  - for different values of l, $\gamma[l]$ is identically zero.[14]

Hence the AC coefficient sequence is

$$\rho[l] = \frac{\gamma[l]}{\sigma_x^2} = \frac{\gamma[l]}{\sigma_e^2(1 + \alpha^2)} = \begin{cases} \rho[0] = 1, \\ \rho[\pm 1] = \frac{\alpha}{1 + \alpha^2}, \\ \rho[l] = 0 \quad \text{for} \quad |l| > 1. \end{cases}$$

The expressions given above clarify that no stationarity constraint actually exists on the value of α: none of the averages of the MA(1) process goes to infinity for some particular value of α. A visual inspection of the plots in Fig. 11.6a–f shows that the mean and the variance remain stable, as expected.

Figure 11.7a–f compares the AC coefficient series (*black dots*), estimated by averaging over 50 realizations of the signals illustrated respectively in Fig. 11.6a–f (each realization being 1024-samples long), with the corresponding theoretical behavior (*gray lines*). Unlike the AR(1) process, the MA(1) process has an AC sequence that goes identically to zero very rapidly. The estimated $\rho[l]$ has absolute value greater than zero at $l = 0$ and $l = 1$. Note that the AC coefficient at lag 1 is positive for $\alpha > 0$ and negative for $\alpha < 0$. Actually, if we observe, for example, the case with $\alpha = 0.8$ in Fig. 11.6a we see sections of the signal in which the values of subsequent samples tend to follow each other for short time intervals, a behavior that suggests some positive AC, even if these sections are shorter than

---

[14]The calculation of the AC is straightforward:

$$E\{x[n]x[n + l]\} = E\{(e[n] + \alpha e[n - 1])(e[n + l] + \alpha e[n + l - 1])\} =$$
$$= E\{e[n]e[n + l]\}$$
$$+ \alpha E\{e[n - 1]e[n + l]\} + \alpha E\{e[n]e[n + l - 1]\} + \alpha^2 E\{e[n - 1]e[n + l - 1]\}.$$

The right-hand terms are all zero, except when the indexes of the noise samples involved are equal. Independently of the index, each term of the kind $E\{(e^2[n + l])\}$ is equal to σ_e^2, so we can write

· for $l = 0$, $E\{x[n]^2\} = \sigma_e^2(1 + \alpha^2)$,
· for $l = 1$, $E\{x[n]x[n + 1]\} = \alpha\sigma_e^2$,
· for $l = -1$, $E\{x[n]x[n - 1]\} = \alpha\sigma_e^2$,
while for different values of l we find zero.

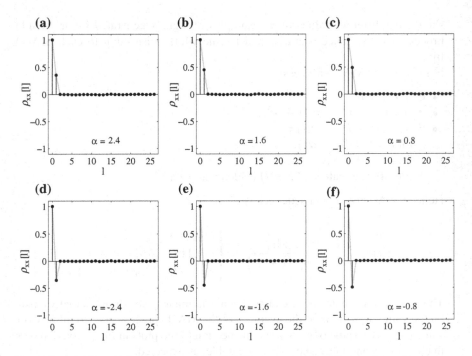

Fig. 11.7 **a–f** AC coefficient sequences of the MA(1) processes with different values of the parameter α illustrated in Fig. 11.6a–f, respectively. The coefficients (*black dots*) are calculated averaging over 50 realization of each process, each realization being $N = 1024$-samples long. The *empty lines* represent the corresponding theoretical values

those observed in the AR(1) signal of Fig. 11.2a. In the case with $\alpha = -0.8$ in Fig. 11.6d the signal tends to oscillate more rapidly, a behavior that suggests some negative AC.

Figure 11.8 shows the zero-pole plots of the MA(1) filters involved in the present example. Note that all poles are in the origin of the z-plane, and the zeros are real and equal to $-\alpha$. For the power spectrum we have

$$P_{xx}(\omega) = \sigma_e^2(1 + \alpha^2 + 2\alpha \cos \omega).$$

This formula can be found in the same manner as the formula for the power spectrum of the AR(1) process, but also directly through the transform of the AC sequence that is particularly simple in this case:

$$P_{xx}(\omega) = \sum_{l=-1}^{1} \sigma_e^2(1 + \alpha^2)\rho[l]e^{-j\omega l} =$$

$$= \sigma_e^2(1 + \alpha^2)\left[1 + \frac{\alpha}{1 + \alpha^2}\left(e^{-j\omega} + e^{j\omega}\right)\right] = \sigma_e^2(1 + \alpha^2 + 2\alpha \cos \omega).$$

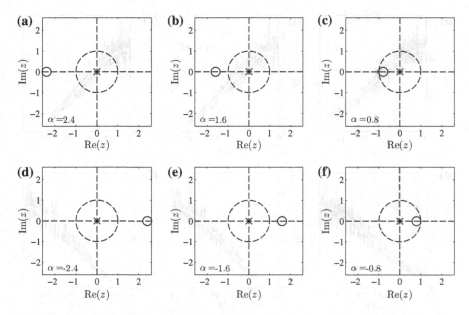

Fig. 11.8 a–f Pole-zero plots of the transfer functions of each of the six MA(1) filters generating the signals that appear in Fig. 11.6a–f, respectively. Zeros are represented by *circles* and poles by *crosses*

Figure 11.9a–f shows the power spectra estimated from 50 realizations of the signals presented respectively in Fig. 11.6a–f and then averaged (*black curves*), in comparison with the theoretical spectra (*gray curves*).

Inspection of these figures tells us that

- if $\alpha > 0$, then the MA(1) process represents *red noise*;
- if $\alpha < 0$, then the MA(1) process represents *blue noise*;
- if $\alpha = 0$, then the MA(1) process represents *white noise*.

However, in our discussion, AR(1) noise is more important than MA(1) red noise, since it typically arises in many stochastic natural processes, in geophysics (Allen and Smith 1996) and other fields in physics, but also in economics (see, e.g., Sella et al. 2013; McConnel and Perez-Quiros 2000; Hess et al. 1997), in medicine and biology (see, e.g., Meltzer et al. 2008), etc. Thus, AR(1) red noise is often assumed as a background spectrum, a null-hypothesis against which spectral peaks are statistically tested to ascertain their significance. We already did this in Sect. 10.8 and will do it again in the following chapters. From now on, when we speak about red noise without specifying anything more, we mean AR(1) red noise.

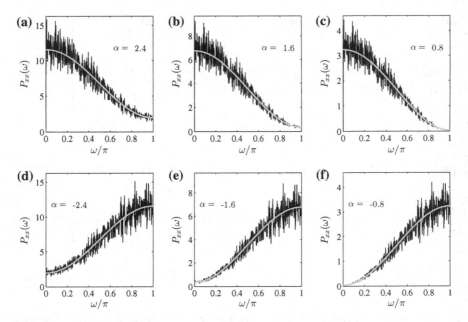

Fig. 11.9 a–f Power spectra of the MA(1) processes with different values of the parameter α illustrated in Fig. 11.6a–f, respectively. These spectra have been calculated via periodograms on 50 realizations of each process (each realization being $N = 1024$-samples long) and then averaged over realizations. The average is represented by the *black curve* in each panel; the *gray curve* shows the theoretical spectrum

11.3.6 Higher-Order AR Models

Here we provide a few graphical examples of AR processes with $p > 1$. This will offer us the possibility of observing some features that are interesting in view of the use of these model for spectral estimation. In Fig. 11.10, 11.11, 11.12 and 11.13 we see, respectively:

- examples of sequences representing a single realization of two AR(2) processes with different parameter values (Fig. 11.10a, b);
- the corresponding AC coefficients (averages over 50 realizations of each process; Fig. 11.11a, b);
- the corresponding poles and zeros of the AR(2) transfer function (Fig. 11.12a, b);
- the corresponding power spectra (averages over 50 realizations of each process), in comparison with their theoretical behavior (Fig. 11.13a, b);
- examples of sequences representing a single realization of two AR(4) processes with different parameter values (Fig. 11.10c, d);
- the corresponding AC coefficients (averages over 50 realizations of each process; Fig. 11.11c, d);
- the corresponding poles and zeros of the AR(4) transfer function (Fig. 11.12c, d);

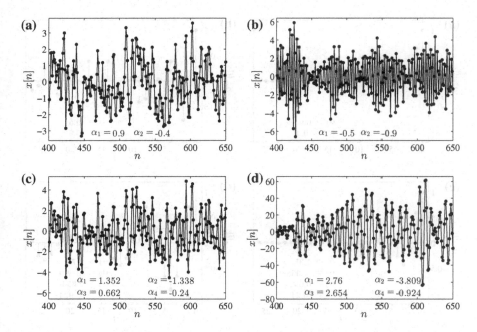

Fig. 11.10 Examples of **a, b** AR(2) signals with different values of the model parameters, and **c, d** AR(4) signals with different values of the model parameters

- the corresponding power spectra (averages over 50 realizations of each process) compared with their theoretical behavior (Fig. 11.13a, b).

All these signals were generated filtering white noise with unit variance and include $N = 1024$ samples.

1. **AR(2) processes**

   For the process of Fig. 11.10a, $\alpha_1 = 0.9$ and $\alpha_2 = 0.4$, i.e., $a_1 = -0.9$ and $a_2 = 0.4$. Thus $a_1^2 - 4a_2$ is positive and the roots of the characteristic polynomial are real and distinct. The AC sequence (Fig. 11.11a) is a mixture of two exponential decays. For the process of Fig. 11.10b, $\alpha_1 = -0.5$ and $\alpha_2 = -0.9$, i.e., $a_1 = 0.5$ and $a_2 = 0.9$. Thus $a_1^2 - 4a_2$ is negative and the roots of the characteristic polynomial are a complex-conjugate pair. The AC sequence behaves like a damped sinusoid (Fig. 11.11b).

   Let us now look at the zero-pole plots (Fig. 11.12a, b). We can see that as required for stationarity, for both AR(2) models the poles lie within the unit circle. In Fig. 11.12a they fall at the complex-plane points $0.4500 \pm j0.4444$, i.e., $0.6325e^{\pm j0.2480\pi}$. In the spectrum, we thus expect a peak at $\omega = 0.2480\pi$ on the positive half-axis. In Fig. 11.12b the poles fall at the complex-plane points $-0.2500 \pm j0.9152$, i.e., $0.9487e^{\pm j0.5849\pi}$. In the spectrum, we thus expect a peak at $\omega = 0.5849\pi$ on the positive half-axis. This is verified in Fig. 11.13a and b,

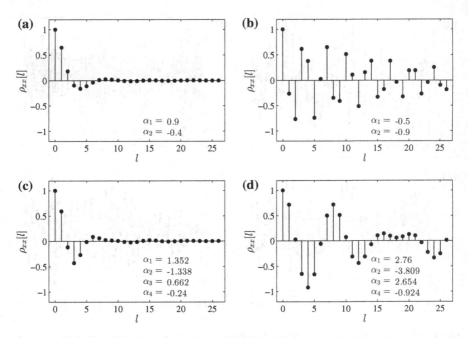

Fig. 11.11 a, b AC coefficient sequences of the AR(2) processes illustrated in Fig. 11.10a, b, respectively. **c, d** AC coefficient sequences of the AR(4) processes illustrated in Fig. 11.10c, d, respectively. The coefficients are calculated averaging over 50 realization of each process, each realization being $N = 1024$-samples long

in which the *vertical dotted lines* identify the frequency at which the peaks are expected to fall for the particular AR(2) processes considered, on the basis of the transfer-function's poles.

2. **AR(4) processes**

For the process of Fig. 11.10c, the model's parameters are such that the AC sequence (Fig. 11.11c) is a mixture of two exponential decays. For the process of Fig. 11.10d, the AC sequence behaves like a damped sinusoid (Fig. 11.11d).

The poles of both AR(4) models (Fig. 11.12c, d) lie within the unit circle. In Fig. 11.12c they fall at the complex-plane points $0.1809 \pm j0.7593$ and $0.4951 \pm j0.4953$. In polar form, the poles are $0.6995e^{\pm j0.4168\pi}$ and $0.7003e^{\pm j0.2500\pi}$. In the spectrum, on the positive frequency half-axis we thus expect two peaks at $\omega = 0.4168\pi$ and $\omega = 0.2500\pi$. In Fig. 11.12d they fall at $0.6231 \pm j0.7593$ and $0.7569 \pm j0.6204$. In polar form, the poles are $0.9822e^{\pm j0.2813\pi}$ and $0.9787e^{\pm j0.2185\pi}$. In the spectrum, on the positive frequency half-axis we thus expect two peaks at $\omega = 0.2813\pi$ and $\omega = 0.2185\pi$. This is verified in Fig. 11.13c, d, in which the *vertical dotted lines* identify the frequency at which the peaks are expected to fall.

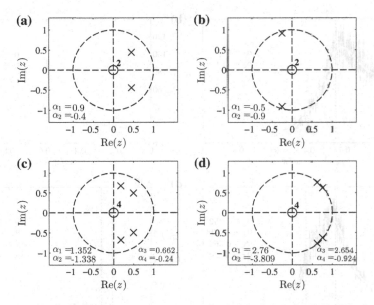

Fig. 11.12 **a, b** Pole-zero plots of the transfer functions of each of the two AR(2) filters generating the signals that appear in Fig. 11.10a, b, respectively. **c, d** Pole-zero plots of the transfer functions of each of the two AR(2) filters generating the signals that appear in Fig. 11.10c, d, respectively. Zeros are represented by *circles* and poles by *crosses*

From Fig. 11.13 we note that the AR power spectrum with order greater than one can exhibit localized and acute peaks. AR spectra can represent both broadband and narrowband features. Indeed, an AR(p) model can represent a wide class of processes, including broadband and narrowband processes, depending on its parameters and on the consequent position of its poles. More precisely, the bandwidth and the peakedness of the spectrum of an AR(p) process depends on the *magnitude r* of its poles, as can be seen comparing Fig. 11.13a, for which $r = 0.6325$, with Fig. 11.13b, for which $r = 0.9487$. A similar argument holds for Fig. 11.13c, for which the pole magnitudes are about 0.70, in comparison with Fig. 11.13d, for which the pole magnitudes are about 0.98: the closer a pole is to the unit circle, the narrower the corresponding spectral peak.

To further illustrate this important point, Fig. 11.14 shows the spectra of several AR(2) processes. The transfer function of each process has a pair of poles at frequencies $\pm 0.4\pi$, but with different magnitudes in the range 0.07–0.97 (*light to dark curves*). We can see that as the poles radii increase towards 1, so that the poles tend to reach the unit circle in the z-plane, the processes gradually pass from being wideband to being narrowband. We can conclude that:

- white noise, a sequence of random uncorrelated variables with zero mean and finite constant variance, has a flat spectrum, so that each frequency equally contributes to the signal variance;

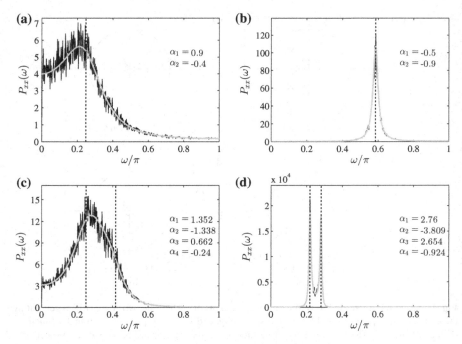

Fig. 11.13 **a, b** Power spectra of the AR(2) processes illustrated in Fig. 11.10a, b, respectively. These spectra have been calculated via periodograms on 50 realizations of each process (each realization being $N = 1024$-samples long) and then averaged over realizations. The average is represented by the *black curve* in each panel; the *gray curve* shows the theoretical spectrum. The *vertical dotted lines* indicate the frequency at which the peak is expected to fall for the particular AR(2) processes being considered, on the basis of the absolute value of the phase angle of its poles. **c, d** Power spectra of the AR(4) processes illustrated in Fig. 11.10c, d, respectively. The *vertical dotted lines* indicate the frequency at which the peaks are expected to fall for the particular AR(4) processes being considered, on the basis of the two different absolute values assumed by the pole's phase angles

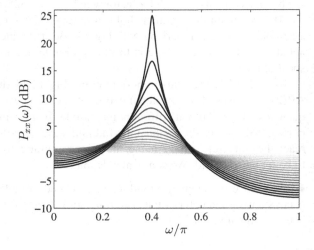

Fig. 11.14 Power spectra of several AR(2) processes with different parameters, chosen in such a way that the magnitudes of the transfer function's poles increase linearly from 0.07 (*light gray curve*) to 0.97 (*black curve*). All processes share the same pole frequencies $\pm 0.4\pi$

- autocorrelated processes have either colored noise spectra, or spectra that are narrowband to a greater or lesser extent. They can be obtained filtering white noise with a filter with proper coefficients. For example, an AR(1) filter has three coefficients: $b_0 = 1$, $a_0 = 1$ and $a_1 = -\rho[1] = -\alpha$, and is the filter that takes a white noise realization as input and transforms it into a realization of an AR(1) process.

11.4 The AR Approach to Spectral Estimation

Rational spectral estimates can be obtained using ARMA(p, q) models, as well as simpler AR(p) models, while the MA(q) approach is related to the non-parametric approach to spectral estimation.

The main difference between all-pole AR(p) and all-zero MA(q) models lies in the fact that the AC sequence of an AR(p) signal decays towards zero only asymptotically for increasing lag, and therefore to compute its DTFT and obtain the power spectrum via correlogram we must truncate the AC sequence. This implies estimation errors. The AC sequence of an MA(q) signal instead has finite length $2q + 1$, and therefore if the available data segment is long enough, we can estimate all the non-zero AC values and use them to estimate the power spectrum via correlogram. We can thus conclude that in principle

- only processes that are truly MA(q) are suitable for spectral estimation via non-parametric methods;
- AR(p) and/or ARMA(p, q) processes naturally require a parametric approach to spectral estimation.

The model most frequently assumed for this purpose is the all-pole model AR(p), for a number of reasons, the main of which are:

- the methods for estimating the parameters of an AR(p) model require the solution of a set of linear equations. Efficient algorithms exist for this task, like the Levinson-Durbin algorithm (Sect. 11.6.3). On the contrary, ARMA methods are rarely used, as they generally result in a set of equations which are nonlinear with respect to the MA parameters;
- any continuous spectrum can be approximated arbitrarily well by an AR(p) model, provided that p is properly chosen: the class of AR processes is rich and flexible enough to represent a wide collection of signal types. In particular, any ARMA(p, q) signal can be approximated satisfactorily by an AR(p) model with sufficiently high order, so that if enough data is available for estimating many parameters, assuming an AR model when a more general ARMA model would be more appropriate is not a severe problem. Therefore we can safely apply AR parametric methods without worrying about the true stochastic nature of the process;
- when high resolution is needed, AR estimators, which correspond to all-pole filters, represent a good choice, in that they are able to reflect spectral behaviors with acute and localized peaks, due to the functional form of their power spectrum.

11.5   AR Modeling and Linear Prediction

We now restrict our attention to AR modeling and show that it can also be interpreted as a strategy for *linear prediction*.

Let us imagine we want to predict the value that the WSS random signal $x[n]$ will assume at the time n, on the basis of p past values, from $x[n - p]$ to $x[n - 1]$, and that we want to apply a *linear forecasting strategy* at some order p:

$$\hat{x}[n] = -\sum_{i=1}^{p} a_i x[n - i],$$

where $\hat{x}[n]$ is the predicted signal sample. The *forward prediction error*, i.e. the error in predicting a signal value from p past values, is

$$\varepsilon_f[n] = x[n] - \hat{x}[n],$$

where $x[n]$ is the measured value at time n. Note that $\varepsilon_f[n]$ is itself a WSS random signal. If this random error is white noise,

$$\varepsilon_f[n] = e[n],$$

then the forecast process is an AR(p) process, satisfying

$$x[n] = -\sum_{i=1}^{p} a_i x[n - i] + e[n] = \sum_{i=1}^{p} \alpha_i x[n - i] + e[n] = \hat{x}[n] + e[n].$$

Therefore the sequence $e[n]$, extracted from a white-noise process with variance unknown a priori, in this approach represents a *modeling error*, a *residual*, i.e., the difference between the signal and the modeled persistence $-\sum_{i=1}^{p} a_i x[n - i] = \sum_{i=1}^{p} \alpha_i x[n - i]$. Note that if the error is white, or nearly white in practical cases, the spectral information on $x[n]$ will be mostly contained in the coefficient a_i. The model is an approximate but close representation of the WSS process $x[n]$, which assumes that the residuals become white or nearly white for reasonably large orders.

In this frame, the corresponding polynomial $A(z)$ is often referred to as the transfer function of the *forward prediction error filter*, which is the reciprocal of the transfer function of the model filter. From previous discussions we know that the forward prediction error filter is stable: it is the FIR filter that responds to the signal $x[n]$ giving in output the forward prediction error, i.e., the filter that transforms the signal into white noise. The white noise samples are referred to as *innovations*.

It can be shown that if we require to minimize the expected value of the forward prediction error variance

$$\sigma_f^2 = \mathrm{E}\left\{\varepsilon_f^2[n]\right\} = \mathrm{E}\left\{e^2[n]\right\} = \sigma_e^2,$$

then we find that the minimization is produced by the same coefficients that are solutions to the YW equations (Stoica and Moses 2005). In other words, the YW equations can be interpreted as the equations that must be satisfied when the optimum linear predictor in the forward-prediction-error least-squares sense is sought. Since $\sigma_e^2 = E\{e^2[n]\}$, this predictor is called the *minimum mean square error predictor* or MMSE predictor. Note that the mean square value, i.e., variance of $x[n]$ equals the mean square value of the prediction plus and the prediction square error:

$$\sigma_x^2 = E\{\hat{x}^2[n]\} + \sigma_e^2.$$

In the frame of different stochastic modeling algorithms, it is also useful to define a *backward prediction error* as the error that occurs when predicting past signal values, i.e., predicting the sample $x[n - p - 1]$ from its future values $x[n - p]$, $x[n - p + 1]$, ... $x[n - 1]$:

$$\hat{x}_b[n - p - 1] = -\sum_{k=1}^{p} a'_k x[n - k].$$

The backward prediction error is

$$\varepsilon_b[n] = x[n - p - 1] - \hat{x}_b[n - p - 1].$$

Both forward and backward predictions use the same samples $x[n - p]$, $x[n - p + 1]$, ... $x[n - 1]$, and in both cases the error is assigned to time n. The backward predictor filter weights are assumed to be different from the forward case, but then it can be shown that they actually are simply an order-reversed version of the forward-case weights:

$$a'[i] = a[p + 1 - i] \quad \text{for} \quad 1 \le i \le p.$$

The FIR filter with these coefficients is known as the *backward prediction error filter*, and its response to the input signal $x[n]$ is the backward prediction error. The mean square values of the forward and backward prediction errors are identical (Vaidyanathan 2008).

An *estimated* AR linear predictor of the same order p can subsequently be written, in which $\hat{a}_i$ are the AR-parameter estimates and $\hat{e}[n]$ are the estimated innovations, obtained solving the YW equations in which the theoretical AC of the signal is substituted with its biased estimate. Here we must clearly distinguish the theoretical AR predictor from the corresponding estimated one: we are now talking about predicting a signal by AR-model-parameter estimation, even if for simplicity we omit the "hat" symbols.

A final remark: the MMSE predictor is defined for *any* WSS discrete-time signal, independently of the process being truly AR, or not, and the corresponding coefficients are always obtained solving the YW equations. But when the considered signal is a true AR(p) signal, then—and only then—the prediction error is white noise.

11.6 AR Modeling Procedure

The AR modeling problem for a discrete-time random WSS signal is the following: given a finite-length part of a sample sequence $(x[n], n = [0, N - 1])$, the set of coefficients of the difference equation that describes the signal, i.e. the set of *model parameters*, must be found under the constraints of stability and minimum phase on the model filter, and in the hypothesis that the input $e[n]$ is a white-noise sequence.

Let us stress once again that when model properties are to be derived, it is assumed that the model holds exactly for the signal to which the model is applied. On the other hand, when the model is applied to a measured random signal, normally the signal is not supposed to necessarily obey the model exactly. The modeling is aimed at providing an approximate description of the signal, i.e., a fit to the signal. The quality of the fit can then be measured according to several criteria, also called *error measures*.

The steps for stochastic AR modeling are the following:

- finding the proper model order;
- estimating the parameters of the model having the chosen order, which provide the "best" approximation to the given signal;
- testing the model.

There are many ways to define what is meant by the "best" approximation to a signal, and, depending upon the definition that is used, there will be different solutions to the modeling problem, along with different techniques for finding the model parameters (Sect. 11.6.3).

As explained above, in practice a signal $x[n]$ that is not truly AR can nevertheless be modeled by an AR process. The fact that $x[n]$ is not really AR means that $e[n]$ (the forward prediction error in the frame of linear prediction) never gets white, but its power spectrum often gets flatter and flatter as the order increases. In this case, the model is the $AR(p)$ approximation to $x[n]$, in the sense that the first $p + 1$ AC samples for the two processes (the model and the measured signal) are equal to each other. So, as the approximation order increases, more and more values of AC are matched to one another. If the measured random process *is* a true AR process of some order, then the model at that order will represent it *exactly*, and the theoretical AC sequence for the random process $x[n]$ and for the model will be exactly equal at all lags l.

11.6.1 Model Order Selection

In modeling and forecasting AR applications, the order is often selected using an iterative procedure that tests the performances of all fitted models from $p = 1$ up, judging for each order the quality of fit according to some criterion and using this judgment to stop the procedure at the order that minimizes the error measure corresponding to that criterion. Usually the fitting process is guided by the principle of parsimony, by which the best model is the simplest possible model—the model with the fewest parameters—that adequately describes the data.

A number of different model-order estimation techniques have been proposed in literature. All of them are somewhat limited in terms of their robustness and accuracy. Selecting the AR model order is equivalent to *model identification*, a more general term that is used when ARMA models are a priori considered, and not just AR models as in the present case.

AR model identification can not only be done by automated iterative procedures, i.e., by fitting many models of increasing orders and using a goodness-of-fit statistic to select the best one, but also by a classical method that consists in looking at plots of *partial autocorrelation* (PAC; the definition is given below).

Since from now on we will be dealing with predictors at different orders, we need to systematically adopt a slightly heavier notation than before: we will use additional subscripts to indicate the order of the model to which a given variable refers.

Identification by visual inspection of partial autocorrelation (PAC)

The classical method of model identification described by Box et al. (2008) for the general ARMA case consists of judging the appropriate model structure and orders from the appearance of the plotted AC and PAC. In the AR case, however, the AC can be a mixture of exponential-decay terms and damped sinusoidal terms. It does not go to zero beyond any finite lag and therefore provides no clue about the appropriate model order. To actually identify the latter, we need a diagram which will have a more distinctive shape when the series is actually AR(p) with some value of p. The PAC plot is such a diagram.

To define PAC, suppose we fit an AR(l) model to our data:

$$x[n] = -\sum_{i=1}^{l} a_i x[n-i] + e[n] = \sum_{i=1}^{l} \alpha_i x[n-i] + e[n].$$

The coefficient α_l is the estimate of the coefficient of $x[n-l)]$. Rewriting the model as

$$x[n] - \sum_{i=1}^{l-1} \alpha_i x[n-i] = \alpha_l x[n-l] + e[n],$$

we see α_l as a plausible estimate of the correlation between $x[n-l]$ and that part of $x[n]$ which cannot be forecast from $x[n-1], \ldots x[n-(l-1)]$. The coefficient α_l is called the *partial autocorrelation* (PAC) between $x[n]$ and $x[n-l]$, i.e., the PAC at lag l. It represents the estimated correlation between $x[n]$ and $x[n-l]$ after the effects of all intermediate x-samples on this correlation have been removed. If the series is actually AR(p), then the (theoretical) PAC will be zero for $l > p$. Thus, we can use the PAC plot to identify the order of the AR model. If the PAC cuts off for lag greater than p, then the appropriate model order is p.

In other words, given a time series $x[n]$, the PAC at lag l is the AC between $x[n]$ and $x[n - l]$ that is not accounted for by lags 1 to $l - 1$, inclusive. The PAC at lags $l = 1, 2, \ldots$ is estimated by fitting a succession of AR models with increasing orders, and retaining the last coefficient of each model. By inspection of the PAC plot and statistical testing of the PAC values we can determine how many AR terms we need to use—what AR order we should choose—to explain the AC pattern in a time series. If the PAC turns out to be significant at lag p and not significant at any higher lags, then this suggests we should fit an AR model of order p to our data.

In practice we will look for the point on the plot where the PAC values for all higher lags are *nearly* zero. Placing on the plot an indication of the uncertainty of the estimated PAC is helpful for this purpose. This uncertainty is usually evaluated as follows: under the hypothesis that the series is a true $AR(p)$ process, the true PAC coefficients at $l > p$ are expected to be zero-mean, independently distributed Gaussian variables. The estimated PAC coefficients are then assumed to be approximately distributed in the same way. In this instance, the standard error of the estimated PAC coefficients of a fitted series with N observations is approximately $1/\sqrt{N}$ for $l > p$. Then, assuming that the true value of the PAC at lag l is zero (null hypothesis), we can establish approximate confidence intervals for the estimated PAC samples. These approximate upper and lower bounds are useful if the record length is not too short (say $N > 30$).

The identification of the proper AR model order from these plots is not easy and requires a lot of experience.

Automated identification by goodness-of-fit criteria

A different way of identifying ARMA models is by trial and error and goodness-of-fit evaluation. In this approach, a suite of candidate models are estimated, and goodness-of-fit statistics are computed that penalize appropriately for excessive model complexity.

Goodness-of-fit statistical measures are functions of the residual (unmodeled) variance and of the number of estimated parameters. The rationale behind this approach is the fact that, in general, the fit improves with model complexity, as the residual variance decreases, but including enough parameters we can *force* a model to nearly perfectly fit any data set. This perfect fit would likely be an *overfitting* to insignificant, minute details of the particular sequence considered, rather than a valid approximation of the underlying random process: it would be an artificial improvement due to increasing model complexity. Overfitting can be understood remembering that with a fixed number of data samples, the larger the number of estimated parameters, the larger the estimation errors.[15]

---

[15] In statistics and machine learning, overfitting occurs when a statistical model ends up describing random errors or noise instead of the underlying relation. Overfitting generally occurs when a model is excessively complex, such as having too many parameters relative to the number of observations. A model that has been overfit will generally have poor predictive performance, as it can exaggerate minor fluctuations in the data. In other words, in case of overfitting, the estimator is too flexible

Focusing on AR modeling, we mention the Akaike's Final Prediction Error (FPE) and the Information Theoretic Criterion (AIC) as two closely related alternative statistical goodness-of-fit measures.

FPE (Akaike 1970) is given by

$$\text{FPE}[p] = \frac{N + p + 1}{N - p - 1}\sigma_{e,p}^2$$

where $\sigma_{e,p}^2$ is the variance of model residuals at order p, N is the length of the time series, and p is the number of estimated coefficients. So FPE is a "corrected" version of $\sigma_{e,p}^2$. FPE is computed for various candidate models, and the model with the lowest FPE is selected as the best-fit model.[16] The estimated $\sigma_{e,p}^2$ is guaranteed to decrease or stay the same as the model order increases for all the AR parameter estimation methods commonly used. Hence we cannot simply monitor the decrease in error power $\sigma_{e,p}^2$ as a means of determining the model order, but we must also take into account the increase in estimation errors that occurs when the amount of data is fixed and we estimate an increasing number of AR parameters. FPE and the following AIC criteria adhere to this philosophy. We can see from the FPE formula that the factor $(N + p + 1)/(N - p - 1)$ is a sort of penalty that increases with increasing order, thus accounting for the growing inaccuracies in the coefficient estimates.

AIC (Akaike 1974) is another widely used goodness-of-fit measure, which is based on information-theory entropy and offers a relative estimate of the information lost when a given AR model is used to represent the process that generates the data. AIC quantifies the compromise between model accuracy and complexity—more precisely, the compromise between model bias and variance—and is defined as

$$\text{AIC}[p] = N \log(\sigma_{e,p}^2) + 2p,$$

where $2p$ represents and additive penalty term for increasing order, i.e., for any extra coefficients that do not significantly reduce the modeling error. As with FPE, the best-fit model has the minimum value of AIC. From the spectral perspective, it is interesting to mention that AIC is an estimate of a properly-defined distance between a modeled spectral density and the true spectral density of the data record (Percival and Walden 1993). It has, however, been shown that it tends to overestimate the true model order as N increases (see, e.g., Rissanen 1983).

(Footnote 15 continued)
and captures illusory trends in the data. These illusory trends are often the result of the noise in the observations. The contrary phenomenon, underfitting, occurs when an estimator is not flexible enough to capture the underlying trends in the observed data, usually because of an insufficient number of parameters.

[16]Recall that we are dealing with centered data, which in practice means that the sample mean has been subtracted from each data sample. If the process true mean value were known, we should write $\text{FPE}[p] = [(N + p)/(N - p)]\sigma_{e,p}^2$.

The performances of FPE and AIC criteria are similar for large data records—
they are actually functionally related to each other, as explained, for example, in
Percival and Walden (1993)—but for short data records, the use of the AIC is
recommended (see Ulrych and Ooe 1979). Neither FPE nor AIC directly address
the question of the model residuals being without AC, as they ideally should be
if the chosen model is able to remove all the persistence in the series. A strategy
for model identification by the FPE or AIC criteria is to find the model that
approximately minimizes FPE or AIC, and then apply diagnostic checking (see
the next subsection) to verify if the model does a good job at producing random
residuals.

A measure similar to AIC is the *minimum description length* (MDL) criterion,
proposed by Rissanen (1978), in which the penalty term $2p$ is substituted by
$(\log N)\, p$, which increases with N, in order to reduce the tendency to overestima-
tion:

$$\text{MDL}[p] = N \log(\sigma_{e,p}^2) + (\log N)\, p.$$

Many other criteria are available, such as the *criterion autoregressive transfer
function* (CAT) proposed by Parzen (1976). Few guidelines as to their use in
practical applications—and in particular in spectral analysis applications—are
however provided in literature.

11.6.2 Model Testing

This step is also called diagnostic checking, or verification (Anderson 1976). Two
important elements of checking are to ensure that the residuals of the model are
random, and that the estimated parameters are statistically significant. In other words,
we must ask ourselves the following questions.

Does the model produce random residuals?
Can the residuals be classified as white noise? The model should effectively
describe the AC persistence in the examined WSS process. If it is so, then the
residuals should be random, i.e., uncorrelated, and the AC function of residuals
should be zero at all lags except lag zero. Of course, the AC estimated from a
finite-length data record will not be exactly zero, but should fluctuate near zero.
The AC of the residuals can be examined in two ways. First, the AC can be scanned
to see if any individual coefficients fall outside some specified confidence interval
around zero, exactly as explained about the PAC in the previous subsection. For the
true residuals, which are unknown, the AC at lag l is normally distributed with zero
mean and standard deviation $1/\sqrt{N}$. An approximate confidence interval for the
AC of residuals at lag k can thus be found by referring to a normal distribution.[17]

---

[17]For example, the 0.975 probability point of the standard normal distribution is 1.96. This means
that for a normally-distributed variable, 95 % of the data lies within $1.96 \approx 2$ standard deviations

A subtle point that should be mentioned is that the AC of the estimated residuals of a fitted AR model has somewhat different properties than that of the true residuals. As a result, the above approximation of standard deviation $(1/\sqrt{N})$ overestimates the width of the confidence interval at small lags when applied to the AC of the residuals of a fitted model (Chatfield 2004). There is, consequently, some bias towards concluding that the model has effectively removed persistence. At large lags, however, the approximation is close.

A different approach to evaluating the randomness of the residuals is to look at their AC sequence—being $\hat{\gamma}_{ee}[k]$—as a whole, rather than at the individual samples separately (Chatfield 2004). The related test is called the *portmanteau lack-of-fit test*, or Q-test, or Box-Pierce test (Box et al. 2008), and the test statistic is

$$Q = N \sum_{k=1}^{K} \hat{\gamma}_{ee}[k]^2.$$

The null hypothesis is that all the true ACs of residuals for lags 1 to K are zero, where the choice of K is up to the user. The Q statistic follows a χ^2 distribution with $K - p$ DOF[18]: if the computed Q does not exceed the value assumed by χ^2 for some chosen probability level, the null hypothesis that the residuals represent a white noise series is accepted at that level.

The estimated parameters are significantly different from zero?

Besides the randomness of the residuals, we are concerned with the statistical significance of the model coefficients. The estimated coefficients should be "significantly" different from zero. If not, the model should probably be simplified by reducing the model order. For example, an AR(2) model for which the second-order coefficient is not significantly different from zero might be discarded in favor of an AR(1) model. But what does "significantly" mean here? Significance of the AR (or ARMA) coefficients can be evaluated by comparing estimated parameters with their standard deviations. For an AR(1) model, the estimated first-order AR coefficient, $\hat{a}_1$, is normally distributed with variance equal to $\text{var}(\hat{a}_1) = (1 - \hat{a}_1^2)/N$. The approximate 95 % confidence interval for $\hat{a}_1$ is therefore two standard deviations around $\hat{a}_1$, and so on.

(Footnote 17 continued)
of the mean. The 95 % confidence interval for the AC at lag l is therefore $\pm 1.96/\sqrt{N}$. For the 99 % confidence interval, the 0.995 probability point of the normal distribution is 2.57, so, 99.5 % of the data lies within $2.57 \approx 3$ standard deviations of the mean. The 99 % confidence interval for the AC is thus $\pm 2.57/\sqrt{N}$. A value outside this confidence interval is evidence that the model residuals are not random at the chosen probability level. If a residual-AC plot for a given series shows that none of the AC samples of the chosen model residuals fall outside the 99 % confidence interval around zero, then the modeling explains the persistence and yields random residuals at that confidence level.

[18] The DOF would be $K - p - q$ in the ARMA case.

Does the model explain a remarkable part of the signal's variance?

For long time series (e.g., many hundreds of observations), AR modeling may yield a model with estimated parameters that are significantly different from zero in the statistical sense, but very small. The persistence described by such a model might actually account for a tiny percentage of the variance of the original time series. A measure of the practical significance of the AR model we have found is then the percentage of the series' variance that is removed by fitting the model itself to the series, i.e., the percentage of original variance *explained* by the model. If the variance of the model residuals is much smaller than the variance of the original series, then the AR model accounts for a large fraction of the original variance, and a large part of the variance of the series is due to persistence. On the contrary, if the variance of the residuals is almost as large as the original variance, then little variance has been removed by AR modeling, and the variance due to persistence is small. A simple measure of the fraction of variance that can be attributed to the signal persistence modeled by an AR(p) model is

$$R_p^2 = 1 - \frac{\sigma_{e,p}^2}{\sigma_x^2}.$$

If R_p^2 is large, then a large part of the signal variance has been explained by the model; if it is small, then very little of the signal's variability has been modeled (and very likely there was little to model at all); the rest is seen as noise. Whether any found value of R_p^2 is practically significant is a matter of subjective judgment and depends on the application. For example, in a noisy time series of paleoclimatological data, 40% would likely be considered practically significant, while 1% might well be dismissed as practically insignificant; with different series and in other fields, even a value as high as 70% might be considered as unsatisfactory.

11.6.3 AR Parameter Estimation

We will now discuss different methods for estimating the AR parameters.

11.6.3.1 Yule-Walker or Autocorrelation Method

In this method, the AR coefficients are obtained by solving the YW equations. We already said (Sect. 11.3.2) that if the AC values were known from lag zero up to lag p, we could solve the YW equations for the coefficients, $\mathbf{a}_p = -\boldsymbol{\Gamma}_p^{-1}\boldsymbol{\gamma}_p$, and then obtain σ_e^2 from the first row of the augmented YW equations. The YW method, also called AC method, is based directly on this approach.

Given N data, we assume that $x[n]$ is zero outside the range $n = [0, N-1]$, and write the data in matrix form. Each matrix column contains the data sequence lagged by l of time steps, from 0 to a maximum lag p. Just to make an example, if $N = 5$ and we take $p = N - 3 = 2$, the data matrix $\mathbf{X}$ is an $(N + p) \times (p + 1) = 7 \times 3$ rectangular matrix:

$$\mathbf{X} = \begin{bmatrix} x[0] & 0 & 0 \\ x[1] & x[0] & 0 \\ x[2] & x[1] & x[0] \\ x[3] & x[2] & x[1] \\ x[4] & x[3] & x[2] \\ 0 & x[4] & x[3] \\ 0 & 0 & x[4] \end{bmatrix}.$$

The estimate of the AC matrix, which is a square matrix $\hat{\boldsymbol{\Gamma}}$ with $p + 1$ rows and columns, can then be obtained from the equation

$$\hat{\boldsymbol{\Gamma}} = \frac{1}{N} \mathbf{X}^T \mathbf{X}.$$

The elements of $\hat{\boldsymbol{\Gamma}}$ are indicated as $\hat{\gamma}(i, k)$, where i is row number and k is column number: so, $\hat{\gamma}(i, k) = \hat{\gamma}(i, i + l)$. Note that the number of samples of $x[n]$ used in the estimate of the elements $\hat{\gamma}(i, i + l)$ of $\hat{\boldsymbol{\Gamma}}$ decreases as l increases. In our example, while $\hat{\gamma}(1, 1)$ corresponding to lag 0 contains 5 product terms, $\hat{\gamma}(1, 3)$ belonging to lag 2 contains only 3 product terms: $x[2]x[0] + x[3]x[1] + x[4]x[2]$. So, the estimates are of good quality only if l is small compared with the number of available data samples N.

The formula for $\hat{\boldsymbol{\Gamma}}$ corresponds to using for AC the biased estimator $c_{xx}[l]$. These estimates are inserted, in place of the true ones, into the YW equations, and these are solved for the coefficients and the noise variance, as explained above in the known-AC case. The use of biased AC estimates ensures the existence of a solution, since the AC matrix is guaranteed to be positive-definite, hence non-singular. The filter poles will lie inside the unit circle, i.e., the method will always produce stable filters.

In summary, the YW method is based on the YW equations, with the true AC elements replaced by their biased estimates. This guarantees the existence of a solution, with model coefficients corresponding to an estimated $\hat{A}_p(z)$ which is stable for any p in the range $1 \leq p \leq N - 1$. From the perspective of linear prediction, the Yule Walker method minimizes the *forward prediction error* in the least squares sense, based on all available observations. Since the AC matrix is a Toeplitz matrix, the so-called Levinson-Durbin recursion introduced below can be applied to solve the equations for the model coefficients, thus speeding up calculations.

11.6.3.2   Levinson-Durbin Recursion

The numerical solution of a set of p linear equations in p unknowns requires $\approx p^3$ floating point operations (flops). In modeling applications, where a priori information about the best order p is usually lacking, AR models with different orders have to be calculated and tested, hence the YW system of equations has to be solved for $p = 1$ up to some specified maximum order p_{max}. By using a general solving method, this task requires a number of flops of the order of $\sum_{p=1}^{P_{max}} p^3 \approx p_{max}^4/4$ flops. This may be a remarkable computational burden if the maximum order is large: this is, for example, the case in applications dealing with narrowband signals, where values of 50 or even 100 for p_{max} are not uncommon.[19] In such applications, it may be important to reduce the number of flops required to determine the model parameters, i.e., the coefficients and σ_e^2.

A method that is widely used to solve YW equations in a fast way and reduce the number of flops is the *Levinson-Durbin (LD) algorithm* (Levinson 1947) that exploits the special algebraic structure of the matrix $\boldsymbol{\Gamma}_{p+1}$ in the augmented YW system of equations. This matrix is highly structured: it is a real symmetric matrix[20] and is also a Toeplitz matrix. This allows for reducing the number of flops for a given p to about p^2. Thus, the number of flops required by the LD algorithm is about p times smaller than that required by a general linear equation solver to determine the parameters of a fixed order model—an efficiency improvement that is already remarkable—and about p_{max}^2 times smaller than that required by a general linear equation solver to determine the parameters of all models from $p = 1$ to p_{max}. The only requirement in the LD algorithm is that the matrix be positive definite, and as we saw, this condition is satisfied not only when true AC values are used, but also when they are replaced by their biased estimates.

The basic idea of the LD algorithm (see, for instance, Vaidyanathan 2008; Stoica and Moses 2005) is to solve the matrix equation

$$\boldsymbol{\Gamma}_{p+1} \begin{bmatrix} 1 \\ \mathbf{a}_p \end{bmatrix} = \begin{bmatrix} \sigma_{e,p}^2 \\ \mathbf{0} \end{bmatrix}$$

—which is unique for any p—recursively in p, starting from the solution for $p = 1$, which is easily determined, and increasing at each iteration the order by one. The LD algorithm is in fact based on the observation that if the solution to the YW equations is known for some order p, then the solution for order $p + 1$ can be obtained by a simple updating process. In this way, we obtain not only the solution to the problem of the selected order, but also those for all the lower orders. To demonstrate the mechanism, let us consider a case in which the chosen order is $p = 3$, so that we have four coefficients. The augmented YW equations become a set of four equations: in matrix form,

---

[19]Narrowband processes have AC values slowly decreasing with increasing lag, which produce large values of p_{max} in the AR-spectrum estimation procedure. Broadband processes have ACs decreasing faster with lag. For an example of this behavior, compare Fig. 11.11a, which is relative to a broadband AR(2) process (as shown in Fig. 11.13a), with Fig. 11.11a, which is relative to a broadband AR(2) process (as shown in Fig. 11.13b).

[20]This matrix would be Hermitian in the case of complex signals.

$$\begin{bmatrix} \gamma[0] & \gamma[1] & \gamma[2] & \gamma[3] \\ \gamma[1] & \gamma[0] & \gamma[1] & \gamma[2] \\ \gamma[2] & \gamma[1] & \gamma[0] & \gamma[1] \\ \gamma[3] & \gamma[2] & \gamma[1] & \gamma[0] \end{bmatrix} \begin{bmatrix} 1 \\ a_{3,1} \\ a_{3,2} \\ a_{3,3} \end{bmatrix} = \begin{bmatrix} \sigma_{e,3}^2 \\ 0 \\ 0 \\ 0 \end{bmatrix}. \tag{A}$$

Let us denote the AC matrix in Eq. (A) by $\mathbf{\Gamma}_4$, since it has $p+1=4$ rows and columns. We will now show how we can pass from the third-order to the fourth-order case. For this, note that we can append a fifth equation to the above set of four and write

$$\mathbf{\Gamma}_5 = \begin{bmatrix} \gamma[0] & \gamma[1] & \gamma[2] & \gamma[3] & \gamma[4] \\ \gamma[1] & \gamma[0] & \gamma[1] & \gamma[2] & \gamma[3] \\ \gamma[2] & \gamma[1] & \gamma[0] & \gamma[1] & \gamma[2] \\ \gamma[3] & \gamma[2] & \gamma[1] & \gamma[0] & \gamma[1] \\ \gamma[4] & \gamma[3] & \gamma[2] & \gamma[1] & \gamma[0] \end{bmatrix} \begin{bmatrix} 1 \\ a_{3,1} \\ a_{3,2} \\ a_{3,3} \\ 0 \end{bmatrix} = \begin{bmatrix} \sigma_{e,3}^2 \\ 0 \\ 0 \\ 0 \\ \mu_3 \end{bmatrix} \tag{B}$$

where

$$\mu_3 = \gamma[4] + a_{3,1}\gamma[3] + a_{3,2}\gamma[2] + a_{3,3}\gamma[1].$$

The matrix $\mathbf{\Gamma}_5$ is symmetric and Toeplitz. It can be verified that its elements $\gamma(i,k)$ satisfy

$$\gamma(4-i, 4-k) = \gamma(i,k), \quad \text{for} \quad 0 \le i, k \le 4.$$

This means that if we take the matrix, reverse the order of all rows, and then reverse the order of all columns, the result is the same matrix. As a consequence, we can also write

$$\begin{bmatrix} \gamma[0] & \gamma[1] & \gamma[2] & \gamma[3] & \gamma[4] \\ \gamma[1] & \gamma[0] & \gamma[1] & \gamma[2] & \gamma[3] \\ \gamma[2] & \gamma[1] & \gamma[0] & \gamma[1] & \gamma[2] \\ \gamma[3] & \gamma[2] & \gamma[1] & \gamma[0] & \gamma[1] \\ \gamma[4] & \gamma[3] & \gamma[2] & \gamma[1] & \gamma[0] \end{bmatrix} \begin{bmatrix} 0 \\ a_{3,3} \\ a_{3,2} \\ a_{3,1} \\ 1 \end{bmatrix} = \begin{bmatrix} \mu_3 \\ 0 \\ 0 \\ 0 \\ \sigma_{e,3}^2 \end{bmatrix}. \tag{C}$$

If we are able to find a linear combination of Eqs. (A) and (C) such that the last element of the right-hand side vector in Eq. (B) becomes zero, we obtain the equations governing the fourth-order process. To this purpose we perform the following operation: Eq. (A) $+ k_4 \times$ Eq. (C), where k_4 is a constant that we define as

$$k_4 = -\frac{\mu_3}{\sigma_{e,3}^2}.$$

After some algebra we get

$$\begin{bmatrix} \gamma[0] & \gamma[1] & \gamma[2] & \gamma[3] & \gamma[4] \\ \gamma[1] & \gamma[0] & \gamma[1] & \gamma[2] & \gamma[3] \\ \gamma[2] & \gamma[1] & \gamma[0] & \gamma[1] & \gamma[2] \\ \gamma[3] & \gamma[2] & \gamma[1] & \gamma[0] & \gamma[1] \\ \gamma[4] & \gamma[3] & \gamma[2] & \gamma[1] & \gamma[0] \end{bmatrix} \begin{bmatrix} 1 \\ a_{4,1} \\ a_{4,2} \\ a_{4,3} \\ a_{4,4} \end{bmatrix} = \begin{bmatrix} u \\ 0 \\ 0 \\ 0 \\ 0 \end{bmatrix} \tag{D}$$

where u indicates an unknown quantity and where we set

$$a_{4,1} = a_{3,1} + k_4 a_{3,3},$$
$$a_{4,2} = a_{3,2} + k_4 a_{3,2},$$
$$a_{4,3} = a_{3,3} + k_4 a_{3,1},$$
$$a_{4,4} = k_4.$$

By comparison with Eq. (A), we conclude that the element denoted by u on the right-hand side of Eq. (D) is $\sigma_{e,4}^2$, which is related to $\sigma_{e,3}^2$ by

$$\sigma_{e,4}^2 = \sigma_{e,3}^2 + k_4 \mu_3 = \sigma_{e,3}^2 - k_4^2 \sigma_{e,3}^2 = \left(1 - k_4^2\right) \sigma_{e,3}^2.$$

If we know the coefficients $a_{3,i}$ and the mean square error for the third-order model, from the above equations we can find the corresponding quantities for the fourth-order model.

The recursion demonstrated above for the third-order case can be easily generalized into arbitrary orders. Adopting the notational convention that given a column vector $v = [v_1 \ldots v_n]^T$ we set $\tilde{v} = [v_n \ldots v_1]^T$, from

$$\mathbf{\Gamma}_{p+1} \begin{bmatrix} 1 \\ \mathbf{a}_p \end{bmatrix} = \begin{bmatrix} \sigma_{e,p}^2 \\ \mathbf{0} \end{bmatrix}$$

it can be easily shown that we can pass to

$$\mathbf{\Gamma}_{p+2} \begin{bmatrix} 1 \\ \mathbf{a}_{p+1} \end{bmatrix} = \begin{bmatrix} \sigma_{e,p+1}^2 \\ \mathbf{0} \end{bmatrix}$$

by the recursive-in-order equations

$$\mathbf{a}_{p+1} = \begin{bmatrix} \mathbf{a}_p \\ 0 \end{bmatrix} + k_{p+1} \begin{bmatrix} \tilde{\mathbf{a}}_p \\ 1 \end{bmatrix}$$

and

$$\sigma_{e,p+1}^2 = \left(1 - k_{p+1}^2\right) \sigma_{e,p}^2.$$

These recursive-in-p equations are initialized and then used for growing order values.

In summary, the LD algorithm comprises of the following two steps:

1. initialization:

$$a_1 = -\frac{\gamma[1]}{\gamma[0]} = k_1,$$

$$\sigma_{e,1}^2 = \left(1 - a_1^2\right) \gamma[0];$$

2. for $p = 1, \ldots p_{max} - 1$, compute

$$k_{p+1} = -\frac{\gamma[p+1] + \tilde{\gamma}[p]\tilde{a}_p}{\sigma_{e,p}^2},$$

$$\sigma_{e,p+1}^2 = \sigma_{e,p}^2 \left(1 - k_{p+1}^2\right),$$

$$\mathbf{a}_{p+1} = \begin{bmatrix} \mathbf{a}_p \\ 0 \end{bmatrix} + k_{p+1} \begin{bmatrix} \tilde{\mathbf{a}}_p \\ 1 \end{bmatrix}.$$

The coefficients k_{p+1} are known as *reflection coefficients*. They can never exceed 1; unity is actually an upper bound for which $\sigma_{e,p+1}^2 = 0$. Now, suppose we compute models of increasing orders using the LD recursion. We know that the error variance decreases with order. Assume that after the order has reached some value p, the error variance does not decrease any further. This represents a stalling condition: increasing the number of included past samples does not help to increase the model accuracy. Whenever such a stalling occurs, it turns out that the error noise is white, and $x[n]$ is the output of an all-pole filter with white-noise input, i.e., $x[n]$ is truly AR(p). Since, in general, increasing the order and its complexity implies a greater ability to model the signal and a decrease of the error variance, for a true AR(p) the LD algorithm would stop spontaneously after p steps, because for larger orders the error variance and the model coefficients would no longer change. If instead the signal is not a true AR process, the algorithm does not reach a point in which to stop naturally. In the limit case in which at some order the error vanishes and the signal turns out to be exactly predictable on the basis of a finite number of past values, the signal can be classified as deterministic; but most real-world signals are never exactly predictable, independently of the predictor order.

11.6.3.3 Covariance and Modified Covariance Methods

Another approach to the estimation of AR parameters is the covariance method, also called least squares (LS) method (see Stoica and Moses 2005). In this case, the data matrix $\mathbf{X}$ is formed differently: in the example with $N = 5$ and maximum lag $p = 2$ it is defined as

$$\mathbf{X} = \begin{bmatrix} x[2] & x[1] & x[0] \\ x[3] & x[2] & x[1] \\ x[4] & x[3] & x[2] \end{bmatrix}$$

and contains no zeros. In this case the AC estimate in matrix form is defined as

$$\hat{\mathbf{\Gamma}} = \frac{1}{N-p} \mathbf{X}^T \mathbf{X}.$$

Its element $\hat{\gamma}(i, j)$ related to lag $l = k - i$ coincides with $c'_{xx}[l]$. In the covariance method, each AC sample estimate is an average of $N - p$ product terms, unlike the autocorrelation method, where the number of data samples used in the estimate of an individual AC sample decreases as the lag increases. Thus the AC estimates, in general, tend to be better than in the autocorrelation method, but the matrix $\hat{\Gamma}$ in this case is not necessarily Toeplitz, and the LD recursion cannot be applied for solving the YW equations (although similar fast algorithms for this method have been developed). Another problem with non-Toeplitz AC matrices is that the AR filter is not guaranteed to have all its poles inside the unit circle. However, instability only appears infrequently and when it occurs, there are simple means to stabilize the model, for instance reflecting the instability-generating poles inside the unit circle. Therefore, possible instability does not represent a serious drawback of the covariance method. From the perspective of linear prediction, the covariance method is also based on the minimization of the forward prediction error in the least squares sense. What then, is the difference between YW and covariance methods?

They represent two distinct finite-sample approximate solutions of the minimization problem for $\mathrm{E}\{e^2[n]\} = \sigma_e^2$. Given a finite set of measurements, we can look at YW and covariance methods in parallel, observing that the minimization of σ_e^2 is expressed in both cases through a finite-sample *cost function* of the AR coefficients that assumes all non-observed signal samples to be zero:

$$f(\boldsymbol{a}_p) = \sum_{n=N_1}^{N_2} e^2[n] = \sum_{n=N_1}^{N^2} \left(x[n] - \sum_{i=1}^{p} a_i x[n - i] \right)^2.$$

The vector $\boldsymbol{a}_p$ that minimizes the cost function depends on the choice of the summation limits N_1 and N_2:

- setting $N_1 = 0$ and $N_2 = N + p - 1$ leads to the YW method, which uses the biased estimator $c_{xx}[l]$ for the AC;
- setting $N_1 = p$ and $N_2 = N - 1$ corresponds to eliminating all the arbitrary zero values of the signal in the pertinent equations and thus to adopting the biased estimator $c'_{xx}[l]$ for the AC; this choice leads to the covariance method.

As N increases while p remains fixed, the difference between the covariance estimates used by the two methods diminishes; consequently, for long records the two methods nearly coincide with one another. Moreover, while the YW-estimated AR model is guaranteed to be stable, the covariance method may lead to unstable models. When applying this method it is advisable to select orders not exceeding $N/2$ to avoid numerical issues, such as singularity of the estimated AC matrix.

The modified covariance method is similar to the covariance method. However, instead of finding the AR model that minimizes the sum of squares of the forward prediction error, the modified covariance method minimizes the sum of the squares of the forward and backward prediction errors. As with the covariance method, the AC matrix involved is not Toeplitz, so the LD recursion cannot be applied. A stable IIR filter is not guaranteed. To avoid singularity of the estimated AC matrix in numerical computations it advisable to choose orders not exceeding $2N/3$. This method is more computationally demanding than the other two methods.

11.6.3.4 Burg's Method

In the early 1960s, geophysicist John P. Burg developed a method for spectral estimation based on AR modeling that he named the "maximum entropy method" or MEM (see Sect. 11.9). As a part of this method, Burg developed an approach to the estimation of the AR model parameters directly from the data, without the intermediate step of computing an AC matrix and solving the YW equations. Burg's method (Burg 1967, 1968) relies on the LD algorithm and is a recursive procedure where, at each step of the recursion, a single reflection coefficient—relating the AR-parameter estimates for a given order to those for the immediately higher order—is estimated. This reflection coefficient is chosen in such a way as to minimize the sum of the squared forward and backward prediction errors: since they are statistically identical, according to Burg (1967, 1968) there is no reason to favor one over the other in fitting the model, so that both should be included in the minimization criterion. In this sense, Burg's method thus resembles the modified covariance method, but it only minimizes the sum of squares of forward and backward prediction error with respect to a single reflection coefficient (see Kay and Marple 1981), while the modified covariance method minimizes it with respect to all of the model coefficients. The models found by this method are always stable. Describing in detail the development of this method would require introducing the so-called "lattice representation" for the prediction error filter, a topic that is beyond the scope of the book.

11.7 AR Spectral Estimation

We saw that the AR(p) approximation to a signal $x[n]$ is such that its AC samples match those of the original process $x[n]$ for the first $p + 1$ values of lag. As p increases, we therefore expect the power spectrum of the AR approximation to resemble the power spectrum of the signal more and more. The spectrum of the AR approximation, being $P_{xx}^{AR}(\omega)$, is called an AR-model-based estimate of $P_{xx}(\omega)$. In essence, $P_{xx}^{AR}(\omega)$ is obtained by extrapolating the finite AC segment we can compute from the data to larger lags, using an AR(p) model, and is nothing but the Fourier transform of the extrapolated AC.

We know that the AR(p) model approximates the power spectrum $P_{xx}(\omega)$ with the all-pole power spectrum

$$P_{xx}^{AR}(\omega) = \frac{1}{N} \frac{|E(e^{j\omega})|^2}{|A(e^{j\omega})|^2}$$

If $P_{xx}(\omega)$ has sharp peaks, then $A(z)$ must have zeros (poles of $P_{xx}^{AR}(z)$) close to the unit circle to reproduce these peaks. If, on the other hand, $P_{xx}(\omega)$ has zeros or sharp dips, then $P_{xx}^{AR}(\omega)$ cannot approximate them very well, because $A(z)$ cannot have poles on the unit circle. Thus, the AR model can be used to obtain a good match of the power spectrum $P_{xx}(\omega)$ near its peaks, but not near its valleys.

The selection of the order p, which is an issue of primary importance in stochastic modeling, is somewhat less central to the problem of practical AR spectral estimation, where the best model order is chosen in an empirical way. Starting from an a priori reasonable minimum value of p—the order should never be smaller than twice the number of cyclical or pseudo-cyclical components that are supposed to exist in the analyzed process—some higher values are also considered, up to a reasonable maximum comprised, as a rule of thumb, between $N/3$ and $N/2$ (Ulrych and Bishop 1975; Ulrych and Clayton 1976). The AR spectra of different orders thus obtained are compared among them and with the spectra obtained for the same record using different methods. Since the model order is also the number of poles of the filter's transfer function, low orders will produce smooth spectra with few peaks, while high orders will produce a number of peaks and a very detailed spectral behavior, regardless of the real spectral content of the time series (Ulrych and Bishop 1975). In the second case, many peaks can be spurious—i.e., not corresponding to actual features present in the data—but other peaks can reflect components actually present in the data record. Comparison is, as usual, the key to understanding the robustness of the results obtained by spectral estimation: which peaks are also present, for example, in an MTM spectrum of the same data? Which peaks survive a reduction of the order?

Note that the issue of AR model-order selection is analogous to the issue of window-width selection in classical spectral estimation. It represents a trade-off between high frequency resolution (high p) and low variance (low p). If the order is too low, the estimate will be too smooth and some real peaks will possibly be missed; if it is too large, spurious peaks and statistical instability will result. The appearance of spurious peaks is due to estimation errors, leading to non-zero AR coefficient estimates even though the true model order may be much lower than the selected p.

Why are FPE, AIC and other similar criteria not routinely applied for order selection in parametric spectral analysis? The main reason is that any order selection method should be appropriate for what we intend to do with the fitted AR model. When we are interested in fitting AR models for spectral analysis, we must take into account that most of the commonly used order-selection criteria are instead geared toward selecting AR models that perform well for one-step-ahead predictions. These criteria are thus not completely appropriate for parametric spectral analysis. We would actually need an assessment of the *spectral estimation performance* of the various AR estimators, in the case in which the model order must also be estimated, in addition to the AR parameters. Moreover, there are some subtle interactions between the various order selection criteria and the different AR parameter estimators. For example, the FPE criterion was originally derived based on the statistical properties of forward least-squares estimators; Ulrych and Bishop (1975) found that, when used in conjunction with Burg's algorithm, this criterion tends to pick out spuriously high-order models. Nevertheless, FPE, AIC and similar tools may serve as model-order indicators in spectral applications.

Each of the parameter estimation methods discussed above corresponds to a different method of parametric spectral estimation, and gives it its name. Once the estimated parameters of the model having the chosen order have been found by any of the above mentioned methods, these parameters are inserted into the AR spectral formula, to get the desired estimate of the signal's PSD:

$$\hat{P}_{xx}^{AR}(\omega) = \frac{\hat{\sigma}_e^2}{\left|\hat{A}(e^{j\omega})\right|^2},$$

where the "hats" represent estimates.

Although parametric AR methods can provide improved resolution with respect to classical methods, it is important to realize that, unless the chosen model is at least approximately appropriate for the process being analyzed, inaccurate results may be obtained. A typical case is a true MA(q) process, for which the spectral estimate obtained by the AR approach can be misleading.

Consider the two spectral estimates shown in Fig. 11.15a that have been obtained for $N = 64$ samples of a process consisting of two sinusoids in unit variance white noise,

$$x[n] = A_1 \sin(\omega_1 n + \vartheta_1) + A_2 \sin(\omega_2 n + \vartheta_2) + e[n],$$

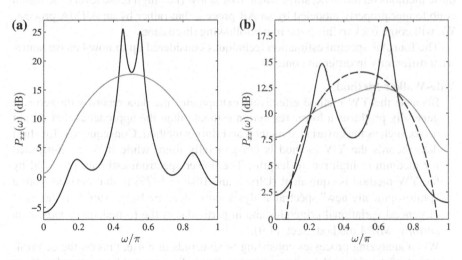

Fig. 11.15 Illustration of how an inappropriate model may lead to an inaccurate spectral estimate. Using a parametric spectrum estimator that assumes an all-pole model for the process (the YW method; *black curves*) and the Blackman-Tukey non-parametric method (with $M = N/10$ and a Blackman window; *gray curves*), the spectral estimates are shown for **a** two sinusoids in noise (with $p = 8$ in the YW method) and **b** a true second-order MA process (with $p = 5$ in the YW method). $N = 64$ samples of each signal have been used. In panel (**b**), the *dashed curve* depicts the true power spectrum of the MA(2) process

with $A_1 = A_2 = 5$, $\omega_1 = 0.45\pi$, and $\omega_2 = 0.55\pi$, and with ϑ_1 and ϑ_2 representing two random initial phases, uniformly distributed between 0 and 2π. The spectrum drawn as a *black curve* was computed using a method that assumes an all-pole model (the YW method with $p = 8$), whereas the one shown by the *gray curve* was computed using the Blackman-Tukey method with a maximum lag of $N/10$ and a Blackman window. Clearly, the estimate produced by the model-based approach provides much better resolution than the Blackman-Tukey method.

On the other hand, consider the MA(2) process having $b_0 = 1$, $b_1 = 0$, $b_2 = -1$, $\sigma_e^2 = 1$, i.e.,

$$x[n] = e[n] - e[n - 2].$$

Figure 11.15b shows the spectral estimates that are produced by the same two approaches (the AR spectrum in this case was estimated by the YW method adopting $p = 5$; see the *black curve*). For this true MA(2) process, the model-based approach inaccurately represents the underlying true spectrum (*dashed curve*), and indicates two non-existent narrowband components. The Blackman-Tukey method, which makes no assumptions about the process, produces a more reasonable estimate (*gray curve*) of the true spectrum.

$$P_{xx}(\omega) = \sigma_e^2 \left| 1 - e^{-j2\omega} \right|^2 = 2\sigma_e^2 [1 - \cos(2\omega)].$$

The utility of AR methods is, moreover, severely limited by the dependence of these methods on the SNR, since when SNR is low (i.e., high noise level), the signal is no longer properly modeled by an AR process, but rather by an ARMA process. We will come back to this issue in the following discussion.

The four AR spectral estimation techniques considered up to now behave somewhat differently in different contexts.

Yule-Walker Method

Because the YW method effectively extrapolates the data record with zeros, it generally produces a lower resolution estimate than the approaches that do not do so, such as the covariance method and Burg's method. Consequently, for short data records the YW method is not generally used, while it is recommended for medium-to-high record lengths. The power spectrum estimate provided by the YW method is equivalent (Ulrych and Bishop 1975) to that derived from a "philosophically new" spectral analysis, introduced by Burg (1967) on the basis of general variational principles and in particular of the formalism of *maximum entropy* (MEM method; Sect. 11.9).

When analyzing processes consisting of sinusoids in noise, bias on the determination of the sinusoids' frequencies can occur. *Frequency bias* means that if we analyze synthetic data containing one or more sinusoids in noise, one or more spectral peaks will fall at frequencies that do not exactly coincide with the frequencies adopted to construct the synthetic data, but are somewhat shifted with respect to their true positions. The method is also subject to spectral line splitting.

Line splitting is the occurrence of two or more peaks in the estimated spectrum, where only one peak should be present.

Covariance Method

The covariance method has been found to be more accurate than the YW method, in the sense that the estimated parameters of the former are on the average closer to the true ones than those of the latter (Marple 1987; Kay 1988). It also can produce higher resolution spectra. This behavior can be explained heuristically observing that the arbitrary zero signal values outside the available measurement interval that are included in the YW-method calculations result in bias in the estimates of the AR parameters. When N is not much greater than the order p, this bias can be significant.

Also with the covariance method, in the case of sinusoids in noise, frequency bias can occur. The peak locations tend to be slightly dependent on the initial phases of the sinusoids. Spectral line splitting may be present.

Modified Covariance Method

The modified covariance method appears to give statistically stable spectral estimates with high resolution, even for short data records.

Furthermore, in the spectral analysis of sinusoids in noise, although also the modified covariance method is affected by a shifting of the peaks from their true locations due to the noise, this shifting appears to be less pronounced than with other AR estimation techniques. In addition, the peak locations tend to be less sensitive to the initial phases of the sinusoids. Finally, unlike the previous methods, it appears that the modified covariance method is not subject to spectral line splitting.

Burg's method

The primary advantage of Burg's method consists of the high resolution provided for short data records, for which Burg's algorithm produces better estimates than the YW method. The accuracy of Burg's method is lower for long data records.

In spectral analysis of sinusoids in noise, Burg's method is able to resolve closely spaced sinusoids when the SNR is not too low. It can, however, be affected by line splitting and frequency bias, with spectral peak locations highly dependent on the initial phase of noisy sinusoidal signals (see, e.g., Chap. 10 in Percival and Walden 1993). An application of the Burg spectral method to a real-world signal will be provided in Sect. 12.3.

A few examples of spectral line splitting and frequency bias for sinusoid in noise can be useful at this point. We will use Burg's method for illustration, since it is particularly prone to these unfavorable effects.

Incidentally, we point out that line splitting can occur even with true AR processes, when p is too large with respect to the available amount of data. To illustrate the phenomenon we report an example proposed by Hayes (1996). We consider (Fig. 11.16) $N = 64$ samples of the AR(2) signal $x[n] = 0.9\,x[n-2] + e[n]$, with noise unit variance, and estimate the power spectrum by Burg's method with $p = 4$ and $p = 12$.

Fig. 11.16 Spectral line splitting in the case of a true AR(2) process analyzed by Burg's method adopting $p = 4$ (*gray curve*), and overmodeling by using $p = 12$ (*black curve*). The *vertical dotted line* identifies the single frequency that characterizes this AR(2) process, which is given by the absolute value of the phase angle of the model's poles

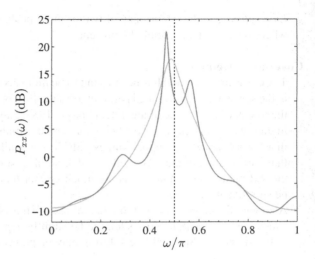

The true spectrum has a single spectral peak at $\omega = \pi/2$ (*vertical dotted line* in Fig. 11.16), and with $p = 4$ this peak is fairly well detected (*gray curve*). However, when overmodeling using $p = 12$, this peak is split into two peaks (*black curve*). The phenomenon will not always be observed: whether or not line splitting occurs depends on the specific white noise realization that is filtered to produce $x[n]$.

Returning to the case of sinusoids in noise, the drawbacks of the Burg algorithm for AR power-spectrum estimation have been studied in depth using this kind of synthetic data. They were reported by Chen and Stegen (1974), who studied frequency bias in particular. Fougère et al. (1976) and Fougère (1977) documented line splitting in detail and saw that the worst case of line splitting occurs when the amount of data is such that an odd number of quarter sinusoid's cycles is contained in the record length, in association to an initial phase of $\pi/4$. Other authors (de Waele and Broersen 2003) documented how an excessive model order may lead to line splitting and/or spurious spectral peaks. Actually, the phenomena of frequency bias and line splitting are influenced by many factors, including data length, SNR, model order, accuracy of parameter estimates and initial phase of the sinusoids in the data. Let us consider a couple of examples.

Spectral line splitting

Here we take $N = 48$ data samples generated by a sinusoid with amplitude $A = 1$, initial phase of $\pi/4$ and frequency $\omega_0 = 0.4\pi$, and include white noise adopting two dramatically different SNRs, namely 5 and 50, corresponding to $\sigma_e = 0.3976$ and $\sigma_e = 0.0022$, respectively. We select the relatively high order $p = 12$. Figure 11.17 illustrates the results of the simulation. The line splitting is clearly evident in the expanded view of Fig. 11.17b. Spectral line splitting is most likely to occur when the SNR and the model order are both high. Moreover, in agreement with what was suggested by Fougère et al. (1976), splitting here occurs for an initial phase of the sinusoid equal to $\pi/4$. Line splitting in this case is favored also by the fact that the number of AR estimated parameters is a large

Fig. 11.17 Spectral line splitting in the case of a sinusoid in noise ($N = 48$ samples) analyzed by Burg's method with $p = 12$. *Black curve*: SNR=5; *gray curve*: SNR=50. The *vertical dotted line* identifies the true sinusoid's frequency. **a** The whole spectrum; **b** the spectrum in the vicinity of the true frequency

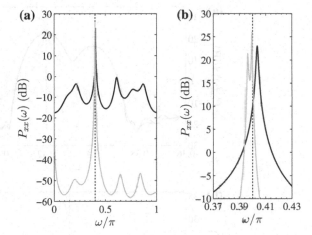

percentage of the number of data samples used for the estimation. It is also evident in Fig. 11.17b that neither of the two spectral peaks coincides with the true frequency, and this is frequency bias. Spurious frequency components appearing in the whole spectrum shown in Fig. 11.17a are due to some overmodeling.

Frequency bias and resolution

We now consider $N = 48$ samples of a signal containing two sinusoids in noise. The amplitudes are $A_1 = A_2 = 1$, the frequencies are 0.4π and 0.41π, the initial phases are chosen at random between 0 and 2π, the SNR is kept fixed at 30 and

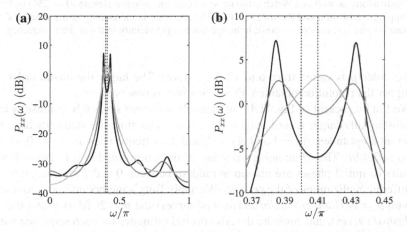

Fig. 11.18 Frequency bias in the case of two sinusoids in noise ($N = 48$ samples) analyzed by Burg's method, adopting p values of 6 (*light gray curve*), 8 (*gray curve*), and 12 (*black curve*). The *vertical dotted lines* identify the true sinusoids' frequencies. **a** The whole spectrum; **b** the spectrum in the vicinity of the true frequencies

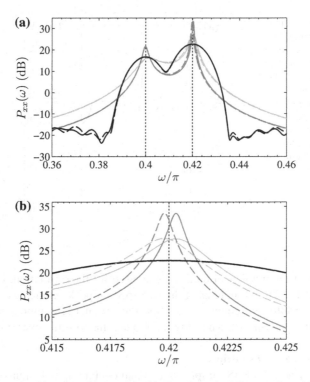

Fig. 11.19 Example of frequency bias produced by Burg's method. For two sinusoids with different amplitudes in noise, two different pairs of initial phases have been chosen at random to generate two different realizations of the process ($N = 1024$). Burg spectral estimates with $p = 14$ (*light gray solid and dashed curves*) and $p = 20$ (*dark gray solid and dashed curves*) are produced for both realizations, as well as a Welch estimate with Hamming window (length $M = 256$) and 50 % overlapping (*black solid and dashed curves*). **a** The spectrum in the vicinity of the sinusoids' true frequencies (see *vertical dotted lines*); **b** the spectrum in the vicinity of the highest frequency

the order p is varied from 6 to 12 (Fig. 11.18). The larger the model order, the higher the resolution obtained. Frequency bias is observed.

We finally present in Fig. 11.19 an example with two sinusoids in noise, with a substantially larger record length, $N = 1024$. This time the sinusoids have different amplitudes, $A_1 = 1$ and $A_2 = 2$, and their frequencies are $\omega_1 = 0.4\pi$ and $\omega_2 = 0.42\pi$. The white noise has standard deviation equal to 0.1. Two different pairs of initial phases are chosen at random between 0 and 2π to generate two different realizations of the process. We apply Burg's method to both sequences, with $p = 14$ (*light gray solid and dashed curves*) and $p = 20$ (*dark gray solid and dashed curves*), and compare the AR spectral estimates for each sequence with a Welch estimate computed adopting a 256-samples-long Hamming window, with overlapping of 128 samples (*black solid and dashed curves*). *Vertical dotted lines* indicate true frequencies. Figure 11.19a shows the region of the spectral peaks. In Fig. 11.19b, which provides an expanded view of the highest-frequency peak, we can see that with both realizations, in Burg's estimates the highest-frequency sinusoid appears shifted with respect to its true value, in a way that changes with changing initial phase.

11.8 Examples of AR Spectral Estimates

A complete set of examples illustrating the properties of the four AR spectrum estimation techniques presented above, for different types of random processes and in comparison with non-parametric spectral methods, would be too long to provide in an introductory book like the present one. Nevertheless, we will try to compare the effectiveness of each approach in estimating the spectra of narrowband and broadband AR processes, though we will not investigate the effects of different AR orders and different record lengths. We will also examine the performances of these techniques when applied to other types of processes, like sinusoids in noise, with varying values of N. More precisely, we will consider

- two medium-length data records, both derived from a low-order AR(4) process but with coefficients chosen in such a way to represent the contrasting possibilities of a broadband process in the first case, and of a narrowband process in the second case;
- one record with sinusoids in Gaussian white noise, with a varying number N of samples.

11.8.1 Broadband and Narrowband AR Processes

We take white noise with unit variance and filter it using the two AR(4) all-pole filters that we already used for Fig. 11.10c, d. We thus generate an AR(4) broadband signal and an AR(4) narrowband signal. We generate 50 realizations of each process, but this time we select a shorter length for each realization ($N = 256$). Each realization is then analyzed using the YW method, the covariance method, the modified covariance method and Burg's method, with $p = 4$ in all cases. The resulting spectral estimates are shown in Figs. 11.20 and 11.21:

- Figure 11.20a is a Burg spectral estimate of the broadband AR(4) process,
- Figure 11.20b is a Burg spectral estimate of the narrowband AR(4) process,
- Figure 11.20c is a YW spectral estimate of the broadband AR(4) process,
- Figure 11.20d is a YW spectral estimate of the narrowband AR(4) process,
- Figure 11.21a is a covariance spectral estimate of the broadband AR(4) process,
- Figure 11.21b is a covariance spectral estimate of the narrowband AR(4) process,
- Figure 11.21c is a modified covariance spectral estimate of the broadband AR(4) process,

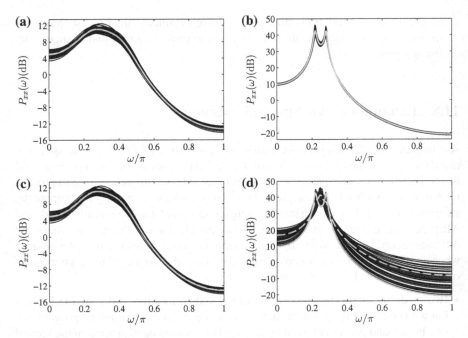

Fig. 11.20 **a** Burg spectral estimates for 50 realizations of a broadband AR(4) process. **b** Burg spectral estimates for 50 realizations of a narrowband AR(4) process. **c** Yule-Walker spectral estimates for 50 realizations of the same broadband AR(4) process of panel **a**. **d** Yule-Walker spectral estimates for 50 realizations of the same narrowband AR(4) process. *Black lines* in each panel show the spectra of 50 individual realizations in an overlaid fashion, *dashed gray curves* show the spectral estimate averaged over realizations, and *solid gray curves* depict the true spectrum. In all cases, $N = 256$ and $p = 4$

- Figure 11.21d is a modified covariance spectral estimate of the narrowband AR(4) process.

Black lines in each panel show the spectra of 50 individual realizations in an overlaid fashion, to indicate the variability of a given type of spectral estimate, *dashed gray curves* show the spectral estimate averaged over realizations, and *solid gray curves* depict the true spectrum. Observe how stable these estimates are if compared to the periodogram (observe Fig. 11.13c). All the estimators except the YW method are unbiased, so that the ensemble-average spectra (*dashed gray curves*) are indistinguishable from true spectra *solid gray curves*, and have comparable variances. The YW method shows a large bias in Fig. 11.20b, being unable to resolve the spectral peaks, and also exhibits a larger variance than the other parametric methods.

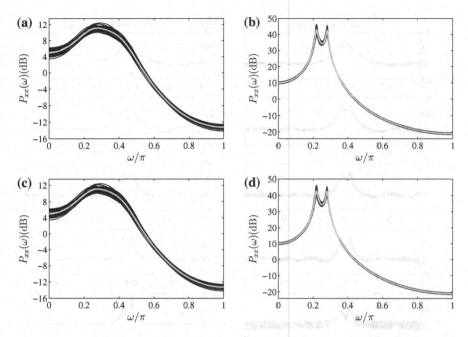

Fig. 11.21 **a** Covariance spectral estimates for 50 realizations of a broadband AR(4) process. **b** Covariance spectral estimates for 50 realizations of a narrowband AR(4) process. **c** Modified covariance spectral estimates for 50 realizations of the same broadband AR(4) process of panel **a**. **d** Modified covariance spectral estimates for 50 realizations of the same narrowband AR(4) process of panel **b**. *Black lines* in each panel show the spectra of 50 individual realizations in an overlaid fashion, *dashed gray curves* show the spectral estimate averaged over realizations, and *solid gray curves* depict the true spectrum. In all cases, $N = 256$ and $p = 4$

11.8.2 Sinusoids in Noise

We consider the process with two sinusoids in Gaussian white noise that we already used for some previous examples (see Fig. 10.9). In Figs. 11.22, 11.23 and 11.24, the parametric estimates of the power spectrum of such a signal ($N = 512$, $A_0 = A_1 = 5$, $\omega_0 = 0.4\,\pi$, $\omega_1 = 0.45\,\pi$; unit noise variance) are shown. Figure 11.22 is relative to the YW method, while Fig. 11.23 is relative to Burg's method; the order of the model varies from 8 to 64 and the record length is $N = 512$ in all cases. Figure 11.24 is still relative to Burg's method, but here the order is always $p = 18$, while the record length progressively decreases from 512 to 40 samples. Figures 11.22 and 11.23 show that as the model order increases, the peaks corresponding to the sinusoids become higher and narrower with both the YW and Burg methods, so that resolution improves. In Sect. 10.3.3 we verified that for the same random process considered here, with a very small record length ($N = 40$), the periodogram's resolution is insufficient (Fig. 10.9). The lowest panels of Fig. 11.24 show that at a reasonable order, even with this very short record, Burg's method provides an acceptable resolution. In conclusion, AR

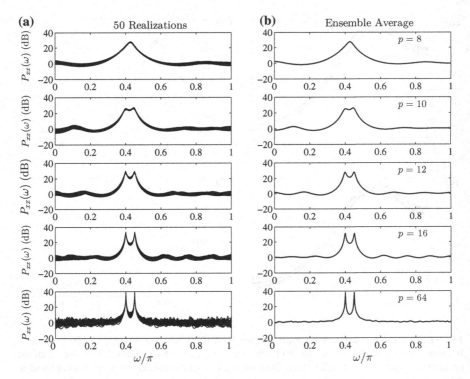

Fig. 11.22 Power spectrum estimates obtained via the YW method for two sinusoids in white noise. The model order varies from 8 to 64; the record length is 512

methods can help with resolution, especially for shorts records: for example, they can deal with sequences of 40–200 samples, for which Fourier methods would perform poorly. However, if the process is not authentically AR, keeping the order low we will overly smooth the spectrum, while increasing it we will probably generate spurious peaks.

11.9 Maximum Entropy Method (MEM)

As we mentioned in Sect. 11.7, the YW method is closely related to a different philosophical approach to the spectral estimation problem: the *maximum entropy method* (MEM).

One of the limitations with the classical approach to power spectrum estimation is that, for a data record of length N, the AC sequence can only be estimated for lags $|l| < N$. Consequently, the AC is set to zero for $|l| \geq N$. Since many signals of interest actually have ACs that are non-zero for $|l| \geq N$, this assumption may severely limit the resolution and the accuracy of the spectral estimate. This is particularly true

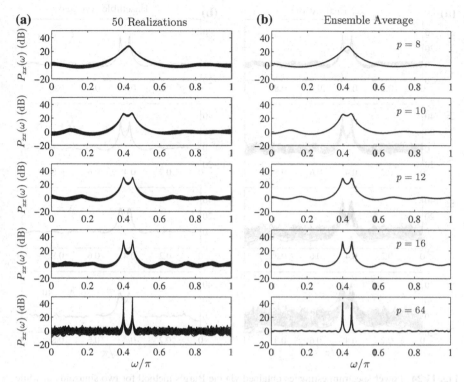

Fig. 11.23 Power spectrum estimates obtained via Burg's method for two sinusoids in white noise. The model order varies from 8 to 64; the record length is 512

in the case of narrowband processes that have ACs decaying slowly with increasing lag. As an example of this behavior, see Fig. 11.11b, d, representing the AC sequences of an AR(2) narrowband process and of an AR(4) narroband process, respectively. Observe these AC sequences in comparison with those illustrated in Fig. 11.11a, c, which represent broadband processes.

The question thus arises as to whether a better extrapolation of the AC sequence may be devised that can improve spectral estimation. This is nothing but the issue already discussed in connection to AR spectral estimation, but the MEM was developed focusing on this issue. The principle on which the MEM relies states that any method aimed at estimating the power spectrum of a process starting from a record of finite length N should do so in such a way that the estimate be consistent with the measured samples, but neutral about the part of the signal that has not been measured, on which no assumptions should be made.

Suppose we are given the AC of a WSS process for lags up to a certain lag p. The values of $\gamma_{xx}[l] \equiv \gamma[l]$, for $|l| \leq p$, have been accurately estimated or determined in some way—they are known. We wish to extrapolate for $|l| > p$. First of all, the extrapolation must lead to a valid power spectrum. If we write (neglecting the

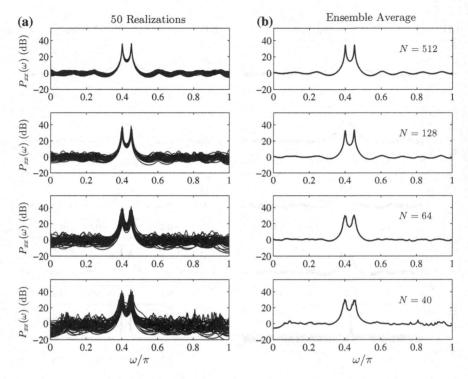

Fig. 11.24 Power spectrum estimates obtained via the Burg's method for two sinusoids in white noise. The model order is 18; the record length varies from 512 to 40 samples

estimation hat symbols on both the spectrum and the AC)

$$P_{xx}(\omega) = \sum_{l=-p}^{+p} \gamma[l]\mathrm{e}^{-jl\omega} + \sum_{|l|>p} \psi[l]\mathrm{e}^{-jl\omega},$$

where $\psi[l]$ are the extrapolated AC values, then $P_{xx}(\omega)$ should be real-valued and non-negative at all frequencies. But this is not sufficient to guarantee a unique extrapolation: additional constraints must be imposed.

Burg's approach was to try to estimate the spectrum of the process generating the observed AC samples $\gamma[l]$ without making any a priori assumptions about the unknown AC values for lags $p+1$ and beyond. He began by noting that there are an infinite number of power spectra which can be inversely Fourier-transformed to yield AC samples identical to the given set of $\gamma[l]$ samples over the range of lags for which the data is available. Beyond that range, of course, all these AC sequences will differ. The problem then becomes: which one of this infinite set of power spectra should we select? To answer this question, Burg used the criterion of *information entropy* as defined for power spectra by Shannon's information theory (Shannon and Weaver 1959). *For a Gaussian random process* $x[n]$ with true power spectrum $P_{xx}(\omega)$, the information entropy $\mathcal{H}_x$ is defined to within a scale factor by (Shannon 1948; Shannon and Weaver 1959; Cover and Thomas 1991)

$$\mathcal{H}_x = \frac{1}{2\pi} \int_{-\pi}^{\pi} \log P_{xx}(\omega) d\omega.$$

Let us assume that all the power spectra under consideration are constrained to have the same average power

$$\sigma_x^2 = P_0 = \frac{1}{2\pi} \int_{-\pi}^{\pi} P_{xx}(\omega) d\omega.$$

Then if $P_{xx}(\omega) = P_0$, the spectrum $P_{xx}(\omega)$ is flat, i.e., it is a white-noise power spectrum, and the entropy assumes a value that we will indicate as $\mathcal{H}_x^w$,

$$\mathcal{H}_x^w = \log(P_0).$$

It is relatively straightforward to convince ourselves that any non-white spectrum will have smaller entropy than a white spectrum having the same average power. We can start by noting that any possible candidate power spectrum must have a mean value of P_0. Then $P_{xx}(\omega)$ can be expressed in the form

$$P_{xx}(\omega) = P_0 \left[1 + P_{xx}'(\omega) \right],$$

where the average of $P_{xx}'(\omega)$ must be zero. In this way we have

$$\mathcal{H}_x = \mathcal{H}_x^w + \frac{1}{2\pi} \int_{-\pi}^{\pi} \log \left[1 + P_{xx}'(\omega) \right] d\omega.$$

For the sake of simplicity we assume that the magnitude of $P_{xx}'(\omega)$ is always much less than one, so that we can expand the log term as

$$\log \left[1 + P_{xx}'(\omega) \right] = P_{xx}'(\omega) - \frac{1}{2} \left[P_{xx}'(\omega) \right]^2 + \cdots.$$

Then we can write

$$\mathcal{H}_x - \mathcal{H}_x^w = \frac{1}{2\pi} \int_{-\pi}^{\pi} \left\{ P_{xx}'(\omega) - \frac{1}{2} \left[P_{xx}'(\omega) \right]^2 + \cdots \right\} d\omega \simeq$$

$$\simeq -\frac{1}{4\pi} \int_{-\pi}^{\pi} \left[P_{xx}'(\omega) \right]^2 < 0,$$

since the integral of $P'_{xx}(\omega)$ is zero. Thus we see that $\mathscr{H}_x < \mathscr{H}_x^w$: for a given $\sigma_x^2 = P_0$, a white noise signal has maximum entropy. Although we have only proven this result under the assumption $\left|P'_{xx}(\omega)\right| \ll 1$, it is true in general.

This is the criterion that Burg used to select one of the infinite possible power spectra: in the absence of any further information about the process generating the data, a reasonable choice for the power spectrum would be the one which both satisfied the constraints imposed by the given AC samples and at the same time had the maximum entropy. White noise has the greatest possible entropy; it is the most random process and its AC vanishes for all non-zero lags. By inference, we can deduce that choosing a power spectrum which is constrained by the known AC samples and yet has the maximum possible entropy, we will choose the spectrum that corresponds to the most random possible process consistent with the observed data, i.e., the process characterized by the fewest assumptions.

Finding the MEM spectral estimator means devising a maximum entropy extrapolation for the AC sequence. For the Wiener-Khinchin theorem, the maximum entropy process and the observed process will share the same spectrum, since they share the same AC sequence.

It can be shown that for a Gaussian random process $x[n]$ with given AC samples $\gamma[l]$, $|l| \leq p$, the MEM power spectrum is the spectrum of an AR(p) process. Indeed, the latter maximizes $\mathscr{H}_x$, subject to the constraint that the inverse Fourier transform of $P_{xx}(\omega)$ equals the given set of AC samples,

$$\gamma[l] = \frac{1}{2\pi} \int_{-\pi}^{\pi} P_{xx}(\omega) e^{jl\omega} d\omega, \qquad |l| \leq p.$$

In order to prove this, we must actually construct the MEM estimator, and in order to do so, we need to find the values $\psi[l]$ of extrapolated AC that maximize the entropy. One way to find the MEM estimator subject to fixed values of AC for $|l| \leq p$ is by use of the so-called Lagrange multipliers (see, e.g., Arfken 1985). We can instead find it by setting the derivative of the entropy with respect to each extrapolated value equal to zero (Robinson 1982):

$$\frac{\partial \mathscr{H}_x}{\partial \psi[l]} = \frac{\partial}{\partial \psi[l]} \frac{1}{2\pi} \int_{-\pi}^{\pi} \log P_{xx}(\omega) d\omega = \frac{1}{2\pi} \int_{-\pi}^{\pi} \frac{e^{-jl\omega}}{P_{xx}(\omega)} d\omega = 0$$

for $|l| > p$. The last equality derives from the fact that from the definition of $P_{xx}(\omega)$ as the DTFT of the known and extrapolated parts of the AC sequence, we can write

$$\frac{\partial}{\partial \psi[l]} \{\log P_{xx}(\omega)\} = \frac{e^{-jl\omega}}{P_{xx}(\omega)} \equiv e^{-jl\omega} Q_{xx}(\omega),$$

where we defined $Q_{xx}(\omega)$ as the reciprocal of the spectrum we are looking for, i.e., $Q_{xx}(\omega) = 1/P_{xx}(\omega)$. $Q_{xx}(\omega)$ is certainly non-negative, and we assume that it is bounded and integrable, i.e., that it is a well-behaved power spectrum in its own right. In this way we can write the condition for entropy maximization as

$$\frac{1}{2\pi} \int_{-\pi}^{\pi} Q_{xx}(\omega) e^{-jl\omega} d\omega = 0, \qquad |l| > p.$$

Being a well-behaved power spectrum, $Q_{xx}(\omega)$ can be associated with a real and even AC sequence $q[l]$, so that the following equations hold:

$$q[l] = \frac{1}{2\pi} \int_{-\pi}^{\pi} Q_{xx}(\omega) e^{jl\omega} d\omega \quad \text{for} \quad -\infty < l < \infty,$$

$$Q_{xx}(\omega) = \sum_{l=-\infty}^{\infty} q[l] e^{-jl\omega}.$$

Comparing the second equation with the maximum-entropy constraint we see that in order to satisfy the latter, $q[l]$ must vanish for $|l| > p$. This means that we can write

$$Q_{xx}(\omega) = \sum_{l=-p}^{p} q[l] e^{-jl\omega}$$

and form the MEM estimator as

$$P_{xx}^{\text{MEM}}(\omega) = \frac{1}{\sum_{l=-p}^{p} q[l] e^{-jl\omega}}.$$

Passing to the z-transform, with

$$P_{xx}^{\text{MEM}}(\omega) \equiv P_{xx}^{\text{MEM}}\left(e^{j\omega}\right) = \left. P_{xx}^{\text{MEM}}(z) \right|_{z=e^{j\omega}},$$

we can write

$$P_{xx}^{\text{MEM}}(z) = \frac{1}{\sum_{l=-p}^{p} q[l] z^{-l}} = \frac{1}{Q(z)},$$

so that the poles of $Q(z)$ are the zeros of $P_{xx}^{\text{MEM}}(z)$ and vice-versa. But since $q[l]$ is a real and even sequence, its z-transform has the following property (Sect. 9.9):

$$Q(z) = Q^*(z^*) = Q(1/z) = Q^*(1/z^*).$$

This implies the possibility of factorizing $Q(z)$, up to a non-negative constant, as $A(z)A(1/z)$, with

$$A(z) = 1 + a_1 z^{-1} + a_2 z^{-2} + \cdots + a_p z^{-p}.$$

being stable and minimum-phase, and with a_0 normalized to 1 (Robinson 1980). So we can write

$$Q(z) = \frac{1}{\sigma^2} A(z)A(1/z),$$

where σ^2 is a non-negative constant. Therefore

$$P_{xx}^{\mathrm{MEM}}(z) = \frac{\sigma^2}{A(z)A(1/z)} = \frac{\sigma^2}{A(z)A^*(1/z^*)}.$$

The corresponding power spectrum on the unit circle is

$$P_{xx}^{\mathrm{MEM}}(\omega) = \frac{\sigma^2}{\left| A\left(\mathrm{e}^{\mathrm{j}\omega}\right) \right|^2} = \frac{\sigma^2}{\left| 1 + \sum_{l=1}^{p} a_l \mathrm{e}^{-\mathrm{j}l\omega} \right|^2}$$

and therefore is a well-behaved, real and non-negative AR(p) spectrum.

Having determined the form of the MEM spectrum, all that remains is to find the parameters appearing in this formula, including σ. Due to the constraints we set on $P_{xx}(\omega)$, these coefficients must be chosen in such a way that the IDFT of the estimated $P_{xx}^{\mathrm{MEM}}(\omega)$ produces an AC sequence that matches the given $\gamma[l]$ for $|l| \le p$. If the parameters are the solution of the augmented YW equations, with $\sigma^2 = \sigma_e^2$, then the AC-matching constraint will be satisfied.

In summary, we will compute the MEM spectrum as follows: we will

- solve the augmented YW equations and find the a_i and σ_e^2;
- incorporate the estimated parameters in $P_{xx}^{\mathrm{MEM}}(\omega)$.

Note that choosing this MEM estimator, then $\gamma[l]$ for any l satisfies the YW equations

$$\gamma[l] = - \sum_{i=1}^{p} a_i \gamma[l - i] \quad \text{for} \quad l > 0,$$

and thus actually the AC is extrapolated according to this recursion. The MEM is thus equivalent to the YW method. The only difference between the two methods lies in the assumptions that are made about the process, since here it is assumed that the process is Gaussian, while with the YW method it is assumed that $x[n]$ is an AR process. The MEM thus boils down to looking for an AR process that "mimics" the original time series. This is why it is classified as a parametric method (see, e.g., Ghil and Taricco 1997).

The properties of the MEM have been studied extensively and, as a spectrum analysis tool, the MEM is subject to different interpretations. In the absence of any information or constraints on the process, and given a finite set of AC values, taking the Fourier transform of the AC sequence formed from the available data values, along with an extrapolation that imposes the least assumptions on what is not observed, appears very reasonable. On the other hand, it may be argued that MEM imposes an all-pole model on the data, and unless the process is known to be consistent with this model, then the estimated spectrum may be inaccurate. This, of course, is a criticism that applies to any AR estimate. Whether or not the MEM estimate—or any other AR estimate—is "better" than an estimate obtained from classical approaches depends critically on what type of process is being analyzed and how closely it may be modeled as an AR process.

The MEM is very efficient in detecting spectral lines and narrowband features of a WSS time series. The art of using MEM, as in any AR method, resides in a reasonable choice for the model order (see, for instance, Ghil and Taricco 1997). The behavior of the spectral estimate depends on this choice: as we already remarked in previous sections, the number of poles (and consequently the number of spectral maxima) depends on the order, so that, for a given time series, the number of peaks will increase with order. Therefore, a trade-off between a good resolution (high order) and few spurious peaks (low order) has to be found for the MEM too. These weaknesses can be remedied partly by

- determining which peaks survive reductions in the order,
- comparing MEM spectra to those produced by the BT method and the MTM, which generally should not share spurious peaks with MEM, and
- using SSA (Chap. 12) to pre-filter the series and thus decompose the original sequence into several components, each of which contains only a few harmonics and is substantially noise-free, so that small order values can be chosen (Penland et al. 1991). Many simulations have actually shown that the resolution of the MEM estimator decreases as SNR decreases (Lacoss 1971; Marple 1977). SNR enhancement using a data-adaptive filter based on SSA improves the resolution of low-order MEM estimates. Of course, the same considerations apply to the other AR spectral estimation methods.

References

Akaike, H.: Statistical predictor identification. Ann. Inst. Statist. Math. **22**, 203–217 (1970)

Akaike, H.: A new look at the statistical model identification. IEEE Trans. Autom. Control **AC19**(6), 716–723 (1974)

Allen, M.R., Smith, L.A.: Monte Carlo SSA: detecting irregular oscillations in the presence of coloured noise. J. Climate **9**(12), 3373–3404 (1996)

Anderson, T.W.: The Statistical Analysis of Time Series. Wiley, New York (1971)

Anderson, O.: Time Series Analysis and Forecasting: The Box-Jenkins Approach. Butterworths, London (1976)

Arfken, G.B.: Mathematical Methods for Physicists. Academic Press, Orlando (1985)

Bartlett, M.S.: An Introduction to Stochastic Processes. Cambridge University Press, Cambridge (1955)

Box, G.E.P., Pierce, D.A.: Distribution of residual autocorrelations in autoregressive-integrated moving average time series models. J. Am. Statist. Assoc. **65**, 1509–1526 (1970)

Box, G.E.P., Jenkins, G.M., Reinsel, G.C.: Time Series Analysis: Forecasting and Control, 4th edn. Wiley, Hoboken (2008)

Burg, J.P.: Maximum entropy spectral analysis. In: 37th Annual International Meeting., Soc. Explor. Geophys., Oklahoma City, OK, USA (1967)

Burg, J.P.: A new analysis technique for time series data. In: NATO Advanced Study Institute on Signal Processing with Emphasis on Underwater Acoustics, Enschede, The Netherlands (1968)

Burg, J.P.: Maximum Entropy Spectral Analysis. Ph.D. Dissertation, Stanford University, Stanford, CA, USA (1975)

Chen, W.Y., Stegen, G.R.: Experiments with maximum entropy power spectra of sinusoids. J. Geophys. Res. **79**, 3019–3022 (1974)

Cover, T.M., Thomas, J.A.: Elements of Information Theory. Wiley, New York (1991)

Chatfield, C.: The Analysis of Time Series. CRC Press, Boca Raton (2004)

de Waele, S., Broersen, P.M.T.: Order selection for vector autoregressive models. IEEE Trans. Signal Process. **51**(2), 427–432 (2003)

Fougère, P.F., Zawalick, E.J., Radoski, H.R.: Spontaneous line splitting in maximum entropy power spectrum analysis. Phys. Earth Planet. In. **12**, 201–207 (1976)

Fougère, P.F.: A solution to the problem of spontaneous line splitting in maximum entropy power spectrum analysis. J. Geophys. Res. **82**, 1051–1054 (1976)

Ghil, M., Taricco, C.: Advanced Spectral Analysis Methods. In: G. Cini Castagnoli and A. Provenzale (eds.) Past and Present Variability of the Solar-Terrestrial System: Measurement, Data Analysis and Theoretical Models, Società Italiana di Fisica, Bologna, & IOS Press, Amsterdam (1997)

Hayes, M.H.: Statistical Digital Signal Processing and Modeling. Wiley, New York (1996)

Haykin, S. (ed.): Nonlinear methods of spectral analysis. Topics in Applied Physics, vol. 34. Springer, New York (1983)

Hess, G.D., Iwata, S.: Measuring and comparing business-cycle features. J. Bus. Econ. Stat. **15**(4), 432–444 (1997)

Lacoss, R.T.: Data-adaptive spectral analysis methods. Geophysics **36**, 661–675 (1971)

Levinson, N.: The Wiener RMS error criterion in filter design and prediction. J. Math. Phys. **25**, 261–278 (1947)

Lysne, D., Tjøstheim, D.: Loss of spectral peaks in autoregressive spectral estimation. Biometrika **74**, 200–206 (1987)

Kay, S.M.: Modern Spectral Estimation: Theory and Applications. Prentice-Hall, Englewood Cliffs (1988)

Kay, S.M., Marple, S.L.: Spectrum analysis-a modern perspective. Proc. IEEE **69**(11), 1380–1419 (1981)

Kolmogorov, A.N.: Interpolation and extrapolation of stationary random sequences. Izv. Akad. Nauk SSSR Ser. Mat. **5**, 3–14 (1941)

Marple, S.L.: Resolution of conventional fourier, autoregressive and special ARMA methods of spectral analysis. In: Proceedings of the 1977 IEEE International Conference on Acoustics, Speech and Signal Process, pp. 74–77 (1977)

Marple, S.L.: Digital Spectral Analysis: With Applications. Prentice-Hall, Upper Saddle River (1987)

Mac Lane, S., Birkhoff, G.: Algebra. AMS Chelsea Publishing, New York (1999)

McConnell, M.M., Perez-Quiros, G.: Output fluctuations in the United States: what has changed since the early 1980s? Am. Econ. Rev. **90**(5), 1464–1476 (2000)

Meltzer, J.A., Zaveri, H.P., Goncharova, I.I., Distasio, M.M., Papademetris, X., Spencer, S.S., Spencer, D.D., Constable, R.T.: Effects of working memory load on oscillatory power in human intracranial EEG. Cerebral Cortex **18**, 1843–1855 (2008)

Montgomery, D.C., Jennings, C.L., Kulahci, M.: Introduction to Time Series Analysis and Forecasting. Wiley, New York (2008)

Parzen, E.: An Approach to Time Series Modeling and Forecasting Illustrated by Hourly Electricity Demands. Technical Report Statistical Science Division, State University of New York, 37, NY, USA (1983)

Penland, C., Ghil, M., Weickmann, K.M.: Adaptive filtering and maximum entropy spectra, with application to changes in atmospheric angular momentum. J. Geophys. Res. **96**, 22659–22671 (1991)

Percival, D.B., Walden, A.T.: Spectral Analysis for Physical Applications. Cambridge University Press, Cambridge (1993)

Powell, M.J.D.: Approximation Theory and Methods. Cambridge University Press, Cambridge (1981)

Priestley, M.B.: Spectral Analysis and Time Series. Academic Press, London (1994)

Rissanen, J.: Modeling by shortest data description. Automatica 14, 465–471 (1978)

Rissanen, J.: A universal prior for the integers and estimation by minimum description length. Ann. Statist. 11, 417–431 (1983)

Robinson, E.A.: Physical Applications of Stationary Time Series. McMillan, New York (1980)

Robinson, E.A.: A historical perspective of spectrum estimation. Proc. IEEE 70(9), 885–907 (1982)

Sella, L., Vivaldo, G., Groth, A., Ghil, M.: Economic Cycles and Their Synchronization: A Survey of Spectral Properties. Nota di Lavoro 105.2013, Fondazione ENI Enrico Mattei, Milan, Italy (2013)

Shannon, C.E.: A mathematical theory of communication. Bell Syst. Tech. J. 27(3), 379–423 (1948)

Shannon, C.E., Weaver, W.: The Mathematical Theory of Communication. The University of Illinois Press, Urbana (1959)

Stoica, P., Moses, R.L.: Spectral Analysis of Signals. Prentice Hall, Upper Saddle River (2005)

Ulrych, T.J., Bishop, T.N.: Maximum entropy spectral analysis and autoregressive decomposition. Rev. Geophys. Space Phys. 13, 183–200 (1975)

Ulrych, T.J., Clayton, R.W.: Time series modelling and maximum entropy. Phys. Earth Planet. In. 12(2–3), 188–200 (1976)

Ulrych, T.J., Ooe, M.: Autoregressive and mixed autoregressive-moving average models and spectra. In: Nonlinear Methods of Spectral Analysis. Springer, New York (1979)

Vaidyanathan, P.P.: The Theory of Linear Prediction. Morgan & Claypool, San Rafael (2008)

Walker, G.: On periodicity in series of related terms. Proc. R. Soc. Lond. A 131, 518–532 (1931)

Wei, W.: Time Series Analysis. Addison-Wesley, New York (1990)

Whittle, P.: Hypothesis Testing in Time Series Analysis. Almquist & Wiksell, Uppsala (1951)

Wold, H.O.: A Study in the Analysis of Stationary Time Series. Almqvist & Wiksell, Uppsala (1938)

Yaglom, A.M.: An Introduction to the Theory of Stationary Random Functions. Courier Dover Publications, New York (2004)

Yule, G.U.: On a method for investigating periodicities in disturbed series with special reference to Wolf's sunspot numbers. Philos. Trans. R. Soc. Lond. A 26, 267–298 (1927)

Chapter 12
Singular Spectrum Analysis (SSA)

12.1 Chapter Summary

This chapter is devoted to an approach of extracting periodic or quasi-periodic components from a random signal. Singular spectrum analysis (SSA) is not a conventional spectral analysis method: it is a technique aimed at representing the signal as a linear combination of elementary variability modes that are not necessarily harmonic components, but, more in general, are data-adaptive functions of time. Thus SSA does not provide an estimate of the power spectrum, but rather is a powerful denoising filter, able to separate autocoherent features—e.g., anharmonic oscillations, quasi-periodic phenomena—from random features. It is also an ally of conventional spectral analysis, because the individual modes extracted via SSA are very "clean" signals for which high-resolution, stable spectra can be obtained that are much more easily interpretable than those that could be estimated analyzing the original series. Finally, SSA is also a very efficient gap filling method, i.e., a method to fill gaps in data records, which is soundly based from a theoretical point of view.

SSA is a non-parametric method: it does not make any assumption about the generation of the observed signal. It does not require particular properties of stationarity or ergodicity. After a brief theoretical introduction, we will focus on real-world application examples illustrating the possibilities offered by SSA.

12.2 Elements of SSA Theory

Singular spectrum analysis (SSA) (Vautard and Ghil 1989)—see also Ghil and Taricco (1997), Ghil et al. (2002) and references therein—allows for a systematic description, quantification, and extraction of long-, medium- and short-term components of time series. It is not, in a strict sense, a simple spectral method, since it is aimed at representing the signal as a linear superposition of elementary *variability*

© Springer International Publishing Switzerland 2016
S.M. Alessio, *Digital Signal Processing and Spectral Analysis for Scientists*, Signals and Communication Technology,
DOI 10.1007/978-3-319-25468-5_12

modes that are not necessarily harmonic components. More in general, these modes are *data-adaptive* functions of time. SSA does not provide an estimate of the power spectrum but rather is a powerful de-noising filter, able to separate autocoherent features, such as anharmonic oscillations and quasi-periodic phenomena, from random features. Spectral analysis of the individual modes extracted via SSA can thus be performed by the Maximum Entropy Method (MEM) or via Burg's method adopting low orders (see Sects. 11.9 and 11.7), and leads to high-resolution spectra that are stable, simply structured and therefore much more easily interpretable than those that could be estimated analyzing the original series. Finally, SSA is also a very efficient *gap filling method* (Kondrashov and Ghil 2006), i.e., a method to fill gaps in data records, which is soundly based from a theoretical point of view.

SSA is a non-parametric method: it does not make any assumption about the generation of the observed signal. It does not require particular properties of stationarity or ergodicity (Ghil et al. 2002). To apply SSA and other methods, such as the BT method and the MTM, a freeware Toolkit has been created (Vautard et al. 1992; Dettinger et al. 1995; Allen and Smith 1996; Allen and Robertson 1996; Mann and Lees 1996; Ghil et al. 2002), which can be downloaded from http://www.atmos.ucla.edu/tcd/ssa/ and can be installed in many Unix-based systems, such as Mac OS X and Linux environments. The webpage also offers a User's Guide, and an interesting review paper (Ghil et al. 2002).

From the mathematical point of view, SSA stems from two classical methodologies: the Karhunen-Loéve spectral decomposition of time series and random fields (Karhunen 1946, 1947; Loéve 1978) and the MañéTakens embedding theorem (Mañé 1981; Takens 1981), thus by combining elements of different fields such as, statistics and probability theory, dynamical systems and signal processing. SSA is structured in the following way:

- after centering the data series (length N), the square and real $(M \times M)$ *lag-covariance matrix* $\boldsymbol{R}$ is built for some choice of $M < N$:

$$\boldsymbol{R} = \{r_{lm}\} \qquad l = [1, M], \qquad [m = 1, M],$$

in which each matrix element represents the correlation between the original series lagged by $l - 1$ time steps and the original series lagged by $m - 1$ time steps, so that the first column of $\boldsymbol{R}$ is the AC sequence of the original sequence ($l - 1 = 0$; $m - 1 = 0$). It must be noted that here l is a position index and not a lag; the corresponding lag in number of samples is $l - 1$. The same is true for m. The matrix is real, square and symmetrical.

There are actually distinct methods to define $\boldsymbol{R}$. In one method, by Broomhead and King (1986a), a window of length M is moved along the time series, producing a set of $N' = N - M + 1$ sequences (vectors). This set is used to obtain the $N' \times M$ *trajectory matrix* $\boldsymbol{D}$, where the ith row is the ith view of the time series through the window. In this approach, the lag-covariance matrix is defined as

$$\boldsymbol{R} = \frac{1}{N} \boldsymbol{D} \boldsymbol{D}^T.$$

Hereafter we will refer to this method as the BK method. In a second method, proposed by Vautard and Ghil (1989), R is estimated directly from the data as a Toeplitz matrix[1]; this method will be referred to as the VG method;

• diagonalizing the matrix R, its M real and non-negative eigenvalues and its M mutually orthogonal eigenvectors are computed. The linear algebra procedure employed to diagonalize the matrix is called *singular value decomposition* (SVD).[2] The reader who is not acquainted with basic elements of linear algebra is invited to consult, as a first approach to this topic, the summary appearing in Hayes (1996), or refer to MacLane and Birkhoff (1999) or similar books presenting linear algebra from its basic principles.

The eigenvectors are called *empirical orthogonal functions* or EOFs. Each eigenvalue λ_i, with $i = [1, M]$, characterizes a different elementary variability mode in

---

[1] Recall that in linear algebra, a Toeplitz matrix is a matrix in which each descending diagonal from left to right is constant, i.e., all elements in a diagonal are equal.

[2] We must underline that SVD is different from the so-called *eigen-decomposition* used to perform diagonalization of a square matrix. Focussing on real matrices, matrix diagonalization is the process of taking a square matrix and converting it into a special type of matrix—a so-called diagonal matrix—that shares the same fundamental properties of the underlying matrix. Matrix diagonalization is equivalent to transforming the underlying system of equations into a special set of coordinate axes in which the matrix takes this diagonal form. Diagonalizing a matrix is also equivalent to finding the matrix's eigenvalues, which turn out to be precisely the entries of the diagonalized matrix. Similarly, the eigenvectors make up the new set of axes corresponding to the diagonal matrix. The relationship between a diagonalized matrix, eigenvalues, and eigenvectors follows from an equation according to which a square matrix R can be decomposed into the form $R = A \Lambda A^{-1}$, where A is a matrix composed of the eigenvectors of R arranged by columns, Λ is the diagonal matrix constructed from the corresponding eigenvalues, and A^{-1} is the matrix inverse of A. It is obvious that if a square matrix R has a matrix of eigenvectors A that is not invertible, then R does not have an eigen-decomposition. However, in the case of the lag-covariance matrix, the eigenvectors are orthogonal to one another. The matrix A having the eigenvectors of R as its columns is thus an orthogonal matrix, so that its inverse is equal to its transpose: $A^{-1} = A^T$. Then R can be written using a so-called SVD of the form $R = A \Lambda A^T$.

More generally, in linear algebra the SVD is a factorization of a real or complex matrix, which is not necessarily square—it can also be rectangular. Even if linear algebra is beyond the scope of the book, we may mention that some key differences between SVD and eigen-decomposition are the following:

– the vectors forming the columns of the eigen-decomposition matrix A are not necessarily orthogonal. On the other hand, the vectors that in the SVD factorization play a similar role are orthonormal; therefore the corresponding matrices—two different matrices—involved in SVD are orthogonal;

– these matrices involved in SVD are not necessarily the inverse of one another. They are usually not related to each other at all. In the eigen-decomposition, the matrices involved are inverses of each other, i.e., they are A and A^{-1};

– in the SVD, the entries in the diagonal matrix Λ are all real and non-negative. In the eigen-decomposition, the entries of the corresponding matrix can be any complex number—negative, positive, imaginary, whatever;

– the SVD always exists for any sort of rectangular or square matrix, whereas the eigen-decomposition only exists for square matrices, and even among square matrices sometimes it doesn't exist.

the signal; the set of the λ_i is referred to as the *eigenvalue spectrum*, or the *eigen-spectrum*, and contains information about the contribution given by the individual ith mode to the total signal's variance: more precisely, the total variance σ_x^2 of the series is equal to $(1/M) \sum_{i=1}^{M} \lambda_i$, and each mode explains a percentage of the total variance given by

$$\text{pct}_i = 100 \frac{\lambda_i}{M\sigma_x^2} = 100 \frac{\lambda_i}{\sum_{i=1}^{M} \lambda_i}.$$

The square roots $\sqrt{\lambda_i}$ of the eigenvalues are referred to as *singular values* and the set they form is the *singular spectrum*; hence the name of the SSA method;

• the eigenvalues and the eigenvectors (EOFs) are then ordered by decreasing eigenvalue importance. Plotting the eigenvalues, thus arranged in decreasing order, versus their order number—the number that identifies the mode, i.e., versus their *rank*—a *SSA spectrum* is obtained, in which often an initial steep descent appears, representing the signal content, followed by a more or less flat plateau, representing the noise in the record (Vautard and Ghil 1989). Error bars are displayed in this plot, showing an ad-hoc range of estimation errors for eigenvalues. These error bars are based on the estimated *decorrelation time* of the time series, according to:

$$\Delta\lambda_i = \left(\frac{2kL}{N}\right)^{1/2},$$

where L is a typical decorrelation time for the time series, and k is a user-supplied decorrelation weight between 1 and 2 (typically, 1.5). The decorrelation length L is normally estimated to be the inverse of the logarithm of the lag-one autocorrelation coefficient of the time series.

As for the use of the VG or the BK method for creating the lag-covariance matrix, in practice, the VG method may suffer from numerical instabilities (yielding negative eigenvalues) when pure oscillations are analyzed. The BK method is somewhat less prone than the VG method to problems with highly non-stationary time series (Allen et al. 1992), although the VG method seems untroubled by all but the most extreme non-stationarities. The VG method also imposes symmetries on the EOF shapes, whereas the BK method does not. However, the VG method appears to yield more stable results under the influence of minor perturbations of the time series (Allen and Smith 1996).

In order to correctly separate signal from noise, visual inspection of the eigenspectrum must be supported by application of proper significance tests, which will soon be described. The first eigenvalues—i.e., the dominant ones—usually come in pairs with nearly equal values, with error bars having a common range. The corresponding eigenvectors are in phase quadrature. Vautard and Ghil (1989) found that, provided the significance tests mentioned above give positive results, these pairs of modes can be interpreted as the nonlinear counterpart of a sine/cosine couple in linear Fourier analysis and therefore represent an oscillation mode, generally non-sinusoidal, detected in the signal. Thus a non-sinusoidal waveform, like a square wave or a sawtooth wave, can be represented by a single pair of modes that capture its periodicity, instead of the many harmonics that would be necessary in linear Fourier analysis that uses base functions of fixed form.

The *trend*, i.e., the long-term series behavior, may be represented by a single mode. More precisely, the trend is usually captured by either 1 or 2 components. If the trend-related EOFs are a pair, then normally one presents a zero-crossing, while the other one does not. If the trend is very strong, it can be analyzed and possibly removed by directly inspecting the shapes of the EOFs. When the trend is not very pronounced, a de-trending test supplied by the SSA-MTM toolkit may give a guidance in trend recognition. This test is based on Kendall's τ nonparametric method for trend detection (Freas and Sieurin 1977; Hirsch and Slack 1984; Hirsch et al. 1982):

- once the visual inspection of the eigenspectrum and the statistical significance tests have detected which eigenvalues represent the useful signal and which represent undesired noise, a selection of modes can be performed, discarding the noise-related modes. In this way the background noise in the signal is eliminated, and the most significant variability modes are retained;
- the time behavior of each selected mode or pair of modes can be reconstructed, and its spectrum analyzed adopting low-order MEM estimators, thus obtaining stable spectra less prone to exhibiting spurious peaks with respect to the direct analysis of the original series (see Sect. 11.9). Such *reconstructed components* or RCs allow us to compare the time behavior of the significant modes with that of possible external forcings acting on the system that generated the signal. Summing all the significant RCs, a de-noised version of the signal can be obtained.

SSA is particularly useful in case of short and noisy records; it is computationally efficient and converges rapidly, meaning that few modes can represent the significant variability contained in a data record.

The parameter that must be fixed in SSA is M, the *window length*, a.k.a. the *embedding dimension*. The choice of M is based on a trade-off between two considerations: quantity of information extracted versus the degree of statistical confidence in that information (Ghil et al. 2002). By and large, it is advisable to choose $M < N/5$, where N is the number of available data samples. Moreover M should be greater than the number of sequence samples contained in one period of the oscillations that must be detected: in fact, SSA is typically efficient in detecting cyclicities with periods in the range between $M/5$ and M (Vautard et al. 1992). M should also be smaller than the number of data samples contained in the typical duration of the intervals in which a possibly intermittent oscillation remains active. Let us make an example: if we want to analyze data with a sampling interval of 1 year and want to study oscillations up to 500 years of period in a 2500 year long record ($N = 2500$), and if we moreover know that typically these oscillations appear intermittently and remain active each time for durations of the order of 1000 years, then $M = 500$ is a reasonable choice. However, it is better to try different values of M, and the robustness of the SSA results to changes of the window length is an important test of their validity.

We still must understand how the significant modes are selected: how is the significance of each dominant mode tested? To this purpose, no hypotheses about the probability distribution of the data are made—for example, they are not assumed to be Gaussian—and therefore no conventional statistical tests can be performed. The significance of the modes is established or rejected on the basis of *Monte Carlo* simulations, according to a method called MC-SSA, introduced by Allen (1992), Allen and Smith (1996). For the test, a null hypothesis of background AR(1) red noise is adopted, with parameter $\alpha = \rho[1]$ and variance σ_e^2 estimated from the signal by maximum likelihood criteria. A high number of realizations of this red-noise process is then generated. This data is called *surrogate data* and typically consists of a few thousands of sequences. For each realization, the lag-covariance matrix C_R is computed and projected onto the basis formed by the eigenvectors of the matrix R derived from the original data. In the case of R, this projection leads to a diagonal matrix $\Lambda = A^T R A$, where A is the matrix having the EOFs as its columns, A^T is the corresponding transposed matrix and Λ has the eigenvalues λ_i on its main diagonal, while the remaining matrix elements are zero.

In the case of C_R, which is derived from surrogate data, since $\Lambda_R = A^T C_R A$ is not the SVD of C_R, the matrix Λ_R is not necessarily diagonal, but its greater or lesser distance from diagonality allows quantifying the similarity of a given surrogate series with the original one. Over the whole collection of surrogate data, the similarity between red noise and the original series can be quantified examining the statistical distribution of the elements of Λ_R. In this way, confidence intervals are obtained at a fixed probabilistic level, outside which a modal contribution to $x[n]$, i.e., a RC reconstructed using one mode or a pair of modes, can be considered significantly different from red noise. The test is bilateral, so that a pair of confidence levels must be specified. For example, setting 0.05 and 0.95 we would plot noise error bars spanning the 5th to 95th percentiles of the noise distribution, and would have a 90 % confidence level (c.l.); setting 0.01 and 0.99 we would plot noise error bars spanning the 1th to 99th percentiles, and would have a 98 % c.l.; setting 0.005 and 0.995 we would plot noise error bars spanning the 0.5th to 99.5th percentiles of the noise distribution, and would have a 99 % c.l. The default in the SSA-Toolkit is 95 %, from 2.5th to 97.5th percentiles.

Let us stress that comparing M data eigenvalues with M confidence intervals computed from the surrogate ensemble, we expect, e.g., $M/10$ to lie above the 90th percentile even if the null hypothesis is true. Thus a small number of excursions above a relatively low percentile should be interpreted with caution. Allen and Smith (1996) discuss this problem in detail, and propose a number of possible solutions.

The MC-SSA algorithm described above can be adapted to eliminate known periodic components and test the residual against noise. This adaptation provides sharper insight into the dynamics captured by the data, since known periodicities (like orbital forcing on the Quaternary time scale or seasonal forcing on the intraseasonal-to-interannual one, in climate studies and other fields) often generate much of the variance at the lower frequencies manifest in a time series and alter the rest of the spectrum. Allen and Smith (1996) describe this refinement of MC-SSA, which consists in restricting the projections to the EOFs that do not account for known periodic behavior. This refinement is implemented in the SSA-MTM toolkit, where it is referred to as the *hybrid basis*.

This procedure can be also used to repeat the MC-SSA test after recognizing some serie's components as significant in a first run of MC-SSA with pure red-noise null-hypotheis: the significant RCs recognized in the first MC-SSA run can be subsequently included in the null hypothesis and the remaining series can be tested again. This should not be done too many times, because each time the MC-SSA test becomes more and more "liberal".

It must be finally mentioned that another version of SSA exists: the so-called "Caterpillar" methodology that was developed in the former Soviet Union, independently of the mainstream SSA work in the West. This methodology became known in the rest of the world more recently (e.g., Golyandina et al. 2001; Golyandina and Zhigljavsky 2013; Zhigljavsky 2010). Caterpillar-SSA emphasizes the concept of separability, a concept that leads, for example, to specific recommendations concerning the choice of SSA parameters. Several practical aspects of SSA and its application to time series are covered in Golyandina et al. (2001) and Golyandina and Zhigljavsky (2013).

Another information that can be useful is that in addition to the freeware SSA-MTM Toolkit, a software named Kspectra exists (http://www.spectraworks.com/web/products.html), working on Mac OS X, that offers some possibilities which are not present in the open-source version. One outstanding feature of this software is the possibility of an interesting approach to time-series forecasting. A combined SSA-AR method is applied, as introduced by Keppenne et al. (1992, 1993) and further described in Ghil et al. (2002). The basic idea is that the RCs in SSA are data-adaptively filtered signals, each of which is dominated by a narrow spectral peak (Penland et al. 1991). Such narrowband signals allow much more accurate predictions than broadband ones. Thus, the individual significant components of a series are predicted separately, and the final prediction results from the sum of the single predictions. This allows for prediction of the SSA-denoised signal, in order to forecast its significant dominant oscillatory modes while reducing noise disturbing effects. The SSA-AR method turns out to be most reliable and robust when the order M_{AR} of the autoregressive model fitted to the data is similar to the SSA embedding dimension M. This result follows from the way the SSA lag-covariance matrix and the AR coefficients are computed in the combined SSA-AR algorithm. For an example of application of this method, see Alessio et al. (2012).

12.3 SSA Application Examples

SSA has been applied extensively in geophysics, especially for the study of climate variability (see, e.g., Feliks et al. 2013; Keppenne et al. 1992, 1993; Kondrashov et al. 2005; Robertson and Mechoso 2000; Taricco et al. 2009), as well as to other areas in the physical sciences such as astrophysics (Greco et al. 2011), in life and biomedical sciences (for instance, Brawanski et al. 2002; Colebrook 1978; Grigorov

2006; Mineva et al. 1996; Rodó et al. 2002), and in socio-economic sciences (e.g., de Carvalho et al. 2012; Hassani and Thomakos 2010; Hassani et al. 2011; Sella 2008; Sella and Marchionatti 2012; Sella et al. 2013; Thomakos et al. 2002).

We will now present four examples of SSA application, two concerning climatology, one concerning astrophysics, and one regarding economics, to illustrate the power and the capabilities of the method.

12.3.1 A Paleoclimatological Application

Climatological applications of SSA include the analysis of paleoclimatic time series, i.e., records useful for the study of past climate variability. As an example, here we report the SSA analysis (Taricco et al. 2009) of a high-resolution ($T_s = 3.87$ y) record of the Oxygen isotope composition $\delta^{18}O$ of planktonic foraminifera shells, aimed at investigating climatic variability on decadal to millennial time scales. The record, covering the last two millennia (200 BC–1979 AD), was measured in a shallow-water sediment core extracted from the Gallipoli Terrace in the Gulf of Taranto, Italy (Ionian Sea, Central Mediterranean).

Planktonic foraminifera are unicellular organisms living in sea water and possessing a calcareous shell, i.e., a shell made of calcite ($CaCO_3$). In particular, the study by Taricco et al. (2009) was performed on the shells of *Globigerinoides ruber*, an organism dwelling in surface waters. When these organisms die, they settle onto the sea bottom. Drilling sediment cores, accurately dating each mud layer, and extracting fossil shell samples from each layer allows for measuring the abundance of stable isotopes of Oxygen and Carbon in the calcite of which the shells are made. This provides time series embedding information on the sea and climate conditions at the time when the foraminifera lived and their shells were built.

The stable isotopes of Oxygen used in paleo-oceanographic studies are ^{16}O and ^{18}O, comprising 99.63 and 0.1995 % of the Oxygen on Earth, respectively. The relative calcite-^{18}O abundance with respect to ^{16}O established during the shells formation is influenced by environmental factors like temperature and seawater Oxygen isotopic composition (Shackleton and Kennett 1975). Note that the isotopic composition of sea water is not considerably affected by temperature changes, since the quantity of water is very large compared to the amount of $CaCO_3$ equilibrated with it. Thus, the $\delta^{18}O$ of a foraminiferal calcite sample, defined with respect to a proper standard[3] as

$$\delta^{18}O = \frac{\left(^{18}O/^{16}O\right)_{\text{calcite sample}} - \left(^{18}O/^{16}O\right)_{\text{standard}}}{\left(^{18}O/^{16}O\right)_{\text{standard}}} \times 1000,$$

---

[3]The reference standard for the Oxygen isotopic composition in carbonates is the PDB standard, which is based on the CO_2 produced from Cretaceous belemnites of Pee Dee formation in South Carolina (Faure, G: Principles of isotope geology, Second Edition, John Wiley and Sons (1986)).

is a *proxy*, i.e., an indicator, of temperature and seawater Oxygen isotopic composition and allows for the study of past climatic variations. The units of $\delta^{18}O$ are per thousand (permil) relative to the standard; for example, $\delta^{18}O = 1$ permil means that the sample has a $^{18}O/^{16}O$ ratio 0.1 % greater than the standard.

One critical factor determining the reliability of climatic reconstructions based on proxy records is the accuracy of the absolute dating of each sample. In the case of the cores from the Gallipoli Terrace, high accuracy of the dating is made possible by the closeness of the drilling site to the volcanic Campanian area, a region that is unique in the world by its detailed historical documentation of volcanic eruptions over the last two millennia. The markers of these eruptions were identified along the core as peaks of the number density of clinopyroxene crystals, carried by the prevailing westerly winds from their source into the Ionian Sea and deposited there as part of the marine sediments. This dating method is known as tephroanalysis. The time-depth relation for the cores retrieved from the Gallipoli Terrace by the authors of this paper, obtained by tephroanalysis, confirmed, improved and extended to the deeper part of each core the dating obtained in the upper 20 cm by the ^{210}Pb method (see, for example, Bonino et al. 1993). This relation turned out to be highly linear, demonstrating that the sedimentation rate at the site remained constant, to a very good approximation, over the last two millennia.

SSA, performed adopting the VG method for constructing the lag-covariance matrix, allowed for detecting in this climatic record, comprising $N = 560$ samples, highly significant oscillatory components with periods of roughly 600, 350, 200, 125 and 11 years (given by RCs 2–3, 4–5, 6–7–8, 9–10, and 11–12, respectively), together with a highly significant long-term trend (represented by RC 1). The authors chose an SSA-window length of $M = 150$ samples, corresponding to about 580 years, in order to be able to detect climatic oscillations with periods as long as 500–600 y, while maintaining sufficient statistical significance; they however obtained coherent results for a fairly wide range of M values, from 120 to 200 points. Taricco et al. (2009) also applied the Continuous Wavelet Transform (CWT; Chap. 13) to the same data. Comparison with other climatic records allowed for identifying the long-term trend and the 200 year oscillation as temperature-driven components, having a dominant role in describing temperature variations over the last two millennia in the Central-Mediterranean area.

We show in detail the results of this analysis, as a good example of how SSA works. The data was kindly provided by C. Taricco. Figure 12.1 shows the raw data and two reconstructed versions of the record: in Fig. 12.1a we can see the SSA reconstruction obtained using all significant components detected by MC-SSA (RCs 1–12); in Fig. 12.1b we can see the SSA reconstruction on centennial–millennial scales (RCs 1–10), compared with the reconstruction obtained by CWT on the basis of significant contributions in the same range of periods (see Chap. 13). Indeed, CWT did not detect the 11 year oscillation as significant, due to the poor frequency resolution that characterizes CWT at high frequency. Figure 12.1 highlights the capabilities of SSA of extracting significant oscillatory modes from short, noisy series. Figure 12.1b also proves that SSA and CWT can depict long-term behaviors that are in very good agreement. Figure 12.2 shows the eigenspectrum, while Fig. 12.3

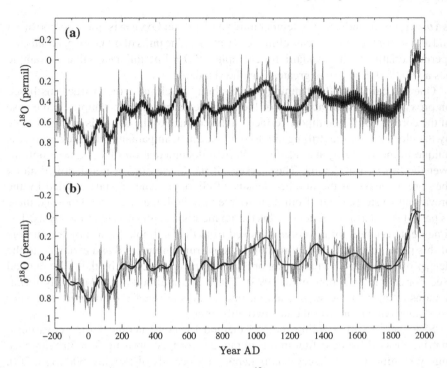

Fig. 12.1 Time series of the Oxygen isotope ratio $\delta^{18}O$ measured by Taricco et al. (2009) in foraminifera shells extracted from the sediments of a Mediterranean Sea core. Isotopic ratios are plotted "upside down", to agree in tendency with environmental temperature. Raw data is shown in *gray solid* in both panels. **a** Signal reconstruction by SSA, using RCs 1–12 *black solid curve*; **b** signal reconstruction by Inverse-CWT, based on the significant periods higher than 100 y (*black dashed curve*), and SSA long-term reconstruction, using RCs 1–10 (*black solid curve*). Window size: $M = 150$. See text for details

gives the same spectrum in terms of percentage of variance explained by each mode (*black dots*), and in terms of cumulative percentage of variance (*empty circles*). The *horizontal dashed line* separates the region of the plot showing percentages of explained variance from the region showing cumulative percentages of variance. The inset of Fig. 12.2 shows how the first 12 modes, recognized as significant in a subsequent MC-SSA test at the 99 % c.l., are grouped together. The inset of Fig. 12.3 shows the percentages of variance explained by each group, i.e., by mode n.1, by the pairs 2–3 and 4–5, by the triplet 6–7–8 and by the pairs 9–10 and 11–12. The first mode in the SSA spectrum represents the long-term trend (RC 1). The other dominant modes, grouped as RCs 2–3, RCs 4–5, RCs 6–7–8, RCs 9–10, and RCs 11–12 (see the inset of Fig. 12.2), are interpreted as oscillatory modes (Ghil et al. 2002), since

- the two members of a pair are associated with very similar eigenvalues (note that the corresponding error bars in Fig. 12.2 overlap)—i.e., they explain more or less the same variance in the series;

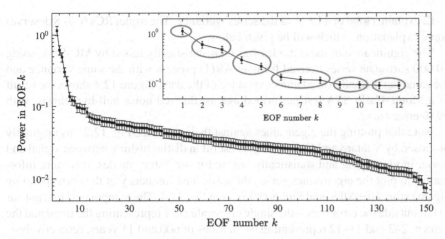

Fig. 12.2 SSA spectrum of the $\delta^{18}O$ record: eigenvalues are plotted versus mode number (*rank*). *Error bars* are displayed, showing an ad hoc range of estimation errors for eigenvalues. *Inset* grouping of the first 12 modes, which in a subsequent MC-SSA test are recognized as significant at the 99 % c.l. (see text). Window size: $M = 150$

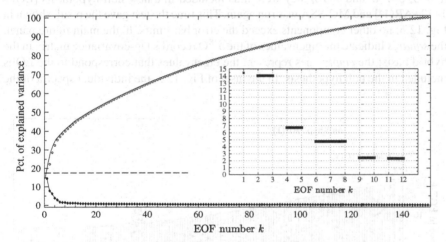

Fig. 12.3 SSA spectrum of the $\delta^{18}O$ record in terms of percentage of variance explained by each mode, and in terms of cumulative percentage of variance (*black diamonds* and *empty circles*, respectively). The *horizontal dashed line* separates the region of the plot showing percentages of explained variance from the region showing cumulative percentages of variance. *Inset* percentages of variance separately explained by mode n.1, by the pairs 2–3 and 4–5, by the triplet 6–7–8 and by the pairs 9–10 and 11–12 (see text). Window size: $M = 150$

- their EOFs (not shown) are nearly in phase quadrature and are approximately associated with the same characteristic frequency.

The corresponding percentages of explained variance are 14.1, 6.7, 4.8, 2.4, and 2.3 % respectively; the trend captures 14.5 % of total variance. Together, modes

1–12 explain a total of 44.8 % of the series' variance. The triplet RCs 6–7–8 deserves some explanation, which will be given below.

The significance of these modes was then statistically tested by MC-SSA, using 10,000 surrogate series derived from an AR(1) process with the same variance and the same lag-1 standardized autocovariance of the data. Figure 12.4 shows the result of a Monte-Carlo SSA test performed against this red noise null-hypothesis, with 99 % error bars.

Note that plotting the eigenvalues against their rank (see Fig. 12.2), as originally proposed by Vautard and Ghil (1989), is useful in distinguishing between signal and noise. In identifying and statistically testing for oscillatory modes, it is more informative to plot the eigenvalues versus the associated frequency of the corresponding eigenvectors (see Allen and Smith 1996), as in Fig. 12.4. The components that appear to fall outside the error bars—the single eigenvalue n. 1 representing the trend and the pairs n. 2–3 and 11–12 representing oscillations of 600 and 11 years, respectively— are distinguished from red noise at the 99 % c.l. Including these components in the null-hypothesis, the Monte-Carlo test was repeated. Other components turned out to exceed the error bar limits in this second MC-SSA test, namely, eigenvalues n. 4–5, 6–7–8, and 9–10; they were thus included in a new null-hypothesis [EOFs 1–12+AR(1)] and MC-SSA was run again. This time the test gave the result shown in Fig. 12.5: no other components exceed the error bar limits. In the main figure panel, the *squares* indicate the eigenvalues of the $\delta^{18}O$ record's lag-covariance matrix in the hybrid basis; the *empty* ones represent the eigenvalues that correspond to the EOFs included in the null hypothesis. In the inset of Fig. 12.5, the individual spectra of the

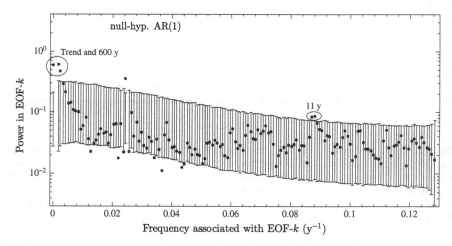

Fig. 12.4 Results of the Monte-Carlo SSA test for the $\delta^{18}O$ record, performed adopting AR(1) as the null-hypothesis, with error bars representing the interval between the 0.05th and 99.5th percentiles. Eigenvalues that lie outside this range are significantly different at the 99 % c.l. from those generated by a red-noise process against which they are tested by using 10,000 surrogate series. They are: the single eigenvalue n. 1 representing the trend and the pairs n. 2–3 and 11–12 representing oscillations of 600 and 11 years, respectively. Window size: $M = 150$

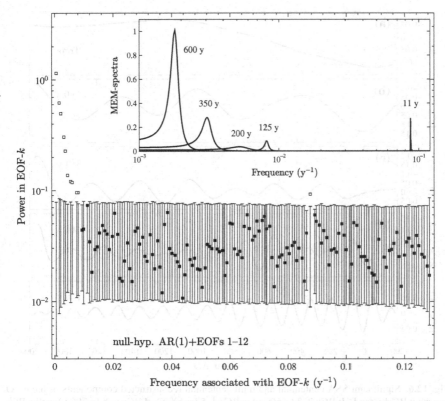

Fig. 12.5 Spectral properties of the $\delta^{18}O$ time series. The main panel shows the Monte-Carlo SSA test using the EOFs 1–12+AR(1) null-hypothesis, with 99 % error bars. The *squares* indicate the eigenvalues of the $\delta^{18}O$ record's lag-covariance matrix in the hybrid basis; the *empty* ones represent the eigenvalues that correspond to the EOFs included in the null hypothesis. *Inset* individual spectra of the RCs, obtained by Burg's method (Sect. 11.7) with the relatively low order $N/28 = 20$. Window size: $M = 150$

significant oscillations are also shown. These spectral estimates were obtained by Burg's method (Sect. 11.7) with an order of 20, which is relatively low compared with the number of available samples ($p \sim N/30$). The dominant periods of the detected oscillations are: ~600 y (RCs 2–3), ~350 y (RCs 4–5), ~200 y (RCs 6–7–8), ~125 y (RCs 9–10) and ~11 y (RCs 11–12).

In Fig. 12.6 we can see the reconstructed significant components in the centennial and multicentennial range. They appear as amplitude- and frequency-modulated oscillations, practically noise-free. Figure 12.7a shows the reconstructed 11 year component (RCs 11–12) obtained from the analysis with $M = 150$. In spite of the closeness to the Nyquist period (7.74 years), this high-frequency component appears very well identified. It is clean and has appreciable amplitude over the whole record duration. It must be noted, in relation to this high-frequency component, that the experimental procedure used to obtain the $\delta^{18}O$ series rules out the issue of frequency

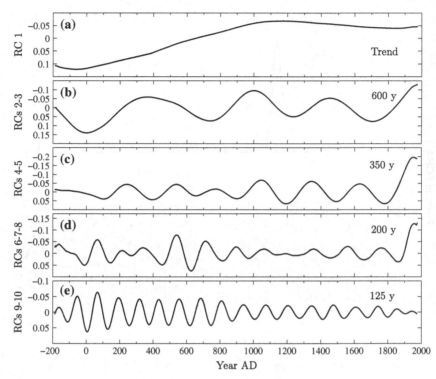

Fig. 12.6 Significant SSA centennial and multicentennial reconstructed components of the $\delta^{18}O$ record: **a** RC 1 (trend), **b** RCs 2–3 (∼600 y), **c** RCs 4–5 (∼400 y), **d** RCs 6–8 (∼200 y), and **e** RCs 9–10 (∼125 y). Window size: $M = 150$

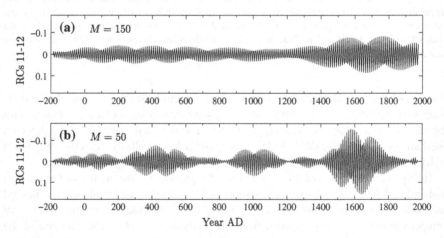

Fig. 12.7 Reconstructed 11 year component of the $\delta^{18}O$ record: **a** reconstructed RCs 11–12 from the analysis with a window of $M = 150$ samples; **b** reconstructed RCs 5–6 from the analysis with a window of $M = 50$ samples

aliasing related to the sampling process. Indeed, discretization of the underlying analog signal here is not due to a punctual sampling, but it derives from measurements in consecutive sediment slices, performed after mixing the material contained in each slice. This mixing cancels out any possible contribution by frequencies higher than the Nyquist frequency, thus acting as an anti-aliasing lowpass filter.

About the triplet formed by EOFs 6–7–8, Fig. 12.8 shows the individual reconstructed components and their sum.

Normally, we would look for EOF pairs, not triplets. This is a case of degeneracy and mode mixing. Note in Fig. 12.2 how the corresponding three eigenvalues are nearly equal: they form an almost completely degenerate triplet. A similar case was reported by Allen and Smith (1997) in the SSA of the 1901–1990 global Earth's temperature record. These authors observed that high-variance components of the colored noise that contaminates the signal in a climatic series can confuse the singular value decomposition. They suggested that when the noise properties can be estimated reliably from the available data, the application of a pre-whitening operator could significantly enhance the signal separation capabilities of SSA (Allen and Smith 1997; Ghil et al. 2002), thus solving this issue. To this aim, Allen and Smith (1997) proposed to pre-process the time series itself or, equivalently but often more efficiently, the lag-covariance matrix, such that the colored noise becomes uncorrelated, i.e., white, in this new representation. SSA could then be performed on the transformed data or covariance matrix and the results transformed back to the original representation for inspection. In Taricco et al. (2009), no such pre-processing was

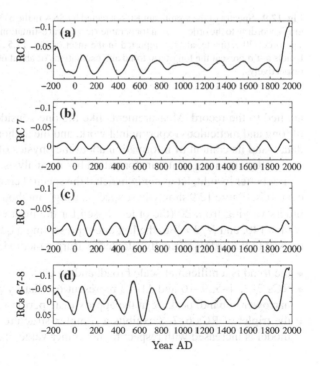

Fig. 12.8 The triplet formed by RCs 6–7–8 in the SSA of the $\delta^{18}O$ record: **a–c** individual RCs and **d** their sum. See text for details

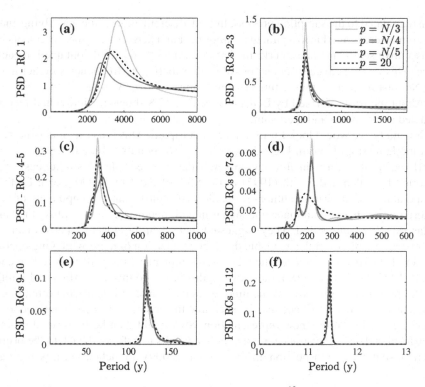

Fig. 12.9 Spectra of the significant RCs detected by SSA in the δ^{18}O record, in different shades of gray according to the order chosen for parametric spectral estimation via Burg's method (Sect. 11.7). The order-20 estimate, already reported in the inset of Fig. 12.5, is represented as a *dotted line*. Panels **a–f** represent the long-term trend and oscillations of about 600, 350, 200, 125 and 11 years, respectively

applied to the record. Measurements like the one considered here are the outcome of long and meticulous experimental work, and the authors preferred not to alter the data—and its content of information about the physics of the natural process under study—by any numerical transformation before analysis.

For better insight, let us study in detail the spectral content of the detected significant RCs. Figure 12.9 shows these spectra. Burg's method (Sect. 11.7) was used, with orders varying from 20 (the order chosen for the inset of Fig. 12.5) to $N/5 = 112$, $N/4 = 140$, and $N/3 = 186$, the last one representing a quite high fraction of N with respect to the orders normally adopted in Burg's method. We can see that

- the trend is a millennial-scale broadband process;
- RCs 2–3, 4–5, 9–10 and 11–12 represent oscillatory processes which are much narrower in band, peaking at the approximate period values given above;
- the triplet of RCs 6–7–8 exhibits a complex structure when the order of the AR model is increased with respect to the former value of 20.

Figure 12.10 shows the spectra of individual components forming the triplet. RC 6 is a process peaking at ∼220 years of period; RC 7 contains power at ∼170 and ∼220 years; RC 8 is a quite narrowband process peaking at ∼170 years; the sum of these three modes thus contains both contributions, at 170 and 220 years, approximately. It is clear that modes are scrambled due to degeneracy. From the point of view of the underlying physics, however, it makes no sense to distinguish between a cycle with a period of 170 years and one with a period of 220 years in this climatic record. For this reason, the three components were collected together into the modulated oscillation shown in Figs. 12.6 and 12.8d that, when analyzed by low-order Burg's method, indicates a period of about 200 years (see the *dotted lines* in Figs. 12.9d and 12.10d and the inset of Fig. 12.5).

Another point that must be made clear is that even when in essence the results of an SSA are robust to M changes, of course the significant components are distributed differently for different values of M: for example, if we increase M we can get a trend formed by two modes. Moreover, when we perform SSA we choose one value of M, but when oscillations of periods which differ in their order of magnitude are present

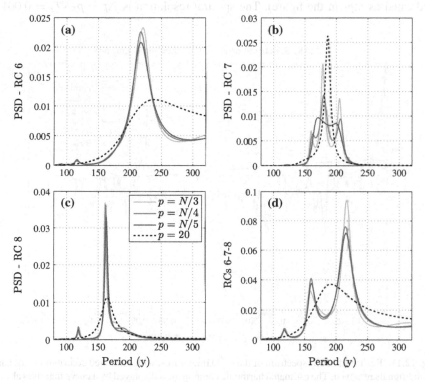

Fig. 12.10 The triplet formed by RCs 6–7–8: spectra of **a–c** the individual RCs and **d** their sum, in different shades of gray according to the order chosen for parametric spectral estimation via Burg's method (Sect. 11.7). The order-20 estimate, used for the triplet in the inset of Fig. 12.5, is represented as a *dotted line*

in the record, the choice of M that has been made in order to be able to detect the longer oscillations may not be optimal for detecting the shorter ones. In the present case, the window of about 600 years selected to detect multicentennial oscillations may not be the best for the 11 year oscillation, which would be better studied using a much narrower window, e.g., 150–200 y. But with this shorter window, we would see the longer multicentennial components merge with the trend. The reconstruction of the 11 year oscillation with a window of $M = 50$ (about 200 years) is shown in Fig. 12.7b. In this case the oscillation is carried by RCs 5–6, since all centennial and multicentennial components are contained in the first four modes when this shorter window is employed. The two reconstructions obtained with different values of M (Fig. 12.7a for $M = 150$ and b for $M = 50$) show an overall agreement, but the oscillation seen through the shorter window exhibits more detailed amplitude modulation.

Other advanced spectral methods were applied to this series by Taricco et al. (2009). As an example, Fig. 12.11 shows the results obtained applying MTM to the $\delta^{18}O$ series. The time-bandwidth parameter p was set to 3 (the time-bandwidth parameter is indicated as npi in the figure), so that the number of tapers is $K = 2p - 1 = 5$ (indicated as ntpr in the figure). The spectral resolution is $\Delta f = p/NT_s = 0.0014$

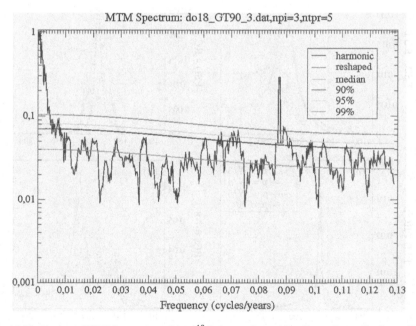

Fig. 12.11 Reshaped MTM spectrum of the $\delta^{18}O$ time series, with estimated contributions of harmonic signals removed. The estimated harmonic component is displayed by a curve that rises above the reshaped spectrum, to indicate the portion of the spectrum above the continuous background which is associated with a harmonic signal. The harmonic 11 year component (0.088 cycles/year) is shown as a narrow spike of breadth equal to the spectral estimation bandwidth (0.0014 cycles/year). Significance curves are also shown, for the confidence levels of 90, 95 and 99 %

cycles/year. Note that Δf depends directly on p and inversely on the record length NT_s, so that for a fixed record length, the lower p, the sharper the resolution, corresponding to a smaller Δf. Longer records can thus be analyzed using higher values of p and K, with the latter providing more substantial variance reduction without resolution degradation, with respect to shorter records. Figure 12.11 was produced directly by the MTM–SSA toolkit and shows the estimated continuous MTM spectrum or *reshaped spectrum*, i.e., the spectrum with estimated contributions of harmonic signals removed, if any. Any estimated harmonic component of the spectrum is displayed by the toolkit as a curve that rises above the reshaped spectrum, to indicate the portion of the spectrum above the continuous background that is associated with a harmonic signal. The latter component is shown as a narrow spike of breadth equal to the spectral estimation bandwidth. Significance curves are also shown, for the confidence levels of 90, 95 and 99 %. It may be seen that in this case, the 11 year component is detected as a harmonic process with frequency of about 0.088 cycles/year.

The study of Taricco et al. (2009) has recently been extended to cover 707 BC– 1979 AD (Taricco et al. 2015).

12.3.2 A Gap-Filling Application

The capabilities of SSA can also be exploited to fill gaps in data records (Kondrashov and Ghil 2006).

The majority of data sets that are obtained from observations and measurements of natural systems, rather than in the laboratory, are often full of gaps, due to the conditions under which the measurements are made. Missing data gives rise to various problems, for example in spectral estimation, and a reliable gap-filling technique is highly needed.

Gap-filling methods proposed in literature are numerous. They can be model-based, i.e., using parameter-dependent models, or relying on the data alone and being nonparametric. SSA and its multivariate counterpart, known as M-SSA, provide a nonparametric approach to gap filling that is particularly useful for data sets that exhibit relatively long, continuous gaps.

A modified, iterative SSA algorithm can self-consistently fill missing points, using the leading oscillatory modes of the time series. This iterative form of SSA can be used to deal with datasets with uneven sampling or missing observations. Gaps are filled-in by utilizing temporal and possibly spatial correlations in the dataset. This gap-filling feature is available in the SSA-MTM Toolkit for univariate and multivariate data. For this task, the window size must be large enough to cover the longest temporal correlations; for a guidance, it can be the largest period deemed to be contained in the examined record(s). The number of SSA components that must be used depends on the dataset, and in particular on the amount of noise that is present. The main idea is to discard higher-ranked components corresponding to noise, and use only "smooth" components, representing true "signal" components. The SSA-based gap-filling method in essence is the following:

- estimates of missing data points are iteratively produced, which are then used to compute a self-consistent lag-covariance matrix, its EOFs and principal components (Ghil et al. 2002);
- cross-validation is used to optimize the window width and the number of dominant SSA modes to be used to fill the gaps.

For a univariate record, the original data is first centered by subtracting the unbiased estimate of the mean. The missing-data values are set to zero. An inner-loop iteration is started, which computes the leading EOF of the centered, zero-padded record (EOF 1). Then the SSA algorithm is performed again on a new time series, in which the RC 1 corresponding to EOF 1 is used to substitute non-zero values in place of the missing points. The record's mean is computed again, as well as the covariance matrix and the EOFs. The reconstruction of the missing data is repeated with a new estimate of RC 1 and tested against the previous one, until a convergence test has been satisfied, i.e., a pre-fixed tolerance is met. Next, an outer-loop begins, in which a second EOF is added for reconstruction (EOF 2). The outer loop starts from the solution with data filled in by RC 1, and the inner iteration is repeated with these two modes until convergence is achieved.

For many climatic and geophysical records, a few leading EOFs correspond to the record's dominant oscillatory modes, while the rest is noise (Ghil et al. 2002); therefore only a few EOFs are expected to be necessary for a good reconstruction. The optimum number of EOFs and the optimum window width to fill the gaps of a given series are found from a set of cross-validation experiments: for each such experiment, a fixed fraction of available data (e.g., 5 %) is left out, and the root-mean-square error in reconstruction is computed as a function of the number of retained EOFs and of the SSA-window size. The absolute minimum of the error function, averaged over all experiments, corresponds to the required optimum parameters, and provides an estimate of the actual error in the reconstructed data set. Considering the window length as fixed, this means computing the estimation error for each EOF being added to the reconstruction; in practice, for a given value of M this error starts to increase once noise-related EOFs are added. The optimal parameters of the procedure depend on number and distribution in time of the missing data, as well as on the variance distribution between oscillatory modes and noise; once they have been estimated, to obtain the actual reconstruction, the inner and outer-loop iterations are repeated, using the optimal parameters found by cross-validation, but with all the available points now being included in the process.

This procedure for a univariate record uses only temporal correlations in the data to fill in the missing points. For a multivariate record, the same methodology can be applied using *multi-channel singular spectrum analysis* (M-SSA).

M-SSA is a natural extension of SSA to a set of vectors or maps, such as, for instance, time-varying temperature or pressure given on a grid of points over the globe. This set is still called a "time series", in a generalized sense. The trajectory matrix of multi-channel time series consists of stacked trajectory matrices of separate times series; the size of different univariate series does not have to be the same. The rest of the algorithm is the same as in the univariate case. The use of M-SSA for such multivariate time series was proposed theoretically, in the context of nonlinear dynamics, by Broomhead and King (1986b) (see also Ghil et al. 2002). M-SSA has many applications, including forecasting (Golyandina and Stepanov 2005). It is especially popular in analyzing and forecasting economic and financial time series with short and long series length (Hassani et al. 2013; Hassani and Mahmoudvand 2013; Patterson et al. 2011). Advanced Monte-Carlo significance tests can be performed in M-SSA, which are based on the so-called *varimax rotation*. The classical varimax rotation is a statistical technique suggested to facilitate the interpretation of the results of principal component analysis (Kaiser 1958). In M-SSA, in order to reduce mode-mixture effects and to improve the physical interpretation, Groth and Ghil (2011) proposed a varimax rotation of the spatio-temporal EOFs (ST-EOFs) obtained from the M-SSA; however, to avoid a loss of spectral properties (Plaut and Vautard 1994), they suggested a slight modification of the common varimax rotation that takes the spatio-temporal structure of ST-EOFs into account. Time-series forecasting according to the SSA-AR approach is possible also in this multivariate setting (this possibility is offered by KSpectra).

In gap-filling applications, M-SSA can be used to take advantage of both spatial and temporal correlations in a data set, to fill-in gaps in time series. This provides a substantial improvement in cases in which data points are often missing in one time series, but not in the other. Using this gap-filling tool (available in the Kspectra software), Kondrashov et al. (2005) and Kondrashov and Ghil (2006) were able to analyze the historical records of the low- and high-water levels of the Nile River, which are among the longest climatic records having near-annual resolution.

The two time series of the Nile flood levels (Popper 1951) represent a human historical record of climate variability for more than 1000 years, which reflects water intake from the Blue and the White Nile in Ethiopia and equatorial Africa. Pharaonic and medieval Egypt depended solely on winter agriculture and hence on the summer floods. Between July and November, the reaches of the Nile running through Egypt would burst their banks and cover the adjacent flood plain. When the waters receded, they left behind a rich alluvial deposit of exceptionally fertile black silt over the croplands.

The rise of the waters of the Nile was measured regularly from the earliest times. Nilometers, i.e., structures for measuring the Nile River's clarity and water level during the flood season, were built along the river. These structures served to predict how satisfactory the flood would be. If the flood was scarce, there would be famine; if it was too abundant, it would be destructive. There was a specific mark that indicated how high the flood should be if the fields were to get good soil. A nilometer (Fig. 12.12), housed in an elaborate and ornate stone structure, can still be seen on the southern tip of Rodah (or Rhoda, or Rawdah) Island in central Cairo. While this structure dates only as far back as AD 861, when the Abbasid caliph al-Mutawakkil ordered its construction, it was built on a site occupied by an earlier nilometer.

Several authors compiled the annual maxima and minima of the water level recorded at nilometers in the Cairo area, in particular at Rodah Island, from A.D.

Fig. 12.12   The nilometer on Rodah Island, Cairo, Egypt

622 to 1922. The most complete records for the time since the Arab invasion in A.D. 622 (Ghaleb 1951; Hurst 1952; Toussoun 1925) were corrected by Popper (1951) to account for changes in the unit of length used (the cubit), for the rise of the bed of the Nile through siltation,[4] and for the differences in lunar and solar calendars. They were subsequently further revised and edited by T. De Putter and D. Percival (see De Putter et al. 1998; Whitcher et al. 2002). Climate researchers extensively studied the resulting multicentury, annually resolved records and demonstrated the significant association between the rainfall in the catchment area of the Nile tributaries, the Indian monsoon, and ENSO (Quinn 1992; Walker 1910).

These records, however, had missing data. The work by Kondrashov et al. (2005) provided instead a complete 1300 year record of Nile River floods (A.D. 622-1922) with annual resolution, of remarkable climatic interest.

The original gappy records, largely based on the above-mentioned compilations of Toussoun (1925), Ghaleb (1951) and Popper (1951), revised and edited by De Putter et al. (1998) and Whitcher et al. (2002), had few gaps in their first part (A.D. 622–1470), where a few missing data points were just linearly interpolated, and larger gaps later (A.D. 1471–1922). The large gaps were caused by social and economic upheavals during the Ottoman rule. The low- and high-water records were found to be strongly correlated in the low-frequency range (periods of 50 years and longer).

---

[4]Siltation, in general, is the pollution of water by fine particulate terrestrial clastic material, with a particle size dominated by silt or clay. Here it refers to the increased accumulation of fine sediments on the river bottom.

Kondrashov et al. (2005) used both iterative SSA and M-SSA on these records, in order to exploit both temporal and spatial correlations in the data set to fill the gaps in either record. Given the fact that high- and low-water levels were not always missing the same year, the M-SSA of both records finally turned out to be the best approach. Kondrashov et al. (2005) found that using the nine leading EOFs of the two-channel SSA and a window width $M = 100$ years minimizes the estimation error of 50 independent cross-validation experiments; these nine EOFs capture a slight nonlinear, data-adaptive upward trend, possibly due to siltation (Popper 1951), accompanied by a very low-frequency oscillation with a 256 year period, and a 64 year mode. Independent information on the signal-to-noise separation was obtained (see Kondrashov and Ghil 2006) by inspecting the slope break in the plot of the M-SSA spectrum of the filled Nile River records with $M = 100$ years, showing the passage from modes representing the signal to modes representing mainly the noise. This plot (not shown) indicated a clear separation between the first nine "signal" EOFs that were used in the reconstruction and the remaining modes, representing the discarded "noise".

The extended low- and high-water records, with the gaps filled in, are reported in Fig. 12.13. Digital Nile River gauge data for drawing this figure were kindly provided by D. Kondrashov, based on the gap filling of Kondrashov et al. (2005). Figure 12.13a shows the original gappy data; Fig. 12.13b shows data with gaps filled in by M-SSA using $M = 100$ years and two channels (low- and high-water). The time series have been centered on the relevant mean, and the amplitudes have been normalized by the standard deviation of the original time series (excluding missing data points).

Analyzing the full extent of the available water-level records, with the missing points filled in, Kondrashov and Ghil (2006) were able to perform an accurate spectral analysis, in which SSA and MTM were applied to identify inter-annual and inter-decadal periodicities, and Monte Carlo significance tests were conducted with improved reliability with respect to the shorter original records. The gap-filling moreover allowed for the study of the evolution of the Nile-river oscillatory modes over the entire 13 centuries covered by the measurements. It must be mentioned that Kondrashov and Ghil (2006) examined not only the high- and low-water records, but also the difference between them, and concluded that the time series of the annual difference between the maxima and minima better represents the Nile floods than the high-level time series alone.

Data sets with red-noise-like spectra, where noisy modes contribute significantly to, or even dominate, the spectrum's low frequencies, present special challenges in SSA. In such cases, it may be beneficial to skip the noisy modes associated with low frequencies and large amplitudes, and study only oscillatory modes in a higher-frequency band. According to this idea, Kondrashov and Ghil (2006) first removed the lowest-frequency component (i.e., the trend and the associated 256 year oscillation) from the gap-filled data. This combination is captured in both the original, short (A.D. 622–1470) records and the long (A.D. 622–1922) records by the two leading eigenmodes of the SSA analysis with a 100 year window. Next, they applied Monte Carlo SSA with a c.l. of 95 % and a window of $M = 75$ years to the detrended time

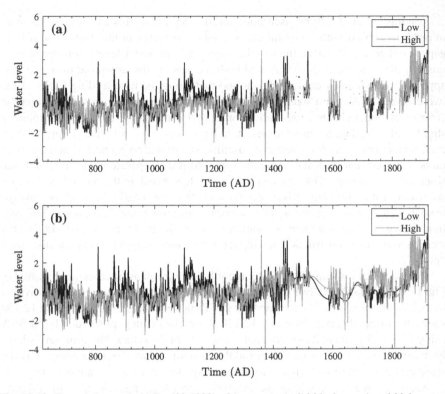

Fig. 12.13 Extended records (A.D. 622-1922) of low-water (*solid black curve*) and high-water (*solid gray curve*) levels of the Nile. **a** Original data with gaps and **b** data with gaps filled in by M-SSA. The time series have been centered on the relevant mean, and the amplitudes have been normalized by the standard deviation of the original time series, excluding missing data points

series. Their analysis revealed several statistically significant features of the records, including:

- quasi-quadriennial and quasi-biennial modes that support the long-established connection between the Nile River discharge and the ENSO phenomenon in the Indo-Pacific Ocean,
- longer periods that might be of astronomical origin, and
- a 7 year periodicity possibly due to North Atlantic influences over North Africa and the Middle East, persistent over the last millennium-and-a-half, which may extend all the way into the Nile River's source area. This suggested a previously undocumented source of interannual climatic variability for tropical East Africa, namely changes in the North Atlantic Ocean circulation.

The authors remarked how it would be tempting to identify the 7 year peak in the Nile River records with the cycle of lean and fat years mentioned in Joseph's biblical story. The story, though, may refer just to a near-regularity of several years, rather than to an exact periodicity.

We may also mention that Kondrashov et al. (2010) and Kondrashov et al. (2014) applied the same SSA-based gap-filling methodology to time series of solar wind parameters.

12.3.3 An Application in Astrophysics

A study by Greco et al. (2011) provides an example of SSA of a series of astrophysical interest. These authors studied γ-ray bursts (GRBs), which are the most instantaneously powerful cosmic explosions known in the universe since the Big Bang. They are identified as brief, intense, and completely unpredictable flashes of high energy γ-rays in the sky, occurring approximately once per day. They come from all different directions of the sky and last from a few milliseconds to a few hundred seconds.

The prompt γ-ray emissions from GRBs exhibit a vast range of extremely complex temporal behaviors, with a typical variability time-scale of the order of the millisecond (i.e., a time scale considerably shorter than the typical overall duration of the burst) and related flux variations of up to 100 %. The analysis of such variability is ultimately aimed at getting clues about the physical mechanisms driving the internal engine of GRBs, which remains hidden from direct observation.

Greco et al. (2011) collected high-quality GRB records including more than 3000 samples each, with a homogeneous sampling time of 64 ms and an SNR level above 50. The list of prompt γ-ray emission candidates was taken from the daily updated compilation provided by the NASA Swift team.[5] In the time interval from January 2005 to September 2010, Greco et al. (2011) identified two events matching their selection criteria, namely GRB 050117 and GRB 100814. One of the approaches used by the authors to analyze these records was SSA, with the aim of investigating the apparent randomness of the GRB series. SSA is particularly indicated in this case, since it allows to separate signal from noise even when SNR varies during the series' duration—a typical feature of GRBs' prompt emissions.

The shape of the SSA spectrum was examined in search of possible evidence of deterministic activity in the prompt emission from GRBs. The authors focused on the selection of low-frequency SSA components, representing nonlinear slow trends. The Kendall's τ nonparametric test for trend detection (Freas and Sieurin 1977; Hirsch and Slack 1984; Hirsch et al. 1982) was also applied in order to reliably identify those components that are significantly non-stationary over the length of the time series, at the 99 % c.l. This is illustrated in Fig. 12.14. This figure and the next two concern the event GRB 050117; the analysis of GRB 100814 event gave very similar

---

[5]Swift is a NASA mission with international participation. Within seconds of detecting a burst, Swift relays its location to ground stations, allowing both ground-based and space-based telescopes around the world the opportunity to observe the burst's afterglow. Swift is part of NASA's medium explorer (MIDEX) program and was launched into a low-Earth orbit on a Delta 7320 rocket on November 20, 2004.

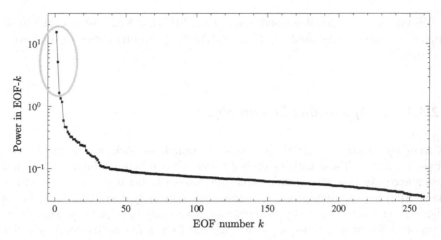

Fig. 12.14 SSA spectrum of prompt emission from GRB 050117. The three modes *circled in gray* were found to be significantly non-stationary over the length of the time series at the 99 % c.l., according to Kendall's τ trend test

results. The data for this figure and the following two figures was kindly provided by G. Greco.

The prompt light curve (Fig. 12.15a, *gray curve*) was subsequently detrended subtracting from it the reconstructed low-frequency oscillation (Fig. 12.15a, *black curve*). The detrended prompt light curve (Fig. 12.15b, *gray curve*) was then analyzed again by SSA. A Monte Carlo SSA test against an AR(1) null-hypothesis was run using 10,000 surrogate realizations of the record and a c.l. of 99.8 %. The first dominant eigenvalues were found to lie outside the intervals that define a purely stochastic behavior: deterministic oscillations with lower variance were thus found to be superimposed to the trend. Figure 12.16 shows the results of the Monte Carlo test. The corresponding significant oscillation reconstructed by SSA is shown in Fig. 12.15b as a *black curve*. This oscillation represents an SSA-de-trended and de-noised version of the GRB 050117 time series. A deterministic signal clearly emerges, which is active during the whole record, out of a remarkable amount of noise. A detailed analysis of the background regions pre- and post-GRB explosion (not shown) confirmed the meaningfulness of these results: indeed, the background was found to have a red-noise behavior, with no exceptions. Thanks to the capabilities of SSA, with this analysis Greco et al. (2011) were able to demonstrate that the temporal variability of GRBs does not follow a pure random behavior, as previously assumed.

12.3.4 An Application in Economics

As an example of SSA application in quantitative economics, here we briefly describe the paper by Sella et al. (2013), who analyzed macroeconomic fluctuations in three European countries. These authors explored five fundamental indicators of the real

Fig. 12.15 SSA
Reconstruction of significant
oscillations (*black curves*) in
the **a** original and
b detrended GRB 050117
series (*gray curves*). The
SSA-detrended series in
panel b revealed a
deterministic signal, clearly
emerging from noise

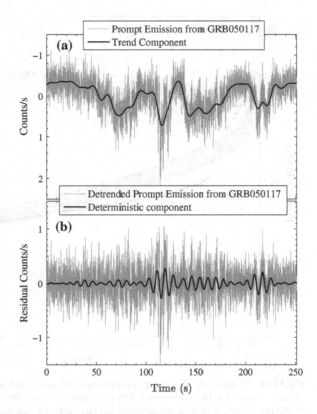

aggregate economy—namely gross domestic product (GDP) at market prices, final
consumption expenditure, gross fixed investments, exports and imports of goods and
services—in a univariate as well as multivariate setting. Here we will focus on the
univariate SSA of the individual series of macroeconomic indicators analyzed by
Sella et al. (2013).

Although it is widely acknowledged that business cycles are multi-country phenomena, showing common characteristics across countries, there is still no accor-
dance about the characterization of co-movements, the existence of supranational
(e.g., European) cycles, and the determinants of economic synchronization. On the
other hand, business cycle synchronization is of great interest in macroeconomics,
e.g., for the study of systems like the Euro area: one reason, for instance, is that if
several countries delegate on some supranational institution the power to perform
a common monetary (and/or fiscal) policy, then they lose this policy stabilization
instrument. If countries have asymmetric business cycles, then applying the same
decision to every country will not be optimal. Business cycle synchronization is a
necessary condition for any monetary union: a country with an asynchronous business cycle will face several difficulties in a monetary union, because of the "wrong"
stabilization policies.

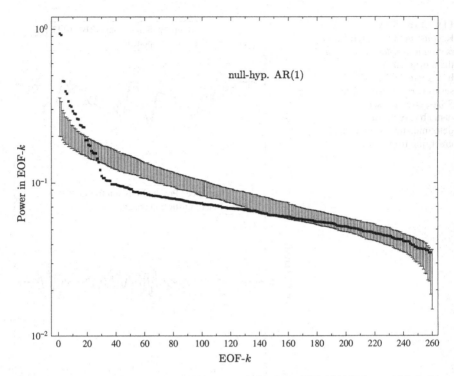

Fig. 12.16 Monte Carlo SSA spectrum of prompt emission from GRB 050117 from which the low-frequency trend has been subtracted. The error bars represent the interval between the 0.01th and 99.9th percentiles: eigenvalues that lie outside this range are significantly different at the 99.8 % c.l. from those generated by a red-noise process against which they are tested by using 10,000 surrogate series

Sella et al. (2013) undertook a quantitative inquiry about economic cycles and synchronization by spectral analysis of quarterly macroeconomic indicators from Italy (henceforth IT), the United Kingdom (UK), and The Netherlands (NL), in the attempt to shed light on supranational synchronization of business cycles. The three countries were selected because they are European economies with different magnitudes and socioeconomic characteristics. All series were expressed in constant euros (base year 2000). They cover different time spans: 54 years for UK (from the first quarter of 1955 to the third quarter of 2008; $N = 216$), 32 years for NL (from the first quarter of 1977 to the third quarter of 2008; $N = 128$), and 28 years for IT (from the first quarter of 1981 to the third quarter of 2008; $N = 112$).

For non-economists, we must first explain that data of this kind is always pre-processed before analysis (see also the appendix of Chap. 9). Pre-processing includes removing seasonality (for example, because of Christmas holidays, in January production always drops a lot), correcting by working days, and removing the typical upward long-term trend. In Sella et al. (2013) this was done applying a Hodrick-Prescott (HP) filter.

In the decomposition of a time series like those of the macroeconomic indicators considered here, one can usually identify, apart from random disturbances, a trend component, one or more cyclical components, and the seasonal component. The HP filter is a mathematical tool (Hodrick and Prescott 1997) that usually takes as an input a data record from which seasonality has already been removed, and decomposes it into the two remaining components—trend and cyclicities. The adjustment of the sensitivity of the trend to short-term fluctuations is achieved by modifying a parameter, usually indicated as λ. The filter was popularized in the field of economics in the 1990s by economists Robert J. Hodrick and Nobel Memorial Prize winner Edward C. Prescott. However, it was first proposed much earlier by Whittaker (1923).

As an example, Fig. 12.17a shows the raw GDP series for the three countries (*solid lines*) and the corresponding trends (*dotted lines*), obtained by applying an HP filter with the recommended smoothing parameter for quarterly time series, i.e., $\lambda = 1600$. The data for these graphs was kindly provided by L. Sella and G. Vivaldo.

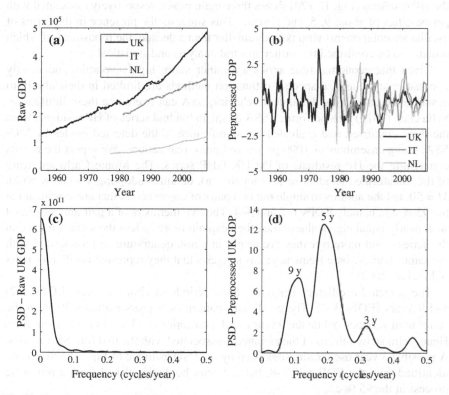

Fig. 12.17 **a** Raw time series of GDP for the UK, Italy, and The Netherlands (*solid lines*, with data expressed in constant euros with base year 2000), and Hodrick-Prescott trends (*dotted lines*). **b** HP Residuals (see text). An underlying cyclical structure is quite evident, which is the subject of the analysis by Sella et al. (2013). **c–d** Blackman-Tukey power spectra for the raw UK GDP series and for the corresponding HP residuals, respectively. Both estimates were obtained using a Hanning window with length $M = 100$

The raw data is an elaboration of EUROSTAT data. The subsequent analysis was performed on the so-called HP residuals, meaning that the trend was extracted from each raw time series and then subtracted from the latter; for normalization purposes, the corresponding residuals were divided by the trend, and finally the relative residuals were standardized to the same variance. GDP residuals for the three countries are shown in Fig. 12.17b.

The same pre-processing was applied to all macroeconomic indicators from the three countries. The pre-processing guarantees reliable estimates of the underlying periodicities and ensures that the results of the analysis are not distorted by the different magnitudes of the trends in the individual series. Figure 12.17c–d illustrates the issue in the case of the UK quarterly GDP series. The pervasive trends that characterize economic variables determine the shape of their estimated power spectra. In fact, they typically show a large bump in the lowest frequency band, the power of which is scattered into the neighboring frequency bands due to leakage (see Fig. 12.17c, illustrating this fact for the raw UK GDP). Conversely, the estimated spectrum of the HP residuals (Fig. 12.17d) shows three main peaks, respectively associated with periodicities of about 9, 5, and 3 years. This suggests the presence in the series of oscillatory components with periods smaller than a decade, the robustness of which remains to be confirmed by further spectral analysis and statistical tests.

Given that economic time series are rather short, highly volatile, and mostly non-stationary, classical spectral estimation methods are limited in their ability to describe the underlying dynamical behavior; SSA can overcome these limitations. Sella et al. (2013) thus performed SSA on the individual series of HP residuals from the three countries, and evaluated the significance of the detected modes by MC-SSA, using ensembles of 1000 proper red noise realizations. We report the results concerning the HP residuals of the UK GDP series. The Monte Carlo spectrum of the transformed residual series (not shown), obtained adopting a window width $M = 50$, led the authors to single out two pairs of eigenvectors that are significant at the 95 % c.l., namely EOFs 1–2 and 3–4. The two members of a pair are associated with nearly equal eigenvalues—i.e., they explain more or less the same variance in the series—and moreover they are nearly in phase quadrature and associated with the same characteristic frequency; this suggests that they represent oscillatory pairs (Ghil et al. 2002).

The detected oscillatory behaviors have periods of about 5 years (EOFs 1–2) and 9 years (EOFs 3–4). The oscillatory pattern of 5 years explains 39 % of the series total variance, while the 9 year oscillation captures 23 % of the total variance. Hence, almost two-thirds of the variance is associated with the first four SSA modes. A possible 3 year oscillation, suggested by the PSD estimate of Fig. 12.17d, can be identified with the EOF-pair 7–8, but it cannot be distinguished from a red noise process at the 95 % c.l.

Extending the same univariate exercise to all time series, similar oscillatory patterns were recurrently detected by Sella et al. (2013). The results of the complete study suggested quite homogeneous cyclical behavior among the indicators of IT, UK and NL. In spite of the different time spans covered and the peculiarities of each national economy, three similar business fluctuations emerged:

- an oscillation of about 9 years, which is statistically significant at the 95 % level only in some of the time series in the sets;
- a fluctuation of about 5 years, significantly isolated in nearly all series;
- a shorter oscillation of about 3 years, which generally explains a low percentage of the variance and often falls in the noisy floor of the spectrum; it is especially noticeable in IT and NL.

These similar patterns are evidence of common features in the dynamics of the economies of the three countries.

As we saw in previous examples, all significant time-domain patterns can be reconstructed by SSA, to inspect both amplitude and phase modulations. In Fig. 12.18, the reconstructions of the 5 year oscillations detected in the pre-processed GDPs of the three countries are plotted as *solid curves* (the *dashed curves* represent pre-pocessed data). This fluctuation is shared by all countries, and is dominant the UK series (Fig. 12.18a). Again, the data for these graphs is courtesy of L. Sella and G. Vivaldo. Characteristic features of the system's dynamic can be recognized observing the reconstructed series: for example, in Fig. 12.18a the reconstruction for UK shows an increase in amplitude (i.e., in the variance of the pattern) in correspondence with the huge energetic shocks of 1973 and 1979. This ~5 year oscillation that for UK is a dominant feature was found to agree in character with the ~5 year oscillatory mode detected in the U.S. economy by Groth et al. (2012): the close connection of the

Fig. 12.18 SSA-reconstructed 5-year oscillations in pre-processed GDPs (*solid curves*), superimposed to pre-processed data (*dashed curves*). The pattern is represented by the first two eigenmodes in each series, explaining **a** the 39 % of the total variance for the UK, **b** the 44 % for the Netherlands, and **c** the 40 % for Italy. Window width is $M = 50$, 45, and 34, respectively

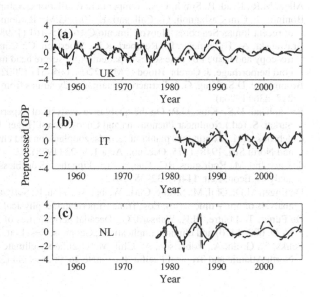

U.K. and the U.S. economy was thus supported. Finally, Sella et al. (2013) observed that their results support the predictions made by Hallegatte et al. (2008), using a simple non-equilibrium dynamic model named NEDyM: namely the presence of an endogenous business cycle with a period of roughly 5–6 years. The model's 5–6 year periodicity is shared by all the indicators examined in this study, and is consistent with the mean business-cycle period found not only by Sella et al. (2013) but also by other authors: see, e.g., the CWT analysis by Aguiar-Conraria and Soares (2011) reported in Sect. 13.6.

This example was meant to illustrate the capability of SSA to automatically identify oscillations in noisy series and to capture the statistically significant dynamical behavior by using just a few components. This parsimonious description of the time series by SSA substantially simplifies the comparison of multiple indicators.

References

Aguiar-Conraria, L., Soares, M.J.: Business cycle synchronization and the Euro: a wavelet analysis. J. Macroecon. **33**, 477–489 (2011)

Alessio, S., Vivaldo, G., Taricco, C., Ghil, M.: Natural variability and anthropogenic effects in a central mediterranean core. Clim. Past **8**, 831–839 (2012)

Allen, M.R.: Interactions between the atmosphere and the oceans on time-scales of weeks to years. Ph.D. Dissertation, Clarendon Laboratory, Oxford (1992)

Allen, M.R., Robertson, A.W.: Distinguishing modulated oscillations from coloured noise in multivariate datasets. Clim. Dyn. **12**, 775–784 (1996)

Allen, M.R., Smith, L.A.: Monte Carlo SSA: detecting irregular oscillations in the presence of colored noise. J. Climate **9**, 3373–3404 (1996)

Allen, M.R., Smith, L.A.: Optimal filtering in singular spectrum analysis. Phys. Lett. A **234**(6), 419–428 (1997)

Allen, M.R., Read, P., Smith, L.A.: Temperature oscillations. Nature **359**, 679 (1992)

Bonino, G., Cini Castagnoli, G., Callegari, E., Zhu, G.M.: Radiometric and tephroanalysis dating of recent Ionian Sea cores. Nuovo Cimento C **16**, 155–161 (1993)

Brawanski, A., Faltermeier, R., Rothoerl, R.D., Woertgen, C.: Comparison of near-infrared spectroscopy and tissue Po_2 time series in patients after severe head injury and aneurysmal subarachnoid hemorrhage. J. Cerebr. Blood F. Met. **22**(5), 605–611 (2002)

Broomhead, D.S., King, G.P.: Extracting qualitative dynamics from experimental data. Phys. D **20**, 217–236 (1986a)

Broomhead, D.S., King, G.P.: On the qualitative analysis of experimental dynamical systems. In: Sarkar, S. (ed.) Nonlinear Phenomena and Chaos. Adam Hilger, Bristol (1986b)

Colebrook, J.M.: Continuous plankton records, zooplankton and environment, North-East Atlantic and North Sea, 1948–1975. Oceanol. Acta **1**, 9–23 (1978)

de Carvalho, M., Rodrigues, P.C., Rua, A.: Tracking the US business cycle with a singular spectrum analysis. Econ. Lett. **114**(1), 32–35 (2012)

Dettinger, M.D., Ghil, M., Strong, C.M., Weibel, W., Yiou, P.: Software expedites singular-spectrum analysis of noisy time series. Eos, Trans. American Geophysical Union **76**(2), 12, 14, 21 (1995)

De Putter, T., Loutre, M.F., Wansard, G.: Decadal periodicities of Nile River historical discharge (A.D. 622–1470) and climatic implications. Geophys. Res. Lett. **25**, 3193–3196 (1998)

Feliks, Y., Groth, A., Robertson, A., Ghil, M.: Oscillatory climate modes in the Indian monsoon, North Atlantic and Tropical Pacific. J. Climate **26**, 9528–9544 (2013)

Freas, W.A., Sieurin, E.: A nonparametric calibration procedure for multi-source urban air pollution dispersion models. In: Fifth Conference on Probability and Statistics in Atmospheric Sciences. American Meteorological Society (AMS), Las Vegas (1977)

Ghaleb, K.O.: Le Mikyas ou Nilomètre de l'Île de Rodah. Mem. Inst. Egypte, 54 (1951)

Ghil, M., Taricco, C.: Advanced spectral analysis methods. In: Cini Castagnoli, G. Provenzale, A. (eds.) Past and Present Variability of the Solar-Terrestrial System: Measurement, Data Analysis and Theoretical Models. Società Italiana di Fisica, Bologna, Italy & IOS Press, Amsterdam (1997)

Ghil, M., Allen, M.R., Dettinger, M.D., Ide, K., Kondrashov, D., Mann, M.E., Robertson, A.W., Saunders, A., Tian, Y., Varadi, F., Yiou, P.: Advanced spectral methods for climatic time series. Rev. Geophys. 40(1), 1.1–1.41 (2002)

Golyandina, N., Nekrutkin, V., Zhigliavsky, A.: Analysis of Time Series Structure: SSA and Related Techniques. Chapman & Hall/CRC, Boca Raton (2001)

Golyandina, N., Stepanov, D.: SSA-Based approaches to analysis and forecast of multidimensional time series. In: Proceedings of the 5th St. Petersburg Workshop on Simulation, June 26-July 2, 2005, pp. 293-298. St. Petersburg State University, St. Petersburg (2005)

Golyandina, N., Zhigljavsky, A.: Singular Spectrum Analysis for Time Series. Springer, Berlin (2013)

Greco, G., Rosa, R., Beskin, G., Karpov, S., Romano, L., Guarnieri, A., Bartolini, C., Bedogni, R.: Evidence of deterministic components in the apparent randomness of GRBs: clues of a chaotic dynamic. Sci. Rep. 1, 91 (2011)

Grigorov, M.: Global dynamics of biological systems from time-resolved omics experiments. Bioinformatics 22(12), 1424–1430 (2006)

Groth, A., Ghil, M.: Multivariate singular spectrum analysis and the road to phase synchronization. Phys. Rev. E 84(036206), 1–10 (2011)

Groth, A., Ghil, M., Hallegatte, S., Dumas, P.: The Role of Oscillatory Modes in U.S. Business Cycles. Working Paper 26. Fondazione ENI Enrico Mattei (FEEM), Milan (2012)

Hallegatte, S., Ghil, M., Dumas, P., Hourcade, J.: Business cycles, bifurcations and chaos in neoclassical model with investment dynamics. J. Econ. Behav. Organ. 67, 57–77 (2008)

Hassani, H., Thomakos, D.: A review on singular spectrum analysis for economic and financial time series. Stat. Interf. 3(3), 377–397 (2010)

Hassani, H., Soofi, A., Zhigljavsky, A.: Predicting daily exchange rate with singular spectrum analysis. Nonlinear Anal.: Real World Appl. 11, 2023–2034 (2011)

Hassani, H., Heravi, S., Zhigljavsky, A.: Forecasting UK industrial production with multivariate singular spectrum analysis. J. Forecast. 32(5), 395–408 (2013)

Hassani, H., Mahmoudvand, R.: Multivariate singular spectrum analysis: a general view and new vector forecasting approach. Int. J. Energy Stat. 1(1), 55–83 (2013)

Hayes, M.H.: Statistical Digital Signal Processing and Modeling. Wiley, New York (1996)

Hirsch, R.M., Slack, J.R.: A nonparametric trend test for seasonal data with serial dependence. Water Resour. Res. 20(6), 727–732 (1984)

Hirsch, R.M., Slack, J.R., Smith, R.A.: Techniques of trend analysis for monthly water quality data. Water Resour. Res. 18(1), 107–121 (1982)

Hodrick, R., Prescott, E.C.: Postwar U.S. business cycles: an empirical investigation. J. Money, Credit, Banking 29(1), 1–16 (1997)

Kaiser, H.F.: The varimax criterion for analytic rotation in factor analysis. Psychometrika 23(3), 187–200 (1958)

Hurst, H.E.: The Nile. Constable, London (1952)

Karhunen, K.: Zur Spektraltheorie stochastischer prozesse. Annales Academiae Scientiarum Fennicae, Ser. A1, Math.-Phys., 34, 1–7 (1946)

Karhunen, K.: ber lineare Methoden in der Wahrscheinlichkeitsrechnung. Annales Academiae Scientiarum Fennicae, Ser. A1, Math.-Phys., 37, 1–79 (1947)

Keppenne, C.L., Ghil, M.: Adaptive filtering and prediction of the Southern Oscillation index. J. Geophys. Res. 97, 20449–20454 (1992)

Keppenne, C.L., Ghil, M.: Adaptive filtering and prediction of noisy multivariate signals: an application to subannual variability in atmospheric angular momentum. Intl. J. Bifurcat. Chaos **3**, 625–634 (1993)

Kondrashov, D., Ghil, M.: Spatio-temporal filling of missing points in geophysical data sets. Nonlin. Process. Geophys. **13**, 151–159 (2006)

Kondrashov, D., Feliks, Y., Ghil, M.: Oscillatory modes of extended Nile River records (A.D. 622–1922). Geophys. Res. Lett. **32**, L10702 (2005)

Kondrashov, D., Shprits, Y.Y., Ghil, M.: Gap filling of solar wind data by singular spectrum analysis. Geophys. Res. Lett. **37**, L15101 (2010)

Kondrashov, D., Denton, R., Shprits, Y.Y., Singer, H.J.: Reconstruction of gaps in the past history of solar wind parameters. Geophys. Res. Lett. **41**, 2702–2707 (2014)

Loéve, M.: Probability Theory, Vol. II. Graduate Texts in Mathematics, vol. 46. Springer, Berlin (1978)

Mac Lane, S., Birkhoff, G.: Algebra. AMS Chelsea Publishing, New York (1999)

Mañé, R.: On the Dimension of the Compact Invariant Sets of Certain Non-Linear Maps. Dynamical Systems and Turbulence. Lecture Notes in Mathematics, vol. 898, pp. 230–242. Springer, Berlin (1981)

Mann, M.E., Lees, J.M.: Robust estimation of background noise and signal detection in climatic time series. Clim. Change **33**(3), 409–445 (1996)

Mineva, A., Popivanov, D.: Method of single-trial readiness potential identification, based on singular spectrum analysis. J. Neurosci. Methods **68**, 91–99 (1996)

Patterson, K., Hassani, H., Heravi, S., Zhigljavsky, A.: Multivariate singular spectrum analysis for forecasting revisions to real-time data. J. Appl. Stat. **38**(10), 2183–2211 (2011)

Penland, C., Ghil, M., Weickmann, K.M.: Adaptive filtering and maximum entropy spectra, with application to changes in atmospheric angular momentum. J. Geophys. Res. **96**, 22659–22671 (1991)

Plaut, G., Vautard, R.: Spells of low-frequency oscillations and weather regimes in the northern hemisphere. J. Atmos. Sci. **51**(2), 210–236 (1994)

Popper, W.: The Cairo Nilometer. University of California Press, Berkeley (1951)

Quinn, W.H.: A Study of Southern Oscillation-Related Climatic Activity for A.D. 622–1900 Incorporating Nile River Flood Data. In: Diaz, H.F., Markgraf, V. (eds.) El Niño: Historical and Paleoclimatic Aspects of the Southern Oscillation. Cambridge University Press, New York (1992)

Robertson, A.W., Mechoso, C.R.: Interannual and interdecadal variability of the South Atlantic convergence zone. Mon. Wea. Rev. **128**, 2947–2957 (2000)

Rodó, X., Pascual, M., Fuchs, G., Faruque, A.S.G.: ENSO and cholera: a nonstationary link related to climate change? Proc. Nat. Acad. Sci. USA **99**(20), 12901–12906 (2002)

Sella, L.: Old and New Spectral Techniques for Economic Time Series. Cognetti de Martiis Working Papers Series, N. 09/2008. Department of Economics. Torino University, Torino (2008)

Sella, L., Marchionatti, R.: On the cyclical variability of economic growth in Italy, 1881–1913: a critical note. Cliometrica **6**(3), 307–328 (2012)

Sella, L., Vivaldo, G., Groth, A., Ghil, M.: Economic Cycles and Their Synchronization: A Survey of Spectral Properties. Working Paper 105.2013. Fondazione ENI Enrico Mattei (FEEM), Milan (2013)

Shackleton, N.J., Kennett, J.P.: Palaeo-temperature history of the cenozoic and the initiation of Antarctic Glaciation: oxygen and carbon isotope analysis in DSDP Sites 277, 279 and 281. In: Kennett, J.P., Houtz, R.E. (eds.) Initial Reports of the Deep Sea Drilling Project, vol. 5, pp. 743–755. US Government Printing Office, Washington (1975)

Takens, F.: Detecting Strange Attractors in Turbulence. Dynamical Systems and Turbulence. Lecture Notes in Mathematics, vol. 898, pp. 366–381. Springer, Berlin (1981)

Taricco, C., Ghil, M., Alessio, S., Vivaldo, G.: Two millennia of climate variability in the central mediterranean. Clim. Past **5**, 171–181 (2009)

Taricco, C., Vivaldo, G., Alessio, S., Rubinetti, S., Mancuso, S.: A high-resolution δ^{18}O record and Mediterranean climate variability. Clim. Past **11**, 509–522 (2015)

Thomakos, D.D., Tao Wang, T., Wille, L.T.: Modeling daily realized futures volatility with singular spectrum analysis. Phys. A, Stat. Mech. Appl. **312**(3–4), 505–519 (2002)

Toussoun, O.: Mémoire sur l'histoire du Nil. Mem. Inst. Egypte **18**, 366–404 (1925)

Vautard, R., Ghil, M.: Singular spectrum analysis in nonlinear dynamics, with applications to paleoclimatic time series. Phys. D **35**, 395–424 (1989)

Vautard, R., Yiou, P., Ghil, M.: Singular-spectrum analysis: a toolkit for short, noisy. Chaotic Signals. Phys. D **58**, 95–126 (1992)

Walker, G.T.: Correlation in seasonal variations of weather: II. Indian Meteorol. Mem. **21**, 1–21 (1910)

Whitcher, B.J., Byers, S.D., Guttorp, P., Percival, D.B.: Testing for homogeneity of variance in time series: long memory, wavelets and the Nile River. Water Resour. Res. **38**(5), 12-1–12-16 (2002)

Whittaker, E.T.: On a new method of graduation. Proc. Edinburgh Math. Assoc. **41**, 63–75 (1923)

Zhigljavsky, A. (Ed.): Special issue on theory and practice in singular spectrum analysis of time series. Stat. Interf. **3**(3) (2010)

Chapter 13
Non-stationary Spectral Analysis

13.1 Chapter Summary

Any time we analyze the spectral content of a signal, we must ask ourselves if assuming stationarity is correct, or at least satisfactory. Very often we may be interested in investigating if, how and to what extent the spectral features of the signal, i.e., its composition in terms of single oscillatory contributions of different amplitude and frequency, vary in time, particularly during the time span over which the segment of a sample sequence we are considering was measured. The techniques of evolutionary spectral analysis address this need.

The traditional approach to this problem, introduced by Gabor (1946), is known as the short-time Fourier transform (STFT) and consists of dividing the signal into segments short enough as to be able to assume that within each segment, spectral characteristics do not vary significantly. Each segment is then analyzed separately. A more modern approach is provided by the continuous wavelet transform (CWT), in which the signal in its entirety is not compared with infinitely-long sinusoids, but with waveforms that are concentrated in time. In the next section we will briefly describe the STFT approach, while in the rest of the chapter we will exhaustively discuss CWT as a tool for spectral analysis of non-stationary random signals.

First, we will introduce the concept of scale, which in this method replaces the concept of period (inverse of a frequency). Using the language of continuous time-signals that allows us to avoid some mathematical difficulties, we will describe what a wavelet is, and how a signal can be analyzed in time and scale via wavelet transform, also investigating the resolution properties of this technique in both domains. We will then establish a relation between scale and frequency, so as to be able to give a spectral description of the signal in "usual" terms. We will also write the inverse continuous wavelet transform (ICWT), and discuss the conditions under which it exists; real and complex wavelets suited for CWT/ICWT will be introduced, and their characteristics described.

© Springer International Publishing Switzerland 2016
S.M. Alessio, *Digital Signal Processing and Spectral
Analysis for Scientists*, Signals and Communication Technology,
DOI 10.1007/978-3-319-25468-5_13

The language of continuous time is useful theoretically, but for practical application the CWT must be made discrete in time and scale. We will discuss what discretization scheme can be adopted.

An average power spectrum estimate, the global power spectrum (GWS), can be derived from the CWT by averaging over time. Moreover, guidelines for performing significance tests for the spectral features detected by the CWT and GWS will be given. Finally, we will present some extensions of wavelet analysis, like wavelet filtering and cross-wavelet spectral analysis. During all the discussion, we will constantly compare the CWT with the STFT. Some real-world applications of the CWT technique conclude the chapter.

13.2   Short-Time Fourier Transform (STFT)

The short-time Fourier transform (STFT) is a classical evolutionary spectral tool also known as *short-time spectral analysis, time-dependent spectral analysis*, and *windowed Fourier transform*. It employs a rectangular or tapered window with length smaller than the signal length. The window, starting from the time origin, frames a part of the signal, with length equal to the window width. The spectrum of this segment is estimated by some spectral method. Then the window moves by a certain number of samples in the direction of increasing discrete time; the spectrum of the framed data is estimated, and so on. For the stationarity assumption to hold, the typical time scale of the variations in spectral characteristics that the signal undergoes must be longer than the length of the window applied to the data in order to obtain the various segments that are analyzed.

This procedure provides a set of spectra, each being relative to a given time-location k of the moving window: $P_{xx}(\omega, k)$. This set of spectra is then displayed in two-dimensional plots having time on the abscissa and frequency on the ordinate, and representing spectral values as contour lines. This representation is referred to as a *spectrogram*. Sometimes, a three-dimensional plot may be preferred. Note that if the window displacement is equal to the window width, then adjacent spectra will be independent of one another. If the displacement of the window is smaller, adjacent spectra will not be independent, but at the same time, spectral features will vary more continuously and smoothly from one spectrum to the following one. The faster the variations of the spectral characteristics of the signal, the shorter the window will have to be, and this will generally imply reduced frequency resolution. On the other hand, as the window width decreases, there will be an increase in the ability to resolve in time the changes of spectral behavior, and changes on shorter time scales will become visible. Thus the choice of the window width ultimately represents a compromise between frequency resolution and time resolution.

The estimate of the power spectrum of each segment can, in principle, be obtained by any method. If we choose a non-parametric method, the window shape—that does not need to be rectangular—is also important: the window transform should ideally have a narrow main lobe with respect to the typical frequency variations of the typical data-segment's spectrum. Recall that the ability to resolve two closely-spaced frequency components depends on the main lobe width of the window transform, while the amount of leakage of a given component in the neighborhood of other components depends on the relative height of the sidelobes.

In summary, the relevant parameters in SFTF are:

- the window shape; Gabor (1946) proposed a Gaussian window (see Fig. 5.7f);
- the window width in number of samples (M);
- the window displacement in number of samples (M_s);
- the number of frequency samples for each DFT (N_{FFT}).

Obviously, we must have $N_{FFT} \geq M \geq M_s$. If a tapered window is used, more external samples of a data segment have a smaller statistical weight than samples located near the center. In this case, a certain overlapping of the segments is advisable ($M_s < M$).

If we apply the STFT to a signal that is actually made of sinusoidal components with constant frequency and amplitude, we might be tempted to expect an evolutionary spectrum exactly constant with time. But this would only be true for a periodic signal with period N_p, for which M and M_s were chosen as integer multiples of N_p. In fact, in this case a segment would include exactly an integer number of periods N_p, and also the window would move by an integer number of periods. In general, even if the signal is exactly periodic, the variable phase relations that result as different segments of the waveform enter the window span produce variations in the transform from one window position to the other. However, for stationary signals the transform amplitude varies only a little from one segment to the other, since the main part of the transform's temporal variability concerns the phase, which does not influence the power spectrum.

We may finally note that AR parametric methods are also indicated for the STFT, especially if the segments are short ($M \approx 60 \div 200$ samples). In this case, the window is always rectangular. More details on the STFT, including its formal definition, are given in the following sections, in that they are useful for comparison with some facets of CWT analysis.

As a simple STFT application, we can consider spectral analysis of some music signals. Time-frequency tools like the STFT representation are often used in the analysis of audio signals. Here we take a look to a very simple case: a single musical note produced by different instruments playing one at a time.

Signal processing techniques are widely applied to music signals. Even neglecting the field of electronic music synthesis, and limiting the attention to the analysis of existing music signals, the vastness of the field is enormous. The information that can be extracted is important for many applications and modeling activities, an example of which—illustrating the complexity of the tasks that may be involved—is the inverse problem of recovering a score-level description, given only the audio. Many techniques in this area were initially borrowed from speech signal processing, another extremely variegated, large and mature field, but the unique properties and stringent demands of music signals often led to independent solutions. Music signals possess specific acoustic and structural characteristics that distinguish them from spoken language or other non-musical audio signals. Signal analysis techniques were designed to specifically address musical dimensions such as pitch, melody, rhythm, and timbre (see Müller et al. 2011).

Pitch is a ubiquitous feature of music: individual notes with distinct pitches are sound oscillations with well-defined fundamental periods. Sequences of pitches create melodies—the "tune" of a piece of music. Most musical instruments—including string-based instruments such as guitars, violins, and pianos, as well as instruments based on vibrating air columns such as flutes, clarinets, horns and trumpets—are explicitly constructed to allow performers to produce sounds with easily controlled, locally stable fundamental periods. Such a signal is well described as a harmonic series of sinusoids at multiples of a fundamental frequency, and results in the perception of a musical note in the mind of the listener. With the exception of unpitched instruments like drums, and a few inharmonic instruments such as bells, the periodicity of individual musical notes is rarely ambiguous, and thus equating the perceived pitch with fundamental frequency is common.

Music exists for the pleasure of human listeners, and thus its features reflect specific aspects of human auditory perception. In particular, humans perceive two signals whose fundamental frequencies fall in a ratio 2:1 (an octave) as highly similar; this fact is sometimes known as "octave equivalence". A sequence of notes—a melody—performed at pitches exactly one octave displaced from an original will be perceived as largely musically equivalent. We note that the sinusoidal harmonics of a fundamental at frequency f_0, falling at frequencies $2f_0$, $3f_0$, $4f_0$, etc., are a proper superset of the harmonics of a note with fundamental frequency $2f_0$ (i.e. $4f_0$, $6f_0$, $8f_0$), and this is presumably the basis of the perceived similarity. Other pairs of notes with frequencies in simple ratios, such as f_0 and $3f_0/2$, will also share many harmonics, and are also perceived as similar, although not as close as in the octave-equivalence case.

Even if different cultures have developed different musical conventions, a common feature is the musical scale, a set of discrete pitches that repeats every octave, from which melodies are constructed. For example, contemporary western music is based on the *equal tempered scale*, which, by a happy mathematical coincidence, allows the octave to be divided into twelve equal steps on a logarithmic axis, while still (almost) preserving intervals corresponding to the most pleasant note combinations. The equal division makes each frequency larger than its predecessor, an interval known as a semitone. The coincidence is that it is even possible to divide the octave uniformly into such a small number of steps, and still have these steps give close, if not exact, matches to the simple integer ratios that result in consonant harmonies, e.g., $\left(2^{1/12}\right)^7 = 1.498 \approx 3/2$, and $\left(2^{1/12}\right)^5 = 1.335 \approx 4/3$.

The western major scale spans the octave using seven of the twelve steps—the "white notes" on a piano, denoted by C or *do*, D or *re*, E or *mi*, F or *fa*, G or *sol*, A or *la*, and B or *si*. The spacing between successive notes is two semitones, except for E/F and B/C, which are only one semitone apart. The "black notes" in between are named in reference to the note immediately below (e.g., A♯), or above (B♭), depending on musicological conventions. The octave degree denoted by these symbols is sometimes known as the pitch's chroma, and a particular pitch can be specified by the concatenation of a chroma and an octave number, where each numbered octave spans C to B. The lowest note on a piano is A0 (27.5 Hz), the highest note is C8 (4186 Hz), and middle C is C4 (262 Hz). Different instruments, of course, span different ranges of frequencies.

Different instruments playing the same note have different timbres. Timbre is defined as the attribute of auditory sensation in terms of which a listener can judge two sounds similarly presented and having the same loudness and pitch as dissimilar. The concept is closely related to sound source recognition: for example, the sounds of the violin and the flute may be identical in their pitch and loudness, but are still easily distinguished. Timbre may be examined in the time domain, looking at the difference between typical waveforms—for example, the waveforms of a flute and a guitar both playing a certain note—and in the frequency domain, looking at the different spectral content of the two sounds. Another feature that distinguishes one instrument from the other is the envelope of the sound amplitude, i.e., the different sound length, the time delay from the onset to the maximum amplitude, and the sound decay in time. For example, a guitar that produces a sound when a string is strummed or plucked produces a sound with a very different envelope with respect to a wind instrument like a flute, or a string instrument in which the sound is produced using a bow, as a cello or a violin.

To illustrate these concepts, Figs. 13.1 and 13.2 show the envelopes and the typical waveforms of an A (*la*) played on a cello, a flute, a French horn and a classical guitar, respectively. These records were downloaded from ftp://ftp.wiley. com/public/college/math/matlab/bporat. They were made available by B. Porat, who analyzed them in his book (Porat 1996).

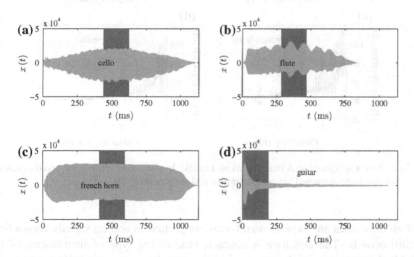

Fig. 13.1 Envelopes of an A note played on **a** a cello, **b** a flute, **c** a French horn, and **d** a classical guitar. The *dark-gray-shaded rectangles* mark sub-intervals of about 186 milliseconds (ms) each, including $N = 8192$ samples, that are analyzed in Figs. 13.3 and 13.4

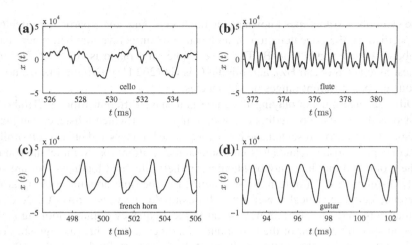

Fig. 13.2 Waveforms of an A note played on **a** a cello, **b** a flute, **c** a French horn, and **d** a classical guitar

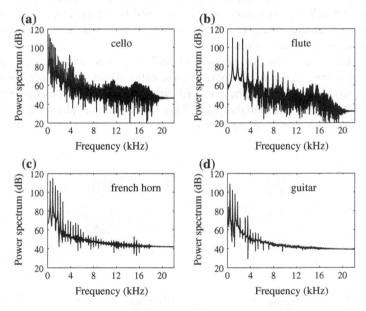

Fig. 13.3 Power spectra of an A note played on **a** a cello, **b** a flute, **c** a French horn, and **d** a classical guitar. Spectra were estimated via periodograms, and are plotted in dB

Both envelopes and typical waveforms are plotted as analog signals versus time in milliseconds. The envelope is characteristic of the type of instrument, of the individual instrument, of the note and of the way the note is played, exactly as the harmonic structure. One obvious difference between bow instruments such as the violin and cello, or wind instruments such as the flute and horn, and plucked string

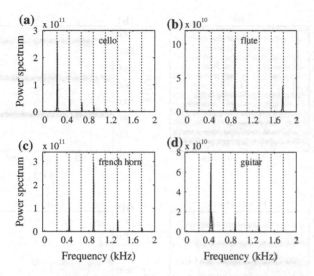

Fig. 13.4 Low-frequency part of the power spectra of an A note (Fig. 13.3) played on **a** a cello, **b** a flute, **c** a French horn, and **d** a classical guitar. Spectra were estimated via periodograms and are plotted on a linear scale

instruments like the guitar, or keyboard instruments like the piano, lies in the time for which the note can be sustained. So in Fig. 13.1 the cello has a gradual rise of the amplitude, followed by a gradual decay. The flute has a characteristic low-amplitude modulation. The French horn has a nearly constant amplitude during the whole note duration. The sound of the guitar rises steeply when the string is plucked, and decays steeply after the string is released. The sounds analyzed here were recorded at 44.1 kHz for about one second.[1] In Fig. 13.1, the *dark-gray-shaded rectangles* mark sub-intervals of about 186 milliseconds each, including $N = 8192$ samples, that are analyzed via periodograms (see Figs. 13.3 and 13.4, showing each spectrum on a dB scale and the low-frequency region of each spectrum on a linear scale, respectively). Each instrument shows the harmonic series at integer multiples of the fundamental, but the energy of the note is distributed differently over the harmonics in the four cases. The cello (Figs. 13.3a and 13.4a) exhibits a very regular series of harmonics, with amplitude decreasing in an exponential way with increasing frequency. The pitch is at 220 Hz. The flute (Figs. 13.3b and 13.4b) exhibits two dominant peaks, at 880 and $2 \times 880 = 1760$ Hz. The French horn is also pitched at 880 Hz, and exhibits, in addition to the 440 Hz fundamental component, higher harmonics (Figs. 13.3c and 13.4c). The guitar (Figs. 13.3d and 13.4d) is pitched at 440 Hz, with a couple of small higher harmonics.

---

[1] When it is necessary to capture audio covering the entire 20–20,000 Hz range of human hearing, such as when recording music or many types of acoustic events, audio waveforms are typically sampled at 44.1 (Compact Disks), 48, 88.2, or 96 kHz.

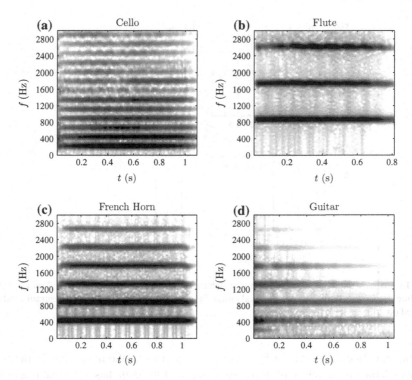

Fig. 13.5 Spectrograms of an A note played on **a** a cello, **b** a flute, **c** a French horn, and **d** a classical guitar, as squared magnitudes of the corresponding windowed transforms in dB

What this picture cannot reveal is the possible presence of amplitude and frequency modulations: it does not offer an evolutionary view of the spectral behavior. At this point, the spectrograms of the four signals (Fig. 13.5) can step in, providing the desired insight into evolutive features of the signals. These spectrograms were obtained adopting a 1024-point Gaussian window and 50 % overlap. We simply plot the squared magnitude of the windowed Fourier transform in dB. The grayscale map has been reversed with respect to the normal convention, which represents maxima with light gray or white and minima with dark gray or black. With this reversal, any regularity or irregularity in the position of maxima in the time-frequency plane is more clearly visible. The cello's spectrogram (Fig. 13.5a) shows that for each harmonic, frequency oscillates regularly in time, so that we see a wavy pattern, especially on higher harmonics. This slight frequency modulation is typical of the cello and other string instruments like the violin. Vertical stripes, like those visible in the spectrogram of the flute (Fig. 13.5b) in the low-frequency frequency band ($f < 400$ Hz), indicate amplitude modulation, which we already detected observing Fig. 13.1b. In the guitar's spectrogram (Fig. 13.5d), the rapid amplitude decay with time observed in Fig. 13.1d is evident; the spectrogram also reveals that high-frequency components decay faster than low-frequency ones. This can be understood thinking of the mechanics of the decay. The vibration of the guitar's string is a standing wave with two nodes at the ends: the elements of the string located outside the nodes move transversely to the line connecting the nodes. If we call x a longitudinal coordinate (distance from one end node) and y a transverse coordinate, the oscillation with a certain wavelength λ can be written as $y(x, t) \propto \sin(2\pi x/\lambda) \cos(2\pi f t)$. The fundamental wavelength for these standing waves, $\lambda = \lambda_0$, is twice the length of the string. The velocity dy/dt is thus $\propto f \sin(2\pi f t)$ and therefore grows with frequency. Damping of the vibration is mainly due to air friction; air is a viscous medium, and the friction on an object moving in a viscous medium is proportional to the speed of the object (Stokes' law, 1851). In conclusion, this simple explanation of the phenomenon leads us to understand that higher harmonics correspond to higher values of transverse velocity and therefore to greater friction and faster damping.

13.3 Wavelet Transform

Wavelets constitute an alternative approach with respect to traditional signal processing methods. They are waveforms concentrated in time, or even with compact temporal support, i.e., localized in the time domain, and concentrated in frequency, which means they are characterized by a passband spectrum and are therefore localized also in the frequency domain.[2] They are, in essence, localized waves: instead of oscillating forever, they start, attain a maximum amplitude and then drop to zero.

When we use wavelets to analyze the spectral content of a signal, we get a map in the *time-scale plane*. The concept of *scale* replaces, in wavelet spectral analysis, the concept of frequency: it is a typical duration and for many wavelets is related to the inverse of a frequency, i.e., to a period. The map in the time-scale plane allows us to study

---

[2] A wavelet is, in general, a finite-energy, zero-mean waveform that in the wavelet transform is stretched or compressed in an auto-similar way, exactly like the Gaussian function in Fig. 4.14 (which has an unbounded time support) and the boxy function of Fig. 4.15 (which has a compact temporal support). For example, we will introduce, on one hand, the historical real Haar wavelet, a waveform with compact time support, and the analytic Morlet wavelet (shown in its real and imaginary parts in Fig. 13.15) that, being a complex exponential oscillation modulated by a Gaussian envelope, never vanishes identically at finite time values, and therefore has an unbounded time support. However, in numerical applications using finite-precision arithmetic, the distinction among the two cases can be considered to some extent as a mathematical nuance. E.g., a Gaussian function that we compute numerically over an extended range of values of the independent variable will end up, as we go farther and farther from its center, assuming such a small value to be practically indistinguishable from zero. For instance, the minimum positive number representable in double precision is $2.22507e-308$ and the maximum is $1.79769e+308$; positive numbers that are smaller than $2.22507e-308$ are treated as zero and positive numbers greater than $1.79769e+308$ are treated as $+\infty$. For this very practical reason, talking about wavelets we will often neglect the conceptual difference between compact support and concentration, and will loosely use expressions like "waveform localized in time", etc. In a similar way, in the frequency domain we will speak about the spectra of the wavelet functions using terms like "passband spectrum, waveform localized in the frequency domain", and so on.

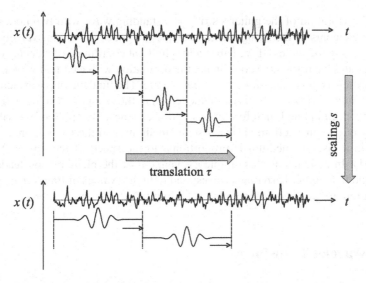

Fig. 13.6 How the CWT is computed: a signal is compared with a wavelet at a given scale s. The wavelet is shifted along the signal (translation or delay τ) and for each wavelet position the cross-correlation between the wavelet and the signal is computed, quantifying the similarity of the two waveforms. Then, using a scaling operation, the wavelet is stretched (or compressed). The wavelet at the new scale is again shifted along the signal, and new correlation values are computed. A correlation matrix as a function of τ and s is thus obtained

non-stationary features of the signal, such as changes in periodicity, as well as isolated events, trends, intermittency, etc.

The analysis is performed starting from a prototype, called *mother wavelet*, that is then scaled (dilated or shrunk) in such a way that the wavelet shape remains unaltered, while the duration—more precisely, the scale—changes. Each *scaling* operation provides a *daughter wavelet* at some scale s. This scaled wavelet is translated along the entire signal's length, and cross-correlated (hereafter, simply correlated) with the signal at each temporal position τ. Then another scaling takes place, followed by a new set of shifts and correlations. This procedure is qualitatively illustrated in Fig. 13.6. The result is a matrix of correlation values that describe the similarity between the signal and the daughter wavelet, at all considered scales and around each temporal location. Choosing the mother wavelet properly, an evolutionary spectral representation of the signal can be obtained that has several advantages with respect to STFT.

13.3.1 Analysis in Time and Scale

In wavelet theory it is convenient use the formalism of continuous-time signals. In addition to time t, we use also the *time translation* or *delay* τ that indicates a

delay with respect to the origin of the time axis, and the *scale s*, which is related to
the duration of the daughter wavelet. However, the analysis will ultimately concern
sampled signals with sampling step T_s; so it is reasonable to formulate the theory
independently of T_s, using a *continuous adimensional time* θ, an *adimensional delay*
b and an *adimensional scale* or *scale factor*[3] a, such that

$$\theta = \frac{t}{T_s}, \qquad b = \frac{\tau}{T_s}, \qquad \text{and} \qquad a = \frac{s}{T_s}.$$

In other words, thinking that we will eventually analyze sampled signals, we directly
use the sampling step to normalize the continuous temporal quantities of interest.

The theory thus deals with an analog signal $x(\theta)$ that is analyzed in time (delay)
b and scale a through the *continuous wavelet transform* (CWT). Let us stress the
fact that a and b are continuous variables, like θ. In Fig. 13.6 we now must imagine
infinitesimal variations of a and b, leading to a continuous correlation function depending on time delay and scale. This function of two independent variables is the
CWT. Note that for the purpose of spectral analysis, scale will have to be translated
into frequency. We will discuss this issue later; for the moment, it is sufficient to think
of frequency as a variable inversely proportional to the scale. Only later, the CWT
is adapted to discrete-time signals $x[n]$, choosing also a discrete set of values for a
and b. The rationale behind this approach is that in the continuous-time domain all
is simpler: concepts like wavelet dilation, or other concepts that will be introduced
later—for example the concept of "regularity"—are not defined in the discrete-time
domain. The continuous-time theory must thus be seen as a useful asymptotic theory
with respect to the practical discrete-time case.[4]

The domain of wavelet techniques and related applications is quite wide. Here
we focus on the discretized CWT, since it provides a tool for evolutionary spectral
analysis of random signals. The elementary functions used for wavelet analysis are
not fixed like the sines and cosines of Fourier analysis, but are functions for which a
remarkable freedom of choice exists, which can accommodate for different types of
signals and different analysis goals. This gives the technique a great versatility. However, the signal representation that is obtained by the discretization scheme usually
adopted for CWT is largely redundant, meaning that the transform contains more
information than would be strictly necessary to reconstruct the signal by inverting the
transform. This redundancy, which sometimes is also referred to as oversampling, is
advantageous for spectral analysis because it makes the analysis more robust—less
sensitive to outliers in the data or to small deviations from basic assumptions.

Redundancy becomes undesirable when the signal must be represented economi-
cally, i.e., retaining the minimum information required for signal reconstruction. This
need occurs, for example, in compression and de-noising applications. In wavelet

---

[3]In the following discussion, for brevity we will often write simply "scale" instead of "scale factor",
provided the context does not allow any ambiguity.
[4]The signals analyzed by CWT may be mono- or bi-dimensional, possibly with space (instead of
time) as the independent variable, as in the case of image processing; in principle, the signals can
even be multi-dimensional. In this book we will only deal with wavelet analysis of a time series.

theory, the issues related to a *critical sampling* of the transform in time and scale are thus examined, as well as the characteristics that are consequently required from the set of functions to be used to represent a signal through its wavelet transform. This leads to the *discrete wavelet transform* (DWT). The DWT is presented in Chap. 14.

The wavelet transform, in its various shades (including the DWT), has been applied in numerous fields, such as

- astronomy,
- climatology, atmospheric physics, environmental physics, oceanography,
- seismology,
- oil and mining exploration,
- turbulence theory,
- signal and image processing,
- study and reproduction of human vision,
- neurosciences in general,
- medical sciences, e.g., cardiology,
- music,
- production of cartoons,
- optics,
- study of fractals and various branches of mathematics,
- economical and social sciences,
- data compression,
- signal de-noising, etc.

13.3.2   Multi-resolution Property of the Wavelet Transform

CWT allows for localization in time of the spectral components in a signal and is an alternative to STFT that avoids a serious drawback of the latter. The point is that in the STFT the window width is constant and therefore if the data contains a certain oscillation,

- at low frequency, so few periods of the oscillation are contained within the window, that frequency cannot be determined accurately;
- at high frequency, so many periods of the oscillation can be seen through the window, that if the oscillation is transient, its precise localization in time is impossible.

On the contrary, in CWT the width of the "window", which later will be identified as the scaled wavelet, varies for each spectral component, since frequency is related to scale and scale determines the temporal support of the wavelet; this allows

- determining accurately the scale (hence the frequency) of large-scale (low-frequency) oscillatory components;
- localizing accurately in time rapidly transient events that, as such, are composed of high-frequency oscillations.

This is a reasonable approach when, as normally occurs, the signal presents high frequency components over short time intervals and low-frequency components that persist for long times. Introducing an analogy related to space rather than to time, it can be said that the CWT allows for "seeing the forest and the (individual) trees", i.e.,

- the repetitive, unlocalized background pattern (the forest), and
- the isolated characteristics at a small scale (the trees).

We may prefer a time-related analogy: we could think of an engine that, on the background of a regular humming, emits time-localized different noises because it speeds up, then decelerates, then misfires, etc. This facet of the CWT is described as *multiresolution analysis* (MRA). Actually, MRA is a property of all wavelet-based techniques, and not of the CWT only. The term MRA expresses the fact that resolution in time and in frequency, meaning the ability to distinguish closely spaced features in any domain, varies with frequency. Resolution is closely related to localization in time and frequency of any particular event in the signal. We will further discuss MRA after formally defining the CWT.

13.4 Continuous Wavelet Transform (CWT)

The CWT is defined in relation to a non-stationary analog signal $x(\theta)$ having *finite energy*, i.e., such that

$$\mathscr{E} = \int_{-\infty}^{+\infty} |x(\theta)|^2 \mathrm{d}\theta < \infty.$$

In all cases of interest in this book, $x(\theta)$ is a random real signal, and mathematically we can express the fact that its energy is finite by saying that $x(\theta)$ belongs to the set $\mathrm{L}^2(\mathbb{R})$ of square-integrable functions (see the appendix to Chap. 3).

The signal is formally of infinite length, as the bounds of the definition integral indicate. Yet, its energy is assumed to be finite. This means that we suppose the signal's fluctuations decay after some finite distance from the time origin in either direction. We will consider centered signals with zero mean, so we can also state that signal variations about its (zero) mean are active only over a certain time interval. On the other hand, we will always be measuring the signal for a finite amount of time, and since we ignore how the signal behaves outside the measuring interval, we can assume that the signal is actually zero outside that interval.

The choice of assuming finite energy for the signal conceptually differentiates this analysis method from those previously discussed. In fact, while in discrete-time random-signal spectral analysis based on the DTFT, employing infinitely persistent waveforms (sines and cosines), we consider an infinitely persistent stationary process with finite average power, here we start with the intention of representing finite-energy signals using finite-energy, time-concentrated waveforms (wavelets). We thus get a distribution of the signal energy in time and scale (frequency). As we will see, we will nonetheless also derive from CWT analysis an average estimate of the power spectrum of the signal, the so-called *global wavelet spectrum* (GWS). This global spectrum is obtained by integrating the energy distribution with respect to time delay b—which transforms it into an energy distribution versus scale or frequency—and then dividing the result by the record length—which transforms energy into average power. If $x(\theta)$ is actually stationary, then the theoretical notion of GWS coincides with the notion of power spectrum we are already familiar with, so that we are allowed to compare the GWS computed from a finite-length sequence with the estimates obtained by Fourier or parametric methods. If it is not so, the GWS describes what *on average* the power spectrum of the sequence looks like over the observation time interval.

Assuming finite energy for the signal is, in a sense, a way to construct a theory which, unlike Fourier analysis that is based on infinite-length periodic functions, is closer to real-world signals. These assumptions of finite or infinite energy are, after all, theoretical nuances that though conceptually important, in applications eventually boil down to the same numbers, because we will always deal with finite-length data records anyway.

The CWT is a linear operation defined as

$$W_x^\psi(a, b) = \int_{-\infty}^{+\infty} x(\theta) \frac{1}{\sqrt{a}} \psi_0^* \left(\frac{\theta - b}{a} \right) d\theta,$$

a formula that expresses the *analysis relation* of the CWT.

$W_x^\psi(a, b)$ indicates the CWT of $x(\theta)$, performed using the wavelet $\psi(\theta)$, as a function of scale factor $a \in \mathbb{R}^+ - \{0\}$ and of adimensional time (shift, delay) b with respect to signal origin. The symbol $\psi_0(\theta)$ indicates the *prototype or mother wavelet*, which can be real or complex, and is a localized-in-time oscillatory function with zero mean. It is a wave-like function with an amplitude that starts from zero, increases, and then decreases back to zero. The expression $(1/\sqrt{a}) \, \psi_0 \, (\theta - b/a)$ represents the daughter wavelet at scale a, i.e., the scaled prototype, normalized to energy independent of scale through the factor $1/\sqrt{a}$, and shifted in time by b time steps with respect to the origin. If we set

$$\psi(\theta) = \frac{1}{\sqrt{a}} \psi_0(\theta),$$

the daughter wavelet $\psi_a(\theta)$ at scale a can be written as

$$\psi_a(\theta) = \psi\left(\frac{\theta}{a}\right) = \frac{1}{\sqrt{a}} \psi_0\left(\frac{\theta}{a}\right).$$

Furthermore, it is normal practice to introduce the adimensional variable

$$\eta = \frac{\theta}{a},$$

so that we can write

$$\psi_a(\theta) = \psi(\eta) = \frac{1}{\sqrt{a}}\psi_0(\eta).$$

For $a = 1$, we get $\eta = \theta$ and $\psi_a(\theta) = \psi(\theta) = \psi_0(\theta)$. We then consider all possible shifts of the daughter wavelet $\psi_a(\theta)$, by writing

$$\psi_{a,b}(\theta) = \psi_a(\theta - b) = \psi\left(\frac{\theta - b}{a}\right) = \frac{1}{\sqrt{a}}\psi_0\left(\frac{\theta - b}{a}\right).$$

As we stated above, the factor $1/\sqrt{a}$ in the definition of the daughter wavelet implies that all daughter wavelets have the same norm in $L^2(\mathbb{R})$, i.e., the same energy independently of a. The norm of the mother wavelet is usually set to 1,

$$\|\psi_0(\theta)\| = \sqrt{\mathscr{E}\{\psi_0(\theta)\}} = \sqrt{\int_{-\infty}^{+\infty}|\psi_0(\theta)|^2\,d\theta} = 1,$$

so that

$$\left\|\psi\left(\frac{\theta}{a}\right)\right\| = \sqrt{\mathscr{E}\left\{\psi\left(\frac{\theta}{a}\right)\right\}} = \sqrt{\int_{-\infty}^{+\infty}\left|\psi\left(\frac{\theta}{a}\right)\right|^2\,d\theta} =$$

$$= \sqrt{\int_{-\infty}^{+\infty}\left|\frac{1}{\sqrt{a}}\psi_0\left(\frac{\theta}{a}\right)\right|^2\,d\theta} = \sqrt{\int_{-\infty}^{+\infty}\left|\psi_0\left(\frac{\theta}{a}\right)\right|^2\,d\left(\frac{\theta}{a}\right)} =$$

$$= \sqrt{\int_{-\infty}^{+\infty}|\psi_0(\theta)|^2\,d\theta} = \|\psi_0(\theta)\| = 1.$$

The scaling that from $\psi(\theta)$ leads to $\psi(\theta/a)$ implies (see Fig. 13.7):

- for $a < 1$, a compression of the original waveform,
- for $a > 1$, a dilation of the original waveform.

Using the notation introduced above, the CWT definition equation can also be written as

$$W_x^{\psi}(a, b) = \int_{-\infty}^{+\infty} x(\theta)\frac{1}{\sqrt{a}}\psi_0^*\left(\frac{\theta - b}{a}\right)d\theta = \int_{-\infty}^{+\infty} x(\theta)\psi^*\left(\frac{\theta - b}{a}\right)d\theta =$$

$$= \int_{-\infty}^{+\infty} x(\theta)\psi_a^*(\theta - b)d\theta = \int_{-\infty}^{+\infty} x(\theta)\psi_{a,b}^*(\theta)d\theta.$$

$W_x^{\psi}(a, b)$ expresses the correlation—that in general will be a complex cross-correlation, as the wavelet is complex in general—between $x(\theta)$ and $\psi_a(\theta) = \psi(\theta/a)$ in the neighborhood of $\theta = b$. It measures the similarity between the scaled wavelet and the signal in that neighborhood.

Fig. 13.7 An example of
scaling: **a** generic waveform
at scale $a = 1$,
b the same waveform at scale
$a = 2$, i.e., stretched, and
c the same waveform at scale
$a = 1/2$, i.e., compressed

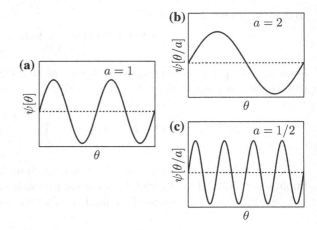

In the wavelet transform, the daughter wavelet plays the role of the transforming
function, but also the role of a window. This window gets wider or narrower as a
varies, so that the number of oscillations that a wave packet-like wavelet presents
does not change, as illustrated in Fig. 13.8. All daughter wavelets have the same
shape as the prototype; they are simply stretched or compressed. This means that
by varying the scale, we change the frequency of the signal's oscillations which the
wavelet resembles, but it also means that the wavelet, i.e., the window, "frames" the
same number of periods of the signal's oscillation, whatever the scale (frequency) of
the oscillation. This is in contrast with the behavior of the window in the STFT: as
shown in Fig. 13.9, in the STFT the window frames many oscillation periods when
the signal varies with high frequency, and few periods when the signal varies with
low frequency.

Fig. 13.8 A wavepacket-
like wavelet is shown at three
different scales: as the
duration of the wavelet
increases, the number of
oscillations in the packet
remains constant

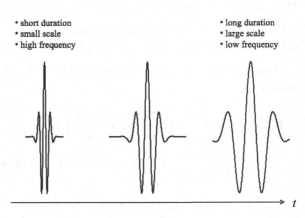

Fig. 13.9 The behavior of the window in the STFT: signal oscillations of high and low frequency are framed by a window with constant width, so that as frequency decreases, a smaller and smaller number of oscillation periods are framed by the window

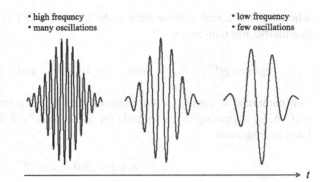

Note that in the CWT, the notion of scale is the same we use for maps:

- large scale means few details, global information about the whole signal, overall *approximation*, overview;
- small scale means many *details* over short signal segments.

Observing that by the substitution $\theta = a\theta'$ we get

$$W_x^{\psi}(a, b) = \frac{1}{\sqrt{a}} \int_{-\infty}^{+\infty} x(\theta)\psi_0^* \left(\frac{\theta - b}{a}\right) d\theta = \sqrt{a} \int_{-\infty}^{+\infty} x(a\theta')\psi_0^* \left(\theta' - \frac{b}{a}\right) d\theta',$$

we can not only state that as a increases, a more and more stretched version of the wavelet is compared with the signal, but also that as a increases, a more and more compressed version of the signal is compared with a merely shifted mother wavelet. This is analogous to looking at a set of smaller-and-smaller-scale maps from the same distance.

13.4.1 Resolution and Coverage of the Time-Frequency Plane

We will now discuss resolution and localization in time and scale (frequency) in the CWT and compare them with resolution and localization in time and frequency provided by the STFT.

In order to be able to formally compare the STFT and CWT, we write the STFT in analog version[5] as

$$S_x(\omega, b) = \int_{-\infty}^{+\infty} x(\theta)g(\theta - b)e^{-j\omega\theta} d\theta = \int_{-\infty}^{+\infty} x(\theta)g_c^*(\theta, b, \omega) d\theta,$$

---

[5]Of course, in digital applications $x(\theta)$ will become a sampled signal $x[n]$ and the continuous variables ω and b will be made discrete.

where $g(\theta)$ is a real window that in the original STFT (Gabor 1946) is Gaussian, symmetric and non-causal,

$$g(\theta) = g(0)e^{-\frac{\theta^2}{2\sigma^2}}, \quad \text{with} \quad g(0) = 1 \quad \text{and} \quad \theta \in (-\infty, +\infty).$$

The parameter σ regulates the width of the Gaussian bell. The *complex window* $g_c(\theta, b, \omega)$ appearing in the formula for $S_x(\omega, b)$ is both shifted in time and modulated in frequency,

$$g_c(\theta, b, \omega) = g(\theta - b)e^{+j\omega\theta},$$
$$g_c^*(\theta, b, \omega) = g(\theta - b)e^{-j\omega\theta},$$

and is introduced so as to make the definition of $S_x(\omega, b)$ formally similar to that of $W_x^\psi(a, b)$ and to allow for identifying the window function that in the STFT corresponds to the daughter wavelet used for the CWT.

Let us recall that in continuous-time Fourier analysis, an impulse in the time domain has a transform containing contributions from all frequencies, with phase features that give, in the reconstruction of the signal via inverse transform, cancellation everywhere, except in the neighborhood of the instant θ_0 at which the impulsive event occurs: a total lack of frequency localization corresponds to a perfect time localization. On the other hand, a spectral line corresponds to an infinitely long sinusoid: a total lack of temporal localization corresponds to a perfect frequency localization.

When we perform an STFT using a Gaussian real window $g(\theta)$, time resolution depends on the window width $\Delta\theta$, which is dictated by the parameter σ, while frequency resolution depends on the bandwidth $\Delta\omega$ of the window's transform[6] $G(\omega)$. The smaller $\Delta\theta$, the greater the time localization, and two rapidly transient events can be distinguished if they are separated in time by more than $\Delta\theta$. Time resolution is thus inversely dependent on $\Delta\theta$. Similar reasoning applies in the frequency domain: both $\Delta\theta$ and $\Delta\omega$ and the corresponding resolutions depend neither on b, nor on ω. We can illustrate this subdivision of the time-frequency plane graphically by arbitrarily discretizing b and ω: the sampling structure inherent in the STFT thus emerges as one being represented by uniform rectangular "tiles", called *Heisenberg boxes*, or *time-frequency atoms*, or *resolution cells* (Fig. 13.10). Note that the precise definition of $\Delta\theta$ and $\Delta\omega$ for a Gaussian $g(\theta)$ is a matter of convention. Usually a mean-square definition is adopted (see Sect. 4.5):

$$\Delta\theta = \sqrt{\frac{\int_{-\infty}^{+\infty} \theta^2 |g(\theta)|^2 \, d\theta}{\int_{-\infty}^{+\infty} |g(\theta)|^2 \, d\theta}}, \qquad \Delta\omega = \sqrt{\frac{\int_{-\infty}^{+\infty} \omega^2 |G(\omega)|^2 \, d\omega}{\int_{-\infty}^{+\infty} |G(\omega)|^2 \, d\omega}}.$$

---

[6]Here we indicate angular frequencies by ω, even if we are formally working in the continuous-time domain: this is justified because we are using an adimensional continuous time and therefore angular frequency is adimensional as well. However, as long as we do not apply some discretization scheme, ω is not constrained by the inequality $-\pi \le \omega < \pi$.

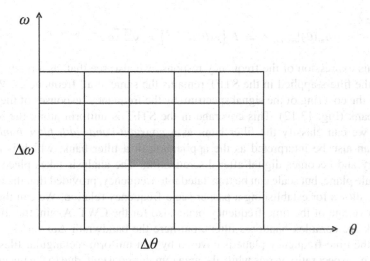

Fig. 13.10 Coverage of the time-frequency plane in the STFT: resolution in time and resolution in frequency are the same at any point and have been represented by uniform rectangular tiles. For clarity, tiles located at different discrete times are drawn juxtaposed, even if in reality they could partially overlap or be separated by empty spaces, according to the spacing adopted for delay values (centers of the horizontal tile sides). Also along the frequency axis the tiles are juxtaposed, but could as well overlap or be separated, according to the difference between frequency values (centers of the vertical tile sides), which in the absence of zero-padding is inversely dependent on the number N of signal samples

These two quantities are inversely related due to the uncertainty principle.

The STFT can also be interpreted as the output of a *bank of analog filters* applied to $x(\theta)$: its definition is actually a convolution integral of the signal with a *complex, analog, frequency-dependent bandpass filter*[7] with impulse response

$$h_\omega(\theta) = g_c^*(-\theta, b = 0, \omega) = g(-\theta)e^{-j\omega\theta} = g(\theta)e^{-j\omega\theta},$$

where the last equality follows from the even symmetry of the Gaussian function. As ω varies, the various filters of the bank are subsequently applied to the signal. The frequency response of each filter is a Gaussian function peaking at some $\omega = \omega_c$ (Fig. 13.11). This can be seen taking the Fourier transform of $h_\omega(\theta)$ for $\omega = \omega_c$: since

$$g(\theta) = e^{-\frac{\theta^2}{2\sigma^2}} \Longleftrightarrow F\{g(\theta)\} \equiv G(\omega) = \sqrt{2\pi}\sigma e^{-\frac{\sigma^2}{2}\omega^2}, \qquad (13.1)$$

---

[7] A complex system is defined as a system with complex-valued impulse response. In the frequency domain, real-valued signals/systems always have even-symmetric amplitude-spectrum/response and odd-symmetric phase-spectrum/response with respect to the zero frequency. Complex signals/systems do not need to have any spectral symmetry properties in general: e.g., the spectral support (region of non-zero amplitude spectrum) can basically be anything. This book focuses on real signals/systems, and for further discussion on the STFT and CWT we need no more information about complex filters. Note also that the analog filter bank related to the STFT will be a digital filter bank in sampled-signal applications.

we have

$$h_\omega(\theta)\big|_{\omega=\omega_c} \Longleftrightarrow F\left\{g(\theta)e^{-j\omega_c\theta}\right\} = \sqrt{2\pi}\,\sigma e^{-\frac{\sigma^2}{2}(\omega-\omega_c)^2}. \tag{13.2}$$

From this expression of the frequency response we also see that, as already stated, $\Delta\omega$ of the filters applied in the STFT remains the same at all frequencies. We can observe the covering of the signal spectrum by the frequency responses of the filters of the bank (Fig. 13.12). This coverage in the STFT is uniform along the ω-axis, so that we can classify the filter bank as a *constant-bandwidth filter bank*. The CWT can also be interpreted as the application of a filter bank, which is analog in theory and becomes digital after discretization. The analysis takes place in the time-scale plane, but scale can be translated into frequency, provided that the mother wavelet allows for establishing a proper scale-frequency relation. We can thus talk about coverage of the time-frequency plane also for the CWT. Again, the filters of the bank are complex bandpass filters, but here the bandwidth $\Delta\omega$ varies with ω, so that the time-frequency plane is covered by non-uniform rectangular tiles: their shape, i.e., aspect ratio, varies while the area remains constant, due to the uncertainty principle. Multiresolution means exactly this: the wavelets used in CWT are subject to the uncertainty principle, so that given a signal with some event in it, one cannot assign simultaneously an exact time and an exact scale (frequency) to that event. The product of the uncertainties on time and scale (frequency) has a lower bound. Thus, such an event marks an entire region in the time-scale plane, instead of just one point. Figure 13.13 shows the CWT coverage of the time-frequency plane by variable tiles.

The definition of the CWT actually appears as the application of a filter bank to the signal $x(\theta)$ if we observe that

$$W_x^\psi(a,b) = \int_{-\infty}^{+\infty} x(\theta)\frac{1}{\sqrt{a}}\psi_0^*\left(\frac{\theta-b}{a}\right)d\theta = \int_{-\infty}^{+\infty} x(\theta)\psi^*\left(\frac{\theta-b}{a}\right)d\theta$$

is the convolution integral of $x(\theta)$ with $\left(1/\sqrt{a}\right)\psi_0^*\left(-\theta/a\right) = \psi^*\left(-\theta/a\right)$. Thus, the CWT can be viewed as the result of the application to $x(\theta)$ of a bank of *complex, analog, scale-dependent bandpass filters* with impulse response

$$h_a(\theta) = \frac{1}{\sqrt{a}}\psi_0^*\left(-\frac{\theta}{a}\right) = \psi^*\left(-\frac{\theta}{a}\right),$$

and with frequency response

$$\Psi^*(a\omega) = \sqrt{a}\Psi_0^*(a\omega).$$

The last statement can be understood as follows. Starting from the Fourier transform

$$\mathcal{F}\left\{\psi_0\left(\frac{\theta}{a}\right)\right\} = a\Psi_0(a\omega),$$

Fig. 13.11 a The real
Gaussian window $g(\theta)$ used
in the STFT, **b** its Fourier
transform, and **c** the Fourier
transform shifted in
frequency by ω_c, as a
consequence of the
modulation applied to $g(\theta)$.
Both transforms are real.
Considering ω_c as a variable
rather than a single value of
peak frequency, panel c
represents the frequency
response of the filter bank
used in STFT, with each
filter associated with a
particular ω_c value

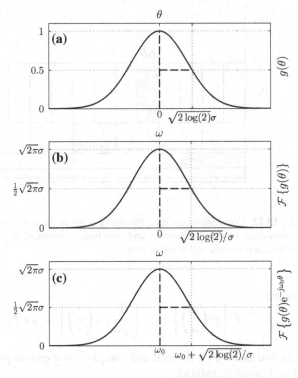

Fig. 13.12 Coverage of the
frequency axis in the STFT:
constant-bandwidth filter
bank. For this sketch, the
center frequency of the
frequency response of the
filter bank has been
arbitrarily discretized, so that
it increases linearly
according to integer
multiples k of the lowest
peak-frequency considered,
which here is simply
indicated as ω_c

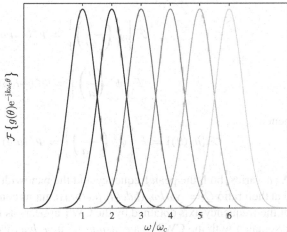

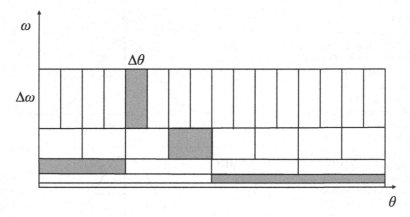

Fig. 13.13 Coverage of the time-frequency plane in the CWT: resolution in time and resolution in frequency vary with frequency and have been represented by juxtaposed rectangular tiles with variable aspect ratio and constant area

we get

$$\mathcal{F}\left\{\psi\left(\frac{\theta}{a}\right)\right\} = \mathcal{F}\left\{\frac{1}{\sqrt{a}}\psi_0\left(\frac{\theta}{a}\right)\right\} = \sqrt{a}\Psi_0\left(a\omega\right) = \Psi\left(a\omega\right).$$

But due to the time-reversal and complex-conjugation properties of the continuous-time Fourier transform,

$$\mathcal{F}\left\{\psi\left(-\frac{\theta}{a}\right)\right\} = \Psi^*(-a\omega),$$

$$\mathcal{F}\left\{\psi^*\left(\frac{\theta}{a}\right)\right\} = \Psi^*(-a\omega),$$

hence

$$\mathcal{F}\{h_a(\theta)\} = F\left\{\psi^*\left(-\frac{\theta}{a}\right)\right\} = \Psi^*(a\omega) = \sqrt{a}\Psi_0^*(a\omega).$$

As a varies, both the peak frequency and the bandwidth $\Delta\omega$ are multiplied by a, so that their ratio $Q = \Delta\omega/\omega_c$ (*fidelity factor*) remains constant. Therefore the coverage of the frequency axis obtained by the CWT appears as in Fig. 13.14. This filter bank associated with the CWT is a *constant relative-bandwidth filter bank*, also referred to as a *constant-Q filter bank*, and the coverage of the ω-axis is logarithmic rather than linear.

We can summarize these results—the essence of MRA—as follows:

• at small scale a—high frequency ω, a short wavelet corresponds to a broadband filter that provides

Fig. 13.14 Coverage of the frequency axis in the CWT: constant-relative-bandwidth filter bank. The analytic Morlet wavelet with $\omega_0 = 6$ (see Sect. 13.4.4) was used to draw this plot. For this sketch, the center frequency of the frequency response of the filter bank has been arbitrarily discretized, so that, starting from the ω_c of the filter associated with the mother wavelet, it increases according to multiples of ω_c that are integer powers of 2

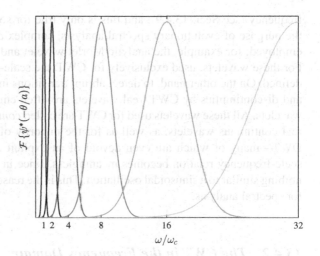

- low frequency resolution,
- high temporal resolution;

• at large scale a—low frequency ω, a long wavelet corresponds to a narrowband filter that provides

- high frequency resolution,
- low temporal resolution.

In principle, two impulsive events separated by an arbitrarily small time interval can always be resolved in CWT, provided we choose a sufficiently small scale. Similarly, two oscillatory persistent patterns with frequencies arbitrarily close to one another can always be resolved, provided we choose a sufficiently large scale. In practice, with sampled signals and with the CWT discretization scheme normally adopted, the minimum and maximum scales are constrained in some way, as we will see later. Considering that

• for the uncertainty principle we have $\Delta\theta\,\Delta\omega =$ constant, while
• for MRA we have $\Delta\omega/\omega_c =$ constant$'$, which for brevity we will write as $\Delta\omega/\omega =$ constant$'$

we see that

$$\Delta\theta = \frac{\text{const.}}{\Delta\omega} = \frac{\text{const.}}{\text{const.}'\omega} = \frac{\text{const.}''}{\omega}:$$

• temporal resolution, which is inversely related to $\Delta\theta$, is inversely proportional to a and directly proportional to ω (assuming $\omega \propto 1/a$);
• frequency resolution, which is inversely related to $\Delta\omega$, is inversely proportional to ω and directly proportional to a.

Up to now we reasoned assuming $\omega \propto 1/a$. This assumption is meaningful if, and only if, the wavelet in use allows us to establish a precise relation between scale and

frequency (see Sect. 13.4.9), and this is only true for some kinds of wavelets. For
the purpose of evolutionary spectral analysis, complex *analytic wavelets* are often
employed: for example, the analytic Morlet wavelet and the analytic Paul wavelet.
For these wavelets, used exclusively for CWT, the scale-frequency relation is well-
defined. On the other hand, to detect abrupt transitions in the signal features, spikes
and discontinuities by CWT, real wavelets are often employed, such as real DOG
wavelets. All these wavelets used for CWT are called *continuous wavelets*. For some
real continuous wavelets, as well as for the majority of *discrete wavelets* used in
DWT—many of which are even devoid of an explicit analytical expression—the
scale-frequency relation becomes meaningless, since in the wavelet shape there is
nothing similar to a sinusoidal oscillation. This is the reason why they cannot be used
for spectral analysis.

13.4.2 The CWT in the Frequency Domain

A convolution integral corresponds to a product in the Fourier transform domain,
so we can compute CWT taking the inverse transform of the product of the Fourier
transforms of $x(\theta)$ and $\psi^*\,(-\theta/a)$:

$$W_x^{\psi}(a, b) = \int_{-\infty}^{+\infty} x(\theta)\psi^*\left(\frac{\theta - b}{a}\right) d\theta = \frac{1}{2\pi} \int_{-\infty}^{+\infty} X(\omega)\Psi^*(a\omega)e^{jb\omega} d\omega.$$

This formula is useful for the computation of the CWT in applications. Note that
since we assumed $\sqrt{\int_{-\infty}^{+\infty} |\psi\,(\theta/a)|^2\, d\theta} = 1$, for Parseval's theorem in the analog
domain we have

$$\frac{1}{2\pi} \int_{-\infty}^{+\infty} |\Psi(a\omega)|^2\, d\omega = 1.$$

This equation assigns unitary energy to wavelets of any scale in the frequency domain too.

13.4.3 Admissibility and Other Constraints on Wavelets

In order to be suitable to be used as a continuous mother wavelet, the function $\psi_0(\theta)$
must satisfy some conditions.

- It must have finite energy.
- It must have zero mean:

$$\int_{-\infty}^{+\infty} \psi_0(\theta)d\theta = 0,$$

and therefore it must be a function that oscillates around zero. This condition is known as the *admissibility condition*. Zero mean implies that $\Psi_0(0)$ must vanish, i.e., that the wavelet must have a bandpass spectrum. As we shall see in Sect. 13.4.5, this condition ensures the existence of an inverse CWT, which in turn guarantees the possibility of reconstructing the signal from its CWT elements.

- It must be non-causal and centered on $\theta = 0$.
- It must satisfy some regularity conditions. *Regularity* is quite a complex concept: we will try to explain it in simple terms using the concept of *vanishing moments in discrete wavelet systems* in the next chapter, in the more general frame of the DWT. For the moment, we just define the wavelet moments, indicated with $m_1[k]$, as

$$m_1[k] = \int_{-\infty}^{+\infty} \psi_0(\theta)\theta^k \mathrm{d}\theta,$$

where k is the moment order, and make a few remarks.

We already know that admissibility requires at least the 0th order moment to vanish, but it may be convenient in some wavelet applications to require that all the first low-order moments vanish up to a certain order $k = k_m$. The number of vanishing moments is related to

- the smoothness of the wavelet function, since generally we can use differentiability to measure smoothness, and if a wavelet function is $(k_m + 1)$-times differentiable, then the first k_m wavelet moments vanish (Daubechies 1992);
- the elimination from the analysis of the most regular (polynomial) part of the signal (e.g., the long-term trend), thus allowing us to study other, more abrupt types of fluctuations in the signal and possible singularities in some high-order derivatives. In this case, the CWT values will be very small in the regions of the time-scale plane where the function is smooth to a certain degree dictated by the number of vanishing moments, and the wavelet transform will only react to higher-order variations of the signal (Farge 1992). The admissibility condition simply implies the removal of the signal mean value. In practice, the signal is usually centered *before* applying the CWT;
- the speed of convergence to zero of the CWT with decreasing scale/increasing frequency: if moments vanish up to the k_mth one, even approximately, then the larger the k_m, the faster the speed with which $W_x^\psi(a, b)$ of a smooth signal $x(\theta)$ decreases with decreasing scale/increasing frequency (see, e.g., Poularikas 2000);
- the locality (concentration) of the wavelet in the time and frequency domains that in turn are related to the extent to which the CWT acts as a local operator in both domains and therefore to resolution in time and frequency.

It must be observed, however, that the wavelets described in this Chapter, like the analytic Morlet and Paul wavelets, as well as the real DOG—that are used for CWT only—satisfy only minimal properties, such as the admissibility condition. The topic of vanishing moments is much more important for those wavelets that are typically used for DWT (even if they can also be adopted for CWT), like orthogonal wavelet families (Daubechies, symlets, coiflets etc.; see Sect. 14.10).

13.4.4 The CWT with Analytic Wavelets

Wavelets employed in the CWT may be analytic or real. For the purpose of evolutionary spectral analysis, it is often advisable to choose an *analytic wavelet*.

When a real mother wavelet is adopted, its Fourier transform has even symmetry around $\omega = 0$. As a consequence, the frequency response of that wavelet filter that is hypothetically centered at $\omega = 0$ also has even symmetry. Thus the filter passband includes both positive and negative frequencies, but the information relative to frequencies $\omega < 0$ is redundant, and can be univocally deduced from the part pertaining to $\omega \geq 0$. This would not be true with general complex mother wavelets, for which the Fourier transform has no particular symmetry properties. Complex analytic mother wavelets avoid this issue, since they have a frequency response which is non-identically-zero only over $\omega \geq 0$. These wavelets have a purely real Fourier transform. The signals with a spectrum covering only $\omega \geq 0$ were introduced by Gabor in its fundamental paper on telecommunication theory (Gabor 1946). He called them *analytic signals*. From the symmetry properties of the Fourier transform of an analog complex signal, it can be shown that for the Fourier transform of a mother wavelet to be real and to vanish at negative frequencies, the real and imaginary part of the complex mother wavelet must be closely related: more precisely, the imaginary part must be obtained taking the real part and shifting all its frequency components by $\pi/2$. This operation is called the *Hilbert transform* (see also Sect. 7.3.2).

We will see later that in the CWT we perform time-frequency analysis looking at the (squared) modulus of the wavelet transform. When we use a real wavelet, this modulus presents the same oscillations as the analyzing wavelet, and it may be difficult to sort out features belonging to the signal or to the wavelet. In the case of analytic wavelets, the quadrature between the real and imaginary parts of the wavelet transform—which is a direct consequence of the phase quadrature between the real and imaginary parts of the wavelet—eliminates these spurious oscillations.[8]

As an example of analytic wavelet, we consider the *analytic Morlet wavelet*:

$$\psi_0(\eta) = \frac{1}{\sqrt[4]{\pi}} e^{j\omega_0 \eta} e^{-\frac{\eta^2}{2}},$$

for which the daughter wavelets are

$$\psi(\eta) = \frac{1}{\sqrt[4]{\pi}} \frac{1}{\sqrt{a}} e^{j\omega_0 \eta} e^{-\frac{\eta^2}{2}},$$

---

[8]Complex wavelets also provide phase information concerning the signal's elementary components, even is this information is not often easy to interpret. For this reason, we will not discuss CWT phase plots.

where ω_0 is the *parameter* of the analytic Morlet wavelet. This wavelet does not respect the admissibility condition in a strict sense, but for $\omega_0 > 5$ its mean value vanishes up to computers' round-off errors. It is usual practice to take $\omega_0 = 6$.

This is the wavelet which is more similar to a wave packet: it is a complex exponential function modulated by a Gaussian bell with $\sigma = 1$. The real and imaginary parts of the analytic Morlet wavelet with $\omega_0 = 6$ are shown in the upper panel of Fig. 13.15a. It can be seen that ω_0 also approximately represents the number of oscillations contained inside the wave packet.

The Fourier transform of the analytic Morlet wavelet, shown in the upper panel of Fig. 13.15b, has the expression

$$\Psi(a\omega) = \sqrt{2}\,^4\sqrt{\pi}\sqrt{a}\,e^{-\frac{(a\omega - \omega_0)^2}{2}}\,U(\omega),$$

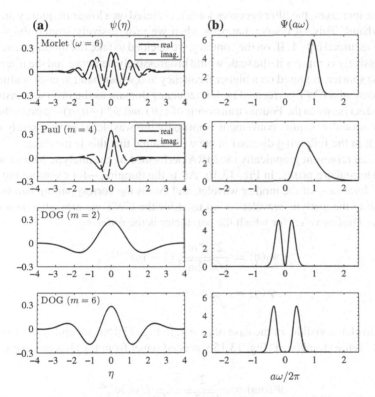

Fig. 13.15 Four different wavelets used for CWT analysis. These are daughter wavelets presented at scale $a = 10$. The plots in column **a** show the real part (*solid curves*) and the imaginary part, if present, (*dashed curves*) of the wavelets in the time domain, while the plots in column **b** show the same wavelets in the frequency domain. From the upper to the lower panel we see two analytic wavelets (the analytic Morlet wavelet with parameter $\omega_0 = 6$ and the analytic Paul wavelet with parameter $m = 4$) and two real wavelets (the DOG wavelet with $m = 2$, also known as Mexican hat wavelet, and the DOG wavelet with $m = 6$)

where $U(\omega)$ indicates the analog unit step, $U(\omega) = 0$ for $\omega < 0$ and $U(\omega) = 1$ for $\omega \geq 0$. Thus the wavelet spectrum is real and vanishes at negative frequencies. The shape of $\Psi(a\omega)$ is that of the spectrum of the Gaussian envelope, but its content is shifted to the neighborhood of $a\omega = \omega_0$: the maximum of the spectral bell falls at $a\omega/2\pi = \omega_0/2\pi = 0.955 \approx 1$. So, the daughter wavelet at scale a is a filter with frequency response centered on $\omega \approx 2\pi/a$. As can be approximately evaluated by visual inspection of the upper panel in Fig. 13.15b, the bandwidth is $\Delta\omega \approx 2\pi/a$, even if the precise value of the bandwidth depends on the definition we adopt for it. Therefore at all scales we can verify that the property of constant relative-bandwidth filter banks holds, the constant being nearly 1 in this case:

$$\frac{\Delta\omega}{\omega} = \text{constant} \approx 1.$$

As scale increases, the filter becomes longer, centered on a lower frequency and more narrowband. This, of course, happens when we progressively stretch the wavelet, having assumed $a \geq 1$. If, on the contrary, we decided to take the mother wavelet and progressively compress it, the scale would progressively decrease and the filter would become shorter, centered on a higher frequency and wider-band at each scaling step. In the case of a CWT performed in the frequency domain as the inverse transform of the product between the Fourier transforms of $x(\theta)$ and $\psi^*(-\theta/a)$—a procedure that computationally is quite convenient—normally the wavelet is progressively dilated ($a \geq 1$); in the following discussion we will assume that this is the case.

We can represent graphically the MRA performed by the analytic Morlet wavelet using Heisenberg boxes. In Fig. 13.16, $\Delta\theta$ is the duration—for example the mean-square duration—of the mother wavelet, and $\Delta\omega$ is the corresponding bandwidth.

Among the analytic wavelets often used for the CWT we may also mention the *analytic Paul wavelet*, for which the parameter is the *order m*:

$$\psi_0(\eta) = \frac{2^m j^m m!}{\sqrt{\pi(2m)!}} (1 - j\eta)^{-m-1},$$

$$\psi(\eta) = \frac{1}{\sqrt{a}} \psi_0(\eta).$$

This wavelet is visible, in the case of $m = 4$, in Fig. 13.15a (second panel from top). Its real Fourier transform (Fig. 13.15b, second panel from top) has the expression

$$\Psi(a\omega) = \frac{2^m \sqrt{a}}{\sqrt{m(2m-1)!}} U(\omega) e^{-a\omega}.$$

With respect to the analytic Morlet wavelet, the analytic Paul wavelet is more concentrated in time and less concentrated in frequency, scale being equal.

Other analytic wavelets sometimes used for the CWT are analytic Shannon wavelets, analytic derivatives of Gaussian (DOG) wavelets and frequency B-spline

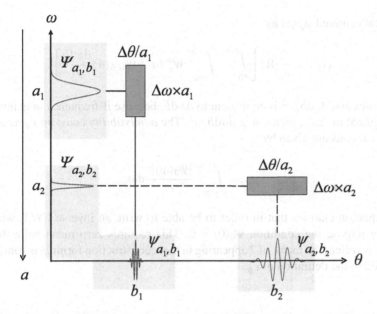

Fig. 13.16 Time-frequency-plane representation of the MRA performed by CWT using the analytic Morlet wavelet with parameter $\omega_0 = 6$: $\Delta\theta$ is the duration—for example the mean-square duration—of the mother wavelet, and $\Delta\omega$ is the corresponding bandwidth; the curves on the horizontal axis represent the shape of the daughter wavelet at two different scales a_1 and a_2, with $a_1 < a_2$; the two daughter wavelets are plotted as if they were applied to a hypothetical signal (not shown) at two different time shifts b_1 and b_2, with $b_1 < b_2$; the curves on the vertical axis represent the shape of the wavelet spectrum; the rectangular Heisenberg boxes, shaded in gray, change their shape as scale varies

(fbsp) wavelets (see, e.g., Mallat 1999; see also the appendix of Chap. 14). The Shannon wavelet (Sinc wavelet) is actually a particular case of fbsp wavelet.

We finally mention that the analytic Morlet wavelet can written with an additional parameter that defines the width of the Gaussian bell and allows for regulating the mother wavelet width in time and frequency (see the appendix in Chap. 14). For example, this additional parameter allows us to choose, at any scale, a better time resolution with respect to that obtainable with a unit parameter; of course this occurs at the expense of a poorer frequency resolution. This is the case, for example, in MATLAB.

13.4.5 Inverse CWT

The CWT performed using an analytic admissible wavelet is *complete*, meaning that any signal in $L^2(\mathbb{R})$ can be represented as a linear combination of scaled/translated wavelets, and preserves the energy of a zero-mean signal. It is then possible to write the inverse CWT (ICWT; *synthesis relation* or *reconstruction formula* of the CWT) for a real centered signal as

$$x(\theta) = \frac{2}{c_\psi} \text{Re} \left[\int_0^{+\infty} \int_{-\infty}^{+\infty} W_x^\psi(a,b) \psi_{a,b}(\theta) \frac{\mathrm{d}a\mathrm{d}b}{a^2} \right].$$

The expression $\mathrm{d}a\mathrm{d}b/a^2$ is equivalent to $\mathrm{d}\omega\mathrm{d}b$, because if frequency ω is inversely proportional to scale, then $\mathrm{d}\omega \propto \mathrm{d}a\mathrm{d}b/a^2$. The *admissibility constant* c_ψ is a finite numerical constant given by

$$c_\psi = \int_{-\infty}^{+\infty} \frac{|\Psi_0(\omega)|^2}{\omega} \mathrm{d}\omega.$$

This equation clarifies that in order to be able to write an inverse CWT, we must actually impose the condition $\Psi_0(0) = 0$. This requires zero mean value for the mother wavelet. The factor of 2 appearing in the reconstruction formula is sometimes absorbed in the definition of c_ψ.

13.4.6   Wavelet-Based Energy and Power Spectra

The total energy contained in the centered signal $x(\theta)$ can be reconstructed using the following equation (*Parseval's theorem* for the CWT):

$$\mathscr{E}[x(\theta)] = \int_{-\infty}^{+\infty} |x(\theta)|^2 \, \mathrm{d}\theta = \frac{2}{c_\psi} \int_0^{+\infty} \int_{-\infty}^{+\infty} \left| W_x^\psi(a,b) \right|^2 \frac{\mathrm{d}a\mathrm{d}b}{a^2}.$$

Observing the energy reconstruction formula, $\left| W_x^\psi(a,b) \right|^2$ appears as an energy density associated with the adimensional measure $(2/c_\psi)\, \mathrm{d}a\mathrm{d}b/a^2 \propto \mathrm{d}\omega\mathrm{d}b$, i.e., a *wavelet energy density function* or *wavelet energy spectrum*:

$$E_w(a,b) = \left| W_x^\psi(a,b) \right|^2.$$

$E_w(a,b)$ represents (Addison 2002) the relative contribution to the signal energy provided by a specific scale factor a and a specific time shift b. The map of $E_w(a,b)$ in the (a,b)-plane (or the corresponding three-dimensional surface plotted above the same plane) is called a *scalogram* or *wavelet spectrogram*.[9] The scalogram allows us to study the temporal evolution of the signal's spectral features, since it highlights the time position and the scale of dominant energetic features within the signal. We may note that in practice, all functions which differ from $\left| W_x^\psi(a,b) \right|^2$ by a constant numerical multiplicative factor only are also scalograms, so that we might prefer

---

[9]Recall that the term spectrogram indicates instead a STFT plot versus time and frequency.

to plot the quantity $(2/c_\psi) \left| W_x^\psi (a, b) \right|^2$, which is associated with the adimensional measure $\mathrm{d}a \mathrm{d}b/a^2$. However, the plotted quantity is usually $\left| W_x^\psi (a, b) \right|^2$. Its units are those of the signal's energy, i.e., the square of the units of data.[10]

Scalogram maps are normally drawn as contour lines of $\left| W_x^\psi (a, b) \right|^2$, and if colors are used, warm colors correspond to high energy-density values, while cool colors correspond to low values. Filled contour lines may be used. Also the use of a gray scale is common. An example of the appearance of a scalogram drawn by gray-scale contour lines—over a finite time interval and a finite scale range, of course—is shown in Fig. 13.17. On the ordinates, scale is reported, or, more often, the corresponding period is used, computed from the inverse of the frequency deduced from the scale-frequency relation holding for the specific wavelet in use: $P = 1/f$.

An alternative way of plotting the scalogram is to draw a three-dimensional surface above the time-period-plane, as in Fig. 13.18. This kind of visualization, though useful for getting a qualitative idea of the scalogram's behavior, is definitely less useful than the scalogram map drawn by contour lines, when quantitative inspection of the resuts of a CWT analysis is needed.

The relative contribution to the total energy contained within the signal at a specific scale a is given by the scale-dependent energy distribution, obtained integrating $E_w(a, b)$ with respect to b:

$$E_w(a) = \frac{2}{c_\psi} \int_{-\infty}^{+\infty} \left| W_x^\psi (a, b) \right|^2 \mathrm{d}b.$$

Fig. 13.17 An example of a scalogram plotted by contour lines in gray scale

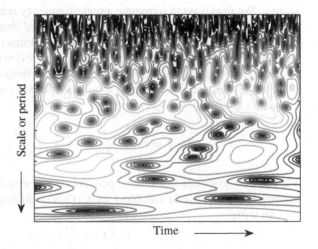

Scale or period

Time

---

[10]We should speak about energy per squared unit of time, but time—delay, scale factor—is adimensional here.

Fig. 13.18 An example of a scalogram plotted as a three-dimensional surface above the time-period-plane

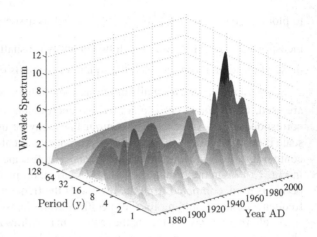

Peaks in $E_w(a)$ highlight the dominant energetic scales within the signal. The total energy in the signal can then be written as

$$\mathscr{E}[x(\theta)] = \int_0^{+\infty} E_w(a)\frac{da}{a^2}.$$

We may want to convert the scale-dependent wavelet energy spectrum $E_w(a)$ into a frequency-dependent wavelet energy spectrum $E_w(f)$ that can be compared directly with the Fourier energy spectrum of the (finite-energy) signal. To do this, we must convert from scale a to frequency—a characteristic frequency of the wavelet at scale a. We thus need to introduce a scale-frequency relation. As we will discuss in greater detail later, one of the most commonly used characteristic frequencies used in practice is the peak frequency of the wavelet's Fourier transform. We will use this frequency here. Other choices equally valid in the following discussion can be found in Sect. 13.4.9. If $f_c = \omega_c/2\pi$ is the analog peak frequency of the mother wavelet ($a = 1$, $s = T_s$), and if we assume that $f \propto 1/a$, we can associate with each scale $s = aT_s$ a *Fourier pseudo-frequency*

$$f = \frac{f_c}{s} = \frac{f_c}{T_s a} = \frac{\omega_c}{2\pi s} = \frac{1}{2\pi T_s}\frac{\omega_c}{a}.$$

We can now associate the scale-dependent energy distribution $E_w(a)$ with this frequency. By the change of variable $f = f_c/(T_s a)$, which implies $da/a^2 = -T_s df/f_c$, we can write

$$E_w(f) = \frac{T_s}{f_c}E_w(a),$$

which defines the *global wavelet energy spectrum* $E_w(f)$. This allows us to express the total energy also as

$$\mathcal{E}\left[x(\theta)\right] = \int_0^{+\infty} E_w(f)\mathrm{d}f = \frac{2T_s}{c_\psi f_c} \int_0^{+\infty} \int_{-\infty}^{+\infty} \left| W_x^\psi(f,b) \right|^2 \mathrm{d}f\mathrm{d}b, \quad (13.3)$$

where we set $\left| W_x^\psi(f,b) \right|^2 = \left| W_x^\psi(a,b) \right|^2$ for $f = f_c/(T_s a)$. If we furthermore define the *energy density in the time-frequency plane* by

$$E_w(f,b) = \frac{2T_s}{c_\psi f_c} \left| W_x^\psi(f,b) \right|^2,$$

we can finally write

$$\mathcal{E}\left[x(\theta)\right] = \int_0^{+\infty} \int_{-\infty}^{+\infty} E_w(f,b)\mathrm{d}f\mathrm{d}b.$$

This concerns the use of frequency as the ordinate in scalogram plots. If we decided to use period $P = 1/f$ as the ordinate, we should make another change of variable, and define new quantities corresponding to those presented above. In practice, however, for our plots we will always use $\left| W_x^\psi(a,b) \right|^2$, irrespective of what independent variable we adopt, among scale, frequency and period. We are allowed to do so, provided we are aware that the contour-line plot (or the three-dimensional plot) of $E_w(a,b)$ versus a and b, as well as the bi-dimensional representation of $E_w(a)$ versus a, do not enclose, respectively, volumes and areas proportional to the energy of the signal, whereas their time-frequency counterparts, $E_w(f,b)$ and $E_w(f)$, would do so. However, the peaks in $E_w(a,b)$ and $E_w(a)$ do correspond to the most energetic parts of the signal, as do the peaks in $E_w(f,b)$ and $E_w(f)$, and therefore the two representations both visualize the signal's energy distribution. Scalograms are normally plotted with a logarithmic y-axis. Given that in $f = f_c/(aT_s)$ both T_s and f_c are constant for a given sampled signal and a given mother wavelet, the plot of $\left| W_x^\psi(f,b) \right|^2$ using a logarithmic frequency axis is simply a shifted, inverted plot of $\left| W_x^\psi(a,b) \right|^2$ using a logarithmic a axis. Similarly, the plot of $\left| W_x^\psi(P,b) \right|^2$ using a logarithmic period scale is simply a shifted plot of $\left| W_x^\psi(a,b) \right|^2$ using a logarithmic a axis, and so on. If we orient increasing a scales and periods towards the bottom of the plot, the plot of $\left| W_x^\psi(a,b) \right|^2$ can also be interpreted as a plot of $\left| W_x^\psi(P,b) \right|^2$ with decreasing P periods towards the top of the plot, or as a plot of $\left| W_x^\psi(f,b) \right|^2$ with increasing frequencies towards the top of the plot. In literature, all three representations are commonplace, but in this book we will adopt period P as the ordinate, with decreasing values of P towards the top of the plot.

Having clarified these issues, we can drop the related distinctions and focus on the substance. The values of $\left| W_x^{\psi}(a, b) \right|^2$ at a fixed value b_0 of b and with variable a (or f, or P), which we would encounter by cutting the scalogram plane along a single vertical line, represent a *local wavelet spectrum* and are representative of the spectral content of the non-stationary signal in the neighborhood of b_0.

Now, imagine we neglected for a moment the non-stationarity of the random process from which our signal is drawn, and wanted to derive from CWT analysis an estimate of its *power* spectrum. This may seem, at first glance, a contradiction and a conceptual jump: we want to establish a connection between the spectral theory of infinite-energy, finite-power random signals and CWT theory, which deals with zero-power, finite-energy signals. However, in practice experimental signals *always* have finite length—assumed to be long enough for the pertinent (stationary) statistics of the underlying process to settle down sufficiently for analysis. We can thus treat our finite-length data record exactly as we did in the theory of stationary random power signals: we consider it as a portion of a typical sample sequence. In the frame of CWT, the effective finite length of the measured signal is in agreement with the assumption of finite energy; in the frame of classical spectral analysis of a random process, energy can be conceptually infinite over the entire time axis, and power can be finite, allowing us to meaningfully use the notion of power spectrum. After all, as we already pointed out, in any case we have a finite-length data record, and the theoretical frame we adopt for its analysis is essentially a matter of convenience. In conclusion, to get the power spectrum of our record from CWT analysis we just need to divide the energy spectrum by the duration of the record, so that the area under the spectral curve we obtain gives the average energy per unit time, i.e., the average power. At each scale, we will sum up all instantaneous scalogram values, as if we were cutting the scalogram plane along a horizontal line corresponding to that scale, and then divide by the record length; we will do this for all scales of interest. We will thus get the *global wavelet spectrum* (GWS) that we will be able to compare with power spectrum estimates of the same record obtained by Fourier or parametric methods, ignoring non-stationarity and aiming at a spectral estimate averaged over the record length. GWS is a very smooth spectral estimate, and it has been shown to be consistent. We will discuss the GWS in greater detail in Sect. 13.5.5, after adapting the theoretical CWT to real-world sampled signals.

13.4.7 The CWT with Real Wavelets

The real wavelets that are most often used for the CWT are the real DOG, for which the parameter is the *order* m of the derivative. The most popular has $m = 2$ and is referred to as DOG($m = 2$) or *Marr wavelet* or *Mexican hat*. The expression of the real DOG(m) wavelet is

$$\psi_0(\eta) = \frac{(-1)^{m+1}}{\sqrt{\Gamma\left(m+\frac{1}{2}\right)}} \frac{d^m}{d\eta^m}\left(e^{-\frac{\eta^2}{2}}\right)$$

that for $m = 2$ reduces to

$$\psi_0(\eta) = \frac{1)}{\sqrt{\Gamma(2.5)}}\left(1-\eta^2\right)\left(e^{-\frac{\eta^2}{2}}\right).$$

The daughter wavelet is, as usual,

$$\psi(\eta) = \frac{\psi_0(\eta)}{\sqrt{a}}.$$

This wavelet is shown in Fig. 13.15a (third panel from top: $m = 2$; lowest panel: $m = 6$). Its Fourier transform is

$$\Psi(a\omega) = \frac{-j^m\sqrt{a}}{\sqrt{\Gamma\left(m+\frac{1}{2}\right)}}(a\omega)^m e^{-\frac{(a\omega)^2}{2}}$$

that for $m = 2$ becomes

$$\Psi(a\omega) = \frac{\sqrt{a}}{\sqrt{\Gamma(2.5)}}(a\omega)^2 e^{-\frac{(a\omega)^2}{2}}.$$

The cases of $m = 2$ and $m = 6$ are illustrated in the third and fourth panels from top in Fig. 13.15b. For even m, the transform is real; however, unlike the Fourier transforms of analytic wavelets, the transform of the real DOG wavelet is different from zero also at negative frequencies. This wavelet is sometimes also used in its "upside down" form. As m increases, the wavelet shrinks in time and widens in frequency.

An inverse transform can also be written for the CWT with real admissible wavelets:

$$x(\theta) = \frac{1}{c_\psi}\int_0^{+\infty}\int_{-\infty}^{+\infty} W_x^\psi(a,b)\psi_{a,b}(\theta)\frac{dadb}{a^2}.$$

The energy reconstruction formula (Parseval's theorem for the CWT) is then

$$\mathcal{E} = \int_{-\infty}^{+\infty}|x(\theta)|^2\,d\theta = \frac{1}{c_\psi}\int_0^{+\infty}\int_{-\infty}^{+\infty}\left|W_x^\psi(a,b)\right|^2\frac{dadb}{a^2},$$

where the scalogram was defined as for analytic wavelets, $E_w(a,b) = \left|W_x^\psi(a,b)\right|^2$, and where

$$c_\psi = \int_{-\infty}^{+\infty}\frac{|\Psi_0(\omega)|^2}{\omega}\,d\omega < \infty.$$

The difference between the synthesis formulas for real and analytic wavelets can be understood as follows:

- a formula identical to the one given for real wavelets reconstructs the analytic signal $x_a(\theta)$ associated with $x(\theta)$, and built by setting $\mathrm{Re}\,[x_a(\theta)] = x(\theta)$ and $\mathrm{Im}\,[x_a(\theta)] = \hat{x}(\theta)$, where $\hat{x}(\theta)$ indicates the Hilbert transform of $x(\theta)$. Thus, assuming we used an analytic wavelet,

$$x_a(\theta) = \frac{1}{c_\psi} \int_0^{+\infty} \int_{-\infty}^{+\infty} W_{x_a}^\psi(a, b)\psi_{a,b}(\theta)\frac{dadb}{a^2};$$

- it can be shown that the CWT of $x_a(\theta)$ is, the mother wavelet being equal, twice the CWT of $x(\theta)$:

$$W_{x_a}^\psi(a, b) = 2W_x^\psi(a, b).$$

Therefore, for an analytic wavelet we have

$$x(\theta) = \frac{2}{c_\psi}\mathrm{Re}\left[\int_0^{+\infty} \int_{-\infty}^{+\infty} W_x^\psi(a, b)\psi_{a,b}(\theta)\frac{dadb}{a^2}\right],$$

that leads us back to the reconstruction formula given in Sect. 13.4.5.

Other real wavelets can be used for CWT analysis: the real Morlet wavelet, the real Meyer wavelet and the compact-support orthogonal wavelets systems used for the DWT (Daubechies wavelets, symlet, coiflets. See the appendix in Chap. 14 for a description of these wavelets).

13.4.8 Morlet's Empirical Reconstruction Formula

We already mentioned that $W_x^\psi(a, b)$ contains the information required to reconstruct $x(\theta)$ in a redundant way. This offers, among many advantages, the possibility of reconstructing the signal using a different synthesis wavelet with respect to the analysis wavelet. In particular, Morlet found an empirical reconstruction formula in which the synthesis wavelet is as simple as a Dirac δ (see Farge 1992):

$$x(\theta) = \frac{1}{c_\delta} \int_0^{+\infty} W_x^\psi(a, \theta)\frac{da}{a^{3/2}},$$

with

$$c_\delta \propto \int_{-\infty}^{+\infty} \frac{\Psi_0(\omega)}{|\omega|}d\omega.$$

We may note how using a Dirac δ makes the integral over delay disappear from the reconstruction formula. This expression is widely used in software implementations.

Fig. 13.19 An example of a sinusoid that captures the dominant oscillation of a wavepacket-like wavelet; the wavelet is DOG($m = 6$) for $a = 10$

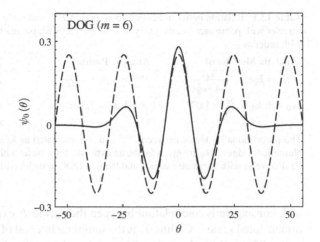

13.4.9 Scale-Frequency Relation

The precise relation between scale and frequency depends on the wavelet shape and on the conventional criterion adopted to define the relation itself. For the analytic Morlet wavelet, for example, it is natural to assume as wavelet frequency at a given scale the reciprocal of the period of the oscillations of the daughter wavelet at the considered scale, but in other cases the choice is much less obvious.

We already used the Fourier pseudo-frequency at scale $s = a T_s$,

$$f = \frac{\omega_c}{2\pi s},$$

where ω_c is the peak frequency of the modulus of the mother wavelet transform, i.e., the non-negative angular frequency that corresponds to the maximum of the transform magnitude. Another possible choice is the passband central frequency, which would be equally valid.

The basic idea is to associate with the mother wavelet a sinusoidal signal with frequency $\omega_c/(2\pi T_s)$, able to capture the dominant wavelet oscillation; when later $\psi_0(\theta)$ is scaled to $\psi(\eta)$, the characteristic frequency becomes $\omega_c/(2\pi a T_s) = \omega_c/(2\pi s)$. This pseudo-frequency is well-defined only at sufficiently large scales, where the wavelet spectrum is narrowband enough. In Fig. 13.19 an example of this association is given for the real wavelet DOG($m = 6$).

Alternatively, we can analyze by CWT a sinusoidal signal with angular frequency ω and compute the scale a at which the maximum absolute value of the CWT is found, as a function of delay b; we will then average these scale factors over delay. If this can be done analytically, it leads to a univocal relation between a and ω, i.e., $\omega = \omega(a)$. This relation can then be translated into the relation between analog frequency and scale, $f = f(s)$. More frequently, we will prefer to derive in this way the relation between adimensional period $P^{ad} = 2\pi/\omega$ and scale factor a, i.e., $P^{ad} = P^{ad}(a)$,

Table 13.1 Relation between adimensional period P^{ad} and scale factor a for the analytic Morlet wavelet with parameter ω_0, the analytic Paul wavelet with parameter m and the real DOG wavelet with order m

Analytic Morlet(ω_0)	Analytic Paul(m)	Real DOG(m)
$P^{ad} = k_0 a = \dfrac{4\pi a}{\omega_0 + \sqrt{\omega_0^2 + 2}}$	$P^{ad} = k_0 a = \dfrac{4\pi a}{2m+1}$	$P^{ad} = k_0 a = \dfrac{2\pi a}{\sqrt{m+\frac{1}{2}}}$
$\omega_0 = 6$: $k_0 \approx \frac{2\pi}{\omega_0} = 1.033$	$m = 4$: $k_0 = 1.396$	$m = 2$: $k_0 = 3.974$
		$m = 6$: $k_0 = 2.465$

The proportionality factor between P^{ad} and a is indicated as k_0 and is called the Fourier factor. Numerical values of k_0 are given for the analytic Morlet wavelet with parameter $\omega_0 = 6$, the analytic Paul wavelet with parameter $m = 4$ and the real DOG wavelet with orders $m = 2$ and 6

and, consequently, the relation between the period P expressed in seconds and the dimensional scale s. Calling T_s is the sampling interval of the data record in the same units, $P^{ad} = P/T_s$. Note that $P^{ad} = 1/\nu$, where ν is adimensional frequency.

This method of defining the relation is particularly useful for continuous wavelets, for which very often the analytical calculation is possible. If, on the contrary, we cannot compute the scale-frequency/scale-period relation by analytical means, we can nevertheless construct a numerical calibration curve, by computing the CWT of many sinusoidal signals with different frequencies.

In Table 13.1, the function $P^{ad} = P^{ad}(a)$ computed by analytical means for three popular continuous wavelet types—analytic Morlet, analytic Paul and real DOG—is given. In all cases, P^{ad} is simply proportional to scale: $P^{ad} = k_0 a$, where the constant k_0, called *Fourier factor*, depends on wavelet type and on the value of its parameter. The remarkable differences in the scale-period relation from one wavelet to another bear no particular meaning, and are simply due to the different wavelet shapes. Note that for the analytic Morlet wavelet, period and scale are essentially equal. The proportionality between P^{ad} and a justifies our previous reasoning based on $f \propto 1/a$.

13.4.10 Cone of Influence (COI) and Locality of the CWT

The CWT synthesis relation shows that $x(\theta = \theta_0)$ cannot be reconstructed on the basis of $W_x^{\psi}(a, b = \theta_0)$ only: the formula contains an integral over time delay b. It also contains an integral over a, meaning that in principle all variability scales contribute to $x(\theta = \theta_0)$. Therefore, in order to reconstruct the signal in the neighborhood of θ_0 we must take into account all the values of $W_x^{\psi}(a, b)$ that belong to a certain area in the time-scale plane . In particular, at each scale a we need all the values of $W_x^{\psi}(a, b)$ included in a certain time interval that is referred to as the *cone of influence* (COI). We can look at the cone of influence also in the transform domain, from the definition of CWT: $W_x^{\psi}(a, b = \theta_0)$ does not contain only $\theta = \theta_0$, but is an integral over time θ. This is related to the temporal locality of the CWT that exists—the time interval centered on θ_0 within which the values of the signal significantly influence $W_x^{\psi}(a, b = \theta_0)$ is limited—but to an extent depending on the scale. As scale increases, temporal locality diminishes. The COI is important also for what concerns the edge effects in the CWT of a finite-length data record.

To understand the concept of the COI, it is actually convenient to focus on the real-world case of a finite-length sampled signal. We will discuss the discretization of the CWT in the next section; for the moment let us imagine two sequences, the data record and a discretized wavelet at some scale a. In the time domain, the convolution integral that defines the CWT at the scale a becomes a discrete-time linear convolution. For continuous wavelets, the analytical wavelet expression is known as a function of time and scale, so that at a given scale we only need to sample it using the same T_s of the data, over an arbitrary time span. The wavelet can therefore be expressed as an arbitrarily long sequence. On the contrary, the signal values are known only as a fixed number N of samples. In the convolution, the daughter wavelet is shifted along the signal, from $b = 0$ up to the total signal length, and at each position b the signal is correlated with the wavelet. If b occupies a central position with respect to the signal's time span, we will have no problems taking a sufficiently long part of the data sequence and computing all the term-by-term products between the data and the wavelet required to compute linear convolution. But if b is close to the beginning or to the end of the record, some data will be missing and we will have to "invent" them. This issue in the CWT is normally solved by zero-padding the signal, so as to extend it. These added zeros constitute a signal discontinuity that influences the values of the CWT at the scale a at both edges. Where this influence is present, the CWT values are reduced with respect to their hypothetical "true" value, i.e., the one we could compute if we had more data—hypothetically, an infinite amount of data. The larger the scale, the wider the time intervals adjacent to the edges in which the CWT is disturbed. The width of each time interval is exactly the COI at the considered scale.[11]

As we will see when discretizing the CWT, in software implementations it may be convenient to compute the wavelet transform in the frequency domain, via FFT. The issue of edge effects then comes into play as in any circular convolution: the linear convolution of two sequences, computed as the product of two DFTs, can be free from time-domain aliasing only if we properly augment the two sequences by zero-padding. For instance, if the data record has N_1 samples and the discretized daughter wavelet at a given scale has N_2 samples, we must

- bring both sequences to a length $M \geq N_1 + N_2 - 1$, according to the length $N_1 + N_2 - 1$ of their linear convolution; we may want to set M equal to the nearest integer power of 2 for FFT efficiency, i.e., $M = N_{\text{FFT}} = 2^{\nu}$, where ν is integer;
- compute both DFTs via FFT;
- multiply them sample by sample;

---

[11]We may note that if the data were cyclical, as in the case of a meteo-climatic data sequence measured at fixed latitude and time as a function of longitude, there would be no need to add zeros and the COI would not exist.

- perform the inverse FFT of the product, thus getting the discretized CWT at the considered scale.

So, the issue of COI-related edge effects is evident in both frequency and time domains.

When the scalogram is plotted in the time-scale or time-period plane, in order to recall that inside the COI relative to the record edges the scalogram values are artificially reduced with respect to their true values, it is advisable to draw the COI curve. We will describe how the equation of this curve can be obtained after discussing the discretization scheme for the CWT, but we can immediately discuss how the COI curve looks in the time-period plane—this also explains why we call it a "cone"— and how it is connected to the locality of CWT analysis, and in particular to locality in the time domain.

Let us first reconsider Fourier analysis. The Fourier transform at a given frequency depends on the whole record, and if, for example, this contains a noise spike, all of the spectrum is affected by this impulsive event. The same is true for the STFT: a noise spike $x(\theta) = \delta(\theta - \theta_0)$, occurring at some instant θ_0, affects the STFT in the neighborhood of θ_0 at all frequencies, and in the time-frequency plane this impulse (Fig. 13.20a) appears as a belt of constant width $\Delta\theta$ (Fig. 13.20b). In the CWT, the impulse only affects the points of the time-scale or time-period plane contained inside the COI, the amplitude of which is proportional to the wavelet support $\Delta\theta_i$ at the scale $a = a_i$. The wavelet transform $W_x^\psi(a_i, b = \theta_0)$ will depend on the values of $x(\theta)$ for $\theta \in \theta_0 - (\Delta\theta_i)/2, \theta_0 + (\Delta\theta_i/)2$. The interval $\Delta\theta_i$ decreases with decreasing a_i (Fig. 13.20c). Therefore the influence of the spike on scalogram values is more and more local in time as the scale decreases. Globally, the time locality of CWT is greater when the largest scales are excluded from the analysis; the very existence of time locality in CWT is due to the use of basis functions with concentrated or compact time support, as opposed to the infinitely-long sines and cosines employed in Fourier analysis. Fig. 13.20c also clarifies the use of the word "cone": speaking with propriety we should call it a "triangle".

The precise definition of the COI width is a matter of convention. Usually it is defined as follows:

- an impulsive disturbance at some θ_0 is considered: $x(\theta) = \delta(\theta - \theta_0)$;
- the wavelet transform $W_x^\psi(a, b)$ is computed;
- at some scale a_0, the value of $\left| W_x^\psi(a_0, b) \right|^2$ will descend to $1/e^2$ times the value it assumes at θ_0 at some time delay b_{a_0} after θ_0. Then, b_{a_0} is assumed as the COI width at the scale a_0;
- if $\left| W_x^\psi(a_0, b) \right|^2$ oscillates and crosses the level $\left| W_x^\psi(a_0, \theta_0) \right|^2 /e^2$ more than once, the maximum delay is assumed as b_{a_0};
- the procedure is repeated at all scales of interest, so that the relation $b_a = b_a(a)$ is found.

Fig. 13.20 Locality of CWT in comparison to STFT: **a** an impulse, **b** the region of the time-frequency plane that is affected by it in the STFT, which has equal width $\Delta\theta$ at all frequencies, and **c** the region of the time-scale plane that is affected by the impulse in the CWT. This region has a width $\Delta\theta$ proportional to scale

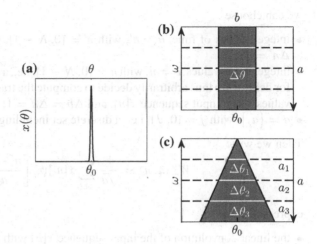

For the analytic Morlet wavelet the result of this procedure is $b_a = \sqrt{2}a$. In terms of dimensional quantities we could equivalently write $\tau_s = \sqrt{2}s$, where τ_s is called *e-folding time*.

So far, we have focused on time locality, but it is clear that the CWT is local not only in time but also in scale. The wavelets do not only have a concentraded or compact time support but also a bandpass spectrum. Even if in the reconstruction formula an integral over scale appears, which is equivalent to an integral over frequency, every oscillatory component of the signal at some scale a will contain significant contributions only from a limited interval of frequencies, centered on the peak frequency of the frequency response of the wavelet filter at the considered scale. The bandwidth of the wavelet filter decreases with increasing scale, according to the MRA property. In Fourier analysis, we have a perfect frequency locality (a sinusoid has a Dirac-δ spectrum) and a non-existent time locality; in the CWT at each scale we have a different trade-off between the two characteristics.

13.5 CWT Discretization

Numerical computation of the CWT requires discretizing it in time and scale, so that

- a finite-length sequence $x[n]$ can be analyzed in place of an analog signal $x(\theta)$ of infinite length;
- a discrete set of scale is considered;
- the definition integral is transformed into a finite sum.

Recalling that

$$W_x^\psi(a,b) = \frac{1}{\sqrt{a}} \int_{-\infty}^{+\infty} x(\theta) \psi_0^* \left(\frac{\theta - b}{a} \right) d\theta = \int_{-\infty}^{+\infty} x(\theta) \psi^* \left(\frac{\theta - b}{a} \right) d\theta,$$

we can choose

- integer values of time, $\theta = n'$, with $n' = [0, N - 1]$, so that $d\theta$ becomes $\Delta\theta = \Delta n' = 1$;
- integer delay values, $b = n$, with $n = [0, N - 1]$, i.e., a maximally dense sampling for delays. We thus arbitrarily decide to compute the transform for all discrete time values in the input sequence $x[n]$, and $\Delta b = \Delta n = 1$;
- $a = \{a_j\}$, with $j = [0, J]$, i.e., a discrete set including $J + 1$ scale factors.

Then we write

$$W_x^{\psi}(a, n) \approx \frac{1}{\sqrt{a}} \sum_{n'=0}^{N-1} x[n']\psi_0^* \left[\frac{n' - n}{a} \right],$$

expressing

- the linear convolution of the input sequence $x[n]$ with $\psi^*(-n/a)$,
- a filtering of the data with a digital bandpass filter with frequency response $\Psi^*(a\omega)$,
- the correlation between the signal and the daughter wavelet at the scale a and at lag n.

The wavelet normalization in the discretized case is expressed by the equation

$$\sum_{n=0}^{N-1} \left| \psi\left(\frac{n}{a}\right) \right|^2 = 1 = \frac{1}{N_{FFT}} \sum_{k=0}^{N_{FFT}-1} |\Psi(a\omega_k)|^2 .$$

The impulse response of the discrete-time wavelet filter is

$$h_a[m] = \psi^* \left[-\frac{m}{a} \right] = \psi^* \left[-k_{dec}m' \right],$$

where k_{dec} is a decimation factor inversely proportional to the scale. Note that in order to have an integer k_{dec} we must necessarily follow the approach in which $a \leq 1$ and the wavelet is progressively contracted, rather than dilated. Therefore if we want to operate in the time domain we must build the mother wavelet, which will represent the maximum scale, over a dense set of points, and then decimate it with factors k_{dec} progressively increasing with decreasing scale. This is, for example, the approach of the function that implements CWT in the Wavelet Toolbox of the popular MATLAB software.

It is, however, very convenient to operate in the frequency domain. For the DFT convolution theorem, the DFT of the discretized CWT is equal to the term-by term product of the DFT of the data over N_{FFT} points, indicated by $X[k]$, with $\Psi^*(a\omega_k)$, with $\omega_k = 2\pi k/N_{FFT}$. The function $\Psi^*(a\omega_k)$ can be viewed as

- the DFT of the impulse response of the discrete-time wavelet filter computed over N_{FFT} points,
- the frequency response of the wavelet filter sampled at the discrete frequencies ω_k,

Fig. 13.21 An example of how a dyadic set of scales is built: the *dark arrow* represents one octave and the *light arrows* indicate voices. The minimum scale factor is 2, the total number of scales is 29, there are 4 voices per octave and the total number of octaves is 7

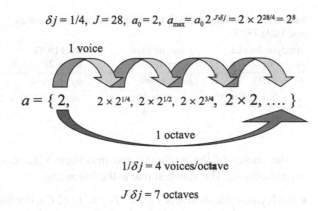

$$\delta j = 1/4, \quad J = 28, \quad a_0 = 2, \quad a_{max} = a_0 2^{J\delta j} = 2 \times 2^{28/4} = 2^8$$

1 voice

$$a = \{\, 2, \quad 2 \times 2^{1/4}, \; 2 \times 2^{1/2}, \; 2 \times 2^{3/4}, \; 2 \times 2, \; .... \,\}$$

1 octave

$$1/\delta j = 4 \text{ voices/octave}$$

$$J \, \delta j = 7 \text{ octaves}$$

- the result of the sampling performed on the continuous-time Fourier transform of the continuous-time wavelet over the same ω_k-set used for $X[k]$. For this approach to $\Psi^*(a\omega_k)$, the analytical expression of $\Psi(a\omega)$ must be known. This is guaranteed if the continuous wavelets introduced in Sect. 13.4.1 are used.

Subsequently we can get $W_x^\psi(a, n)$ by inverse DFT of $X[k]\Psi^*(a\omega_k)$. As a last step, the redundant samples will be eliminated: if the data has N samples and the DFTs have been computed over N_{FFT} points, the samples of $W_x^\psi(a, n)$ with $n = [N + 1, N_{FFT}]$ will be discarded.

As for the set of discrete scales, in principle it can be chosen arbitrarily, but usually a *dyadic set of scales* is constructed:

$$a = a_0 2^{j\delta j}, \qquad j = [0, J],$$

where a_0 is the minimum scale factor and δj is the reciprocal of the number of scales per octave. An *octave* is a jump by a factor of 2 in scale factors: for example, from 2 to 4, or from 8 to 16, etc. The parameter δj is set to a value small enough to have a smooth discretized CWT. Its optimal value varies according to the bandwidth of the mother wavelet; for instance, with the analytic Morlet wavelet it is advisable to set $\delta j < 0.5$, but if the computational burden allows it, often we will choose smaller values, like $\delta j = 0.1$, or even 0.05. The maximum scale factor is thus $a_{max} = a_0 2^{J\delta j}$, and therefore $\log_2(a_{max}) = \log_2(a_0) + J\delta j$. Hence

$$J = \frac{1}{\delta j} \log_2 \frac{a_{max}}{a_0}.$$

The total number of scales is $J + 1$, the octaves are $J\delta j$ and the number of jumps from one scale to the nearest one inside a given octave is $1/\delta j$: these jumps are called *voices*. Figure 13.21 shows an example of a set of dyadic scales.

Table 13.2 Minimum scale factor a_0 for different mother wavelets, in relation to Fourier factor k_0 (see Table 13.1)

Analytic Morlet ($\omega_0 = 6$)	Analytic Paul ($m = 4$)	Real DOG ($m = 2$)	Real DOG ($m = 6$)
$k_0 = 1.033$	$k_0 = 1.396$	$k_0 = 3.974$	$k_0 = 2.465$
$a_0 = 2/1.033 \approx 2^1$	$a_0 = 2/1.396 \approx 2^1$	$a_0 = 2/3.974 \approx 2^{-1}$	$a_0 = 2/2.465 \approx 2^0$

The choice of the minimum and maximum scale can be done on the basis of several criteria. The simplest one is the following:

- the Nyquist theorem sets $f_{\max} = f_{Ny} = 1/(2T_s)$. For the scale-frequency relation, $f_{\max}$ corresponds to some $s_{\min}$ and therefore to some $a_{\min} \equiv a_0$. More precisely, from $f_{\max} = 1/(2T_s)$ we deduce $P_{\min}^{ad} \equiv P_0^{ad} = 2$ and thus $a_0 = P_0^{ad}/k_0$. The value of a_0 obtained in this way is then rounded up to the nearest higher integer power of 2. Table 13.2 gives the minimum scale factor for different mother wavelets.
- In a similar way, we can set $f_{\min} = 1/(NT_s)$ as in Fourier analysis, where the Rayleigh frequency $\Delta f = 1/(NT_s)$ sets to this value the frequency that is closest to the zero frequency. From $f_{\min}$ we can then derive $s_{\max}$ and $a_{\max}$.

In this way, for a given wavelet, the minimum scale $s_{\min}$ is dictated by T_s, while the maximum scale $s_{\max}$ is determined by both N and T_s, i.e., by the record's duration. As a consequence, $a_{\min} \equiv a_0$ is constant and $a_{\max}$ depends on the number N of available data points.

This scale discretization scheme is arbitrary and not binding, and implies, in association with the choice of maximally dense delays, a remarkable redundance of the wavelet transform. Moreover, the criteria for selecting a_0 and $a_{\max}$ given above are simple but can be subject to criticism.

For instance, it may seem more sensible to determine a_0 by requiring the wavelet at scale a_0 (i.e., the one having the highest center frequency and the largest bandwidth) to be approximately bandlimited to $\omega = \pi$, in the sense that only a negligible percentage of the continuous-wavelet energy in the frequency domain should belong to frequencies $\omega > \pi$. If we call a_0^* the value chosen on the basis of this criterion, and if we set $a_0 < a_0^*$, scales will exist for which the product $X[k]\Psi(a\omega_k)$ will use only a part of the bell-shaped curve of $\Psi(a\omega_k)$, because the rest of the bell protrudes beyond the upper bound π of the frequencies ω_k over which $X[k]$ is calculated.

Another criterion, connected to the previous one, could be based on requiring for the wavelet at scale a_0 a duration—quantitatively defined in some way—equal to at least $2T_s$, which is different from requiring that $k_0 a_0 = 2$. Considering that evaluating the minimum scale dictated by these criteria for all the different wavelets we may want to use is not immediate, and, above all, observing that the precise choice of a_0 is not crucial in applications, we hereafter will adopt the approximate criterion that leads to Table 13.2.

Also the maximum scale factor $a_{\max}$ can be set on the basis of different criteria: for instance, we can require the duration of the wavelet not to exceed the record duration, or maybe two times as much, etc.; we could also require its spectrum to cover at least two intervals between the discrete frequencies over which the data spectrum is sampled. The simple criterion based on the Rayleigh frequency gives a reasonable approximate idea of what $a_{\max}$ should be. In applications, we will also be able to judge the quality our choice looking at the results of the analysis, i.e., looking at

- the appearance of the discretized scalogram $\left| W_x^{\psi}(a, n) \right|^2$ and of the related COI plot: if most of the scalogram lies outside the COI relative to record edges, we will be sure we did not go too high with the maximum scale;
- the appearance of the GWS plot: if the GWS drops off at low frequencies, we will be confident we did not neglect any long-wave features in the data;
- the comparison between the original data and the data reconstructed by inverse CWT based on all scales of variability: if the reconstructed signal is similar to the original one, we will be confident about having included all relevant scales in the analysis.

13.5.1 Scalogram Plot and Edges-Related COI Curve

Figure 13.22b shows a typical example of a scalogram plot. This is the same scalogram appearing in Fig. 13.17, but this time it is drawn by filled gray-scale contour lines. The wavelet is the analytic Morlet wavelet with $\omega_0 = 6$. The data is the same SST series used for Fig. 10.19, with $T_s = 1/4$ year. On the abscissa, $t = n T_s$ in years is reported; the acronym AD stands for *Anno Domini* and indicates the years of the Christian Era. On the ordinate, we see the dimensional period P expressed in years. The P-axis reversed, so that the associated frequency axis would be upright. Moreover, the ordinate is reported on a logarithmic base-2 scale, in agreement with the dyadic discretization of scale factors. The regions of the plane in which the energy of the signal concentrates are clearly visible.[12] The COI curve related to edge effects is overlaid to the map (*white cup-shaped curve*). Significance contour lines for scalogram values at the 95 % c.l. are also drawn as *black curves* (see Sect. 13.5.2)

The relation between the COI width and the scale factor, i.e., $b_a = b_a(a)$, is given in Table 13.3 for the most commonly adopted continuous wavelets. The relation is linear. The proportionality coefficient between b_a and a, the numerical value of which is reported in Table 13.3, will hereafter be indicated by k_{COI}. Now, to draw the edges-related COI curve in the time-period plane we actually need the equation

$$P_{\text{COI}} = P_{\text{COI}}(t), \qquad \text{with} \qquad t = n T_s,$$

---

[12]The plot should also include a colorbar telling the scalogram value associated with each shade of gray. Since this is only a qualitative example, the colorbar has been omitted.

where P_{COI} is the ordinate on the *white cup-shaped curve* that corresponds to the abscissa t. This equation can be derived as follows.

If $T_d = (N - 1)T_s$ is the duration of the data record, then

- for the time interval $0 \le t \le T_d/2$ we imagine an impulsive disturbance at $t = 0$. Recalling that $P = k_0 s$ and $\tau_s = T_s b_a = T_s k_{COI} a = k_{COI} s$, we observe that imposing $t = \tau_s$ we get

$$P = P_{COI} = k_0 s_{COI}, \quad t = k_{COI} s_{COI}.$$

Then we have

$$P_{COI} = \frac{k_0}{k_{COI}} t,$$

representing a curve we can draw in the time-period plane. The curve delimits the COI region related to the impulse at $t = 0$; this region, enclosed between the ordinate axis and the curve, would be a triangle if we used a linear P-axis, but

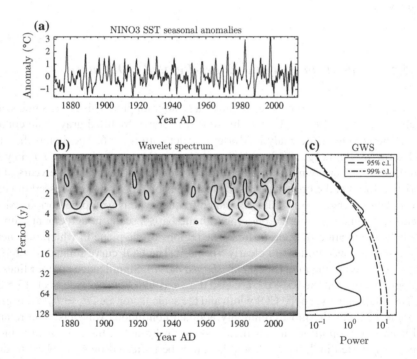

Fig. 13.22 A typical graph summarizing the results of the CWT analysis of a record: **a** the series (the same analyzed in 13.17 and previously used for Fig. 10.19), **b** its scalogram with special *black contours* enclosing significant areas at the 5 % confidence level and with the COI curve (*white line*), and **c** the GWS (*solid curve*) with its significance curves at the significance levels of 5 and 1 % (*dashed and dashed-dotted curves*, respectively). The wavelet is analytic Morlet with $\omega_0 = 6$.

Table 13.3 Linear relation between scale factor and cone-of-influence (COI) width, $b_a = b_a(a)$, for the analytic Morlet wavelet with parameter $\omega_0 = 6$, the analytic Paul wavelet with parameter $m = 4$, and the real DOG wavelets with orders $m = 2$ and $m = 6$

Analytic Morlet(ω_0)	Analytic Paul(m)	Real DOG($m = 2$)	Real DOG($m = 6$)
$b_a = \sqrt{2}a$	$b_a = 0.7013a$	$b_a = 2.1243a$	$b_a = 1.5829a$

The numerical proportionality factor between b_a and a is called k_{COI} in the text

becomes a curvilinear triangle with concavity facing upwards when we use the base-2-logarithmic P-axis;

- for the remaining part of the time axis, we will place the disturbance at $t = T_d$ and will obtain in a similar way another COI curve delimiting a curvilinear triangle that will be a mirror image of the first one.

This is illustrated in Fig. 13.23: in this example, the analytic Morlet wavelet with $\omega_0 = 6$ has been used, for which $P_{COI} = (k_0/k_{COI})\, t = \left(1.033/\sqrt{2}\right) t = 0.73t$.

We can see that the larger the delay τ_s corresponding to a given P_{COI}, the more extended in time the area of the map affected by edge effects. This is the reason why when k_{COI} is calculated, if the equation

$$\left|W_x^\psi(a_0, b_{a_0})\right|^2 = \left|W_x^\psi(a_0, \theta_0)\right|^2 / e^2$$

has more than one solution for b_{a_0}, the maximum delay b_{a_0} is conservatively assumed, which corresponds to the widest COI.

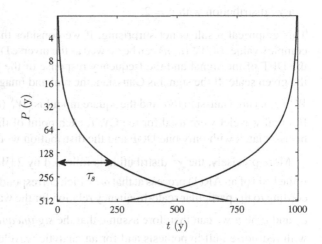

Fig. 13.23 Cone-of-influence (COI) at the edges of the time-period plane, found applying two impulsive disturbances, one in the time origin and the other one at the end of the data record. A analytic Morlet wavelet with $\omega_0 = 6$ has been used; the record is composed by $N = 1000$ samples and the sampling interval has been assumed to be 1 year. As a consequence, times and periods on the axes are expressed in years too

13.5.2 Scalogram Significance Levels

In order to evaluate the statistical significance of the spectral features detected in CWT analysis, a test similar to that used for the periodogram is applied to every local wavelet spectrum. We may observe that in our discretized CWT, with $\Delta b = \Delta n = 1$, every local spectrum represents the frequency distribution of *energy per unit time*, and therefore is itself a power spectrum estimate, exactly as a periodogram—it is the local-in-time power spectrum estimate for our series. The GWS can now simply be seen as the *average of all local-in-time power spectra*.

The significance test for each local spectrum is the same we used for the periodogram. This test was devised for WSS random processes; thus, in principle, the test is not relevant to the case of non-stationary signals. However, it can still provide useful guidelines in CWT analysis. The background spectrum is assumed to be white or red noise with variance σ_x^2 estimated from the data. *Monte Carlo simulations* have shown that for an analytic wavelet, $\left| W_x^{\psi}(a, n) \right|^2$ for a given $n = n_0$ follows a χ^2 distribution with $\nu = 2$ DOF. The simulations were performed in this way:

- a high number, e.g., 10,000, of Gaussian red-noise sequences was generated, all with the same length $N = 512$;
- the scalogram of each sequence was computed;
- from each scalogram, a "vertical slice", i.e., a local wavelet spectrum, was taken far from the edges, at $n_0 = 256$;
- at each scale, the 10,000 spectral samples thus obtained were ordered by increasing value and plotted as a function of order number, to see the amplitude distribution of the local spectrum at each scale. In this way, Torrence and Compo (1998) found a χ^2 distribution with $\nu = 2$.

This empirical result is not surprising, if we consider that at any given scale, the complex values of $W_x^{\psi}(a, n)$ can be viewed as the inverse DFT of the product between the DFT of the signal and the frequency response of the bandpass wavelet filter at the given scale. If the signal is Gaussian, the real and imaginary parts of such values $W_x^{\psi}(a, n)$ are Gaussian too, and the square modulus $\left| W_x^{\psi}(a_j, n) \right|^2$ is χ_2^2-distributed. If a real wavelet were used for the CWT, each point of the time-scale plane would be associated with only one DOF and the distribution would be χ_1^2.

More precisely, the χ^2 distribution is followed by $2 \left| W_x^{\psi}(a_j, n) \right|^2 / P_j$, where P_j is the PSD of an AR(1) process at that ω_j which corresponds to the scale factor a_j, according to the particular scale-frequency relation for the wavelet in use. At each scale a_j and time n we can therefore assume that the *significance level* for $\left| W_x^{\psi}(a_j, n) \right|^2$, with red-noise null-hypothesis and for an analytic wavelet—also referred to as the $p\%$-c.l. scalogram—is

$$\left| W_x^{\psi}(a_j, n) \right|_{\text{signif}}^2 (a_j) = P_j \frac{\chi_{1-p,2}^2}{2}$$

at the $p\%$ confidence level (c.l.). In the scalogram map, the areas in which $\left|W_x^\psi(a_j,n)\right|^2$ exceeds the significance level are highlighted by drawing a particular and clearly visible set of contour lines representing scalogram values equal to the significance level. These contour lines will enclose significant scalogram areas, as in Fig. 13.22b, where we can see them as *black curves*. Here the c.l. is 95 %.

13.5.3 Interpretation of the Scalogram

In evolutionary spectral analysis, maps like the one shown in Fig. 13.22b allow for observing what band of periods—or frequencies, or scales—is particularly active in contributing to the signal's varibility in the neighborhood of every time value, and allow for deciding if such activity can be considered significant at some confidence level against a properly chosen null-hypothesis for the noise background.

For example, the scalogram values in Fig. 13.22b are high in the period range from 2 to 4 years from the beginning of the series to the 1920s and again from 1960 AD to 2000 AD, while the interval 1920–1960 exhibits lower values in this period range. The Nino3 SST anomalies analyzed in Fig. 13.22b are indicators of interannual climatic variability in the Pacific Ocean, which is characterized by the El Niño–La Niña phenomenon (Sect. 10.8). The scalogram allows us to appreciate the temporal variations of the frequency with which warm events (El Niño) and cold events (La Niña) occurred in the last one-and-a-half centuries. It is known that between 1875 and 1920 AD many remarkable events took place, while between 1920 and 1960 only a few events were identified; they did not occur in a strictly periodic way, but typically took place every 2–7 years. The scalogram of Fig. 13.22b tells us exactly this, and describes the spectral evolution of the phenomenon. From 1875 to 1915 AD, a shift of significant energy density from about 4 years of period to about 2 years of period is evident; from 1960 to 2000 AD an opposite shift can be observed, from shorter to larger periods (more precisely, from a dominant period of about 3 years in 1960–1965 to a dominant period of about 5 years from 1980 AD on). Periods up to about 6–7 years are associated with significant scalogram values in the most recent decades. Note how localized events in the signal appear as single, narrow scalogram maxima where time resolution is high, i.e., at small scale/period.

We can also observe that the maximum scale considered for drawing Fig. 13.22b, which corresponds to a period of about 90 years, has been chosen in a reasonable way: this record covers 1871–2014 AD, i.e., about 140 years, so that the maximum period is about 65 % of the record length. The interval of scales for which the scalogram points lie completely inside the COI is a small fraction of the whole vertical axis.

If we count the number of scalogram points that exceed the significance level at the 95 % c.l., we find about 5 % of the total. On this basis, the series cannot be distinguished from noise. However, other scalogram features can inform us about the*randomness* of the series. Imagine if we analyzed a pure-noise sequence. What would the scalogram look like? In case of white noise with variance σ_x^2, we might expect the scalogram to exhibit a constant value everywhere. In reality it would show a random distribution of maxima and minima around the constant value σ_x^2. Similarly, the scalogram of a red noise signal would show, on average, a gradual increase of scalogram values with increasing scale/period and decreasing frequency, with a random distribution of maxima and minima around this average behavior. Now imagine we analyzed a generic stationary signal characterized by a true energy spectrum of some kind. The scalogram would show random fluctuations around this spectrum. Apart from these fluctuations, the scalogram would appear uniform along the time axis. All local spectra would be similar to one another, and the global wavelet spectrum would echo their shape. At last, the scalogram of a non-stationary signal, like the one shown in Fig. 13.22b, is recognizable from the presence of organized features in time and scale. The scalogram of a red noise sequence with parameter equal to the lag-1 autocorrelation of the Nino3 anomaly series would display an average gradual increase of scalogram values with period, with maxima and minima randomly distributed around this average trend; in our actual scalogram, on the contrary, significant regions are clearly organized in time and scale, which indicates a lesser randomness of the stochastic process underlying the generation of the data record, with respect to red noise.

Scalograms also allow us to see if passing from one scale to the other, repetitive patterns appear that might indicate fractality of the process. To see an example of this behavior, we can consider the von Koch curve (Fig. 13.24a), a mathematical curve which is one of the earliest fractal curves that has been described in literature (von Koch 1904). If we analyze this signal by CWT we actually see (Fig. 13.24b) a regular pattern through scales. The CWT of the von Koch curve has been computed using a "nearly-symmetric orthogonal wavelet" named coiflet of order 3, a wavelet that is normally employed for DWT (see Chap. 14) but can be used for CWT too, and is particularly fit for the task of fractality detection. Figure 13.24b simply shows the absolute values of the CWT-matrix elements: we do not need to refer to concepts like energy density and scalogram in this case, since we are just looking for repetitive patterns. Indeed, it is intuitive that since the CWT is a "resemblance index" between the signal and the wavelet, if a signal is similar to itself at different scales, then the resemblance index also will be similar to itself at different scales, and this self-similarity will generate a characteristic pattern. Note that to make the regular pattern more visible, in Fig. 13.24b only a part of the time axis is shown: the signal extends over some 8000 samples, while the CWT magnitude is shown only for time indexes from 3200 to 4200.

The choice of the mother wavelet is important to put one or another facet of the data spectral behavior into evidence. We just pointed out this fact in the case of fractality detection. By and large, we can say that the wavelet should "look like" the feature of the signal that we want to detect: for example, if a signal contains transient bursts and we want to detect them, the Mexican hat wavelet can be a good choice. This may be the case of an ECG signal in which we want to locate the onset and demise of eachheartbeat, or the case of a seismologic signal in which we want to

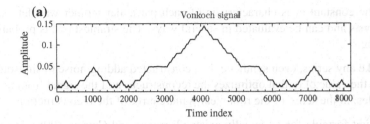

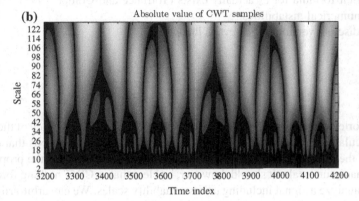

Fig. 13.24 a The fractal von Koch curve over $N = 8192$ points, and **b** an expanded view of the absolute value of its CWT in the time-scale plane. A coiflet wavelet of order 3 was used (see the text and Chap. 14)

detect earthquake tremors. For spectral analysis, the best choice is a wave-packet-like wavelet like the analytic Morlet wavelet, but even in this case we may desire to privilege time resolution, or frequency resolution. If we want a finer time resolution at all scales with respect to what we can obtain with the analytic Morlet wavelet, we may want to use the analytic Paul wavelet that, being narrower in time, can produce the desired result; or we may want to modify the analytic Morlet wavelet (see Chap. 16).

In normal applications, the redundancy of the CWT discretizaton scheme is not too heavy in terms of computational costs, and makes the scalogram maps smoother and easier to interpret than those that could be obtained with a discretization made paying attention to "economy".

13.5.4 Signal Reconstruction from Discretized CWT Samples

The discrete version of Morlet's reconstruction formula that uses a Dirac δ is

$$x[n] = \frac{\delta j}{c_\delta \psi_0(0)} \sum_{j=0}^{J} \frac{\text{Re}\left\{W_x^\psi(a_j, n)\right\}}{\sqrt{a_j}},$$

where the constant c_δ is characteristic of each particular mother wavelet used for the analysis and can be evaluated in several ways. The simplest one is probably the following:

- we take any series, even synthetic, i.e., constructed adding noise to sinusoids;
- we fit the value of c_δ that optimizes the reconstruction of the signal from its CWT samples, in the sense of the root mean square error of the reconstruction.

An explicit formula for c_δ actually exists (Torrence and Compo 1998), but it may lead to numerical instabilities.

The discrete version of Parseval's theorem is

$$\sigma_x^2 = \frac{\delta j}{N c_\delta} \sum_{n=0}^{N-1} \sum_{j=0}^{J} \frac{\left| W_x^\psi(a_j, n) \right|^2}{a_j}.$$

Both Morlet's and Parseval's formulas are useful to check the accuracy of the numerical calculation of the CWT and find the interval of scales or periods that contains most of the signal's energy. They help with deciding which is the proper value of the maximum scale. It should, however, be clear that in CWT nothing forces us to always analyze a signal including *all* its variability scales. We can arbitrarily decide which scales are of interest (see also Sect. 13.5.6).

13.5.5   Global Wavelet Spectrum (GWS) and Significance Levels

The *global wavelet spectrum* (GWS) is computed as the time average of all local wavelet spectra:

$$\overline{W}^2(a) = \frac{1}{N} \sum_{n=0}^{N-1} \left| W_x^\psi(a, n) \right|^2 .$$

The GWS of the SST record appears in Fig. 13.22c. Significance levels for the GWS at the 95 and 99 % c.l. are also drawn, which are introduced below. Figure 13.22 is an example of how the final result of the CWT analysis of a data record can appear. In a single figure, three plots are drawn: the series (Fig. 13.22a), its scalogram (Fig. 13.22b) with special contour lines enclosing areas with energy spectral density that is significant at the some c.l. (here, 95 % c.l.) and with the COI curve, and the GWS (Fig. 13.22c) with its significance curves at different probability levels (here, 95 % c.l. and 99 % c.l.).

The GWS power spectrum estimate is consistent. It is comparable with the periodogram $|X[k]|^2 / N$ and other classical and parametric estimates but is typically very smooth, even if its smoothness varies with the mother wavelet and is always greater at small periods. A characteristic of the GWS is in fact the remarkable smoothing undergone by possible high-frequency peaks: for example, if we take a signal with two sinusoids of different frequencies and equal amplitudes, possibly immersed in noise, in the periodogram we would see two peaks of equal heights, while in the GWS we will find that the higher-frequency peak is much smaller than the lower-frequency one (see the example of Fig. 13.25). Therefore, if we expect to detect peaky spectral features, the GWS is not a good tool to determine their relative heights correctly. The reason for this bias is the bandwidth of the wavelet filter, which increases with frequency according to the multiresolution property. It is directly related to the fact that the GWS has poor frequency resolution at high frequency.

If we smooth the periodogram of a record by a running-average filter, the result becomes more similar to the GWS, but the amount of smoothing required for similarity increases with increasing frequency. This is illustrated in Fig. 13.26 that shows the GWS of the same record analyzed in Fig. 13.22b (*black curve* in all panels) versus period, superimposed to smoothed versions of the periodogram. The smoothing has been repeated four times, progressively averaging over an increasing number q of

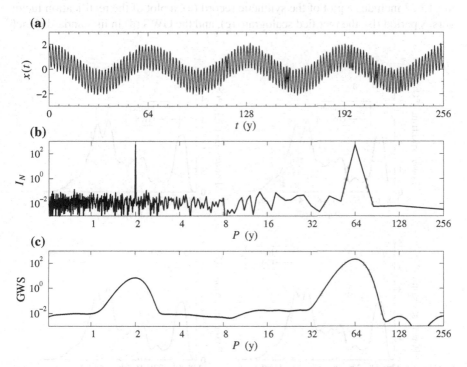

Fig. 13.25 Comparison between periodogram and GWS for a synthetic signal with hypothetical sampling interval of 1 year, composed of a sinusoid with high frequency and a sinusoid of low frequency, with the same amplitude, immersed in white noise. **a** The signal; **b** the periodogram, showing two peaks of equal height; **c** the GWS obtained using the analytic Morlet wavelet with $\omega_0 = 6$, which exhibits a remarkable smoothing of the high frequency peak; on the abscissa of both spectral plots, dimensional period P in years is reported

spectral values (*gray curve* in all panels). It may be seen that at high period, GWS is similar to the periodogram smoothed adopting a value of q as small as 3 (Fig. 13.26a). At lower period values, GWS is more similar to the periodogram smoothed adopting larger values of q, i.e., 7 and 19 (Fig. 13.26b and c, respectively). In the lowest period range, similarity requires a value of q as high as 71.

The bias described above has been discussed by Liu et al. (2007). These authors suggested that since computing the scalogram implies, at each scale, integrating the energy of the signal over a time interval, the length of which increases with scale, a consistent definition of the wavelet spectrum should include dividing each value by the scale to which the value refers. In essence, according to Liu et al. (2007) the traditional definition of wavelet spectrum confuses energy with energy integrated with respect to time. If we apply this normalization (the authors speak about "rectifying the spectrum"), the analysis of a signal containing three sinusoids with equal unit amplitudes leads to a GWS with peaks of nearly equal height, in contrast with what happens with the traditional definition of wavelet spectrum (Fig. 13.27). An analytic Morlet wavelet with $\omega_0 = 6$ has been used for this comparison. Figure 13.27 includes a plot of the synthetic record (a), a plot of the rectification factor versus period (b), the rectified scalogram (c), and the GWS (d) in its standard (*black*

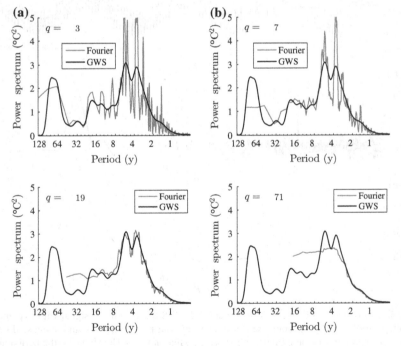

Fig. 13.26 GWS (*black curves*) and periodogram smoothed by a running-average filter over an increasing number q of values (*gray curves*) for the same record analyzed in Fig. 13.22. The GWS was obtained using the analytic Morlet wavelet with $\omega_0 = 6$. **a** $q = 3$, **b** $q = 7$, **c** $q = 19$, and **d** $q = 37$. See text for details

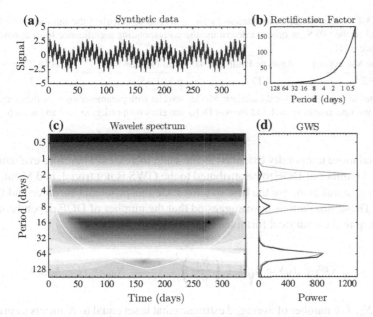

Fig. 13.27 CWT analysis of a signal containing three sinusoids with equal unit amplitudes, and periods of 2, 8, and 64 days; the sampling interval is assumed to be one hour and the record length is $N = 8192$. The standard wavelet spectrum, computed using the analytic Morlet wavelet with $\omega_0 = 6$, has been rectified dividing it by a factor proportional to scale. **a** Time series; **b** rectification factor versus period; **c** rectified scalogram; **d** the GWS in its standard (*black curve*) and rectified (*gray curve*) versions. In the rectified GWS, the heights of the peaks are similar; only the peak corresponding to the 64 days sinusoid is slightly lower, because it is reduced with respect to its true value by the zero-padding edge effects (COI). In the standard GWS, the peaks have very different heights, so different that on the linear axis adopted to plot the GWS, the peak at 2 days of period is barely visible

curve) and rectified (*gray curve*) versions. In order to leave to the discussed effect all its natural evidence, the GWS is plotted onto a linear axis, rather than onto a logarithmic one. The rectifying procedure actually leads to the desired result (*gray curve* in Fig. 13.27d): the rectified GWS allows for estimating the relative sinusoids' amplitudes on the basis of peaks' heights. However, this rectification produces an unpleasant effect, namely the rectified spectrum of a white noise record would no longer be flat. For this reason, this correction is seldom applied.

A *significance level for the GWS* at some $p\%$ c.l. can be established. It is also referred to as the $p\%$-c.l. GWS. The time average contained in the definition of the GWS makes the number of DOF increase with respect to the minimum value of 2 that characterizes the periodogram and all local wavelet spectra. The amount of increase in the number of DOF depends on the number of independent estimates $\left| W_x^\psi (a, n) \right|^2$ that are averaged to produce $\overline{W}^2(a)$. These estimates are, however, correlated among themselves in time and scale, and the correlation in time becomes

Table 13.4 Values of the decorrelation factor γ needed to calculate the number of DOF to be attributed to the GWS, in order to determine the corresponding significance level at some given confidence level

Analytic Morlet(ω_0)	Analytic Paul($m = 4$)	Real DOG($m = 2$)	Real DOG($m = 6$)
$\gamma = 2.32$	$\gamma = 1.17$	$\gamma = 1.43$	$\gamma = 1.37$

The values of γ are shown for the analytic Morlet wavelet with parameter $\omega_0 = 6$, the analytic Paul wavelet with parameter $m = 4$ and the real DOG wavelets with orders $m = 2$ and $m = 6$

more and more temporally extended as the scale increases. Therefore evaluating the effective number of DOF to be attributed to the GWS is not trivial, and Monte Carlo numerical simulations are the best way to tackle this problem (Torrence and Compo 1998). These simulations have suggested that the number of DOF depends on scale according to the empirical formula

$$\text{DOF}_{\text{GWS}}(a) = 2\sqrt{1 + \left(\frac{N_{av}}{\gamma a}\right)^2}, \qquad N_{av} = N - k_{\text{COI}}a$$

where N_{av} is a number of averaged estimates that is set equal to N minus a correction to take into account the fact that the points located inside the COI have a statistical weight that is about half the statistical weight of the others. The constant γ is a parameter deduced from the simulations and is known as the *decorrelation factor*. Table 13.4 shows the value of γ for the analytic wavelets of Morlet and Paul and for DOG(2) and DOG(6) real wavelets.

Once the number of DOF has been determined as a function of a, the usual formula is applied: the significance upper bound for the GWS at the c.l. of $p\%$ is a function of scale given by

$$\text{GWS}_{\text{signif}}(a_j) = P_j \frac{\chi^2_{1-p,\text{DOF}(a)}}{\text{DOF}(a)},$$

where P_j is again the PSD of an AR(1) process at that ω_j which corresponds to the scale factor a_j, according to the particular scale-frequency relation for the considered wavelet. Through the scale-frequency relation, this upper bound can be translated into a $\text{GWS}_{\text{signif}}(f)$, or a $\text{GWS}_{\text{signif}}(P)$. The plot of this function, for one or more values of p, is superposed to the GWS curve, so that the peaks detected as significant at the c.l. of $p\%$ can be visually identified (see Fig. 13.22c).

13.5.6 Extensions of Wavelet Analysis

We conclude presenting two developments of the CWT that are very useful in the analysis of time series.

Wavelet Filtering

By reconstructing the signal on the basis of a subset of scales only, say $\{a_j\}$, $j = [j_1, j_2]$, we can obtain a filtered series containing only the contributions of frequency components falling inside the passband of the overall wavelet filter. We can write

$$x'[n] = \frac{\delta j}{c_\delta \psi_0(0)} \sum_{j=j_1}^{j_2} \frac{\mathrm{Re}\left\{W_x^\psi(a_j, n)\right\}}{\sqrt{a_j}}.$$

The overall filter has a frequency response given by the sum of all $\Psi^*(a_j \omega)$ for $j = [j_1, j_2]$. An example of CWT-based signal reconstruction is given in Chap. 16.

Wavelet cross spectrum (WCS)

The evolutionary *wavelet cross spectrum* (WCS) of two contemporary time series—being $x[n]$ and $y[n]$—can be estimated as

$$W_{xy}^\psi(a, n) = \left[W_x^\psi(a, n)\right]^* W_y^\psi(a, n).$$

When, as is common practice, complex-valued wavelets are used, the WCS is complex and bears both magnitude and phase information. A *cross-scalogram* is then defined as $\left|W_{xy}^\psi(a, n)\right|$. More often, however, the concept of *wavelet coherence* is used. The reader may recall that in Chap. 10, when dealing with classical Fourier-based cross-spectral analysis, we defined the magnitude square coherence as the square of the cross-spectrum normalized by the individual power spectra. This gives a quantity between 0 and 1, and measures the cross-correlation between two time series as a function of frequency. Unfortunately, as noted by Liu (1994), coherence defined in this way in wavelet analysis would be identically one at all times and scales. In Fourier analysis, moreover, coherence is normally estimated smoothing the cross-spectrum and the individual spectra before forming the coherence estimate (e.g., using Welch estimates). In wavelet analysis, it is not immediately clear what sort of smoothing (presumably in time) should be done to arrive at a useful measure of coherence. The smoothing would prevent wavelet coherence to be one everywhere in the time-scale plane, but in a sense would also seem to defeat the very purpose of wavelet analysis by decreasing the localization in time (Torrence and Compo 1998). If we denote the smoothing operator by $\mathscr{S}$, the definition of wavelet coherence adopted in literature is

$$\text{wavelet coherence} = \frac{\mathscr{S}\left[W_{xy}^\psi(a, n)\right]}{\sqrt{\mathscr{S}\left[\left|W_x^\psi(a, n)\right|^2\right]}\sqrt{\mathscr{S}\left[\left|W_y^\psi(a, n)\right|^2\right]}}.$$

This is a complex quantity that is then decomposed into a modulus, which we will indicate as *wavelet coherence magnitude*, and a phase angle,[13] called *wavelet coherence phase*. The smoothing operator is often applied in time only, and can be a simple running mean; for a more detailed discussion see Torrence and Webster (1999) and Grinsted and Chan (2011).

Also the wavelet cross spectrum is often smoothed before plotting, so that it becomes

$$\hat{\text{WCS}} = \mathscr{S}\left[W_{xy}^{\psi}(a, n)\right].$$

The WCS or the wavelet coherence of two contemporary time series can reveal localized similarities in time and scale. In the coherence magnitude map, areas in the time-period plane where the two time series co-vary with elevated shared energy can emerge. The phase map visualizes the relative phase behavior of the two series, i.e., the phase difference between the two signals at the various scales or periods, and at different time instants. The phase map is meaningful only in those regions of the time-scale plane in which the values of coherence magnitude are not negligibly low.

A typical example of application of these concepts may be given using a noisy sine and a Doppler signal. The first time series is a 4-Hz sine wave with additive Gaussian noise that is sampled on a grid of 1024 points over the interval [0,1]. The second time series is a Doppler signal with decreasing frequency over time. Consider (Fig. 13.28) the CWT of the two individual signals computed using the analytic DOG wavelet of order 2, for integer scales from 1 to 512. The figure shows the modulus and phase angles of each CWT; the scale axis is upright in this case.

The analysis of the noisy sine function on the left exhibits the scale associated with the period, which is equal to 1024/8 = 128. The analysis of the Doppler signal on the right shows a typical time-scale pattern in which the dominant scale increases—the dominant frequency decreases—with increasing time. The corresponding smoothed wavelet cross spectrum is shown in Fig. 13.29. Smoothing was performed in time by a running average over 21 points. After convolution of the WCS with the running average filter, only the central part of the convolution was retained over the same time-length of the WCS (see Sect. 16.4.1.1 for the description of the precise way in which this operation is performed). The magnitude plot (upper panel of Fig. 13.29b) is the most instructive. It shows the similarity of the local frequency behavior of the two time series in the time-scale plane. Both signals have a similar contribution around scale 128 over the time-index interval [300, 700]. This is consistent with the behavior observed by visual inspection of the time-domain plot in Fig. 13.29a. Additional interesting information is discernible in the wavelet coherence plot (Fig. 13.30), in which filled gray-scale *contour lines* represent the magnitude and *arrows* have been drawn to show the phase angle.

---

[13]Recall that the phase angle of a complex quantity is always obtained by taking the arc tangent of the ratio of the imaginary and real parts.

Smoothing was again performed in time by a 21-point running average of the CWS and of the individual scalograms before forming the coherence. The phase information can be interpreted by locating the regions of the time-scale plane that highlight possible coherent behaviors. Some transient minor contributions to the variability of the two coupled time series occur at small scales at the beginning of the series, where the Doppler signal exhibits rapid oscillations. The behavior is not coherent and the phase changes very quickly. However, at values of the time index greater than about 150 and scales greater than about 130, numerous coherent regions can be easily detected.

For focusing on the phase of wavelet coherence, another representation can be adopted (Fig. 13.31). The phase information here is coded both by the orientation of the *arrows* and by the background color. The background color is associated with a mapping onto the interval $[-\pi, \pi]$.

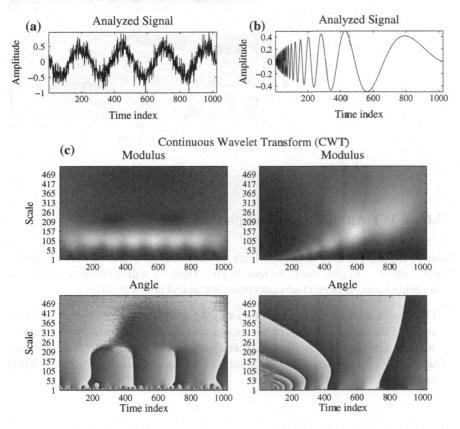

Fig. 13.28 Individual CWTs of a noisy sinusoidal signal and a Doppler signal, obtained using the analytic DOG wavelet of order 2 at integer scales from 1 to 512. **a, b** The noisy sinusoid and Doppler series, respectively; **c** the modulus and the phase angle of the two wavelet transforms

Fig. 13.29 Smoothed
wavelet cross spectrum
between a noisy sinusoidal
signal and a Doppler signal,
computed using the analytic
DOG wavelet of order 2 for
integer scales from 1 to 512.
a The noisy sinusoid and
Doppler series superposed to
one another; **b** the modulus
and the phase angle of the
smoothed wavelet cross
spectrum

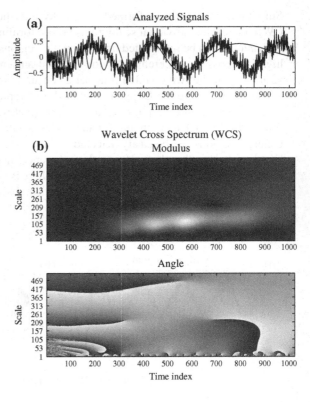

13.6 CWT Application Examples

Real-world applications of the CWT are so numerous in literature that a comprehensive set of examples would be hard to give. Here we will show three cases: one
concerning economics and two concerning social and political sciences.

1. Our first example is taken from a paper by Aguiar-Conraria and Soares (2011): a
   study of business cycle synchronization across the Euro countries performed by
   wavelet analysis. In substance, Aguiar-Conraria and Soares (2011) approached
   using CWT an investigation very similar to the one for which Sella et al. (2013)
   resorted to SSA (see Sect. 12.3).

Fig. 13.30 Wavelet coherence between a noisy sinusoidal signal and a Doppler signal, computed using the analytic DOG wavelet of order 2 at integer scales from 1 to 512. **a** The noisy sinusoid and Doppler series superposed to one another; **b** wavelet coherence plotted using gray-scale filled *contour lines* for the magnitude and *arrows* for the phase angle

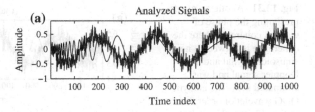

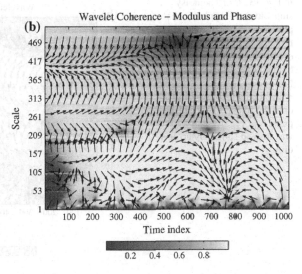

Aguiar-Conraria and Soares (2011) analyzed monthly data of the industrial production (IP) index for the countries in the EU-15, i.e., the twelve countries that first joined the Euro—Austria, Belgium, Finland, France, Germany, Greece, Ireland, Italy, Luxembourg, Netherlands, Portugal and Spain: EU-12 hereafter—and the three countries that were part of the European Union in 1999, but chose not to join the monetary union—United Kingdom, Sweden, and Denmark. Here we focus on the EU-12 results. IP-index wavelet spectra were compared to one another to investigate the degree of synchronization among countries. Then, cross-wavelet analysis was used to study in detail whether in some time and frequency intervals a given country has behaved in synchronism with the other ones.

For non-economists, we must first recall from Sect. 12.3 that data of this kind is always pre-processed before analysis. Pre-processing includes removing seasonality, and the typical upward long-term trend. In Aguiar-Conraria and Soares (2011) this is done applying a wavelet bandpass filter to preliminary remove these confounding variability components. Also, it is common in economics to take the logarithm of the data (see, for instance, Gujarati and Porter 2009), and this paper is no exception.[14]

---

[14]Time-series modeling is common practice in economics. Logarithms possess properties that assist with model-building and with the visual display of models and data in graphs. In essence,

Fig. 13.31 Another way of presenting the phase of wavelet coherence: here the coherence is between a noisy sinusoidal signal and a Doppler signal and was computed using the analytic DOG wavelet of order 2, as in Fig. 13.30. **a** The noisy sinusoid and Doppler series plotted onto one another; **b** phase as coded both by the orientation of the *arrows* and by the background color. The background color is associated with a mapping onto the interval $[-\pi, \pi]$. In panel b of Fig. 13.30, phase was instead shown by arrows only

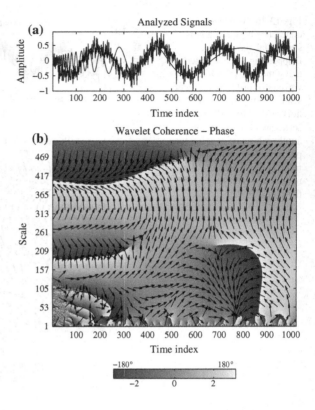

Using the International Financial Statistics database of the International Monetary Fund (IMF), Aguiar-Conraria and Soares (2011) gathered non-seasonally-adjusted IP data from July 1975 to May 2010. They also derived an Euro-12 IP index by calculating a weighted average of the industrial productions of the twelve individual countries in EU-12. As weights, the authors used the GDP of each country in the year 2000. They removed seasonal effects from the series of the individual countries and from the series of the Euro-12 IP index, and then estimated wavelet spectra between 1.5 and 8 year periods, since business-cycle periodicities are commonly believed to lay in this range. In Fig. 13.32a we can see the Euro-12 IP index series; Fig. 13.32b shows its scalogram. The data and the software for producing Fig. 13.32 and the subsequent four figures (Figs. 13.33, 13.34, 13.35 and 13.36) were kindly provided by L. Aguiar-Conraria. The wavelet spectra for

---

(Footnote 14 continued)

log-linearization is a solution to the problem of reducing computational complexity in systems of numerically specified equations that need to be solved simultaneously. Log-linearization converts a nonlinear equation into an equation that is linear in terms of the log-deviations of the associated variables from their steady-state values. For small deviations from the steady state, log-deviations have a convenient economic interpretation: they are approximately equal to the percentage deviations from the steady state. Log-linearization can greatly simplify the computational burden and, therefore, help solve a model that may otherwise be intractable.

Fig. 13.32 a Euro-12
Industrial Production Index,
and **b** corresponding
scalogram obtained adopting
the analytic Morlet wavelet
with parameter $\omega_0 = 6$. The
cone of influence is shown as
a *light-gray cup-shaped
curve*. The *black segments*
indicate the position of
scalogram's maxima

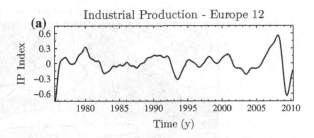

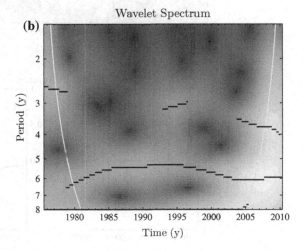

the individual countries (not shown) revealed the same patterns that are apparent
in the scalogram of the European aggregate (Fig. 13.32b): first, with the exception
of Greece, every country showed a persistent scalogram maximum at a period of
about 6 years. Second, the maximum at a period of about 3 years during the 1990s
that is observed for the Euro-12 aggregate was found to be common to several
countries, although not all of them. Finally, after 2005 volatility increased at all
frequencies and across all countries.

Aguiar-Conraria and Soares (2011) then studied the wavelet coherence between
each country's IP index and the rest of Europe, and tested the significance of their
results by applying proper statistical tests. They found that one feature which is
shared by the majority of the countries is a high-coherence region during the late
2000s. This is not surprising: the global crisis hit many countries simultaneously
and, as a consequence, they started behaving in a highly-synchronized way. The
analysis, however, revealed that Portugal, Greece, Finland—and to a lesser extent Ireland too—do not exhibit many regions of high coherence and therefore
can be classified as economies that do not follow closely the Euro-cycle. Conversely, the European core was identified as being formed by Germany and France.
Aguiar-Conraria and Soares (2011) found, perhaps surprisingly, that France
shows more regions of high coherence with the whole of Europe than Germany.
Moreover, in the shorter run (1.5–4.5 years of period) both France and Germany are

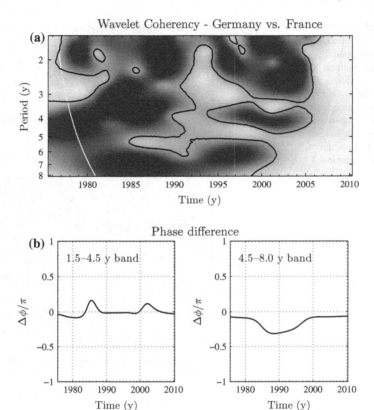

Fig. 13.33 **a** IP-index wavelet coherence, Germany versus France. The cone of influence is shown as a *light-gray cup-shaped curve*. The *black contours* enclose regions of significant coherence at the 95 % c.l. **b** Phase difference between the two countries, averaged over the range of periods 1.5–4.5 y (on the left) and 4.5–8 y (on the right). The wavelet is analytic Morlet with $\omega_0 = 6$

remarkably in phase with the rest of Europe, but on the longer run (4.5–8 years of period) it is France, and not Germany, that has been leading the European cycle. This result is shown in Fig. 13.33, where we see (Fig. 13.33a) the wavelet coherence of Germany versus France. In Fig. 13.33b, the phase-difference between the two countries is shown in the period bands 1.5–4.5 years (on the left) and 4.5–8 years (on the right). A zero phase-difference at the specified period would indicate that the two time series move together at the corresponding frequency. A phase-difference between 0 and $\pi/2$ means that the Germany leads France; a phase-difference between $-\pi/2$ and zero means that France leads Germany. In the right-hand panel of Fig. 13.33b we actually observe negative phase-difference values, indicating a delay in the variations of the German IP index with respect to France.

Aguiar-Conraria and Soares (2011) thus defined through CWT analysis an Euro-core and a Euro-periphery in terms of business cycles synchrony. From this study, Germany and France turned out to form the Euro-core around which the other

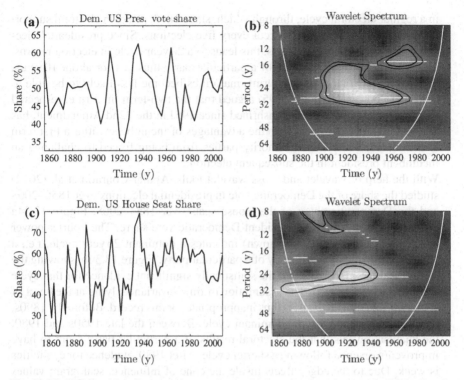

Fig. 13.34 U.S. President Democratic vote share: **a** time series, **b** scalogram obtained using the analytic Morlet wavelet with $\omega_0 = 6$. The cone of influence is shown as a *white cup-shaped curve*. The *black contours* enclose regions wich are significant at the 95 % c.l. *White segments* highlight the positions of maxima. Democratic House Seat share: **c** time series, **d** scalogram obtained using the same wavelet as in panel c

countries gravitate; all the other Euro-12 countries were found to be synchronized at 95 % c.l., with the exception of Portugal, Greece, Finland and Ireland that were identified as belonging to the Euro-area periphery, with their cycles out of synchrony with the rest of the Euro-12 group.

2. The second example is a study by Aguiar-Conraria et al. (2012) in political sciences.

   These authors tackled by CWT the study of two lingering puzzles in the political science literature: the existence of cycles in election returns in the United States and in the severity of great-power wars. Wavelet analysis, with its ability to detect transient, irregular cycles and structural breaks in the periodicity of cycles, is the ideal tool for these investigations.

   About election returns, previous studies have shown that the presidential vote in the United States can be modeled as an AR(2) process: Norpoth (1995) estimated the parameters of this AR(2) model fitting the Republican presidential share of the vote from 1860 to 1992. The AR(2) nature of the U.S. presidential vote results in a regular electoral cycle, through which, since 1860, peaks in electoral support for a particular party tend to occur every five elections. Since presidential elections take place every four years, this leads to a 20 year cycle in election returns. The half-cycles of ascendancy of a particular party thus last for about 10 years, which suggests that after two terms, majorities become less likely to be able to hold on to power. Norpoth (1995) argued that the two-term limit rule, respected by all presidents but one and enshrined since 1951 in the 22nd Amendment, has allowed majority parties to reap the advantages of incumbency after a first term is completed, while saving minority parties from being forced to challenge an incumbent president in the subsequent election.

With the help of wavelet and cross-wavelet tools, Aguiar-Conraria et al. (2012) studied the share of the Democratic vote in presidential elections over 1856–2008 and the Democratic share of the House seats over 1854–2008. Figure 13.34a displays the series of U.S.-President Democratic vote share. The Fourier power spectrum of this record (not shown) indicates a dominant 26 year cycle, i.e., a 13 year half-cycle of ascendancy of a particular party. Figure 13.34b, showing the series' scalogram, reveals that the statistically significant evidence for that cycle is limited to 1900–1970. The assumption of time invariance, lying at the heart of stationary spectral analysis, is thus inappropriate for this record. Before the 1890s, there is no evidence for any dominant cycle. Between the late 1950s and 1980, the analysis reveals that the electoral results of a particular party seem to have improved/worsened following a shorter cycle. After 1980, evidence for cyclicities is weak. Due to the edge effects inside the cone of influence, scalogram values may be underestimated at the beginning and the end of the series, before 1875 and after 1985; nevertheless, the results point to the fact that significant cyclicities are temporally localized. The scalogram of Fig. 13.34b also suggests the existence of longer, multidecadal cycles, which are not, however, statistically significant at the 95 % c.l.—and not even at the 75 % c.l. They would, anyway, be poorly represented by only 150 years of available data. Figure 13.34c shows the series for the House of Representatives. While the Fourier power spectrum of this record (not shown) indicates a dominant 25 year cycle, the scalogram reported in Fig. 13.34d demonstrates that such a cycle is not dominant throughout the entire series: for the House of Representatives, evidence for such a 25 year cycle is present from the beginning of the series in mid-19th century to the 1940s, but disappears after that decade.

Aguiar-Conraria et al. (2012) also estimated the wavelet coherence and the phase difference between these two series. Figure 13.35a shows the wavelet coherence of Presidential Election versus House of Representatives. Two main regions of high coherence between the two series are detected in Fig. 13.35a: the first one between the 1890s and 1940 in the 20–30 year period band, and the second one between the 1960s and 2000 in the 10–16 year period band. However, only the first region corresponds to a portion of the time-period plane where relevant cycles in individual scalograms are present (see Fig. 13.34a, c). Indeed, the region between the 1890s and 1940 in the 20–30 year band in Fig. 13.35a roughly corresponds to the intersection of the highest-energy density regions in Fig. 13.34a and c. The phase information for that time interval and period band (Fig. 13.35b) tells us that the series were in phase, i.e., their cycles were highly coordinated. Aguiar-Conraria et al. (2012) remarked that wavelet analysis allowed them to discover that the dynamics in the presidential and House elections are less similar than conventional methods would lead to believe. While the 26 year cycle for presidential elections started in the late 1890s and lasted until the late 1960s, a cycle of about the same period characterized the House elections only until the 1940s. Synchrony between the presidential and House series turned out to be a time-localized phenomenon, existing only for a 50 year interval between the late nineteenth century and the mid-twentieth century, with the House of Representatives vote share displaying an increasing lag since then, with respect to presidential elections (Fig. 13.35).

3. The third example concerns a study on the severity of great-power wars from 1495 to 1975. It is contained in same paper by Aguiar-Conraria et al. (2012). This analysis was performed on well-established data from literature (Goldstein 1988; Levy 1983). The question of whether wars occur cyclically has intrigued observers at least since the sixteenth century, but the answer has remained extremely elusive for modern social scientists. On one hand, variations in data sources and in the dimensions of the phenomenon of war—outbreak, duration, and magnitude, or severity—make data analysis difficult. On the other hand, solid theories that might explain why cyclical rhythms should be expected in wars are scarce. The data on war severity examined in Aguiar-Conraria et al. (2012) was taken from Goldstein (1988). They cover 1495–1975 and measure battle fatalities in great-power wars (GPWAR) in units of thousands, transformed according to

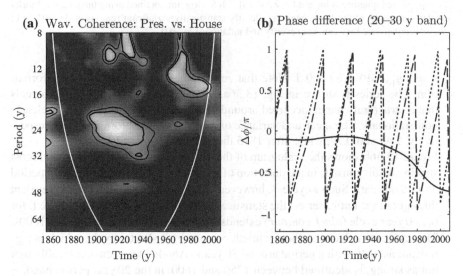

Fig. 13.35 Presidential Election versus House of Representatives: **a** wavelet coherence, and **b** phase averaged over the 20–30 year period band (*dotted line*: presidential election; *dashed line*: House of Representatives). Also shown is the related phase difference (*solid line*). The wavelet is analytic Morlet with $\omega_0 = 6$

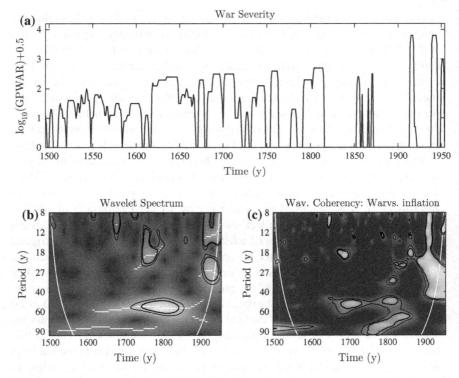

Fig. 13.36 **a** War severity series: GPWAR = battle fatalities from great-power wars, in thousands. The represented quantity is logged GPWAR + 0.5. **b** Scalogram obtained using the analytic Morlet wavelet with $\omega_0 = 6$; the *black contours* identify statistically significant regions at the 90% c.l.; **c** wavelet coherence between war severity and inflation (see text)

$x = \log_{10}(\text{GPWAR}) + 0.5$. Note that zeros remain zeros after this transformation. The series is shown in Fig. 13.36a. Visual inspection of the series reveals that a structural change occurred around 1815, so that the time series is clearly non-stationary. The mean and variance of the time series before 1815 are, respectively, 1.45 and 0.72, while after 1815 they become 0.46 and 1.28.

Figure 13.36b shows the scalogram of the war-severity series. The first and most striking result consists in the detection of a statistically significant cycle of period around 60 years. Such a cycle is, however, localized in time, rather than prevalent throughout the entire series: the statistically significant region at the 90% c.l. for the 60 year cycle (*black contour*) extends from the early 1700s to the mid-1800s. This cycle is replaced in the twentieth century by a shorter cycle, which is significant as well, with a period around 30 years. Another transient cycle, although not as strong, is identified between 1750 and 1800, in the 20 year period band.

Aguiar-Conraria et al. (2012) investigated a hypothesis according to which inflation could provide the explanation for these patterns, since a precise relation between inflation and war would exist (see Goldstein 1988, 2003). Wavelet analysis allows testing this hypothesis even in the presence of irregular cycles. Figure 13.36c shows the wavelet coherence between war severity and British inflation. Inflation was computed from the yearly South English Consumer Price Index (Goldstein 1988)—let us call it $p(i)$ where i is year—as

$$\text{Inflation}[i] = \frac{p[i] - p[i - 1]]}{\text{mean}\{p[i - 1], p[i]\}}.$$

The two series appear to be significantly coherent after 1875 in the 18–45 year period band. Before 1875, high coherence is observed at 60–100 year periods between the mid-18th and the mid-19th centuries, as well as in small regions around 1700 and 1800 in the 45–55 year period band. These, however, are not regions of high energy density in the war series, as can be seen in Fig. 13.36b. Therefore, the wavelet coherence results do not support the widespread notion that cycles in war severity consistently precede cycles in inflation.

The data and software for the plots related to this example are courtesy of L. Aguiar-Conraria.

References

Addison, P.S.: The Illustrated Wavelet Transform Handbook—Introductory Theory and Applications in Science, Engineering, Medicine and Finance. CRC Press (Taylor & Francis Group), Boca Raton (2002)

Aguiar-Conraria, L., Soares, M.J.: Business cycle synchronization and the Euro: a wavelet analysis. J. Macroecon. **33**, 477–489 (2011)

Aguiar-Conraria, L., Magalhães, P.C., Soares, M.J.: Cycles in politics: wavelet analysis of political time series. Am. J. Polit. Sci. **56**(2), 500–518 (2012)

Daubechies, I.: Ten Lectures on Wavelets. CBMS-NSF Regional Conference Series in Applied Mathematics. SIAM—Society for Industrial and Applied Mathematics, Philadelphia (1992)

Farge, M.: Wavelet transforms and their application to turbulence. Annu. Rev. Fluid Mech. **24**, 395–457 (1992)

Gabor, D.: Theory of communication. J. Inst. Electr. Eng. **93**, 429–457 (1946)

Goldstein, J.: Long Cycles: Prosperity and War in the Modern Age. Yale University Press, New Haven (1988)

Goldstein, J.: War and economic history. In: Mokyr, J. (ed.) The Oxford Encyclopedia of Economic History. Oxford University Press, New York (2003)

Grinsted, J.C., Chan, A.K.: Fundamentals of Wavelets: Theory, Algorithms, and Applications. Wiley, New York (2011)

Gujarati, D.N., Porter, D.C.: Basic Econometrics. McGraw-Hill/Irwin, New York (2009)

Levy, J.S.: War in the Modern Great Power System, 1495–1975. University Press of Kentucky, Lexington (1983)

Liu, Y.G., Liang, X.S., Weisberg, R.H.: Rectification of the bias in the wavelet power spectrum. J. Atmos. Ocean. Tech. **24**, 2093–2102 (2007)

Liu, P.C.: Wavelet spectrum analysis and ocean wind waves. In: Foufoula-Georgiou, E., Kumar, P. (eds.) Wavelets in Geophysics, pp. 151–166. Academic Press, San Diego (1994)

Mallat, S.G.: A Wavelet Tour of Signal Processing. Academic Press, Burlington (1999)

Müller, M., Ellis, D.P.W., Klapuri, A., Richard, G.: Signal processing for music analysis. IEEE J. Sel. Top. Sign. Proces. **5**(6), 1088–1110 (2011)

Norpoth, H.: Is Clinton doomed? An early forecast for 1996. PS. Polit. Sci. Polit. **28**(2), 201–207 (1995)

Porat, B.: A Course in Digital Signal Processing. Wiley, New York (1996)

Poularikas, A.D.: Transforms and Applications Handbook. CRC Press (Taylor & Francis Group), Boca Raton (2000)

Sella, L., Vivaldo, G., Groth, A., Ghil, M.: Economic cycles and their synchronization: a survey of spectral properties. Working Paper 105.2013. Fondazione ENI Enrico Mattei (FEEM), Milan (2013)

Torrence, C., Compo, G.P.: A practical guide to wavelet analysis. B. Am. Meteor. Soc. **79**(1), 61–78 (1998)

Torrence, C., Webster, P.J.: Interdecadal changes in the ENSO-monsoon system. J. Clim. **12**, 2679–2690 (1999)

von Koch, H.: Sur une courbe continue sans tangente, obtenue par une construction géométrique élémentaire. Archiv för Matemat., Astron. och Fys., 1, 681–702 (1904)

Part IV
Signal Decomposition, De-noising and Compression

Chapter 14
Discrete Wavelet Transform (DWT)

14.1 Chapter Summary

The wavelet transform can be seen as a wavelet-based expansion (decomposition) of a finite-energy signal. Orthogonality of the basis set of functions employed for the expansion is the key point in the discrete wavelet transform (DWT), in that it leads to economy in the representation of the signal through its DWT coefficients (signal decomposition), together with the possibility of perfect signal reconstruction. These are crucial features in many DWT applications, and in order to obtain them, the choice of the mother wavelet and of the discretization scheme become central issues. The basis set that can give DWT appealing features is not unique: there are many different wavelets systems that can be used effectively. The simplest formulation of the DWT problem includes two types of functions for the basis set: the scaling function and the wavelet function. We will describe this formulation, examining how an ideal, infinite-length but finite-energy signal can be decomposed from the point of view of function spaces, and how this decomposition, leading to a unique set of DWT coefficients, can be obtained through the application of a filter bank to the signal. An analysis (decomposition) filter bank is a set of filters, often comprising two types of filters (a lowpass and a highpass) that iteratively separate the input signal into disjoint frequency bands. This kind of operation is known as subband coding. At any time, the original signal can be recombined using a corresponding synthesis (reconstruction) filter bank.

The description of Mallat's algorithm, a fast wavelet decomposition and reconstruction scheme introduced in 1988 by Stéphane Mallat, will subsequently lead us to the practical implementation of the DWT in the real-world case of a finite-length, sampled input signal. We will focus on orthonormal (ON; orthogonal and properly normalized), compact-support wavelet systems for this discussion. Mallat's algorithm is a tree-shaped structure that starting from the crown—the input signal—leads us to the DWT coefficients—the roots. It also allows for decomposing the signal into an approximation, describing its coarse features, and a number of

© Springer International Publishing Switzerland 2016
S.M. Alessio, *Digital Signal Processing and Spectral*
Analysis for Scientists, Signals and Communication Technology,
DOI 10.1007/978-3-319-25468-5_14

details, describing its finer structure, in full agreement with the concept of multiresolution analysis (MRA). If then we want to re-synthesize the signal from its DWT coefficients, we just have to climb the tree from the roots to the crown. Under the assumptions made above, the filters involved in the process are four FIR filters: a lowpass/highpass pair for decomposition and another pair for reconstruction. The lowpass/highpass reconstruction filter coincides, to within a time folding, with the corresponding analysis filter. The properties of the lowpass and highpass decomposition/reconstruction filters are strictly related to those of the scaling and wavelet function, respectively. Perfect reconstruction (PR) realizable filters are needed for the filter bank to work properly: these filters must satisfy a number conditions, which are reflected in constraints on the scaling and wavelet functions.

Nevertheless, even in the presence of these constraints, a number of degrees of freedom remain available to design different wavelet systems. Scaling and wavelet functions satisfying the above-mentioned conditions may be extraordinarily irregular, even fractal in nature. This may be an advantage in analyzing rough or fractal signals, but it is likely to be a disadvantage for the analysis of most signals. We will thus investigate the smoothness (differentiability) of the scaling and wavelet functions through the concept of vanishing moments. The most popular compact-support ON wavelet systems (Daubechies wavelets, symlets and coiflets) will then be described. Biorthogonal wavelet systems also exist, which behave as an orthogonal basis, except for the presence of a dual basis set. In a biorthogonal DWT, the relation between analysis and synthesis filters is not a simple time-reversal.

Orthogonality (or possibly biorthogonality) is optimal for many applications, but DWT theory also considers cases in which the basis system dictated by the problem at hand cannot, or should not, be made orthogonal/biorthogonal. We will briefly touch upon these more general approaches, in which a generalization of the concept of basis is introduced that is known as a frame. A frame still allows representing a signal as a wavelet expansion, but the coefficients are not necessarily unique. In some cases, the lack of uniqueness can be advantageous because it opens up the possibility of choosing the coefficients that fit a certain application best, and also makes the representation of the signal less sensitive to noise.

A real-world example of signal DWT decomposition will be provided. The chapter ends with an appendix in which the various wavelet systems used for the DWT, or the CWT, or both transforms, are reviewed.

14.2 Wavelet Expansion Sets: Bases and Frames

DWT is a complex topic related on one side to the mathematical facets of functional analysis, and on the other side to engineering theories of subband coding and perfect reconstruction filter banks. We first approach the DWT from the functional analysis point of view.

In the appendix to Chap. 3 (Sect. 3.8) we provided a general description of the way in which a function in $L^2(\mathbb{R})$, i.e., a continuous-time energy signal, can be expanded. We will now focus on *wavelet expansions*. In the case of wavelets, the functions forming the expansion set are wavelet functions; actually, in many cases they are both *wavelet and scaling functions*, discretized according to a pair of indexes j and k.

The wavelet expansion of a signal $x(\theta)$ can be expressed as

$$x(\theta) = \sum_k \sum_j \alpha_{jk} f_{jk}(\theta),$$

where both j and k are integer, the $\{f_{jk}(\theta)\}$ form the expansion set and the $\{\alpha_{jk}\}$ form the set of expansion coefficients, which is called the *discrete wavelet transform* (DWT) of $x(\theta)$ (Daubechies 1992; Meyer 1993; Mallat 1989). The expansion is the *inverse discrete wavelet transform* (IDWT). We will assume the functions of the expansion set to belong to a suitable function space equipped with a finite norm: for instance,[1] $L^1(\mathbb{R})$. The set may correspond to an orthogonal basis, a biorthogonal basis, or a frame.

In the discretized-CWT analysis discussed in the previous chapter, orthogonality is unnecessary and even undesirable. The continuous wavelets used for the CWT usually lead, after discretization, to frames that are close to being tight. This is enough for the purposes for which the CWT is employed: it allows for writing a synthesis relation (inverse transform) and therefore being able to reconstruct the signal from its CWT samples; it also means that a generalized Parseval's theorem holds. This behavior is referred to as *quasi-orthogonality*. The conditions for quasi-orthogonality are that the wavelet $\psi(\theta)$ must be concentrated in time and must be an oscillating function with zero mean and a bandpass spectrum (Chap. 13). Nevertheless, an arbitrary discretization choice for scale factors, with several voices per octave, and a maximally dense discretization of time delays lead to a remarkable redundancy in the signal's representation: we are very far from that "economy" that a true orthogonal basis can guarantee. We have to store and process much more than the minimum number of coefficients that would be strictly necessary for a proper reconstruction.

This economy becomes important in typical DWT-related techniques like signal compression and de-noising. Orthogonality (possibly biorthogonality) of the basis is an important point in the DWT. Sampling of the time delay-scale continuum is normally much sparser than in the CWT, and the sparser the sampling, the more severe the constraints that must be imposed on the scaling and wavelet functions. The key questions here are the following: if we discretize in time and scale in a very "economic" way (*critical sampling*), can we get a true orthogonal basis? What are the characteristics that the basis functions must possess to achieve the desired DWT features, i.e., *economy* in the signal's representation coupled with the possibility of *perfect reconstruction* of the signal from its DWT coefficients?

---

[1] Note that $L^1(\mathbb{R})$ is more restrictive than $L^2(\mathbb{R})$, because an absolutely integrable function is also square-integrable, but the converse is not necessarily true.

Actually, the last question must be reformulated. Scaling and wavelet functions are completely implicit in the DWT: we do not design the functions but two digital lowpass filters and two digital highpass filters associated to the scaling and wavelet functions, respectively. These filters dictate the shape of $\varphi(\theta)$ and $\psi(\theta)$. Iterated application of a lowpass/highpass pair of filter constitutes a filter bank able to decompose the signal and produce its DWT coefficients. A distinct filter bank allows for reconstructing the signal. A perfect signal reconstruction cannot be achieved arbitrarily choosing $\varphi(\theta)$ and $\psi(\theta)$, but only starting from the filters and designing them properly. Given the filters, $\varphi(\theta)$—provided it exists—and $\psi(\theta)$ can univocally be found by numerical procedures. In most cases, the functions $\varphi(\theta)$ and $\psi(\theta)$ involved in the DWT do not even possess an explicit functional form: they have no analytical expression. So, how must we design DWT filters to achieve the desired DWT features?

The answer to the above questions is that we can get an orthogonal DWT, provided that we choose filters that satisfy very restrictive conditions. However, even doing so, some freedom remains: the wavelet expansion set is not unique, and there are many different scaling/wavelet function sets—often collectively referred to as *wavelet systems*—that can be constructed and used, and each of them has its own features, its advantages and drawbacks in applications.

Both the continuous wavelets used in the CWT and the discrete wavelets used in the DWT are wavelet systems; however, one important difference between them lies in the greater or lesser severity of the constraints that are imposed. These constraints are mild in the CWT, and much more stringent in the DWT. Another difference is that the discrete wavelets used for the DWT are naturally associated with *digital filter banks*, while the continuous wavelets used for the CWT are not. The STFT and the CWT can be related to filters banks, but these are analog in theory, and become digital only after discretization, related to sampled-signal applications. The DWT filter banks stem from theory as digital entities, and this naturally leads to the practical implementation of the DWT of a sampled signal as the application of digital filter banks to the signal; this is not the case in the CWT. CWT and DWT are characterized by this complete difference in the respective implementation schemes. The reader is referred to Sect. 14.9 and to the appendix of this chapter for the description of the most popular wavelet systems.

Virtually all wavelet systems and related expansions have the following characteristics:

- a wavelet system is a set of building blocks to construct or represent a signal. It is a two-dimensional expansion set (usually a basis) for some class of signals; the wavelet expansion of a signal maps it into a two-dimensional array of coefficients;
- the wavelet expansion gives a time-frequency localization of the signal, and most of the energy of the signal is well represented by a (more or less) modest number of expansion coefficients; this feature is referred to as the greater or lesser *sparseness* of the wavelet representation;

- the calculation of the expansion coefficients from the signal can be done efficiently. In the DWT, the total complexity of the transform is linear with respect to the size of the data, with a constant term that grows linearly with respect to the length of the filters used (see, for instance, Misiti et al. 2007). This is really remarkable, since the DWT complexity is lower than that of the FFT.

Wavelet expansions and transforms have proven to be very effective in analyzing and processing a very wide class of signals and phenomena. What are the properties that give this effectiveness? We have already discussed, while describing the CWT, the very appealing MRA property of wavelet techniques: we can now explicitly extend our remarks to the DWT expansion, saying that it allows an accurate local description and separation of signal characteristics.[2] Two more properties are, however, particularly important in the DWT:

- wavelet expansions of many signals have coefficients $\{\alpha_{jk}\}$ that drop off rapidly with j and k for a large class of signals. Therefore the signal can be efficiently represented by a small number of them. This *sparseness* property is related to the wavelet system being an *unconditional basis* (Sect. 3.8), and it is why wavelets are so effective in signal compression and de-noising (Burrus et al. 1998);
- wavelets are adjustable and adaptable; since there is not just one wavelet system, they can be designed to fit individual applications.

14.3 Elements of DWT Theory

In the discretized CWT (Chap. 13) we adopted arbitrarily dense dyadic scales. For the DWT it seems natural to adopt "economic" scaling operations of the type $a = a_0^{\pm j}$, with $j \in \mathbb{Z}$. Obviously we may write $a = a_0^{+j}$ or $a = a_0^{-j}$ indifferently, covering anyway the same range of scales. Most often $a_0 = 2$ is chosen, so that we get dyadic scales with octaves only and no voices.[3] The choice $a = a_0^{-j}$ means *scanning scales from the largest to the smallest*, as j increases from $-\infty$ to $+\infty$. This choice is the standard one in discussing MRA from the point of view of function spaces. Instead, the choice $a = a_0^{+j}$ means scanning scales from the smallest to the largest and

---

[2] To be precise, we must remark that while the continuous wavelets involved in the CWT are subject to the uncertainty principle of Fourier analysis, the discrete wavelet systems involved in the DWT, often defined through the associated digital filters, do not. However, the DWT does share the MRA property of all forms of wavelet transform: discrete wavelet bases can be shown to possess the MRA property by introducing the concept of nested spanned function spaces (Sect. 14.3).

[3] There is an interesting correspondence between wavelet notation and musical notation. In a musical score, each note specifies a frequency and a position in time by its vertical and horizontal placements, respectively, in a way that closely resembles a wavelet signal representation, except that it has fractional jumps in frequency. An article written in 1994 by G. Strang for a nontechnical audience (American Scientist, vol. 82, pp. 250–255) clearly explains this similarity.

is the option that we originally selected for the CWT, and also the standard one for describing how filter banks work (Sect. 14.5). Since we will start our discussion from function spaces, we now set

$$a = a_0^{-j}, \qquad j \in \mathbb{Z}.$$

As for the discretization in time delay b, the sampling theorem (Sect. 4.2) suggests broader and broader steps as the scale increases, exactly as we would do passing from a microscope (tiny steps to observe details of a small object) to a spyglass (large steps to observe the general features of an extended object). Therefore we set

$$b = ka = ka_0^{-j},$$

where k is an integer. We can thus write a generic *scaled and shifted discrete wavelet* as

$$\psi_{jk}(\theta) = a_0^{j/2} \psi \left(\frac{\theta}{a_0^{-j}} - k \right) = a_0^{j/2} \psi \left(a_0^j \theta - k \right) =$$
$$= a_0^{j/2} \psi \left(\frac{\theta - b}{a_0^{-j}} \right) = a_0^{j/2} \psi \left[a_0^j \left(\theta - b \right) \right],$$

where $\psi(\theta)$ is some *prototype wavelet function*, θ is continuous adimensional time, j is the scale index, k is the time-delay index, and the factor $a_0^{j/2}$ ensures wavelet energy normalization, i.e., equal energy for wavelets of any scale. To take into account scaling operations only we would write

$$\psi_j(\theta) = a_0^{j/2} \psi \left(\frac{\theta}{a_0^{-j}} \right) = a_0^{j/2} \psi \left(a_0^j \theta \right).$$

With a sampling scheme of this type, we must find a $\{\psi_{jk}(\theta)\}$ set ensuring, for a finite energy signal $x(\theta)$, a satisfactory reconstruction from the transform coefficients, which for the moment we indicate by W_{jk}:

$$x(\theta) = \sum_j \sum_k W_{jk} \psi_{jk}(\theta).$$

This is the general inverse wavelet transform definition, which is valid not only for the DWT, but also for the discretized CWT, while $\{W_{jk}\}$ are the elements of the wavelet transform matrix. The functions $\psi_{jk}(\theta)$ are a finite or infinite set that may constitute an orthogonal basis, a biorthogonal basis, or a frame.

Orthogonality of the basis would require that all (real) $\psi_{jk}(\theta)$ satisfy the condition on inner products[4]

$$\langle\psi_{jk}(\theta)\psi_{j'k'}(\theta)\rangle = \int \psi_{jk}(\theta)\psi_{j'k'}(\theta)\mathrm{d}\theta = 0$$

for $j \neq j'$ and $k \neq k'$. The DWT coefficients could then be calculated as inner products:

$$W_{jk} = \langle x(\theta)\psi_{jk}(\theta)\rangle.$$

Normality would further require that all $\psi_{jk}(\theta)$ satisfy the condition on inner products

$$\langle\psi_{jk}(\theta)\psi_{j'k'}(\theta)\rangle = \int \psi_{jk}(\theta)\psi_{j'k'}(\theta)\mathrm{d}\theta = 1$$

for $j = j'$ and $k = k'$. *Orthonormality* would thus require

$$\langle\psi_{jk}(\theta)\psi_{j'k'}(\theta)\rangle = \int \psi_{jk}(\theta)\psi_{j'k'}(\theta)\mathrm{d}\theta = \delta(j - j')\delta(k - k').$$

Wavelet theory leads to the following results:

1. if the set of scales is dense, and if the minimum Δb is small, the information in the transform domain is redundant, and the reconstruction takes place under non-restrictive conditions on the form of the wavelet system; this is the case of the discretized CWT;
2. if the sampling of scales is sparse and close to, or coincident with, the critical one, as in the DWT, a true orthogonal basis is obtained only for very special choices of the wavelet system. Provided that a proper wavelet system is adopted, the critical sampling scheme corresponds to setting $a_0 = 2$.

The theory of *wavelet frames* (see, e.g., Mallat 1989) covers these two extreme cases and all the intermediate situations, allowing us to balance redundancy and restrictions on the wavelet form, under the constraint that the reconstruction be possible and therefore $x(\theta)$ be adequately represented by the coefficients $\{W_{jk}\}$, i.e., that the information embedded in them be sufficient to reconstruct $x(\theta)$ with good accuracy.

The simplest formulation of the DWT problem includes *two types of functions for the basis set*: the *scaling function* $\varphi(\theta)$ and the *wavelet function* $\psi(\theta)$. We will start by defining the scaling function and later will define the wavelet function in terms of the former. In the final expansion formula, only integer translations of the scaling function, indicated by $\varphi(\theta - k)$, will come into play, while for the wavelet function

---

[4]The signals we are considering are real, and in the DWT they are normally expanded using real functions. For this reason, from now on we will use the notation which is appropriate for real functions, thus writing the inner product without any conjugation sign.

both translation and scaling operations, denoted by $\psi\,[(\theta/a) - k]$, will be used.[5] We will restrict ourselves to the critical sampling scheme ($a_0 = 2$).

Let $L^2(\mathbb{R})$ be the ensemble of all signals that can be represented by the considered expansion, i.e., the span of the expansion set. We then take a *scaling set*

$$\varphi_k(\theta) = \varphi(\theta - k), \quad k \in \mathbb{Z}, \quad \text{with} \quad \varphi(\theta) \in L^1(\mathbb{R}), \quad k = (-\infty, +\infty),$$

where $\varphi(\theta)$ is some *prototype scaling function* of which we consider shifts only. We introduce V_0 as the subspace of $L^2(\mathbb{R})$ spanned by this scaling set and write

$$x(\theta) = \sum_k c_k \varphi_k(\theta) \quad \text{for any} \quad x(\theta) \in V_0 = \text{Span}_k\,\{\varphi_k(\theta)\},$$

with c_k as expansion coefficients.

At this stage of the discussion it is, however, convenient to include both translations and scalings of $\varphi(\theta)$. Only later will the scalings of $\varphi(\theta)$ disappear. We thus generate the *two-dimensional family of scaling functions*

$$\varphi_{jk}(\theta) = 2^{j/2}\varphi\left(2^j\theta - k\right).$$

Note that pure scalings of $\varphi(\theta)$ would be indicated by

$$\varphi_j(\theta) = 2^{j/2}\varphi\left(2^j\theta\right).$$

The factor $2^{j/2}$ ensures energy normalization for the scaling functions. The members of this family correspond to subspaces

$$V_j = \text{Span}_k\,\{\varphi_{jk}(\theta)\} = \text{Span}_k\,\{\varphi_k\left(2^j\theta\right)\},$$

so we can write, for any specific V_j,

$$x(\theta) = \sum_k c_{jk}\varphi_{jk}(\theta) \quad \text{for any} \quad x(\theta) \in V_j.$$

Observe that the expansion coefficients c_{jk} now bear two indexes, j and k.

For $j > 0$, V_j is wider than V_0, because $a_j = 2^{-j}$ is smaller than the scale factor $2^0 = 1$ corresponding to V_0. The function $\varphi_{jk}(\theta)$ is narrower in time and translated

---

[5] The prototype (unscaled) wavelet function is what we called the "mother wavelet" in the previous chapter. The prototype scaling function is sometimes referred to as the "father wavelet". Note that in order to simplify the notation, here we dropped the subscript adopted in the previous chapter to indicate the mother wavelet: we wrote $\psi(\theta)$ instead of $\psi_0(\theta)$.

Fig. 14.1 The subspaces V_j of $L^2(\mathbb{R})$, with $j \in (-\infty, +\infty)$, are nested spanned spaces

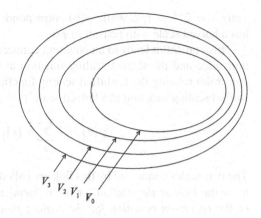

V_3 V_2 V_1 V_0

by smaller steps, so that it can represent signals $x(\theta)$ exhibiting finer-scale details.[6] Thus we have infinite spaces V_j, becoming wider and wider as j increases from $-\infty$ to $+\infty$, while scales 2^{-j} become smaller and smaller; $V_{+\infty}$ is as wide as the whole $L^2(\mathbb{R})$; $V_{-\infty}$ is empty, i.e., $V_{-\infty} = \emptyset$. Signals with finer and finer details can be represented as j increases.

Of course, if a signal with a given degree of detail can be represented in some V_j, then all the signals that only exhibit coarser details can also be represented in V_j. Thus in the DWT the MRA of the space $L^2(\mathbb{R})$ relies on the requirement that the V_j are *nested spanned spaces*, as illustrated in Fig. 14.1:

$$\ldots \subset V_{-2} \subset V_{-1} \subset V_0 \subset V_1 \subset V_2 \ldots L^2(\mathbb{R}),$$

where the symbol $\subset$ means, for example, that V_1 is contained in V_2: in general, $V_j \subset V_{j+1}$ for all integer j, with $V_{-\infty} = \emptyset$ and $V_{+\infty} = L^2(\mathbb{R})$.

Since the space that contains signals with fine-scale details also contains less detailed signals, i.e., signals with lower *time resolution*, if the *time support* of the functions of the set is halved, and therefore translations take place at steps of halved width, the set will become able to represent exactly a wider class of signals than before, or to represent any signal with better accuracy. In this general discussion, the expression "support" can be meant in a strict sense, i.e., we can assume the time support of the functions of the set to be compact, but also in a wider sense, meaning that most of the energy of the functions is confined in a time interval, outside of which the variability of the functions drops to zero.

The subspaces V_j must therefore satisfy a natural scaling condition, according to which if $x(\theta) \in V_j$, then $x(2\theta) \in V_{j+1}$. This nesting of spaces is obtained by requiring that if $\varphi(\theta) \in V_0$, then $\varphi(\theta) \in V_1$ be also true, V_1 being the space spanned by $\varphi(2\theta)$. Indeed, $\varphi(2\theta)$ is a compressed version of $\varphi(\theta)$; $\varphi(2\theta)$ corresponds to a

---

[6]For $j < 0$, the contrary would be true: only coarser features could be represented, and V_j would be narrower than V_0.

scale $a = 2^{-1} = 1/2$, while $\varphi(\theta)$ corresponds to a scale $a = 2^0 = 1$. Thus, $\varphi(2\theta)$ has a halved scale with respect to $\varphi(\theta)$.

This constraint leads to an equation connecting the unshifted scaling function at unit scale and the shifted scaling function at the halved scale. In reality, this is a constraint relating the unshifted scaling function at a given, unspecified scale to the shifted scaling function at a halved scale[7]:

$$\varphi(\theta) = 2 \sum_k h_s[k]\varphi(2\theta - k).$$

The two scales connected by this link are called *twin scales*. The equation is referred to as the *twin-scale relation*, or the *refinement equation*, or the *dilation equation*, or the *recursion equation for the scaling function*. This equation states that $\varphi(\theta)$ is obtained by a convolution of $\varphi(2\theta)$ with the impulse response $h_s[n]$ of a filter called the *scaling filter*. Given a scaling function $\varphi(\theta)$, $h_s[n]$—provided it exists—is a unique sequence, and is a series of real or complex numbers; conversely, designing $h_s[n]$ properly we can get—up to normalization—a unique function $\varphi(\theta)$ with some desired properties.

As an example of twin-scale relation, we may consider the Haar scaling function, which is defined as the real function

$$\varphi(\theta) = \begin{cases} 1 & \text{for } 0 < \theta < 1, \\ 0 & \text{elsewhere.} \end{cases}$$

The recursion equation in this case reduces to $\varphi(\theta) = \varphi(2\theta) + \varphi(2\theta - 1)$, since for the Haar system,

$$h_s[k] = \frac{1}{2} \text{[1 1]}, \quad h_{LR}[k] = \frac{1}{\sqrt{2}} \text{[1 1]},$$

where [1 1] and [1 -1] represent row vectors.

The Haar system is such that integer translates $\varphi(\theta - k)$ of the basic scaling function span the space of piecewise-constant functions over integers. This scaling function satisfies the twin-scale relation, since $\varphi(\theta)$ can be built using the scaling function at halved scale, as qualitatively shown in Fig. 14.2.

At this point in the discussion it is convenient to call upon the set of wavelet functions $\psi_{jk}(\theta)$. Their spans are the *differences* between contiguous V_j subspaces. For convenience, we will now *assume orthogonality for both scaling and wavelet sets*, though keeping in mind that this requirement is not essential. Let us call W_j the *orthogonal complement* to V_j in V_{j+1}: in symbols,

$$V_{j+1} = V_j \oplus W_j,$$

---

[7]Note that in this equation we should actually write $\eta = \theta/a$ in place of θ, since the larger scale is unspecified. However, the one presented here is the standard way in which the relation is written in literature.

Fig. 14.2 The Haar scaling
function $\varphi(\theta)$ can be built
using the Haar scaling
function at halved scale

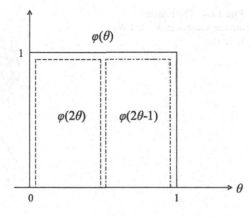

Fig. 14.3 W_j is the
orthogonal complement
to V_j in V_{j+1}

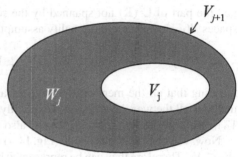

as shown in Fig. 14.3. All functions $\in V_j$ are orthogonal to those $\in W_j$, so that

$$\langle \varphi_{jk}(\theta)\psi_{jl}(\theta)\rangle = \int \varphi_{jk}(\theta)\psi_{jl}(\theta)\mathrm{d}\theta = 0$$

for all the appropriate values of the integers j, k, and l.

Now, what values of j do we need to consider for the expansion of a signal $x(\theta) \in L^2(\mathbb{R})$? We will start from a minimum value $j = j_0$, typically $j_0 = 0$, and then let j increase, so as to begin with the maximum scale of interest and subsequently descend from large to small scales. In practice the subspace V_0 is chosen in such a way to represent the coarsest features of interest in the signal. All the features corresponding to scales larger than this upper bound, i.e., to subspaces V_j with $j < 0$, are not considered in an explicit way. The nested spanned spaces $\ldots \subset V_0 \subset V_1 \subset V_2 \subset \ldots L^2(\mathbb{R})$, with $V_1 = V_0 \oplus W_0$, $V_2 = V_1 \oplus W_1 = V_0 \oplus W_0 \oplus W_1$, etc., allow us to write, as illustrated in Fig. 14.4,

$$L^2(\mathbb{R}) = V_0 \oplus W_0 \oplus W_1 \oplus W_2 \oplus W_3 \ldots,$$

Fig. 14.4 The relation
among subspaces V_j and W_j
in $L^2(\mathbb{R})$

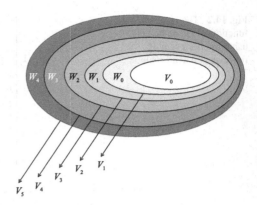

i.e., the part of $L^2(\mathbb{R})$ not spanned by the scaling space V_0 is spanned by wavelet
spaces W_j. Under the orthogonality assumption we have

$$V_0 \perp W_0 \perp W_1 \perp W_2 \perp W_3 \ldots,$$

meaning that all the mentioned subspaces are orthogonal to each other.[8] Since V_0
includes all the wavelet spaces not explicitly considered when stating that $L^2(\mathbb{R}) =
V_0 \oplus W_0 \oplus W_1 \oplus W_2 \oplus W_3 \ldots$, we can also write $V_0 = W_{-\infty} \oplus \cdots \oplus W_{-2} \oplus W_{-1}$.

Now, wavelet functions reside (Fig. 14.4) in the next wider scaling subspace, i.e.,
$W_0 \subset V_1$. Therefore they can be represented by a weighted sum of shifted versions of
$\varphi(2\theta)$. A second recursion equation, the *twin-scale relation for the wavelet function*,
can thus be written, connecting the unshifted wavelet function at unit scale and the
shifted scaling function at halved scale. Again, this is actually a constraint relating
the unshifted wavelet function at a given, unspecified scale to the shifted scaling
function at a halved scale:

$$\psi(\theta) = 2 \sum_k h_w[k]\varphi(2\theta - k).$$

---

[8]Note that the choice $j_0 = 0$ is arbitrary, and we might as well decide to chose a smaller starting
scale—a larger degree of detail, e.g. $j_0 = 10$, $a = 2^{-10}$—and write

$$L^2(\mathbb{R}) = V_{10} \oplus W_{10} \oplus W_{11} \oplus W_{12} \ldots,$$

or we might prefer a larger starting scale—a smaller degree of detail, e.g. $j_0 = -5$, $a = 2^5$—and
write

$$L^2(\mathbb{R}) = V_{-5} \oplus W_{-5} \oplus W_{-4} \oplus W_{-3} \ldots.$$

We could even start from $j_0 = -\infty$, i.e., from an infinitely large scale: since $V_{-\infty} = 0$, we would
then write

$$L^2(\mathbb{R}) = \cdots \oplus W_{-2} \oplus W_{-1} \oplus W_0 \oplus W_1 \oplus W_2 \oplus \cdots.$$

In this way we would eliminate the scaling function and would get an expansion of the signal on
the basis of wavelets solely. In the following discussion we will set $j_0 = 0$, unless explicitly stated
otherwise.

Fig. 14.5 The Haar wavelet function $\psi(\theta)$ may be built using the Haar scaling function at halved scale

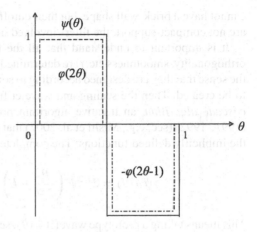

This equation states that $\psi(\theta)$ is obtained by convolution, using the shifted scaling function at halved scale and a new filter $h_w[n]$, which is called the *wavelet filter*. A qualitative example of the twin-scale wavelet relation using the Haar wavelet is given in Fig. 14.5. The Haar wavelet function is defined as

$$\psi(\theta) = \begin{cases} 1 & \text{for } 0 < \theta < \frac{1}{2}, \\ -1 & \text{for } \frac{1}{2} < \theta < 1, \\ 0 & \text{elsewhere.} \end{cases}$$

The recursion equation in this case is is $\psi(\theta) = \varphi(2\theta) - \varphi(2\theta - 1)$, since since for the Haar wavelet,

$$h_w[k] = \frac{1}{2} \, [1 \; -1], \quad h_{\text{LR}}[k] = \frac{1}{\sqrt{2}} \, [1 \; -1].$$

Note how in DWT theory, continuous-time quantities (functions) and discrete-time quantities (sequences) are mixed and deeply interconnected: in discrete time, we have digital filters and their impulse responses, $h_s[n]$ and $h_w[n]$; in continuous time, we have scaling and wavelet functions.

The scaling and wavelet filters introduced above can be shown to be strictly connected to one another. They are a lowpass and a highpass filter, respectively.[9] When the functions of the set are compact-support (in a strict sense, as in the case of the Haar system), all these filters are LTI stable and causal FIR systems with equal even lengths, which we will denote by $2M$. Moreover, they are *half-band filters*. If they were ideal brick-wall filters, their passband would occupy exactly half of the interval $[0, \pi]$. Realizable, causal FIR filters employed in the computational DWT

---

[9]This statement is not in contradiction with what we said in the previous chapter about the frequency response of the filters associated with complex analytic wavelets. Those wavelets are used for CWT only and do not have a corresponding scaling function. They actually act as passband filters.

cannot have a brick-wall shape, but their cutoff is $\pi/2$. When the functions of the set are not compact-support, the filters involved in the DWT may be IIR.

It is important to understand that all the features of $\varphi(\theta)$ and $\psi(\theta)$ (support, orthogonality, smoothness etc.) are determined by the filters. The filters come first, in the sense that they are designed according to some requirement on the wavelet system to be created. Then the scaling and wavelet functions are obtained using so-called *cascade algorithm*, an iterative algorithm proposed by Daubechies and Lagarias (1991, 1992) (see, e.g., Misiti et al. 2007) that provides excellent *approximations* of the implicitly defined functions. The complete set of wavelet functions is

$$\psi_{jk}(\theta) = 2^{j/2}\psi\left(\frac{\theta}{2^{-j}} - k\right) = 2^{j/2}\psi(2^j\theta - k).$$

This means taking a prototype wavelet $\psi(\theta)$, scaling time by $\theta \to \theta/a$ with $a = 2^{-j}$, and shifting time by k. Keep in mind that in our present approach

- large j means small scale, fine detail, i.e., *elevated time resolution* of the DWT, high frequency, and small translation steps;
- small j means large scale, coarse features, i.e., *poor time resolution* of the DWT, low frequency, and wide translation steps.

Moreover, recall that the factor $1/\sqrt{a} = 2^{j/2}$ is for normalization: when we scale the wavelet we also multiply it by $2^{j/2}$ to preserve its energy. If the prototype is normalized to unit energy, wavelets at all scales have unit energy. Under the present assumption of orthogonality, the system thus becomes ON.

The DWT that we desire must allow for *decomposition* (analysis) of a signal and subsequent satisfactory *reconstruction* (synthesis). Indeed, *perfect reconstruction* (PR) is the most crucial property for the DWT. We need an *invertible transform*; PR is equivalent to *exact invertibility*. The "gap" between decomposition and reconstruction may involve some manipulation of the DWT coefficients, for the purpose of denoising, compression etc. However, in this chapter we assume that this gap can be closed, and focus on the conditions under which PR can be achieved in the absence of any manipulation of the DWT coefficients. The two opposite procedures of analysis and synthesis actually involve, as we will see, not two but *four filters*. These four filters are iteratively applied to the signal, thus becoming the essential components of two distinct *two-band filter banks*: the analysis filter bank that operates a decomposition of $x(t)$, and the synthesis filter banks that performs reconstruction.

PR means that the reconstructed signal must be equal to the original one, except for a possible time delay connected to the use of causal filters. We will discuss later how the filters work on the signal and which characteristics they must have in order to make invertibility possible. For the moment, we just introduce them through their relation with the scaling and wavelet filters. We will distinguish the four filters by the subscripts LD, HD, LR, and HR, referring to "Lowpass for Decomposition", "Highpass for Decomposition", "Lowpass for Reconstruction", and "Highpass for Reconstruction", respectively.

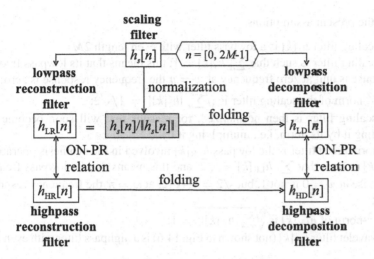

Fig. 14.6 The four PR filters used in the DWT for signal decomposition and reconstruction, in the case of LTI real and causal FIR filters corresponding to a real, compactly-supported wavelet system forming an ON basis, chosen in such a way that all filters have the same length. The relation of each of the four filters with the scaling filter and with the other three filters is shown. The subscripts LD, HD, LR, and HR refer to "Lowpass for Decomposition", "Highpass for Decomposition", "Lowpass for Reconstruction", and "Highpass for Reconstruction", respectively. ON-PR indicates perfect reconstruction in the selected orthonormal basis (see text)

These filters and their mutual relations are shown in Fig. 14.6. For simplicity, in Fig. 14.6 these relations are illustrated referring to *LTI real, causal PR FIR filters that correspond to a real, compactly-supported wavelets system forming an ON basis. The basis chosen in such a way that all filters have the same length.* These systems are commonly adopted for the so-called *fast wavelet transform* (FWT), a.k.a. *Mallat's algorithm*, presented in Sect. 14.5. They include Daubechies wavelets, symlets and coiflets (Sect. 14.9). Indeed,

- it can be shown that $\varphi_{jk}(\theta)$ and $\psi_{jk}(\theta)$ can form an orthogonal basis while having, at the same time, a *compact support*, i.e., they can be different from zero over a bounded and closed interval of the θ axis and be zero elsewhere;
- a property referred to as the *compact support property* establishes that compact-support scaling functions correspond to FIR filters. More precisely, a theorem (Burrus et al. 1998) states that if $\varphi(\theta)$ has compact support on an integer interval, which must have even length $2M$ (i.e., if $\varphi(\theta)$ is compactly supported on $0 \leq \theta \leq 2M-1$), and its integer translates $\varphi(\theta-k)$ are linearly independent, then $h_s[n]$, and $h_{LR}[n]$ that is nothing but a normalized version of $h_s[n]$, also have compact support over $0 \leq n \leq 2M-1$. Thus $h_{LR}[n]$ is an FIR filter with length $2M$. Orthogonality can further be required, in the form of additional constraints leading to the same length $2M$ for the remaining filters, $h_{HR}[n]$, $h_{LD}[n]$ and $h_{HD}[n]$. If the translates $\varphi(\theta-k)$ are not independent, or do not satisfy some equivalent restriction, $h_{LR}[n]$ can be IIR.

Under the present assumptions

- the scaling filter $h_s[k]$ is a lowpass filter with even length $2M$;
- the scaling filter is such that $\sum_k h_s[k] = 1$; this means that its lowpass frequency response is unit at zero frequency at $w = n$ the frequency response is zero;
- the ℓ^2-norm of the scaling filter is $\sqrt{\sum_k |h_s[k]|^2} = 1/\sqrt{2}$;
- the scaling filter is then normalized, for reasons that will soon become clear, dividing it by its norm, i.e., multiplying it by $\sqrt{2}$;
- the normalized filter is the lowpass $h_{LR}[k]$ involved in the synthesis operation;
- $h_{LR}[k]$ is such that $\sum_k h_{LR}[k] = \sqrt{2}$, and this means that its lowpass frequency response at $\omega = 0$ is not 1, but $\sqrt{2} = 1.4142$; at $\omega = \pi$ the frequency response is zero;
- the ℓ^2-norm of $h_{LR}[k]$ is $\sqrt{\sum |h_w[k]|^2} = 1$;
- the wavelet filter $h_w[k]$ (not shown in Fig. 14.6) is a highpass filter with even length $2M$;
- the wavelet filter has $\sum h_w[k] = 0$, as expected for a highpass filter whose frequency response must vanish at $\omega = 0$; at $\omega = \pi$ the frequency response is 1;
- its ℓ^2-norm is $\sqrt{\sum |h_w[k]|^2} = 1/\sqrt{2}$;
- the wavelet filter is then normalized multiplying it by $\sqrt{2}$, to give a new highpass filter $h_{HR}[k]$ involved in the synthesis operation;
- $h_{HR}[k]$ is such that its highpass frequency response at $\omega = \pi$ is not 1, but $\sqrt{2} = 1.4142$; at $\omega = 0$ the frequency response is zero, since $\sum h_{HR}[k] = 0$;
- $h_{HR}[k]$ has ℓ^2-norm $\sqrt{\sum_k |h_{HR}[k]|^2} = 1$. However, we do not need to explicitly consider the wavelet filter, because
- the reconstruction highpass filter $h_{HR}[k]$ can be directly derived from $h_{LR}[k]$, exploiting the fact that the two filters are a pair of ON-PR filters (see next section);
- the lowpass/highpass decomposition filters, which we denote by $h_{LD}[k]$ and $h_{HD}[k]$ respectively, are related to the corresponding reconstruction filters by a folding operation, i.e., a time reversal. Moreover, the highpass decomposition filter $h_{HD}[k]$ could be derived directly from $h_{LD}[k]$, because these two filters also are a pair of ON-PR filters.

PR filters are special realizable filters that ensure invertibility of the DWT when a signal $x(\theta)$ is synthesized from its DWT coefficients. They are such that the magnitude frequency response of $h_{HR}[n]$ is a mirror image of that of $h_{LR}[n]$ with respect to $\omega = \pi/2$; for the DWT algorithm to work properly, the filters must satisfy precise constraints, which will be described in Sect. 14.4. The fundamental filter—the one that is designed first—is $h_{LR}[n]$; $h_{HR}[n]$ then follows by the ON-PR-filters relation, which is (see, e.g., Qian 2001; Misiti et al. 2007)

$$h_{HR}[n] = (-1)^n h_{LR}[2M - 1 - n], \qquad n = [0, 2M - 1],$$

where the $(-1)^n$ term (modulation term) provides lowpass-to-highpass conversion. The inverse relation is

$$h_{\mathrm{LR}}[n] = -(-1)^n h_{\mathrm{HR}}[2M - 1 - n], \qquad n = [0, 2M - 1].$$

The corresponding decomposition filters are finally obtained by time folding. Note, however, that folding without delay would lead to anticausal filters; a proper delay is therefore applied, leading to the relations

$$h_{\mathrm{LD}}[n] = h_{\mathrm{LR}}[2M - 1 - n], \qquad n = [0, 2M - 1],$$
$$h_{\mathrm{HD}}[n] = h_{\mathrm{HR}}[2M - 1 - n], \qquad n = [0, 2M - 1].$$

The decomposition filters are also related by the ON-PR-filters relation:

$$h_{\mathrm{HD}}[n] = -(-1)^n h_{\mathrm{LD}}[2M - 1 - n], \qquad n = [0, 2M - 1],$$
$$h_{\mathrm{LD}}[n] = (-1)^n h_{\mathrm{HD}}[2M - 1 - n], \qquad n = [0, 2M - 1].$$

In summary,

- the lowpass reconstruction filter derives from the scaling filter via normalization;
- the highpass reconstruction filter is obtained from the lowpass reconstruction filter by modulation and folding;
- the decomposition filters are obtained as folded versions of the corresponding reconstruction filters.

At this point, it is useful to see a couple of examples of these filters. We select the Daubechies wavelet system, indicated by the acronym db, and the coiflets, indicated by coif. These wavelets can have different *orders* (Sect. 14.9), and we choose an order of 3 for db and an order of 2 for coif. Figures 14.7 and 14.8 show the impulse responses of the four filters used for signal decomposition and reconstruction in the two cases. For db3, $M = 6$ was chosen, so the plots extend over $n = [0, 2M - 1] = [0, 11]$; for coif2, $M = 12$ was selected, so the plots extend over $n = [0, 2M - 1] = [0, 23]$. Moreover, in the lower panels the magnitude and phase frequency responses of the filters are shown. Note that

$$\left| H_{\mathrm{LR}}(e^{j\omega}) \right| = \left| H_{\mathrm{LD}}(e^{j\omega}) \right|, \qquad \left| H_{\mathrm{HR}}(e^{j\omega}) \right| = \left| H_{\mathrm{HD}}(e^{j\omega}) \right|,$$

since a folding operation in the time domain leaves the magnitude frequency response unchanged, due to the properties of the DTFT. Note also that we are mainly interested in the passband phase response of each filter. For this reason, passband phase responses are drawn as *black curves*, while stopband phase responses are drawn as *gray curves*. For both wavelets, the LD filter is identified by *solid lines*, the HD filter by *dashed lines*, the LR filter by *dot-dashed lines*, and the HR filter by *dotted lines*. Observe how the frequency responses of the filters are *flat* near $\omega = 0$ and $\omega = \pi$. This is a typical feature of these filters. More details about them can be found in

Fig. 14.7 Impulse-response stem plots of the four filters used in the DWT for signal decomposition and reconstruction, in the case of the Daubechies wavelet of order 3. The magnitude and phase frequency responses are also shown. Passband phase responses are drawn as *black curves*, while stopband phase responses are drawn as *gray curves*. LD: *solid lines*; HD: *dashed lines*; LR: *dot-dashed lines*; HR: *dotted lines*

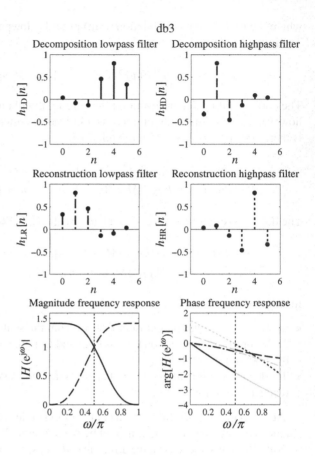

Strang and Nguyen (1996). Note also that the phase response is nearly linear in the coif2 case for all the four filters.

Our examples concerned an ON compactly supported wavelet system associated to equal-length FIR filters. Biorthogonal bases are also used in some DWT applications. Biorthogonal wavelet systems with compact support correspond to FIR filters and exact inversion of the transform is possible with them, but the filters involved in decomposition and reconstruction have different lengths, and the relation between the impulse responses of lowpass/highpass decomposition and reconstruction filters is not a simple time reversal.

We will now consider an example of decomposition of the $L^2(\mathbb{R})$-space using the Haar wavelet, which is nothing but a Daubechies wavelet with order 1. The space V_0 is spanned by $\varphi(\theta - k)$ and therefore represents the space of all $L^2(\mathbb{R})$ functions that are constant over intervals of unit length. V_1 is spanned by $\varphi(2\theta - k)$, V_2 is spanned by $\varphi(4\theta - k)$, and so on. In general, V_j is spanned by $\varphi(2^j\theta - k)$ and as j increases, arbitrary signals with shorter and shorter step-function behaviors can be approximated. Haar demonstrated that when $j \to \infty$, $V_j \to L^2(\mathbb{R})$. Now, suppose

Fig. 14.8 Impulse-response stem plots of the four filters used in the DWT for signal decomposition and reconstruction, in the case of the Coifman wavelet of order 2. The magnitude and phase frequency responses are also shown. Passband phase responses are drawn as *black curves*, while stopband phase responses are drawn as *gray curves*. LD: *solid lines*; HD: *dashed lines*; LR: *dot-dashed lines*; HR: *dotted lines*

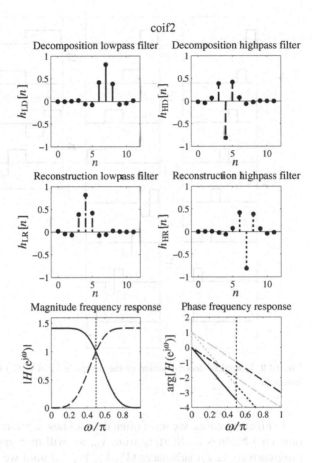

that for a particular purpose we need V_3 as the widest scaling subspace, i.e., suppose we are considering a signal $x(\theta) \in V_3$. Then, there are seven spaces involved in our decomposition, namely, V_0, W_0, V_1, W_1, V_2, W_2, V_3:

$$V_0 = \text{Span}_k \{\varphi(\theta - k)\}, \qquad W_0 = \text{Span}_k \{\psi(\theta - k)\};$$
$$V_1 = V_0 \oplus W_0 = \text{Span}_k \{\varphi(2\theta - k)\} = \text{Span}_k \{\varphi(\theta - k)\} \oplus \text{Span}_k \{\psi(\theta - k)\},$$
$$W_1 = \text{Span}_k \{\psi(2\theta - k)\};$$
$$V_2 = V_1 \oplus W_1 = \text{Span}_k \{\varphi(4\theta - k)\} = \text{Span}_k \{\varphi(2\theta - k)\} \oplus \text{Span}_k \{\psi(2\theta - k)\},$$
$$W_2 = \text{Span}_k \{\psi(4\theta - k)\};$$
$$V_3 = V_2 \oplus W_2 = \text{Span}_k \{\varphi(8\theta - k)\} = \text{Span}_k \{4\varphi(\theta - k)\} \oplus \text{Span}_k \{\psi(4\theta - k)\}.$$

The complete decomposition of V_3 using scaling and wavelet Haar functions is shown in Fig. 14.9, and can be expressed as $V_3 = V_0 \oplus W_0 \oplus W_1 \oplus W_2$. Had we desired to represent a function with finer details, e.g., $x(\theta) \in V_4$, we would have considered $V_4 = V_0 \oplus W_0 \oplus W_1 \oplus W_2 \oplus W_3$, and so on.

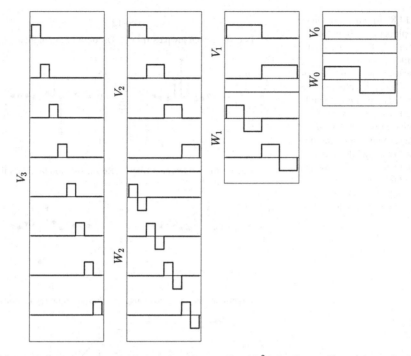

Fig. 14.9 Complete decomposition of the subspace V_3 of $L^2(\mathbb{R})$ using scaling and wavelet Haar functions

In practical cases, we most often do not know a priori the subspace a given function $x(\theta)$ belongs to. Starting from V_0, we will then approximate our function by progressively larger subspaces (V_1, V_2, V_3, ...) until we attain a good signal representation at some $j = j_{\max}$. For example, in Fig. 14.10a, a synthetic signal $x(\theta)$—a sinusoid with two discontinuities—is compared with its successive *approximations* obtained using the Haar set in the scaling spaces V_0–V_6. The approximation to $x(\theta)$ in V_6 is defined as the signal in the right-hand side of the following equation:

$$x(\theta) \approx \sum_k c_0[k]\varphi_{0k}(\theta) + \sum_{j=0}^{5}\sum_k d_j[k]\psi_{jk}(\theta);$$

similar expressions can be written for approximations in the narrower scaling subspaces, V_5–V_0. Observe the notation we adopted here for the coefficients. We are using both scaling and wavelet functions and need to distinguish *approximation coefficients* related to the scaling function (a.k.a. *scaling coefficients*), which we denote by $c_j[k]$, from *detail coefficients* related to the wavelet function (a.k.a. *wavelet coefficients*), which we indicate by $d_j[k]$. This notation stresses the fact that all coefficients involved in the expansion are treated as sequences, i.e., vectors: one sequence of approximation coefficients $c_0[k]$ and 6 sequences of detail coefficients $d_j[k]$, $j = [0, 5]$. Of

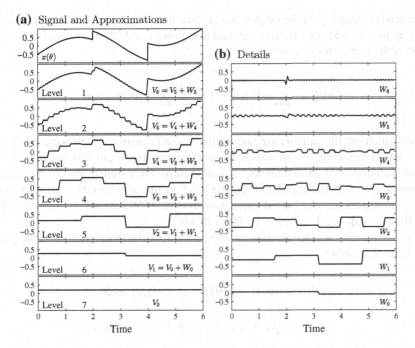

(a) Signal and Approximations   **(b)** Details

Fig. 14.10 **a** Successive approximations of a synthetic signal $x(\theta)$ containing a sinusoid with two discontinuities. The approximations are obtained using the Haar scaling/wavelet set in the scaling spaces V_0–V_6. Also shown are **b** the corresponding details in W_0–W_6. The level numbering on the *left-hand side* of the panels of column a, which holds for column b too, will prove useful in Sect. 14.5

course, also the approximation coefficients $c_1[k]$–$c_5[k]$ exist; we simply do not need them to form the approximation in V_6.

The accuracy of the approximation to $x(\theta)$ improves passing from V_0 to V_6. The coarsest approximation is in V_0, and represents the signal's mean value. The difference between one approximation and the following finer one represents a *detail* that, being neglected in one approximation, is then included in the subsequent one. The details of the synthetic signal are shown in Fig. 14.10b: for example, the detail in W_3 is the difference between the approximation in V_4 and the approximation in V_3. We can also see that the approximation in V_6 does not reproduce the signal $x[n]$ in a satisfactory way. If this were a real-world application, we would probably want to increase by 1 the subspaces included in the expansion (provided that this is possible; see Sect. 14.5). The inaccuracy of the approximation in V_6 is witnessed also by the fact that the detail in W_6 does not appear insignificant; it is evidently needed to reproduce the sharp transition that $x(\theta)$ exhibits near time 2. In terms of subspaces, the decomposition we operated is

$$V_6 = V_0 \oplus W_0 \oplus W_1 \oplus W_2 \oplus W_3 \oplus W_4 \oplus W_5.$$

If instead of stopping the decomposition at some finite j_{max} we consider all the values of j up to $j = +\infty$, i.e., all the (finer and finer) *time resolutions* with which we may look at the signal, we can write

$$x(\theta) = \sum_k c_0[k]\varphi_{0k}(\theta) + \sum_{j=0}^{+\infty} \sum_k d_j[k]\psi_{jk}(\theta).$$

In this equation, the equality sign holds, because all possible details were taken into account. The first right-hand sum is the approximation in V_0, to which all details are added (second right-hand term with a double sum). Only translations of the scaling function appear in the expansion: $\varphi_{0k}(\theta)$ has k as the only variable subscript. This equation represents the formal expression for the IDWT of the signal $x(\theta)$. The coefficients $c_0[k]$ and $d_j[k]$ are the DWT of $x(\theta)$.

In the present case, in which we assumed an ON basis, the coefficients are given by the inner products

$$c_0[k] = \langle x(\theta), \varphi_{0k}(\theta)\rangle = \int x(\theta)\varphi_{0k}(\theta)d\theta,$$

$$d_j[k] = \langle x(\theta), \psi_{jk}(\theta)\rangle = \int x(\theta)\psi_{jk}(\theta)d\theta.$$

The DWT coefficients completely describe the signal and can be used to approximate, filter, de-noise, compress the signal. Moreover, with an ON basis, Parseval's theorem holds, which subdivides the signal's energy among the DWT coefficients:

$$\mathcal{E} = \int |x(\theta)|^2 d\theta = \sum_{k=-\infty}^{+\infty} |c_0[k]|^2 + \sum_{j=0}^{+\infty}\sum_{k=-\infty}^{+\infty} |d_j[k]|^2.$$

This is equivalent to stating that the L^2-norm of $x(\theta)$ is equal to the ℓ^2-norm of the DWT coefficients.

In practice, the DWT coefficients must be calculated from a sampled version $x[n]$ of the signal $x(\theta)$, with $n = [0, N-1]$. This implies the existence of an upper limit to resolution: in a sampled signal we cannot see details falling between one sample and the next one. Now, if the scaling set $\varphi_j(\theta) = 2^{j/2}\varphi\left(2^j\theta\right)$ is "well chosen", then at a sufficiently high value of j—at sufficiently small scale—the scaling function becomes narrow enough to be approximated by a Dirac δ. At this particular j-value, the inner product simply samples the signal at all available values of discrete time, and $c_j[k] \approx x[k]$.

This observation suggests that in order to arrive at a computational implementation of the DWT decomposition, we will probably have to reverse our perspective and *start from the finest scale*. We will first define a j-th level of scaling coefficients using the samples of the input sequence. These j-th-level coefficients will serve as inputs to a recursive procedure with which to *scan, going backwards with decreasing values of j, larger and larger scales.* As we will see, this is actually the Mallat approach to a realizable DWT and IDWT (Sect. 14.5). The procedure uses only the impulse responses of the four filters introduced above, and the discrete-time signals representing the approximation and detail coefficients. The underlying functions $\varphi(\theta)$ and $\psi(\theta)$ remain completely implicit.

We must now get an idea of how the DWT decomposition can be seen as the application of a *filter bank*. For this purpose, we start again from the scaling recursion equation and assume that we know the impulse response $h_s[n]$ of the scaling filter (in the sense that we assume that the filter has already been designed according to some specifications). We thus look at $\varphi(\theta)$ as the solution of the equation, which will be unique up to normalization.

From the twin-scale relation for the scaling function, changing θ into $2^j\theta - k$ (i.e., scaling and shifting the time variable), adopting n as the summation index and recalling that $h_s[n] = (1/\sqrt{2})h_{\mathrm{LR}}[n]$, we can build the scaling set

$$\varphi(2^j\theta - k) = \sqrt{2}\sum_n h_{\mathrm{LR}}[n]\varphi\left[2\left(2^j\theta - k\right) - n\right] = \sqrt{2}\sum_n h_{\mathrm{LR}}[n]\varphi\left(2^{j+1}\theta - m\right),$$

with $m = 2k + n$. We can also write, eliminating the index n by using $n = m - 2k$,

$$\varphi(2^j\theta - k) = \sqrt{2}\sum_m h_{\mathrm{LR}}[m - 2k]\varphi\left(2^{j+1}\theta - m\right).$$

Now we consider the expansion of a signal $x(\theta)$ belonging to some scaling space, which we will call V_{j+1}. We can get the approximation in $V_{j+1} = \mathrm{Span}_k\left\{\varphi_{j+1\ k}(\theta)\right\}$ using scaling functions only:

$$x(\theta) = \sum_k c_{j+1}[k]2^{\frac{j+1}{2}}\varphi\left(2^{j+1}\theta - k\right).$$

If instead we use scaling functions at the next coarser j-th scale, we need to use also wavelets functions for the expansion, so as to provide the j-th detail that is not available if just scaling functions at the j-th scale are employed:

$$x(\theta) = \sum_k c_j[k]2^{\frac{j}{2}}\varphi\left(2^j\theta - k\right) + \sum_k d_j[k]2^{\frac{j}{2}}\psi\left(2^j\theta - k\right).$$

In the case of an orthogonal basis the coefficients are obtained by inner products as

$$c_j[k] = \langle x(\theta), \varphi_{jk}(\theta)\rangle = \int x(\theta)2^{\frac{j}{2}}\varphi\left(2^j\theta - k\right)d\theta.$$

Inserting $\varphi\left(2^j\theta - k\right) = \sqrt{2}\sum_m h_{\mathrm{LR}}[m - 2k]\varphi\left(2^{j+1}\theta - m\right)$ in the expression of $c_j[k]$ and considering scaling functions $\varphi(\theta) \in L^1(\mathbb{R})$, so as to be allowed to interchange the sum and the integral, we can write

$$c_j[k] = \int x(\theta) 2^{\frac{j+1}{2}} \sum_m h_{LR}[m - 2k] \varphi \left(2^{j+1}\theta - m\right) d\theta$$

$$= \sum_m h_{LR}[m - 2k] \int x(\theta) 2^{\frac{j+1}{2}} \varphi \left(2^{j+1}\theta - m\right) d\theta.$$

But the integral is the inner product of $x(\theta)$ with the scaling function at the $(j+1)$-th scale, so it equals $c_{j+1}[m]$. Working on the expression of $d_j[k]$ in a similar way we can obtain the following two equations (see Burrus et al. 1998):

$$c_j[k] = \sum_m h_{LR}[m - 2k] c_{j+1}[m],$$

$$d_j[k] = \sum_m h_{HR}[m - 2k] c_{j+1}[m],$$

or

$$c_j[k] = \sum_m h_{LD}[2k - m] c_{j+1}[m],$$

$$d_j[k] = \sum_m h_{HD}[2k - m] c_{j+1}[m],$$

where we have taken into account that the decomposition filters here are folded versions of the reconstruction filters. Note that for the sake of simplicity we neglected causality and simply indicated folding of $h_{LR}[m - 2k]$ as $h_{LR}[2k - m]$, etc.

Recalling that sharp time resolution means small scale, so that resolution is $\propto 2^j$ and increases with j, these equations tell us that the coefficients at the level of poorer resolution can be calculated from the coefficients at the next level of higher resolution. They are recursive relations leading to a branched-tree algorithm, i.e., the filter bank that implements the DWT.

Better insight into the operations appearing in the above equations can be gained by explicitly writing down the linear convolution

$$G_j[l] \equiv h_{LD}[l] * c_{j+1}[l] = \sum_m h_{LD}[l - m] c_{j+1}[m].$$

If we take only every other sample of $G_j[l]$ and set it equal to $c_j[k]$, i.e., if we set $c_j[k] = G_j[l]\big|_{l=2k}$, thus operating a downsampling by a factor of 2 of the result of the convolution, we arrive at the equation for $c_j[k]$ written above. Similar reasoning can explain the equation for $d_j[k]$. We thus understand that in the DWT decomposition, the two decomposition filters work on the signal by *convolution and subsequent downsampling by a factor of 2*. The application of each decomposition filter is immediately followed by downsampling. If it were not so, we would start with one input sequence and end up with two (longer) sequences, i.e., with over two times as many numbers to process. This would not be very practical. Downsampling by 2 is the solution, but in this way we would halve the energy. Energy must be preserved, and this is the reason why each impulse response is multiplied by $\sqrt{2}$ with respect to the original scaling and wavelet filters: in the energy, this gives a factor of 2 that restores the lost energy. This normalization is reflected in the lowpass and highpass frequency responses being $\sqrt{2}$ at $\omega = 0$ and π respectively, rather than being 1.

The fact that lower-resolution coefficients can be calculated from the higher-resolution ones in the way described above allows for a very efficient calculation of all the coefficients that are needed in any practical application. But once a decomposition has been carried out, how do we invert the transform? We need similar formulas to compute higher-resolution coefficients from lower-resolution ones. It can be shown that

$$c_{j+1}[k] = \sum_m c_j[m]h_{LR}[k - 2m] + \sum_m d_j[m]h_{HR}[k - 2m].$$

Let us examine the first sum of the equation written above, for the element $k = 0$:

$$\sum_m c_j[m]h_{LR}[-2m] = \cdots + c_j[-1]h_{LR}[+2] + c_j[0]h_{LR}[0] + c_j[1]h_{LR}[-2] + \cdots.$$

If we insert zeros between adjacent samples of $c_j[m]$ we obtain

$$\sum_m c_j[m]h_{LR}[-2m] = \cdots + c_j[-1]h_{LR}[+2] + 0 \times h_{LR}[1]$$
$$+ c_j[0]h_{LR}[0] + 0 \times h_{LR}[-1] + c_j[1]h_{LR}[-2] + \cdots.$$

If we define

$$\tilde{c}_j[m] = \ldots c_j[-1],\ 0,\ c_j[0], 0, c_j[-1] \ldots, \text{ i.e., } \tilde{c}_j[2m] = c_j[m] \text{ and } \tilde{c}_j[2m+1] = 0,$$

we get

$$\sum_m c_j[m]h_{LR}[-2m] = \sum_m \tilde{c}_j[2m]h_{LR}[-2m] = \sum_m \tilde{c}_j[m]h_{LR}[-m].$$

By extension to other values of k we can write the general formula

$$\sum_m c_j[m]h_{LR}[k - 2m] = \sum_m \tilde{c}_j[m]h_{LR}[k - m],$$

representing the convolution of $\tilde{c}_j[m]$ with the filter $h_{LR}[m]$. The operations needed in each reconstruction step thus become clear: they are *an upsampling by factor of 2 and a subsequent filtering*. Of course, to obtain $c_{j+1}[m]$ we must add the output of the $d_j[m]$ branch to the result of these two processing steps performed on $c_j[m]$.

In Sect. 14.5 we will describe in detail how to computationally realize the DWT decomposition through the efficient application of a filter bank, i.e., by iterating the single-step operations described above.

14.4 Perfect Reconstruction (PR) Filters

The theoretical DWT is applied to signals that are defined on an infinite length time interval. The actual, computationally realizable DWT is applied to sampled signals that are defined on a finite-length time interval. Moreover, while in theory lowpass and bandpass filters can have ideal top-hat frequency responses, in a computational environment we must employ realizable filters. Perfect reconstruction, which would be ensured in the theoretical case, requires some care when realizable filters are employed to operate on finite-length data sequences. In fact, a real-world DWT decomposition is always affected by *distortions related to filtering* and by *aliasing caused by downsampling*. These negative effects must be compensated by appropriately designed PR synthesis filters. Focusing on PR filters we make sure that the output of the synthesis bank is equal to the input signal, possibly up to some time delay; but filters satisfying restrictive conditions are needed to attain this result.

The properties of PR filters described above derive from a number of theorems that determining the necessary and/or sufficient conditions that the four filters $h_{LR}[n]$, $h_{HR}[n]$, $h_{LD}[n]$ and $h_{HD}[n]$ must satisfy for the solution $\varphi(\theta)$ of the recursion equation to exist and to possess, together with $\psi(\theta)$, some given properties, leading to a biorthogonal basis, an orthogonal basis, or a tight frame (see the appendix of Chap. 3). The conditions on PR filters can be expressed in the time domain, i.e., they can be specified in terms of constraints on the impulse responses of the filters, or in the frequency domain, which leads to constraints on the frequency responses/transfer functions. The mathematical properties of PR filters constitute a quite complex topic. For a rigorous mathematical discussion, the reader is referred to Burrus et al. (1998), Daubechies (1992). We will only mention a few fundamental results.

1. It can be shown that the filter $h_{LR}[n]$ must satisfy the so-called *linear admissibility condition*

$$\sum_n h_{LR}[n] = \sqrt{2}.$$

This is the weakest condition on $h_{LR}[n]$. No assumption of orthogonality of the basis functions is made to derive this constraint, nor are any other properties of $\varphi(\theta)$ other than a non-zero integral required. This equation shows that, unlike LCCDEs, not just any set of filter coefficients will support a solution to the scaling recursion equation.

The equivalent frequency-domain condition is

$$H_{LR}(e^{j0}) = \sqrt{2},$$

i.e., the filter's frequency response at zero frequency must equal $\sqrt{2}$, and therefore the filter cannot be highpass or bandpass.

2. It can be shown that if we require orthogonality, the following *frequency-domain orthogonality condition* on the lowpass reconstruction filter must hold:

$$\left|H_{\text{LR}}\left(e^{j\omega}\right)\right|^2 + \left|H_{\text{LR}}\left[e^{j(\omega+\pi)}\right]\right|^2 = 2.$$

In the time domain the same condition appears as the so-called *quadratic admissibility condition*, which characterizes *PR orthogonal filters*:

$$\sum_n h_{\text{LR}}[n]h_{\text{LR}}[n-2k] = \delta[k].$$

Indeed, it can be shown that in order for a solution $\varphi(\theta)$ of the scaling recursion equation to be orthogonal under integer translations, it is necessary that the coefficients of the recursive equation be orthogonal themselves after downsampling by a factor of 2. Note that the norm of the impulse response $h_{\text{LR}}[n]$ is set to 1 by this condition, and that this does not depend on any particular normalization of $\varphi(\theta)$. Not only must the sum of $h_{\text{LR}}[n]$ be $\sqrt{2}$, but for orthogonality of the solution, the sum of squares of $h_{\text{LR}}[n]$ must be 1, both independent of any normalization on the scaling function. If $\varphi(\theta)$ is normalized by dividing by the square root of its energy, then integer translates of $\varphi(\theta)$ are not only orthogonal, but also orthonormal. The quadratic admissibility condition also implies

$$\sum_n h_{\text{LR}}[2n] = \sum_n h_{\text{LR}}[2n+1] = 1/\sqrt{2},$$

i.e., not only must the sum of $h_{\text{LR}}[n]$ equal $\sqrt{2}$, but for orthogonality of the solution, the individual sums of the even and odd terms in $h_{\text{LR}}[n]$ must be $1/\sqrt{2}$, independent of any normalization of $\varphi(\theta)$.

3. If $h_{\text{LR}}[n]$ satisfies the linear and quadratic admissibility conditions, then it is also true (Burrus et al. 1998) that

$$\sum_n h_{\text{HR}}[n] = H_{\text{HR}}(e^{j0}) = 0,$$

$$\left|H_{\text{HR}}(e^{j\omega})\right| = \left|H_{\text{LR}}(e^{j[\omega+\pi]})\right|,$$

$$\left|H_{\text{LR}}(e^{j\omega})\right|^2 + \left|H_{\text{HR}}(e^{j\omega})\right|^2 = 2,$$

under the condition $\int \psi(\theta)d\theta = 0$. Note that the second equation implies

$$\left|H_{\text{HR}}(e^{j\pi})\right| = \left|H_{\text{LR}}(e^{j2\pi})\right| = \left|H_{\text{LR}}(e^{j0})\right| = \sqrt{2},$$

i.e., the highpass frequency response is $\sqrt{2}$ at $\omega = \pi$.

We will now try to explain the conceptual meaning of the quadratic admissibility condition for orthogonality. For convenience, we will actually start from a general biorthogonal basis set and only later will we impose orthogonality. Figure 14.11 illustrates a two-channel decomposition filter bank followed by a two-channel reconstruction filter bank. The scheme depicts a single stage of application of the two filters (lowpass and highpass) in each bank. In the decomposition, the input signal $x[n]$ is separated in the frequency domain into two subbands: the high-band $(\pi/2 \le |\omega| < \pi)$ and the low-band $(0 \le |\omega| < \pi/2)$. The corresponding output signals are $g[n]$ and $f[n]$, respectively. Each band is then downsampled by a factor of 2; let us call $v[n]$ the resulting signal in the low-band. The *vertical dashed line* symbolizes the gap between analysis and synthesis, in which simple transmission, or manipulation of the signals, can take place. Here we assume simple transmission. Reconstruction is performed via upsampling by a factor of 2, which generates $u[n]$ on the low-band channel. Finally, a further filtering in each band, followed by addition of the corresponding output signals, produces the reconstructed signal $\hat{x}[n]$. This structure was introduced in the 1980s by Stéphane Mallat, who discovered the connection between wavelets and filter banks, and is nothing but a single stage of what is known as *subband coding* (Vetterli and Kovačević 1995). The filters of Fig. 14.11 cannot be ideal brick-wall filters. They are realizable filters and their cutoff is not sharp. Moreover, recall that the input signal is a sampled, finite-length signal.

Perfect reconstruction (PR), apart from a possible time delay, can be achieved only if no information is lost. In order to understand the related issues, consider that according to the formulas reported in Sect. 6.9.1, a downsampling by a factor of 2 of a signal $f[n]$ leads to a signal

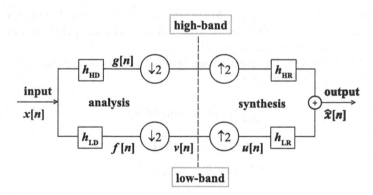

Fig. 14.11 Two-band analysis- and synthesis-filter banks in a general biorthogonal DWT. The scheme depicts a single stage of application of two filters (lowpass and highpass) in each bank. Decomposition is performed via filtering and subsequent downsampling by a factor of 2 in each band. The *vertical dashed line* symbolizes the gap between analysis and synthesis, in which we assume simple transmission. Reconstruction is performed via upsampling by a factor of 2 and subsequent filtering in each band

$$v[n] = f[2n] \qquad n = 0, 1, 2\ldots,$$

whose z-transform is

$$V(z) = \frac{1}{2} \left[F\left(z^{1/2}\right) + F\left(e^{j\pi}z^{1/2}\right) \right] = \frac{1}{2} \left[F\left(z^{1/2}\right) + F\left(-z^{1/2}\right) \right].$$

An upsampling by a factor of 2 of a signal $f[n]$ produces a signal given by

$$w[2n] = f[n], \qquad n = 0, 1, 2\ldots,$$
$$w[2n+1] = 0, \qquad n = 0, 1, 2\ldots,$$

whose z-transform is

$$W(z) = F(z^2).$$

On the unit circle we have

$$V\left(e^{j\omega}\right) = \frac{1}{2} \left\{ F\left(e^{j\omega/2}\right) + F\left[e^{j(\omega/2+\pi)}\right] \right\}$$

and

$$W\left(e^{j\omega}\right) = F\left(e^{j2\omega}\right).$$

The effects of these operations are visible in Fig. 14.12.

Figure 14.12a shows the spectrum $F(e^{j\omega})$ of the low-band signal $f[n]$, which is has band limits of $\pm\pi/2$, meaning that most of its energy is in $|\omega| < \pi/2$. The spectrum has a period of 2π. Due to downsampling, $F(e^{j\omega})$ is stretched in frequency by a factor of 2, so that its bandwidth doubles as $F(e^{j\omega})$ becomes $F(e^{j\omega/2})$; moreover, a shifted spectral image $F[e^{j(\omega/2+\pi)}]$ is added to $F(e^{j\omega/2})$; the resulting sum is multiplied by 1/2 to give the spectrum $V(e^{j\omega})$ of the downsampled signal (Fig. 14.12b). Since adjacent spectral copies overlay, *aliasing* occurs, which is represented by *gray triangles*.

If we applied an upsampling by a factor of 2 to the same signal $f[n]$ whose spectrum is shown in Fig. 14.12a, thus obtaining an upsampled signal $w[n]$, the spectrum $F\left(e^{j\omega}\right)$ would be compressed, and images of the compressed spectrum would appear, centered on $\pm\pi$. This is illustrated in Fig. 14.12c, where these images are represented by *dashed lines*. This effect is called *imaging*, and causes the period of the resulting spectrum $W(e^{j\omega})$ to be π rather than 2π.

Note that imaging is the opposite of aliasing. In aliasing, two input frequencies ω and $\omega+\pi$ give the same output. In imaging, one input frequency ω gives two outputs, one at frequency $\omega/2$ and another one at $\omega/2+\pi$. *Upsampling causes imaging while downsampling causes aliasing.*

However, the process described above is not what happens in the bank of Fig. 14.11: in the actual bank we first apply a downsampling by a factor of 2, thus obtaining $v[n]$, and then immediately upsample $v[n]$ by 2, generating $u[n]$. The

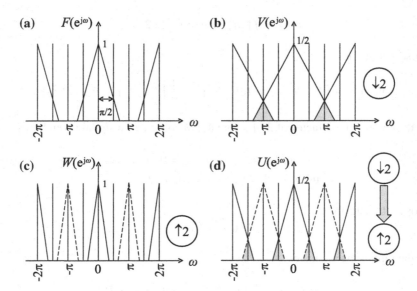

Fig. 14.12 **a** The spectrum $F(e^{j\omega})$ of a signal $f[n]$ that has most of its energy in the frequency range $|\omega| < \pi/2$. The period of the spectrum is 2π. **b** Aliasing produced by a downsampling of the signal $f[n]$ by a factor of 2. The downsampled signal is $v[n]$ and its spectrum is $V(e^{j\omega})$ (see text). Aliasing is highlighted by *gray triangles*. **c** Imaging produced by upsampling: the spectrum $W(e^{j\omega})$ of a signal $w[n]$ obtained upsampling $f[n]$ by a factor of 2. *Dashed lines* represent images of the compressed spectrum (see text). **d** The combined effects of downsampling and upsampling: the spectrum $U(e^{j\omega})$ of the signal $u[n]$ obtained downsampling $f[n]$ and then immediately upsampling the result. The spectrum is stretched by downsampling and compressed by upsampling, so that its shape is as in panel a, but the spectral images generated by upsampling (symbolized by *dashed lines*) produce aliasing, indicated by *gray triangles*

signal $u[n]$ has a spectrum (Fig. 14.12d) that also shows aliasing, since stretching due to decimation and compression due to upsampling compensate each other, but the spectral images generated by upsampling are also stretched and overlap. The corresponding z-transform and DTFT of $u[n]$ are

$$U(z) = \frac{1}{2}\left[V(z) + V\left(e^{j\pi}z\right)\right] = \frac{1}{2}\left[V(z) + V(-z)\right],$$
$$U\left(e^{j\omega}\right) = \frac{1}{2}\left\{V\left(e^{j\omega}\right) + V\left[e^{j(\omega+\pi)}\right]\right\}.$$

Downsampling and upsampling are also present along the high-band channel, and therefore similar considerations hold in that case.

It is worth noting that the operations described above represent what would happen in a single low-band channel of a hypothetical "two-channel filter bank without filters". This strange concept is useful to understand the effects of downsampling and upsampling, but in reality such a hypothetical bank would produce a very undesirable result: it would simply eliminate the odd-indexed samples of the original signal, and

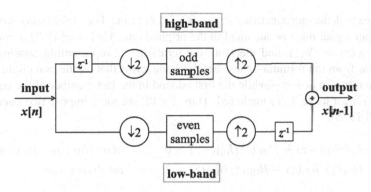

Fig. 14.13 A hypothetical "two-channel filter bank without filters": in order to retain the odd-indexed samples of the input signal in the high-band channel and the even-indexed ones in the low-band channel, we must insert delays in our scheme (see text)

with them, the possibility of reconstructing it later. The scheme can work if we insert delays by one time step, in the positions indicated in Fig. 14.13. With these delays, the low-band channel downsamples and upsamples; it produces a signal with all odd components of $x[n]$ replaced by zeros. This signal is delayed by one time step before output. The high-band channel delays $x[n]$ at the start, so that downsampling followed by upsampling removes the even-number components of $x[n]$, which have already been coded in the low-band channel. The combined output is just a delayed version of the input signal.

Going back to Fig. 14.11, we now know that there is aliasing in each channel. There will also be amplitude distortion and possibly phase distortion due to filtering. Therefore, the synthesis filters must be specially adapted to the analysis filters, in order to cancel the errors introduced by the analysis bank. The goal of this section is to discover the conditions for exact signal reconstruction. We need a synthesis bank which is the inverse of the analysis bank.

We can extend the downsampling and upsampling z-transform expressions written above to include the filtering operations appearing in Fig. 14.11: with $\hat{x}[n]$ indicating the output, and recalling that in the z-domain a filtering operation (convolution) becomes a multiplication, we can write

$$\hat{X}(z) = \frac{1}{2} H_{\mathrm{LR}}(z) \left[H_{\mathrm{LD}}(z) X(z) + H_{\mathrm{LD}}(-z) X(-z) \right] +$$

$$+ \frac{1}{2} H_{\mathrm{HR}}(z) \left[H_{\mathrm{HD}}(z) X(z) + H_{\mathrm{HD}}(-z) X(-z) \right] =$$

$$= \frac{1}{2} \left[H_{\mathrm{LR}}(z) H_{\mathrm{LD}}(z) + H_{\mathrm{HR}}(z) H_{\mathrm{HD}}(z) \right] X(z) +$$

$$+ \frac{1}{2} \left[H_{\mathrm{LR}}(z) H_{\mathrm{LD}}(-z) + H_{\mathrm{HR}}(z) H_{\mathrm{HD}}(-z) \right] X(-z),$$

where we took into account that $F(z) = H_{LD}(z)X(z)$, etc. For perfect reconstruction, the output signal must be identical to the original one: $\hat{x}[n] = x[n]$. Thus we must require $\hat{X}(z) = X(z)$, and therefore the terms in $-z$ representing aliasing must disappear from the formula—these are the same terms that on the unit circle would produce terms in $\omega + \pi$—while the first addend in the last member of the equation for $\hat{X}(z)$ must leave $X(x)$ unaltered. Thus, for PR we must impose two conditions (Vetterli 1986):

$$H_{LR}(z)H_{LD}(-z) + H_{HR}(z)H_{HD}(-z) = 0 \qquad \textit{aliasing cancellation,}$$
$$H_{LR}(z)H_{LD}(z) + H_{HR}(z)H_{HD}(z) = 2 \qquad \textit{no distortion}$$

(see also Qian 2001; Misiti et al. 2007; Mallat 2008).

Since we desire each individual filter to be causal, should we include an overall delay i.e., should we write $\hat{x}[n] = x[n - l]$ with some value of l? This would affect the way in which we write the no-distortion condition. The answer would be yes if we decided to impose causality at this point of the discussion. However, it is more convenient to impose causality later, when establishing the conditions for orthogonality. Thus we leave the no-distorion condition as it is.

Furthermore, we must observe that the scheme in Fig. 14.11 is a theoretical scheme. In the computational DWT, the situation is more complicated. Figure 14.14 shows an example of the operations actually included in a one-stage DWT decomposition employing causal filters. The number of samples that exist after each operation is also shown. The example refers to a signal which is $N = 256$ samples long. We imagine to decompose the signal at level 1 using a coiflet of order 2 (coif2; see Sect. 14.9) and to immediately reconstruct it without any intermediate manipulation of the coefficients. The causal FIR filters associated with coif2 have length $2M = 12$.

First, the input sequence is extended before entering the low- and high-branches (extension is indicated by the symbol E). This is because each filtering in the bank requires linear convolution between the input or some intermediate signal, which we will call $y[n]$, and the impulse response of a filter. But linear convolution asks for $y[-1]$, $y[-2]$, ... $y[-(2M-1)]$ that do not exist. Signal-extension schemes are applied on the signal boundaries to solve this issue (see Sect. 14.5.1). The extension inserts $2M - 1$ samples at the beginning and end of the record. Since the signal is treated like a vector of numbers, without any association with a corresponding pre-defined vector of discrete times, this produces what can be seen as a delay (a shift to the right) of $2M - 1$ samples. This is symbolized in Fig. 14.14 by the term $z^{-(2M-1)}$. The sequence's length becomes $N_E = N + 2(2M - 1)$. We assume that the extension is performed according to *half-point symmetry* (Sect. 14.5.1): $x[-1] = x(0)$, $x[-2] = x[1]$, etc.

We now focus on the low-branch, starting with decomposition. The extended sequence is filtered by the lowpass-decomposition filter via linear convolution. This produces a sequence with the length of a *full convolution*, i.e., $N_{LD} = N_E + 2M - 1$. It also produces some delay z^{-1}. Then, the initial and final transients ($2M - 1$ samples each) are eliminated: the resulting "clean convolution" is sometimes referred to as the *valid* part of the convolution. This operation is indicated by the symbol V. We are

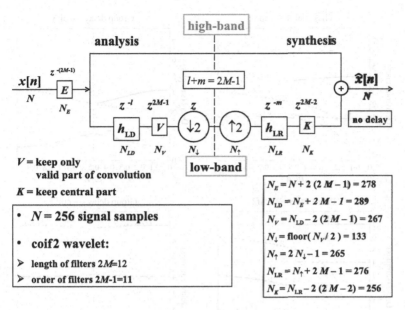

Fig. 14.14 Two-band analysis- and synthesis-filter banks in a one-stage DWT decomposition employing causal filters. The *vertical dashed line* symbolizes the gap between analysis and synthesis, in which we assume simple transmission. Reconstruction is performed via upsampling by a factor of 2 and subsequent filtering in each band

now left with $N_V = N_{LD} - 2(2M - 1)$ samples, and can look at this shift to the left of the vector, due to the elimination of its head, as an advance by $2M - 1$ samples, as described by the term z^{2M-1} in the scheme. Downsampling by a factor of 2 follows: since the first preserved sample is the second one, this gives a shift to the left of 1 sample, i.e., an advance by 1 step ($z^1 = z$). The signal length is essentially halved, but since we had an odd number of samples, the last sample is discarded. Now we have a vector of length $N_\downarrow = \text{floor}(N_V/2)$, where floor means rounding up $N_V/2$ to the nearest lower integer.

Immediately after downsampling, the signal is upsampled by a factor of 2, inserting a zero between the first and second sample, between the second and third sample, etc., but not after the last one. A vector with length $N_\uparrow = 2N_\downarrow - 1$ is obtained. This vector is filtered by the lowpass-reconstruction filter via linear convolution. The output is a sequence with the length of a full convolution, i.e., $N_{LR} = N_\uparrow + 2M - 1$. Filtering also causes some delay z^{-m}. Finally, only the central part of this output sequence is kept (see the symbol K in the scheme). This is done by discarding $2M - 2$ samples on each side, for a total of $2(2M - 2)$ samples. We end up with N_K samples. The shift to the left by $2M - 2$ samples (advance) produces a term z^{2M-2}.

If we progressively substitute N_{LR} in N_K, then $N_\uparrow$ in the resulting expression, etc., and keep into account that $2N_\downarrow = 2\text{floor}(N_V/2) = N_V - 1$, we find $N_K = N$: the output sequence has the same length as the input sequence. If we add together all the exponents of the delay terms, we get an overall delay of $l + m - (2M - 1)$ time steps.

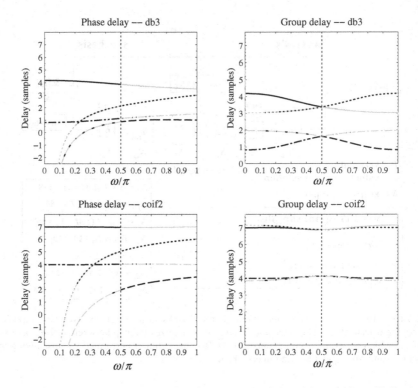

Fig. 14.15 Phase and group delays of the four filters related to db3, and the same for coif2. Passband delays are drawn as *black curves*, while stopband delays are drawn as *gray curves*. LD: *solid lines*; HD: *dashed lines*; LR: *dot-dashed lines*; HR: *dotted lines*

But for coiflets, $l \simeq 4M/3 - 1$ and $m \simeq 2M/3$, i.e., 7 and 4 for coif2, which has $M = 6$. This can be seen in Fig. 14.15, were, for db3 and coif2 wavelets, the phase and group delays are plotted , which are the negative ratio of the phase response of the filter to ω and the negative first derivative of the phase response with respect to ω, respectively. Note that in general, phase and group delays are not constant with PR filters: they vary with frequency. In Fig. 14.15, passband delays are drawn as *black curves*, while stopband delays are drawn as *gray curves*, to highlight that what matters is the filter's behavior in the passband. For both wavelets, the LD filter is identified by *solid lines*, the HD filter by *dashed lines*, the LR filter by *dot-dashed lines*, and the HR filter by *dotted lines*.

This means that we can set $l + m = 2M - 1$, and the overall delay produced by the bank vanishes. It is clear from the previous discussion that, on the contrary, if we adopted causal filters in the "theoretical bank" depicted in Fig. 14.11, we should expect a delay of $l + m = 2M - 1$ samples of the reconstructed signal with respect to the input one. We should then write the no-distortion condition as $H_{\mathrm{LR}}(z)H_{\mathrm{LD}}(z) + H_{\mathrm{HR}}(z)H_{\mathrm{HD}}(z) = 2z^{-(2M-1)}$. Nevertheless, for the moment we do not make this correction.

Now we will examine the no-aliasing and no-distortion constraints to deduce their consequences. They define biorthogonal filter banks; orthogonality is a special case of biorthogonality that requires imposing additional constraints. This will be done later.

1. The no-aliasing constraint can be satisfied by letting

$$H_{\text{LR}}(z) = H_{\text{HD}}(-z) \quad \text{and} \quad H_{\text{HR}}(z) = -H_{\text{LD}}(-z).$$

Recalling the modulation property of the z-transform (Sect. 3.2.5), in the time domain these equations imply

$$h_{\text{LR}}[n] = (-1)^n h_{\text{HD}}[n], \qquad h_{\text{HR}}[n] = -(-1)^n h_{\text{LD}}[n],$$

i.e., the reconstruction filters can be obtained from the decomposition filters by *alternating the signs* of the elements.

2. To see the consequences of the no-distortion condition, let us rewrite the above z-domain relations as

$$H_{\text{HD}}(z) = H_{\text{LR}}(-z) \qquad H_{\text{HR}}(z) = -H_{\text{LD}}(-z),$$

and substitute them into the no-distortion relation. This leads to

$$H_{\text{LR}}(z)H_{\text{LD}}(z) - H_{\text{LD}}(-z)H_{\text{LR}}(-z) = P_L(z) - P_L(-z) = 2,$$

where we defined $P_L(z) = H_{\text{LD}}(z)H_{\text{LR}}(z)$, which is the product of the transfer functions of two lowpass filters. It can be shown (see Qian 2001) that the above equation contrains odd-indexed samples of the corresponding impulse response $p_L[n]$, but leaves the even-indexed samples arbitrary. If we design some $P_L(z)$, we can later factorize it into $H_{\text{LD}}(z)$ and $H_{\text{LR}}(z)$. Obviously, not only there are many ways to design $P_L(z)$, but also there are many ways to factor it. As a consequence, the choice of the four filters not unique. Let us stress that the conditions we imposed do not ensure that the resulting filters form orthogonal filter banks and that the resulting wavelets are orthogonal wavelets. In general, the bank will be biorthogonal, and there will be two sets of scaling and wavelet functions.

As the above discussion indicates, there exist many possible designs for biorthogonal filter banks, even if we restrict our attention to compactly supported wavelet systems and consequently to FIR filters. Historically, the first system of biorthogonal wavelets, which was made popular by Ingrid Daubechies, is the so-called *Cohen-Daubechies-Feauveau* or CDF biorthogonal wavelet system (Daubechies et al. 1990), a.k.a. *B-spline biorthogonal wavelet system*, whose construction is based on B-splines (basis splines).[10] B-spline biorthogonal systems have a pair of

---

[10] A B-spline is a piecewise polynomial function in one independent variable, exhibiting knots or break-points. The number of internal knots is equal to the degree of the polynomial if there are no

analysis/synthesis wavelet functions and a pair of analysis/synthesis scaling functions. Associated filters have different length, which can also be odd. However, in computation these filters are padded with zeros, to augment them to equal even lengths. They have a nice property, namely, filter symmetry: the impulse responses of the filters are symmetric/antisymmetric around their mid-point. Symmetry can be an important property for wavelet systems, especially in image-related applications. As Daubechies explains in Chap. 8 of Daubechies (1992), for some applications symmetry does not matter at all, but in image coding, for example, "quantization errors will often be most prominent around edges in images; it is a property of our visual system that we are more tolerant of symmetric errors than asymmetric ones. In other words, less asymmetry would result in greater compressibility [of the image] for the same perceptual error. Moreover, symmetric filters make it easier to deal with the boundaries of the image. [...omissis...] Symmetric filters are often called *linear phase filters* in the subband-coding engineering literature. If a filter is not symmetric, then its deviations from symmetry is judged by how much its phase deviates from a linear function." So, *symmetry leads to linear phase*.

Given that two constraints on four filters do not determine them in a unique way, we can narrow the field by imposing *additional conditions for orthogonality*. We have already satisfied the no-aliasing condition, so we proceed with the no-distortion equation. We saw that the time-domain condition for orthogonality is the quadratic admissibility condition

$$\sum_n h_{LR}[n]h_{LR}[n - 2k] = \delta[k], \qquad k = 0, 2M - 1,$$

representing M bilinear or quadratic equations. It can be verified that the ON *causal* filters adopted for Mallat's algorithm, presented in Sect. 14.3 (see the discussion concerning Figs. 14.6, 14.7 and 14.8), satisfy this condition. Do they also satisfy the no-distortion condition in the z-domain? To be more precise, how does the no-distortion condition in the z-domain look like for these particular filters?

In order to see this, we z-transform three relations given in Sect. 14.3, making use of the z-transform properties of time reversal, time shift and modulation (Sect. 3.2.5):

---

knot multiplicities. A B-spline is a continuous function at the knots. For any given set of knots, the B-spline is unique, hence the name, B being short for Basis.

we obtain, with $n = [0, 2M - 1]$,

$$
\begin{aligned}
h_{\text{HR}}[n] = (-1)^n h_{\text{LR}}[2M - 1 - n] &\iff H_{\text{HR}}(z) = (-z)^{2M-1} H_{\text{LR}}(-z^{-1}), \\
h_{\text{LD}}[n] = h_{\text{LR}}[2M - 1 - n] &\iff H_{\text{LD}}(z) = z^{2M-1} H_{\text{LR}}(z^{-1}), \\
h_{\text{HD}}[n] = h_{\text{HR}}[2M - 1 - n] &\iff H_{\text{HD}}(z) = z^{2M-1} H_{\text{HR}}(z^{-1}).
\end{aligned}
$$

Substituting these equations into the no-distortion constraint leads to an expression in terms of $H_{\text{LR}}(z)$ alone:

$$
\begin{aligned}
H_{\text{LR}}(z) H_{\text{LD}}(z) &- H_{\text{LD}}(-z) H_{\text{LR}}(-z) = \\
H_{\text{LR}}(z) z^{2M-1} H_{\text{LR}}(z^{-1}) &- (-z)^{2M-1} H_{\text{LR}}(-z^{-1}) H_{\text{LR}}(-z) = 2.
\end{aligned}
$$

But since $2M - 1$ is odd, $-(-z)^{2M-1} = z^{2M-1}$ and we get the following particular form of the no-distortion constraint:

$$
\left[H_{\text{LR}}(z) H_{\text{LR}}(z^{-1}) + H_{\text{LR}}(-z^{-1}) H_{\text{LR}}(-z) \right] = 2z^{-(2M-1)}.
$$

Note how assuming causality together with orthogonality actually leads to the delay term $z^{-(2M-1)}$ that we did not insert before in the no-distortion equation. The conclusion we can draw is that if the filter $H_{\text{LR}}(z)$ is designed in such a way to satisfy

1. one linear equation for admissibility, and
2. M bilinear or quadratic equations for orthogonality (quadratic admissibility), implying the above condition on $H_{\text{LR}}(z)$,

then PR will be ensured with an orthogonal basis corresponding to causal filters. Passing to frequency responses after dropping the causality-related delay factor to return to the theoretical setting we have

$$
\begin{aligned}
H_{\text{LR}}\left(e^{j\omega}\right) H_{\text{LR}}\left(e^{-j\omega}\right) &+ H_{\text{LR}}\left(e^{-j\pi} e^{-j\omega}\right) H_{\text{LR}}\left(e^{j\pi} e^{j\omega}\right) = \\
= H_{\text{LR}}\left(e^{j\omega}\right) H_{\text{LR}}^*\left(e^{j\omega}\right) &+ H_{\text{LR}}\left[e^{-j(\omega+\pi)}\right] H_{\text{LR}}\left[e^{j(\omega+\pi)}\right] = \\
= \left| H_{\text{LR}}\left(e^{j\omega}\right) \right|^2 &+ H_{\text{LR}}^*\left[e^{j(\omega+\pi)}\right] H_{\text{LR}}\left[e^{j(\omega+\pi)}\right] = \\
= \left| H_{\text{LR}}\left(e^{j\omega}\right) \right|^2 &+ \left| H_{\text{LR}}\left[e^{j(\omega+\pi)}\right] \right|^2 = 2.
\end{aligned}
$$

This is exactly the frequency-domain orthogonality condition on the lowpass reconstruction filter that we reported when talking about the theorems on PR filters at the beginning of this section.

So, *the ON-PR solution is not unique.* Indeed, the length of $h_{\text{LR}}[n]$ is $2M$. After satisfying $M + 1$ conditions, the number of degrees of freedom (DOF) in choosing these $2M$ coefficients is $M - 1$. This freedom is used in the design of different wavelet system having some desired features.

14.5   Mallat's Algorithm

In the late 1980s, Stéphane Mallat proposed a fast DWT decomposition and reconstruction algorithm (Mallat 1989) in the form of a *two-channel subband coder* using
PR filters (see, e.g., Qian 2001; Misiti et al. 2007). This algorithm is sometimes
referred to as the *fast wavelet transform* (FWT). We will now describe this algorithm, referring to a *finite-length sampled input signal* of length N. *We assume the
basis set to be ON and compactly supported; the filters involved are FIR.*

For biorthogonal wavelets, the same algorithm holds, but the decomposition filters
on one hand and the reconstruction filters on the other hand are obtained from two
distinct scaling functions associated with two multiresolution analyses in duality. In
this case, the filters for decomposition and reconstruction are, in general, of different
lengths. By zero-padding, the four filters can be extended in such a way that they
will have the same even length.

14.5.1   Edge Effects and Extension Modes

Each stage of Mallat's algorithm implies filtering in the time domain via linear convolution. When a linear convolution $z[n] = \sum_k s[n - k]h[k]$ is performed on a
finite-length signal $s[n]$, $n = [0, N - 1]$, border distortions arise, since the computation of $z[n]$ will ask, e.g., for $s[-1]$, which does not exist. Signal-extension
schemes are applied on the signal boundaries to solve this issue; their purpose is
actually to define $s[-1]$ and the other samples possibly needed outside the record's
time support.

Moreover, classically the computational DWT is defined for data records with
length of some integer power of 2, and ways of extending records of other sizes
are needed. However, extension related to convolution is needed *at each stage* of
the decomposition process. Therefore, by "signal" in this subsection we mean the
original signal itself in the first stage, or the input to any other stage.

Different signal-extension schemes can be adopted to deal with edge effects in
the computational DWT. They are referred to as *extension modes*, and include:

1. *zero-padding*: extending the signal by zeros over the time span required by the
   convolution, which in turn depends linearly on the length of the filter. In this case
   we suppose that the signal is zero everywhere outside of the original support. The
   disadvantage of zero-padding is that discontinuities are created at the signal's
   ends;
2. *symmetrization*: extending the signal by reflection, i.e., by symmetric boundary-
   value replication over the span required by the convolution. In other words, signals
   are extended outside of their support by repeating near edge values by symmetry. Symmetrization has the disadvantage of creating discontinuities in the first
   derivative at the borders, but this is better than creating jumps in the signal itself;

3. *regular extrapolation*: a low degree polynomial extrapolation of the signal outside the support is performed. A typically case is linear extrapolation. This method works poorly for noisy signals;
4. *periodization*: extending the signal by transforming it into its periodic extension over the time span required by the convolution. Thus, this method consists of assuming that the signal is periodic. This procedure may include a pre-processing, consisting of the fact that if the signal's initial length is odd, then the signal is first extended by adding an extra-sample equal to the last value on the right; then a minimal periodic extension is performed on each side. This extension also creates artificial discontinuities at the edges.

How can we choose the best extension mode for a given application? This is a rather technical issue about which we cannot enter into details. The rationale behind each extension scheme is explained in Chap. 8 of Strang and Nguyen (1996). However, we can observe that the heart of the finite-length problem in the DWT concerns the downsampling operation by a factor of 2, which can give very different results depending on the extension choice that is made.

For example, let us focus on symmetrization. There are two symmetric ways to extend a signal that starts with $x[0]$: everything depends on the choice of $x[-1]$. We may take $x[-1] = x[1]$, and then continue by $x[-2] = x[2]$, etc. The center of symmetry, also called point of symmetry, of this extension is $x[0]$. When we do this at both ends, $n = 0$ and $n = N - 1$, the period of the resulting signal is $2N - 2$. Neither the sample $x[0]$, nor the sample $x[N - 1]$ are repeated; this is called whole-point symmetry. The other possibility is to choose $x[-1] = x[0]$, $x[-2] = x[1]$, etc. The point of symmetry is halfway between two samples, at $n = -0.5$. This half-point symmetry produces a signal with period $2N$. Which choice is better depends on the filter being used, namely, on its length and symmetry. In the DWT cascade, after downsampling we still desire to have a symmetric extension. Now, suppose we are working with symmetric/antisymmetric filters of even length, as those associated to B-spline biorthogonal wavelets. An even-length FIR filter can be symmetric or antisymmetric around a half-integer value of the index. A signal downsampled by a factor of 2 will still be periodic anyway, but it will be symmetric only if, in association with an even-length symmetric/antisymmetric FIR filter, we adopt an analogous extension mode—half-point symmetry, etc. This is just to give an idea of the complexity of the extension-mode issue.

Conceptually, extension makes it possible to associate a finite-length sequence with an infinite-length signal, for which the DWT algorithm is theoretically justified (Misiti et al. 2007). Provided that the assumption on which an extension mode relies is reasonable, the extended signal should behave as the infinitely-long signal we consider in theory. Of course, any extension mode is just a way to "invent" data samples that we do not posses. The properties of

- preservation of the square norm (energy) by decomposition,
- orthogonality,
- perfect reconstruction,

verified for infinitely long signals, in practice are lost to a lesser or greater extent for a finite-length sequence. Since perfect invertibility of the transform is the main goal in DWT, only PR is always strictly preserved in computation. In all the DWT examples in the present chapter, the extension mode has been set to half-point symmetry.

14.5.2   Signal Decomposition: Subband Coding

To describe Mallat's algorithm we must adopt a "computational" perspective, as opposed to the theoretical perspective presented in Sect. 14.3. We thus need to change our notation.

Talking about scaling and wavelet subspaces we used j_0 (typically 0) and j_{max} to denote the initial and final subspaces in a decomposition; j_{max} could go to infinity, or be finite. In the finite case, the number of j-values involved, called *levels of decomposition* from now on, was $N_{lev} = j_{max} - j_0 + 1$. With $j_0 = 0$, $N_{lev} = j_{max} + 1$.

Now we must consider that

- in any computational environment (as opposed to the theoretical one) indexes run from 1 rather than from zero. It is normal practice to number the decomposition stages from level 1 to a maximum level of decomposition that we will call J: $j = [1, J]$. For each signal length and each wavelet (type and order), the total number J of levels cannot exceed some upper bound N_{lev}^{max};
- the Mallat decomposition algorithm includes, in general, several stages, each of them corresponding to a certain level of decomposition. It is common practice to number the stages *inversely* with respect to our previous discussion on scaling and wavelet subspaces.

In our new level numbering convention, $j = 0$ does not exist. However, we can imagine to start from a *non-existent* level 0 in which the hypothetical scaling coefficient sequence $c_0[k]$ is well approximated by the signal $x[k]$. This level corresponds to the smallest possible scale, i.e., the highest time resolution, and to the scaling subspace V_J. The first actual level ($j = 1$) corresponds to V_{J-1}; this level receives the signal's samples as the input and produces a pair of coefficient vectors $c_1[k]$ and $d_1[k]$. Stopping the procedure at this point would produce what is called a *single-level decomposition* of the signal. If we proceed further, then we perform a *multi-level decomposition*, in which the recursive relations for the scaling and wavelet coefficients are exploited to compute all the coarser-resolution (larger scale) scaling and wavelet coefficients, until a proper maximum level J is attained (coarsest detail, largest scale considered). Level J corresponds to subspace V_0. The approximation and detail coefficients $c_{j+1}[k]$ and $d_{j+1}[k]$ for some intermediate level $j + 1 \in [1, J]$ are obtained by convolution of the preceding $c_j[k]$ with the impulse responses of the two $2M$-samples-long lowpass/highpass decomposition filters, respectively, coupled with downsampling/upsampling operations.

Note that in spite of the adoption of this "computational-style" level numbering, we will continue to use the symbol j to indicate a generic level, since this notation is consolidated in practice. Note also that passing from the "old" j to the "new" one corresponds to substituting the old j with a new index $J - j$. As a consequence, we are also changing the definition of scale: in the computational environment scale varies with j as 2^j, with the minimum scale being dictated by the fact that the signal is sampled, and the maximum scale being upperly bounded essentially by the number of available samples in the data record. This also clarifies why the "old" j, corresponding to the "new" $j = J$, can always be assigned a value of 0. We start from the minimum scale, determined by the signal's sampling rate, and proceed towards larger scales by repeated downsamplings. If the sampling of the underlying analog signal has been correctly performed according to Nyquist's criterion, then this minimum scale actually identifies the scaling subspace to which the signal belongs. If enough signal samples are available, i.e., if N is large, the decomposition can proceed to V_0, which will contain the signal's mean value; if the amount of data is not large enough, the decomposition will necessarily stop earlier. In any case, in terms of the "old" j that serve to number subspaces, a label J will be assigned to the subspace related to the starting level of minimum scale, and a label 0 will be assigned to the subspace related to the final level of maximum scale. The level related to V_0 will not necessarily correspond to the signal's mean level, but will correspond anyway to the coarsest features of the signal that we are able to study using N samples.

A decomposition up to some level J produces, in the end, one approximation coefficient vector ($c_J[k]$ in V_0) and J detail coefficient vectors $d_j[k]$, with $j = [1, J]$. The tree-structured subband-coding algorithm includes J stages, or levels. The first level takes $x[n]$ as the input and outputs $c_1[k]$ and $d_1[k]$ by filtering and downsampling (Fig. 14.16). To attain level J, we must go through $J - 1$ more subsequent stages, each of them being structured as in Fig. 14.17. In any intermediate stage, a $c_j[k]$ produces a $c_{j+1}[k]$ and a $d_{j+1}[k]$. Iterating the single-stage processing along the approximation-coefficient (low-band) branch depicts a tree-shaped structure with three stages, and so on, until the last stage produces $c_J[k]$ and $d_J[k]$. The DWT decomposition thus appears as a tree that we descend from the crown towards the roots, always following the approximation-coefficients branch (Fig. 14.18). In the frequency domain, each stage divides the spectrum of the input signal—which can be $x[k]$, as in Fig. 14.16, or some $c_j[k]$, as in Fig. 14.17—into one lowpass and one highpass half, since the filters are half-band filters. In other words, each stage is a two-band processor.

Fig. 14.16 Starting stage of the DWT decomposition of a finite-length sampled signal (see text for details). Note that this scheme depicts a process progressing from left to right, as indicated by the *gray arrow*

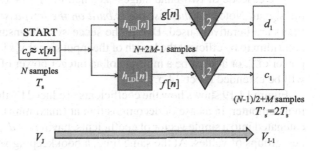

Fig. 14.17 An intermediate
two-band stage in the
tree-structured DWT
decomposition of a
finite-length sampled signal.
Note that this scheme depicts
a process progressing from
right to left, as shown by the
gray arrow

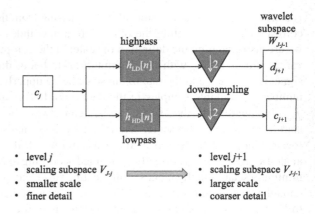

- level j
- scaling subspace $V_{J\text{-}j}$
- smaller scale
- finer detail

- level j+1
- scaling subspace $V_{J\text{-}j\text{-}1}$
- larger scale
- coarser detail

Fig. 14.18 Subband coding:
scheme for the DWT
decomposition of a
finite-length sampled signal
of length Nequal to an
integer power of 2

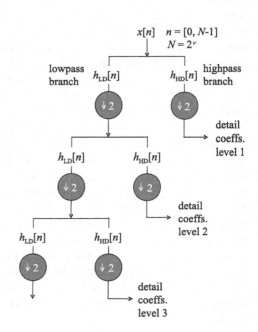

The cascade of two-band stages sketched in Fig. 14.18 is the *subband coding* of
the signal. Note that *the cascade is built on the lowpass branch*, and that only two
filters are iteratively used. Due to the successive downsamplings by 2, for Mallat's
algorithm to be efficient the length of the input signal $x[k]$ must be equal to an integer
power of 2, or at least be a multiple of an integer power of 2. If it is not so, the signal
will be extended (Sect. 14.5.1).

Figure 14.19 shows how the coefficients produced by the DWT cascade are stored
in a computer, in a case of decomposition at (maximum) level $J = 3$: they are concatenated into a single vector of coefficients, having c_3, d_3, d_2, and d_1 as components,
i.e., groups of values. At the same time, a bookkeeping vector is created, giving the

Fig. 14.19 The way the DWT coefficients are stored in a computer: an example of decomposition at level $J = 3$

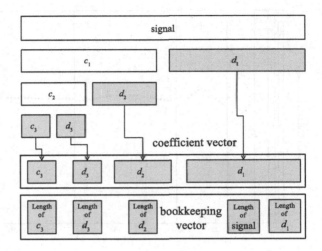

length of each component. The last element of the bookkeeping vector reports the length of the signal. We denote by $L[i]$ the elements of the bookkeeping vector, with $i = 1, J+2$. The vector concatenating all sequences of DWT coefficients for a given decomposition is denoted by $C[l]$, $l = 1, N_{DWT}$, with $N_{DWT} = \sum_{i=1}^{J+1} L(i)$. Inside $C[l]$,

- the part of the vector $C[l]$ containing the approximation coefficients can be indicated as $C_a[m]$, $m = [1, N_a]$, with $N_a = L(1)$;
- the part of the vector $C[l]$ containing the detail coefficients of all levels can be indicated as $C_d[m]$, $m = [1, N_d]$, with $N_d = \sum_{i=2}^{J+1} L(i)$;
- the part of the vector $C[l]$ containing the detail coefficients of level j can be indicated as $C_d^{(j)}[m]$, $m = [1, N_d^{(j)}]$, with $N_d^{(j)} \equiv L(j+1)$.

An example of a vector $C[l]$ of concatenated coefficients appears in Fig. 14.20. The coefficients plotted in Fig. 14.20 result from the decomposition that we saw in Fig. 14.10: the decomposition of a sine function with two discontinuities, performed using an ON compactly-supported wavelet (the Haar wavelet). It may be useful, at this point, to go back to that example. We now understand that we pushed the decomposition up to level $J = 7$. If we observe the column showing the original signal and the seven reconstructed approximations (Fig. 14.10a), we can read not only the names of the scaling subspace which each plotted signal belongs to, but also the corresponding decomposition-level numbering, according to the convention we have now adopted for the index j. We can see that the level-1 approximation is in V_6, the level-2 approximation is in V_5, ..., and the level-7 approximation is in V_0.

We can now see what the cascade looks like in the frequency domain. The filters involved in the cascade have *constant relative bandwidth*, as the MRA property of any shade of wavelet transform implies. The cascade represents an iterated application of the same two filters at different scales. We thus see the DWT as the application of

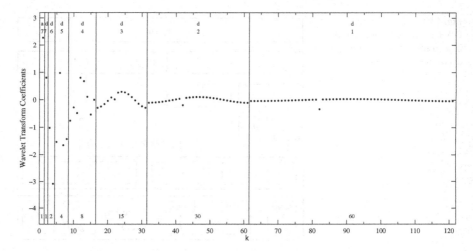

Fig. 14.20 How the DWT coefficients resulting from the decomposition of a signal are concatenated into a single vector. The coefficients for this example have been taken from the decomposition at level $J = 7$ of the sine function with two discontinuities shown in Fig. 14.10

a *constant-Q filter bank*, with $Q = \Delta\omega/\omega_c$ (*fidelity factor*) being equal for all filters in the bank. In the tree-structured algorithm,

- the first stage divides the spectrum of the original signal $x[k] \equiv c_0[k]$ into two bands of equal width, a lowpass and a highpass band;
- the second stage further divides the lowpass band into two bands of equal width, a lowpass and a highpass band;
- the third stage further divides the new lowpass band into two bands of equal width, a lowpass and a highpass band, and so on.

This is illustrated in Fig. 14.21, depicting the progressive (ideal) subdivision of the input signal bandwidth $[0, \pi]$ in the DWT cascade. The half-band lowpass filter associated with V_0 corresponds to the last scaling coefficient sequence, i.e., the sequence $c_J[k]$ that will not be filtered/downsampled any further, since j has reached its maximum value: the coarsest resolution has been attained. The value of J is dictated by the length N of the record because it is the level at which, after repeated downsampling operations that progressively reduce the length of the sequences involved, we are left with only two coefficient samples: one sample in $c_J[k]$ and one sample in $d_J[k]$. The maximum possible value of J can thus be estimated as $N_{\text{lev}}^{\max} = \log_2 N$. Actually, the maximum decomposition level does not only depend on the signal's length N, but also on the wavelet we adopt and on its order. This will be explained in Sect. 14.5.5. Of course, the cascade can be stopped earlier.[11] As for the highpass filter associated with wavelet coefficients, in this example, the rightmost one is associated with the

---

[11]Returning for a moment to Fig. 14.10, at this point we may wonder if we would be in a position to proceed further to a more accurate level-8 approximation in V_{-1}. The answer is no; in the example of Fig. 14.10, the signal length is $N = 120$, and after repeated downsampling operations, the

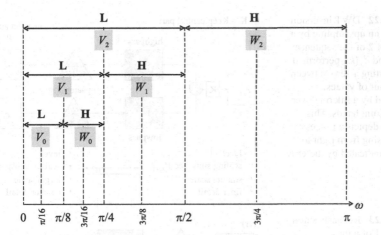

Fig. 14.21 The progressive subdivision of the input signal bandwidth $[0, \pi]$ in the DWT cascade, in a case with $J = 3$. The subspaces of $L^2(\mathbb{R})$ associated with each subdivision are also shown

wavelet subspace W_2, operating on the interval $\pi/2 - \pi$. The remaining half-band, i.e., $0 - \pi/2$, is again divided into two parts, the highpass one being associated with the wavelet subspace W_1. In the subsequent stage, the highpass band is related to the wavelet subspace W_0, and the remaining part of the signal's spectrum is covered by V_0. Thus, this example describes the following decomposition of the signal's space: $V_0 \oplus W_0 \oplus W_1 \oplus W_2$.

14.5.3 Signal Reconstruction from DWT Coefficients

In order to re-synthesize the signal from its DWT coefficients, we must climb the tree from the roots to the crown: we must progressively reconstruct the scaling coefficients at a finer scale from the scaling and wavelet coefficients at the adjacent coarser scale, by upsampling by a factor of 2 at each stage and then applying the two reconstruction filters $h_{LR}[k]$ and $h_{HR}[k]$.

Figure 14.22 illustrates a generic stage of the reconstruction process. Note that this scheme depicts a process progressing from right to left (see the *gray arrow*). Iterating the single-stage reconstruction we get the complete reconstruction tree, which begins with $j = J$ and stops at $j = 1$, the level at which the original signal is obtained as the final output of the PR filter bank.

(Footnote 11 continued)
coarsest-resolution approximation and detail coefficients ($c_7[k]$, $d_7[k]$) are made by just one sample each. The decomposition was pushed up to the maximum possible level. The remaining detail coefficients up to d_1 are progressively longer, since they represent less downsampled signals. Now, suppose we had $N = 240$ samples. We would then be able to attain $J = 8$, and ($c_7[k]$, $d_7[k]$) would have one sample each. However, in Fig. 14.10 we would see level 1 in V_7, since $J - j = 8 - 1 = 7$, and so on, up to level 8 in V_0: we would always label the last subspace as V_0.

Fig. 14.22 DWT inversion implies an upsampling by a factor of 2 of the sequences $c_j[k]$ and $d_j[k]$, performed by inserting a zero between every pair of values, followed by a filtering over two disjoint bands. This scheme depicts a process progressing from right to left, as indicated by the *gray arrow*

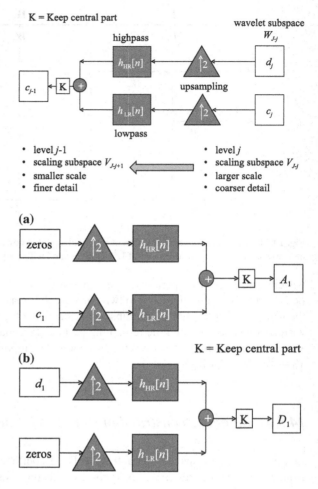

Fig. 14.23 Reconstruction at level 1 of **a** the approximation signal and **b** the detail signal

Adopting a similar scheme, we can reconstruct an *approximation signal* $A_1[k]$ at level 1 from the coefficients $c_1[k]$, by using a sequence of zeros as input to the high-band branch of a single reconstruction stage, as shown in Fig. 14.23a. In the same way, the scheme of Fig. 14.23b provides a *detail signal* $D_1[k]$ at level 1. Adding level-1 approximation and detail signals gives back the original signal.[12]

Extending the reconstruction of approximations and details at all levels, we can obtain all the N-length signals into which $x[n]$ may be decomposed (Fig. 14.24). We can re-assemble $x[n]$ in many different ways, depending on the level J at which we

---

[12]Note that the coefficient vectors $c_1[k]$ and $d_1[k]$ cannot be directly combined to reproduce the signal. The coefficients are produced by downsampling and are only half the length of the original signal. It is necessary to reconstruct the approximations and details before combining them.

Fig. 14.24 All the N-length
signals into which a signal
$x[n]$ of length N may be
decomposed

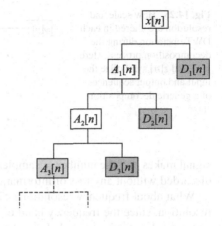

stop. We can write $x[n] = A_1[n] + D_1[n] = A_2[n] + D_2[n] + D_1[n] = A_3[n] + D_3[n] + D_2[n] + D_1[n]$, etc.; in general,

$$x[n] = A_J[n] + \sum_{j=1}^{J} D_J[n].$$

Do not miss that reconstructed detail and approximation signals have the same length and the same sampling interval $\Delta n = 1$ as the original signal, and must be clearly distinguished from approximation (scaling) and detail (wavelet) coefficients.

14.5.4 Multiresolution in Subband Coding

In a DWT decomposition, scale, time resolution and frequency resolution of the signals change at each level. To see how and where this occurs, let us simplify the picture and reason on the basis of ideal half-band filters. We can refer to Fig. 14.21.

After passing the signal through a half-band lowpass filter, half of the samples can be eliminated according to Nyquist's rule, since the signal now has a bandlimit of $\pi/2$ instead of π. Downsampling the signal by a factor of 2 then becomes harmless, and leaves us with a signal with half the number of samples and a doubled sampling interval. This means that the scale is doubled. Note that the lowpass filtering removes the high frequency information, but leaves the scale unchanged; only the downsampling process changes the scale. Time resolution, on the other hand, is related to the amount of information in the signal, and therefore it is affected by the filtering operation. Halfband lowpass filtering removes half of the frequencies, which can be interpreted as losing half of the information. Therefore, time resolution is halved by the filtering operation. However, the downsampling operation after filtering does not affect time resolution, since removing half of the spectral components from the

Fig. 14.25 How scale and
resolution are altered in each
DWT operation during the
decomposition process. Here
$y[n]$ and $z[n]$ symbolize the
input and output sequences
of a generic decomposition
stage

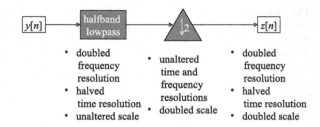

- doubled
  frequency
  resolution
- halved
  time resolution
- unaltered scale

- unaltered
  time and
  frequency
  resolutions
- doubled scale

- doubled
  frequency
  resolution
- halved
  time resolution
- doubled scale

signal makes half the number of samples redundant anyway; half the samples can be
discarded without any loss of information.

What about frequency resolution? The half-band filtering doubles the frequency
resolution, since the frequency band of the signal now spans only half the previous
frequency band, effectively halving the uncertainty on frequency.

In summary, at each level in the decomposition cascade, the lowpass filtering
halves the time resolution and doubles the frequency resolution, but leaves the scale
unchanged. The signal is then subsampled by 2 since half of the number of samples
are redundant, and this doubles the scale because it doubles the sampling interval
and halves the number of samples. These concepts are illustrated in Fig. 14.25.
With reference to Fig. 14.24, we can say that all the approximation and detail signals
into which $x[n]$ may be decomposed share the same sampling interval and therefore
belong to the same scale of $x[n]$, but have different time resolutions, meaning that
they show with different degrees of detail the time behavior of the original signal.
To memorize this fact we can think of a set of pictures of the same object, taken
with progressively smaller resolution: a detail signal corresponds to the difference
between one picture and the next one.

14.5.5 Maximum Decomposition Level

We saw in the previous discussion that DWT decomposition of a signal is limited
to a maximum level, determined by the requirement that at that level, at least one
scaling/wavelet coefficient must be correctly computable. The maximum decomposition level depends on the signal's length N and on the wavelet system we adopt, as
illustrated by Fig. 14.26, which shows N_{lev}^{max} for compactly supported ON wavelets:
Daubechies wavelets, symlets and coiflets (Sect. 14.9). It can be seen that the maximum level for all these wavelets increases linearly with $\log_2 N$. At fixed N, the higher
the order of the ON wavelet, the lower the maximum allowed level. This feature is
simply due to the varying length of the filters involved.

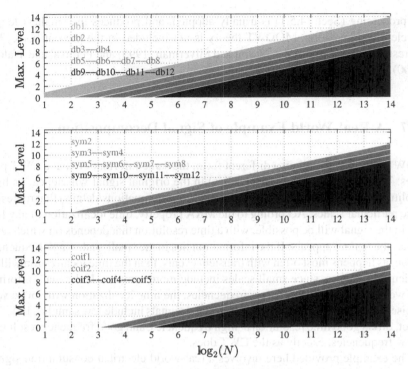

Fig. 14.26 Maximum level of DWT decomposition for Daubechies wavelets, symlets and coiflets of different orders; their names are abbreviated as db, sym and coif, respectively, with an attached number that represents the order

14.6 Another Approach to DWT Implementation

The DWT can be implemented using different approaches. The most popular ones are Mallat's algorithm described above, and the *algorithme á trous*.[13]

The á trous is a non-orthogonal, shift-invariant, dyadic, undecimated, redundant DWT algorithm introduced by Holschneider et al. (1989) (see also Shensa 1991) that implements the so-called *maximal overlap discrete wavelet transform* (MODWT), also known as the *stationary wavelet transform* (SWT), the undecimated DWT, and so on. The MODWT algorithm was designed to overcome the lack of translation invariance that characterizes the DWT, and that is related to the fact that downsampling is a linear operation, but is not time-invariant. Translation invariance, which may be desirable in some applications, is achieved by avoiding downsampling: the impulse responses of the filters at each level of the algorithm are upsampled instead. The name "algorithme á trous" actually refers to inserting zeros in the impulse responses for the purpose of upsampling. The choice of the extension mode depends on the algorithm adopted. For the MODWT, the periodic padding extension mode with

---

[13]The word "trous" means "holes" in English.

pre-processing (Sect. 14.5.1) is usually adopted, and produces the smallest length wavelet decomposition. MODWT theory is not included in the present book. The interested reader is referred to Nason and Silverman (1995) and Percival and Walden (2000).

14.7 A Real-World Example of Signal Decomposition

A DWT conveys, though in a different form, the same information that a CWT provides. The dominant scales (frequencies) in the original signal will appear as high amplitudes in the DWT-reconstructed signals (approximations/details) that represent those particular scales. According to the MRA property, time localization of any feature in the signal will be possible, with a time resolution that depends on which level (scale) the feature appears. If the information lies in the small-scale (high frequency) region, as happens most often with transient events, then the time localization will be particularly precise, since small scales include more samples. If the main information we are looking for lies at low frequencies, the time localization will not be very precise, since the corresponding reconstructed signals include few samples. DWT, in effect, offers good time resolution at high frequencies, and good frequency resolution at low frequencies, exactly as the CWT does.

The example provided here involves a real-world electrical consumption signal, originally measured and analyzed by Misiti et al. (1994) over 5 weeks, and that we will analyze over the course of 3 days (namely days 9–11 of the original record). This signal is particularly interesting because of noise introduced by a defect developed in the monitoring equipment as the measurements were being made. We will use this feature to illustrate de-noising in Sect. 15.2. The data consists of measurement of a complex, highly aggregated plant: the electrical load consumption, sampled minute by minute. The number of samples is 4320 (4320 min = 72 h = 3 days). This load curve is the aggregation of hundreds of sensors measurements, generating measurement errors, and is plotted in Fig. 14.27. Time is in minutes and starts from 0:00 of day 9 of the original record.

Roughly speaking, 50 % of the consumption is accounted for by industry, and the rest by more than 10 million individual consumers. The component of the load curve produced by industry has a rather regular profile and exhibits low-frequency changes. On the other hand, the consumption of individual consumers is highly irregular, leading to high-frequency components. The fundamental period is a daily cycle linked to economic rhythms. Daily consumption patterns also change according to rate changes at different times (e.g., relay-switched water heaters to benefit from special night rates). A long-term trend is also evident during the examined 3-day interval, connected to longer time-scale variations. For the observations from minute 2400 to minute 3400, the measurement errors are unusually high, due to sensor failures. Figure 14.27 highlights in two shades of gray two small time intervals, one relative to the end of the night (*dark gray*) and another one relative to midday (*light gray*). In the second interval, the signal structure is complex, while in the first one it is

Fig. 14.27 A part of the electrical consumption signal measured and analyzed by Misiti et al. (1994). Two small time intervals, one relative to the end of the night (shaded in *dark gray*) and another one relative to midday (shaded in *light gray*) are highlighted, since they are discussed in the text

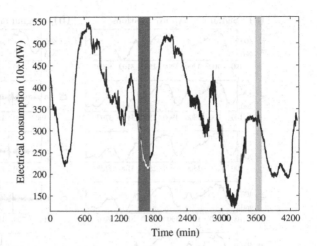

much simpler. The midday period has a complicated structure because the intensity of the electricity consumer activity is high and presents very large changes. At the end of the night, the activity is low and the signal changes slowly.

We focus on the components of the signal for which the period is less than 30 min. Therefore we carry out the decomposition up to the level $J = 5$, because $2^5 = 32$ is very close to 30 units of T_s, i.e., 30 min. The analyzing wavelet used here is the Daubechies wavelet of order 3. We reconstruct the details and the approximations. Of course, details are orthogonal with respect to approximations and among themselves. In order to avoid edge effects we suppress edge values and plot (Fig. 14.28).

Visual inspection of Fig. 14.28 suggests that the approximation A_3 might provide a satisfactory smooth representation of this signal. However, in order to make sure that we do not need to include more details to correctly represent the signal in all its facets, let us zoom on the two time intervals highlighted in Fig. 14.27.

First, we zoom in on the neighborhood of the minimum at end of the night (minutes 1551–1750, i.e., from 01:50 to 5:10 of day 10; Fig. 14.29). The shape of the curve during the end of the night is a slow descent, globally smooth, but locally highly irregular. One can hardly distinguish two successive local extrema in the vicinity of time t = 1600 min and t = 1625 min: a tiny local maximum and a small local minimum, respectively. The approximations A_4 and A_5 are piecewise linear and appear to provide a good signal representation, except at these two points. This means that this portion of the signal is a low-frequency signal corrupted by noises. The massive and simultaneous changes of personal electrical appliances that would give the signal a more complex structure are absent. The details D_1, D_2, and D_3 contain local short-period irregularities caused by noises, and the inspection of D_2 and D_3 allows for detecting the local minimum around t = 1625 min. The details D_4 and D_5 exhibit the slope changes of the regular part of the signal. In conclusion, none of the high-level details provide essential information on this portion of the signal: the two local extrema can be considered as unessential features. We could just retain approximation A_4 or A_5 without any further correction. The approximation A_3,

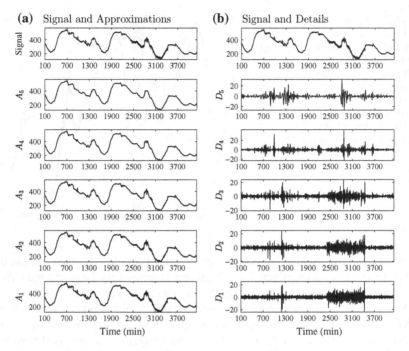

Fig. 14.28 **a** Approximations and **b** details from the decomposition of the electrical consumption signal plotted in Fig. 14.27, performed using the Daubechies wavelet of order 3 up to the level $J = 5$

selected on the basis of the overall signal behavior (that is less smooth than A_4 or A_5) thus appears more than adequate to ensure we do not neglect any signal-related feature in this night time interval.

Next we focus on the neighborhood of the midday maximum (minutes 3601–3700, i.e., from 12:00 to 13:40 of day 11; Fig. 14.30), in which the signal has a complex structure. It shows a peak between $\sim$12:30 p.m. and 1:00 p.m. (minutes 3631–3660), preceded and followed by a hollow off-peak, and a second smoother peak around 1:10 p.m. (minutes 3665–3675). The approximation A_5, corresponding to the time scale of 32 min, is a very crude signal approximation, particularly for the central peak: it exhibits a peak time-lag and an underestimation of the maximum value. So at this level, the most essential information is missing. We have to look at a lower scale, 4 for instance. Let us examine the corresponding details. The details D_1 and D_2 have small values and may be considered as local short-period discrepancies caused by the high-frequency components of sensor, and state, noises. In this frequency band, these noises are essentially due to measurement errors and fast variations of the signal induced by millions of state changes of personal electrical appliances. The detail D_3 exhibits high values mainly at times corresponding to the 12:40 p.m. signal maximum (minute $\simeq$ 3640). But it is the detail D_4 which contains the main patterns: three successive modes (minimum-maximum-minimum) synchronized with the original signal. It is remarkably close to the shape of the original curve. The amplitude of the

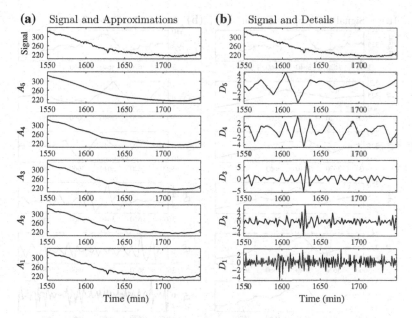

Fig. 14.29 **a** Approximations and **b** details from the decomposition of the electrical consumption signal performed using the Daubechies wavelet of order 3 up to the level $J = 5$, in the neighborhood of the minimum electrical consumption at end of the night

oscillations in this detail is much higher than that of the other details. The detail D_5 does not bear much additional information. So the contribution of level 4 is the highest one, both from the qualitative and the quantitative points of view: it captures the shape of the curve in the examined period. In conclusion, with respect to the approximation A_4, the detail D_4 is the main correction needed. The components with period of 8–16 min, belonging to level 4, contain the crucial dynamics of the analyzed process. The approximation A_3 we selected on the basis of the overall signal behavior should thus ensure that signal-related features are preserved also in this time interval, characterized by complex signal structure.

This example was meant to show how a DWT decomposition can separate the variability of the signal belonging to different scales into the corresponding approximation and detail signals.

14.8 Wavelet Packets

Developing the DWT cascade along both lowpass and highpass branches, rather than on the lowpass branch only, a technique called *wavelet packet decomposition* is obtained. This method allows for great freedom in coding the signal. The tree for wavelet packet decomposition is shown in Fig. 14.31.

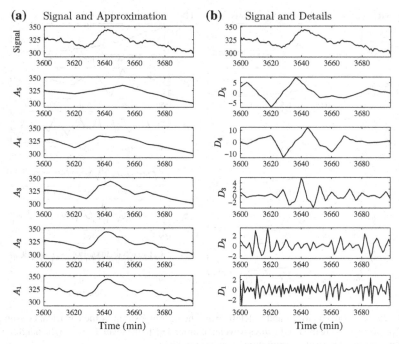

(a) Signal and Approximation   **(b)** Signal and Details

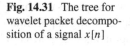

Fig. 14.30 **a** Approximations and **b** details from the decomposition of the electrical consumption signal performed using the Daubechies wavelet of order 3 up to the level $J = 5$, in the neighborhood of the midday maximum of electrical consumption

Fig. 14.31 The tree for wavelet packet decomposition of a signal $x[n]$

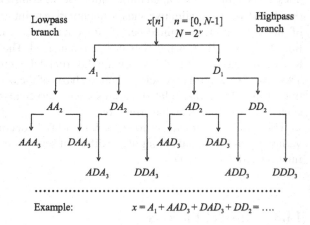

The complete binary tree is then pruned according to criteria for optimal decomposition. In order to do so, the information gain related to each tree node, i.e., split point, is quantified, and a decision is made about proceeding further, or not. In other words, at each node one can judge if it is worthwhile to go ahead or if it is better to stop at that node. The analysis performed with wavelet packets corresponds to virtually arbitrary tilings of the time-scale plane. Wavelet packets are not discussed in this book.

14.9 Regularity, Moments, and Wavelet System Design

The $M + 1$ conditions $\sum_n h_{\mathrm{LR}}[n] = \sqrt{2}$ and $\sum_n h_{\mathrm{LR}}[n]h_{\mathrm{LR}}[n - 2k] = \delta[k]$ ensure the existence and orthogonality/orthonormality of the scaling and wavelet functions. After satisfying these constraints, $M - 1$ DOF remain available to design the scaling filter and determine the $2M$ samples of the impulse response $h_{\mathrm{LR}}[n]$. How can this freedom be usefully employed?

These basic conditions do not imply any particular property of smoothness of the scaling and wavelet functions. Indeed, they may be highly irregular, even fractal in nature. This may be an advantage if we want to analyze rough or fractal signals, but it is likely to be a disadvantage for the analysis of most signals we may encounter in the real world. The $M - 1$ DOF can thus be exploited to build smooth functions, if this is advisable for a particular application. This topic in wavelet theory involves introducing the concepts of *regularity* and *vanishing moments*—two expressions that often recur in discussions about wavelet properties.

The continuous-time k-th moments of $\varphi(\theta)$ and $\psi(\theta)$ are defined as

$$m[k] = \int \theta^k \varphi(\theta)\mathrm{d}\theta, \qquad m_1[k] = \int \theta^k \psi(\theta)\mathrm{d}\theta,$$

and the discrete-time k-th moments of $h_{\mathrm{LR}}[n]$ and $h_{\mathrm{HR}}[n]$ are defined as

$$\mu[k] = \sum_n n^k h_{\mathrm{LR}}[n], \qquad \mu_1[k] = \sum_n n^k h_{\mathrm{HR}}[n].$$

Recall that $m[0]$ cannot vanish because $\varphi(\theta)$ must have a non-zero integral, an correspondingly, $\mu[0]$ cannot vanish because $\mu[0] = \sum_n h_{\mathrm{LR}}[n] = \sqrt{2}$. On the other hand, $m_1[0] = 0$ because $\psi(\theta)$ must have zero-mean, and $\mu_1[0] = 0$ because $\mu_1[0] = \sum_n h_{\mathrm{HR}}[n] = 0$.

These moments are not independent: using the basic recursion it can be shown that

$$m[k] = \frac{1}{(2^k - 1)\sqrt{2}} \sum_{l=1}^{k} \binom{k}{l} \mu[l] m[k - l],$$

$$m_1[k] = \frac{1}{2^k \sqrt{2}} \sum_{l=0}^{k} \binom{k}{l} \mu_1[l] m[k - l].$$

In essence,

- the number of vanishing moments of $h_{HR}[n]$ and $\psi(\theta)$ is related to the smoothness of $\varphi(\theta)$ and $\psi(\theta)$, while
- the number of vanishing moments of $h_{LR}[n]$ and $\varphi(\theta)$ is related to the quality of the approximation of high-resolution scaling coefficients by samples of the signal, as well as to the symmetry and concentration in time of the scaling and wavelet functions.

What does *smoothness* mean here? Is it related to *differentiability*? What about the word *regularity*?

In complex analysis, a regular function (also called holomorphic function) is an infinitely differentiable function. Wavelet analysis uses function of real variables, so there is no concept of "regular function" involved in wavelet theory. We are still interested in the differentiability of the functions, because it is related to their smoothness. The higher the differentiability, i.e., the existence of derivatives of high order, the greater the smoothness. This topic is mathematically involute and is beyond the scope of the book; we can just give some general ideas on the subject.

The smoothness of the scaling function is related to the so-called *K-regular unitary scaling filters* (Burrus et al. 1998). First, we define a *unitary scaling filter* to be an FIR filter with coefficients $h_{LR}[n]$ from the basic recursion, satisfying the admissibility condition $\sum_n h_{LR}[n] = \sqrt{2}$ and the orthogonality condition $\sum_n h_{LR}[n] h_{LR}[n - 2k] = \delta[k]$. Nothing new up to now, just a new name for $h_{LR}[n]$. Second, a unitary scaling filter is said to be *K-regular*[14] if its transfer function $H_{LR}(z)$ has K zeros at $z = e^{j\pi} = -1$, i.e., if we can write

$$H_{LR}(z) = \left(\frac{1 + z^{-1}}{2}\right)^K Q(z),$$

where $Q(z)$ is a polynomial assumed to have no poles or zeros at $z = e^{j\pi} = -1$. The length of the scaling filter is $2M$, which means that $H_{LR}(z)$ is a $2M - 1$ degree polynomial. Since the multiple zero at $z = e^{j\pi} = -1$ is order K, the degree of the polynomial $Q(z)$ is $2M - 1 - K$.

Any unitary scaling filter is at least $K = 1$ regular, since the scaling filter must be lowpass, hence $H_{LR}(e^{j\pi}) = 0$. This corresponds to having $M - 1$ DOF for choosing the coefficients of $h_{LR}[n]$. Requiring $K > 1$ corresponds to having $M - K \geq 0$ DOF; therefore K is constrained by $1 \leq K \leq M$.

---

[14]Note that that here we are presenting a definition of regularity of the scaling filter, not of the scaling function or of the wavelet.

Requiring some regularity is a way to employ the DOF that remain available in wavelet design. For example, Daubechies used the DOF to obtain maximum regularity for a given M, or to obtain the minimum M for a given regularity. Other authors allowed for a smaller regularity and used the resulting extra DOF for other design purposes.

Regularity is defined in terms of zeros of the transfer function $H_{LR}(z)$, hence in terms of the frequency response $H_{LR}(e^{j\omega})$. Up to normalization, $h_{LR}[n]$ determines the scaling function. Now, the differentiability of a function is tied to how fast its Fourier-series coefficients drop off as their index goes to infinity, or how fast the Fourier-transform magnitude drops off as frequency goes to infinity. Fourier-transforming the twin-scale scaling relation gives

$$\Phi(\omega) = \frac{1}{\sqrt{2}} H_{LR}\left(e^{j\omega/2}\right) \Phi\left(\frac{\omega}{2}\right),$$

meaning that when time is rescaled multiplying it by 2, so that scale is halved, frequency is divided by 2 and thus shifts upwards by an octave. Iterating over scaling operations, it can be shown (see, e.g., Burrus et al. 1998) that the continuous Fourier transform of the scaling function is related to the frequency response of the FIR filter $h_{LR}[n]$ by the infinite product

$$\Phi(\omega) = \Phi(0) \prod_{j=1}^{\infty} \left[\frac{1}{\sqrt{2}} H_{LR}\left(e^{j\omega/2}\right)\right].$$

Since $h_{LR}[n]$ is lowpass, we can expect that if its frequency response has a high order zero at $\omega = \pi$, as it happens if its transfer function has a high order zero at $z = -1$, $\Phi(\omega)$ should drop off rapidly and, therefore, $\varphi(\theta)$ should be smoother and smoother as K increases. This turns out to be true.

K-regular scaling filters, which we defined as having K zeros at $z = e^{j\pi} = -1$, can actually be characterized in several equivalent ways. A variety of equivalent characteristics for the K-regular scaling filter can be specified, which relate to the smoothness of scaling and wavelet functions. They also relate to the possibility of representing and approximating polynomial signals by the considered wavelet system. Since many signals exhibit polynomial behavior, this feature is important.

A theorem states that a unitary scaling filter is K-regular if and only if the following equivalent statements are true:

- all moments of the wavelet filter are zero, $\mu_1[k] = 0$ for $k = [0, K - 1]$;
- all moments of the wavelet function are zero, $m_1[k] = 0$ for $k = [0, K - 1]$;
- $H_{LR}\left(e^{j\omega}\right)$ has a zero of order K at $\omega = \pi$; this is related not only to the smoothness of $\varphi(\theta)$, but also to the flatness of $\left|H_{LR}\left(e^{j\omega}\right)\right|$ at $\omega = \pi$;
- the k-th derivative of $\left|H_{LR}\left(\frac{\omega}{2^j}\right)\right|^2$ is zero at $\omega = 0$ for $k = [1, 2K - 1]$; thus $\left|H_{LR}\left(e^{j\omega}\right)\right|$ is flat at $\omega = 0$;

• all polynomial sequences up to degree $K - 1$ can be expressed as a linear combination of shifted versions of $h_{LR}[n]$;
• all polynomials of degree up to $K - 1$ can be expressed as a linear combination of shifted versions of $\varphi(\theta)$ at any scale.

The properties of K-regular filters tie the number of vanishing moments of $h_{HR}[n]$ and $\psi(\theta)$ to the smoothness of the scaling function, to the smoothness of $H_{LR}\left(e^{j\omega}\right)$ at $\omega = 0$ and π, and to the degree of polynomials that can be exactly expressed as a sum of weighted and shifted scaling functions. Other theorems exist, which relate vanishing wavelet moments to the smoothness of the wavelet function.

Daubechies (1988) proposed the following design criterion for orthogonal wavelet systems: given a number of DOF, maximize the number of vanishing wavelet moments. The primary motivation of the criterion was to obtain smooth wavelets. Such a criterion was used in the design of several wavelet systems, including orthogonal Daubechies wavelet systems (Daubechies 1988), biorthogonal spline wavelet systems (Cohen et al. 1992; Vetterli and Herley 1992), and biorthogonal quincunx spline wavelet systems (Cohen and Daubechies 1993; Kovačević and Vetterli 1992).

Orthogonal Daubechies wavelets have compact support and the maximum number of vanishing moments $m_1[k]$, $\mu_1[k]$. Their construction procedure is based on defining the frequency response of the scaling filter, and therefore the corresponding impulse response; the impulse response of the wavelet filter then follows. The desired regularity K determines, via $1 \leq K \leq M$, the minimum order $2M_{min} - 1$ of the scaling filter. For example, if $K = 4$, $2M_{min} - 1 = 8$. Therefore for a given K we do not have only one pair $\{h_{LR}[n], h_{HR}[n]\}$, but an infinite number of possible pairs, from order $2M_{min} - 1$ up. If all the available $M - 1$ DOF are used to make the moments $m_1[k]$ and $\mu_1[k]$ vanish, then $K = M$, and this is the case of Daubechies wavelets. The filters $h_{LR}[n]$ and $h_{HR}[n]$ implicitly determine the shape of the scaling and wavelet functions, which have no explicit expression, and can only be obtained numerically. An exception is the case with $K = 1$, $2M = 2$, that coincides with the Haar wavelet.

The Daubechies wavelet system[15] is illustrated in Figs. 14.32 and 14.33. The wavelet order is set equal to the number K of vanishing wavelet moments $m_1[k]$, $\mu_1[k]$. Daubechies wavelets are only very approximately symmetric (except for the Haar wavelet, i.e., the Daubechies wavelet of order 1); the smoothness of the wavelet increases with order. When the order is small, the wavelet is discontinuous. They are indicated by db$K$, which stands for "Daubechies wavelet of order K". Do not confuse the wavelet "order" with the order of the related scaling filter. The order of the scaling filter and the wavelet-support width are $2K - 1$, so the regularity of the scaling filter corresponds to half the filters' length. This is the most known discrete wavelet system: Ingrid Daubechies invented what are called compactly supported ON wavelets, thus making discrete wavelet analysis practicable.

---

[15]Daubechies (1992) defined two classes of wavelets, via criteria that select a particular scaling filter. One criterion leads to "extremal phase" (minimum phase) Daubechies wavelets, i.e., the ones illustrated here. Another criterion leads to "least asymmetric" Daubechies wavelets, also called "symlets".

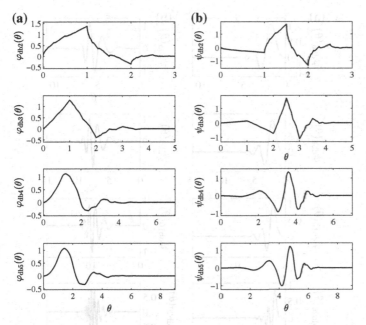

Fig. 14.32 Daubechies **a** scaling functions and **b** wavelet functions. The order is equal to number of vanishing moments K; here $K = 2$–5. Smoothness increases with K

Modifications to the Daubechies family give the symlets, shown in Figs. 14.34 and 14.35. Again, the order of a symlet is equal to the number K of vanishing moments. The order of the scaling filter and the wavelet-support width are $2K - 1$, therefore the regularity of the scaling filter corresponds to half the filters' length. These wavelets are indicated as sym$K$. These wavelets are approximately symmetric, and their symmetry increases with order.

Motivated by wavelet-based numerical analysis, Coifman proposed a different criterion: given a number of DOF, maximize both the number of wavelet *and* scaling vanishing moments. Daubechies used such a criterion to construct orthogonal Coiflet wavelet systems (coiflets) of even orders (Daubechies 1993), while Tian and Wells (1995) constructed odd-ordered ones (Fig. 14.36). Coifman wavelets are designed so as to have, in addition to a number of vanishing wavelet moments $m_1[k]$, $\mu_1[k]$, also a number (not necessarily equal) of vanishing scaling moments $m[k]$, $\mu[k]$. Here we consider the case in which the number of vanishing scaling moments is $K - 1$ and the number of vanishing wavelet moments is K, i.e., they are essentially equal, except for the φ-related moments $\mu[0]$, $m[0]$ that cannot vanish. Thus the total number of vanishing moments is $2K - 1$. The order of a coiflet is defined as $O = K/2$. The length of the four associated filters is $2M = 6O = 3K$, so that $M = 3K/2 = 3O$. The order of the scaling filter and the wavelet-support width are $2M - 1 = 6O - 1 = 3K - 1$. Coiflets are denoted coif$O$; thus, coif4 will have filters with length 24, coif5 will have filters with length 30, etc. These wavelets

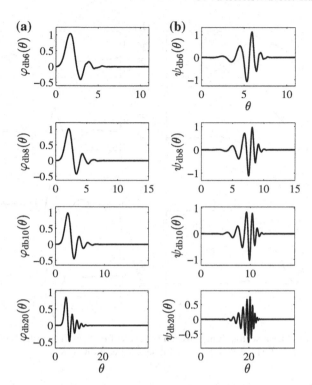

Fig. 14.33 Daubechies **a** scaling functions and **b** wavelet functions. The order is equal to number of vanishing moments K; here $K = 6, 8, 10,$ and 20. Smoothness increases with K

are more symmetrical than Daubechies wavelets, which can be important in certain applications, and for them the approximation $c_0[k] \approx x[n]$ is particularly good. We finally mention that biorthogonal coiflets also exist.

The property of vanishing wavelet moments, related to the smoothness of $\varphi(\theta$ and $\psi(\theta)$, makes the DWT of a smooth signal a *sparse representation* of the signal, i.e., only a small portion of the expansion coefficients are needed to approximate the signal accurately. In fact, this is the fundamental reason for the success of wavelet representations in certain applications where information in the transform domain is selectively discarded, such as data compression and noise reduction.

Vanishing scaling moments are also important. The first moment of the scaling function is unity for normalization reasons, but if the next $K - 1$ scaling moments vanish, as in coiflets, the starting scaling coefficients can be accurately approximated by the samples of the signal. The property implies that we can apply Mallat's algorithm directly on the samples of $x(\theta)$ to generate a valid wavelet representation. This property is extremely important for wavelet-based applications, in which only signal samples rather than continuous-time functions are available. Generally, in applications it is not advisable to treat signal samples as the starting scaling coefficients without proper filtering applied before they enter the filter bank (Strang and Nguyen

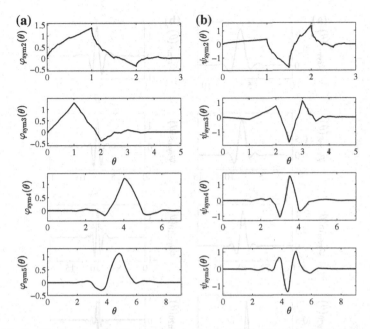

Fig. 14.34 Symlet **a** scaling functions and **b** wavelet functions. The order is equal to number of vanishing moments K; here $K = 2$–5. Smoothness increases with K

1996). However, if the above property holds, i.e., if the wavelet system has a sufficient number of vanishing scaling moments, pre-processing becomes unnecessary, and the combination of Mallat's algorithm and properties related to vanishing scaling moments yields a valid, purely discrete-domain, and fast method for computing scaling and wavelet coefficients.

It can also be shown that vanishing scaling moments are related to another interesting feature, namely the fact that PR filters are linear or nearly-linear phase. This produces a scaling function that is exactly or nearly symmetric.

14.10 Appendix: Wavelet Systems

In this appendix we will review the most popular *wavelets systems*. The review is meant as a reference to the reader.

Many wavelet systems have been proposed in literature. Some wavelets can be used for both continuous and discrete analysis. Others are suitable for CWT only, like analytic Morlet and Paul wavelets; others are mainly associated with discrete analysis, such as orthogonal and biorthogonal wavelets. The choice of the wavelet is dictated by the signal characteristics and by the nature of the application. Understanding

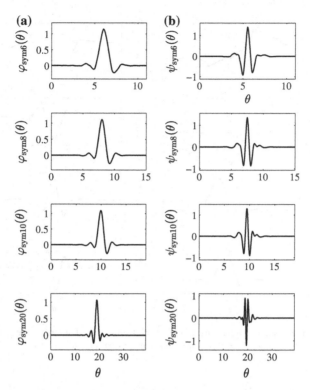

Fig. 14.35 Symlet **a** scaling functions and **b** wavelet functions. The order is equal to number of vanishing moments K; here $K = 6, 8, 10$, and 20. Smoothness increases with K

the properties of the analysis and synthesis basis functions is therefore crucial for choosing a wavelet system that is optimized for a particular application.

While in the CWT we typically use *continuous wavelets that do not posses a scaling function and are not associated to digital filters*, so that almost any zero-integral function can be admissible and appropriate, in the DWT we must restrict ourselves to *discrete wavelets associated with digital filters*, which allow for multiresolution orthogonal or biorthogonal analyses via FWT. In nearly all cases, orthogonal or biorthogonal wavelets, which have a compact support and allow discrete decompositions using the FWT, are defined by their associated filters. These wavelet do not possess a closed-form expression. Nonetheless, by using an iterative procedure deduced from the Mallat reconstruction algorithm and known as the *cascade algorithm* (see Burrus et al. 1998), very good *approximations* of the implicitly defined wavelet can be obtained.

Wavelets are grouped in systems, or *families*. Each family groups similar wavelets that are characterized by particular values of one or more parameters. Wavelet systems can be distinguished according to important properties, such as:

Fig. 14.36 Coifman
a scaling functions and
b wavelet functions of orders
$O = 2$–5 (see text)

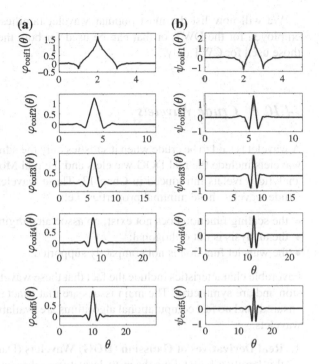

- the existence of a scaling function;
- the orthogonality or biorthogonality of the analysis stemming from it, together with the possibility of exact signal reconstruction;
- the support of the wavelet in time and frequency and the rate of decay;
- the smoothness of the wavelet. Smoother wavelets provide sharper frequency resolution; moreover, iterative algorithms for wavelet construction converge faster;
- the number of vanishing moments. Wavelets with increasing numbers of vanishing moments result in sparse representations for a large class of signals, which is advantageous for applications like signal de-noising and compression;
- the symmetry or antisymmetry of the wavelet, implying that the accompanying PR filters have linear phase; recall that in discrete wavelet analysis, the analysis and synthesis filters are of greater interest than the associated scaling and wavelet functions.

Perfect signal reconstruction (PR) is ensured for orthogonal and biorthogonal wavelets used in the DWT and related applications. However, all wavelets allow for reconstructions that are accurate enough for the applications in which they are employed: for example, a CWT performed with an analytic Morlet wavelet does not ensure a perfect signal reconstruction in a subsequent inverse CWT, but provides an approximate reconstruction that is satisfactory for all facets of evolutionary spectral analysis.

We will now list the most popular wavelet families, including not only those employed for the DWT, or that can be used for both the DWT and CWT, but also those used for CWT only.

14.10.1   Crude Wavelets

A wavelet is said to be crude when it satisfies only the admissibility condition. Crude wavelets include the real DOG wavelets and the real Morlet wavelet, as well as the analytic wavelets introduced in Chap. 13. These wavelets are used for CWT only. Crude wavelets have minimal properties, i.e.:

- the scaling function does not exist; no associated digital filters exist;
- the analysis is not orthogonal;
- the wavelet function is not compactly supported.

Favorable characteristics include the fact that these wavelets have an explicit expression and are symmetric. The main issues are that exact signal reconstruction is not ensured, and no fast computational algorithms are available in association with these wavelets.

1. **Real Derivatives of Gaussian (DOG) Wavelets** (Daubechies 1992):
   this family is built taking the m-th derivative of the Gaussian function. The integer m is the parameter of this family. Real DOGs can be symmetric (m even) or antisymmetric (m odd). Symmetric real DOGs can be used for the CWT in the frequency domain. A famous DOG wavelet is the Mexican Hat real wavelet, having $m = 2$.
2. **Real Morlet Wavelet** (Daubechies 1992):
   this wavelet is derived from a function that is proportional to the second derivative function of the Gaussian probability density function. The real Morlet wavelet is defined as $\psi(\theta) = Ce^{-\theta^2}\cos(5\theta)$. The constant C is used for normalization in view of reconstruction. This wavelet does not technically satisfy the admissibility condition. It is symmetrical.
3. **Analytic Wavelets**: see Sect. 14.10.5

14.10.2   Infinitely Regular Wavelets

This family includes the Meyer wavelets (Abry 1997), which are real and have the following properties:

- the scaling function exists;
- the analysis is orthogonal;
- the scaling and wavelet functions are indefinitely derivable;
- the scaling and wavelet functions are not compactly supported.

The Meyer wavelet and scaling functions are defined in the frequency domain, starting with an auxiliary function; by changing the auxiliary function, a family of different wavelets can be obtained. Meyer wavelets can be used for CWT and also for DWT, but in this case the corresponding filters are not FIR. Favorable characteristics include symmetry and infinite regularity. The main issue is that no fast algorithms are available. The wavelet function does not have compact support, but decreases to 0 when $\theta \to \infty$ faster than any inverse polynomial. This property also holds for the derivatives; the wavelet is infinitely derivable. A compactly supported variant of these wavelets exists: the *discrete Meyer wavelets*, corresponding to FIR filters that can be used in the DWT.

14.10.3 Orthogonal Compactly Supported Wavelets

This family includes Daubechies wavelets, symlets and coiflets. These wavelets are real and have the following general properties:

- the scaling function exists;
- the analysis is orthogonal;
- the scaling and wavelet functions are compactly supported;
- the wavelet function has vanishing moments.

They can be used for the DWT via FWT, and also for the CWT. Nice properties of these wavelets include compact support, vanishing moments, and the fact that the related filters are FIR. Regularity can be poor, and this is a disadvantage. Specific properties are:

- for Daubechies wavelets: asymmetry;
- for symlets: near-symmetry;
- for coiflets: near-symmetry; both the scaling and the wavelet functions have vanishing moments.

1. **Haar wavelet**:
   this is the simplest wavelet. It is the only real-valued wavelet that is compactly supported, symmetric and orthogonal. Its symmetry ensures that the wavelet filter has linear phase characteristics, meaning that when a wavelet filtering operation is performed on a signal with this wavelet, there will be no phase distortion in the filtered signal. It is however discontinuous: it looks like a step function. The scaling function is $\varphi(\theta) = 1$ on $[0, 1]$ and 0 otherwise. The wavelet function is $\psi(\theta) = 1$ on $[0, 0.5]$, $\psi(\theta) = 1$ on $[0.51]$ and 0 otherwise. The support width is thus 1; the associated filters have length 2; the number of vanishing moments for $\psi(\theta)$ is 1. The Haar wavelet is the same as the Daubechies wavelet of order 1.

2. **Daubechies wavelets** (Daubechies 1992):
   the appearance of Daubechies compactly-supported orthogonal wavelets made discrete wavelet analysis practicable. Daubechies wavelets are indicated by db$K$, where "db" stands for Daubechies and K is the order, a strictly positive integer

equal to the number of vanishing wavelet moments. Associated filters have length $2M = 2K$. The Daubechies wavelets are designed so as to have the highest number of *vanishing wavelet moments* for a given support width, which does not imply the best smoothness for a given support width, and the associated scaling filters are minimum-phase filters (a.k.a extremal phase filters). For this reason, Daubechies wavelets are also known as the Daubechies extremal phase wavelets. They are not symmetric. They are not defined in terms of the scaling and wavelet functions, but in terms of the associated filters: in fact, these functions cannot be written down in closed form, and are derived numerically using the cascade algorithm.

3. **Symlets** (Daubechies 1992):
   the symlets are nearly-symmetrical, compactly supported orthogonal wavelets with least asymmetry and highest number of *vanishing wavelet moments* for a given support width, proposed by Daubechies as modifications to the Daubechies family. They are more symmetric than the extremal phase wavelets, and are also known as Daubechies least-asymmetric wavelets. The properties of the two wavelet families are similar. Symlets are usually indicated as sym$K$, where K is the order ($K = 2, 3, \ldots$), which is equal to the number of vanishing wavelet moments. Associated filters are near-linear-phase filters with length $2M = 2K$.

4. **Coiflets** (Daubechies 1992):
   also called *Coifman wavelets*, these wavelets were designed by Daubechies after suggestion by Ronald R. Coifman, to allow for *both the scaling function and the wavelet function to have a number of vanishing moments*. These are compactly supported orthogonal wavelets with the highest number of vanishing moments for both $\varphi(\theta)$ and $\psi(\theta)$ for a given support width. Their name is abbreviated as coif$O$, where O is the order ($O = 1, \ldots, 5$). The order is equal to $K/2$, where K is the number of vanishing wavelet moments. The number of vanishing scaling moments is $K - 1$. The total number of vanishing moments is $2K - 1$. The length of the four associated filters is $2M = 6O = 3K$, so that $M = 3K/2 = 3O$. The order of the scaling filter and the wavelet-support width are $2M - 1 = 6O - 1 = 3K - 1$. These wavelets are more symmetrical than Daubechies wavelets, which can be important in certain applications, and for them the approximation $c_0[k] \approx x[n]$ is particularly good.

14.10.4 Biorthogonal Compactly Supported Wavelet Pairs

This family includes B-spline biorthogonal wavelets (Daubechies et al. 1990; Daubechies 1992; Cohen et al. 1992). These wavelets are real and have the following properties:

- the scaling function exists;
- the analysis is biorthogonal;

- the scaling and wavelet functions both for decomposition and reconstruction are compactly supported;
- the scaling and wavelet functions for decomposition have vanishing moments.

They can be used for both the DWT via FWT, and the CWT. The main disadvantage lies in that orthogonality is lost. Positive features include:

- the fact that desirable properties for decomposition and reconstruction are split, allowing flexibility. Biorthogonal wavelets feature a pair of scaling functions and associated scaling filters, one for analysis and one for synthesis; there is also a pair of wavelets and associated wavelet filters, one for analysis and one for synthesis. Synthesis and analysis highpass/lowpass filters therefore are not related by a simple folding relation, as in the case of orthogonal wavelets;
- symmetry and association with linear-phase FIR filters. Perfect symmetry and PR are incompatible when the same filters are used for decomposition and reconstruction, as it happens in orthogonal systems, except for the Haar wavelet. On the contrary, biorthogonal B-spline wavelets are compactly supported wavelets for which symmetry and exact reconstruction are possible with FIR filters.

Biorthogonal B-spline wavelets are characterized by two orders, O_R and O_D, where "R" stands for reconstruction and "D" for decomposition. Usual values of the orders are: $O_R = 1$, $O_D = 1, 3, 5$; $O_R = 2$, $O_D = 2, 4, 6, 8$; $O_R = 3$, $O_D = 1, 3, 5, 7, 9$; $O_R = 4$, $O_D = 4$; $O_R = 5$, $O_D = 5$; $O_R = 6$, $O_D = 8$. Their name is bior$O_R.O_D$. The analysis and synthesis wavelets can have different numbers of vanishing moments and differenty properties. Associated filters have different length, which can also be odd: for example, for bior6.8 the lowpass decomposition and highpass reconstruction filters have length 17, while the highpass decomposition and lowpass reconstruction filters have length 11. However, in computation these filters are padded with zeros, to augment them to equal even lengths.

Reverse biorthogonal wavelets also exist, derived from the previous ones.

14.10.5 Analytic Wavelets

This family includes complex DOG wavelets, the complex Morlet wavelet, the complex Shannon wavelet and complex frequency B-spline wavelets, i.e., complex wavelets the spectrum of which are splines. These wavelets are used for CWT only, and exhibit minimal properties:

- the scaling function does not exist;
- the analysis is not orthogonal;
- the wavelet function is not compactly supported.

The main favorable properties are symmetry, and the fact that the wavelets have explicit expressions. The main disadvantage is that exact reconstruction is not ensured and fast algorithms are unavailable.

1. **Complex Derivatives of Gaussian (DOG) Wavelets**:
   this family, used for the CWT, is built starting from the complex Gaussian function

   $$f(\theta) = C_m e^{-j\theta} e^{-\theta^2}$$

   and then taking the m-th derivative of $f(\theta)$. The parameter of this family is m; the normalization constant C_m is such that the squared norm of the m-th derivative of $f(\theta)$ is 1. Values of m normally considered go from 1 to 8. The symmetry properties are the same as with real DOGs.

2. **Complex Morlet Wavelet** (Teolis 1998):
   the complex Morlet wavelet is defined, in its more general form, as

   $$\psi(\theta) = \frac{1}{\sqrt{\pi T_p}} e^{2j\pi f_c \theta} e^{-\frac{\theta^2}{T_p}}.$$

   where T_p is a time parameter regulating the width of the Gaussian bell and f_c is the center frequency of the wavelet's spectrum.[16]

3. **Complex Shannon Wavelet** (Teolis 1998):
   the main characteristic of the Shannon wavelet is that its spectrum is constant over some interval of frequencies excluding the origin and zero elsewhere. The frequency interval of support is described in terms of a desired center frequency f_c and a desired bandwidth f_b, with the condition $f_c > f_b/2$. The Shannon wavelet is thus defined primarily in the frequency domain, and then obtained via inverse Fourier transform. Its expression is

   $$\psi(\theta) = \sqrt{f_b} \text{Sinc} \, (f_b\theta) e^{2j\pi f_c \theta}.$$

   Because the envelope of the Shannon wavelet is a Sinc function, the time decay is poor (inversely proportional to time). This wavelet can be obtained from the complex frequency B-spline wavelets by setting $m = 1$.

4. **Complex Frequency B-Spline Wavelets** (Teolis 1998):
   as in the Shannon case, the frequency B-spline wavelets are defined directly in the frequency domain on a compact frequency interval with support described in terms of a desired center frequency and a desired bandwidth. A complex frequency B-spline wavelet is defined by

---

[16]Note that in the previous chapter we gave a different and less general definition of the complex Morlet wavelet,

$$\psi_0(\theta) = \frac{1}{\sqrt[4]{\pi}} e^{j\omega_0 \theta} e^{-\frac{\theta^2}{2}},$$

that corresponds to fixed values of T_p and f_c. More precisely, it corresponds to $T_p = 2$ and $f_c = 2\pi/\omega_0$. In the standard expression of the Gaussian probability density function, we would have $T_p = 2\sigma^2$, where σ is the standard deviation of the Gaussian distribution. Thus, $T_p = 2$ means $\sigma = 1$. Note, however, that the constant factor $1/\sqrt{\pi T_p}$, that for $T_p = 2$ becomes $1/\sqrt{2\pi}$, does not coincide with the factor $1/\sqrt[4]{\pi}$.

$$\psi(\theta) = \sqrt{f_b} \left\{ \mathrm{Sinc}\left(\frac{f_b\theta}{m}\right) \right\}^m e^{2j\pi f_c\theta},$$

an expression depending on three parameters: m is the integer order-parameter ($m \geq 1$), f_b is the bandwidth parameter, and f_c is the wavelet center frequency. The frequency B-spline wavelets are an entire family of wavelets indexed by an integer order parameter m. These wavelets are a generalization of the Shannon wavelet, in the sense that the complex frequency B-spline wavelet with $m = 1$ is the Shannon wavelet.

References

Abry, P.: Ondelettes et turbulence, Diderot edn. France, Paris (1997)

Burrus, C.S., Gopinath, R., Guo, H.: Introduction to Wavelets and Wavelet Transforms: A Primer. Prentice-Hall, Upple Saddle River (1998)

Cohen, A., Daubechies, I.: Non-separable bidimensional wavelet bases. Rev. Mat. Iberoam. 9(1), 51–137 (1993)

Cohen, A., Daubechies, I., Feauveau, J.-C.: Biorthogonal bases of compactly supported wavelets. Commun. Pure Appl. Math. 45, 485–560 (1992)

Daubechies, I.: Orthonormal bases of compactly supported wavelets. Commun. Pure Appl. Math. 41, 909–996 (1988)

Daubechies, I.: Ten lectures on wavelets. In: CBMS-NSF Regional Conference Series in Applied Mathematics, Society for Industrial and Applied Mathematics (SIAM), Philadelphia, PA, USA (1992)

Daubechies, I.: Orthonormal bases of compactly supported wavelets II. Variations on a theme. SIAM J. Math. Anal. 24(2), 499–519 (1993)

Daubechies, I., Lagarias, J.C.: Two-scale difference equations I. Existence and global regularity of solutions. SIAM J. Math. Anal. 22(5), 1388–1410 (1991)

Daubechies, I., Lagarias, J.C.: Two-scale difference equations. II. Local regularity, infinite products of matrices, and fractals. SIAM J. Math. Anal. 23(4), 1031–1079 (1992)

Daubechies, I., Cohen, A., Feauveau, J.-C.: Biorthogonal bases of compactly supported wavelets. Commun. Pure Appl. Math. XLV, 485–560 (1990)

Holschneider, M., Kronland-Martinet, R., Morlet, J., Tchamitchian, P.: A real-time algorithm for signal analysis with the help of the wavelet transform. In: Combes, J.-M., Grossmann, A. (eds.) Wavelets, Time-Frequency Methods and Phase Space (Inverse Problems and Theoretical Imaging). Springer, Berlin (1989)

Kovačević, J., Vetterli, M.: Nonseparable multidimensional perfect reconstruction filter banks and wavelet bases for R^n. IEEE Trans. Inform. Theory 38(2) 533–555 (1992)

Mallat, S.: A theory for multiresolution signal decomposition: the wavelet representation. IEEE Trans. Pattern Anal. Machine Intell. 11(7), 674–693 (1989)

Mallat, S.: A Wavelet Tour of Signal Processing. Academic Press, Waltham (2008)

Meyer, Y.: Wavelets and Operators. Cambridge University Press, Cambridge (1993)

Misiti, M., Misiti, Y., Oppenheim, G., Poggi, J.M.: Décomposition en ondelettes et méthodes comparatives: étude d'une courbe de charge électrique. Revue de Statistique Appliquée XLII(2), 57–77 (1994)

Misiti, M., Misiti, Y., Oppenheim, G., Poggi, J.-M. (eds.): Wavelets and Their Applications. Wiley-ISTE, New York (2007)

Nason, G.P., Silverman, B.W.: The stationary wavelet transform and some statistical applications. In: Antoniadis, A., Oppenheim, G.: Wavelets and Statistics. Lecture Notes in Statistics, vol. 103, pp. 281–300. Springer, New York (1995)

Percival, D.B., Walden, A.T.: Wavelet Methods for Time Series Analysis. Cambridge University Press, New York (2000)

Qian, S.: Introduction to Time-Frequency and Wavelet Transforms. Prentice Hall, Upper Saddle River (2001)

Shensa, M.J.: Discrete Wavelet Transforms: The Relationship of the á trous and Mallat algorithms. Treizième Colloque GRETSI—Juan Les Pins, France (1991)

Strang, G., Nguyen, T.: Wavelets and Filter Banks. Wellesley-Cambridge Press, Wellesley (1996)

Tian, J., Wells, R.O. Jr.: Vanishing Moments and Wavelet Approximation. Technical Report CML TR95-01, Computational Mathematics Laboratory, Rice University, Houston, TX, USA (1995)

Teolis, A.: Computational Signal Processing with Wavelets (Applied and Numerical Harmonic Analysis). Birkhauser, Boston (1998)

Vetterli, M.: Filter banks allowing perfect reconstruction. Signal Process. **10**, 219–244 (1986)

Vetterli, M., Herley, C.: Wavelets and filter banks: theory and design. IEEE Trans. Acoust. Speech Signal Process. **40**(9), 2207–2232 (1992)

Vetterli, M., Kovačević, J.: Wavelets and Subband Coding. Prentice Hall PTR, Englewood Cliffs (1995). Reissued in 2013 by CreateSpace Independent Publishing Platform

Chapter 15
De-noising and Compression by Wavelets

15.1 Chapter Summary

This chapter offers an introduction to the problem of signal de-noising, a task that can be effectively accomplished wavelets. A related issue is that of signal compression, and we will briefly touch upon the main ideas concerning it. Real-world examples will be provided for both de-noising and compression.

The term "signal de-noising" indicates the recovery of a digital signal that has been contaminated by additive noise; it is a particularly interesting specialization of a more general problem known in statistics as function estimation, in which an unknown function must be recovered from some corrupted version. The noise contaminating the "true" signal is most often assumed to be Gaussian and white. In theory, the performance of a de-noising algorithm is measured using the mean square error (MSE) between the true signal and its reconstructed version. In practice, since the true signal is unknown, the MSE can only be estimated.

The orthogonal DWT is particularly suited for de-noising, for a number of reasons, including the fact that the decomposition is additive, and consequently, the analysis of the noisy signal is equal to the sum of the analyses of the true signal and of the additive noise. Moreover, if noise is supposed to be white, then the detail coefficients on all scales are essentially white noises with the same variance. The basic idea in wavelet-based de-noising consists of thresholding the detail coefficients of the noisy signal, preserving only those that are larger than the characteristic amplitude of the noise, which is another parameter that must be estimated. Approximation coefficients are normally left untouched, since they represent low-frequency terms that usually contain important components of the signal, and are less affected by noise. The interest of wavelets, within this framework, stems from their capacity to represent the true signal using very few significantly non-zero coefficients. Real-world signals are, in many cases, fairly smooth, except in rare locations like the beginning and the end of transitory phenomena, or ruptures. This renders the decomposition of the true signal by wavelets very sparse, so that the signal is very well represented by the

© Springer International Publishing Switzerland 2016   715
S.M. Alessio, *Digital Signal Processing and Spectral*
Analysis for Scientists, Signals and Communication Technology,
DOI 10.1007/978-3-319-25468-5_15

coefficients of a rather rough approximation, to which some large detail coefficients are added. The threshold can generally be a function of level and time, but usually is function of level only, or even a scalar. There are many variants of de-noising procedures that differ primarily by strategy of thresholding (global, i.e., performed on all levels at the same time; level by level; local in time; by data blocks, etc.), rule for thresholding (hard or soft), and threshold selection method (Square2log, SURE, HeurSURE, Minimax, etc.). An essential ingredient in the recipe for a threshold is the estimate of the standard deviation of the white noise contaminating the signal. This estimate can be obtained from the level-1 detail coefficients; in cases in which the presence of non-white noise is suspected, the noise standard deviation can be estimated level-by-level.

Signal compression is a process aimed at retaining only the information necessary to reconstruct significant features of the original signal, for reasons of storage saving. The relation with the de-noising issue is obvious, but in compression the focus is on the extent to which the number of DWT coefficients to be stored to make signal reconstruction possible can be reduced with respect to the complete set, while preserving a substantial amount of the signal's variability. While in de-noising we seek the best strategy to extract the true signal, irrespective of how many coefficients we will have to retain, the threshold and the thresholding method chosen for compression most often come from external constraints, such as frequency bandwidth of interest, available memory, prescribed compression performance, etc. The compression procedure contains the same steps as de-noising, i.e., computing the wavelet decomposition and thresholding the detail coefficients. The retained coefficients and their positions in the original sequence are memorized, so as to be able to reconstruct the compressed signal. The main differences from the de-noising procedure lie in the choice of the threshold and in the rule adopted for applying it, i.e., for thresholding the detail coefficients.

In this chapter, wavelet families will be indicated with their abbreviated names introduced in Sect. 14.10.

15.2 Signal De-noising by DWT

It is well known to any scientist and engineer who work with real-world data that signals do not exist without noise, which may be negligible (i.e., high SNR) under certain conditions. However, there are many cases in which the noise corrupts the signals in a significant manner, and it must be removed from the data before proceeding with further analysis. The process of noise removal is generally referred to as *signal de-noising*. Although the term "signal de-noising" is general, it usually indicates the recovery of a digital signal that has been contaminated by *additive* noise. Moreover, most practical cases are covered by the assumption of additive *Gaussian* white noise. We will now describe how de-noising can be tackled by DWT.

15.2.1 Theory of De-noising by Wavelets

The model for the estimation of an unknown discrete-time signal corrupted by additive noise is (see, e.g., Johnstone and Silverman 1997)

$$x[n] = s[n] + \sigma_e e[n], \qquad\qquad N = [0, N-1].$$

The $e[n]$ are normally-distributed random variables with zero mean and unit variance; the factor σ_e allows for noise with non-unit variance. The goal is to recover the underlying signal $s[n]$ from the noisy data $x[n]$ with little error, i.e., to suppress $e[n]$ and to recover the "true signal" $s[n]$.

In theory, the performance of a de-noising algorithm is usually measured using the mean square error (MSE) between the true signal and its reconstructed version $\hat{s}[n]$, or using its square root, indicated by RMSE. On the basis of this simple criterion, we look for a sequence $\hat{s}[n]$ (estimated true signal; de-noised signal) minimizing the MSE,

$$R(s, \hat{s}) = \frac{1}{N} \left\{ \sum_{n=0}^{N-1} |s[n] - \hat{s}[n]|^2 \right\}.$$

MSE measures the degree of similarity between the de-noised signal and the true signal. The smaller the MSE, the more faithful is the reconstruction of $s[n]$ provided by the de-noised signal. $R(s, \hat{s})$ can be seen as a form of *risk function*, meaning a measure of the "risk" associated with approximating $s[n]$ by $\hat{s}[n]$. Note that de-noising is a particularly interesting specialization of a more general problem known in statistics as *function estimation*, in which an unknown function must be recovered from its corrupted version.

In practice, since the true signal $s[n]$ is unknown, the MSE is impossible to evaluate directly, and cannot be used as an objective criterion to optimize the parameters of the de-noising procedure. The risk function itself must be properly estimated, and several ways to do so have been proposed. The unbiased risk estimator tools, among which SURE (see next subsection) is a well-known representative, aim at tackling this issue.

As Misiti et al. (2007) pointed out, two of the greatest successes of wavelets are signal de-noising and compression, which are often regarded as particularly difficult tasks. As we will see, the issue of signal compression is conceptually strictly related to that of de-noising. The de-noising and compression procedures based on an orthogonal DWT are simple and powerful algorithms that are often easier to fine-tune than the traditional methods of function estimation. This is mainly due to the following reasons:

- due to the orthogonality of the transformation, the orthogonal DWT decomposition of a noisy process is additive, and consequently, the analysis of the noisy signal is equal to the sum of the analyses of the true signal and of the additive noise;

- most true signals we may be interested in recovering from their noisy versions are smooth enough to have very *sparse DWT representations*, i.e., to be very well represented using only a few significantly non-zero coefficients: if we were able to analyze the true signal by an orthogonal DWT, we would find that only a small fraction of the detail coefficients are non-zero, and the position where they are non-zero cluster around the discontinuities and abrupt variations of the true signal. We could thus represent it in a satisfactory manner by the coefficients of a rather rough approximation, to which some large detail coefficients were added. More precisely, the energy in the DWT coefficients for most real-world smooth signals decays in an approximately exponential way toward smaller scales (Cheng 1997);
- a white noise process with given variance yields a set of detail coefficients which at all scales are white noises with the same variance. If we were able to analyze the white noise alone, we would find the same white noise at all levels. Thus, in the DWT of a signal corrupted by additive white noise, the energy of the noise component spreads equally across the scales. Since the energy of a smooth true signal lies mainly in the coarsest approximation coefficient vector, the energy of the additive noise component mainly affects the detail coefficients[1];
- provided that we can suitably estimate the *noise level*,[2] i.e., the amount of noise, represented by its standard deviation, if the irregularities of the signal generate detail coefficients that are larger than the amplitude scale σ_e of noise fluctuations, the de-noising process can focus on selecting and retaining the detail coefficients related to the signal and discarding the rest; the selection takes place through a thresholding;
- the DWT analysis is local and, consequently, thresholding leads to a local de-noising of the signal.

On the basis of these observations, the problem of recovering the unknown signal can be reformulated as the problem of selecting those few detail coefficients of the noisy signal that are significantly non-zero, against a Gaussian white noise background. This approach leads to estimating a threshold below which the detail coefficients of the noisy signal can be assumed to represent only noise, and can thus be discarded. The threshold value is determined based on certain statistical assumptions about the noise and/or the target signal. The procedure is efficient in reducing noise effectively, while preserving possible sharp features of the underlying true signal.

Wavelet thresholding methods for noise removal were first introduced by Donoho in 1993 (see Donoho 1993a, b, c; Donoho and Johnstone 1994a, b; Donoho 1995; Donoho and Johnstone 1995; Johnstone and Silverman 1997; Donoho and Johnstone 1998). These methods do not require any particular assumptions about the nature of

---

[1]Note that for Gaussian white noise, an orthogonal DWT will also preserve the Gaussian nature of the noise: the histogram of the noise in the transform domain will be a symmetric bell-shaped curve about its mean value, with the same width at half height that the noise amplitude distribution has in the time domain. In other words, the noise and the details have the same probabilistic properties, and the details inherit the Gaussian and centered nature of the noise.

[2]Often σ_e is referred to as the noise level. When using this term, we should be aware that the word level here has nothing to do with the word level as it is used in DWT theory, as in the expression "level of decomposition".

the signal, are undisturbed by the presence of discontinuities in the signal, and exploit the multiresolution properties of the wavelet transform. As a consequence, these techniques offer more possibilities than any frequency-selective filtering, because they can remove noise at any frequency. Last but not least, if the presence of non-white noise is suspected, DWT allows for analyzing the noise level separately at each wavelet scale, and adapting the de-noising algorithm accordingly.

Starting with a noisy signal, we will first decompose it in an orthogonal (orthonormal) wavelet basis using the DWT. We will use a wavelet whose form is similar to the true signal features we want to detect: the wavelet must be able to capture the transient spikes of the original signal that we desire to preserve. Then we will select those detail coefficients that are larger than the characteristic amplitude of the noise, and discard the others. We will keep the approximation coefficients vector of a suitably chosen level intact: indeed, the approximation coefficients represent low-frequency terms that usually contain important components of the signal, and are less affected by the noise.

The selection will be performed by thresholding the detail coefficients. Actually, it can be shown (Misiti et al. 2007) that thresholding is the optimal selection strategy. Each coefficient is compared with a threshold, denoted by λ, in order to decide whether it constitutes a desirable part of the signal $x[n]$, or not. The threshold can be global or level-dependent; it may be different in different sections of the signal, or be independent of time. Thus, the threshold will generally be a function of the decomposition/resolution level j and of the time index k, i.e., $\lambda = \lambda(j, k)$, but usually it will be a function of j only, i.e., $\lambda = \lambda(j)$, or even a scalar. Note that the value of a threshold is always positive. The threshold value(s) may be estimated according to different criteria (Sect. 15.2.3).

Also the thresholding rule can vary: it can be hard or soft.

- *Hard* thresholding is defined, for a generic real sequence $y[n]$ to be thresholded and for a threshold λ independent of time over the duration of $y[n]$, as

$$\delta_{hard,\,\lambda}[n] = \begin{cases} y[n] & \text{if } |y[n]| > \lambda, \\ 0 & \text{otherwise,} \end{cases}$$

where $\delta_{hard,\lambda}[n]$ indicates the output of the thresholding operation performed on the sample $y[n]$.

- *Soft* thresholding *shrinks* the kept coefficients by setting

$$\delta_{soft,\,\lambda}[n] = \begin{cases} y[n] - \text{sign}(y[n])\lambda & \text{if } |y[n]| > \lambda, \\ 0 & \text{otherwise.} \end{cases}$$

The soft thresholding rule can be also written as

$$\delta_{soft,\,\lambda}[n] = \frac{1}{2}\text{sign}\,(y[n])\,(|y[n]| - \lambda + ||y[n]| - \lambda|).$$

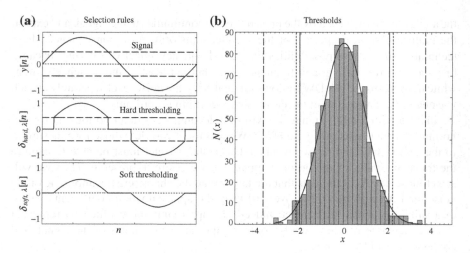

Fig. 15.1 **a** The two rules for thresholding a sequence $y[n]$: hard and soft. This is a case of a threshold λ independent of time; values of $y[n]$ equal to $\pm\lambda$ are indicated by *dashed horizontal lines*. **b** Thresholds selected, for a series of Gaussian *white* noise with unit variance, according to three different criteria (see text): Square2log (*dashed vertical line*), SURE (*solid vertical line*), and Minimax (*dotted vertical line*)

All else being equal, soft thresholding, also called *shrinkage*, leads to smoother true-signal estimators, and is the most common choice in de-noising. Hard thresholding is preferred for compression. Figure 15.1a is an illustration of these two rules. Note how the hard thresholding sets to zero the values of $y[n]$ that are below the threshold in absolute value, while the soft one pulls them towards the origin by an amount equal to the threshold (observe the *dashed horizontal lines*, representing $y[n] = \pm\lambda$).

Once the strategy, the threshold(s), and the thresholding rule have been selected, only one essential ingredient remains to be found: the noise level. How noisy is the signal? How does the noice level affect the threshold? After answering these question, we will be able to actually threshold the detail coefficients; later, using the thresholded detail coefficients and the untouched approximation coefficients, we can reconstruct the signal by IDWT. The signal obtained in this manner is the de-noised signal.

15.2.2 Estimation of Noise Level

The basic signal model assumes we know σ_e, but in reality the noise standard deviation is unknown a priori and must be estimated. The level-1 detail coefficients (the finest scale; the least decimated detail coefficients, i.e., the longest sequence among all the $d_j[k]$) are essentially noise coefficients with variance σ_e^2. In most cases, only some large level-1 detail coefficients are not ascribable to noise. In fact, the orthogonality of the DWT has the fundamental statistical consequence we already mentioned—which is not present in the biorthogonal transform—according to which the DWT transforms white noise into white noise. Hence, for the additive noise case, the transformed noise is also a white noise process with variance σ_e^2, and once we can identify the wavelet coefficients that correspond to the noise, estimating their variance is equivalent to estimating the variance of the actual noise. It is then enough to use a robust estimator, i.e., an estimator not sensitive to outliers, of the standard deviation of the level-1 detail coefficients to estimate σ_e. In the Gaussian-distribution case, the estimator has the form:

$$\hat{\sigma}_e = \text{median}\left\{\left|C_d^{(1)}[m]\right|\right\}/0.6745.$$

Here we exploit the relation between the standard deviation and the median[3] for the Gaussian distribution. Indeed, for L independent Gaussian centered random variables $(x_0, \ldots x_{L-1})$ with a standard deviation σ, the ensemble average of the median of the variables taken in absolute value is (Staudte and Sheather 1990) E $[\text{median}(|x_i|)] \approx 0.6745 \sigma$ for $i = [0, L-1]$. Thus the constant 0.6745 makes the estimate of the standard deviation unbiased for the normal distribution.

If we assume that noise is white, we will thus estimate σ_e from level-1 detail coefficients and use it for all levels. However, the assumption of white noise is sometimes too stringent, and we may need to relax it. If we need to account for non-white noise, a level-dependent (i.e., frequency dependent) estimation of σ_e^2 is necessary. A similar formula will then be used to estimate σ_e level by level.

15.2.3 Threshold Estimation

Several threshold selection methods have been proposed. Two main families are: *Donoho-Johnstone methods* (Square2log, SURE, HeurSURE, and Minimax), and *parametric methods*, like Birgé-Massart threshold and Birgé-Massart penalized threshold. Among the Donoho-Johnstone methods, SURE and HeurSURE typically serve to obtain level-by level threshold estimates, while Square2log and Minimax provide global thresholds. The Birgé-Massart threshold is level-dependent; the Birgé-Massart penalized threshold is global.

We now examine each of these threshold selection approaches. We will list both approaches normally adopted for de-noising (which are thus associated with the soft-

---

[3]The median is the numerical value separating the higher half of a data sample, a population, or a probability distribution, from the lower half. The median of a finite list of numbers can be found by arranging all the observations from lowest value to highest value and picking the middle one. If there is an even number of observations, then there is no single middle value; the median is then usually defined to be the mean of the two middle values.

thresholding rule) and approaches usually adopted for compression (which are thus associated with the hard-thresholding rule).

Threshold estimation and noise-level estimation must be clearly distinguished. It is simpler, from now on, to refer to the white noise case (we have already discussed how to deal with non-white noise). It is also convenient, for the moment, to focus on Donoho-Johnstone methods. We may then think of expressing the (global or level-dependent) threshold as

$$\lambda = \sigma_e \lambda_0,$$

where λ_0 is the threshold chosen according to one of the previously mentioned methods, for a signal model including white noise with $\sigma_e = 1$. λ_0 thus appears as a threshold for unit-variance noise that later is scaled multiplying it by the estimated noise level for the data we are actually dealing with. Depending on the threshold estimation method, λ_0 can be constant or level-dependent. The approaches to threshold estimation described below serve to define λ_0, from which λ is derived using the other essential ingredient in the recipe, namely the estimate of the noise level σ_e.

In practice, starting with a given data record, we will decompose it at a suitably selected level J. Then

- if we are using Square2log or Minimax, we will compute the scalar λ_0 and multiply it by σ_e to get the scalar λ;
- if we are using SURE or HeurSURE, we will reduce the detail coefficients of the jth level ($j = [1, J]$) by a factor equal to the estimated σ_e, i.e., we will scale them by the standard deviation of noise, thus turning to the unit-variance noise case. We will then compute $\lambda_0(j)$. Finally we will form $\lambda(j) = \sigma_e \lambda_0(j)$ for use with the original, unscaled coefficients.

Let us now examine the most popular threshold selection criteria.

1. **Square2log Criterion**

   The simplest thresholding method is to perform a DWT of the data and then use a constant threshold, i.e., a single threshold for all the detail coefficients in the expansion, independently of the level. This criterion is often employed for soft thresholding in de-noising.

   The threshold that is typically used is

   $$\lambda = \sigma_e \lambda_0 = \sigma_e \sqrt{2 \log N_{DWT}}$$

   where N_{DWT} is the size of the DWT coefficient vector $C[l]$ which contains all approximation and detail coefficients for a decomposition at a given level J. Note that N_{DWT} is of the order of data size N; note also that the threshold λ_0 depends only on N_{DWT}. The Square2log criterion is illustrated in Fig. 15.1b (*dashed vertical line*), where it has been applied for threshold selection to a sequence $x[n]$ of Gaussian white noise with unit variance.

   The theoretical argument in favor of this expression is the following: as we already pointed out, if the noise samples $e[n]$ are normally distributed random variables with zero mean and variance σ_e^2, then the detail coefficients on all the scales are

Gaussian white noises with the same variance. Moreover, in spite of the fact that the noise values $e[n]$ are not bounded, it can be shown that

$$\lim_{N \to \infty} P \left\{ \max_{n=[0,N-1]} |e[n]| > \sigma_e \sqrt{2 \log N} \right\} = 0,$$

where P stands for probability. Observe how the upper noise bound, $\max |e[n]|$, is just inflated by a multiplicative factor when we pass from the case with $\sigma_e = 1$ to the general case ($\sigma_e \neq 1$). For a wide range of N values (say, from 2^6 to 2^{19}), statistical calculations reveal that only in about one-tenth of the realizations will pure noise variables exceed the threshold. In the DWT domain, the DWT detail coefficients will behave in the same way, with the only difference being that they are precisely N_{DWT} in number. This result ensures that if we choose the threshold in the DWT domain according to the Square2log criterion, then, in the limit, no pure-noise detail coefficient will be let through the threshold.

2. **Stein's Unbiased Risk Estimate (SURE) Criterion**
This is an unbiased threshold estimator proposed by Stein (1981). It may be employed for a level-by level threshold estimation, or for a global threshold estimation. We will describe the first, more general approach. The SURE criterion constructs a threshold which is meant to be used for soft thresholding in de-noising applications.[4] The related threshold $\lambda(j)$ can be evaluated as follows.
First, the $N_d^{(j)}$ detail coefficients of level j, with $j = [1, J]$, are divided by σ_e and then are squared and arranged in *ascending order* of squared value. This operation is indicated by a sort operator:

$$q_j^2[m] = \text{sort}_{k=\left[1,N_d^{(j)}\right]} \left| C_d^{(j)}[k] \right|^2, \qquad m = \left[1, N_d^{(j)} \right].$$

A risk vector for level j is then computed as

$$r_j[m] = \frac{N_d^{(j)} - 2m + \sum_{i=1}^{m} q_j^2[i] + (N_d^{(j)} - m)q_j^2[m]}{N_d^{(j)}}.$$

The index $m_{\text{best}}^{(j)}$ is found, corresponding to the smallest element of $r_j[m]$, i.e., to the minimum risk. The threshold is finally defined as

$$\lambda(j) = \sigma_e \lambda_0(j) = \sigma_e \sqrt{q_j^2[m_{\text{best}}^{(j)}]}.$$

The rationale behind SURE is that the estimation problem for MSE actually leads to the above-reported threshold as the value of minimum risk. It may be useful to remark that SURE is not a surrogate for MSE, but minimizing SURE is a surrogate

---

[4]For this reason, the corresponding de-noising procedure is referred to as *SUREShrink*.

for minimizing MSE. When choosing a statistical estimator, we often want the one that will minimize MSE, but we cannot compute MSE without knowing the true signal, or the true characteristics of the noise we are trying to suppress. SURE gives us an unbiased estimate of what the unknown MSE is. SURE is still an estimator, so it has a distribution, whereas MSE is a fixed value.

The choice of the threshold in the SURE method being data-adaptive, this algorithm is claimed to be smoothness-adaptive: if the unknown function contains a jump, the reconstruction (de-noised signal) is essentially supposed to do the same. If the unknown function has a smooth piece, the reconstruction is essentially expected to be as smooth as the mother wavelet will allow.

If a single SURE threshold for all levels is desired, then the criterion may be applied to the detail coefficients belonging to level J only.

3. **HeurSURE Criterion**

According to this criterion, the threshold is selected using a combination of Square2log and SURE methods. In fact, if the SNR is very small, the SURE method is found to perform poorly: it turns out to be too conservative, i.e., to give low thresholds prone to preserving part of the noise. In such cases, the Square2log method gives a better threshold estimate. However, the Square2log criterion in this case is used in a level-by level manner. This is obtained by setting

$$\lambda_0^{\text{Square2log}}(j) = \sqrt{2 \log N_d^{(j)}}.$$

Subsequently, the detail coefficients are reduced to the unit-variance noise case (dividing them by σ_e) and the SURE threshold, which we denote by $\lambda_0^{\text{SURE}}(j)$, is computed from the modified coefficients. With the same coefficients, the following two parameters are formed:

$$\eta = \sum_{k=1}^{N_d^{(j)}} \left| C_d^{(j)}[k] \right|^2, \qquad \chi = \frac{\left[\log_2 N_d^{(j)} \right]^{3/2}}{\sqrt{N_d^{(j)}}}.$$

Finally the following threshold is defined:

$$\lambda(j) = \begin{cases} \sigma_e \lambda_0^{\text{Square2log}}(j) & \text{for } \eta < \chi, \\ \sigma_e \min \left[\lambda_0^{\text{Square2log}}(j), \lambda_0^{\text{SURE}}(j) \right] & \text{for } \eta \geq \chi. \end{cases}$$

If a single HeurSURE threshold for all levels is desired, then the criterion will be applied to the detail coefficients belonging to level J only.

Threshold determination is an important problem. An exceedingly small threshold may yield a result which may still be noisy, while a threshold that is too large can cut a significant part of the signal, thus causing loss of important details. Heursure emphasizes the importance of SNR: when SNR is low and the SURE estimator would lead to a noisy result, then the less conservative Square2log algorithm is used.

4. **Minimax Criterion**

   This method finds a global threshold λ using the so-called *minimax principle* that is used in statistics to design estimators. Since the de-noised signal can be assimilated to the estimator of an unknown sequence contaminated by noise, the Minimax estimator is the option that realizes the minimum, over a given set of signals $\hat{s}[n]$, of the related MSE. The threshold is defined as

   $$\lambda = \sigma_e \lambda_0 = \begin{cases} \sigma_e \left(0.3936 + 0.1829 \log_2 N_{DWT} \right) & \text{for} \quad N_{DWT} > 32, \\ 0 & \text{for} \quad N_{DWT} \leq 32. \end{cases}$$

   SURE and Minimax threshold selection rules are particularly conservative (see Fig. 15.1b, *solid* and dotted vertical lines, respectively) and are convenient when small details of the true signal lie in the noise range. Other rules remove the noise more efficiently.

5. **Birgé-Massart Criterion**

   There are two shades of this parametric method.

   - *Birgé-Massart Threshold*: scarce high, scarce medium, scarce low.
     This method is well suited mainly for compression purposes, though it may be used for de-noising. It is a levelwise threshold estimation method suggested by Birgé and Massart (1997). The Birgé-Massart threshold is classified as a parametric method of threshold estimation, because three parameters characterize it: the level of the decomposition J, a positive constant M and a sparsity parameter α, with $1 < \alpha < 5$. The sparsity of the wavelet representation of the de-noised signal grows with α. The level-dependent threshold prescribed by this method is such that at level J the approximation is kept, and for any level j from 1 to J, the n_j largest coefficients are kept, with

     $$n_j = \frac{M}{(J + 2 - j)^\alpha}.$$

     So the Birgé-Massart threshold leads to selecting the highest coefficients in absolute value at each level, with the number of retained coefficients growing with $J - j$. If N_a is the length of vector of the coarsest approximation coefficients (level J), M is set proportional to N_a, and three different choices for the M are proposed: scarce high, $M = N_a$; scarce medium, $M = 1.5 N_a$; scarce low, $M = 2 N_a$. As for the sparsity parameter, typically the choice is $\alpha = 3$ for de-noising, and $\alpha = 1.5$ for compression.
   - *Birgé-Massart Penalized Threshold*: penalized high, penalized medium, penalized low.
     This is a global threshold estimation method, which was suggested by Birgé and Massart (2001, 2006). It is classified as a parametric method since it

includes a sparsity parameter $\alpha > 1$. It is commonly used for de-noising. It applies a scalar threshold λ defined as the one that minimizes

$$\text{crit}(m) = -\sum_{k \leq m} v^2[k] + 2\sigma_e^2 m \left(\alpha + \log \frac{N_d}{m} \right) \qquad m = 1, 2, \ldots N_d,$$

where $v^2[k]$ are the detail coefficients, arranged in *descending order* of magnitude and then squared, and N_d is their number. If we call m_{best} the minimizer of $\text{crit}(m)$, then

$$\lambda = \sigma_e \lambda_0 = \sigma_e \sqrt{v^2[m_{\text{best}}]}.$$

Three different intervals of α-values are proposed: penalized high, $2.5 \leq \alpha < 10$ (typically, 6.25); penalized medium, $1.5 < \alpha < 2.5$ (typically, 2); penalized low, $1 < \alpha < 2$ (typically, 1.5).

15.2.4 De-noising Examples

Now we will illustrate the capabilities of the de-noising procedures presented above, providing a couple of real-world examples.

1. To see a first example of de-noising, we use the same electrical load signal studied in Sect. 14.7 (Misiti et al. 1994). On the basis of the previous discussion, we use db3 and decompose at level 3, since we have already seen that the level-3 approximation represents an overall good smoothing of the signal. We will simply use zero padding to extend the signal (for extension modes see Sect. 14.5). We do not worry too much about the extension mode, since it is expected to mainly affect the edges of the de-noised signal.
   Figure 15.2a, b show the original signal and the A_3 approximation, respectively. Only the most disturbed part of the signal is shown, to avoid cluttering (minutes 1950–3900). The approximation A_3 can be considered as a crude approach to de-noising. A more rigorous approach is to apply soft thresholding to detail coefficients, with a global positive threshold that will be chosen, according to the Square2log criterion, computing the median of the absolute value of the detail coefficients obtained from a level-1 decomposition with db1 (the simplest compactly-supported orthonormal wavelet, having the lowest number of vanishing wavelet moments among all dbK), dividing by 0.6745 and multiplying by $\sqrt{2 \log N}$. The threshold, in this case with $N = 4320$ data samples, turns out to be equal to $\left[\sqrt{2 \log N} \right] \text{median}(|d_1|)/0.6745 = 4.0917*1.4663/0.6745 = 8.8947 \approx 8.90$.
   We may wonder if this estimate would change substantially if we employed another wavelet in place of db1 to estimate the median. Using db3, db4, db8 and db12, for example, we would get thresholds of 7.80, 7.64, 8.09, and 7.85, respectively. Therefore using an order-1 wavelet provides an upper bound for the

Fig. 15.2 Electrical load
signal studied by Misiti et al.
(1994). **a** A noisy portion
of the original signal,
b approximation A_3 as a
crude form of de-noising,
and **c** de-noised signal
obtained decomposing at
level 3 by db3 and
soft-thresholding the detailed
coefficients with a global
positive threshold threshold
selected according (see text
for details)

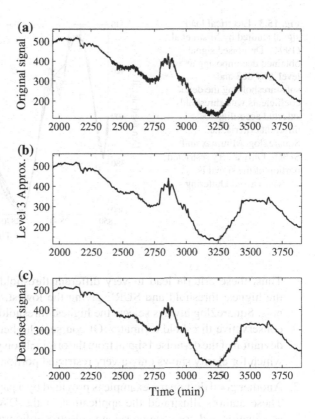

median and therefore for the threshold, though the use of higher-order wavelets
leads to very similar values.

The result of the de-noising procedure with soft thresholding and noise variance
estimate independent of level (i.e., assumption of additive white noise) is shown in
Fig. 15.2c. The result of Fig. 15.2c is quite good, in spite of the time heterogeneity
of the nature of the noise after and before the beginning of a sensor failure that
occurred around minute 2450. Notice how we have removed the noise without
compromising the sharp detail of the original signal (for instance, observe the
abrupt transition around minute 3450). This would not be the case if we used the
cruder A_3 approach (Fig. 15.2b).

Maintaining the white noise assumption, we can repeat the de-noising using different criteria, in order to compare the results. This time we adopt db3 for noise
level estimation, i.e., the same wavelet adopted for signal decomposition, and
apply Square2log, SURE and Minimax criteria. The thresholds thus obtained
are:

- Square2log: $4.0924\,\sigma_e = 7.80$,
- SURE: $0.3657\,\sigma_e = 0.697$,
- Minimax: $2.6032\,\sigma_e = 4.96$.

Fig. 15.3 Electrical load
signal studied by Misiti et al.
(1994). De-noised signal
obtained decomposing at
level 3 by db3 and
soft-thresholding the detail
coefficients with thresholds
selected according to
different criteria, i.e.,
Square2log, Minimax and
SURE. Only a very restricted
portion of the signal is
shown to avoid cluttering

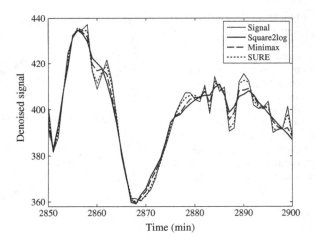

Thus, these criteria lead to very different thresholds, with Square2log giving
the highest threshold and SURE giving the lowest. Actually, for different signals, Square2log always selects the highest thresholds, while SURE is the most
conservative threshold estimator. Of course, a higher threshold means a greater
deviation of the de-noised signal from the original noisy one, as visible in Fig. 15.3,
which for clarity shows only a very restricted portion of the results.

2. Another good de-noising example is provided by a paper by Castillo et al. (2013).
   These authors illustrated the application of the DWT for baseline wandering
   elimination and noise suppression in electrocardiographic (ECG) signals.
   ECG acquisition from the human skin involves the use of high gain instrumen-
   tation amplifiers, and this fact makes the ECG signal prone to be contaminated
   by different sources of noise. This effect is particularly important when the target
   is the measurement of fetal ECG signals acquired over the mother's abdomen.
   Noise makes visual inspection and ECG-feature extraction difficult.
   In general, ECG contaminants can be classified into different categories, including
   power line interference, electrode pop or contact noise, patient-electrode motion
   artifacts, electromyographic (EMG) noise, and baseline wandering. Among these
   noises, the power line interference and the baseline wandering (BW) are the
   most significant. The power line interference is a very narrow-band type of noise
   centered at 50 or 60 Hz, with a bandwidth of less than 1 Hz.[5] BW usually comes
   from respiration, at frequencies varying between 0.15 and 0.3 Hz. Other noise
   components may be wideband, and usually involve a complex stochastic process,
   which also distorts the ECG signal. Usually the ECG signal acquisition analog
   hardware can remove the power line interference, but the BW and other wideband
   noise components are not easily suppressed by analog circuits. De-noising is

---

[5]In software applications, often such narrow-band undesired components are removed by special
filters called *notch filters*. An example of a notch filter is provided in the next chapter.

thus better tackled in software by DWT-based processing, and it is decisive for subsequent parameter extraction in clinic applications.

Castillo et al. (2013) discuss in depth the choice of parameters for this task, such as wavelet type, decomposition level, threshold selection and thresholding rule, as well as noise level estimation, through simulations with synthetic ECG signals and application to real-world data. Here, we focus on the application of the method to an ECG record taken from the DaISy dataset (DeMoor 2010). The data, contributed by Lieven De Lathauwer, contain 8 leads of skin potential recordings of a pregnant woman (Callaerts 1989; De Lathauwer et al. 2000). The lead recordings, three thoracic and five abdominal, were sampled at 250 Hz and are 10 s long (2500 samples each). The sequence analyzed here is from lead 4 (abdominal). The db6 wavelet is chosen, because of its similarity to the basic waveform in an ECG. The extension mode is zero padding.

The processing steps are as follows:

- decomposition up to a high level and visual identification of the approximation that best captures the BW;
- decomposition at the level of the approximation that captures the BW;
- reconstruction of the signal on the basis of details only, i.e., BW suppression; this step leads to a BW-corrected signal;
- de-noising of the BW-corrected signal to obtain the BW-corrected and de-noised signal.

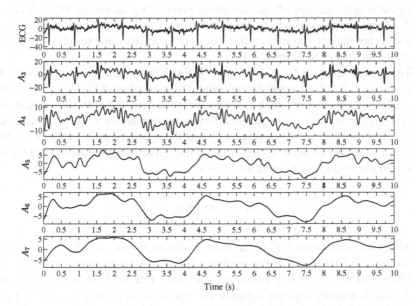

Fig. 15.4 Approximations to a fetal ECG signals acquired over the mother's abdomen, obtained by DWT decomposition at levels 3 to 7 using db6 wavelet

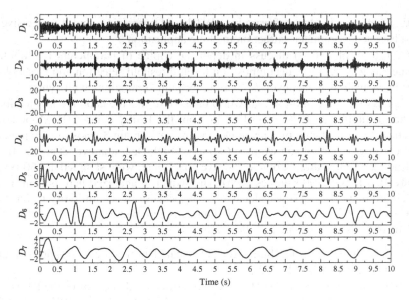

Fig. 15.5 Details of the fetal ECG signal, obtained by DWT decomposition of the ECG signal at level 7 using db6 wavelet

We start from the maximum decomposition level (7) advisable on the basis of signal length and type of wavelet (db6). Figure 15.4 shows the ECG signal and the approximation sequences for levels from 3 to 7. This figure shows that A_6 and A_7 are the approximations that best capture the baseline wander. Figure 15.5 shows the details associated with A_7.

The most important frequencies in BW are below a certain frequency f_c. For example, $f_c \simeq 1$ Hz for wandering coming from respiration (0.15–0.3 Hz). Other wandering components, such as motion of the patient and of the instrument, may have higher-frequency components. To remove wandering, it would be necessary to select the decomposition level such that the approximation captures the ECG components for frequencies lower than this f_c. The decomposition level for BW suppression can be estimated as follows:

$$J_{BW} = \text{int} \left(\log_2 \frac{f_{Ny}}{f_c} \right),$$

where int means round off to the nearest integer. In our case, $f_{Ny} = 125$ Hz, and setting $f_c = 1$ Hz (respiration) we obtain $J_{BW} = 7$. After applying seven low-pass filters and downsampling processes, the DWT tree leads to an approximation A_7 which actually captures frequencies from 0 Hz to about 1 Hz and is a good estimation of BW. This confirms the choice made by visual inspection of the approximations 1–7 in Fig. 15.4. These arguments are illustrated by Figs. 15.6 and 15.7, which show the spectra of the signal and of its various components.

Fig. 15.6 Periodograms of the original fetal-ECG signal and of the approximations obtained by DWT decomposition at level 7 using db6 wavelet

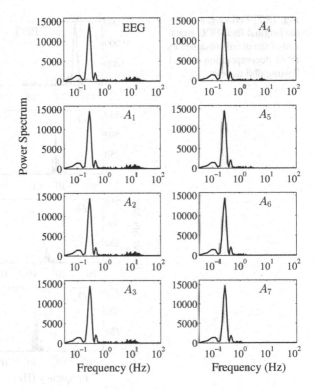

These plots allow us to see how the chosen approximation and the corresponding details cover the spectrum of the original signal. Observe how considering A_6 or A_7 as representative of BW actually corresponds to suppressing the signal's spectral content above 1 Hz.

The result of the process of BW suppression is illustrated in Fig. 15.8a–c, in which the original signal, the estimated baseline (ultimately chosen as A_7) and the BW-corrected signal are shown, respectively.

We must now select a decomposition level for the subsequent de-noising of the BW-corrected signal. De-noising here is supposed to suppress noise-related features also at frequencies not covered by the BW suppression. The high-frequency noise in the original signal is contained mainly in details D_3–D_1 from the decomposition at level 7 (see Figs. 15.5 and 15.7; the latter shows how the spectral content of D_3–D_1 lies above ∼10 Hz). To avoid eliminating clinically important components of the signal, such as PQRST[6] morphologies, only these details must be touched by the de-noising process.

---

[6]The acronym PQRST indicates the pattern of electrical activity of the heart during one cardiac cycle, as recorded by electrocardiography. Typically, an ECG exhibits five deflections, arbitrarily named P to T waves. The Q, R, and S waves occur in rapid succession, do not all appear in all

Fig. 15.7 Periodograms of the original fetal-ECG signal and of the details obtained by DWT decomposition at level 7 using db6 wavelet

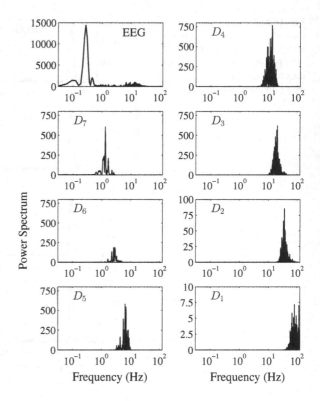

At this point, we decompose the BW-corrected signal up to level 4, still using db6. The resulting approximation is shown in Fig. 15.8d; the details are visible in Fig. 15.8e–h.

The effect of noise is evident in D_2 and D_1. The details higher than level 2 contain most of the significant information for diagnostics. These observations suggest performing de-noising of the BW-corrected signal using db6 and a decomposition at level 3. A global threshold associated with soft thresholding is applied. The estimate of the noise standard deviation is single-level, based on level 1. Figure 15.9 shows the original ECG signal and the BW-corrected-and-de-noised signal. The BW-corrected-and-de-noised abdominal ECG signal still exhibits characteristics

(Footnote 6 continued)

leads, and reflect a single event, so that they are usually considered together. A Q wave is any downward deflection after the P wave. An R wave follows as an upward deflection, and the S wave is any downward deflection after the R wave. The T wave follows the S wave, and in some cases an additional U wave follows the T wave. The QRS complex is the name for the combination of three of the graphical deflections seen on a typical ECG. It is usually the central and most visually obvious part of the tracing; it corresponds to the depolarization of the right and left ventricles of the human heart. In adults, it normally lasts 0.06–0.10 s, while in children and during physical activity, it may be shorter.

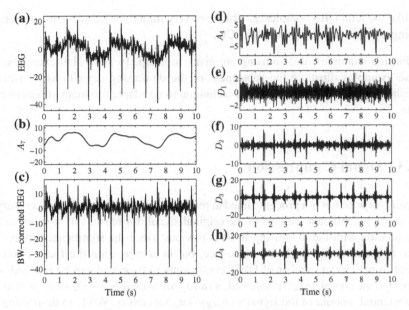

Fig. 15.8 a The original fetal-ECG signal, **b** the baseline extracted by DWT with db6 as the approximation at level 7, and **c** the BW-corrected signal. Subsequent decomposition of the BW-corrected signal using db6 at level 4: **d** approximation, **e–h** details

Fig. 15.9 a Original fetal ECG signal and **b** BW-corrected-and-de-noised signal

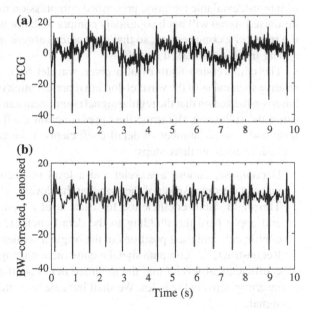

like the fetal QRS complexes, which are very important for subsequent processing, such as separation of fetal ECG.

Further examples of de-noising are given in Chap. 16 as Matlab exercises. All these examples illustrate the robustness of the de-noising algorithms by wavelet thresholding and the capacity of the technique to be adapted to many different contexts.

15.3 Signal Compression

Signal compression is a process aimed at retaining only the information necessary to reconstruct significant features of the original signal, for reasons of storage saving. The relation with the de-noising issue is obvious, since the significant features we want to preserve are evidently noise-free, but in compression the focus is on the extent to which the number of DWT coefficients to be stored can be reduced with respect to the complete set (*compression ratio; compression score*), while preserving a substantial amount of the signal's energy, i.e., variability. While in de-noising we seek the best strategy to extract the true signal, irrespective of how many coefficients we will have to retain, the threshold and the thresholding method chosen for compression most often comes from external constraints, such as frequency bandwidth of interest, available memory, prescribed compression ratio, etc. This is why a well-de-noised signal will not be optimally compressed, nor will de-noising be optimal in the result of a compression, so that fluctuations absent from the original signal may appear in the compressed signal.

The compression features of a given wavelet basis are primarily linked to the relative sparseness of the wavelet domain representation of the signal. Compression is based on the notion that the regular signal component can be accurately approximated using the following elements: the approximation coefficients at a suitably chosen level and a small number of detail coefficients. Like de-noising, the compression procedure contains three steps.

1. Decompose: choose a wavelet and a level of decomposition J. Compute the wavelet decomposition of the signal at that level.
2. Threshold detail coefficients: for each level j from 1 to J, select a threshold and apply hard thresholding to the detail coefficients. Memorize the retained coefficients and their positions in the original sequence.
3. Reconstruct, i.e., compute signal reconstruction using the original approximation coefficients of level J and the modified detail coefficients of levels from 1 to J, inserting zeros in the holes. We shall indicate by x_c the compressed reconstructed signal.

The main differences from the de-noising procedure are found in step 2, i.e., in the choice of the threshold and in performing hard thresholding rather than shrinkage.

There are two compression approaches available. The first consists of taking the wavelet expansion of the signal and keeping the largest absolute value detail coefficients. In this case, we may establish a global-threshold value, or start directly from some desired compression performance, or from some prescribed relative square-norm recovery performance (see below). Only a single threshold needs to be selected in this case. The second approach consists of applying suitably determined level-dependent thresholds.

When compressing using orthogonal (orthonormal) wavelets, the *retained energy* in percentage (a.k.a. *squared-norm recovery*) is defined as

$$\text{RE} = 100 * \frac{\|x_c[n]\|^2}{\|x[n]\|^2}.$$

RE gives the ratio of the ℓ^2-squared-norm of the compressed signal to the ℓ^2-squared-norm of the original signal. If this number is close to 100 %, this means that the energy in the compressed signal is very close to the energy in the uncompressed signal. Note that RE is also equal to the ratio of the ℓ^2-squared-norm of the thresholded wavelet coefficients to the ℓ^2-squared-norm of the original wavelet coefficients.

The *compression score* CS is defined as

$$\text{CS} = 100 * \frac{\text{number of coefficients set to zero}}{\text{total number of coefficients}}.$$

It gives the ratio of the number of thresholded wavelet coefficients that have been set to zero to the total number of wavelet coefficients, expressed as a percentage. If this number is close to 100 %, it means that virtually all the thresholded wavelet coefficients have been zeroed: the signal is being reconstructed based on a very sparse set of wavelet coefficients. Note that normally the approximation coefficients are left untouched: only detail coefficients are processed.

15.3.1 A Compression Example

Compressing applications for sequences are numerous in any field where large amounts of data are collected and then need to be stored. Let us examine an example of compression using a global threshold in association with hard thresholding, for a given and unoptimized wavelet choice, to verify that it can produce a nearly complete squared-norm recovery.

We use the electrical load signal used in Sects. 14.7 and 15.2 and select a very noisy part of it, i.e., minutes 2300–3500 (see Fig. 14.27). Decomposing this signal at level $J = 3$ with a db3 wavelet we obtain the coefficient vector plotted in Fig. 15.10. The inset shows the detail coefficients only, on an expanded scale. The length of the uncompressed signal is 1201 samples; the complete coefficient vector, including approximation and detail coefficients, is 1215 samples-long. The level-3 approxima-

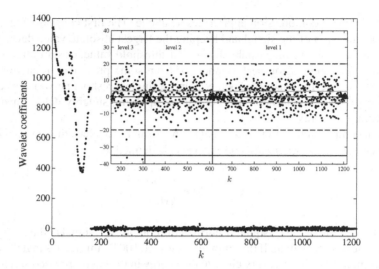

Fig. 15.10 Coefficients produced by decomposing the electric consumption signal of Fig. 14.27 (minutes 2300–3500) at level 3 with db3 wavelet, versus position k in the wavelet coefficients vector (see Sect. 14.5.3). The inset shows the detail coefficients only, on an expanded scale. The *gray*, *dashed* and *black horizontal lines* in the inset mark thresholds of 3.38, 20 and 35, respectively

tion coefficients are the first 154; the level-3 detail coefficients follow, and are 154; the subsequent level-2 and level-1 detail coefficients are 304 and 603, respectively. Compression is thus aimed at saving space with respect to about 1200 numbers to store, while obtaining a reasonably low value of the *root-mean-square deviation* (RMSD), defined as

$$\text{RMSD} = \sqrt{\sum_n |x[n] - x_c[n]|^2} = \sqrt{\sum_k |C^{\text{or}}[k] - C^{\text{co}}[k]|^2}.$$

Here we indicated by $C^{\text{or}}[k]$ and $C^{\text{co}}[k]$ the concatenated detail coefficient vector for the original and compressed signal, respectively.

How can we find a simple but effective threshold value in this case of compression? A basic approach often adopted for compression consists of the following steps.

We decompose again the signal at level 1 by db1 and compute the median of the absolute value of the detail coefficients. If the result is non-zero, we choose this value as the threshold. If the median vanishes, then we just prepare to "kill" the lowest coefficients of the previous decomposition at level J, by setting the threshold to 5 % of the maximum absolute value of these detail coefficients. In the present case, the resulting threshold is 3.3774 ≈ 3.38; this default value is represented by the *gray line* in the inset of Fig. 15.10. We may note that this threshold is quite conservative; many detail coefficients exceed it, and we may expect to obtain a relatively low CS using it. A threshold of 20 (about 6 × 3.3774; see the *dashed line* in the inset of Fig. 15.10) would, on the other hand, lead to retaining only a few high-valued detail coefficients, and getting a higher score. A threshold of 35 (over 10×3.3774; see the *black line* in the inset of Fig. 15.10) would mean killing practically all coefficients, except a couple of them at level 3. Note that the maximum absolute value found among these coefficients is 37.34.

Let us try the highest threshold.

The value of CS is $100 \{1 - [(154 + 2)/1215]\} = 87.16\,\%$, so that only $13.84\,\%$ of the coefficients are retained; RE is $99.96\,\%$ and the RMSD is 3.32. In spite the drastic reduction in the number of coefficients, RE is quite high. We cannot do much more than that in terms of CS without touching the approximation coefficients: indeed, if we neglected all detail coefficients, we would get $100 \{1 - [154/1215]\} = 87.33\,\%$, the retained coefficients would be $12.68\,\%$, and RE would approximately remain $99.96\,\%$, with a RMSD of 3.41.

Figure 15.11 compares the signal, the approximation A_3 that can be considered as the result of a crude de-noising/compression, and the compressed signal obtained using the fixed threshold of 35. The inset presents an expanded view of the region of the maximum, which allows appreciating how the approximation A_3 (*gray curve*) and the compressed signal (*dashed curve*) actually differ only in the restricted time interval in which the signal (*black curve*) undergoes very sharp changes, with the compressed signal following the original one more closely than the approximation. Even the contribution of a couple of coefficients thus turns out to be crucial in determining the behavior of the compressed signal at "critical" points. This can be

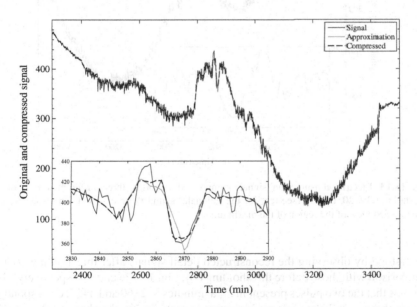

Fig. 15.11 The original electrical load signal, the approximation A_3 as a smooth signal representation, and the compressed signal obtained using hard thresholding with a fixed threshold of 35, i.e., killing all detail coefficients but 2. The inset presents an expanded view of the region of the maximum, in which the original signal undergoes the sharpest changes

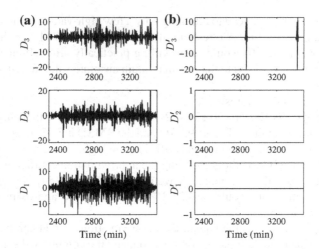

Fig. 15.12 Electric consumption signal of Fig. 14.27 decomposed at level 3 with db3: reconstructed detail signals **a** before and **b** after hard thresholding with a fixed threshold of 35

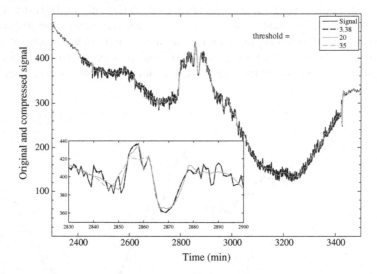

Fig. 15.13 Electric consumption signal, compressed adopting three different values of the threshold: 3.38, 20, and 35 (see text for the rationale behind these choices). The inset presents an expanded view of the region of the maximum

understood by observing the reconstructed detail signals after thresholding (D'_j), in comparison with those before thresholding (D_j; Fig. 15.12a and b, respectively). It is obvious that the two pulses present in D'_3 at minutes ≈ 2860 and 3425, corresponding to two sharp signal changes, are important in reproducing the shape of the original signal at these instants.

The lower the threshold, the closer the compressed signal to the original one, and the more modest the compression performance. This can be seen in Fig. 15.13,

Fig. 15.14 a Compression score (CS) and percentage of retained coefficients; **b** squared-norm recovery (RE); **c** root-mean square deviation (RMSD) of compressed signal to original signal for a range of threshold values from 3.38 to $12\times 3.38 \approx 40$. In all panels, the *dashed* and *solid vertical lines* correspond to two threshold values (20 and 35) discussed in the text. Note that 20 and 35 are about 5.92 and 10.36 times the default threshold of 3.38, respectively. In panel a, the *horizontal lines* represent the asymptotic values of CS and percentage of retained coefficients. In the other two panels, the *dashed* and *solid horizontal lines* indicate the performances for the thresholds 20 and 35, respectively

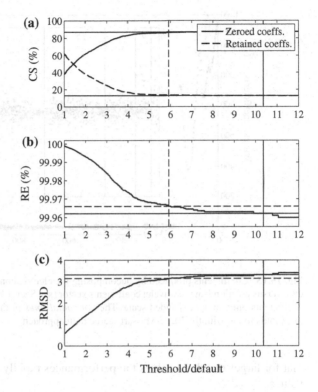

comparing the compressed signals obtained adopting three different threshold values, from the conservative value of about 3.38 to 20, and finally to 35. With the threshold set to 3.38, the original signal (*black solid line*) and the compressed one (*black dashed line*) are almost indistinguishable.

In order to be more quantitative in judging the results of the compression procedure, we can compute CS and percentage of retained coefficients, RE and RMSD of the compressed signal from the original one for a range of threshold values from 3.38 to $12 \times 3.38 \approx 40$, in steps of 0.01. Figure 15.14a–c shows the corresponding plots. The *dashed* and *solid vertical lines* highlight two threshold values (20 and 35) mentioned above. In Fig. 15.14a, CS and percentage of retained coefficients are shown. The *horizontal lines* represent the asymptotic values of CS and percentage of retained coefficients, i.e., the values that would be attained discarding all the detail coefficients; the values for the thresholds 20 and 35 are indistinguishable from the asymptotic ones. In Fig. 15.14b, we can see the values of RE. The *dashed* and *solid horizontal lines* respectively indicate RE values for the thresholds of 20 and 35. Figure 15.14c shows the RMSD of the compressed signal from the original one. The *dashed* and *solid horizontal lines* indicate the RMSD values for the thresholds of 20 and 35, respectively. Observe the rapid variation of all curves for thresholds values below 4–5 times the default threshold of 3.38, which is quite conservative. Note also

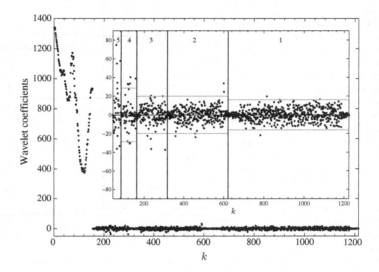

Fig. 15.15 Coefficients produced by decomposing the electric consumption signal at level 5 with db3, versus position k in the wavelet coefficients vector (see Sect. 14.5.3). The inset shows the detail coefficients only, on an expanded scale. The *horizontal lines* in the inset mark the level-by level thresholds found with the Birgé-Massart scarce high approach

that for larger threshold values the performances rapidly approach their asymptotic values.

As a more elaborated approach to compression, we can try the levelwise Birgé-Massart scarce-high strategy. Figure 15.15 shows the coefficients from the decomposition of the electrical load signal using db3 at level 5, which gives 42 level-5 approximation coefficients and 42 level-5 detail coefficients, as well as 79, 154, 304 and 603 detail coefficients for levels 4 to 1, respectively. The length of the DWT is thus 1224. Recall that the signal is 1201-samples long.

Adopting default values for M and α, namely $M = N_a$ and $\alpha = 1.5$, the thresholds that the method provides level by level are 16.05, 19.94, 19.83, 28.65, and 14.42 for levels 1 to 5, respectively, and correspond to keeping 15 detail coefficients at level 5, 8 at level 4, 5 at level 3, 4 at level 2 and 3 at level 1, for a total of 35 kept detail coefficients and 42 kept approximation coefficients out of 1224. The thresholds are shown by *horizontal gray segments* in the inset of Fig. 15.15. Figure 15.16 illustrates the result of the compression. The CS value is very high: $100 \{1 - [(42 + 35)/1224]\}$ = 93.71 % zeroed coefficients (only 6.29 % retained). The RE value is as high as 99.97 %.

Finally, we may mention that for compression some authors prefer to adopt biorthogonal wavelets rather than orthogonal wavelets. Biorthogonal wavelets can be very efficient for compression, because the analysis part (decomposition) and the synthesis part (reconstruction) are treated by two distinct wavelets. Adopting a compactly supported analysis wavelet with many zero moments is convenient because it makes the representation of the signal as sparse as possible. However, such a wavelet

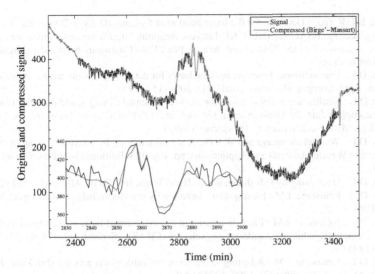

Fig. 15.16 The original signal and the compressed signal obtained using the Birgé-Massart scarce high strategy. The inset presents an expanded view of the region of the maximum

will be rather asymmetric and irregular, and therefore ill-adapted to reconstruction, for which it is desirable to use a regular and symmetric wavelet in order to minimize reconstruction artefacts. The framework of biorthogonal wavelets makes it possible to meet these two requirements simultaneously.

References

Birgé, L., Massart, P.: From model selection to adaptive estimation. In: Pollard, D. (ed.) Festchrift for Lucien Le Cam, Research Papers in Probability and Statistics, pp. 55–88. Springer, New York (1997)

Birgé, L., Massart, P.: Gaussian model selection. J. Eur. Math. Soc. 3(3), 203–268 (2001)

Birgé, L., Massart, P.: Minimal penalties for gaussian model selection. Probab. Theor. Relat. Fields **138**(1–2), 33–73 (2006)

Callaerts, D.: Signal separation methods based on singular value decomposition and their application to the real-time extraction of the fetal electrocardiogram from cutaneous recordings. Ph.D. Thesis, K.U. Leuven, E.E. Dept., Leuven, Belgium (1989)

Castillo, E., Morales, D.P., García, A., Martínez-Martí, F., Parrilla, L., Palma, A.J.: Noise suppression in ECG signals through efficient one-step wavelet processing techniques. J. Appl. Math., Article ID 763903, 13 p (2013)

Cheng, C.L.: High frequency compensation of low sample-rate audio files: a wavelet-based excitation algorithm. In: Rikakis, T. (ed.) Proceedings of the International Computer Music Conference, pp. 458–461. Thessaloniki, Greece, 1997. International Computer Music Association, San Francisco (1997)

De Lathauwer, L., De Moor, B.L.R., Vandewalle, J.: Fetal electrocardiogram extraction by blind source subspace separation. IEEE Trans. Biomed. Eng. **47**(5); Spec. Top. Sect. Adv. Stat. Sig. Process. Biomed. 567–572 (2000)

DeMoor, B.L.R. (ed.): Database for the Identification of Systems (DaISy). Department of Electrical Engineering, ESAT/STADIUS, KU Leuven, Belgium. http://homes.esat.kuleuven.be/~smc/daisy/. Accessed 20 Dec 2014 (Used dataset: [96-012]—Cutaneous potential recordings of a pregnant woman)

Donoho, D.L.: Unconditional bases are optimal bases for data compression and for statistical estimation. Appl. Comput. Harmonic Anal. 1(1), 100–115 (1993)

Donoho, D.L.: Nonlinear wavelet methods for recovery of signals, images, and densities from noisy and incomplete data. In: Daubechies, I. (ed.) Different Perspectives on Wavelets, pp. 173–205. American Mathematical Society, Providence (1993)

Donoho, D.L.: Wavelet shrinkage and W.V.D.: A ten-minute tour. In: Meyer, Y., Roques, S. (eds.) Progress in Wavelet Analysis and Applications, pp. 109–128. Editions Frontières, Gif-sur-Yvette (1993)

Donoho, D.L.: De-noising by soft-thresholding. IEEE Trans. Inf. Theor. 41(3), 613–627 (1995)

Donoho, D.L., Johnstone, I.M.: Ideal spatial adaptation by wavelet shrinkage. Biometrika 81, 425–455 (1994)

Donoho, D.L., Johnstone, I.M.: Ideal de-noising in an orthonormal basis chosen from a library of bases. Comptes Rendus de l'Académie des Sciences (CRAS)—Series I, vol. 319, pp. 1317–1322, Paris (1994)

Donoho, D.L., Johnstone, I.M.: Adapting to unknown smoothness via wavelet shrinkage. J. Amer. Statist. Assoc. (JASA) 90(432), 1200–1224 (1995)

Donoho, D.L., Johnstone, I.M.: Minimax estimation via wavelet shrinkage. Ann. Stat. 26(3), 879–921 (1998)

Johnstone, I.M., Silverman, B.W.: Wavelet threshold estimators for data with correlated noise. J. Roy. Statis. Soc. Ser. B 59(2), 319–351 (1997)

Misiti, M., Misiti, Y., Oppenheim, G., Poggi, J.M.: Décomposition en ondelettes et méthodes comparatives: étude d'une courbe de charge électrique. Revue de Statistique Appliquée XLII(2), 57–77 (1994)

Misiti, M., Misiti, Y., Oppenheim, G., Poggi, J.M.: Wavelets and Their Applications. ISTE, London (2007)

Staudte, R.G., Sheather, S.J.: Robust Estimation and Testing. Wiley, New York (1990)

Stein, C.M.: Estimation of the mean of a multivariate normal distribution. Ann. Stat. 9(6), 1135–1151 (1981)

Part V
Exercises in Matlab Environment

Chapter 16
Exercises with Matlab

16.1 Chapter Summary

This final chapter presents exercises on most of the techniques discussed in the book, including filter design and filtering implementation, stationary and evolutionary spectral analysis, etc.

A set of Matlab scripts and functions is provided to perform the various tasks. They require installation of the Signal Processing, Statistics and Wavelet Toolboxes.

No preliminary tutorial on Matlab is given, since very good books and guides are available on this topic, including the Matlab User's Guide and Reference Guide. Moreover, the programs presented here employ only very simple commands, so that any user with a basic knowledge of the Matlab language can go through them and possibly modify them if desired. Every exercise comes with comments on the results and with the reference to the chapter or section where the considered technique is explained.

Just a warning: do not forget that in the theory we use indexes starting from 0, while in Matlab all indexes start from 1. For example, $[0, N - 1]$ here becomes $[1, N]$. As for the names of the variables appearing in each program, they have been chosen in such a way to be—hopefully—intuitive, in order to avoid the need for detailing their meaning. We will often use the name of these variables instead of their mathematical symbols: for example, we will often write T_s in place of T_s.

16.2 Generation of Synthetic Data

To study features and performances of signal processing algorithms, numerical simulations are often performed on synthetic signals, usually containing an arbitrary number of sinusoids (Sect. 2.3.2) with given amplitudes and frequencies. If we aim at simulating random data from real-world measurements, the initial phase of each

© Springer International Publishing Switzerland 2016
S.M. Alessio, *Digital Signal Processing and Spectral*
Analysis for Scientists, Signals and Communication Technology,
DOI 10.1007/978-3-319-25468-5_16

sinusoidal component will be chosen at random in the interval $0-2\pi$, with a uniform distribution. For the same randomization purpose, we will add Gaussian white noise (see Sects. 9.7, 11.3.5, etc.) with properly chosen variance to the sinusoids.

Samples of Gaussian white noise can be produced using a generator of pseudo-random numbers, and modifying its output so as to get a Gaussian amplitude distribution. Let us suppose we generate a sequence of real numbers according to the following rules:

- all numbers must be in [0,1];
- no discrimination is allowed in favor or against any number or interval of numbers; all numbers must have the same probability of being extracted;
- each number must be chosen independently of the others; we are not allowed to take note of the numbers already generated and change our subsequent choices on the basis of them.

In this way we generate a white noise sequence with a uniform amplitude distribution.

The Gaussian case is different. Imagine an antenna which receives energy from all sources of radio signals in the Universe and adds all these energies together. The resulting tension will be random, but not uniform: most of the time the incoming radio signals will tend to mutually cancel and will produce tensions close to zero; once in a while, they will interfere constructively and will produce higher tensions. The amplitude distribution of these tension values will be Gaussian.

In order to generate Gaussian white noise, we will adopt the following rules:

- the numbers must be extracted from a Gaussian population;
- they must be chosen independently of one another.

More details on pseudo-random number generation, in general and in Matlab in particular, may be found in Moler (2004). Single chapters of the book are also available for download at www.mathworks.com/moler/.

The exercises of this chapter require several sets of data. In some cases, the data is the outcome of real-world measurements; in most of the cases, it is synthetic. Real synthetic sequences are often used to test algorithms in discrete-time signal processing. When we process a sequence, the sampling interval does not play any role. However, even in the case of a synthetic series, here we wish to maintain a mental link to a hypothetical real-world application, about which we suppose that an analog signal exists, from which our sequence derives by sampling at constant rate $1/T\_s$. Therefore we will *arbitrarily* assign a value of T_s to any synthetic data sequence we create. Hereafter, this arbitrary value is assumed to be 5 s, unless stated otherwise.

The synthetic sequences we need are often composed by sinusoids and noise. For example, we will need a sequence with length N, containing Gaussian white noise with standard deviation sigma_wn, possibly superimposed to Nsin sinusoids with some amplitudes included in a vector A, and some analog frequencies included in a vector f_sd.[1] Thus we will start with a script devoted to this purpose. The script must

---

[1]Note that since we assigned a value of T_s to our synthetic data, we can express frequencies in analog terms, as we would do in the real world.

also work for Nsin = 0, i.e., for generating pure noise; if Nsin = 1, A and f_sd will be scalars. Actually, it is convenient to include the possibility of generating not just one sequence, but a set of sequences with different lengths, contained in a vector N.

This first script, like many others presented later, prompts the user for input of parameters. For example, prompt = 'N[] =' requires input of one or more values of N included in square brackets; prompt = 'sigma_wn =' requires input of a single value for sigma_wn, etc. The commands s = rng and rng(s) that appear at the beginning and end of the script reported below (data_Nsin_noise_variable_N) respectively serve to get the random number generator settings at the beginning of the work session and then, after one or more calls to the Matlab functions rand or randn, to restore the original generator settings, so that if the script is run a second time, it will produce exactly the same sequence. This allows for reproducibility of the results. The name sd, chosen for the output data, is nothing but the acronym of "synthetic data". Note that the sequence is centered before output.

```matlab
1   % data_Nsin_noise_variable_N
2   % Generate synthetic data (Nsin sinusoids in noise)
3   % with variable number N of samples
4
5   s = rng;
6   prompt ='N[] =';
7   N = input(prompt);
8   prompt ='Nsin (0 if noise only) =';
9   Nsin = input(prompt);
10  prompt ='T_s (s) =';
11  T_s = input(prompt); % s
12  prompt='A[] =';
13  if Nsin>0
14      A = input(prompt);
15      prompt ='f_sd (as omega/pi)[] =';
16      w_over_pi_sd = input(prompt);
17      f_sd = w_over_pi_sd./(2*T_s); % Hz
18  end
19  prompt ='\sigma_wn =';
20  sigma_wn = input(prompt);
21
22  t_sd=NaN(max(N),length(N)); % initialize time vector
23  sd=zeros(max(N),length(N)); % inizialize data vector
24  for m=1:length(N)
25    t_sd(1:N(m),m)=0:T_s:(N(m)-1)*T_s; % time vector in s
26    if Nsin>0 % build sinusoids
27      for i=1:Nsin
28        rndph=rand(1,1)*2*pi;
29        sd(1:N(m),m)=sd(1:N(m),m)+A(i)*sin(rndph+2*pi*f_sd(i)*t_sd(1:N(m),m));
30      end
31    end % if Nsin=0 data are zeros
32    sd(1:N(m),m)=sd(1:N(m),m)+sigma_wn*randn(N(m),1); % add noise
33    sd(1:N(m),m)=sd(1:N(m),m)-mean(sd(1:N(m),m)); % remove mean
34  end
35  rng(s);
```

One of the exercises requires a data set including sequences with white noise with variable standard deviation $\sigma_e \equiv$ sigma_wn, superimposed to Nsin sinusoids. The length N is fixed in this case. The script data_Nsin_noise_variable_sigma_wn given below generates this data, which is centered before output.

```
1   % data_Nsin_noise_variable_sigma_wn
2   % Generate synthetic data (Nsin sinusoids in noise) with variable sigma_wn.
3
4   s = rng
5   prompt ='N =';
6   N = input(prompt);
7   prompt ='Nsin (0 if noise only) =';
8   Nsin = input(prompt);
9   prompt ='T_s (s) =';
10  T_s = input(prompt); % s
11  if Nsin>0
12      prompt ='A[] =';
13      A = input(prompt);
14      prompt ='f_sd (as omega/pi)[] =';
15      w_over_pi_sd = input(prompt);
16      f_sd = w_over_pi_sd./(2*T_s); % Hz
17  end
18  prompt ='\sigma_wn[]=';
19  sigma_wn = input(prompt);
20
21  t_sd=(0:T_s:(N-1)*T_s)'; % time vector in s
22  sines=zeros(N,1); % initialize sinusoids vector
23  if Nsin>0 % build sinusoids
24    for i=1:Nsin
25      rndph=rand(1,1)*2*pi;
26      sines=sines+A(i)*sin(rndph+2*pi*f_sd(i)*t_sd);
27    end
28  end % if Nsin=0 sines are zeros
29  for m=1:length(sigma_wn)
30      sd(1:N,m)=sines+sigma_wn(m)*randn(N,1); % add noise
31      sd(1:N,m)=sd(1:N,m)-mean(sd(1:N,m)); % remove mean
32  end
33  rng(s);
```

16.3 DFT

We start our exercises by computing the DFT of a real sequence (Sect. 3.5). We will use the function fft of Matlab that implements the FFT (Sect. 3.6), provided that the specified number $N_{FFT} \equiv$ N_f of transform samples is equal to an integer power of 2. For the choice of N_f, the function nextpow2 may be useful. Given an integer that in our case is the number N of data samples, the function computes the exponent of the next higher integer power of two, which can be assumed as the value of N_f. For instance, if N = 500, then N_f $= 512 = 2^9$ is a good choice; of course, 2^{10} or even more would also be acceptable. Before any DFT computation we will reduce the input sequence to zero mean. Then, if N_f > N, the sequence must be zero-padded, and this is done automatically by fft. The output DFT samples correspond to analog frequencies extending from 0 to the frequency that immediately precedes the sampling frequency f_s = 1/T_s = 2 f_Ny. Thus discrete frequencies ω are in [0, 2π). Parseval's theorem allows for checking the result of the DFT computation.

The steps to be performed are the following:

- generating the zero-mean data vector sd to be transformed;
- computing the energy of sd;
- computing N_f samples the DFT of sd, being SD (as usual, we denote a transform with the same name as the data, but capitalized). The syntax of the function fft is SD = fft(sd,N_f);
- taking the modulus of SD: modSD = abs(SD);
- computing the phase angle of SD and unwrapping it, i.e., correcting the phase angle in radians by adding a multiple of $\pm 2\pi$ when the absolute jump between two consecutive phase elements is greater than, or equal to, 2π radians. This eliminates any spurious discontinuity introduced by the arc tangent used for computing the phase angle, and provides a smoother phase plot;
- computing the sum of squares, (modSD)^2, and checking if it equals the data energy as expected;
- computing the analog frequencies f corresponding to the SD samples; they cover [0, 2π);
- plotting the modulus and phase of SD as a function of f. The plot of modSD is usually linear, or in dB, or even double logarithmic: respectively,

```
plot(f,modSD)
```

or

```
plot(f,10*log\_10(modSD))
```

or

```
loglog(f,modSD)
```

The zero frequency f(1) is included only if the f-axis is linear;
- shifting the zero-frequency component to the center of the DFT vector, as it may be preferable in some applications. This can be obtained by the command SDshift = fftshift(SD). The samples of SDshift will correspond to frequencies—being fnegpos—in [-f_Ny,f_Ny), i.e., [-f_Ny,f_Ny-deltf], where deltf = 1/(N_f T_s);
- plotting the modulus of SDshift.

For this exercise,

- generate N = 8192 synthetic data with T_s = 5 s, containing three sinusoids with equal unit amplitudes, random initial phases and discrete angular frequencies 0.15π, 0.35π, and 0.55π, plus white noise with sigma_wn = 1; use the script data_Nsin_noise_variable_N for this task;

- compute and plot the DFT with N_f = N, using the script given below. Note that thanks to the linearity of the DFT, we can extend the signal model (sinusoids + noise) from the time domain to the frequency domain.

The script that follows (DFT) does the job.

```
1   % DFT
2   % Compute via FFT the DFT of a sequence and plot it.
3
4   % Compute data energy
5   E=sum(sd.^2)
6
7   % Compute DFT, modulus and phase angle
8   N_f=N;
9   SD=fft(sd,N_f);
10  modSD=abs(SD);
11  angleSD=angle(SD);
12  uangleSD=unwrap(angleSD);
13
14  % Compute right-hand term
15  % of Parseval's relation
16  E_check=sum(modSD.^2)/N_f
17
18  % Build frequency vector
19  f_Ny=1/(2*T_s);
20  deltaf=1/(N_f*T_s);
21  f=0:deltaf:(2*f_Ny-deltaf);
22
23  % Shift DFT, compute modulus and corresponding frequency vector
24  SDshift=fftshift(SD);
25  modSDshift=abs(SDshift);
26  fnegpos=(-f_Ny:deltaf:f_Ny-deltaf)';
27
28  % Plot results
29  figure(1) % modulus vs. f, linear scale
30  plot(f,modSD)
31  set(gca,'XLim',[0 2*f_Ny])
32  hlabelx=get(gca,'Xlabel');
33  set(hlabelx,'String','$f$ (Hz)','Interpreter','Latex','FontSize',16)
34  hlabely=get(gca,'Ylabel');
35  set(hlabely,'String','$|X|$','Interpreter','Latex','FontSize',16)
36
37  figure(2) % modulus in dB versus f
38  plot(f(2:end),10*log10(modSD(2:end)))
39  set(gca,'XLim',[0 2*f_Ny])
40  hlabelx=get(gca,'Xlabel');
41  set(hlabelx,'String','$f$ (Hz)','Interpreter','Latex','FontSize',16)
42  hlabely=get(gca,'Ylabel');
43  set(hlabely,'String','$10\log_{10}|X|$','Interpreter','Latex','FontSize',16)
44
45  figure(3) % modulus vs. f, log-log scale
46  loglog(f(2:end),modSD(2:end))
47  set(gca,'XLim',[0 2*f_Ny])
48  hlabelx=get(gca,'Xlabel');
```

```
49  set(hlabelx,'String','$f$ (Hz)','Interpreter','Latex','FontSize',16)
50  hlabely=get(gca,'Ylabel');
51  set(hlabely,'String','$|X|$','Interpreter','Latex','FontSize',16)
52
53  figure(4) % phase angle vs. f
54  plot(f,uangleSD/pi)
55  set(gca,'XLim',[0 2*f_Ny])
56  hlabelx=get(gca,'Xlabel');
57  set(hlabelx,'String','$f$ (Hz)','Interpreter','Latex','FontSize',16)
58  hlabely=get(gca,'Ylabel');
59  set(hlabely,'String','$\arg(X)$','Interpreter','Latex','FontSize',16)
60
61  figure(5) % shifted modulus vs. related f. linear scale
62  plot(fnegpos,modSDshift)
63  set(gca,'XLim',[-f_Ny f_Ny])
64  hlabelx=get(gca,'Xlabel');
65  set(hlabelx,'String','$f$ (Hz)','Interpreter','Latex','FontSize',16)
66  hlabely=get(gca,'Ylabel');
67  set(hlabely,'String','$|X|$','Interpreter','Latex','FontSize',16)
```

SD is complex and *symmetrical*, being the DFT of a real sequence. Thus, the useful information is contained in the SD samples indexed from $k = 1$ to $k = N\_f/2+1$. At the zero frequency ($k = 1$) and at the Nyquist frequency ($k = N\_f/2+1$) the DFT is real; the sample at $k = 1$ is equal to the data mean value, and therefore is very small (ideally, it should be zero). We may observe that

- the dB plot appears rather flat, while the loglog plot does not;
- the loglog plot puts into evidence a base of rapidly varying spectral values that extends over the whole f-axis and is very evident especially at high frequency. This base is due to white noise, which has a flat spectrum;
- the peaks corresponding to the sinusoids have a finite width, as expected from windowing effects (Sect. 5.2.1). Do they fall at the expected analog frequencies? To compute analog frequencies corresponding to discrete angular frequencies $0.15\,\pi$, 0.35π, and 0.55π, recall that for sd we assumed a sampling interval of 5 s.

16.4 Digital Filters

Here we will experiment with the filtering of discrete-time signals (Chaps. 2 and 6), with the design of equiripple FIR filters (Chap. 7), and with the classical design method for IIR filters (Chap. 8).

16.4.1   FIR Filters

16.4.1.1   Filtering a Sequence with an FIR Filter

In this exercise we will filter a sequence using a LP FIR-Type I bandpass filter
(Sect. 6.5.2) whose impulse response is known. The impulse response of the filter is
given in Table 16.1. Its length is 55 (order 54) and we will assume it resides in a file
named h.dat. Figure 16.1 shows the time- and frequency-domain characteristics of
this filter. The passband bounds are 0.2150π and 0.464π.

   Filtering can be implemented in different ways in Matlab. We will use the functions
filter, conv, and fftfilt.

- The function filter works with both FIR and IIR filters because it implements
  filtering via the difference equation (Sects. 2.4.6 and 6.6). Its syntax is

```
sdf = filter(b,a,sd)
```

  The name sdf assigned to the output refers to "sd, filtered". When an FIR filter is
  used, b = h (the impulse response vector) and a = 1: therefore the command we
  need here is sdf = filter(h,1,sd).
- The function conv implements linear convolution (Sect. 2.4.1) and can therefore
  be used with FIR filters only. Its syntax is

```
sdf = conv(h,sd)
```

- The function fftfilt filters data using an efficient FFT-based method called the
  overlap-add method (Oppenheim and Schafer 2009). This is a filtering technique
  related to convolution performed in the frequency domain (Sect. 3.5.4) and therefore only works for FIR filters (see Matlab documentation). Its syntax is

```
sdf = fftfilt(h,sd,N_f)
```

Table 16.1 Impulse response of a Linear Phase FIR passband filter (see text)

n	$h[n]$	n	$h[n]$	n	$h[n]$	n	$h[n]$	n	$h[n]$
1	0.0145	12	−0.0054	23	0.0575	34	0.0459	45	0.0229
2	0.0258	13	−0.0144	24	−0.0664	35	0.0005	46	0.0268
3	−0.0052	14	0.0009	25	−0.2051	36	0.0200	47	0.0030
4	−0.0182	15	−0.0053	26	−0.1319	37	0.0509	48	−0.0086
5	−0.0166	16	−0.0376	27	0.1327	38	0.0187	49	−0.0001
6	0.0018	17	−0.0354	28	0.2843	39	−0.0354	50	0.0018
7	−0.0001	18	0.0187	29	0.1327	40	−0.0376	51	−0.0166
8	−0.0086	19	0.0509	30	−0.1319	41	−0.0053	52	−0.0182
9	0.0030	20	0.0200	31	−0.2051	42	0.0009	53	−0.0052
10	0.0268	21	0.0005	32	−0.0664	43	−0.0144	54	0.0258
11	0.0229	22	0.0459	33	0.0575	44	−0.0054	55	0.0145

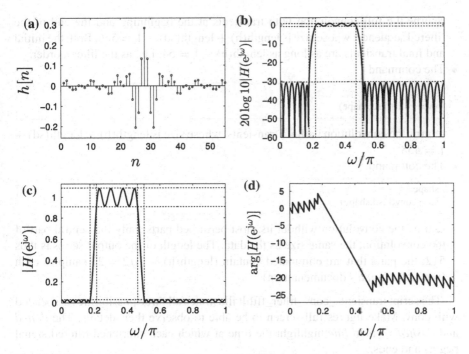

Fig. 16.1 Characteristics of the filter given in Table 16.1: **a** impulse response, **b** square modulus of the frequency response in dB, **c** frequency-response modulus in normal units, and **d** unwrapped phase versus ω/π

Before performing this exercise, it is however instructive to filter, via conv, filter and fftfilt and using the filter given above, a simple sinusoid with frequency falling in the filter's passband. This will offer us the possibility of examining in each case the head and the tail of the output sequence, to see if transients are present, and to determine their length. For the preliminary exercise, the script given below (transients_conv_filter_fftfilt) can be used. First, a synthetic series containing one sinusoid in white noise must be generated using datasint_Nsin_noise_variable_N. It will contain $N = 512$ samples, with $T\_s = 1$ s, of a sinusoid (unit amplitude and frequency 0.35π, falling inside the filter's passband) with a small amount of additive white noise (sigma_wn = 0.2). This data will be filtered by running transients_conv_filter_fftfilt, using the three functions and the impulse response h.

The function conv accepts in input an additional parameter, indicated as shape, that can assume three values: 'full' (default), 'valid' and 'same'.

- The command

```
1  shape='full'
2  yf = conv(h,sd,shape)
```

returns the full convolution with transients at the beginning and the end of the filtered sequence, whose size is length(h) + length(sd) $- 1 = 566$. Both the initial and final transients are as long as length(h) $- 1 = 54$, i.e., as the filter's order.

• The command

```
1  shape='valid'
2  yv = conv(h,sd,shape)
```

returns the convolution without transients, whose size is length(h) $-$ length(sd) $+ 1 = 458$.

• The command

```
1  shape='same'
2  ys = conv(h,sd,shape)
```

returns the convolution without its most perturbed parts: only the central part of the convolution, the same size as the data. The length of the output series is thus 512; the parts that are eliminated contain (length(h) $- 1)/2 = 27$ samples each (see also Matlab's documentation).

The script transients_conv_filter_fftfilt illustrates all three possibilities associated with conv. Make figures full-screen to be able to observe their details. The *dotted* and *dashed vertical line* highlight the time at which each shortened filtered signal begins and ends.

```
1   % transients_conv_filter_fftfilt
2
3   % Load filter
4   load h.dat
5   M=length(h)
6   order=M−1
7
8   % Filter with conv
9
10  % Full − convolution
11      shape ='full';
12  yf = conv(sd,h,shape);
13  Nf=length(yf)
14  tf=(−floor((M−1)/2))*T_s:T_s:(N−1+floor((M−1)/2))*T_s;
15  % Valid − convolution
16      shape ='valid';
17  yv = conv(sd,h,shape);
18  Nv=length(yv)
19  tv=(floor((M−1)/2))*T_s:T_s:(N−1−floor((M−1)/2))*T_s;
20  % Same − convolution
21      shape ='same';
22  ys = conv(sd,h,shape);
23  Ns=length(ys)
```

```
24   ts= 0:T_s:(N−1)*T_s;
25   figure(99)
26   subplot(2,1,1)
27   plot(tv,yv,...
28   'Color',[0.4 0.4 0.4 ],'LineStyle','−','LineWidth',10,'Marker','d',
29   'MarkerFaceColor',[0.4 0.4 0.4 ],'MarkerEdgeColor',[0.4 0.4 0.4 ],'
         MarkerSize',6)
30   hold on
31   plot(ts,ys,...
32   'Color',[0.75 0.75 0.75],'LineStyle','−','LineWidth',5,'Marker','d',
         MarkerFaceColor',[0.75 0.75 0.75],'MarkerEdgeColor',[0.75 0.75
         0.75],'MarkerSize',4.5)
33   plot(tf,yf,...
34   'Color',[0 0 0],'LineStyle','−','LineWidth',1,'Marker','d','MarkerFaceColor'
         ,[0 0 0],'MarkerEdgeColor',[0 0 0],'MarkerSize',2)
35   legend('valid','same','full','Location','NorthWest')
36   plot([tv(1) tv(1)],[−1.1 1.1],'k−−','LineWidth',1.5)
37   plot([ts(1) ts(1)],[−1.1 1.1],'k:','LineWidth',1.5)
38   hold off
39   set(gca,'XLim',[tf(1)−0.36 tf(80)+0.35])
40   set(gca,'XMinorTick','on')
41   set(gca,'YLim',[−1.1 1.1])
42   hlabelx=get(gca,'Xlabel');
43   set(hlabelx,'String','Time (s)','FontName','Times New Roman','Fontsize',14)
44   hlabely=get(gca,'Ylabel');
45   set(hlabely,'String','Output of conv − Head','FontName','Times New
         Roman','Fontsize',14)
46   subplot(2,1,2)
47   plot(tv,yv,...
48   'Color',[0.4 0.4 0.4 ],'LineStyle','−','LineWidth',10,'Marker','d',
49   'MarkerFaceColor',[0.4 0.4 0.4 ],'MarkerEdgeColor',[0.4 0.4 0.4 ],'
         MarkerSize',6)
50   hold on
51   plot(ts,ys,...
52   'Color',[0.75 0.7 0.7],'LineStyle','−','LineWidth',5,'Marker','d',
53   'MarkerFaceColor',[0.75 0.75 0.75],'MarkerEdgeColor',[0.75 0.75 0.75],'
         MarkerSize',4.5)
54   plot(tf,yf,...
55   'Color',[0 0 0],'LineStyle','−','LineWidth',1,'Marker','d','MarkerFaceColor'
         ,[0 0 0],'MarkerEdgeColor',[0 0 0],'MarkerSize',2)
56   plot([tv(end) tv(end)],[−1.1 1.1],'k−−','LineWidth',1.5)
57   plot([ts(end) ts(end)],[−1.1 1.1],'k:','LineWidth',1.5)
58   hold off
59   set(gca,'XLim',[tf(end−79)−0.36 tf(end)+0.35])
60   set(gca,'XMinorTick','on')
```

```matlab
61   set(gca,'YLim',[−1.1 1.1])
62   hlabelx=get(gca,'Xlabel');
63   set(hlabelx,'String','Time (s)','FontName','Times New Roman','Fontsize',14)
64   hlabely=get(gca,'Ylabel');
65   set(hlabely,'String','Output of conv − Tail','FontName','Times New Roman',
         'Fontsize',14)
66
67   % Filter with filter
68   h=h';
69   sd=sd';
70   yfil = filter(h,1,sd);
71   Nfil=length(yfil)
72   tfil = 0:T_s:(Nfil−1)*T_s;
73   yvfil = yfil(M:end);
74   Nvfil=length(yv)
75   tvfil= (M−1)*T_s:T_s:(Nfil−1)*T_s;
76   figure(999)
77   subplot(2,1,1)
78   plot(tvfil,yvfil,...
79   'Color',[0.7 0.7 0.7],'LineStyle','−','LineWidth',3,'Marker','d','
         MarkerFaceColor',[0.7 0.7 0.7],'MarkerEdgeColor',[0.7 0.7 0.7],'
         MarkerSize',5)
80   hold on
81   plot(tfil,yfil,...
82   'Color',[0 0 0],'LineStyle','−','LineWidth',1,'Marker','d','MarkerFaceColor'
         ,[0 0 0],'MarkerEdgeColor',[0 0 0],'MarkerSize',3)
83   legend('no transient','full','Location','NorthWest')
84   plot([tvfil(1) tvfil(1)],[−1.1 1.1],'k−−','LineWidth',1.5)
85   hold off
86   set(gca,'XLim',[tfil(1)−0.36 tfil(80)+0.35])
87   set(gca,'XMinorTick','on')
88   set(gca,'YLim',[−1.1 1.1])
89   hlabelx=get(gca,'Xlabel');
90   set(hlabelx,'String','Time (s)','FontName','Times New Roman','Fontsize',14)
91   hlabely=get(gca,'Ylabel');
92   set(hlabely,'String','Output of filter − Head','FontName','Times New
         Roman','Fontsize',14)
93   subplot(2,1,2)
94   plot(tvfil,yvfil,...
95   'Color',[0.7 0.7 0.7],'LineStyle','−','LineWidth',3,'Marker','d','
         MarkerFaceColor',[0.7 0.7 0.7],'MarkerEdgeColor',[0.7 0.7 0.7],'
         MarkerSize',5)
96   hold on
97   plot(tfil,yfil,...
```

```
 98   'Color',[0 0 0],'LineStyle','−','LineWidth',1,'Marker','d','MarkerFaceColor'
          ,[0 0 0],'MarkerEdgeColor',[0 0 0],'MarkerSize',2)
 99   hold off
100   set(gca,'XLim',[tfil(end−79)−0.36 tfil(end)+0.35])
101   set(gca,'XMinorTick','on')
102   set(gca,'YLim',[−1.1 1.1])
103   hlabelx=get(gca,'Xlabel');
104   set(hlabelx,'String','Time (s)','FontName','Times New Roman','Fontsize',14)
105   hlabely=get(gca,'Ylabel');
106   set(hlabely,'String','Output of filter − Tail','FontName','Times New Roman',
          'Fontsize',14)
107
108   % Filter with fftfilt
109   yfft = fftfilt (h,sd);
110   Nfft=length(yfft)
111   tfft= 0:T_s:(Nfft−1)∗T_s;
112   yvfft = yfft(M:end);
113   Nvfft=length(yvfft)
114   tvfft= (M−1)∗T_s:T_s:(Nfft−1)∗T_s;
115   figure(9999)
116   subplot(2,1,1)
117   plot(tvfil,yvfil,...
118   'Color',[0.7 0.7 0.7],'LineStyle','−','LineWidth',3,'Marker','d',
          MarkerFaceColor',[0.7 0.7 0.7],'MarkerEdgeColor',[0.7 0.7 0.7],'
          MarkerSize',5)
119   hold on
120   plot(tfil,yfil,...
121   'Color',[0 0 0],'LineStyle','−','LineWidth',1,'Marker','d','MarkerFaceColor'
          ,[0 0 0],'MarkerEdgeColor',[0 0 0],'MarkerSize',3)
122   legend('no transient','full','Location','NorthWest')
123   plot([tvfil(1) tvfil(1)],[−1.1 1.1],'k−−','LineWidth',1.5)
124   hold off
125   set(gca,'XLim',[tfil(1)−0.36 tfil(80)+0.35])
126   set(gca,'XMinorTick','on')
127   set(gca,'YLim',[−1.1 1.1])
128   hlabelx=get(gca,'Xlabel');
129   set(hlabelx,'String','Time (s)','FontName','Times New Roman','Fontsize',14)
130   hlabely=get(gca,'Ylabel');
131   set(hlabely,'String','Output of fftfilt − Head','FontName','Times New
          Roman','Fontsize',14)
132   subplot(2,1,2)
133   plot(tvfil,yvfil,...
134   'Color',[0.7 0.7 0.7],'LineStyle','−','LineWidth',3,'Marker','d',
135   'MarkerFaceColor',[0.7 0.7 0.7],'MarkerEdgeColor',[0.7 0.7 0.7],'
          MarkerSize',5)
```

```
136   hold on
137   plot(tfil,yfil,...
138   'Color',[0 0 0],'LineStyle','−','LineWidth',1,'Marker','d','MarkerFaceColor'
         ,[0 0 0],'MarkerEdgeColor',[0 0 0],'MarkerSize',2)
139   hold off
140   set(gca,'XLim',[tfil(end−79)−0.36 tfil(end)+0.35])
141   set(gca,'XMinorTick','on')
142   set(gca,'YLim',[−1.1 1.1])
143   hlabelx=get(gca,'Xlabel');
144   set(hlabelx,'String','Time (s)','FontName','Times New Roman','Fontsize',14)
145   hlabely=get(gca,'Ylabel');
146   set(hlabely,'String','Output of fftfilt − Tail','FontName','Times New Roman'
         ,'Fontsize',14)
```

We know that an LP-Type I FIR filter applies a delay of half the filter's order. What does this mean in practice? Let us see it through an example. Figure 16.2 shows a section of a signal with sampling interval of 1 min, exhibiting two evident pulses (*black curve*), and its filtered versions with and without the initial transient (*dotted* and *dashed lines*, respectively; fftfilt was used). The filter employed here is an LP-FIR lowpass filter with a very wide passband, covering 98 % of the principal interval. The

Fig. 16.2 A section of a signal with sampling interval of 1 min, exhibiting two evident pulses (*black curve*), and its filtered versions with and without the initial transient (*light gray* and *dark gray curves*, respectively; fftfilt was used. The filter employed here is an LP-FIR lowpass filter with a very wide passband and order $M - 1 = 54$. **a** Signals plotted versus the respective time vectors; **b** signals treated as vectors and plotted versus their respective indexes

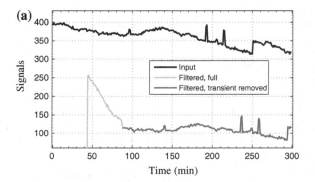

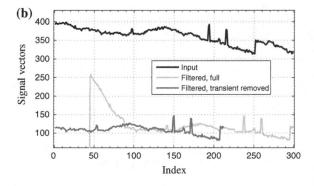

filter order is $M - 1 = 54$. This filter passes nearly all frequencies, and only applies a delay to the input signal. It thus approximates, though crudely, an ideal delayer. In Fig. 16.2a, signals are plotted versus the respective time vectors, which begin at 0 s (original signal, *black curve*), 0 s (full filtered signal, *light gray curve*), and 54 s (filtered signal with initial transient removed, *dark gray curve*). This is equivalent to plotting them versus the *same* index n. A delay of 27 samples is evident in the filtered signal with respect to the original one. In Fig. 16.2b, the signals are treated as vectors and plotted versus their respective indexes. The full filtered vector (*light gray curve*) still appears delayed by 27 samples with respect to the original one (*black curve*), but the filtered vector with the initial transient-phase removed (*dark gray curve*) obviously appears *anticipated* by 27 samples. This example was meant to show that we must be careful: we will see the delay only if we plot the original and the filtered signal versus the same index, or time vector. Removing a transient does not alter delay if we plot the signal versus the same index or time vector as the original signal, but turns a delay into an advance if we just use vectors as sequences of numbers, without remembering that one of them actually had its head cut off, and therefor begins later in real time.

We can now proceed to the main filtering exercise. We will filter the same data used for the DFT exercise (three sinusoids with discrete angular frequencies 0.15π, 0.35π, and 0.55π in Gaussian white noise). Thus, retrieve those data, or generate them again. Which frequencies, among those of the sinusoids contained in sd, fall inside the filter's passband? Which frequencies are expected to be attenuated, since they are external to the filter's passband? The script reported below (filtering_FIR) compares the DFT moduli of the original and filtered sequences to check the result of the filtering operation. The plots are limited to [0, f_Ny]. Run the script once for each filtering function.

```
1   % filtering_FIR
2   % Filter data record using FIR – Type I bandpass filter.
3
4   prompt ='Filtering function? (1 = filter, 2 = conv, 3 = fftfilt)';
5   flag = input(prompt);
6
7   load h.dat % impulse response of the filter
8
9   % Compute DFT and modulus
10  N_f=N;
11  SD=fft(sd,N_f);
12  modSD=abs(SD);
13
14  % Compute frequency vector
15  f_Ny=1/(2*T_s);
16  deltf=1/(length(SD)*T_s);
17  f=0:deltf:(2*f_Ny−deltf);
18
19  % Perform filtering with chosen function
20  if flag == 1
21      sdf=filter(h,1,sd);
22  end
```

```
23   if flag == 2
24       sdf=conv(h,sd);
25   end
26   if flag == 3
27       sdf=fftfilt(h,sd,N_f);
28   end
29
30   N_sdf=length(sdf) % length of filtered data
31   sdf=sdf−mean(sdf); % remove mean from filtered data
32
33   % Compute DFT and modulus
34   % Frequency vector is the same as for original data
35   SDF=fft(sdf,N_f);
36   modSDF=abs(SDF);
37
38   % Plot results
39   figure(6)% Filtered sequence − head
40   plot(1:100,sdf(1:100))
41   set(gca,'XLim',[−1 101])
42   hlabelx=get(gca,'Xlabel');
43   set(hlabelx,'String','$n$','Interpreter','Latex','FontSize',16)
44   hlabely=get(gca,'Ylabel');
45   set(hlabely,'String','$x_{\mathrm{f}}$','Interpreter','Latex','FontSize',16)
46
47   figure(7)% Filtered sequence − tail
48   plot(N−99:N,sdf(end−99:end))
49   set(gca,'XLim',[N−100 N+1])
50   hlabelx=get(gca,'Xlabel');
51   set(hlabelx,'String','$n$','Interpreter','Latex','FontSize',16)
52   hlabely=get(gca,'Ylabel');
53   set(hlabely,'String','$x_{\mathrm{f}}$','Interpreter','Latex','FontSize',16)
54
55   figure(8)% DFT modulus in dB for original and filtered data
56   plot(f(2:N_f/2+1),10*log10(modSD(2:N_f/2+1)))
57   hold on
58   plot(f(2:N_f/2+1),10*log10(modSDF(2:N_f/2+1)),'g')
59   hold off
60   set(gca,'XLim',[0 f_Ny])
61   hlabelx=get(gca,'Xlabel');
62   set(hlabelx,'String','$f$ (Hz)','Interpreter','Latex','FontSize',16)
63   hlabely=get(gca,'Ylabel');
64   set(hlabely,'String','$10\log_{10}|X_{\mathrm{f}}|$','Interpreter','Latex','FontSize',16)
```

   The spectral plots show that the only sinusoidal component transmitted through the filter is the one with discrete angular frequency equal to 0.35π. The other two components are attenuated, since they fall outside the filter passband.

   As for the length of the filtered record and the initial/final transients:

- the function filter computes the output starting from the first time step of the input, assuming rest initial conditions and stopping at the last data sample. Therefore the length of sdf is equal to the length of sd (i.e., 8192) and an initial transient is present, covering the first 54 samples (recall that 54 is the filter order). If we discarded the transient (thus keeping $8192 - 54 = 8138$ samples) and then computed and plotted again modSDF over 8192 frequency values (using zero padding), we would find that those initial 54 samples affected by the transient have little or no effect on the spectrum;

- with conv, the output is longer (8246 samples), and an initial and a final transient are present. Each transient is 54-samples long. In fact, with 55 impulse response samples and $N = 8192$ data, the linear convolution discarding transients contains $8192 - 55 + 1 = 8138$ values, while including transients it contains $8192 + 55 - 1 = 8246$ values (see, e.g., Sect. 2.4.2). In this case, it can also be verified that the transients have little or no effect on the spectrum;

- with fftfilt, the length of sdf is equal to the length of sd, i.e., 8192, and an initial transient is present, which however does not affect the spectrum in a significant way.

16.4.1.2 Filtering a Sequence with an FIR Filter After Downsampling

Imagine that we want to filter a new sd sequence characterized by a sampling interval of $T\_s = 5$ s, and containing three sinusoids with frequencies 0.2π, 0.4π, and 0.8π, corresponding to $2 \times 10^{-3}, 4 \times 10^{-3}$, and 8×10^{-3} Hz, respectively. Imagine that we want to use the same bandpass filter employed above, and that we want to preserve the frequency component 4×10^{-3} Hz, while attenuating the other components. Unfortunately, this frequency is excluded from the filter's passband, which for $T\_s = 5$ s covers 2.15×10^{-2} to 4.64×10^{-2} Hz.

However, if we decimate the data by a factor $K = 8$, the sampling interval T_s becomes 40 s and the Nyquist frequency becomes 1.25×10^{-2} Hz. The filter's passband ($0.2150\pi \leq \omega \leq 0.464\pi$) now covers the interval of analog frequencies from 2.69×10^{-3} to 5.80×10^{-3} Hz, a range that includes 4×10^{-3} Hz.

The cutoff of the ideal anti-aliasing lowpass filter required before downsampling would be $\pi/K = \pi/8 = 0.1250\pi$, but we will actually use a realizable filter with a non-zero-width transition band, and consequently, we will choose a passband limit smaller than this theoretical cutoff—let us say 0.1π, so that at 0.1250π the amplitude response has already attained a sufficiently small value (Sect. 6.9.1). The impulse response of a lowpass filter with the desired characteristics is given in Table 16.2. Its length is again 55 and the tolerances assumed for its design are $\delta_p \equiv$ deltap $= 0.105$ and $r_s \equiv$ rs $= 30$ dB. We will assume it resides in a file named haa.dat. Figure 16.3 shows the time- and frequency-domain characteristics of this filter. The cutoff is 0.1π and at 0.125π the attenuation is about 11 dB; it attains 30 dB at 0.14π. Thus, this filter will efficiently remove all components higher than the Nyquist frequency as it is after downsampling.

In this exercise we will

- generate $N = 8192$ synthetic data (sd) with $T\_s = 5$ s, containing three sinusoids with equal unit amplitudes, random initial phases and discrete angular

Table 16.2 Impulse response of a Linear Phase FIR lowpass anti-aliasing filter designed for a decimation factor of 8 (see text)

n	h[n]	n	h[n]	n	h[n]	n	h[n]	n	h[n]
1	−0.0107	12	−0.0059	23	0.0604	34	0.0415	45	0.0006
2	0.0092	13	−0.0124	24	0.0784	35	0.0234	46	0.0063
3	0.0095	14	−0.0181	25	0.0943	36	0.0072	47	0.0106
4	0.0111	15	−0.0220	26	0.1068	37	−0.0060	48	0.0134
5	0.0129	16	−0.0233	27	0.1147	38	−0.0156	49	0.0145
6	0.0143	17	−0.0214	28	0.1174	39	−0.0214	50	0.0143
7	0.0145	18	−0.0156	29	0.1147	40	−0.0233	51	0.0129
8	0.0134	19	−0.0060	30	0.1068	41	−0.0220	52	0.0111
9	0.0106	20	0.0072	31	0.0943	42	−0.0181	53	0.0095
10	0.0063	21	0.0234	32	0.0784	43	−0.0124	54	0.0092
11	0.0006	22	0.0415	33	0.0604	44	−0.0059	55	−0.0107

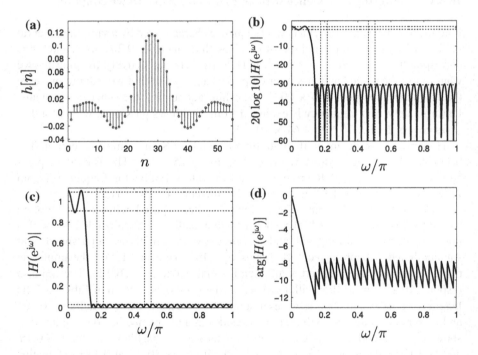

Fig. 16.3 Characteristics of the filter given in Table 16.2: **a** impulse response, **b** square modulus of the frequency response in dB, **c** frequency-response modulus in normal units, and **d** unwrapped phase versus ω/π

frequencies 0.2π, 0.4π, and 0.8π, plus white noise with sigma_wn = 1 (use data_Nsin_noise_variable_N);

- filter sd with the anti-aliasing filter haa, thus obtaining a sequence that we will name sdfaa;
- decimate sdfaa, taking one sample every K = 8 samples, thus obtaining a sequence that we will call sddec;
- filter sddec with the passband filter whose impulse response is h, obtaining a sequence that we will call sdf;
- compute the modulus of the DFT of sdf and plot it together with the modulus of the DFT of the original data. For graphical convenience, we will compensate the factor 1/K on the transform amplitude, introduced by decimation (Sect. 6.9.1). Pay attention to the fact that the two transforms correspond to different frequency vectors. This is because the Nyquist frequency changes when decimation is applied.

In this case we filter twice, and transients that are not eliminated accumulate. If we repeated the process eliminating all transients, we would however notice that their effect is visible in the spectrum but does not substantially alter the results.

The script for this exercise (filtering_FIR_with_decimation) is given below. The user can choose if to use filter, conv or fftfilt.

```
1   % filtering_FIR_with_decimation
2   % Filter data record after decimating;
3   % use FIR − Type I bandpass filter.
4
5   prompt ='Filtering function? (1 = filter, 2 = conv, 3 = fftfilt)';
6   flag = input(prompt);
7
8   load haa.dat % imp. response of anti−aliasing filter
9   K=8; % decimation factor
10  load h.dat % imp. response of bandpass filter
11
12  % Filter data record
13  % with anti−aliasing lowpass filter
14  % using the chosen function
15  if flag == 1
16      sdfaa=filter(haa,1,sd);
17  end
18  if flag == 2
19      sdfaa=conv(haa,sd);
20  end
21  if flag == 3
22      sdfaa=fftfilt(haa,sd,N_f);
23  end
24  N_sdfaa=length(sdfaa) % length of aa−filtered data
25  sdfaa=sdfaa−mean(sdfaa); % remove mean
26
27  % Compute DFT modulus
28  % for original and aa−filtered data
29  N_f=N;
30  SD=fft(sd,N_f);
31  modSD=abs(SD);
```

```
32   SDFAA=fft(sdfaa,N_f);
33   modSDFAA=abs(SDFAA);
34
35   % Build corresponding frequency vector
36   f_Ny=1/(2*T_s);
37   deltf=1/(N_f*T_s);
38   f=0:deltf:(2*f_Ny−deltf);
39
40   % Plot DFT modulus in dB for original
41   % and aa−filtered data
42   figure(flag*10)
43   plot(f(2:N_f/2+1),10*log10(modSD(2:N_f/2+1)))
44   hold on
45   plot(f(2:N_f/2+1),10*log10(modSDFAA(2:N_f/2+1)),'g')
46   hold off
47   set(gca,'XLim',[0 f_Ny])
48   hlabelx=get(gca,'Xlabel');
49   set(hlabelx,'String','$f$ (Hz)','Interpreter','Latex','FontSize',16)
50   hlabely=get(gca,'Ylabel');
51   set(hlabely,'String','$10\log_{10}|X_{\mathrm{faa}}|$','Interpreter','Latex','FontSize',16)
52
53   % Decimate sequence
54   sddec=sdfaa(1:K:end);
55   N_sddec=length(sddec)
56
57   % Filter decimated data with bandpass filter
58   % using chosen function
59   if flag == 1
60       sdf=filter(h,1,sddec);
61   end
62   if flag == 2
63       sdf=conv(h,sddec);
64   end
65   if flag == 3
66       sdf=fftfilt(h,sddec,N_f);
67   end
68   N_sdf=length(sdf) % length of filtered data
69   sdf=sdf−mean(sdf); % remove mean
70
71   % DFT of filtered data and modulus
72   SDF=fft(sdf,N_f);
73   modSDF=abs(SDF);
74
75   % Build corresponding frequency vector
76   % (Nyquist frequency has changed)
77   f_Ny_dec=1/(2*K*T_s);
78   deltf_dec=1/(N_f*K*T_s);
79   f_dec=0:deltf_dec:(2*f_Ny_dec−deltf_dec);
80
81   % Plot DFT modulus in dB for original and filtered data
82   figure(flag*10+1)
83   plot(f(2:N_f/2+1),10*log10(modSD(2:N_f/2+1)))
84   hold on
```

```
85    plot(f_dec(2:N_f/2+1),10*log10(K*modSDF(2:N_f/2+1)),'g')
86    hold off
87    set(gca,'XLim',[0 f_Ny])
88    hlabelx=get(gca,'Xlabel');
89    set(hlabelx,'String','$f$ (Hz)','Interpreter','Latex','FontSize',16)
90    hlabely=get(gca,'Ylabel');
91    set(hlabely,'String','$10\log_{10}|X_{\mathrm{faa}}|$','Interpreter','Latex','FontSize',16)
```

16.4.1.3   Characteristics of a Digital FIR Filter

We will now study, in the dual domains of time and frequency, the characteristics of an FIR filter (Sect. 2.4.5; Chap. 6). We initially suppose that we know the impulse response $h[n]$ of the filter, namely the one given in Table 16.1 that we have used so far.

1. Frequency response from impulse response:
   we start with a stem plot of $h[n]$. Then we compute the frequency response in $[0, \pi)$ (N_f points) and the corresponding vector of ω-values (the variable ω is called w in the following script) using the function freqz, whose syntax is

```
1    [H,w] = freqz(b,a,N_f)
```

for a general IIR filter in difference-equation form, and

```
1    [H,w] = freqz(h,1,N_f)
```

in our FIR case, since the denominator transfer-function coefficients are all zero, except a_0, which is 1 (Chap. 2).

We plot the frequency response amplitude in linear and logarithmic units and verify that it satisfies the following specifications: passband $0.2150\pi \leq \omega \leq 0.464\pi$; transition bandwidth 0.04π; passband ripple 0.09; stopband attenuation 30.5 dB. Note that the graph in dB is useful to read the attenuation in the stopband; the passband ripple and the transition bandwidth are more visible in the linear plot. Then we plot the unwrapped phase response. We also compute the frequency response over the interval $[0, 2\pi)$ (H_whole) and save it for future use in the binary file H.mat. Note that if we do not change N_f, the number of samples will be the same as before. The script for this task (freq_resp) is reported below.

```
1     % freq_resp
2     % Compute frequency response from impulse response
3
4     load h.dat % impulse response
5     n=length(h)−1
6
7     % Compute frequency response in [0,pi) vs. omega
8     N_f=1024
9     [H,w]=freqz(h,1,N_f);
10
```

```
11   % Compute complete H over [0,2 pi)
12   % and save for future use
13   [H_whole,w_whole]=freqz(h,1,N_f,'whole');
14   save H.mat H_whole w_whole N_f
15
16   % Characteristic frequencies
17   % and tolerances of given filter
18   % (guess and then check)
19   deltap=0.09
20   rs=30.5
21   deltas=10^(−rs/20)
22   deltw_over_pi= 0.04
23   w1_over_pi=0.215
24   w2_over_pi=0.464
25
26   figure(40)
27
28   % Plot impulse response
29   subplot(2,2,1)
30
31   stem(h,'MarkerSize',3)
32   set(gca,'XLim',[0 n+1])
33   set(gca,'YLim',[min(h)−0.05 max(h)+0.05])
34   hlabelx=get(gca,'Xlabel');
35   set(hlabelx,'String','$n$','Interpreter','Latex','FontSize',16)
36   hlabely=get(gca,'Ylabel');
37   set(hlabely,'String','$h$','Interpreter','Latex','FontSize',16)
38
39   % Plot amplitude of frequency response in dB vs. omega/pi,
40   % with tolerances
41   subplot(2,2,2)
42   plot(w/pi,20*log10((1−deltap)*ones(1,N_f)),'g')
43   hold on
44   plot(w/pi,20*log10((1+deltap)*ones(1,N_f)),'g')
45   plot(w/pi,20*log10(deltas*ones(1,N_f)),'g')
46   plot([w1_over_pi w1_over_pi],[−60 4],'m')
47   plot([w2_over_pi w2_over_pi],[−60 4],'m')
48   plot([w1_over_pi−deltw_over_pi w1_over_pi−deltw_over_pi],[−60 4],'m')
49   plot([w2_over_pi+deltw_over_pi w2_over_pi+deltw_over_pi],[−60 4],'m')
50   plot(w/pi,20*log10(abs(H)))
51   hold off
52   set(gca,'XLim',[0 1])
53   set(gca,'YLim',[−60 4])
54   hlabelx=get(gca,'Xlabel');
55   set(hlabelx,'String','$\omega/\pi$','Interpreter','Latex','FontSize',16)
56   hlabely=get(gca,'Ylabel');
57   set(hlabely,'String','$20\log10|H|$','Interpreter','Latex','FontSize',16)
58
59   % Plot amplitude of frequency response vs. omega/pi,
60   % with tolerances
61   subplot(2,2,3)
62   plot(w/pi,(1+deltap)*ones(1,N_f),'g')
63   hold on
```

```
64  plot(w/pi,(1−deltap)*ones(1,N_f),'g')
65  plot(w/pi,deltas*ones(1,N_f),'g')
66  plot([w1_over_pi w1_over_pi],[0 1.13],'m')
67  plot([w2_over_pi w2_over_pi],[0 1.13],'m')
68  plot([w1_over_pi−deltw_over_pi w1_over_pi−deltw_over_pi],[0 1.13],'m')
69  plot([w2_over_pi+deltw_over_pi w2_over_pi+deltw_over_pi],[0 1.13],'m')
70  plot(w/pi,abs(H))
71  hold off
72  set(gca,'XLim',[0 1])
73  set(gca,'YLim',[0 1.13])
74  hlabelx=get(gca,'Xlabel');
75  set(hlabelx,'String','$\omega/\pi$','Interpreter','Latex','FontSize',16)
76  hlabely=get(gca,'Ylabel');
77  set(hlabely,'String','$|H|$','Interpreter','Latex','FontSize',16)
78
79  % Plot phase angle of frequency response vs. omega/pi
80  subplot(2,2,4)
81  plot(w/pi,unwrap(angle(H)))
82  set(gca,'XLim',[0 1])
83  set(gca,'YLim',[min(unwrap(angle(H)))−3 max(unwrap(angle(H)))+3])
84  hlabelx=get(gca,'Xlabel');
85  set(hlabelx,'String','$\omega/\pi$','Interpreter','Latex','FontSize',16)
86  hlabely=get(gca,'Ylabel');
87  set(hlabely,'String','$\arg(H)$','Interpreter','Latex','FontSize',16)
```

2. **Impulse Response from Frequency Response:**
to deduce the impulse response from the frequency response of a filter, the inverse fft function (ifft) can be used, but this requires knowing the frequency response over the frequency interval $[0, 2\pi)$, i.e., the *complete* DFT of $h[n]$. This is why we saved H_whole in the file H.mat in the previous exercise. Now we will simply load it and take the inverse transform. We will then compare the stem plots of the original impulse response and of the one deduced from H_whole. The script (imp_resp) is given below.

```
1   % imp_resp
2   % Compute mpulse response from frequency response.
3
4   load H.mat % frequency response
5   load h.dat % impulse response we expect to find
6   M=length(h);
7
8   % Compute impulse response
9   hi=ifft(H_whole,N_f);
10
11  % Get rid of negligible imaginary parts due to numerical errors
12  % and reduce length discarding zeros
13  hi_r=real(hi(1:M));
14
15  % Compare the two impulse responses
16  figure(50)
17      stem(hi_r,'MarkerSize',8,'MarkerFaceColor','w')
18      hold on
```

```
19      stem(h,'MarkerSize',4,'MarkerEdgeColor','g','MarkerFaceColor','g')
20  hold off
21  set(gca,'XLim',[0 M+1])
22  set(gca,'YLim',[−0.25 0.3])
23  hlabelx=get(gca,'Xlabel');
24  set(hlabelx,'String','$n$','Interpreter','Latex','FontSize',16)
25  hlabely=get(gca,'Ylabel');
26  set(hlabely,'String','$h$','Interpreter','Latex','FontSize',16)
```

Note that the impulse response computed from H_whole by ifft (which is called
hi in the script) is N_f = 1024 samples long and is complex, but the imaginary
parts of its elements are negligible, and, moreover, only the first 55 samples of hi
are different from zero within numerical errors. These 55 samples turn out to be,
as expected, identical to the samples of the impulse response given in Table 16.1.

16.4.1.4   Design of Equiripple FIR Filters

We will now design Type-I equiripple FIR filters (Chap. 7) with different frequency
selectivities, using the Matlab functions firpmord and firpm.

1. Bandpass filter:
   reproduce the passband filter used in previous exercises. Therefore, filter specifications in this case include:

   • type of frequency selectivity (bandpass filter with only one passband);
   • even order;
   • passband bounded by $0.2150\pi \leq \omega \leq 0.464\pi$;
   • transition bandwidth equal to 0.04π;
   • linear phase;
   • desired passband amplitude response equal to one and maximum ripple deltap
     $= 0.09$;
   • minimum stopband attenuation rs $= 30.5$ dB (desired stopband amplitude
     response equal to zero).

   The first thing to do is to estimate the minimum filter order required to meet these
   specifications (Sect. 7.8). This is possible using the function firpmord, whose
   syntax is

```
[n0,f0,a0,whts] = firpmord(f,a,dev,f_s)
```

   In input, the frequencies f that bound the passband and the two stopbands can be
   given in different forms, according to the value of f_s that we provide. By default,
   the function firpmord assumes f_s = 2, so that the elements of f are expected
   to be in terms of ω/π. We will provide frequencies in this form. Note, however,
   that since we know the actual T_s of the sequence to which the filter must be
   applied, we could also set f_s = 1/T_s and give the elements of f in terms of
   analog frequencies. The zero frequency and the Nyquist frequency are omitted
   from the vector f, and in this passband case we only need to provide

- the upper limit of the first stopband, i.e., the lower limit of the first transition band, $0.2150 - 0.04 = 0.1750$;
- the upper limit of the first transition band, i.e., the lower passband limit, 0.2150;
- the upper passband limit, i.e., the lower limit of the second transition band, 0.464;
- the upper limit of the second transition band, i.e., the lower limit of the second stopband, $0.464 + 0.04 = 0.5040$.

The corresponding desired amplitudes in vector a must have one element for each band: $a = [0\ 1\ 0]$. In general, the length of f is always twice the length of a, minus 2. The vector dev contains one ripple tolerance for each band. So we first derive deltas from rs using the formula

$$20 \log_{10} \delta_s = -r_s, \qquad \text{i.e.,} \qquad \delta_s = 10^{-r_s/20},$$

thus obtaining deltas $= 0.0299$; then we set dev $=$ [deltas deltap deltas].
In output we get the variables n0, f0, a0, and whts, to be used as inputs to the Matlab function that implements the Parks-McClellan design algorithm (Sect. 7.7), i.e., firpm. Since we provided the elements of f in terms of ω/π, the output frequencies will also be in terms of ω/π. Therefore we call them w_over_pi0. For a general multiband filter, the outputs of firpmord (inputs to firpm) are:

- the minimum filter order n0;
- a frequency vector w_over_pi0 having 0 and 1 at the edges, and including the bounds of all stopbands and passbands in terms of ω/π. In the present case, w_over_pi0 $= [0\ 0.1750\ 0.2150\ 0.464\ 0.5040\ 1]$;
- a desired amplitude vector a0 with one element for each element of f0: a0 $=$ $[0\ 0\ 1\ 1\ 0\ 0]$;
- a vector whts of weights for the minimization of the maximum error, with one element for each band. In the present case, the passband weight is set to 1 and the weights for the stopbands are equal to $\delta_p/\delta_s = 3.01$. Thus, whts $= [3.01$ $1.\ 3.01]$.

Before calling firpm, we will check if n0 is odd, and if it is we will increase it by 1 since we desire a Type-I filter (Sect. 7.4). We will then initialize n as n $=$ n0 $- 2$ and start a while loop for filter design, which is meant to repeat the design process until the result is satisfactory. Indeed, since firpmord often underestimates the minimum order, it is necessary to test if the designed filter meets the specifications, and in particular if its ripples do not exceed the tolerances.
At the start of the loop, a flag named repeat is introduced and initialized to 1 (meaning "yes"). Inside the loop we first set n $=$ n $+ 2$ and then call firpm as follows:

```
h_pm = firpm(n,w_over_pi0,a0,whts)
```

In this way we obtain the impulse response h_pm of the designed filter. Then we compute the amplitude of the frequency response and check visually if the specifications have been met. We will be prompted to set repeat = 1 (meaning "no") if we are not satisfied yet, or repeat = 0 (meaning "yes") to terminate the design process. Another issue to keep into account is that for a bandpass filter, non-equiripple solutions can occur, which satisfy the minimax criterion but are irregular and always discarded in practice (Sect. 7.11). We will also set repeat = 1 in such cases. In this exercise, in which we have a target filter to reproduce, the filter check does not only include the amplitude frequency response plot, but also the comparison between the designed and target impulse responses. The following script (design_FIR_bandpass) performs the steps described above. The final order turns out to be 54, as expected. The designed impulse response coincides with the $h[n]$ of our target filter (Table 16.1).

```
1   % design_FIR_bandpass
2   % Design linear phase FIR – Type I bandpass filter
3   % with rs = 30.5 dB, deltap = 0.09, passband 0.2150 pi – – 0.4640 pi, transition
        bandwidth 0.04 pi.
4
5   % Tolerances
6   rs = 30.5
7   deltas = 10^(−rs/20)
8   deltap = 0.09
9   rp = 20*log10((1+deltap)/(1−deltap))
10
11  % Characteristic frequencies, amplitudes and deviations
12  w1_over_pi = 0.2150
13  w2_over_pi = 0.4640
14  deltw_over_pi = 0.04
15  w_over_pi = [w1_over_pi−deltw_over_pi w1_over_pi w2_over_pi w2_over_pi+
        deltw_over_pi];
16  a = [0 1 0];
17  dev = [deltas deltap deltas];
18
19  % Estimate minimum filter order n0
20  [n0,w_over_pi0,a0,whts]=firpmord(w_over_pi,a,dev);
21
22  % Check if minimum order is odd
23  % and if it is, increase n0 by 1
24  if mod(n0,2)~=0
25      n0=n0+1;
26  end
27
28  % Initialize order for design process
29  n=n0−2;
30
31  repeat=1;
32
33  % Start loop until satisfaction
34  while repeat==1
35
36      % Design filter
37      n=n+2
38      h_pm=firpm(n,w_over_pi0,a0,whts);
```

```
39
40    % Compute frequency response and plot it
41    % with specifications to check performances
42    N_f=1024
43    [H,w]=freqz(h_pm,1,N_f);
44
45    figure(60)
46
47    subplot(1,2,1)
48    plot(w/pi,20*log10(abs(H)))
49    hold on
50    plot(w/pi,-rs*ones(1,N_f),'g')
51    hold off
52    set(gca,'XLim',[0 1])
53    set(gca,'XTick',[0:0.25:1])
54    set(gca,'YLim',[-60 5])
55    set(gca,'YTick',[-60:10:0])
56    hlabelx=get(gca,'Xlabel');
57    set(hlabelx,'String','$\omega/\pi$','Interpreter','Latex','FontSize',16)
58    hlabely=get(gca,'Ylabel');
59    set(hlabely,'String','$20\log_{10}|H|$','Interpreter','Latex','FontSize',16)
60
61    subplot(1,2,2)
62    plot(w/pi,abs(H))
63    hold on
64    plot(w/pi,(1+deltap)*ones(1,N_f),'g')
65    plot(w/pi,(1-deltap)*ones(1,N_f),'g')
66    hold off
67    set(gca,'XLim',[0 1])
68    set(gca,'XTick',[0:0.25:1])
69    set(gca,'YLim',[0 1.15])
70    hlabelx=get(gca,'Xlabel');
71    set(hlabelx,'String','$\omega/\pi$','Interpreter','Latex','FontSize',16)
72    hlabely=get(gca,'Ylabel');
73    set(hlabely,'String','$|H|$','Interpreter','Latex','FontSize',16)
74
75    % Plot impulse response and compare it with target one
76    load h.dat % Target filter
77
78    figure(61)
79    stem(h_pm,'MarkerSize',8,'MarkerFaceColor','w')
80    hold on
81    stem(h,'MarkerSize',4,'MarkerEdgeColor','g','MarkerFaceColor','g')
82    hold off
83    set(gca,'XLim',[0 n+2])
84    set(gca,'YLim',[-0.25 0.3])
85    hlabelx=get(gca,'Xlabel');
86    set(hlabelx,'String','$n$','Interpreter','Latex','FontSize',16)
87    hlabely=get(gca,'Ylabel');
88    set(hlabely,'String','$h$','Interpreter','Latex','FontSize',16)
89    prompt='Is the filter satisfactory? (0 = yes, 1 = no)';
90    repeat = input(prompt);
91  end
```

2. Bandstop filter:

design a linear phase FIR—Type I bandstop filter with rs = 30 dB and deltap = 0.09, suitable to attenuate, in a data sequence with sampling frequency of 150 Hz, all components internal to the interval 5–25 Hz. Specify transition bands 2.5 Hz wide. Express frequencies in Hz in the plots. Note that in this and in the following FIR design exercises we start from characteristic *analog* frequencies. Use the script given below (design_FIR_bandstop).

```
1    % design_FIR_bandstop
2    % Design linear phase FIR — Type I bandpass filter
3    % with rs = 30 dB and deltap = 0.09, suitable to attenuate 5–25 Hz components
4    % in data with sampling frequency of 150 Hz. Specify transition bands 2.5 Hz wide.
5
6    % Tolerances
7    rs = 30
8    deltas = 10^(−rs/20)
9    deltap = 0.09
10   rp = 20*log10((1+deltap)/(1−deltap))
11
12   % Characteristic frequencies, amplitudes and deviations
13   f_s = 150
14   f_Ny = f_s/2
15
16   fs1 = 5
17   fs2 = 25
18   deltf = 2.5
19   fp1 = fs1−deltf
20   fp2 = fs2+deltf
21   f = [fp1 fs1 fs2 fp2];
22   a = [1 0 1];
23   dev = [deltap deltas deltap];
24
25   % Estimate minimum order
26   [n0,w_over_pi0,a0,whts]=firpmord(f,a,dev,f_s) ;
27
28   % Check if minimum order is odd
29   % and if it is, increase n0 by 1
30   if mod(n0,2)~=0
31       n0=n0+1;
32   end
33
34   % Initialize order for design process
35   n=n0−2;
36
37   repeat=1;
38
39   % Start loop until satisfaction
40   while repeat==1
41
42       % Design filter
43       n=n+2
44       h_pm=firpm(n,w_over_pi0,a0,whts);
45
```

```
46    % Compute frequency response and plot it
47    % with specifications to check performances
48    N_f=1024
49    [H,f]=freqz(h_pm,1,N_f,f_s);
50
51    figure(70)
52
53    subplot(1,2,1)
54    plot(f,20*log10(abs(H)))
55    hold on
56    plot(f,−rs*ones(1,N_f),'g')
57    hold off
58    set(gca,'XLim',[0 f_Ny])
59    set(gca,'YLim',[−60 5])
60    set(gca,'YTick',[−60:10:0])
61    hlabelx=get(gca,'Xlabel');
62    set(hlabelx,'String','$f$ (Hz)','Interpreter','Latex','FontSize',16)
63    hlabely=get(gca,'Ylabel');
64    set(hlabely,'String','$20\log_{10}|H|$','Interpreter','Latex','FontSize',16)
65
66    subplot(1,2,2)
67    plot(f,abs(H))
68    hold on
69    plot(f,(1+deltap)*ones(1,N_f),'g')
70    plot(f,(1−deltap)*ones(1,N_f),'g')
71    hold off
72    set(gca,'XLim',[0 f_Ny])
73    set(gca,'YLim',[0 1.15])
74    hlabelx=get(gca,'Xlabel');
75    set(hlabelx,'String','$f$ (Hz)','Interpreter','Latex','FontSize',16)
76    hlabely=get(gca,'Ylabel');
77    set(hlabely,'String','$|H|$','Interpreter','Latex','FontSize',16)
78
79    prompt='Is the filter satisfactory? (0 = yes, 1 = no)';
80    repeat = input(prompt);
81  end
```

3. Highpass filter:
   design a linear phase FIR—Type I highpass filter with rs = 30 dB and deltap =
   0.09, able to attenuate, in a data sequence with sampling frequency of 1 kHz, all
   components with frequency smaller than 200 Hz. Specify transition bands 0.04π
   wide. Express frequencies in Hz in the plots. Use the design_FIR_highpass script
   given below.

```
1    % design_FIR_highpass
2    % Design linear phase FIR − Type I highpass filter
3    % with rs = 30 and deltap = 0.09, suitable to attenuate components with frequency
            smaller than 200 Hz
4    % in data with sampling frequency of 1 kHz. Specify transition bands 0.04 pi wide.
5
6    % Tolerances
7    rs = 30
8    deltas = 10^(−rs/20)
```

```
 9   deltap = 0.09
10   rp = 20*log10((1+deltap)/(1−deltap))
11
12   % Characteristic frequencies, amplitudes and deviations
13   f_s = 10^3
14   f_Ny = f_s/2
15
16   fs = 200
17   deltw_over_pi = 0.04
18   deltf = deltw_over_pi*f_Ny
19   fp = fs+deltf
20   f = [fs fp];
21   a = [0 1];
22   dev = [deltas deltap];
23
24   % Estimate minimum order
25   [n0,w_over_pi0,a0,whts]=firpmord(f,a,dev,f_s) ;
26
27   % Check if minimum order is odd and if it is, increase n0 by 1
28   if mod(n0,2)~=0
29       n0=n0+1;
30   end
31
32   % Initialize order for design process
33   n=n0−2;
34
35   repeat=1;
36
37   % Start loop until satisfaction
38   while repeat==1
39
40       % Design filter
41       n=n+2
42       h_pm=firpm(n,w_over_pi0,a0,whts);
43
44       % Compute frequency response and plot it
45       % with specifications to check performances
46       N_f=1024
47       [H,f]=freqz(h_pm,1,N_f,f_s);
48
49       figure(80)
50
51       subplot(1,2,1)
52       plot(f,20*log10(abs(H)))
53       hold on
54       plot(f,−rs*ones(1,N_f),'g')
55       hold off
56       set(gca,'XLim',[0 f_Ny])
57       set(gca,'YLim',[−60 5])
58       set(gca,'YTick',[−60:10:0])
59       hlabelx=get(gca,'Xlabel');
60       set(hlabelx,'String','$f$ (Hz)','Interpreter','Latex','FontSize',16)
61       hlabely=get(gca,'Ylabel');
```

```
62      set(hlabely,'String','$20\log_{10}|H|$','Interpreter','Latex','FontSize',16)
63
64      subplot(1,2,2)
65      plot(f,abs(H))
66      hold on
67      plot(f,(1+deltap)*ones(1,N_f),'g')
68      plot(f,(1-deltap)*ones(1,N_f),'g')
69      hold off
70      set(gca,'XLim',[0 f_Ny])
71      set(gca,'YLim',[0 1.15])
72      hlabelx=get(gca,'Xlabel');
73      set(hlabelx,'String','$f$ (Hz)','Interpreter','Latex','FontSize',16)
74      hlabely=get(gca,'Ylabel');
75      set(hlabely,'String','$|H|$','Interpreter','Latex','FontSize',16)
76
77      prompt='Is the filter satisfactory? (0 = yes, 1 = no)';
78      repeat = input(prompt);
79    end
```

4. Lowpass filter:

   design a linear phase FIR—Type I lowpass filter with rs = 50 dB and deltap = 0.06, fit to attenuate, in data with sampling interval of 1 min, all components with period higher than 5 min. Specify transition bands 0.04π wide. Express frequencies in Hz in the plots. Use the design_FIR_lowpass script given below.

```
1    % design_FIR_lowpass
2    % Design linear phase FIR - Type I lowpass filter
3    % with rs = 50 and deltap = 0.06, suitable to attenuate components with period higher
          than 5 min
4    % in data with sampling interval of 1 min. Specify transition bands 0.04 pi wide.
5
6    % Tolerances
7    rs = 50
8    deltas = 10^(-rs/20)
9    deltap = 0.06
10   rp = 20*log10((1+deltap)/(1-deltap))
11
12   % Characteristic frequencies, amplitudes and deviations
13   T_s = 60
14   f_s = 1/T_s
15   f_Ny = f_s/2
16
17   fs = 1/300
18   deltw_over_pi = 0.04
19   deltf = deltw_over_pi/(2*T_s)
20   fp = fs-deltf
21   f = [fp fs];
22   a = [1 0];
23   dev = [deltap deltas];
24
25   % Estimate minimum order
26   [n0,w_over_pi0,a0,whts]=firpmord(f,a,dev,f_s) ;
27
28   % Check if minimum order is odd and if it is, increase n0 by 1
```

```
29    if mod(n0,2)~=0
30        n0=n0+1;
31    end
32
33    % Initialize order for design process
34    n=n0-2;
35
36    repeat=1;
37
38    % Start loop until satisfaction
39    while repeat==1
40
41        % Design filter
42        n=n+2
43        h_pm=firpm(n,w_over_pi0,a0,whts);
44
45        % Compute frequency response and plot it
46        % with specifications to check performances
47        N_f=1024
48        [H,f]=freqz(h_pm,1,N_f,f_s);
49
50        figure(90)
51
52        subplot(1,2,1)
53        plot(f,20*log10(abs(H)))
54        hold on
55        plot(f,-rs*ones(1,N_f),'g')
56        hold off
57        set(gca,'XLim',[0 f_Ny])
58        set(gca,'YLim',[-60 5])
59        set(gca,'YTick',[-60:10:0])
60        hlabelx=get(gca,'Xlabel');
61        set(hlabelx,'String','$f$ (Hz)','Interpreter','Latex','FontSize',16)
62        hlabely=get(gca,'Ylabel');
63        set(hlabely,'String','$20\log_{10}|H|$','Interpreter','Latex','FontSize',16)
64
65        subplot(1,2,2)
66        plot(f,abs(H))
67        hold on
68        plot(f,(1+deltap)*ones(1,N_f),'g')
69        plot(f,(1-deltap)*ones(1,N_f),'g')
70        hold off
71        set(gca,'XLim',[0 f_Ny])
72        set(gca,'YLim',[0 1.15])
73        hlabelx=get(gca,'Xlabel');
74        set(hlabelx,'String','$f$ (Hz)','Interpreter','Latex','FontSize',16)
75        hlabely=get(gca,'Ylabel');
76        set(hlabely,'String','$|H|$','Interpreter','Latex','FontSize',16)
77
78        prompt='Is the filter satisfactory? (0 = yes, 1 = no)';
79        repeat = input(prompt);
80    end
```

5. Notch filter:
   design a linear phase FIR—Type I narrowband bandstop filter (notch filter) with
   rs = 30 and deltap = 0.04, suitable to strongly attenuate, in a data sequence with
   1 kHz sampling frequency, the 50 Hz component due to power line interference
   (Europe). Express frequencies in Hz in the plots. Use the design_FIR_notch script
   given below.

```
1   % design_FIR_notch
2   % Design linear phase FIR − Type I narrowband bandstop filter
3   % (notch filter) with rs = 30 and deltap = 0.09,
4   % to eliminate the 50 Hz component due to power line interference (Europe)
5   % from data with 1 kHz sampling frequency.
6
7   % Tolerances
8   rs = 30
9   deltas = 10^(−rs/20)
10  deltap = 0.04
11  rp = 20*log10((1+deltap)/(1−deltap))
12
13  % Characteristic frequencies, amplitudes and deviations
14  f_s = 10^3
15  f_Ny = f_s/2
16
17  fno = 50
18  ww_over_pi = fno/f_Ny
19  deltw_over_pi = 0.012
20  wp1_over_pi = ww_over_pi−deltw_over_pi
21  wp2_over_pi = ww_over_pi+deltw_over_pi
22  deltw_over_pi = 0.006
23  ws1_over_pi = wp1_over_pi+deltw_over_pi
24  ws2_over_pi = wp2_over_pi−deltw_over_pi
25  w_over_pi = [wp1_over_pi ws1_over_pi ws2_over_pi wp2_over_pi];
26  f = w_over_pi*f_Ny;
27  a=[1 0 1];
28  dev=[deltas deltap deltas];
29
30  % Estimate minimum order
31  [n0,w_over_pi0,a0,whts]=firpmord(f,a,dev,f_s) ;
32
33  % Check if minimum order is odd and if it is, increase n0 by 1
34  if mod(n0,2)~=0
35      n0=n0+1;
36  end
37
38  % Initialize order for design process
39  n=n0−2;
40
41  repeat=1;
42
43  % Start loop until satisfaction
44  while repeat==1
45
46      % Design filter
```

```
47      n=n+2
48      h_pm=firpm(n,w_over_pi0,a0,whts);
49
50      % Compute frequency response and plot it
51      % with specifications to check performances
52      N_f=1024
53      [H,f]=freqz(h_pm,1,N_f,f_s);
54
55      figure(100)
56
57      subplot(1,2,1)
58      plot(f,20*log10(abs(H)))
59      hold on
60      plot(f,-rs*ones(1,N_f),'g')
61      hold off
62      set(gca,'XLim',[0 f_Ny])
63      set(gca,'YLim',[-60 5])
64      set(gca,'YTick',[-60:10:0])
65      hlabelx=get(gca,'Xlabel');
66      set(hlabelx,'String','$f$ (Hz)','Interpreter','Latex','FontSize',16)
67      hlabely=get(gca,'Ylabel');
68      set(hlabely,'String','$20\log_{10}|H|$','Interpreter','Latex','FontSize',16)
69
70      subplot(1,2,2)
71      plot(f,abs(H))
72      hold on
73      plot(f,(1+deltap)*ones(1,N_f),'g')
74      plot(f,(1-deltap)*ones(1,N_f),'g')
75      hold off
76      set(gca,'XLim',[0 f_Ny])
77      set(gca,'YLim',[0 1.15])
78      hlabelx=get(gca,'Xlabel');
79      set(hlabelx,'String','$f$ (Hz)','Interpreter','Latex','FontSize',16)
80      hlabely=get(gca,'Ylabel');
81      set(hlabely,'String','$|H|$','Interpreter','Latex','FontSize',16)
82
83      prompt='Is the filter satisfactory? (0 = yes, 1 = no)';
84      repeat = input(prompt);
85  end
```

16.4.2   IIR Filters

16.4.2.1   Design of IIR Filters

We will now turn to the classical design of IIR filters (Chap. 8). The designs proposed below can be tackled using Butterworth, Chebyshev I and II, and elliptic filters. When prompted, input the desired value of the related flag. Repeat for all types of filter. The Matlab functions involved in these exercises are listed below; their syntax is exemplified assuming that all characteristic frequencies are specified in terms of ω/π.

- For filter order estimation on the basis of specifications, use

```
1  [N,wc_over_pi] = buttord(wp_over_pi,ws_over_pi,rp,rs)
2  [N,wc_over_pi] = cheb1ord(wp_over_pi,ws_over_pi,rp,rs)
3  [N,wc_over_pi] = cheb2ord(wp_over_pi,ws_over_pi,rp,rs)
4  [N,wc_over_pi] = ellipord(wp_over_pi,ws_over_pi,rp,rs)
```

where wc_over_pi represents the cutoff angular frequency divided by π (see Chap. 8).
- For filter design, use

```
1  [z,p,k] = butter(N,wn_over_pi,filtertype)
2  [z,p,k] = cheby1(N,rp,wp_over_pi,filtertype)
3  [z,p,k] = cheby2(N,rs,ws_over_pi,filtertype)
4  [z,p,k] = ellip(N,rp,rs,wp_over_pi,filtertype)
```

where filtertype can be 'low', 'high', 'bandpass', or 'stop', and the output filter is given through its zero-pole-gain parameters (z,p,k). Note that also the [b,a] syntax is possible, e.g.,

```
1  [b,a] = butter(N,wn_over_pi,'high')
```

but we will design our IIR filters in terms of zero-pole-parameters, in order to avoid possible numerical instabilities (see the following discussion). Later we will use the function sos = zp2sos(z,p,k) (see the Matlab documentation) to convert the zero-pole-gain filter values into the so-called *second-order sections form* (sos), in order to be able to use the function freqz for computing the frequency response.

1. Bandpass filter:
   design an IIR bandpass filter with rs = 45 dB and rp = 3 dB, useful to attenuate, in data with sampling frequency of 1.5 Hz, all components external to the interval 500–560 Hz. Specify transition bands 10 Hz wide. Express frequencies in Hz in the plots. Note that in this and in the following IIR design exercises we start from characteristic *analog* frequencies. Use the design_IIR_bandpass script given below.

```
1   % design_IIR_bandpass
2   % Design IIR bandpass filter with rs = 45 dB and rp = 3 dB
3   % useful to attenuate components external to the interval 500–560 Hz
4   % in data sampled at 1.5 kHz. Specify transition bands 10 Hz wide.
5
6   prompt='Filter type? (1 = Butterworth, 2 = Cheby_I, 3 = Cheby_II, 4 =
          ellipt)';
7   fflag = input(prompt);
8
9   selectivity ='bandpass'
10
```

```
11   % Tolerances
12   rp = 3
13   rs = 45
14   deltap = 1−10^(−rp/20)
15   deltas = 10^(−rs/20)
16
17   % Characteristic frequencies
18   f_s = 1500
19   T_s = 1/f_s
20   f_Ny = f_s/2
21
22   fp1 = 500
23   fp2 = 560
24   deltf = 10
25   fs1 = fp1−deltf
26   fs2 = fp2+deltf
27   wp_over_pi = [fp1 fp2]/f_Ny;
28   ws_over_pi = [fs1 fs2]/f_Ny;
29
30   % Design filter in zero−pole−gain form
31   if fflag==1
32       [n,wn_over_pi] = buttord(wp_over_pi,ws_over_pi,rp,rs);
33       order = n
34       [z,p,k] = butter(n,wn_over_pi,selectivity);
35   end
36   if fflag==2
37       [n,wp_over_pi] = cheb1ord(wp_over_pi,ws_over_pi,rp,rs);
38       order = n
39       [z,p,k] = cheby1(n,rp,wp_over_pi,selectivity);
40   end
41   if fflag==3
42       [n,ws_over_pi] = cheb2ord(wp_over_pi,ws_over_pi,rp,rs)
43       [z,p,k] = cheby2(n,rs,ws_over_pi,selectivity);
44       order = n
45   end
46   if fflag==4
47       [n,wp_over_pi] = ellipord(wp_over_pi,ws_over_pi,rp,rs)
48       [z,p,k] = ellip(n,rp,rs,wp_over_pi,selectivity);
49       order = n
50   end
51
52   % Convert zero−pole−gain filter parameters
53   % to second−order sections form and compute frequency response
54   sos = zp2sos(z,p,k);
55   N_f=1024
```

```
56   [H,f] = freqz(sos,N_f,f_s);
57   % Compute amplitude response
58   ampl=abs(H);
59   % Compute phase response in units of pi
60   phase=unwrap(angle(H))/pi;
61
62   % Plot frequency response
63   figure(100+fflag*10)
64   % Plot of amplitude response and specifications
65   % Linear scale
66   subplot(3,1,1)
67   plot([fs1 fs1],[−0.1 1.1],'m')
68   hold on
69   plot([fp1 fp1],[−0.1 1.1],'m')
70   plot([fs2 fs2],[−0.1 1.1],'m')
71   plot([fp2 fp2],[−0.1 1.1],'m')
72   plot([f(1) f(end)],[1−deltap 1−deltap],'g')
73   plot([f(1) f(end)],[1 1],'k')
74   plot([f(1) f(end)],[deltas deltas],'g')
75   plot(f,ampl,'LineWidth',1.5)
76   hold off
77   set(gca,'XLim',[0 f_Ny])
78   set(gca,'YLim',[−0.1 1.1])
79   set(gca,'YTick',[0:0.2:1])
80   hlabely=get(gca,'Ylabel');
81   set(hlabely,'String','$|H|$','Interpreter','Latex','FontSize',16)
82
83   % Plot amplitude response and specifications in dB
84   subplot(3,1,2)
85   y1a=−60;
86   y2a=max(20*log10(ampl))+5;
87   plot([fs1 fs1],[y1a y2a],'m')
88   hold on
89   plot([fp1 fp1],[y1a y2a],'m')
90   plot([fs2 fs2],[y1a y2a],'m')
91   plot([fp2 fp2],[y1a y2a],'m')
92   plot([f(1) f(end)],[0 0],'k')
93   plot([f(1) f(end)],[−rp −rp],'g')
94   plot([f(1) f(end)],[−rs −rs],'g')
95   plot(f,20*log10(ampl),'LineWidth',1.5)
96   hold off
97   set(gca,'XLim',[0 f_Ny])
98   set(gca,'YLim',[y1a y2a])
99   hlabely=get(gca,'Ylabel');
100  set(hlabely,'String','$20\log_{10}|H|$','Interpreter','Latex','FontSize',16)
```

```
101
102   % Plot phase response
103   subplot(3,1,3)
104   y1p=min(phase)−0.5;
105   y2p=max(phase)+0.5;
106   plot([fs1 fs1],[y1p y2p],'m')
107   hold on
108   plot([fp1 fp1],[y1p y2p],'m')
109   plot([fs2 fs2],[y1a y2a],'m')
110   plot([fp2 fp2],[y1a y2a],'m')
111   plot(f,phase,'LineWidth',1.5)
112   hold off
113   set(gca,'XLim',[0 f_Ny])
114   set(gca,'YLim',[y1p y2p])
115   hlabelx=get(gca,'Xlabel');
116   set(hlabelx,'String','$f$ (Hz)','Interpreter','Latex','FontSize',16)
117   hlabely=get(gca,'Ylabel');
118   set(hlabely,'String','$\psi/\pi$','Interpreter','Latex','FontSize',16)
```

2. Bandstop filter:

   design an IIR bandstop filter with rs = 30 dB and rp = 3 dB, useful to attenuate, in data sampled at 150 Hz, all components internal to the interval 5–25 Hz. Specify transition bands 2.5 Hz wide. Express frequencies in Hz in the plots. Use the design_IIR_bandstop script given below.

```
 1   % design_IIR_bandstop
 2   % Design IIR bandstop filter with rs = 30 dB and rp = 3 dB
 3   % useful to attenuate components internal to the interval 5−25 Hz
 4   % in data with f_s = 150 Hz. Specify transition bands 2.5 Hz wide.
 5
 6   prompt='Filter type? (1 = Butterworth, 2 = Cheby_I, 3 = Cheby_II, 4 =
            ellipt)';
 7   fflag = input(prompt);
 8
 9   selectivity ='stop'
10
11   % Tolerances
12   rp = 3
13   rs = 30
14   deltap = 1−10^(−rp/20)
15   deltas = 10^(−rs/20)
16
17   % Characteristic frequencies
18   f_s = 150
19   T_s = 1/f_s
```

```
20   f_Ny = f_s/2
21
22   fp1 = 5
23   fp2 = 25
24   deltf = 2.5
25   fs1 = fp1+deltf
26   fs2 = fp2−deltf
27   wp_over_pi = [fp1 fp2]/f_Ny;
28   ws_over_pi = [fs1 fs2]/f_Ny;
29
30   % Design filter in zero−pole−gain form
31   if fflag==1
32       [n,wn_over_pi] = buttord(wp_over_pi,ws_over_pi,rp,rs);
33       order = n
34       [z,p,k] = butter(n,wn_over_pi,selectivity);
35   end
36   if fflag==2
37       [n,wp_over_pi] = cheb1ord(wp_over_pi,ws_over_pi,rp,rs);
38       order = n
39       [z,p,k] = cheby1(n,rp,wp_over_pi,selectivity);
40   end
41   if fflag==3
42       [n,ws_over_pi] = cheb2ord(wp_over_pi,ws_over_pi,rp,rs)
43       [z,p,k] = cheby2(n,rs,ws_over_pi,selectivity);
44       order = n
45   end
46   if fflag==4
47       [n,wp_over_pi] = ellipord(wp_over_pi,ws_over_pi,rp,rs)
48       [z,p,k] = ellip(n,rp,rs,wp_over_pi,selectivity);
49       order = n
50   end
51
52   % Convert zero−pole−gain filter parameters
53   % to second−order sections form
54   % and compute frequency response
55   sos = zp2sos(z,p,k);
56   N_f=1024
57   [H,f] = freqz(sos,N_f,f_s);
58   % Compute amplitude response
59   ampl=abs(H);
60   % Compute phase response in units of pi
61   phase=unwrap(angle(H))/pi;
62
63   % Plot frequency response
64   figure(200+fflag*10)
```

```
65   % Plot of amplitude response and specifications
66   % Linear scale
67   subplot(3,1,1)
68   plot([fs1 fs1],[−0.1 1.1],'m')
69   hold on
70   plot([fp1 fp1],[−0.1 1.1],'m')
71   plot([fs2 fs2],[−0.1 1.1],'m')
72   plot([fp2 fp2],[−0.1 1.1],'m')
73   plot([f(1) f(end)],[1−deltap 1−deltap],'g')
74   plot([f(1) f(end)],[1 1],'k')
75   plot([f(1) f(end)],[deltas deltas],'g')
76   plot(f,ampl,'LineWidth',1.5)
77   hold off
78   set(gca,'XLim',[0 f_Ny])
79   set(gca,'YLim',[−0.1 1.1])
80   set(gca,'YTick',[0:0.2:1])
81   hlabely=get(gca,'Ylabel');
82   set(hlabely,'String','$|H|$','Interpreter','Latex','FontSize',16)
83
84   % Plot amplitude response and specifications in dB
85   subplot(3,1,2)
86   y1a=−60;
87   y2a=max(20*log10(ampl))+5;
88   plot([fs1 fs1],[y1a y2a],'m')
89   hold on
90   plot([fp1 fp1],[y1a y2a],'m')
91   plot([fs2 fs2],[y1a y2a],'m')
92   plot([fp2 fp2],[y1a y2a],'m')
93   plot([f(1) f(end)],[0 0],'k')
94   plot([f(1) f(end)],[−rp −rp],'g')
95   plot([f(1) f(end)],[−rs −rs],'g')
96   plot(f,20*log10(ampl),'LineWidth',1.5)
97   hold off
98   set(gca,'XLim',[0 f_Ny])
99   set(gca,'YLim',[y1a y2a])
100  hlabely=get(gca,'Ylabel');
101  set(hlabely,'String','$20\log_{10}|H|$','Interpreter','Latex','FontSize',16)
102
103  % Plot phase response
104  subplot(3,1,3)
105  y1p=min(phase)−0.5;
106  y2p=max(phase)+0.5;
107  plot([fs1 fs1],[y1p y2p],'m')
108  hold on
109  plot([fp1 fp1],[y1p y2p],'m')
```

```
110   plot([fs2 fs2],[y1a y2a],'m')
111   plot([fp2 fp2],[y1a y2a],'m')
112   plot(f,phase,'LineWidth',1.5)
113   hold off
114   set(gca,'XLim',[0 f_Ny])
115   set(gca,'YLim',[y1p y2p])
116   hlabelx=get(gca,'Xlabel');
117   set(hlabelx,'String','$f$ (Hz)','Interpreter','Latex','FontSize',16)
118   hlabely=get(gca,'Ylabel');
119   set(hlabely,'String','$\psi/\pi$','Interpreter','Latex','FontSize',16)
```

3. Highpass filter:

   design an IIR highpass filter with rs = 26 dB and rp = 3 dB, useful to attenuate, in data with sampling interval of 1 year, all components with period higher
   than 10 years. Specify transition bands 60 years wide in terms of period. Express
   frequencies in y^{-1}. Use the design_IIR_highpass script given below.

```
1    % design_IIR_highpass
2    % Design IIR highpass filter with rs = 26 dB and rp = 3 dB
3    % useful to attenuate components with period higher than 10 y
4    % in data with T_s = 1 y. Specify transition bands 60 y wide in terms of period.
5
6    prompt='Input filter type (1 = Butterworth, 2 = Cheby_I, 3 = Cheby_II, 4 = ellipt) =';
7    fflag = input(prompt);
8
9    selectivity ='high'
10
11   % Tolerances
12   rp = 3
13   rs = 26
14   deltap = 1−10^(−rp/20)
15   deltas = 10^(−rs/20)
16
17   % Characteristic frequencies
18   T_s = 1 % years − express frequencies in years^(−1)
19   f_s = 1/T_s
20   f_Ny = f_s/2
21
22   fp = 1/10
23   deltf = 1/60
24   fs = fp−deltf
25   wp_over_pi = fp/f_Ny;
26   ws_over_pi = fs/f_Ny;
27
28   % Design filter in zero−pole−gain form
29   if fflag==1
30       [n,wn_over_pi] = buttord(wp_over_pi,ws_over_pi,rp,rs);
31       order = n
32       [z,p,k] = butter(n,wn_over_pi,selectivity);
33   end
34   if fflag==2
```

```
35        [n,wp_over_pi] = cheb1ord(wp_over_pi,ws_over_pi,rp,rs);
36        order = n
37        [z,p,k] = cheby1(n,rp,wp_over_pi,selectivity);
38     end
39     if fflag==3
40        [n,ws_over_pi] = cheb2ord(wp_over_pi,ws_over_pi,rp,rs)
41        [z,p,k] = cheby2(n,rs,ws_over_pi,selectivity);
42        order = n
43     end
44     if fflag==4
45        [n,wp_over_pi] = ellipord(wp_over_pi,ws_over_pi,rp,rs)
46        [z,p,k] = ellip(n,rp,rs,wp_over_pi,selectivity);
47        order = n
48     end
49
50     % Convert zero-pole-gain filter parameters
51     % to second-order sections form
52     % and compute frequency response
53     sos = zp2sos(z,p,k);
54     N_f=1024
55     [H,f] = freqz(sos,N_f,f_s);
56     % Compute amplitude response
57     ampl=abs(H);
58     % Compute phase response in units of pi
59     phase=unwrap(angle(H))/pi;
60
61     % Plot frequency response
62     figure(300+fflag*10)
63     % Plot of amplitude response and specifications
64     % Linear scale
65     subplot(3,1,1)
66     plot([fs fs],[−0.1 1.1],'m')
67     hold on
68     plot([fp fp],[−0.1 1.1],'m')
69     plot([f(1) f(end)],[1−deltap 1−deltap],'g')
70     plot([f(1) f(end)],[1 1],'k')
71     plot([f(1) f(end)],[deltas deltas],'g')
72     plot(f,ampl,'LineWidth',1.5)
73     hold off
74     set(gca,'XLim',[0 f_Ny])
75     set(gca,'YLim',[−0.1 1.1])
76     set(gca,'YTick',[0:0.2:1])
77     hlabely=get(gca,'Ylabel');
78     set(hlabely,'String','$|H|$','Interpreter','Latex','FontSize',16)
79
80     % Plot amplitude response and specifications in dB
81     subplot(3,1,2)
82     y1a=−60;
83     y2a=max(20*log10(ampl))+5;
84     plot([fs fs],[y1a y2a],'m')
85     hold on
86     plot([fp fp],[y1a y2a],'m')
87     plot([f(1) f(end)],[0 0],'k')
```

```
88   plot([f(1) f(end)],[−rp −rp],'g')
89   plot([f(1) f(end)],[−rs −rs],'g')
90   plot(f,20*log10(ampl),'LineWidth',1.5)
91   hold off
92   set(gca,'XLim',[0 f_Ny])
93   set(gca,'YLim',[y1a y2a])
94   hlabely=get(gca,'Ylabel');
95   set(hlabely,'String','$20\log_{10}|H|$','Interpreter','Latex','FontSize',16)
96
97   % Plot phase response
98   subplot(3,1,3)
99   y1p=min(phase)−0.5;
100  y2p=max(phase)+0.5;
101  plot([fs fs],[y1a y2a],'m')
102  hold on
103  plot([fp fp],[y1a y2a],'m')
104  plot(f,phase,'LineWidth',1.5)
105  hold off
106  set(gca,'XLim',[0 f_Ny])
107  set(gca,'YLim',[y1p y2p])
108  hlabelx=get(gca,'Xlabel');
109  set(hlabelx,'String','$f$ (Hz)','Interpreter','Latex','FontSize',16)
110  hlabely=get(gca,'Ylabel');
111  set(hlabely,'String','$\psi/\pi$','Interpreter','Latex','FontSize',16)
```

4. Lowpass filter:

   design an IIR lowpass filter with rs = 14 dB and rp = 2 dB, able to attenuate, in data sampled at 10 Hz, all components higher than 2 Hz. Specify transition bands 0.2 Hz wide. Express frequencies in Hz in the plots. Use the design_IIR_lowpass script given below. In the Butterworth case, save the filter in sos form in a binary file named filter_sos_lp.mat for future use.

```
1    % design_IIR_lowpass
2    % Design IIR lowpass filter with rs = 14 dB and rp = 2 dB
3    % useful to attenuate components higher than 2 Hz
4    % in data sampled at 10 Hz. Specify transition bands 0.2 Hz wide.
5
6    prompt ='Filter type? (1 = Butterworth, 2 = Cheby_I, 3 = Cheby_II, 4 =
            ellipt)';
7    fflag = input(prompt);
8
9    prompt ='Save filter? (1 = yes, 2 = no)';
10   saveflag = input(prompt);
11
12   selectivity ='low'
13
14   % Tolerances
15   rp = 2
16   rs = 14
```

```
17   deltap = 1−10^(−rp/20)
18   deltas = 10^(−rs/20)
19
20   % Characteristic frequencies
21   f_s = 10
22   T_s = 1/f_s
23   f_Ny = f_s/2
24
25   fp = 2
26   deltf = 0.2
27   fs = fp+deltf
28   wp_over_pi = fp/f_Ny;
29   ws_over_pi = fs/f_Ny;
30
31   % Design filter in zero−pole−gain form
32   if fflag==1
33       [n,wn_over_pi] = buttord(wp_over_pi,ws_over_pi,rp,rs);
34       order = n
35       [z,p,k] = butter(n,wn_over_pi,selectivity);
36   end
37   if fflag==2
38       [n,wp_over_pi] = cheb1ord(wp_over_pi,ws_over_pi,rp,rs);
39       order = n
40       [z,p,k] = cheby1(n,rp,wp_over_pi,selectivity);
41   end
42   if fflag==3
43       [n,ws_over_pi] = cheb2ord(wp_over_pi,ws_over_pi,rp,rs)
44       [z,p,k] = cheby2(n,rs,ws_over_pi,selectivity);
45       order = n
46   end
47   if fflag==4
48       [n,wp_over_pi] = ellipord(wp_over_pi,ws_over_pi,rp,rs)
49       [z,p,k] = ellip(n,rp,rs,wp_over_pi,selectivity);
50       order = n
51   end
52
53   % Convert zero−pole−gain filter parameters
54   % to second−order sections form
55   % and compute frequency response
56   sos = zp2sos(z,p,k);
57   N_f=1024
58   [H,f] = freqz(sos,N_f,f_s);
59   % Compute amplitude response
60   ampl=abs(H);
61   % Compute phase response in units of pi
```

```
62   phase=unwrap(angle(H))/pi;
63
64   % Plot frequency response
65   figure(400+fflag*10)
66   % Plot of amplitude response and specifications
67   % Linear scale
68   subplot(3,1,1)
69   plot([fs fs],[−0.1 1.1],'m')
70   hold on
71   plot([fp fp],[−0.1 1.1],'m')
72   plot([f(1) f(end)],[1−deltap 1−deltap],'g')
73   plot([f(1) f(end)],[1 1],'k')
74   plot([f(1) f(end)],[deltas deltas],'g')
75   plot(f,ampl,'LineWidth',1.5)
76   hold off
77   set(gca,'XLim',[0 f_Ny])
78   set(gca,'YLim',[−0.1 1.1])
79   set(gca,'YTick',[0:0.2:1])
80   hlabely=get(gca,'Ylabel');
81   set(hlabely,'String','$|H|$','Interpreter','Latex','FontSize',16)
82
83   % Plot amplitude response and specifications in dB
84   subplot(3,1,2)
85   y1a=−60;
86   y2a=max(20*log10(ampl))+5;
87   plot([fs fs],[y1a y2a],'m')
88   hold on
89   plot([fp fp],[y1a y2a],'m')
90   plot([f(1) f(end)],[0 0],'k')
91   plot([f(1) f(end)],[−rp −rp],'g')
92   plot([f(1) f(end)],[−rs −rs],'g')
93   plot(f,20*log10(ampl),'LineWidth',1.5)
94   hold off
95   set(gca,'XLim',[0 f_Ny])
96   set(gca,'YLim',[y1a y2a])
97   hlabely=get(gca,'Ylabel');
98   set(hlabely,'String','$20\log_{10}|H|$','Interpreter','Latex','FontSize',16)
99
100  % Plot phase response
101  subplot(3,1,3)
102  y1p=min(phase)−0.5;
103  y2p=max(phase)+0.5;
104  plot([fs fs],[y1a y2a],'m')
105  hold on
106  plot([fp fp],[y1a y2a],'m')
```

```
107   plot(f,phase,'LineWidth',1.5)
108   hold off
109   set(gca,'XLim',[0 f_Ny])
110   set(gca,'YLim',[y1p y2p])
111   hlabelx=get(gca,'Xlabel');
112   set(hlabelx,'String','$f$ (Hz)','Interpreter','Latex','FontSize',16)
113   hlabely=get(gca,'Ylabel');
114   set(hlabely,'String','$\psi/\pi$','Interpreter','Latex','FontSize',16)
115
116   % Save filter if required
117   if saveflag==1
118       save filter_lp_sos.mat sos
119   end
```

5. Notch filter:
   design a narrowband bandstop Butterworth filter (notch filter) with given tolerances, namely, rs = 30 dB and deltap = 0.04, fit for eliminating, from data with
   1 kHz sampling frequency, the 50 Hz component due to power line interference
   (Europe). The script given below (design_IIR_notch) performs this task, and also
   compares the result with the corresponding notch filter obtained using another
   Matlab function,

```
1   [b,a] = iirnotch(w0_over_pi,bw_over_pi)
```

   that designs a digital notch filter with the notch located at the ω/π value
   specified by w0_over_pi, and with the bandwidth at the -3 dB point set to
   bw_over_pi (the bandwidth is also in terms of ω/π). The bandwidth bw_over_pi
   is related to w0_over_pi through the so-called *Q-factor* of the filter: bw_over_pi
   = w0_over_pi/Q. In the present case, Q is set to 35. The design procedure on
   which the iirnotch function is based can be found in Orfanidis (1996).

```
1   % design_IIR_notch
2   % Design narrowband bandstop IIR filter (notch filter)
3   % for eliminating the 50 Hz component due to power line interference
        (Europe)
4   % from data sampled at 1 kHz.
5
6   selectivity ='stop'
7
8   % Tolerances
9   rp = 3
10  rs = 60
11  deltap = 1−10^(−rp/20)
12  deltas = 10^(−rs/20);
13
```

```
14  % Characteristic frequencies
15  f_s = 1000
16  T_s = 1/f_s
17  f_Ny = f_s/2
18
19  fp1 = 47
20  fp2 = 53
21  deltf = 2.9
22  fs1 = fp1+deltf
23  fs2 = fp2−deltf
24  wp_over_pi = [fp1 fp2]/f_Ny;
25  ws_over_pi = [fs1 fs2]/f_Ny;
26
27  % Design filter in zero−pole−gain form
28  [n,wn_over_pi] = buttord(wp_over_pi,ws_over_pi,rp,rs);
29  order = n
30  [z,p,k] = butter(n,wn_over_pi,selectivity);
31
32  % Convert zero−pole−gain filter parameters
33  % to second−order sections form
34  % and compute frequency response
35  sos = zp2sos(z,p,k);
36  N_f=1024
37  [H,f] = freqz(sos,N_f,f_s);
38  % Compute amplitude response
39  ampl=abs(H);
40  % Compute phase response in units of pi
41  phase=unwrap(angle(H))/pi;
42
43  % Plot frequency response
44  figure(500)
45  % Plot of amplitude response and specifications
46  % Linear scale
47  subplot(3,1,1)
48  plot([f(1) f(end)],[1−deltap 1−deltap],'g')
49  hold on
50  plot([f(1) f(end)],[1 1],'k')
51  plot([f(1) f(end)],[deltas deltas],'g')
52  plot(f,ampl,'LineWidth',1.5)
53  hold off
54  set(gca,'XLim',[0 0.2*f_Ny])
55  set(gca,'YLim',[−0.2 1.1])
56  set(gca,'YTick',[0:0.2:1])
57  hlabely=get(gca,'Ylabel');
58  set(hlabely,'String','$|H|$','Interpreter','Latex','FontSize',16)
```

```
59
60   % Plot amplitude response and specifications in dB
61   subplot(3,1,2)
62   plot([f(1) f(end)],[0 0],'k')
63   hold on
64   plot([f(1) f(end)],[−rp −rp],'g')
65   plot([f(1) f(end)],[−rs −rs],'g')
66   plot(f,20*log10(ampl),'LineWidth',1.5)
67   hold off
68   set(gca,'XLim',[0 0.2*f_Ny])
69   set(gca,'YLim',[−60 10])
70   hlabely=get(gca,'Ylabel');
71   set(hlabely,'String','$20\log_{10}|H|$','Interpreter','Latex','FontSize'
        ,16)
72
73   % Plot phase response
74   subplot(3,1,3)
75   plot(f,phase,'LineWidth',1.5)
76   set(gca,'XLim',[0 0.2*f_Ny])
77   set(gca,'YLim',[−2.2 0.2])
78   hlabelx=get(gca,'Xlabel');
79   set(hlabelx,'String','$f$ (Hz)','Interpreter','Latex','FontSize',16)
80   hlabely=get(gca,'Ylabel');
81   set(hlabely,'String','$\psi/\pi$','Interpreter','Latex','FontSize',16)
82
83   % Use iirnotch function
84
85   % Design filter
86   f0=50
87   w0_over_pi=f0/f_Ny
88   Q=35
89   bw_over_pi=w0_over_pi/Q
90   [b,a] = iirnotch(w0_over_pi,bw_over_pi);
91
92   % Compute frequency response
93    [H1,f1] = freqz(b,a,N_f,f_s);
94   % Compute amplitude response
95   ampl1=abs(H1);
96   % Compute phase response in units of pi
97   phase1=unwrap(angle(H1))/pi;
98
99   % Plot frequency response
100  figure(600)
101  % Plot of amplitude response
102  % Linear scale
```

```
103  subplot(3,1,1)
104  plot(f1,ampl1,'LineWidth',1.5)
105  hold off
106  set(gca,'XLim',[0 0.2*f_Ny])
107  set(gca,'YLim',[−0.2 1.1])
108  set(gca,'YTick',[0:0.2:1])
109  hlabely=get(gca,'Ylabel');
110  set(hlabely,'String','$|H|$','Interpreter','Latex','FontSize',16)
111
112  % Plot amplitude response and specifications in dB
113  subplot(3,1,2)
114  plot(f1,20*log10(ampl1),'LineWidth',1.5)
115  hold off
116  set(gca,'XLim',[0 0.2*f_Ny])
117  set(gca,'YLim',[−15 2])
118  hlabely=get(gca,'Ylabel');
119  set(hlabely,'String','$20\log_{10}|H|$','Interpreter','Latex','FontSize'
         ,16)
120
121  % Plot phase response
122  subplot(3,1,3)
123  plot(f1,phase1,'LineWidth',1.5)
124  set(gca,'XLim',[0 0.2*f_Ny])
125  hlabelx=get(gca,'Xlabel');
126  set(hlabelx,'String','$f$ (Hz)','Interpreter','Latex','FontSize',16)
127  hlabely=get(gca,'Ylabel');
128  set(hlabely,'String','$\psi/\pi$','Interpreter','Latex','FontSize',16)
```

Note that for convenience, only the frequencies 0–100 Hz are plotted.
6. Instabilities in the design of IIR filters:
   in general, it is advisable to use the [z,p,k] syntax to design IIR filters. If the
   filter is designed using the [b,a] syntax, numerical problems can be encountered.
   These problems are due to round-off errors, and can occur even at very low orders.
   For example, the following code (instability_cheby2) illustrates how an unstable
   Chebyshev-II filter can emerge: see the Signal Processing Toolbox User's Guide
   (The MathWorks 2014).

```
1   % instability_cheby2
2   % Provide an example of unstable Chebyshev−II filter.
3
4   % Specifications
5   rs = 80
6   f_s = 1e4
7   f_Ny = f_s/2
8   fs = [25 290]
9   ws_over_pi = fs/f_Ny
10  n = 6
```

```
11
12   % Transfer function design
13   [b,a] = cheby2(n,rs,ws_over_pi,'bandpass');
14   % This filter is unstable
15
16   % Zero−pole−gain design
17   [z,p,k] = cheby2(n,rs,ws_over_pi,'bandpass');
18   sos = zp2sos(z,p,k);
19
20   % Plot and compare the results
21   figure(700)
22   N_f=2^(13)
23   [H,f] = freqz(sos,N_f);
24   [Hu,fu] = freqz(b,a,N_f);
25   semilogx(f/pi,20*log10(abs(H)))
26   hold on
27   semilogx(fu/pi,20*log10(abs(Hu)),'r')
28   hold off
29   set(gca,'XLim',[10^(-4) 1])
30   set(gca,'XTick',[10^(-4) 10^(-3) 10^(-2) 10^(-1) 10^0])
31   set(gca,'YLim',[-110 10])
32   set(gca,'YTick',[-100:10:0])
33   hlabelx=get(gca,'Xlabel');
34   set(hlabelx,'String','$\omega/\pi$','Interpreter','Latex','FontSize',16)
35   hlabely=get(gca,'Ylabel');
36   set(hlabely,'String','$20\log10|H|$','Interpreter','Latex','FontSize',16)
```

16.4.2.2   Filtering a Sequence with an IIR Filter

In this exercise we will filter a data record using the IIR Butterworth lowpass filter generated before and stored in the binary file filter_sos_lp.mat. Recall that the filter was stored in its second-order sections form (sos). As a consequence, we cannot use the function filter for implementing the filtering operation: we must use a dedicated function. This function is sosfilt, which implements second-order (biquadratic) IIR filtering (see the Matlab documentation): for an input vector sd, the output vector sdf of filtered data is obtained by

```
1   sdf = sosfilt(sos,sd)
```

The data record to be generated for this purpose, using data_Nsin_noise_variable_N with a scalar N, has $T\_s = 0.1$ s and N = 8192; it includes three sines with amplitudes A = 5, 2.5, 1, and frequencies f_synth = 1.5, 3, and 4.5 Hz (0.3π, 0.6π, and 0.9π, respectively), plus white noise with sigma_wn = 1. For this exercise use the filtering_IIR script reported below.

```
1   % filtering_IIR
2   % Filter data record using IIR filter in sos form.
3
4   % Load lowpass Butterworth filter
```

```
 5 | load filter_lp_sos.mat
 6 |
 7 | % Compute DFT amplitude for original data
 8 | N_f=N;
 9 | SD=fft(sd,N_f);
10 | modSD=abs(SD);
11 |
12 | % Build frequency vector
13 | f_Ny=1/(2*T_s);
14 | deltf=1/(length(SD)*T_s);
15 | f=0:deltf:(2*f_Ny-deltf);
16 |
17 | % Filter data sequence
18 | sdf = sosfilt(sos,sd);
19 | sdf=sdf-mean(sdf);
20 | N_sdf=length(sdf) % length of filtered data
21 |
22 | % Compute DFT amplitude for filtered data
23 | SDF=fft(sdf,N_f);
24 | modSDF=abs(SDF);
25 |
26 | % Plots
27 | figure(800) % DFT modulus
28 | % for original and filtered data
29 | plot(f(2:N_f/2+1),10*log10(modSD(2:N_f/2+1)))
30 | hold on
31 | plot(f(2:N_f/2+1),10*log10(modSDF(2:N_f/2+1)),'g')
32 | hold off
33 | set(gca,'XLim',[0 f_Ny])
34 | hlabelx=get(gca,'Xlabel');
35 | set(hlabelx,'String','$f$ (Hz)','Interpreter','Latex','FontSize',16)
36 | hlabely=get(gca,'Ylabel');
37 | set(hlabely,'String','$10\log_{10}|X_{\mathrm{f}}|$','Interpreter','Latex','FontSize',16)
```

16.5 Methods of Stationary Spectral Analysis

In this section we will apply the stationary power spectrum estimation methods studied in Chaps. 10 and 11 and see how they are implemented in MATLAB. The exercises include the application of

- non-parametric methods (Chap. 10):

  - simple periodogram (Sect. 10.3) and modified periodogram (Sect. 10.5);
  - Bartlett's method (Sect. 10.4) and Welch's method (Sect. 10.6);
  - Blackman-Tukey (BT) method (Sect. 10.7);
  - MTM (Sect. 10.9);

- parametric methods (Chap. 11):

  - Burg's method (Sect. 11.6.3.4),
  - Yule-Walker AR method (Sect. 11.6.3.1).

16.5.1   Periodogram

Schuster's periodogram can be obtained by simply taking the squared modulus of
the DFT of the data sequence and dividing it by the number of spectral samples
(Sect. 10.3). However, Matlab offers a dedicated function: periodogram, which allows
computing both simple and modified periodograms. The syntax of this function for
a centered input sequence x is

```
[PSD,f] = periodogram(x,window,N_f,f_s,range)
```

where window must have the same length N of the input sequence. Recall that our
sd data have actually been centered. Be careful: if we were analyzing a sequence x
with a mean value different from zero, we should write

```
[PSD,f] = periodogram(x−mean(x),window,N_f,f_s,range)
```

As for the window, we can choose among a number of different window shapes,
including the following:

- rectangular (indicated as boxcar or rectwin in Matlab),
- Hamming (hamming),
- Hanning (hanning),
- Blackman (blackman), etc.

For calculating the periodogram, the window must be rectangular (default). The
range parameter can be 'twosided' or 'onesided'; for real data the default value is
'onesided'. We can omit one or more input variables and use default values, but
any unspecified input variable different from range, and falling between two other
variables that are explicitly provided, must be substituted by the empty vector []. For
example, to specify N_f and f_s, we will write

```
[PSD,f] = periodogram(x,[],N_f,f_s)
```

Note that zero padding is performed automatically if $N_f > N$. In output to periodogram, PSD and f are the power spectral density and the corresponding frequency
vector, respectively.

The nature of the frequency vector f and the spectral normalization varies according to the vaue of f_s that we provide in input.

- If we input a value for f_s (any value), MATLAB assumes we want to refer to the
  notion of spectrum as defined for a continuous-time signal, and therefore, given an
  input sequence $x[n]$ with some sampling frequency, it divides $|X[k]|^2/N$ by f_s,
  i.e., it multiplies it by T_s. Thus, if the frequency unit is Hz, the unit for PSD is
  power/(Hz); power means variance for a zero-mean data record, and its units are
  the square of the data units.
- If we wish to use adimensional frequencies ν, we will set f_s = 1. The spectral
  plot units are then cycles/sample in abscissa, power/(cycles/sample) in ordinate.

- If f_s is not assigned, by default Matlab assumes we want to use adimensional angular frequencies ω—in other words, by default f_s is set to 2π—and thus divides $|X[k]|^2/N$ by 2π. In this case, the spectral plot units are radians/sample in abscissa, power/(radians/sample) in ordinate.

For consistency with the convention followed throughout this book, we will set range = 'twosided' and remove the normalizations listed above, so as to always plot $|X[k]|^2/N$, independently of the frequency variable used in abscissa.

We will now perform some tests with the periodogram. We will analyze sequences with T_s = 5 s and use N_f = 1024, unless otherwise specified.

1. Using data_Nsin_noise_variable_N, generate a set of white noise sequences with sigma_wn = 1 and $N = 64, 256$, and 1024. Compute and plot the power spectra via periodograms. The related script (Periodogram_variable_N) is given below.

```
 1  % Periodogram_variable_N
 2  % Power spectrum estimation via Periodogram
 3
 4  prompt='N_f =';
 5  N_f = input(prompt);
 6
 7  range='twosided';
 8
 9  f_s=1/T_s;
10  f_Ny=f_s/2;
11  [Nmax,M]=size(sd);
12
13  figure(1000)
14  for m=1:M
15      N_samples=N(m);
16      variance=var(sd(1:N(m),m))  % data variance
17
18      [PSD,f]=periodogram(sd(1:N(m),m),boxcar(N(m)),N_f,f_s,range);
19      PSD=PSD*f_s;  % remove normalization
20      subplot(M,1,m)
21      plot(f(2:N_f/2+1),10*log10(PSD(2:N_f/2+1)))
22      set(gca,'XLim',[0 f_Ny])
23      set(gca,'YLim',[-40 40])
24      set(gca,'YTick',[-40:20:40])
25      ht=text(0.005,25,sprintf('$N$ = %d',N(m)));
26      set(ht,'Interpreter','Latex','FontSize',12)
27      hold off
28      if m==M
29          hlabelx=get(gca,'Xlabel');
30          set(hlabelx,'String','$f$ (Hz)','Interpreter','Latex','FontSize',16)
31      end
32      if m==ceil(M/2)
33          hlabely=get(gca,'Ylabel');
34          set(hlabely,'String','PSD (dB)','Interpreter','Latex','FontSize',16)
35      end
36      ave_power=sum(PSD)/N_f  % average power
37  end
```

Let us observe the plots and think about the following questions.

- Theoretically, what should the behavior of the power spectrum be, in each case? How do the estimated spectra deviate from this expected behavior?
- Do the oscillations of the spectrum vary with increasing N? How? Why?
- Does the average value of each spectrum approximately agree with its theoretical constant value? Does the quality of the agreement vary with N?

2. Now add a sinusoid to white noise and compute the periodogram for different values of N. Therefore, using data_Nsin_noise_variable_N generate a set of sequences with N = 40, 64, 256, and 1024, each containing one sinusoid with amplitude A = 5 and frequency 0.4π immersed in white noise with sigma_wn = 1. Compute the power spectrum of each sequence using Periodogram_variable_N (if you are keeping figure windows open, you may want to change the figure number in the script). These plots show clearly that the sinusoid's peak is initially low and wide, and then, as N increases, progressively narrows and becomes taller and taller.

3. Using data_Nsin_noise_variable_sigma_wn, generate a set of sequences with one sinusoid in white noise (the same as above) and N = 64 samples. Consider values of sigma_wn growing from 1 to 10 and to 100. Compute and plot the power spectrum using Periodogram_variable_sigma_wn, to examine the influence of the SNR on the spectrum (Sect. 10.3.3).

```
1    % Periodogram_variable_sigma_wn
2    % Power spectrum estimation via Periodogram
3
4    prompt='N_f =';
5    N_f = input(prompt);
6
7    range='twosided';
8
9    f_s=1/T_s;
10   f_Ny=f_s/2;
11   [N,M]=size(sd);
12
13   figure(1002)
14   for m=1:M
15       sigma_noise=sigma_wn(m)
16       variance=var(sd(:,m)) % data variance
17       SNR_o= N*A(1)^2/(sigma_wn(m)^2) % output SNR
18       [PSD,f]=periodogram(sd(:,m),boxcar(N),N_f,f_s,range);
19       PSD=PSD*f_s;
20       subplot(M,1,m)
21       plot(f(2:N_f/2+1),10*log10(PSD(2:N_f/2+1)))
22       set(gca,'XLim',[0 f_Ny])
23       set(gca,'YLim',[-22 60])
24       set(gca,'YTick',[-20:20:60])
25       ht=text(0.005,43-10*(m-1),sprintf('Output SNR = %g',SNR_o));
26       set(ht,'Interpreter','Latex','FontSize',12)
27       hold off
28       if m==M
```

```
29      hlabelx=get(gca,'Xlabel');
30      set(hlabelx,'String','$f$ (Hz)','Interpreter','Latex','FontSize',16)
31    end
32    if m==ceil(M/2)
33      hlabely=get(gca,'Ylabel');
34      set(hlabely,'String','PSD (dB)','Interpreter','Latex','FontSize',16)
35    end
36    ave_power=sum(PSD)/N_f % average power
37  end
```

When sigma_wn is 1, with as little as 64 data samples the sinusoid's peak is clearly visible. As sigma_wn increases, the peak is progressively submerged by the noise background and ends up disappearing completely. This behavior is due to the decline of the output SNR, which for a rectangular window with processing gain of 1 must satisfy (Sect. 10.3.3)

$$\text{SNR}_o = \frac{NA^2}{4\sigma_{wn}^2} = N \, \text{SNR}_i > 25.$$

With $A = 5$, $N = 64$ and sigma_wn = 1, 10 and 100, SNR_o assumes the values 400, 4 and 0.04, respectively. Thus it initially exceeds the prescribed threshold of 25, but for higher noise variance it becomes insufficient.

4. Now let us study the periodogram's resolution, considering signals with two sinusoids in noise. By data_Nsin_noise_variable_N, generate a set of sequences with two sinusoids of equal amplitudes $A = 5$ and frequencies 0.4π and 0.45π, immersed in white noise with sigma_wn $= 1$. Use $N = 40$, 64 and 256. Compute and plot the power spectrum via periodogram, using Periodogram_variable_N. With $N = 40$, we see that resolution is borderline; with $N = 64$, the resolution is sufficient; with $N = 256$, the two sinusoids are well resolved.

In fact, to resolve two close sinusoids in a periodogram, their frequency separation $\Delta\omega = (2\pi/f_s)\Delta f$ must be larger than the mainlobe width of the spectral window $W_B(e^{j\omega})$ centered on each sinusoid's frequency, which can be defined as the width at half height: about $2\pi/N$ in terms of ω, and $f_s/N = 1/(NT_s)$ in analog terms (Sect. 10.3.3). Therefore, the resolution rule for the periodogram is

$$\Delta\omega > \frac{2\pi}{N} \quad \rightsquigarrow \quad N > \frac{2\pi}{\Delta\omega},$$

$$\Delta f > \frac{f_s}{N} \quad \rightsquigarrow \quad N > \frac{f_s}{\Delta f},$$

which provides an approximate lower limit for the number of samples required for sufficient resolution. In the present exercise, the criterion for resolution is satisfied, since we have $\Delta\omega = 0.05\pi$, corresponding to $\Delta f = 5 \times 10^{-3}$ Hz with T_s $= 5$ s, f_s $= 2\times10^{-1}$ Hz; hence the constraint is $N > 40$. Of course, at the same time, the SNR_o for each sinusoid to be detected must be sufficiently high. Since for our data $A = 5$ and sigma_wn = 1, we have $\text{SNR}_i = 6.25$, $\text{SNR}_o = 250$.

In conclusion, we expect, with N = 40 samples, to be barely able to distinguish the single sinusoids against the noise background and to get a better resolution with N = 64 samples, as it actually occurs.

Note that we sample the DTFT of the sequence over a set of discrete frequencies determined by N_f, and the frequency of each sinusoid contained in the data in general will not coincide with any of them (see Sect. 5.2.2). The corresponding spectral peak will be close to its theoretical height $N\, A^2/4 + sigma\_wn^2$ (see Fig. 10.8) only if the sinusoid's frequency is very close to one of the sampling frequencies, and this is the reason why we perform a substantial zero padding: we need a dense set of sampling frequencies to ensure that this condition is (approximately) satisfied.

16.5.2 Modified Periodogram

This example illustrates the fact that the modified periodogram (Sect. 10.5) offers advantages with respect to the periodogram when the goal is the detection of a weak sinusoid close to a dominant one. Indeed, with the modified periodogram we obtain a reduction of the relative height of the sidelobes of the spectral window.

Generate a sequence with two sinusoids in white noise, with amplitudes 0.1 and 1, and frequencies 0.225π and 0.3π, respectively. Set sigma_wn = 0.05 (low noise level), N = 128, and N_f = 1024. Compute and plot the power spectrum via periodogram and via modified periodogram. The script (ModPeriodogram) listed below does exactly this.

```
1   % ModPeriodogram
2   % Power spectrum estimation via Modified Periodogram
3
4   prompt='N_f =';
5   N_f = input(prompt);
6
7   range='twosided';
8
9   f_s=1/T_s;
10  f_Ny=f_s/2;
11
12  variance=var(sd) % data variance
13
14  figure(1004)
15  for m=1:2 % periodogram or modified periodogram
16      subplot(2,1,m)
17      if m==1
18          [PSD,f]=periodogram(sd,boxcar(N),N_f,f_s,range);
19          ht=text(0.055,10,'Periodogram');
20          set(gca,'Box','on')
21      else
22          [PSD,f]=periodogram(sd,hamming(N),N_f,f_s,range);
23          ht=text(0.055,10,'Modified periodogram');
```

```
24        set(gca,'Box','on')
25     end
26     PSD=PSD*f_s;
27     set(ht,'Interpreter','Latex','FontSize',12)
28     plot(f(2:N_f/2+1),10*log10(PSD(2:N_f/2+1)))
29     set(gca,'XLim',[0 f_Ny])
30     set(gca,'YLim',[-60 20])
31     set(gca,'YTick',[-60:20:20])
32     if m==2
33         hlabelx=get(gca,'Xlabel');
34         set(hlabelx,'String','$f$ (Hz)','Interpreter','Latex','FontSize',16)
35     end
36     hlabely=get(gca,'Ylabel');
37     set(hlabely,'String','PSD (dB)','Interpreter','Latex','FontSize',16)
38     ave_power=sum(PSD)/N_f % average power
39 end
```

In this exercise, the SNR is good for both sinusoidal components, and their frequency separation would be sufficient to resolve them if they had comparable amplitudes. However, with the periodogram, the weaker sinusoid is masked by the dominant one. With the modified periodogram with Hamming window, the weaker sinusoid is clearly visible.

16.5.3 Bartlett's Method and Welch's Method

Both methods (see Sects. 10.4 and 10.6) are implemented in Matlab by the function pwelch:

```
1 [PSD,f] = pwelch(x,window,noverlap,N_f,f_s,range)
```

in which the meaning of the variables and the PSD normalization scheme are the same as explained for the periodogram, except that

- window can be a vector of window samples but also an integer, in which case a Hamming window with length equal to that integer is assumed by default; if window is omitted or specified as empty, a default Hamming window is used to obtain 8 sections of x.
- noverlap is the amount of overlapping in number of samples for Welch's method. Default is 50 % of the window length, and this is the most advisable choice for a moderate taper, such as the triangular or the Hanning one. If the length of x is such that it cannot be divided exactly into an integer number of sections with 50 % overlap, x will be truncated accordingly.

We will now consider a few cases.

1. Start with Bartlett's method (Sect. 10.4). Using data_Nsin_noise_variable_N, generate a white noise sequence with sigma_wn = 1 and N = 1024. Compute and plot the Bartlett estimate of the power spectrum, for subsequence lengths of M =

N = 1024 (periodogram), M = 256, and M = 64. The number of subsequences is
K = 1, 4, and 16, respectively, since M = N/K. The script named Bartlett performs
these operations.

```
1    % Bartlett
2    % Power spectrum estimation via Bartlett method
3
4    prompt='N_f =';
5    N_f = input(prompt);
6
7    range='twosided';
8
9    f_s=1/T_s;
10   f_Ny=f_s/2;
11
12   KB=[1 4 16] % number of subsequences
13   M=N./KB % length of subsequences
14
15   variance=var(sd) % data variance
16
17    figure(1005)
18   for m=1:length(M)
19       [PSD,f]=pwelch(sd,boxcar(M(m)),0,N_f,f_s,range);
20       PSD=PSD*f_s;
21       subplot(3,1,m)
22       plot(f(2:N_f/2+1),10*log10(PSD(2:N_f/2+1)))
23       ht=text(0.005,-24,sprintf('$K$ = %d',KB(m)));
24       set(ht,'Interpreter','Latex','FontSize',12)
25       ht=text(0.022,-24,sprintf('$M$ = %d',M(m)));
26       set(ht,'Interpreter','Latex','FontSize',12)
27       set(gca,'XLim',[0 f_Ny])
28       set(gca,'YLim',[-35 15])
29       set(gca,'YTick',[-30:10:10])
30       if m==3
31          hlabelx=get(gca,'Xlabel');
32          set(hlabelx,'String','$f$ (Hz)','Interpreter','Latex','FontSize',16)
33       end
34       if m==2
35          hlabely=get(gca,'Ylabel');
36          set(hlabely,'String','PSD (dB)','Interpreter','Latex','FontSize',16)
37       end
38       ave_power=sum(PSD)/N_f % average power
39   end
```

A gradual reduction of the spectral variance can be observed as K increases.
2. To apply the more general Welch's method (Sect. 10.6), using data_Nsin_noise_
   variable_N generate N = 512 samples of a sequence with two sinusoids in
   white noise, with equal amplitudes A = 5, and frequencies 0.4π and 0.45π.
   Set sigma_wn = 1. Compute and plot the power spectrum via periodogram, via
   Bartlett's method using K = 4 and 8 (M = 128 and 64), and via Welch's method,
   using a Hanning window of length M = 128 and 50 % overlap. Compare the
   results. Use the following script (Periodogram_Bartlett_Welch).

```
 1    % Periodogram_Bartlett_Welch
 2    % Power spectrum estimation via Bartlett and Welch methods and via Periodogram
 3
 4    prompt='N\_f =';
 5    N_f = input(prompt);
 6
 7    range='twosided';
 8
 9    f_s=1/T_s;
10    f_Ny=f_s/2;
11    K=[4 8]
12    M=N./K
13
14    variance=var(sd) % data variance
15
16    figure(1006)
17    subplot(4,1,1) % Periodogram
18    [PSD,f]=periodogram(sd,boxcar(N),N_f,f_s,range);
19    PSD=PSD*f_s;
20    plot(f(2:N_f/2+1),10*log10(PSD(2:N_f/2+1)))
21    set(gca,'XLim',[0 f_Ny])
22    set(gca,'YLim',[-35 45])
23    set(gca,'YTick',[-20:20:40])
24    text('String','Periodogram, $M = N = 512$, $K = 1$','Position',[0.02 -22],'Interpreter','
            Latex','FontSize',12)
25     ave_power=sum(PSD)/N_f % average power - periodogram
26
27    for ib=1:length(K)
28        subplot(4,1,1+ib) % Bartlett
29        [PSD_B,f]=pwelch(sd,boxcar(M(ib)),0,N_f,f_s,range);
30        PSD_B=PSD_B*f_s;
31        plot(f(2:N_f/2+1),10*log10(PSD_B(2:N_f/2+1)))
32        set(gca,'XLim',[0 f_Ny])
33        set(gca,'YLim',[-35 45])
34        set(gca,'YTick',[-20:20:40])
35        ht1=text(0.02,-22,sprintf('Bartlett, $M$ = %d',M(m)));
36        ht2=text(0.12,-22,sprintf('$K$ = %d',K(m)));
37        set(ht1,'Interpreter','Latex','FontSize',12)
38        set(ht2,'Interpreter','Latex','FontSize',12)
39        if ib==1
40            hlabely=get(gca,'Ylabel');
41            set(hlabely,'String','PSD (dB)','Interpreter','Latex','FontSize',16)
42        end
43        ave_power_B=sum(PSD_B)/N_f % average power - Bartlett
44    end
45
46    subplot(4,1,4) % Welch
47    [PSD_W,f]=pwelch(sd,hanning(M(1)),M(1)/2,N_f,f_s,range);
48    PSD_W=PSD_W*f_s;
49    plot(f(2:N_f/2+1),10*log10(PSD_W(2:N_f/2+1)))
50    set(gca,'XLim',[0 f_Ny])
51    set(gca,'YLim',[-35 43])
52    set(gca,'YTick',[-20:20:40])
```

```
53   text('String','Welch, Hanning, $M = 128$, $K = 7$','Position',[0.02 −22],'Interpreter','
        Latex','FontSize',12)
54   hlabelx=get(gca,'Xlabel');
55   set(hlabelx,'String','$f$ (Hz)','Interpreter','Latex','FontSize',16)
56   ave_power_BW=sum(PSD_W)/N_f % average power − Welch
```

This exercise proves that K being nearly equal, the variance is approximately the same with the Bartlett and Welch methods (see the bottom panels of the figure that has been created, with K = 8 for Bartlett's method, and K = 7 for Welch's method; compare with Fig. 10.14). Moreover, though the Hanning window has a spectral mainlobe larger than that of the rectangular window used for the Bartlett estimate, the resolution that we get by Welch's method is similar to the one that we get by Bartlett's method, or even better, thanks to the 50 % overlap present in the Welch's method. The overlap allows for M being 128, while in the Bartlett estimation we had M = 64. Applying the Welch's method instead of the Bartlett's method we also obtain a leakage reduction, thanks to the reduction of the height of the spectral window's sidelobes.

16.5.4   Blackman-Tukey Method

Take the same sequence with two sinusoids in white noise used in the previous exercise (N = 512). Apply the Blackman-Tukey (BT) method (Sect. 10.7) to this sequence, using both a Bartlett and a Hamming window. For the value of M, chose an integer close to N/6, and also try N/8 and N/10; this means using M = 85, 64 and 50, respectively. Use the BlackmanTukey script given below.

```
1    % BlackmanTukey
2    % Power spectrum estimation via Blackman−Tukey method and Periodogram
3
4    prompt='N_f =';
5    N_f = input(prompt);
6    prompt='maxlag [] =';
7    maxlag = input(prompt);
8
9    flagw = [1 3];
10   %1=bartlett, 3=hamming in function pbt
11   range='twosided';
12
13   f_s=1/T_s;
14   f_Ny=f_s/2;
15
16   variance=var(sd) % data variance
17
18   figure(1007)
19   [PSD,f]=periodogram(sd,boxcar(N),N_f,f_s,range);
20   PSD=PSD*f_s;
21   ave_power=sum(PSD)/N_f % average power − periodogram
22
```

```
23  for im=1:length(maxlag)
24      for iw=1:length(flagw)
25          subplot(3,2,2*(im−1)+iw)
26          plot(f(2:N_f/2+1),10*log10(PSD(2:N_f/2+1)),'g')
27          hold on
28          PSD_BT=pbt(sd,maxlag(im),N_f,flagw(iw));
29          plot(f(2:N_f/2+1),10*log10(PSD_BT(2:N_f/2+1)))
30          hold off
31          if flagw(iw)==1
32              ht1=text(0.006,−34,'Bartlett window');
33          elseif flagw(iw)==3
34              ht1=text(0.006,−34,'Hamming window');
35          end
36          set(ht1,'Interpreter','Latex','FontSize',12)
37          ht2=text(0.06,26,sprintf('$M$ = %d',maxlag(im)));
38          set(ht2,'Interpreter','Latex','FontSize',12)
39          set(gca,'XLim',[0 0.1])
40          set(gca,'XTick',[0:0.02:0.1])
41          set(gca,'YLim',[−45 45])
42          set(gca,'YTick',[−40:20:40])
43          if 2*(im−1)+iw >= 5
44              hlabelx=get(gca,'Xlabel');
45              set(hlabelx,'String','$f$ (Hz)','Interpreter','Latex','FontSize',16)
46          end
47          if 2*(im−1)+iw == 3
48              hlabely=get(gca,'Ylabel');
49              set(hlabely,'String','PSD (dB)','Interpreter','Latex','FontSize',16)
50          end
51      end
52      ave_power_BT=sum(PSD_BT)/N_f  % average power − BT
53  end
```

This script calls the function pbt, reported next.

```
1   function PSD = pbt(x,maxlag,N_f,flagw)
2   % Blackman−Tukey spectral estimation for zero−mean sequence
3
4   rxx = xcorr(x,maxlag,'biased');
5   if flagw == 1
6       w = bartlett(2*maxlag+1);
7   elseif flagw == 2
8       w = hanning(2*maxlag+1);
9   elseif flagw == 3
10      w = hamming(2*maxlag+1);
11  elseif flagw == 4
12      w = blackman(2*maxlag+1);
13  end
14  PSD = abs(fft(w.*rxx,N_f));
```

This example illustrates that for a given window type, the variance of the spectral estimate obtained by the BT method decreases with increasing M. For a given value of M, the Bartlett window allows for a greater variance reduction with respect to the Hamming window. The BT spectral estimate is quite smooth in any case; its resolution is nearly the same for both windows, and increases with M, but is always relatively poor.

16.5.5 MultiTaper Method

In Matlab, the MTM (Sect. 10.9) is implemented by the function pmtm. The command's syntax for a centered input x is

```
[PSD,PSDc,f] = pmtm(x,nw,N_f,f_s,method,p,range)
```

The meaning of most variables is the usual one, but for MTM we have additional parameters:

- nw is the value of the time-bandwidth parameter,
- method specifies how the individual spectral estimates (modified periodograms) are combined: the combination may be linear or not. More precisely, we can choose among setting

  - method = 'adapt' (default), i.e., the nonlinear Thomsom adaptive combination,
  - method = 'unity', i.e., linear combination with unit weights,
  - method = 'eigen', i.e., linear combination with weights equal to the eigenvalues of the variational problem of leakage minimization.

  We will hold to the default value;
- p indicates a probability level. If p is specified, pmtm returns the p×100%-confidence-level interval for the PSD estimate at each frequency. The corresponding two vectors are contained in the matrix PSDc, having N_f rows and two columns; the first column contains the lower confidence bounds, the second column contains the upper bound.

As an example, take the same sequence with two sinusoids in white noise used in the previous two exercises (N = 512), compute and plot the power spectrum via MTM, using nw = 2, 3 and 4. Observe how the variance-versus-resolution tradeoff varies with nw. Use the following MTM script.

```
1    % MTM
2    % Power spectrum estimation via MultiTaper method
3
4    prompt='N_f =';
5    N_f = input(prompt);
6
7    range='twosided';
8
9    f_s=1/T_s;
10   f_Ny=f_s/2;
11
12   nw=[2 3 4]
13   p=0.95;
14   method='adapt'
15
16   variance=var(sd) % data variance
17
18   figure(1008)
19   for m=1:length(nw)
20       subplot(3,1,m)
21       [PSD,PSDc,f]=pmtm(sd,nw(m),N_f,f_s,method,p,range);
22       PSD=PSD*f_s;
23       PSDc=PSDc*f_s;
24       for k=1:2
25           plot(f(2:N_f/2+1),10*log10(PSDc(2:N_f/2+1,k)),'c')
26           hold on
27       end
28       plot(f(2:N_f/2+1),10*log10(PSD(2:N_f/2+1)))
29       hold off
30       set(gca,'XLim',[0 f_Ny])
31       set(gca,'YLim',[-25 35])
32       set(gca,'YTick',[-20:10:30])
33       ht=text(0.06,20,sprintf('$n_w$ = %d',nw(m)));
34       set(ht,'Interpreter','Latex','FontSize',12)
35       ht1=text(0.078,20,sprintf('$p$ = %g',p));
36       set(ht1,'Interpreter','Latex','FontSize',12)
37       if m==3
38           hlabelx=get(gca,'Xlabel');
39           set(hlabelx,'String','$f$ (Hz)','Interpreter','Latex','FontSize',16)
40       end
41       if m==2
42           hlabely=get(gca,'Ylabel');
43           set(hlabely,'String','PSD (dB)','Interpreter','Latex','FontSize',16)
44       end
45       ave_power=sum(PSD)/N_f % average power
46   end
```

16.5.6   Parametric Methods

We will examine two parametric methods (Chap. 11) implemented in Matlab: Burg's
method (Sect. 11.6.3.4), implemented by the function pburg, and the autoregressive
Yule-Walker method (Sect. 11.6.3.1), implemented by the function pyulear. These
functions apply the PSD normalization scheme described for the periodogram and
have similar syntaxes: for example,

```
[PSD,f] = pburg(x,p,N_f,f_s,range)
```

where, in input, p is the AR-model order and x is a centered data sequence.

1. We will now compute with pburg the power spectrum of those 40 signal samples
   for which the periodogram did not allow for resolving the two sinusoids in a
   satisfactory way, in spite of the relatively low noise level. Thus, retrieve those data,
   or generate again $N = 40$ samples of a sequence with two sinusoids in white noise,
   with equal amplitudes $A = 5$ and frequencies 0.4π and 0.45π, and with sigma_w
   $= 1$. To estimate the power spectrum, use both Burg and Yule AR methods, with
   model orders p $= 8, 10, 12$ and 16; compute also the periodogram for comparison.
   Plot the results. Use the Burg_YuleAR_Periodogram script reported below.

```
1    % Burg_YuleAR_Periodogram
2    % Power spectrum estimation via Burg or Yule-AR methods and Periodogram
3
4    prompt='N_f =';
5    N_f = input(prompt);
6    prompt='Method? (1 = Burg, 2 = Yule AR)';
7    flagmeth = input(prompt);
8
9    range='twosided';
10
11   f_s=1/T_s;
12   f_Ny=f_s/2;
13
14   variance=var(sd) % data variance
15
16   figure(1009+flagmeth-1)
17   [PSD,f]=periodogram(sd,boxcar(N),N_f,f_s,range);
18   PSD=PSD*f_s;
19   ave_power=sum(PSD)/N_f % average power - periodogram
20
21   p=[8 10 12 16];
22   for m=1:length(p)
23       subplot(4,1,m)
24       if flagmeth==1
25           [PSD_AR,f] = pburg(sd,p(m),N_f,f_s,range);
26       end
27       if flagmeth==2
28           [PSD_AR,f] = pyulear(sd,p(m),N_f,f_s,range);
29       end
30       plot(f(2:N_f/2+1),10*log10(PSD(2:N_f/2+1)),'g')
```

```
31      hold on
32      PSD_AR=PSD_AR*f_s;
33      plot(f(2:N_f/2+1),10*log10(PSD_AR(2:N_f/2+1)))
34      hold off
35      set(gca,'XLim',[0 f_Ny])
36      set(gca,'YLim',[-20 60])
37      set(gca,'YTick',[-20:20:60])
38      ht=text(0.005,42,sprintf('AR model order = %d',p(m)));
39      set(ht,'Interpreter','Latex','FontSize',12)
40      if m==length(p)
41          hlabelx=get(gca,'Xlabel');
42          set(hlabelx,'String','$f$ (Hz)','Interpreter','Latex','FontSize',16)
43      end
44      if m==3
45          hlabely=get(gca,'Ylabel');
46          set(hlabely,'String','PSD (dB)','Interpreter','Latex','FontSize',16)
47      end
48      ave_power_AR=sum(PSD_AR)/N_f % average power - AR
49  end
```

This exercise shows that even with 40 data samples, if we choose a sufficiently high model order we can resolve the two sinusoidal peaks. Note that the maximum considered order p = 16 is about (2/5) N, so we are within reasonable order limits. An order value of 18 would also be acceptable.

2. Now, try Yule's method again, but this time generate N = 2048 samples of a sequence with two sinusoids in white noise, with equal amplitudes A = 5 and frequencies 0.4π and 0.45π, setting sigma_wn = 1 and N_f = 2048, to verify that Yule's method also works fairly well for longer series.

16.6 Stationary Analysis of Nino3 Historical Series

Here we use, as a case study, the historical time series of the area-averaged sea surface temperature (SST) from 5°S–5°N and 150°W–90°W, available for download at http://www.esrl.noaa.gov/psd/gcos_wgsp/Timeseries/Data/nino3.long.data. The data is given on a monthly basis in a typical matrix format, in which the first column represents year A.D., and the other 12 columns represent SST values in degrees C°. The series covers January 1870–September 2014 (N = 1737 samples). We will assume that this matrix, given below, resides in a file named sst_nino3_monthly.dat.

	Year												
1	1870	24.23	25.35	25.74	26.99	27.74	27.65	27.40	26.19	25.41	25.30	23.94	24.19
2	1871	25.33	25.83	26.65	26.91	26.22	25.93	25.07	24.66	24.59	24.55	24.66	24.52
3	1872	24.82	25.61	26.46	26.62	26.27	25.84	25.19	24.17	23.97	23.81	24.01	24.14
4	1873	24.63	25.17	25.60	26.62	26.25	26.06	25.12	24.65	24.43	23.97	23.94	24.05
5	1874	24.25	24.94	25.63	26.41	26.23	25.67	24.48	23.76	23.76	23.70	23.74	24.03
6	1875	24.81	25.85	26.45	26.68	26.02	25.80	24.73	24.24	24.08	24.01	24.06	24.13
7	1876	24.33	24.89	25.70	25.99	25.62	25.81	25.16	24.77	24.88	25.16	25.22	25.44
8	1877	25.80	26.57	27.32	27.61	27.43	26.96	26.96	26.58	26.97	27.07	27.27	27.71
9	1878	28.15	28.84	28.28	28.32	27.72	27.18	25.86	24.90	24.49	24.33	24.05	24.34
10	1879	25.02	26.09	26.72	26.91	26.06	25.89	24.97	24.43	24.19	24.07	23.95	24.16
11	1880	24.57	25.73	26.43	26.88	26.36	25.80	25.14	24.91	25.15	25.27	25.34	25.20
12	1881	25.69	26.30	27.14	27.55	26.97	26.51	25.35	24.68	24.49	24.50	24.37	24.66
13	1882	24.97	25.61	26.40	27.33	26.82	25.75	24.66	24.29	24.43	24.30	24.09	24.32
14	1883	24.79	25.66	26.57	27.17	26.63	26.37	25.66	24.89	24.66	24.45	24.74	24.80
15	1884	25.25	26.12	27.17	27.79	27.35	26.41	25.84	25.28	25.19	25.29	25.46	25.70
16	1885	25.82	26.47	27.03	27.32	27.03	26.25	25.30	25.03	25.25	25.30	25.49	25.84
17	1886	25.13	25.68	26.39	26.76	25.88	25.17	24.56	23.84	23.81	23.72	23.72	23.70
18	1887	24.65	25.08	25.85	26.37	26.11	26.07	25.24	24.27	24.62	24.67	24.87	25.15
19	1888	25.68	26.62	27.20	27.59	27.64	27.05	26.19	25.74	25.97	26.43	27.17	26.87
20	1889	27.60	27.69	27.84	27.84	27.16	26.46	24.86	24.03	24.01	23.69	23.97	24.10
21	1890	23.58	24.66	26.09	26.34	26.00	25.53	24.71	23.93	23.93	23.85	23.97	24.54
22	1891	25.16	26.04	27.05	27.65	27.33	26.91	25.89	25.10	24.72	24.72	24.91	25.05
23	1892	25.12	25.90	26.45	26.74	26.18	25.77	24.93	24.33	23.85	23.29	23.38	23.74
24	1893	24.01	24.90	25.80	26.24	25.89	25.51	24.59	24.03	24.00	23.90	24.06	24.23
25	1894	24.72	25.41	26.19	26.68	26.16	25.81	24.99	24.34	24.11	24.21	24.14	24.52
26	1895	24.86	25.74	26.53	27.21	26.96	26.31	25.40	25.29	25.37	25.46	25.50	25.69
27	1896	25.96	26.52	27.17	27.48	27.12	26.54	25.95	26.03	25.86	25.86	26.55	26.75
28	1897	27.03	27.41	27.61	27.26	26.87	26.40	25.55	24.81	24.54	24.45	24.45	24.69
29	1898	25.10	25.76	26.29	26.78	26.39	25.94	25.06	24.38	24.60	24.64	24.36	24.37
30	1899	25.02	25.76	26.73	27.21	27.11	26.65	25.47	25.89	25.69	25.92	26.22	26.43
31	1900	26.98	27.70	28.15	28.07	27.57	27.29	26.30	25.58	25.32	25.07	24.87	25.45
32	1901	26.04	26.37	26.86	27.09	26.51	25.99	25.09	24.46	24.46	24.59	24.83	25.07
33	1902	25.48	26.27	27.08	27.70	27.29	27.33	27.10	26.28	26.23	26.48	26.60	26.41
34	1903	26.67	26.77	27.47	27.37	26.76	26.50	25.11	24.39	24.14	24.25	24.17	23.97
35	1904	25.05	25.66	26.16	26.71	26.82	26.69	26.45	25.81	25.07	25.90	25.52	25.90
36	1905	26.24	26.82	28.03	27.62	28.29	27.33	26.39	26.23	26.27	25.92	26.08	26.07
37	1906	26.50	26.90	27.34	27.74	26.89	26.45	24.72	24.14	24.00	24.13	24.47	24.49
38	1907	24.89	25.79	26.55	26.86	26.37	26.13	24.80	23.95	24.99	24.48	24.55	24.93
39	1908	25.28	26.45	26.66	26.73	26.26	26.37	24.83	24.53	24.22	24.46	24.16	23.94
40	1909	25.40	25.36	26.64	26.62	26.37	25.18	24.55	23.84	23.83	23.62	23.42	24.07
41	1910	24.60	25.77	26.32	26.46	25.72	25.62	24.88	24.28	23.89	24.08	24.58	24.52
42	1911	24.98	25.60	26.35	26.41	25.75	25.91	25.59	25.23	25.65	25.40	25.91	26.60
43	1912	27.19	27.20	27.66	28.00	27.15	26.29	25.26	24.58	25.10	24.77	25.06	24.74
44	1913	25.55	26.50	26.99	26.31	26.57	26.30	25.53	25.07	25.01	25.27	25.86	26.10
45	1914	26.39	27.02	27.40	27.90	26.82	26.42	26.16	25.98	25.85	25.43	25.45	25.89
46	1915	26.38	26.80	27.50	28.03	27.71	27.87	25.70	25.01	24.96	24.89	24.56	24.48
47	1916	25.16	25.39	26.27	26.53	26.27	25.64	24.07	23.28	23.41	23.37	23.12	23.42
48	1917	23.96	24.75	26.13	27.12	26.92	26.60	25.85	25.31	24.99	24.86	24.52	24.11
49	1918	24.54	25.40	26.27	27.09	27.20	27.25	26.15	25.60	25.62	26.04	26.38	26.55
50	1919	27.06	27.36	27.70	27.98	27.72	27.13	26.07	25.59	25.43	25.03	24.52	25.33
51	1920	26.30	26.97	27.61	27.54	27.09	26.66	25.24	24.98	24.97	24.86	24.82	25.26
52	1921	25.86	26.07	25.97	26.83	26.17	26.29	25.34	24.40	24.52	24.91	24.08	24.36
53	1922	24.87	26.27	26.98	26.98	26.94	25.82	24.80	23.81	24.14	24.53	24.18	24.41
54	1923	25.11	25.61	26.05	26.92	26.80	26.41	25.68	25.07	25.45	25.35	25.48	25.87
55	1924	25.55	26.51	27.07	27.13	25.92	25.55	24.55	24.20	23.95	24.52	24.02	24.33
56	1925	24.66	26.26	26.98	27.25	26.85	26.79	26.40	25.86	25.73	25.76	25.82	26.65
57	1926	27.07	27.43	28.27	28.55	27.88	26.99	26.29	25.09	24.67	24.69	24.69	25.03
58	1927	25.63	26.68	26.93	27.15	26.46	26.19	25.13	24.80	24.88	25.06	24.89	25.10
59	1928	25.73	26.38	26.91	27.14	26.78	25.92	25.34	24.61	24.69	24.65	24.65	24.94
60	1929	25.25	26.26	26.99	27.57	27.05	26.59	25.45	25.05	25.29	25.19	25.09	25.54
61	1930	25.56	26.30	27.13	27.78	26.94	26.68	26.36	25.93	26.21	26.31	26.78	26.70
62	1931	27.16	27.53	28.27	28.35	27.42	26.59	25.88	24.83	24.40	24.85	24.51	24.75
63	1932	25.34	26.33	27.44	27.73	27.51	26.88	25.66	25.17	24.85	24.85	24.74	24.71
64	1933	25.30	26.30	26.93	27.44	26.55	25.71	24.89	24.15	24.20	23.96	23.95	24.00
65	1934	24.93	25.66	26.44	27.38	26.63	26.27	25.13	24.84	24.76	24.70	24.93	24.71
66	1935	25.06	25.70	26.79	27.36	26.68	25.91	24.77	25.00	25.08	25.00	25.27	25.18
67	1936	26.14	26.72	26.98	27.62	26.69	25.89	25.28	24.65	24.90	25.36	25.01	25.59
68	1937	25.42	26.56	26.88	27.60	26.37	26.05	25.53	24.59	25.18	24.67	24.94	25.20
69	1938	25.21	26.36	26.84	27.14	26.70	25.25	24.03	23.96	24.31	24.22	24.11	24.08
70	1939	24.89	25.44	26.16	27.16	26.90	26.71	25.72	25.18	24.97	24.70	25.35	25.16
71	1940	26.76	27.63	28.20	28.28	27.28	26.87	25.55	25.69	25.01	25.12	25.28	26.26
72	1941	26.83	27.63	28.49	28.76	28.26	27.15	26.06	25.69	25.55	25.84	26.12	26.45
73	1942	26.09	26.55	27.13	27.53	26.77	25.65	24.57	23.73	23.61	23.42	23.31	23.96
74	1943	24.69	25.32	26.03	27.01	26.88	26.48	25.59	25.03	24.53	24.71	24.52	24.67
75	1944	25.05	26.47	26.88	27.63	27.04	26.54	25.69	24.99	24.73	24.58	24.34	24.57
76	1945	25.10	26.03	26.26	26.88	27.00	26.18	25.20	24.45	24.42	24.32	24.73	24.61
77	1946	24.88	25.70	26.50	26.94	26.45	26.01	25.27	24.04	24.61	24.58	24.60	24.95
78	1947	25.81	26.19	27.26	27.34	26.63	26.35	25.20	24.48	24.02	24.22	24.46	25.28
79	1948	25.46	26.67	27.60	27.52	27.05	26.17	25.23	24.97	24.55	24.16	24.65	25.35
80	1949	24.84	26.35	26.59	27.72	26.95	25.61	24.89	24.42	23.97	23.81	23.52	23.81
81	1950	24.25	24.75	26.15	26.49	25.74	25.44	24.55	24.49	23.78	24.05	23.53	24.17
82	1951	24.86	25.91	26.48	27.48	26.95	26.72	26.72	26.26	25.76	25.95	26.00	25.95
83	1952	26.01	26.54	27.22	27.62	26.54	25.72	24.82	24.45	24.27	24.66	24.43	24.45
84	1953	25.65	26.59	27.26	28.27	27.43	26.95	25.81	25.04	25.59	24.94	25.10	25.23
85	1954	25.43	26.22	26.79	26.31	25.81	25.12	24.23	23.67	23.55	23.65	23.59	23.96
86	1955	24.83	25.67	26.22	26.49	25.55	25.04	24.33	24.21	23.46	23.23	22.89	23.50
87	1956	24.29	25.42	26.43	26.73	26.40	25.83	24.82	24.33	24.18	24.32	23.93	24.34
88	1957	24.71	26.03	27.34	27.82	27.62	27.35	26.77	26.14	25.58	25.76	25.99	26.30
89	1958	26.77	27.26	27.67	27.54	26.91	26.49	25.58	25.08	24.69	24.66	24.83	25.06
90	1959	25.56	26.32	27.07	27.60	26.92	26.02	25.21	24.50	24.47	25.03	24.73	25.00

91	1960	25.51	25.98	26.96	27.17	26.89	26.12	25.29	25.03	25.02	24.58	24.23	24.83
92	1961	25.24	26.45	26.98	27.78	27.12	26.54	24.91	24.41	23.96	24.00	24.65	24.82
93	1962	25.41	26.22	26.52	26.71	26.34	25.90	25.20	24.90	24.35	24.43	24.29	24.36
94	1963	25.21	25.98	27.11	27.46	27.13	26.84	26.42	25.90	25.53	25.71	25.64	26.11
95	1964	26.02	26.37	26.75	26.45	25.68	25.22	24.94	23.97	24.13	24.01	24.05	24.01
96	1965	24.88	25.99	26.94	27.58	27.60	27.16	26.53	26.32	26.15	26.30	26.41	26.45
97	1966	26.70	26.86	27.16	27.45	26.46	26.13	25.41	24.50	24.46	24.62	24.27	24.46
98	1967	25.09	25.93	26.64	26.78	26.65	26.33	25.17	24.29	23.95	23.96	24.06	24.29
99	1968	24.42	25.03	25.92	26.78	26.21	26.36	25.85	25.38	25.19	25.19	25.34	25.70
100	1969	25.95	26.67	27.53	27.84	27.72	27.12	25.88	25.60	25.58	25.75	25.83	25.01
101	1970	26.24	26.47	27.01	27.24	26.44	25.59	23.91	23.51	23.83	24.05	23.82	25.89
102	1971	24.27	24.93	25.94	26.70	26.19	25.64	24.97	24.19	24.14	23.85	23.88	24.00
103	1972	25.01	26.03	26.91	27.80	27.55	27.38	26.91	27.02	26.54	26.85	27.12	27.59
104	1973	27.20	27.16	27.42	27.15	26.17	25.49	24.33	23.78	23.72	23.58	23.39	23.54
105	1974	24.00	25.16	26.17	26.91	26.69	26.23	25.36	25.04	24.67	24.30	24.22	24.40
106	1975	25.15	25.85	26.67	27.21	26.14	25.35	24.71	24.37	23.84	23.71	23.75	23.61
107	1976	23.93	25.33	26.54	27.24	27.00	27.06	26.44	25.97	25.84	26.02	26.02	25.91
108	1977	26.54	26.95	27.51	26.97	26.84	26.61	25.74	24.79	24.90	25.48	25.55	25.67
109	1978	26.02	26.57	26.96	26.75	26.26	26.75	24.86	24.28	24.43	24.78	24.95	25.40
110	1979	25.50	26.15	27.02	27.71	27.15	26.75	25.55	25.47	25.70	25.50	25.49	25.72
111	1980	26.01	26.37	26.98	27.51	27.14	26.80	25.49	24.82	24.86	24.66	24.98	25.35
112	1981	24.84	25.69	26.87	27.14	26.85	26.42	25.24	24.68	24.93	25.04	24.88	25.43
113	1982	25.78	26.36	26.99	27.68	27.70	27.34	26.21	26.12	26.70	27.27	27.66	28.22
114	1983	28.48	28.79	29.12	29.21	29.02	28.19	26.61	25.83	25.08	24.44	24.18	24.52
115	1984	24.97	25.91	26.86	27.04	26.30	25.35	25.00	24.50	24.51	24.07	23.99	23.79
116	1985	24.58	25.33	26.16	26.50	25.96	25.60	24.69	24.21	24.16	24.09	24.30	24.47
117	1986	24.76	25.81	26.64	27.03	26.43	26.18	25.62	24.98	25.22	25.70	25.86	25.89
118	1987	26.62	27.32	28.16	28.43	28.13	27.57	26.92	26.43	26.53	26.16	26.13	26.21
119	1988	26.10	26.43	27.21	26.73	25.76	24.60	23.81	23.44	23.50	23.06	23.13	23.39
120	1989	24.19	25.56	26.06	26.60	26.28	26.03	25.31	24.49	24.58	24.56	24.64	24.78
121	1990	25.31	26.49	26.99	27.62	27.30	26.46	25.50	25.00	25.00	24.86	24.85	25.11
122	1991	25.76	26.38	27.11	27.38	27.44	27.13	26.33	25.29	25.00	25.57	26.01	26.46
123	1992	26.97	27.56	28.30	28.80	28.44	26.90	25.57	24.66	24.54	24.58	24.69	24.89
124	1993	25.52	26.59	27.47	28.40	28.12	27.08	25.72	25.01	24.98	25.13	25.08	25.14
125	1994	25.54	26.06	26.75	27.05	26.85	26.43	25.21	24.81	24.91	25.55	25.83	26.09
126	1995	26.42	26.92	27.17	27.25	26.54	26.22	25.49	24.42	24.20	24.20	24.23	24.41
127	1996	25.00	25.82	26.82	26.93	26.53	26.11	25.33	24.72	24.56	24.53	24.49	24.35
128	1997	24.88	25.96	27.17	27.76	28.13	28.13	27.85	27.75	27.74	28.04	28.19	28.39
129	1998	28.61	28.85	29.14	28.98	28.31	26.54	25.41	24.73	24.44	24.19	24.27	24.15
130	1999	24.61	25.58	26.74	26.85	26.48	25.76	24.94	24.26	23.96	23.77	23.55	23.68
131	2000	23.87	25.20	26.28	27.18	26.57	25.70	25.00	24.63	24.59	24.45	24.26	24.42
132	2001	24.87	25.94	27.01	27.44	26.76	26.21	25.42	24.72	24.22	24.39	24.39	24.51
133	2002	25.10	26.15	27.31	27.54	27.22	26.87	25.86	25.34	25.40	25.86	26.21	26.39
134	2003	26.30	26.63	27.21	27.05	26.17	25.95	25.60	25.07	24.92	25.31	25.33	25.67
135	2004	26.01	26.49	27.05	27.43	26.75	26.26	25.56	25.25	25.23	25.40	25.46	25.69
136	2005	25.86	26.24	26.95	27.62	27.31	26.63	25.71	25.09	24.65	24.50	24.05	23.88
137	2006	24.68	25.95	26.53	27.22	26.93	26.43	25.58	25.36	25.61	25.78	25.90	26.21
138	2007	26.22	26.35	26.75	27.03	26.30	25.94	24.95	24.04	23.60	23.44	23.26	23.48
139	2008	23.98	25.10	26.37	27.01	26.87	26.40	25.91	25.54	24.93	24.91	24.83	24.51
140	2009	24.96	25.86	26.35	27.40	27.40	27.16	26.47	25.72	25.66	25.71	26.16	26.62
141	2010	26.68	27.14	27.72	28.01	27.06	25.92	24.75	23.93	23.56	23.22	23.41	23.60
142	2011	24.17	25.51	26.33	27.12	26.84	26.54	25.64	24.60	24.22	23.95	23.93	24.35
143	2012	24.85	26.21	26.96	27.47	27.07	26.95	26.42	25.75	25.29	24.97	25.14	24.87
144	2013	25.00	25.90	27.14	27.37	26.39	25.74	25.02	24.52	24.68	24.83	24.87	25.13
145	2014	25.35	25.74	26.99	27.74	27.65	27.40	26.19	25.41	25.30	NaN	NaN	NaN

16.6.1 Preliminary Data Processing

We start pre-processing the Nino3-SST matrix:

- we transform the monthly SST data matrix into a time series, eliminating the missing data (NaN) in the last year; we plot this monthly time series. A plot of the series should *always* be made before analysis, to get a general idea of the series behavior, of the interval of values involved, as well as to identify visually any constant drifts or trends, any sharp changes in the mean value and/or variance, etc. A plot also allows for singling out possible outliers and/or missing data, which usually are represented by out-of-range values, e.g., -999 for a variable that typically assumes values between -1 and 1. The record considered here is the result of previous data processing and therefore does not contain outliers or missing data, except for the three last months of 2014, which are represented by NaN because we are working in the Matlab environment, which offers this possibility of dealing with what must be considered as "Not-a-Number". In general, a preliminary plot is mandatory;

- from the monthly SST data matrix we compute the annual SST cycle (mean SST vs. calendar month) and plot it;
- we transform the monthly data matrix into a matrix of monthly anomalies, by subtracting from each data sample the average over 1870–2014 of the corresponding calendar month;
- we transform the monthly anomaly matrix into a time series of monthly anomalies and plot it;
- since later on we will also need this data on a seasonal basis, we transform the monthly data matrix into a seasonal data matrix, considering complete years only. In climatological studies, seasons are usually defined as:

  - Spring: March–April–May (MAM),
  - Summer: June–July–August (JJA),
  - Autumn: September–October–November (SON),
  - Winter: December–January–February (DJF). Forming winter data thus involves averaging samples belonging to two successive years;

- we transform the seasonal data matrix into a time series, eliminating the missing data (NaN) in the last year; we plot the seasonal time series;
- we compute the annual SST cycle from seasonal data (mean seasonal SST vs. season);
- we transform the seasonal data matrix into a matrix of seasonal anomalies, by subtracting from each data sample the average over 1870–2014 of the corresponding season;
- we transform the matrix of seasonal anomalies into a time series, eliminating the missing data (NaN) in the last year; we plot the seasonal anomaly time series;
- we save all generated time series and the corresponding time vectors for future use. Be careful: this time the time series are not centered before output. The anomalies have a very small mean value, but the SSTs have a mean value of the order of 26 °C.

These tasks are performed by the following script, which is called preliminary.

```
 1   % preliminary
 2   % Preliminary data processing of sst_nino3_monthly data
 3
 4   load sst_nino3_monthly.dat
 5   [Nyears,M1]=size(sst_nino3_monthly)
 6   data=sst_nino3_monthly(:,2:13);
 7   years=sst_nino3_monthly(:,1);
 8   T_s_m=1/12;
 9   T_s_s=1/4;
10
11   % Transform monthly data matrix into time series
12   % Eliminate missing data in the last year
```

```
13   k=0;
14   for iy=1:Nyears
15      for im=1:12
16         if isfinite(data(iy,im))
17            k=k+1;
18            sst_nino3_m(k)=data(iy,im);
19            time_m(k)=years(1)+(1/24)+(k−1)*T_s_m;
20         end
21      end
22   end
23   Nmonths=length(sst_nino3_m)
24   min_sst_nino3_m=min(sst_nino3_m)
25   max_sst_nino3_m=max(sst_nino3_m)
26
27   % Plot monthly time series
28   figure(2000)
29   plot(time_m,sst_nino3_m)
30   set(gca,'XLim',[time_m(1) time_m(Nmonths)])
31   set(gca,'YLim',[0.99*min_sst_nino3_m 1.01*max_sst_nino3_m])
32   hlabelx=get(gca,'Xlabel');
33   set(hlabelx,'String','Time AD (years)','Interpreter','Latex','FontSize',16)
34   hlabely=get(gca,'Ylabel');
35   set(hlabely,'String','$T (^{\circ}$C)','Interpreter','Latex','FontSize',16)
36   title('Nino3 SST monthly series')
37
38   % Annual cycle from monthly data
39   dim=1;
40   anncycle_m=sum(data(1:Nyears−1,:),dim)/(Nyears−1);
41   min_anncycle_m=min(anncycle_m)
42   max_anncycle_m=max(anncycle_m)
43
44   % Plot annual cycle
45   figure(2001)
46   tmon=1:12;
47   bar(tmon,anncycle_m)
48   set(gca,'XLim',[0 13])
49   set(gca,'XTick',[1:12])
50   set(gca,'YLim',[0.99*min_anncycle_m 1.01*max_anncycle_m])
51   hlabelx=get(gca,'Xlabel');
52   set(hlabelx,'String','Month','Interpreter','Latex','FontSize',16)
53   hlabely=get(gca,'Ylabel');
54   set(hlabely,'String','$T (^{\circ}$C)','Interpreter','Latex','FontSize',16)
55   title('Nino3 SST monthly series − Annual cycle')
56
57   % Transform monthly data matrix into matrix of monthly anomalies
```

```matlab
58   for iy=1:Nyears
59      for im=1:12
60         if isfinite(data(iy,im))
61            a_sst_nino3_monthly(iy,im)=data(iy,im)-anncycle_m(im);
62         else
63            a_sst_nino3_monthly(iy,im)=NaN;
64         end
65      end
66   end
67
68   % Transform matrix of monthly anomalies into time series
69   % Eliminate missing data in the last year
70   k=0;
71   for iy=1:Nyears
72      for im=1:12
73         if isfinite(a_sst_nino3_monthly(iy,im))
74            k=k+1;
75            a_sst_nino3_m(k)=a_sst_nino3_monthly(iy,im);
76         end
77      end
78   end
79   min_a_sst_nino3_m=min(a_sst_nino3_m)
80   max_a_sst_nino3_m=max(a_sst_nino3_m)
81
82   % Plot monthly anomaly time series
83   figure(2002)
84   plot(time_m,a_sst_nino3_m)
85   set(gca,'XLim',[time_m(1) time_m(Nmonths)])
86   set(gca,'YLim',[0.99*min_a_sst_nino3_m 1.01*max_a_sst_nino3_m])
87   hlabelx=get(gca,'Xlabel');
88   set(hlabelx,'String','Time AD (years)','Interpreter','Latex','FontSize',16)
89   hlabely=get(gca,'Ylabel');
90   set(hlabely,'String','$T (^{\circ}$C)','Interpreter','Latex','FontSize',16)
91   title('Nino3 monthly SST anomaly series')
92
93   % Transform monthly data matrix into seasonal data matrix (complete
        years only)
94   i_m_s(1,1:3)=3:5;
95   i_m_s(2,1:3)=6:8;
96   i_m_s(3,1:3)=9:11;
97
98   for iy=1:Nyears-1
99      for is=1:3
100        if sum(isfinite(data(iy,i_m_s(is,:))))==3
101           sst_nino3_seas(iy,is)=mean(data(iy,i_m_s(is,:)));
```

```
102          else
103              sst_nino3_seas(iy,is)=NaN;
104          end
105      end
106          % Winter involves two successive years
107          if sum(isfinite(data(iy,12))+isfinite(data(iy+1,1))+isfinite(data(iy
                 +1,2)))==3
108              sst_nino3_seas(iy,4)=sum(data(iy,12)+data(iy+1,1)+data(iy
                     +1,2))/3;
109          else
110              sst_nino3_seas(iy,4)=NaN;
111          end
112  end
113
114  % Transform seasonal data matrix into time series
115  % Eliminate missing data
116  k=0;
117  for iy=1:Nyears−1
118      for is=1:4
119          if isfinite(sst_nino3_seas(iy,is))
120              k=k+1;
121              sst_nino3_s(k)=sst_nino3_seas(iy,is);
122              time_s(k)=years(1)+(1/4+1/24)+(k−1)*T_s_s;
123          end
124      end
125  end
126  Nseas=length(sst_nino3_s)
127  min_sst_nino3_s=min(sst_nino3_s)
128  max_sst_nino3_s=max(sst_nino3_s)
129
130  % Plot seasonal time series
131  figure(2003)
132  plot(time_s,sst_nino3_s)
133  set(gca,'XLim',[time_s(1) time_s(Nseas)])
134  set(gca,'YLim',[0.99*min_sst_nino3_s 1.01*max_sst_nino3_s])
135  hlabelx=get(gca,'Xlabel');
136  set(hlabelx,'String','Time AD (years)','Interpreter','Latex','FontSize',16)
137  hlabely=get(gca,'Ylabel');
138  set(hlabely,'String','$T (^{\circ}$C)','Interpreter','Latex','FontSize',16)
139  title('Nino3 seasonal SST series')
140
141  % Annual cycle from seasonal data
142  dim=1;
143  anncycle_s=sum(sst_nino3_seas(1:Nyears−1,:),dim)/(Nyears−1);
144  min_anncycle_s=min(anncycle_s)
```

```matlab
145   max_anncycle_s=max(anncycle_s)
146
147   % Transform seasonal data matrix into matrix of seasonal anomalies
148   for iy=1:Nyears−1
149       for is=1:4
150           if isfinite(sst_nino3_seas(iy,is))
151               a_sst_nino3_seas(iy,is)=sst_nino3_seas(iy,is)−anncycle_s(is);
152           else
153               a_sst_nino3_seas(iy,is)=NaN;
154           end
155       end
156   end
157
158   % Transform matrix of seasonal anomalies into time series
159   % Eliminate missing data
160   k=0;
161   for iy=1:Nyears−1
162       for is=1:4
163           if isfinite(a_sst_nino3_seas(iy,is))
164               k=k+1;
165               a_sst_nino3_s(k)=a_sst_nino3_seas(iy,is);
166           end
167       end
168   end
169   Nseas_anom=length(a_sst_nino3_s)
170   min_a_sst_nino3_s=min(a_sst_nino3_s)
171   max_a_sst_nino3_s=max(a_sst_nino3_s)
172
173   % Plot seasonal anomaly time series
174   figure(2004)
175   plot(time_s,a_sst_nino3_s)
176   set(gca,'XLim',[time_s(1) time_s(Nseas_anom)])
177   set(gca,'YLim',[0.99*min_a_sst_nino3_s 1.01*max_a_sst_nino3_s])
178   hlabelx=get(gca,'Xlabel');
179   set(hlabelx,'String','Time AD (years)','Interpreter','Latex','FontSize',16)
180   hlabely=get(gca,'Ylabel');
181   set(hlabely,'String','$T (^{\circ}$C)','Interpreter','Latex','FontSize',16)
182   title('Nino3 seasonal SST anomaly series')
183
184   % Compute mean values
185   musm=mean(sst_nino3_m)
186   muam=mean(a_sst_nino3_m)
187   muss=mean(sst_nino3_s)
188   muas=mean(a_sst_nino3_s)
189
```

```
190   % Save time series
191   sst_nino3_m=sst_nino3_m';
192   a_sst_nino3_m=a_sst_nino3_m';
193   sst_nino3_s=sst_nino3_s';
194   a_sst_nino3_s=a_sst_nino3_s';
195   time_m=time_m';
196   time_s=time_s';
197   save sst_nino3_m.dat sst_nino3_m −ascii
198   save a_sst_nino3_m.dat a_sst_nino3_m −ascii
199   save sst_nino3_s.dat sst_nino3_s −ascii
200   save a_sst_nino3_s.dat a_sst_nino3_s −ascii
201   save time_m.dat time_m −ascii
202   save time_s.dat time_s −ascii
```

16.6.2 Elementary Statistical Analysis

Now we focus on monthly SST data and anomalies, and examine their mean value, standard deviation, minimum and maximum value, and amplitude distribution.

The sample mean and the standard deviation of the data can be computed using the functions mean and std, respectively. However, if we desire also confidence intervals for these parameters, we must use the function normfit:

```
1   [mu,sigma,muci,sigmaci] = normfit(x,alpha)
```

This returns the estimates mu and sigma of the parameters of the normal distribution, given the data in x. Note that in the form reported above, normfit computes mu using the sample mean and sigma using the square root of the unbiased estimator of the variance (Sect. 9.10.2). It also returns the related $100(1 - \text{alpha})\%$ confidence intervals for mu and sigma. The output variables muci and sigmaci contain two values each, representing the lower and upper confidence bounds for mu and sigma, respectively. For alpha we must provide a value in the range [0 1], specifying the width of the confidence interval. By default, alpha is 0.05, which corresponds to 95 % confidence intervals.

Amplitude distributions are plotted as histograms. The Matlab function involved is hist:

```
1   [Nx,xout] = hist(x,nbins)
```

where x is the input sequence of length N, nbins the number of bins, Nx the number of data values falling in each bin, and xout the output vector of bin centers. The proper number of bins can be estimated, as a rule of thumb, by the nearest higher integer to $\sqrt{N}$. It is useful to superimpose to the plot of Nx versus xout a Gaussian curve with the same mean and standard deviation of the data record. A semilogy plot helps observing possible deviations from Gaussianity in the tails of the distribution.

Next we compute and plot the autocorrelation/autocovariance (Sect. 9.10.3) of the Nino3 data. The two Matlab functions involved are xcorr for autocorrelation, and xcov for autocovariance. If the input data is centered, we can use xcorr or xcov indifferently.[2] By default, the output is computed for lags from $-(N-1)$ to $+(N-1)$, with a total of $2N-1$ values; the Nth output sample corresponds to lag zero. In input to xcorr or xcov, we will provide the data sequence and the value of a character flag, specifying a scaling option for the autocorrelation or autocovariance: in fact, in output we can get

- a 'biased' estimate (c_{xx}), or
- an 'unbiased' estimate (c'_{xx}), or
- an autocorrelation coefficient estimate 'coeff' (ρ_{xx}).

We can also specify a desired maximum lag: for example, if we type

```
1   maxlags = 10
2   [rho,lags] = xcov(sst_nino3_m,maxlags,'coeff')
```

we will get the sequence of autocorrelation coefficients of monthly SST data over lags from -10 to $+10$ (21 samples) and the corresponding lag vector, with lag zero at the 11th position. The script given below (statistics_m_SST_anom) performs these tasks. Run it for both monthly SSTs and anomalies.

```
1    % statistics_m_SST_anom
2    % Load Nino3 montly SST data or anomalies, compute mean value and
          standard deviation, plot,
3    % draw histograms to check Gaussianity, compute and plot autocorrelation
          coefficient.
4
5    prompt='Data? (1=SST, 2= Anomalies)';
6    flagdat = input(prompt);
7    if flagdat==1
8        load sst_nino3_m.dat;
9        x=sst_nino3_m;
10   else
11       load a_sst_nino3_m.dat;
12       x=a_sst_nino3_m;
13   end
14   load time_m.dat;
15
16   % Series length, sample mean,sample standard deviation, min and max
17   N=length(x)
18   alpha=0.05
19   [mu,sigma,muci,sigmaci] = normfit(x,alpha)
20   mi=min(x)
```

---

[2] Actually these functions also serve to compute the cross-correlation and cross-covariance between two records.

```
21   ma=max(x)
22
23   % Data plot
24   figure(2100+flagdat−1)
25   plot(time_m,x)
26   hold on
27   plot(time_m,mu+sigma*ones(1,N),'m')
28   plot(time_m,mu+2*sigma*ones(1,N),'g')
29   plot(time_m,mu−sigma*ones(1,N),'m')
30   plot(time_m,mu−2*sigma*ones(1,N),'g')
31   hold off
32   set(gca,'XLim',[time_m(1) time_m(N)])
33   set(gca,'YLim',[floor(min(x)) ceil(max(x))])
34   hlabelx=get(gca,'Xlabel');
35   set(hlabelx,'String','Time AD (years)','Interpreter','Latex','FontSize',16)
36   hlabely=get(gca,'Ylabel');
37   set(hlabely,'String','$T (^{\circ}$C)','Interpreter','Latex','FontSize',16)
38   if flagdat==1
39     title('Nino3 monthly SST series')
40   else
41     title('Nino3 monthly SST anomaly series')
42   end
43
44   % Histogram
45   % xout = centers of bins
46   nbins=ceil(sqrt(N))
47   [Nx,xout]=hist(x,nbins);
48
49   % Related Gaussian curve
50   ampbin=(ma−mi)/nbins
51   limbins=NaN(nbins,2);
52   for k=1:nbins
53       limbins(k,1)=xout(k)−ampbin/2;
54       limbins(k,2)=xout(k)+ampbin/2;
55   end
56   for k=1:nbins
57       area(k)=(limbins(k,2)−limbins(k,1))*Nx(k);
58   end
59   are=sum(area);
60   ny=10^3;
61   y=linspace(limbins(1,1),limbins(nbins,2),ny);
62   ga=are*(1/(sigma*sqrt(2*pi)))*exp(−(y−mu).*(y−mu)/(2*sigma*sigma));
63
64   % Histogram plots
65   %semilogy
```

```
66   figure(2200+flagdat−1)
67   semilogy(xout,Nx,'o','MarkerFaceColor','b','MarkerSize',4);
68   hold on
69   semilogy(y,ga,'g','LineWidth',1.5)
70   semilogy([mu mu],[8*10^(−1) 2*10^2],'r')
71   hwhm=sigma*sqrt(2*log(2));
72   semilogy([mu−hwhm mu+hwhm],[max(ga)/2 max(ga)/2],'r')
73   hold off
74   set(gca,'XLim',[mu−5*sigma mu+5*sigma])
75   if flagdat==1 set(gca,'YLim',[8*10^(−1) 1.2*10^2]);end
76   if flagdat==2 set(gca,'YLim',[8*10^(−1) 2*10^2]);end
77   hlabelx=get(gca,'Xlabel');
78   set(hlabelx,'String','$x$','Interpreter','Latex','FontSize',16)
79   hlabely=get(gca,'Ylabel');
80   set(hlabely,'String','$N$','Interpreter','Latex','FontSize',16)
81     title('Histogram')
82
83   %linear
84   figure(2300+flagdat−1)
85   hist(x,nbins);
86   hold on
87   plot(y,ga,'g','LineWidth',1.5)
88   plot([mu mu],[0 145],'r')
89   plot([mu−hwhm mu+hwhm],[max(ga)/2 max(ga)/2],'r')
90   hold off
91   set(gca,'XLim',[mu−5*sigma mu+5*sigma])
92   if flagdat==1 set(gca,'YLim',[0 95]);end
93   if flagdat==2 set(gca,'YLim',[0 145]);end
94   hlabelx=get(gca,'Xlabel');
95   set(hlabelx,'String','$x$','Interpreter','Latex','FontSize',16)
96   hlabely=get(gca,'Ylabel');
97   set(hlabely,'String','$N$','Interpreter','Latex','FontSize',16)
98     title('Histogram')
99
100  % Autocorrelation coefficients
101  maxlag=100;
102  xc=x−mu;
103  [rho,lags]=xcov(xc,maxlag,'coeff');
104  % lags are in units of adimensional discrete−time
105
106  figure(2400+flagdat−1)
107  plot(lags,rho)
108  set(gca,'XLim',[lags(1) lags(end)])
109  hlabelx=get(gca,'Xlabel');
110  set(hlabelx,'String','Lag','Interpreter','Latex','FontSize',16)
```

```
111   hlabely=get(gca,'Ylabel');
112   set(hlabely,'String','$\rho$','Interpreter','Latex','FontSize',16)
113     title('Aurocorrelation coefficients')
```

Looking at the SST histograms on a linear scale, we can ask ourselves the following questions: is the amplitude distribution of our data Gaussian? Does it seem to have a skewness in one direction or the other? Try a change of nbins: how does the distribution change as we increase/decrease the number of bins? Looking at the histograms of the anomalies, we can ask ourselves: is the distribution remarkably different from that of the SSTs? Does it look more Gaussian?

On the basis of these plots, we can conclude that the assumption of Gaussianity is *approximately* valid for anomalies, though looking at the semilogx anomaly histogram we observe a tail of high values, which is much less evident for the SSTs.

As explained in Sect. 9.11, *pre-whitening*, or simply *whitening*, is a process in which some dominant signals are removed from a data record, in order to obtain an amplitude distribution closer to Gaussianity. Climatological variables like our SSTs contain the annual cycle of the seasons that dominates the power spectrum and influences the amplitude distribution. By removing it, as we did by passing to anomalies, we obtain a signal that is more similar to white noise, in the sense that—as we will soon see—the power spectrum no longer contains the huge peak related to the seasonal cycle. This is a form of whitening, and allows for a more accurate study of possible remaining cyclicities in the data.

What is the behavior of the SST's sequence of autocorrelation coefficients? Does it exhibit regular or irregular oscillations? Does it rapidly or slowly fall off to zero with increasing lag? Passing to anomalies, how does the sequence of autocorrelation coefficients change? Here again, whitening determines a crucial change. SSTs exhibit a ρ_{xx} with marked and persistent oscillations of 12 months = 1 year of period, which are typical of periodic signals: the annual cycle dominates the autocorrelation behavior. By removing it, we obtain an autocorrelation coefficient's sequence that settles around zero at lags of a few tens of time steps. The SST-anomaly random process has finite memory. In the next subsection we will assume assume both stationarity and ergodicity for it (Chap. 9).

16.6.3 Stationary Spectral Analysis

We compute and plot the power spectrum of the monthly SST data and anomalies, including white and red noise background curves and the corresponding significance levels at the 95 and 99.9 % confidence level (c.l.). The following script (periodogram_m_SST_anom) does the job. Frequencies are expressed in years^{-1}. The user can decide if data will be standardized before computing the spectrum, or not. Standardization reduces the data variance to 1 and is sometimes performed for normalization purposes, especially when spectra of different data with dramatically different variances have to be subsequently compared. The spectrum maintains its shape, but its absolute value changes. In our case, the choice between standardizing or not is irrelevant; we can arbitrarily decide to standardize our data. The script periodogram_m_SST_anom given below performs this task. For the significance tests, two functions are provided: AR1_param that estimates the lag-1 AC of the data, and redbckg, which using the output of AR1_param computes the red-noise background power spectrum.

```
1   % periodogram_m_SST_anom
2   % Load Nino3 montly SST data or anomalies, center them, if desired
        standardize them;
3   % compute periodogram as |X|^2/N and plot it versus frequency in years
        ^(-1) on linear and log-log scales.
4   % Then plot periodogram versus period in years on a reversed log2 scale.
5   % Plot white-noise-background spectrum and corresponding significance
6   % levels at 95 and 99.9 % confidence levels.
7   % Plot red-noise-background spectrum and corresponding significance
        levels at 95 and 99.9 % confidence levels.
8
9   prompt='Data? (1 = SST, 2 = Anomalies)';
10  flagdat = input(prompt);
11  prompt='Standardize data? (1 = no, 2 = yes)';
12  flagstand = input(prompt);
13  if flagdat==1
14      load sst_nino3_m.dat;
15      x=sst_nino3_m;
16  else
17      load a_sst_nino3_m.dat;
18      x=a_sst_nino3_m;
19  end
20  load time_m.dat;
21
22  if flagdat==1
23      if flagstand==1
24          tit='Nino3 monthly SST series'
25      else
26          tit='Standardized Nino3 monthly SST series'
27      end
28  else
29      if flagstand==1
30          tit='Nino3 monthly SST anomaly series'
31      else
32          tit='Standardized Nino3 monthly SST anomaly series'
33      end
34  end
35
36  T_s=1/12;
```

```
37   f_Ny=1/(2*T_s);
38   x=x−mean(x); % center data
39   variance=var(x)
40   if flagstand==2
41       x=x/sqrt(variance);
42   end
43   post_variance=var(x)
44   N=length(x)
45
46   N_f=2048;
47   range='twosided';
48   [power,freq]=periodogram(x,boxcar(N),N_f,1/T_s,range);
49   pow=power(1:N_f/2+1)/T_s;
50   f=freq(1:N_f/2+1);
51   pw=pow(2:end);
52   ff=f(2:end);
53   period=1./ff;
54
55   Pwhite=var(x)*ones(1,length(period));
56   signif_fac=chi2inv(.95,2)/2;
57   signif_fac_1=chi2inv(.999,2)/2;
58
59   Pred=redbckg(x,N_f);
60   Pred=Pred(2:end);
61
62   figure(2500+(flagdat−1)*10+flagstand−1)
63   plot(f,pow)
64   set(gca,'XLim',[0 f_Ny])
65   hlabelx=get(gca,'Xlabel');
66   set(hlabelx,'String','$f$ (y$^{−1}$)','Interpreter','Latex','FontSize',16)
67   hlabely=get(gca,'Ylabel');
68   set(hlabely,'String','PSD ($^{\circ}$C$^2$)','Interpreter','Latex','FontSize'
         ,16)
69   title(tit)
70
71   figure(2600+(flagdat−1)*10+flagstand−1)
72   loglog(ff,pw)
73   set(gca,'XLim',[f(2) f_Ny])
74   hlabelx=get(gca,'Xlabel');
75   set(hlabelx,'String','$f$ (y$^{−1}$)','Interpreter','Latex','FontSize',16)
76   hlabely=get(gca,'Ylabel');
77   set(hlabely,'String','PSD ($^{\circ}$C$^2$)','Interpreter','Latex','FontSize'
         ,16)
78   title(tit)
79
```

```
80   figure(2700+(flagdat−1)∗10+flagstand−1)
81   plot(log2(period),Pwhite∗signif_fac_1,'r')
82   hold on
83   plot(log2(period),Pwhite∗signif_fac,'g')
84   plot(log2(period),Pwhite,'k')
85   legend('99.9% c.l.','95% c.l.','w.n. background')
86   plot(log2(period),pw,'LineWidth',1.5)
87   hold off
88   Xticks=2.^(fix(log2(min(period))):fix(log2(max(period))));
89   set(gca,'XDir','reverse')
90   set(gca,'Xlim',log2([min(period),max(period)]));
91   set(gca,'XTick',log2(Xticks(:)));
92   set(gca,'XTickLabel',Xticks);
93   set(gca,'Ylim',[0 50]);
94   hlabelx=get(gca,'Xlabel');
95   set(hlabelx,'String','Period (y)','Interpreter','Latex','FontSize',16)
96   hlabely=get(gca,'Ylabel');
97   set(hlabely,'String','PSD ($^{\circ}$C$^2$)','Interpreter','Latex','FontSize'
        ,16)
98   title(tit)
99
100  figure(2800+(flagdat−1)∗10+flagstand−1)
101  plot(log2(period),Pred∗signif_fac_1,'r')
102  hold on
103  plot(log2(period),Pred∗signif_fac,'g')
104  plot(log2(period),Pred,'k')
105  legend('99.9% c.l.','95% c.l.','r.n. background')
106  plot(log2(period),pw,'LineWidth',1.5)
107  hold off
108  Xticks=2.^(fix(log2(min(period))):fix(log2(max(period))));
109  set(gca,'XDir','reverse')
110  set(gca,'Xlim',log2([min(period),max(period)]));
111  set(gca,'XTick',log2(Xticks(:)));
112  set(gca,'XTickLabel',Xticks);
113  set(gca,'Ylim',[0 50]);
114  hlabelx=get(gca,'Xlabel');
115  set(hlabelx,'String','Period (y)','Interpreter','Latex','FontSize',16)
116  hlabely=get(gca,'Ylabel');
117  set(hlabely,'String','PSD ($^{\circ}$C$^2$)','Interpreter','Latex','FontSize'
        ,16)
118  title(tit)
```

The previous script uses the functions AR1_param and redbckg, given below.

```
1   function [rholag1]=AR1_param(x)
2   % Estimation of lag 1 autocorrelation (rholag1) of data series (x)
3   % for red−noise−background spectrum
4
5   rho=xcov(x,3,'coeff');
6   rho1=rho(3); % lag 1 autocorrelation from xcov
7   rho2=rho(2); % lag 2 autocorrelation from xcov
8
9   % If data are strongly anticorrelated at lag 1 OR 2,
10  % then send a warning message
11  if ((rho1<0 & abs(rho1)>0.1) | (rho2<0 & abs(rho2)>0.1)),
12          error('Data do not resemble white or red noise')
13  % If data are weakly anticorrelated at lag 1 OR 2,
14  % then ASSUME white noise
15  elseif ((rho1<0 & abs(rho1)<=0.1) | (rho2<0 & abs(rho2)<=0.1)),
16          rholag1= 0;          % White noise
17  else
18  % If data are uncorrelated or positively correlated at lag 1 AND 2,
19  % then estimate rolag1; if rolag1 turns out to be zero because ro1=ro2=0,
20  % then the noise is actually white
21          rholag1=(rho1+sqrt(rho2))/2;
22  end
23  return
```

```
1   function [Pred,omred]=redbckg(x,N_f)
2   % Compute red−noise−background power spectrum (Pred)
3   % with parameter alpha estimated from data series (x)
4   % and corresponding frequencies omega (omred) over N_f points from 0 to pi.
5
6   alpha=AR1_param(x);
7   num=1−(alpha*alpha);
8   k=0:1:N_f/2;
9   omred=2*pi*k/N_f;
10  arg=(2*pi/N_f)*k;
11  den=1+alpha*alpha−2*alpha*cos(arg);
12  p=num./den;
13  s2=var(x);
14  Pred=s2*p';
```

Starting with the SST data, let us observe the power spectrum plotted on a linear scale: which is the most prominent peak? Are other peaks observed? Due to the huge amplitude of the annual cycle in comparison with SST fluctuations of different origin, the peak at 1 year of period dominates the spectrum and it is difficult to observe any other peak. The log-log plot evidences another important peak at high frequency that corresponds to $2\,y^{-1}$, i.e., to a period of 0.5 years, or 6 months. This is a higher harmonic of the annual cycle. The remaining spectral features are still difficult to observe. The last type of plot produced by periodogram_m_SST_anom, which reports period instead of frequency in abscissa, on a base-2-logarithmic scale (reversed, so as to agree in direction with the frequency axis of the previous plots) reveals a group of peaks that are significant against both white and red noise.

Now, focus on the last plots that include significance levels:

- does white noise appear a reasonable choice for the background spectrum? Does red noise appear a better choice?
- which peaks or group of peaks pass the significance test at the 95 % c.l. with red noise background? Which ones pass it at the 99.9 % c.l.?

Red noise definitely appears to follow the general spectral shape more closely than white noise, but in the presence of the annual cycle, only the peaks related to it pass the test, even at the lowest confidence level considered (95 %).

Analyzing the anomalies, the issue of the huge annual peak is solved, and the El Niño/Southern Oscillation (ENSO) range (2–8 y) turns out to be the most populated by peaks that pass the red-noise-background test at the 95 % c.l., and sometimes even at the 99.9 % c.l.. Based on the content of Sect. 10.8, we can ask ourselves the following question: if we have a given number n_s of spectral samples that exceed the considered significance level (e.g., the one at the 95 % c.l.), does that mean that all these spectral values are *certainly* due to some process different from red noise? The answer is no; our conclusions are to be interpreted in a probabilistic way. Then, what percentage of the n_s "significant" spectral values can be expected to exceed the significance level *accidentally*, i.e., by chance?

16.6.4   Checking Stationarity

In the previous subsection we assumed stationarity of the data record. To investigate about the quality of this assumption for the Nino3-SST-anomaly monthly series, divide the monthly data sequence of anomalies into three sections. For each section, center the data and standardize it; compute the periodogram as $|X|^2/N$, and plot it versus period in years on a reversed $\log_2$ scale. Plot also the red noise background spectrum and the corresponding significance-level curves at the 95 c.l. and 99.9 % confidence levels. Observe the results to decide if stationarity of the process can be assumed. Use the script reported below.

```
1   % stationarity
2   % Load Nino3 SST anomaly monthly series, divide it into three sections, compute
        Periodogram of each section
3   % and plot it versus period in years on a reversed log2 scale.
4   % Plot red noise background spectrum and corresponding significance levels at 95 and
        99.9 % confidence levels.
5
6   load a_sst_nino3_m.dat;
7   Ntot=length(a_sst_nino3_m);
8   load time_m.dat;
9
10  T_s=1/12;
11  f_Ny=1/(2*T_s);
```

```
12   T_Ny=2*T_s;
13   NN=Ntot/3;
14
15   figure(2900)
16   for i=1:3
17       x=a_sst_nino3_m(1+NN*(i-1):NN*i);
18       x=x-mean(x);
19       N=length(x);
20       N_f=2048;
21       range='twosided';
22       [pow,f]=periodogram(x,boxcar(N),N_f,1/T_s,range);
23       power=pow/T_s;
24       pow=power(2:N_f/2+1);
25       period=1./(f(2:N_f/2+1));
26       signif_fac=chi2inv(.95,2)/2;
27       signif_fac_1=chi2inv(.999,2)/2;
28       [Pred]=redbckg(x,N_f);
29       Pred=Pred(2:end);
30
31       subplot(3,1,i)
32       plot(log2(period),Pred*signif_fac_1,'r')
33       hold on
34       plot(log2(period),Pred*signif_fac,'g')
35       plot(log2(period),Pred,'k')
36       if i==1
37         legend('99.9% c.l.','95% c.l.','r.n. background','Location','NorthWest')
38       end
39       plot(log2(period),pow,'LineWidth',1.5)
40         hold off
41       iniyear=num2str(fix(time_m(1+NN*(i-1))));
42       finyear=num2str(fix(time_m(NN*i)));
43       dash='-';
44       tx=strcat(iniyear,dash,finyear);
45       ht=text(0.2,43,tx);
46       set(ht,'FontSize',12)
47       Xticks=2.^(fix(log2(min(period))):fix(log2(max(period))));
48       set(gca,'XDir','reverse')
49       set(gca,'Xlim',log2([min(period),max(period)]));
50       set(gca,'XTick',log2(Xticks(:)));
51       set(gca,'XTickLabel',Xticks);
52       set(gca,'Ylim',[0 50]);
53       if i==3
54           hlabelx=get(gca,'Xlabel');
55           set(hlabelx,'String','Period (y)','Interpreter','Latex','FontSize',16)
56       end
57       hlabely=get(gca,'Ylabel');
58       set(hlabely,'String','PSD ($^{\circ}$C$^2$)','Interpreter','Latex','FontSize',16)
59   end
```

The plots produced by this code clearly tell us that the series is non-stationary. This naturally leads us towards evolutionary spectral analysis.

16.7 Evolutionary Spectral Analysis of Nino3 Series via CWT

We perform evolutionary spectral analysis via CWT (Chap. 13) on the Nino3 SST anomaly seasonal series ($N = 576$), so as to process a reduced number of samples and speed up numerical calculations, though preserving a representative record duration from a climatological point of view. We will standardize our data before analysis.

The CWT script given below calls a few functions that will be described first. Each of them performs part of the calculations. The CWT is computed in the frequency domain, as the inverse DFT of the product between the DFT of the data and the sampled continuous-time Fourier transform of the daughter wavelet at each scale. The software described and commented here is not only inspired by, but substantially equal to, the Matlab software made available on the web by Christopher Torrence and Gilbert P. Compo at http://paos.colorado.edu/research/wavelets/ (see Torrence and Compo 1998), with only a few changes.

16.7.1 The CWT Using the Complex Morlet Wavelet

The functions providing the building blocks for CWT computation are w_parameters, w_wavfun_fd, w_transform, w_significance, and AR1_param. We will now describe the operations performed by each function (AR1_param has already been described above).

- w_parameters:
  this function provides characteristic parameters of a few complex and real wavelets employed for the CWT: Morlet, Paul, and real DOGs. The function is called by w_transform and w_significance, and also by the main CWT script, which will be called w_nino3. The syntax of w_parameters is

```
[consts] = w_parameters(mother,param)
```

where in input

- mother is the name of the mother wavelet $\psi_0(\theta)$ (θ = continuous adimensional time): 'Morlet, 'Paul', 'DOG';
- param is the parameter that defines the mother wavelet: ω_0 for the complex Morlet wavelet (default $\omega_0 = 6$), the order m in the remaining cases (default: $m = 4$ for Paul, $m = 2$ for DOG; also the choice $m = 6$ is available for DOG).

In output, consts is a vector containing the values of several constants, which are typical of the considered wavelet. We are interested in:

- consts(1) that returns dofmin for significance tests, i.e., the minimum number of DOF, used for the scalogram test (Sect. 13.5.2): 2 for complex wavelets and 1 for real wavelets;

- consts(2) that returns c_δ for signal reconstruction (Sect. 13.4.8);
- consts(3) that returns the *time decorrelation factor* γ, useful for the computation of the number of DOF in the significance test for the global wavelet spectrum or GWS (Sect. 13.5.5);
- consts(5) that returns $\psi_0(0)$ (value of mother wavelet function at time origin), a value which is used for the inverse transform;
- consts(6) that returns the Fourier factor k_0, thus defining the scale-frequency relation (Sect. 13.4.9);
- consts(7) that returns k_{COI}, useful for drawing the COI curve (Sects. 13.4.10 and 13.5.1) on the scalogram plot.

```
1   function [consts] = w_parameters(mother,param)
2   % Scalar parameters that are characteristic of a wavelet
3   % (Morlet, Paul or DOG)
4   %
5   % Outputs:
6   % consts(1) = dofmin (minimum number of degrees of freedom: 2 for complex wavelets, 1
            for real wavelets)
7   % consts(2) = Cdelta (reconstruction constant)
8   % consts(3) = time decorrelation factor
9   % consts(4) = scale decorrelation factor
10  % consts(5) = value of wavelet function at time origin
11  % consts(6) = Fourier factor (ratio between Fourier period and scale)
12  % consts(7) = ratio between e−folding time tau_s and scale
13
14  % Check number of input arguments
15  if (nargin < 2) param=−1; end
16  if(nargin<1)
17      error('Must input a MOTHER')
18  end
19
20  % Default values
21  if (mother == −1), mother ='Morlet'; end
22
23  % Set parameters
24  mother = upper(mother);
25  if (strcmp(mother,'MORLET'))
26      if (param == −1), param = 6.; end
27      om0 = param;
28      consts = [2.,−1,−1,−1,−1,−1,−1];
29      if (om0 == 6)
30          consts(2:5)=[0.771,2.32,0.60,pi^(−1/4)];
31      end
32      if (om0 == 6)
33          consts(6) = (4*pi)/(om0 + sqrt(2 + om0^2));
34      end
35      if (om0 == 6)
36          consts(7) = sqrt(2);
37      end
38  elseif (strcmp(mother,'PAUL'))
39      if (param == −1), param = 4.; end
```

```
40      m = param;
41      consts = [2.,-1,-1,-1,-1,-1,-1];
42      if (m == 4)
43          consts(2:5)=[1.080,1.17,1.5,1.079];
44      end
45      if (m == 4)
46          consts(6) = 4*pi/(2*m+1);
47      end
48      if (m == 4)
49          consts(7) = 0.7013;
50      end
51  elseif (strcmp(mother,'DOG'))
52      if (param == -1), param = 2.; end
53      m = param;
54      consts = [1.,-1,-1,-1,-1,-1,-1];
55      if (m==2)
56          consts(2:5) = [3.431,1.43,1.4,0.867];
57          consts(6) = 2*pi*sqrt(2./(2*m+1));
58          consts(7) = 2.124345;
59      end
60      if (m == 6)
61          consts(2:5) = [1.893,1.37,0.97,0.884];
62          consts(6) = 2*pi*sqrt(2./(2*m+1));
63          consts(7) = 1.582865;
64      end
65  else
66      error('Mother must be one of MORLET,PAUL,DOG')
67  end
68  return
```

- w_wavfun_fd:
  this function is "transparent" to the user; it is called by w_transform for each value
  aa of the scale factor a. The syntax of w_wavfun_fd is

```
1   [daughter] = w_wavfun_fd(om,aa,mother,param)
```

where in input, besides mother and param and the considered scale factor value
(aa), we can see the vector om of the discrete frequencies ω_k corresponding to the
samples of the DFT of the data. In output, the function returns the vector of values
of the continuous Fourier transform of the daughter wavelet at the considered scale,
sampled at the frequencies contained in vector om, i.e., at the same frequencies
ω_k corresponding to the samples of the DFT of the input sequence.

```
1   function [daughter] = w_wavfun_fd(om,aa,mother,param)
2   %Wavelet functions Morlet, Paul, or DOG in the frequency domain.
3
4   % Inputs:
5   % om = vector of adimensional angular frequencies
6   % aa = present value of scale factor
7   % Optional inputs: mother, param
8   %
9   % Output:
```

```
10  | % daughter = the wavelet in the frequency domain at scale factor aa
11  |
12  | % Check number of input arguments
13  | if (nargin < 4), param = −1; end
14  | if (nargin < 3), mother= −1; end
15  | if (nargin < 2),
16  |     error('Must input vector of adimensional angular FREQUENCIES and value of
    |           SCALE FACTOR')
17  | end
18  |
19  | % Default
20  | if (mother==−1), mother='Morlet'; end
21  | mother = upper(mother);
22  | if (strcmp(mother,'MORLET'))
23  |     if (param == −1), param = 6.; end
24  |     om0 = param;
25  |     expnt = −(aa.*om − om0).^2/2.*(om > 0.);
26  |     norm = sqrt(aa)*(pi^(−0.25))*sqrt(2*pi);
27  |     daughter = norm*exp(expnt);
28  |     daughter = daughter.*(om > 0.); % Heaviside step function
29  | elseif (strcmp(mother,'PAUL'))
30  |     if (param == −1), param = 4.; end
31  |     m = param;
32  |     expnt = −(aa.*om).*(om > 0.);
33  |     norm = sqrt(2*pi*aa)*(2^m/sqrt(m*prod(2:(2*m−1))));
34  |     daughter = norm*((aa.*om).^m).*exp(expnt);
35  |     daughter = daughter.*(om > 0.); % Heaviside step function
36  | elseif (strcmp(mother,'DOG'))
37  |     j=sqrt(−1);
38  |     if (param == −1), param = 2.; end
39  |     m = param;
40  |     expnt = −(aa.*om).^2 ./ 2.0;
41  |     norm = sqrt(2*pi*aa/gamma(m+0.5));
42  |     daughter = −norm*(j^m)*((aa.*om).^m).*exp(expnt);
43  | else
44  |     error('Mother must be one of MORLET,PAUL,DOG')
45  | end
46  | return
```

- w_transform:
  this function computes the CWT, by

  - applying the necessary zero padding to the data record, in order to avoid time-domain aliasing;
  - computing the DFT of the padded data via FFT;
  - building the vector a of scale factors as $a = a_0 2^{j\delta_j}$, with $j = [0, J]$, on the basis of the values of a0 $\equiv a_0$, dj $\equiv d_j$ and J $\equiv J$ that are set by the user in the main script w_nino3, which calls w_transform;
  - building the vector om of angular frequencies[3];

---

[3]Angular frequencies here are in $(-\pi, \pi]$. There is no particular reason for this choice, which is perfectly equivalent to the usual one $[-\pi, \pi)$.

– initiating a loop over scale factors, in which at each scale
  · the function w_wavfun_fd is called to compute the daughter-wavelet Fourier transform samples over the frequency set ω_k;
  · the sampled daughter-wavelet Fourier transform is multiplied by the DFT of the data sample by sample;
  · the inverse DFT of the resulting product is taken, to obtain the CWT samples at the considered scale; at the end of the loop, the whole CWT matrix (wave) is ready;
– getting rid of the effects of the padding, by discarding the CWT elements with time indexes exceeding the original number of data;
– computing the wavelet energy density as the square modulus of the wave matrix; this new matrix is called power; note that this is just a name;
– calling w_parameters to retrieve consts(6) and consts(7);
– computing, from the vector a of scale factors and from consts(6) (Fourier factor), the vector of adimensional periods (period);
– computing the adimensional cone-of-influence (coi); for this purpose,
  · the ratio consts(6)/consts(7) is calculated;
  · this ratio is multiplied by the vector of time indexes (i.e., delay indexes). The resulting vector contains the critical value of period ($P_{COI}^{ad} = (k_0/k_{COI})n$) corresponding to each adimensional time value; this allows drawing the cup-shaped curve of the COI in the scalogram plane;
– checking the computed values of wave. An inverse CWT (Sect. 13.4.5) is performed, reconstructing the signal on the basis of all elements of wave. The reconstruction is plotted onto the original series. If all the time scales at which the input signal varies are taken into account when the numbers of octaves and voices are chosen in the main CWT script, the reconstruction performed by w_transform is expected to be approximately equal to the original series. The mean square reconstruction error and the variance of the reconstructed series are also computed; this variance should approximate the variance of the input record, which after standardization is 1. If the user selected a smaller set of scales for the analysis, the reconstruction will lack the largest scales (the lowest frequencies) involved in the signal's variability. Indeed, the minimum scale is usually set according to Nyquist's criterion, on the basis of the scale-frequency relation that holds for the chosen mother wavelet; therefore reducing the number of octaves ultimately leads to neglecting the largest scales. The maximum scale, on the other hand, is constrained by the record length (Sect. 13.5).

The syntax of w_transform is

```
[wave,period,a,coi,power,xcheck] = w_transform(x,dj,a0,J,mother,param)
```

where x represents the input sequence, and dj ($\equiv \delta_j$), a0 ($\equiv a_0$) and J ($\equiv J$) are the user-provided constants needed to build the vector of scale factors. In ouput, xcheck is the reconstructed test-series; the meaning of the other output arguments has already been explained.

```
1    function [wave,period,a,coi,power,xcheck] = w_transform(y,pad,dj,a0,J,
         flag_tests,mother,param)
2    % Wavelet transform of vector y in the frequency domain
3
4    % Optional inputs:
5    % pad = flag for zero padding
6    % dj = spacing between discrete scale factors
7    % a0 = minimum scale factor
8    % J = total number of scales −1
9    % flag_tests = flag for performing tests
10   %              (reconstruction, Parseval's equation, etc.)
11   % mother = mother wavelet name
12   % param = parameter for mother wavelet
13   % See default values.
14   %
15   % Outputs:
16   % wave = complex arry containing the CWT of y
17   % period = adimensional periods
18   % a = scale factors
19   % coi = adimensional cone−of−influence
20   % power = square modulus of CWT
21
22   % Check number of input arguments
23   if (nargin < 8), param = −1; end
24   if (nargin < 7), mother = −1; end
25   if (nargin < 6), flag_test = 0; end
26   if (nargin < 5), J = −1; end
27   if (nargin < 4), a0 = −1; end
28   if (nargin < 3), dj = −1; end
29   if (nargin < 2), pad = 0; end
30   if (nargin < 1)
31       error('Must input a DATA vector y')
32   end
33   N1 = length(y);% original series
34
35   % Default values
36   if (dj == −1), dj = 1./4.; end
37   if (a0 == −1), a0=2; end
38   if (J == −1), J=fix((log(N1/a0)/log(2))/dj); end
39   if (mother == −1), mother ='MORLET'; end
40
41   % Perform zero−padding if required
42   x(1:N1) = y−mean(y);% redundant if y is already centered
43   if (pad == 1)
```

```
44      base2 = nextpow2(N1);% power of 2 nearest to N1
45      x = [x,zeros(1,2^(base2)−N1)];
46   end
47   N = length(x);% padded series
48
49   % Compute FFT of the padded series
50   X = fft(x);
51
52   % Construct SCALE FACTOR array
53   a = a0*2.^((0:J)*dj);
54
55   % Allocate empty complex WAVE array
56   wave = zeros(J+1,N);
57   wave = complex(wave,wave);
58
59   % Construct vector of adimensional angular frequencies
60   % used in transform
61   om = 1:fix(N/2);
62   om = om.*(2*pi/N);
63   om = [0., om, −om(fix((N−1)/2):−1:1)];
64
65   % Loop through all scale factors and compute transform
66   for jj = 1:J+1
67       [daughter]=w_wavfun_fd(om,a(jj),mother,param);
68       wave(jj,:) = ifft(X.*daughter); % wavelet transform
69   end
70
71   % Get rid of padding
72   wave = wave(:,1:N1);
73
74   % Compute wavelet power
75   power = (abs(wave)).^2 ;
76
77   % Get more parameters for the chosen wavelet
78   [consts]=w_parameters(mother,param);
79
80   % Adimensional periods
81   period = consts(6)*a;
82
83   % Adimensional cone of influence
84   coi=consts(6)/consts(7);
85   coi = coi*[1E−5,1:((N1+1)/2−1),fliplr((1:(N1/2−1))),1E−5];
86
87   % Tests
88   summnd=NaN(J+1,N1);
```

```
 89   if(flag_tests==1)
 90       % Reconstruction (inverse transform)
 91       Cdelta=consts(2);
 92       psizerozero=consts(5);
 93       coeffi=dj/(Cdelta*psizerozero);
 94       for jj=1:J+1
 95           summnd(jj,:)=real(wave(jj,:))/sqrt(a(jj));
 96       end
 97       xcheck=sum(summnd);
 98       xcheck=coeffi*xcheck;
 99       figure(9999)
100       plot(x(1:N1),'k')
101       hold on
102       plot(xcheck,'r')
103       title('Reconstruction check: data vs. index');
104       legend('Original data','Reconstructed data')
105       hold off
106
107       % Reconstruction error
108       test=xcheck-x(1:N1);
109       test=test.*test;
110       rec_mse=sum(test)/N
111       rec_rmse=sqrt(rec_mse)
112
113       % Variance of reconstructed data
114       test_var=var(xcheck)
115
116       % Parseval's theorem
117       for jj=1:J+1
118           summnd(jj,:)=(abs(wave(jj,:))).^2/a(jj);
119       end
120       summnd1=sum(summnd);
121       parseval=sum(summnd1);
122       parseval=parseval*dj/(N1*Cdelta)
123   end
124   return
```

- w_significance:
this functions computes the significance levels for the scalogram and GWS. Its syntax is

```
1   [signif,P_red] = w_significance(y,a,sigtest,rolag1,conflev,dof,mother,param)
```

In input,

- y can be the input series, or directly its variance; if it is a scalar, w_significance assumes it is a variance; if it is a vector, the corresponding variance is computed;
- a is the vector of scale factors;
- sigtest can be 0, or 1. If sigtest = 0 (default), a regular χ^2 test is performed on scalogram values; if sigtest = 1, a suitable test for the GWS is applied;
- rholag1 is the parameter α of the AR(1) model for the red-noise background;
- conflev is the confidence level for the significance tests (default 95 %, assumed if conflev = −1);
- dof is the number of DOF, depending on the value of sigtest. If sigtest = 0, then dof is automatically set to consts(1) = dofmin, i.e., 2 for a complex mother wavelet, and 1 for a real one; if sigtest = 1, then dof must be provided in input. It must be set to dof = N − consts(7)*a, where N is the number of time steps averaged together to form an individual GWS sample, and the term consts(7)*a is an empirical correction related to the zero-padding performed at the edges of the series. The term subtracted from N is aimed at taking into account the distortion due to the cone of influence, meaning that the spectral values inside the coi reduce the number of independent estimates that are averaged together, in a way that is more and more statistically important as the scale increases.

The output vector, called signif, is the vector of significance level values (one for each scale). Also the background spectrum P_red can be retrieved, but we will not use it in the present exercise.

```
 1  function [signif,P_red] = w_significance(y,a,sigtest,rholag1,siglvl,dof,mother,param)
 2  % Significance test for wavelet transform
 3
 4  % Inputs:
 5  %
 6  %    y = the time series, or, the variance of the time series.
 7  %       (If this is a single number, it is assumed to be the variance)
 8  %    a = the vector of scale factors.
 9  %
10  % Optional inputs: mother,param
11  %
12  % *** Note *** setting any of the following to −1 will cause the default value to be used
13  %
14  %    sigtest = 0, or 1. If omitted, then assume 0.
15  %
16  %       If 0 (default), then just do a regular chi−square test.
17  %       If 1, then do a "time−average" test.
18  %          In this case, dof should be set to N_a, the number of local wavelet spectra
19  %          that were averaged together.
20  %          For the GWS, this is N_a=N, where N is the number of points in the input
           series.
21  %          minus a correction for the effects of the cone of influence
22  %
23  % rholag1 = lag−1 autocorrelation, used for significance levels. Default is 0.0.
24  %
25  % siglvl = confidence level to use for significance test. Default is 0.95.
```

```
26  %
27  %    dof = degrees−of−freedom for significance test.
28  %        If sigtest=0, then (automatically) dof = 2 (or 1 for mother='DOG')
29  %        If sigtest=1, then dof = N_a, the number of time steps averaged together.
30  %
31  %    Note: If sigtest=1, then dof can be a vector (same length as a=scale factors),
32  %          in which case N_a is assumed to vary with scale.
33  %          This allows one to take into account the effect of the cone of influence.
34  % Outputs:
35  %
36  %    signif = significance level
37  %    P_red = theoretical red−noise spectrum
38
39  % Check number of input arguments
40  if (nargin < 8), param = −1; end
41  if (nargin < 7), mother = −1; end
42  if (nargin < 6), dof = −1; end
43  if (nargin < 5), siglvl = −1; end
44  if (nargin < 4), rholag1 = −1; end
45  if (nargin < 3), sigtest = −1; end
46  if (nargin < 2)
47      error('Must input a DATA vector y, and SCALE FACTOR vector a')
48  end
49
50  % Default values
51  if (sigtest == −1), sigtest = 0; end
52  if (rholag1 == −1), rholag1 = 0.0; end
53  if (siglvl == −1), siglvl = 0.95; end
54  if (mother == −1), mother = 'Morlet'; end
55
56  % Check
57  if dof==−1
58      if sigtest==1
59          error('Must input dof if sigtest is 1')
60      end
61  end
62  N1 = length(y);
63  J = length(a) − 1;
64  dj = log(a(2)/a(1))/log(2.);
65  if (N1 == 1)
66      variance = y;
67  else
68      variance = std(y)^2;
69  end
70
71  % Get appropriate parameters
72  [consts]=w_parameters(mother,param);
73  period = a.*consts(6); % adimensional periods
74  dofmin = consts(1); % degrees of freedom with no smoothing
75  Cdelta = consts(2); % reconstruction factor
76  gamma_fac = consts(3); % time−decorrelation factor
77  dj0 = consts(4); % scale−decorrelation factor
78
```

```
79   % Red noise spectrum and significance levels for scalogram
80   nu = 1 ./ period;  % normalized frequency
81   P_red = (1−rholag1^2) ./ (1−2*rholag1*cos(nu*2*pi)+rholag1^2);
82   P_red = variance*P_red;  % include time−series variance
83
84   signif=NaN(1,J+1);  % initialize signif
85
86   if (sigtest == 0)   % no smoothing, dof=dofmin: for scalogram
87       dof = dofmin;
88       chisquare = chi2inv(siglvl,dof)/dof;
89       signif = P_red*chisquare ;
90
91   elseif (sigtest == 1)  % time−average significance: for global wavelet spectrum
92       if (length(dof) == 1), dof=zeros(1,J+1)+dof; end
93       truncate = find(dof < 1);
94       dof(truncate) = ones(size(truncate));
95       dof = dofmin*sqrt(1 + (dof/gamma_fac ./ a).^2 );
96       truncate = find(dof < dofmin);
97       dof(truncate) = dofmin*ones(size(truncate));
98       for jj = 1:J+1
99           chisquare = chi2inv(siglvl,dof(jj))/dof(jj);
100          signif(jj) = P_red(jj)*chisquare;
101      end
102
103  else
104      error('sigtest must be either 0, or 1')
105  end
106  return
```

Now we can put all together in the main script w_nino3, to perform the following steps.

- Choose a mother wavelet (mother) and its parameter.
- Choose the minimum scale factor, a0. This depends on the wavelet, since it is chosen according to the scale-frequency relation (see Sect. 13.4.9).
- When requested to state if the data have to be standardized, or not, choose standardization, for coherence with the previous exercises.
- Set confidence levels for significance tests: a single one for the scalogram (95 %), which is also used for the GWS, and four additional levels for the GWS; for example, choose 90 %, 92 %, 98 % and 99 %;
- Load the data record (a_nino3_SST_s.dat) and the corresponding time values (time_s.dat) and place them in vectors z and time, respectively.
- Compute the length N of z.
- Compute the mean value, standard deviation and variance of z.
- Center the data by subtracting the mean (even if the mean is already very close to zero in the present case). Standardize them dividing by their standard deviation. Place the pre-processed data in vector x. The variance of x in case of standardization will be 1. This leads to a standardized scalogram.
- Specify the value of the sampling interval for the data record ($T\_s = 0.25$ years).

- Set the total number of octaves to be included in the analysis (n_oct = 8 in the present case) and the inverse of the number of voices per octave ($\delta_j \equiv$ dj $= 0.05$). Compute the corresponding total number of scales, minus 1 ($J \equiv$ J). The choice of the values n_oct = 8 and dj = 0.05 can be justified as follows. Since $f_{\min} = 1/(NT_s)$, $\nu_{\min} = 1/N$; then $P_{\max}^{ad} = k_0 a_{\max} = 1/\nu_{\min} = N$, hence $a_{\max} = N/k_0$. At the same time, $a_0 = 2/k_0$; therefore, the ratio of the maximum scale factor to the minimum is

$$\frac{a_{max}}{a_0} = \frac{N}{2} = \frac{576}{2} = 288,$$

which is comprised between $2^8 = 256$ and $2^9 = 512$. But

$$\frac{a_{max}}{a_0} = 2^{J\delta_j} = 2^{n_{oct}},$$

where n_{oct} is the number of octaves. Hence,

$$n_{oct} = \log_2 \frac{a_{max}}{a_0}.$$

We choose $n_{oct} = 8$. Adopting dj $= 0.05$ for a smooth scalogram plot, we get

$$J = \frac{1}{\delta_j} \frac{\log\left(\frac{a_{max}}{a_0}\right)}{\log 2} = 160.$$

A smaller value of $\delta_j \equiv$ dj would give the scalogram a finer scale structure, but would also imply a greater computational burden. We will later check the quality of these choices observing the results of the reconstruction test and the scalogram and GWS plots (see Sect. 13.5).
- Compute the CWT of x by calling the function w_transform, thus obtaining as outputs the matrices wave and power, and the vectors a, period and coi.
- Compute the parameter $\alpha \equiv$ rholag1 of the AR(1) model for the red-noise background, by calling AR1_param according to the syntax

```
rholag1 = AR1_param(z)
```

- Compute the significance level for the scalogram (signif), as a function of scale or period, by calling w_significance. Provide in input

  - the variance of the data after standardization, i.e., 1;
  - the scale-factor vector a;
  - sigtest = 0;
  - the parameter rholag1;
  - the confidence level, conflev;
  - dof = dofmin, or -1 (default);
  - mother and param.

The output vector (signif) has one element for each scale; to get a quantity that can be plotted in the time-period plane, expand signif, transforming it into a matrix (sig) with dimensions $(J + 1) \times N$. Then redefine sig as the ratio between power and sig, so that in the scalogram contour plot, special contour lines can be drawn where sig attains 1. The areas included inside these special contour lines have sig ≥ 1 and thus exhibit significant power.

- Compute the GWS (global_ws) averaging power over time at each scale.
- Compute the significance level for the GWS at the same confidence level chosen for the scalogram (global_signif), as well as at a set of four different confidence levels (glob_sign), performing new calls to w_significance.[4] At each call, provide in input

  - the variance of the data after standardization, i.e., 1;
  - the scale-factor vector a;
  - sigtest $= 1$;
  - the parameter rholag1;
  - the confidence level conflev;
  - dof. This variable must contain the number of independent estimates that were averaged to obtain the GWS at each scale. So, call w_parameters to get consts and set dof $= N -$ consts(7)*a. This statement also turns dof into a vector of length $J + 1$, which is useful in w_significance, where the dof will be further processed to produce a variable representing the actual number of DOF for the GWS test at each scale.
  - mother and param.

- Translate, using T_s, the adimensional coi and period vectors into the corresponding values in years.

The main script (w_nino3) is listed below.

```
 1  % w_nino3.m
 2  % Perform CWT analysis of Nino3 seasonal SST anomalies
 3
 4  clear
 5
 6  % Chose mother wavelet
 7  prompt='Mother wavelet? (1 = Morlet, 2 = Paul(m=4), 3 = DOG(m=2), 4=
          DOG(m=6))';
 8  flagmother = input(prompt);
 9
10  if flagmother == 1
```

---

[4]We could decide to always include the c.l. used for the scalogram in this set of four probability values and avoid the calculation of global_signif. However, we prefer to keep the choice of the additional c.l. values for the GWS separate from the choice made for the scalogram. At the same time, it is evident that if we used the 95 % for the scalogram, for instance, we must also use it for the GWS. So, global_signif is directly related to the choice made for the scalogram significance test, while glob_sign is completely free for what concerns the probability levels.

```
11        mother = 'Morlet'
12        param = 6
13        % mother and param are Morlet, 6 by default
14        % in functions called later.
15        % Therefore this statement could be omitted
16        a0 = 2;     % minimum scale factor
17    elseif flagmother == 2
18        mother = 'Paul'
19        param = 4
20        a0 = 2;
21    elseif flagmother == 3
22        mother='DOG'
23        param = 2
24        a0 = 1/2;
25    elseif flagmother == 4
26        mother='DOG'
27        param=6
28        a0 = 1;
29    else
30      error('Must input 1, or 2, or 3, or 4')
31    end
32
33    % Decide if data must be standardized
34    prompt='Standardize data? (1 = yes, 0 = no)';
35    istand = input(prompt);
36
37    % Decide if tests in w_transform are of interest
38    prompt='Perform tests in w_transform? (1 = yes, 0 = no)';
39    flag_tests = input(prompt);
40
41    % Confidence levels for tests
42    cl=NaN(1,4);% for the GWS
43    cl(1)=0.90;
44    cl(2)=0.92;
45    cl(3)=0.98;
46    cl(4)=0.99;
47    conflev=−1; % 0.95 for the scalogram and the GWS
48
49    pad = 1;   % pad the time series with zeroes
50    %            (recommended)
51
52    % Input data
53    z = load('a_sst_nino3_s.dat');
54    time=load('time_s.dat');
55    N=length(z)
```

```matlab
56  T_s = 0.25 ; % sampling interval in years (seasonal data)
57
58  % Octaves, voices & scales
59  n_oct=8;        % number of octaves
60  dj = 0.05;      % inverse of n. of intervals per octave
61  J = n_oct/dj;   % n. of powers-of-two (octaves)
62  %                    with 1/dj intervals each
63
64  % Computation
65  n_scales=J+1
66
67  glob_sign=NaN(n_scales,4); % initialize glob_sign
68
69  % Compute variance of data
70  variance = var(z)
71
72  % Remove mean -- variance remains unchanged
73  mu=mean(z);
74  x = z - mu;
75
76  % If required, standardize data:
77  % variance becomes 1
78  if istand==1
79      x = x/sqrt(variance);
80  end
81
82  % Compute variance again
83  variancex = var(x)
84
85  % Wavelet transform
86  [wave,period,a,coi,power] = w_transform(x,pad,dj,a0,J,flag_tests,mother,
         param);
87
88  % Significance levels for scalogram
89  [rholag1] = AR1_param(z) % Compute AR(1) parameter
90  %                        for red noise background
91  [signif] = w_significance(variancex,a,0,rholag1,conflev,-1,mother,param);
92  % signif is a (J+1)-long vector
93  sig = (signif')*(ones(1,N)); % expand signif --> (J+1)x(N) array
94  sig = power./ sig;           % where ratio > 1, power is significant
95
96  % Global wavelet spectrum & significance levels
97  global_ws=sum(power')/N;
98  [consts]=w_parameters(mother,param);
99  dof = N - consts(7)*a;
```

```
100  % where the last term is an empirical correction
101  % for padding at the edges of the series
102  [global_signif] = w_significance(variancex,a,1,rholag1,conflev,dof,mother,
          param);
103  for icl=1:4
104     [glob_sign(:,icl)] = w_significance(variancex,a,1,rholag1,cl(icl),dof,
             mother,param);
105  end
106
107  % Turn to dimensional cone-of-influence and period
108  coi=coi*T_s;
109  period=period*T_s;
110
111  % Save results in .mat file
112  if flagmother==1
113     save w_nino3_Morl.mat
114  elseif flagmother==2
115     save w_nino3_Paul.mat
116  elseif flagmother==3
117     save w_nino3_DOG2.mat
118  else
119     save w_nino3_DOG6.mat
120  end
```

Run the main script, using the Morlet wavelet, choosing to standardize the data and to perform the reconstruction test in w_transform. The test-reconstruction plot will appear in a figure numbered 9999. After running the main script, use the following code (plot_w_nino3) to plot the input series, the scalogram with its single sigificance level, and the GWS with its various significance levels. Make figure full-screen to be able to observe its details.

```
1   % plot_w_nino3
2   % Plot results of w_nino3
3
4   % Chose mother wavelet
5   prompt= 'Mother wavelet? (1 = Morlet, 2 = Paul(m=4), 3 = DOG(m=2), 4= DOG(m=6))';
6   flagmother = input(prompt);
7
8   % Load computation results from mat file
9   if flagmother==1
10     load w_nino3_Morl.mat
11  elseif flagmother==2
12     load w_nino3_Paul.mat
13  elseif flagmother==3
14     load w_nino3_DOG2.mat
15  else
16     load w_nino3_DOG6.mat
17  end
```

```matlab
18
19   label1y='Anomaly ($^{\circ}$C)';
20   title1='NINO3 SST seasonal anomalies';
21   label2x='Year AD';
22   label2y='Period (y)';
23   title2='Wavelet spectrum';
24   label3x='Power';
25   title3='GWS';
26   xlim = [time(1),time(end)];
27   ylim1= [min(z) max(z)];
28   xti=1880:20:2000;
29   gws_lim1=0.05;
30   gws_lim2=50;
31   xti_gws=[10^(-1) 10^0 10^1 10^2];
32   levels=linspace(log2(min(power(:))),log2(max(power(:))),64);
33
34   figure(3000+flagmother-1)
35   %--- Plot time series
36   subplot('position',[0.1 0.73 0.62 0.18])
37   plot(time,z)
38   set(gca,'XLim',xlim(:))
39   set(gca,'YLim',ylim1(:))
40   set(gca,'XTick',xti)
41   set(gca,'YTick',[-2:4])
42   hlabely=get(gca,'Ylabel');
43   set(hlabely,'String',label1y,'Interpreter','Latex','FontSize',12)
44   title(title1,'FontSize',10,'FontWeight','Bold')
45
46   %--- Contour plot wavelet power spectrum
47   subplot('position',[0.1 0.12 0.62 0.48])
48   pow_lev=2.^levels
49   nlev=length(pow_lev)
50   [c,H]=contourf(time,log2(period),log2(power),levels);
51   colormap(jet)
52   set(H,'LineStyle','none')
53   set(gca,'XLim',xlim(:))
54   set(gca,'XTick',xti)
55   Yticks = 2.^(fix(log2(min(period))):fix(log2(max(period))));
56   set(gca,'YLim',log2([min(period),max(period)]),'YDir','reverse','YTick',log2(Yticks(:)),'
              YTickLabel',num2str(Yticks'),'layer','top')
57   hlabelx=get(gca,'Xlabel');
58   set(hlabelx,'String',label2x,'Interpreter','Latex','FontSize',12)
59   hlabely=get(gca,'Ylabel');
60   set(hlabely,'String',label2y,'Interpreter','Latex','FontSize',12)
61   title(title2,'FontSize',10,'FontWeight','Bold')
62   hold on
63   % significance contour, levels at -99 (fake) and 1 (signif)
64   [c,h] = contour(time,log2(period),sig,[-99,1],'k','LineWidth',1.5);%95%
65   % cone-of-influence, anything "below" is dubious
66   plot(time,log2(coi),'w','LineWidth',1.5)
67   hold off
68
69   %--- Plot global wavelet spectrum
```

```
70   subplot('position',[0.74 0.12 0.23 0.48])
71   semilogx(global_signif,log2(period),'k','LineWidth',1.5)%95%
72   hold on
73   semilogx(glob_sign(:,1),log2(period),'g','LineWidth',1.5)%90%
74   semilogx(glob_sign(:,2),log2(period),'m','LineWidth',1.5)%95%
75   semilogx(glob_sign(:,3),log2(period),'b','LineWidth',1.5)%98%
76   semilogx(glob_sign(:,4),log2(period),'r','LineWidth',1.5)%99%
77   semilogx(global_ws,log2(period),'k','LineWidth',1.5)
78   hold off
79   Yticks = 2.^(fix(log2(min(period))):fix(log2(max(period))));
80   set(gca,'YLim',log2([min(period),max(period)]),'YDir','reverse','YTick',log2(Yticks(:)),'
         YTickLabel','')
81   set(gca,'XLim',[gws_lim1,gws_lim2])
82   set(gca,'XTick',xti_gws)
83   hlabelx=get(gca,'Xlabel');
84   set(hlabelx,'String',label3x,'Interpreter','Latex','FontSize',12)
85   title(title3,'FontSize',10,'FontWeight','Bold')
```

The interpretation of these graphs has been given in Sect. 13.5.3.

16.7.2 The CWT with Other Wavelets

We will now repeat the analysis with the other three wavelets (Paul and real DOGs with orders 2 and 6) to see if changing the mother wavelet modifies the results.

1. Run again the CWT and plotting scripts with the Paul wavelet with parameter $m = 4$. This wavelet is narrower in time than the Morlet wavelet, and this produces a better temporal localization of transient events, at the cost of a poorer frequency resolution. Anyway, the main features detected in the series by the previous analysis are visible in this new scalogram. The GWS is much smoother. On the other hand, the cone of influence is less extended.

2. Run again the CWT and plotting scripts with the DOG wavelet with parameter $m = 2$ (Mexican Hat wavelet). The fine scale structure observed in this case is due to the fact that the wavelet is real, and therefore captures positive and negative variations of the centered input sequence as separate peaks in the scalogram. The Morlet wavelet, on the contrary, is a complex analytic function containing six oscillations under the Gaussian dome, and therefore combines negative and positive local extrema into a single peak, whose width depends on the scale. If we decided to plot the real and imaginary parts of the Morlet-based CWT separately, we would obtain a picture very similar to the one obtained using the Mexican Hat. Moreover, DOG($m = 2$) is relatively narrow in time and broadband in frequency with respect to the Morlet wavelet, and therefore in the corresponding scalogram the high-power areas are narrow in time and quite elongated in period. The main features revealed by the previous scalograms remain unaltered, though the scalogram is overall less easily interpretable than with complex wavelets.

3. Run again the CWT and plotting scripts with the DOG wavelet with parameter
   $m = 6$ and observe the changes with respect to the DOG($m = 2$) case.

16.7.3 Global Wavelet Spectrum Versus Periodogram

It is interesting to compare the GWS obtained using the Morlet wavelet with the
periodogram. The GWS must be compared with a *twosided* periodogram of the
same data. This comparison is performed by the script named periodogram_gws. In
this script we take the running average over n_ra samples of the PSD computed via
periodogram for the centered and standardized seasonal anomalies ($n\_ra = 3, 7, 21$,
and 71), and compare these smoothed periodograms with the GWS.

```
 1   % periodogram_gws
 2   % Compare GWS with smoothed periodograms obtained by running averages
 3   % over different numbers of periodogram samples
 4
 5   % Chose mother wavelet
 6   prompt='Mother wavelet? (1 = Morlet, 2 = Paul(m=4), 3 = DOG(m=2), 4= DOG(m=6))';
 7   flagmother = input(prompt);
 8
 9   % Load computation results from mat file
10   if flagmother==1
11       load w_nino3_Morl.mat
12   elseif flagmother==2
13       load w_nino3_Paul.mat
14   elseif flagmother==3
15       load w_nino3_DOG2.mat
16   else
17       load w_nino3_DOG6.mat
18   end
19
20   load a_sst_nino3_s.dat;
21
22   % Set number of values for periodogram's running average
23   n_ra=[3 7 21 71];
24
25   % Compute periodogram
26   T_s=1/4;
27   f_s=1/T_s;
28   N=length(z);
29   N_f=1024;
30   range='twosided';
31   [PSD,f]=periodogram(x,boxcar(N),N_f,f_s,range);%use same data as CWT, centered only
         , or centered and standardized
32   PSD=PSD*f_s; % remove normalization
33   pow=PSD(2:end);% eliminate zero frequency
34   % corresponding to infinite period
35   per=1./f(2:end);
36
```

```
37   % Smooth periodogram with running average filter
38   % over different numbers of points and plot
39   % together with GWS
40   figure(3100)
41   for ia=1:length(n_ra)
42       n_rav=n_ra(ia);
43       % Generate filter weights
44       a = 1;
45       b = ones(n_rav,1)*1/n_rav;
46       % Filter periodogram and corresponding periods
47       per_ra0 = filter(b,a,per);
48       pow_ra0 = filter(b,a,pow);
49       % Discard transients
50       per_ra=per_ra0(n_rav:end);
51       pow_ra=pow_ra0(n_rav:end);
52       % Plot
53       subplot(2,2,ia)
54       plot(log2(per),pow,'g')%periodogram
55       hold on
56       plot(log2(per_ra),pow_ra,'r')%smoothed periodogram
57       text('String','$q=$','Interpreter','Latex','Position',[log2(100) 9],'FontSize',16)
58       n_rav_str=num2str(n_rav);
59       text('String',n_rav_str,'Position',[log2(30) 9],'FontSize',12)
60       plot(log2(period),global_ws)%GWS
61       hold off
62       Xticks=2.^(fix(log2(min(period))):fix(log2(max(period))));
63       set(gca,'XDir','reverse')
64       set(gca,'Xlim',log2([min(period),max(period)]));
65       set(gca,'XTick',log2(Xticks(:)));
66       set(gca,'XTickLabel',Xticks);
67       set(gca,'YLim',[0 10])
68       set(gca,'YTick',[0:2:10]);
69       hlabelx=get(gca,'Xlabel');
70       set(hlabelx,'String','Period (y$^{-1}$)','Interpreter','Latex','FontSize',16)
71       hlabely=get(gca,'Ylabel');
72       set(hlabely,'String','PSD ($^{\circ}$C$^2$)','Interpreter','Latex','FontSize',16)
73   end
```

Observe how the Fourier spectrum tends to become very similar to the GWS as the smoothing applied to the periodogram increases. The amount of smoothing required to obtain this similarity increases with decreasing period (increasing frequency; see Sect. 13.5.5).

16.7.4 Reconstruction of Significant Oscillations

Our goal here is to reconstruct the significant oscillation detected in the anomaly series. This oscillation is in the period range that is typical of the ENSO phenomenon. In order to be able to do so, we must decide how to select the range of periods involved in the reconstruction, and we will base our decision on the GWS shape. We plot the

GWS with its significance levels and decide to include the complete GWS peaks found in this range, i.e., periods from 2 to 8 years, represented by *vertical lines* in the figure produced by the following script, plot_gws_nino3.

```
1   % plot_gws_nino3.m
2   % Plot GWS from CWT analysis of Nino3 SST anomalies.
3
4   % Chose mother wavelet
5   prompt='Mother wavelet? (1 = Morlet, 2 = Paul(m=4), 3 = DOG(m=2), 4= DOG(m=6))';
6   flagmother = input(prompt);
7
8   % Load computation results from mat file
9   if flagmother==1
10      load w_nino3_Morl.mat
11  elseif flagmother==2
12      load w_nino3_Paul.mat
13  elseif flagmother==3
14      load w_nino3_DOG2.mat
15  else
16      load w_nino3_DOG6.mat
17  end
18
19  pow=global_ws;
20  gsig=global_signif;
21  gsi=glob_sign;
22
23  figure(3200)
24  plot(log2(period),pow,'LineWidth',1.5)
25  hold on
26  plot(log2(period),gsi(:,1),'c')
27  plot(log2(period),gsi(:,2),'g')
28  plot(log2(period),gsi(:,3),'m')
29  plot(log2(period),gsi(:,4),'r')
30  Xticks=2.^(0:3.5);
31  set(gca,'XDir','reverse')
32  set(gca,'XLim',[0 3.5]);
33  set(gca,'XTick',log2(Xticks(:)));
34  set(gca,'XTickLabel',Xticks);
35  set(gca,'YLim',[0 6.5]);
36  hlabelx=get(gca,'Xlabel');
37  set(hlabelx,'String','Period (y)','Interpreter','Latex','FontSize',16)
38  hlabely=get(gca,'Ylabel');
39  set(hlabely,'String','GWS','Interpreter','Latex','FontSize',16)
40  legend('Power','90%','95%','98%','99%')
41  plot(log2(2)*ones(2,1),[0 6.5],'k')
42  plot(log2(8)*ones(2,1),[0 6.5],'k')
43  hold off
```

Then we reconstruct the oscillation using the script called w_recostr_nino3, in which we select the contributions of cyclicities with periods in the range 2–8 years. For reconstruction we avoid data standardization.

```
1   % w_recostr_nino3
2   % Perform CWT analysis and series reconstruction.
3   % Plot reconstruted series and list included periods and scales.
4
5   % Chose mother wavelet
6   prompt='Mother wavelet? (1 = Morlet, 2 = Paul(m=4), 3 = DOG(m=2), 4= DOG(m=6))';
7   flagmother = input(prompt);
8
9   if flagmother == 1
10     mother = 'Morlet'
11     param = 6
12     a0 = 2;
13  elseif flagmother == 2
14     mother = 'Paul'
15     param = 4
16     a0 = 2;
17  elseif flagmother == 3
18     mother='DOG'
19     param = 2
20     a0 = 1/2;
21  elseif flagmother == 4
22     mother='DOG'
23     param=6
24     a0 = 1;
25  else
26     error('Must input 1, or 2, or 3, or 4')
27  end
28
29  % Choose range of periods on which the reconstruction will be based
30  prompt='Minimum and maximum dimensional period of interest []?';
31  incl_per = input(prompt);
32
33  % Do not standardize data and pad the time series with zeroes
34  istand=0;
35  pad = 1;
36
37  % Load data
38  z = load('a_sst_nino3_s.dat');
39  time=load('time_s.dat');
40  N=length(z)
41  T_s = 0.25 ;% sampling interval
42  incl_adim_per = incl_per/T_s; % adimensional periods
43
44  dj = 0.05; % inverse of n. of intervals per octave
45  J = 8/dj;   % n. of powers−of−two (octaves)
46  %              with 1/dj intervals each
47  n_scales=J+1
48
49  % Remove mean − variance remains unchanged
50  mu=mean(z);
51  x = z − mu;
52
53  % Wavelet transform and reconstruction
```

```
54  [rec,a,period] = w_transform_rec(x,pad,dj,a0,J,mother,param,incl_adim_per);
55  % adimensional periods
56
57  % Plot reconstruction
58  figure(3300)
59
60  plot(time,z)
61  hold on
62  plot(time,rec+mu,'r')
63  hold off
64  set(gca,'XLim',[time(1) time(end)])
65  set(gca,'YLim',[min(z) max(z)])
66  hlabelx=get(gca,'Xlabel');
67  set(hlabelx,'String','Time AD (y)','Interpreter','Latex','FontSize',16)
68  hlabely=get(gca,'Ylabel');
69  set(hlabely,'String','$T$ anomaly ($^{\circ}$C)','Interpreter','Latex','FontSize',16)
70
71  period=period*T_s ;% dimensional periods
72  included_periods = period(find((period >= incl_per(1)) & (period <= incl_per(2))))
```

The reconstruction script calls a modified version of w_transform, reported below
(w_transform_rec).

```
1   function [rec,a,period] = w_transform_rec(y,pad,dj,a0,J,mother,param,incl_adim_per);
2
3   % Wavelet transform of vector y in the frequency domain
4   % and data reconstruction on the basis of scale factors in a given range
5   %
6   % Optional inputs:
7   % pad = flag for zero padding
8   % dj = spacing between discrete scale factors
9   % a0 = minimum scale factor
10  % J  = total number of scales -1
11  % mother = mother wavelet name
12  % param = parameter for mother wavelet
13  % See default values.
14  %
15  % Outputs:
16  % wave = complex arry containing the CWT of y
17  % period = adimensional periods
18  % a     = scale factors
19  % coi   = adimensional cone-of-influence
20  % power = square mod. of CWT
21  % rec   = reconstructed series
22
23  % Check number of input arguments
24  if (nargin < 8), ia=[1 J+1]; end
25  if (nargin < 7), param = -1; end
26  if (nargin < 6), mother = -1; end
27  if (nargin < 5), J = -1; end
28  if (nargin < 4), a0 = -1; end
29  if (nargin < 3), dj = -1; end
30  if (nargin < 2), pad = 0; end
31  if (nargin < 1)
```

```
32        error('Must input a data vector y')
33     end
34     N1 = length(y);% original series
35
36     % Default values
37     if (dj == −1), dj = 1./4.; end
38     if (a0 == −1), a0=2; end
39     if (J == −1), J=fix((log(N1/a0)/log(2))/dj); end
40     if (mother == −1), mother = 'Morlet'; end
41
42     % Perform zero−padding if required
43     x(1:N1) = y−mean(y);% redundant if y is already centered
44     if (pad == 1)
45        base2 = nextpow2(N1);% power of 2 nearest to N1
46        x = [x,zeros(1,2^(base2)−N1)];
47     end
48     N = length(x);% padded series
49
50     % Compute FFT of the padded series
51     X = fft(x);
52
53     % Construct scale factor array
54     a = a0*2.^((0:J)*dj);
55
56     % Allocate empty complex wave array
57     wave = zeros(J+1,N);
58     wave = complex(wave,wave);
59
60     % Construct vector of adimensional angular frequencies
61     % used in transform
62     om = 1:fix(N/2);
63     om = om.*(2*pi/N);
64     om = [0., om, −om(fix((N−1)/2):−1:1)];
65
66     % Loop through all scale factors and compute transform
67     for jj = 1:J+1
68        [daughter]=w_wavfun_fd(om,a(jj),mother,param);
69        wave(jj,:) = ifft(X.*daughter); % wavelet transform
70     end
71
72     % Get rid of padding
73     wave = wave(:,1:N1);
74
75     % Get more parameters for the chosen wavelet
76     [consts]=w_parameters(mother,param);
77
78     % Adimensional periods
79     period = consts(6)*a;
80
81     % Find indexes corresponding to adimensional period bounds
82     eps=10^(−1);
83     for jj=1:J+1
84        for i=1:2
```

```
85        if abs(period(jj)−incl_adim_per(i))<eps
86            ia(i)=jj;
87        end
88     end
89 end
90 % Reconstruction (inverse transform)
91 Cdelta=consts(2);
92 psizerozero=consts(5);
93 coeffi=dj/(Cdelta*psizerozero);
94 summnd=NaN(J+1,N1);
95 for jj=ia(1):ia(2)
96    summnd(jj,:)=real(wave(jj,:))/sqrt(a(jj));
97 end
98 rec=nansum(summnd);
99 rec=coeffi*rec';
100 return
```

In order to show that our results make sense, we compare the reconstructed ENSO oscillation from Nino3 data with the behavior of the Nino3.4 SST index and Southern Oscillation Index (SOI). The Nino3.4 index is one of several ENSO indicators based on SSTs. Nino3.4 is the average SST anomaly in the region from 5°N to 5°S, and from 170°W to 120°W. This region has large SST variability on El Niño time scales, and is close to the region where changes in local SST are important for shifting the large region of rainfall typically located in the far western Pacific Ocean. Recall that the region Nino3 is somewhat smaller, being the area averaged SST from 5°S–5°N and 150°W–90°W. Nino3.4 neglects explicit atmospheric processes; by adding the SOI, which expresses the pressure difference between Tahiti and Darwin, these processes are more directly included. El Niño/La Niña events can be identified using those

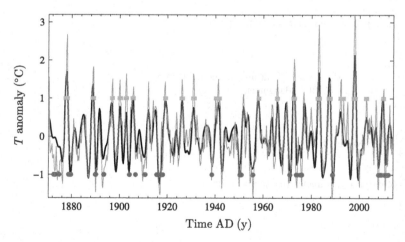

Fig. 16.4 Comparison between the reconstructed ENSO oscillation, based on CWT analysis of Nino3 data and inclusion of periods of 2–8 years, and the behavior of the Nino3.4 SST index coupled with Southern Oscillation Index(SOI). The months classified as as El Niño/La Niña months because both the Nino3.4SST and the SOI exceeded the 20th percentile are indicated by *squares* and *circles*, respectively

months that correspond to both high Nino3.4 valuess and low SOI values, or vice versa (Smith and Sardeshmukh 2000). The months where both the Nino3.4 index and the SOI exceeded the 20th percentile (± 1.28) are classified as as El Niño/La Niña months. The list of these months can be dowloaded from http://www.esrl.noaa. gov/psd/people/cathy.smith/best/table.txt. Figure 16.4 shows the satisfactory result of this comparison.

Finally, we may mention that the CWT can also be implemented using the Matlab function cwt. This function performs the transform in the time domain via convolution. The syntax is

```
coefs = cwt(x,a,wname)
```

where x is the signal vector, the vector a expresses scale factors, and wname is the wavelet name. The output coefs is an la-by-lx matrix, where la is the length of vector a and lx is the length of the vector x. The output coefs is a real or complex matrix, depending on the wavelet type. Available wavelet types have been described in Chap. 14. Alternatively, Matlab also offers a function to compute the CWT in the frequency domain, cwtft, which uses an FFT algorithm, similarly to our previous approach. The analytic Morlet wavelet is available as 'morl' (default). The corresponding inverse function icwtft allows signal reconstruction. For the cwtft and icwtft syntax the reader is referred to the Wavelet Toolbox User's Guide (The Math-Works 2014), since notions on structure and cell arrays in Matlab would be needed to describe it, which are beyond the scope of these exercises.

16.8   DWT

16.8.1   Signal Decomposition

The main Matlab functions for DWT calculation (see Chap. 14 for the theory) are the following.

- dwtmode:
  this function sets the signal extension mode for the DWT. Extension modes represent different ways of handling the problem of border distortion in signal analysis.

```
st = dwtmode
```

displays the current mode and returns it in st;

```
dwtmode('mode')
```

sets the DWT extension mode according to the value of 'mode':

- 'sym' or 'symh': symmetric-padding (half-point symmetry): boundary value symmetric replication—default mode;
- 'symw': symmetric-padding (whole-point symmetry): boundary value symmetric replication;

– 'asym' or 'asymh': antisymmetric-padding (half-point): boundary value antisymmetric replication;
– 'asymw': antisymmetric-padding (whole-point): boundary value antisymmetric
  replication;
– 'zpd': zero-padding;
– 'spd' or 'sp1': smooth-padding of order 1 (first derivative interpolation at the
  edges);
– 'sp0': smooth-padding of order 0 (constant extension at the edges);
– 'ppd': periodic-padding (periodic extension at the edges);
– 'per': periodization.

Details on the rationale of the mentioned schemes can be found in Strang and
Nguyen (1996).

● wmaxlev:
  this function returns the maximum wavelet-decomposition level for signal x, given
  the adopted wavelet (wname). Note that normally, a smaller value is adopted in
  applications.

```
lev_max = wmaxlev(x,wname)
```

● dwt:
  this function performs a single-level signal decomposition (Daubechies 1992; Mallat 1989; Meyer 1993). The command

```
[cA,cD] = dwt(x,wname)
```

computes the approximation coefficients vector cA and detail coefficients vector
cD, obtained by wavelet decomposition. Starting from a signal x, in a single-level
decomposition two sequences of coefficients are computed: approximation coefficients (cA at level 1), and detail coefficients (cD at level 1). Available orthogonal or
biorthogonal wavelet names, indicated by wname, are Daubechies wavelets ('db1'
or 'haar', 'db2', …,'db10', …, 'db45'), coiflets ('coif1', …, 'coif5'), symlets
('sym2', ldots , 'sym8', …, 'sym45'), biorthogonal wavlets ('bior1.1', 'bior1.3',
'bior1.5', 'bior2.2', 'bior2.4', 'bior2.6', 'bior2.8', 'bior3.1', 'bior3.3', 'bior3.5',
'bior3.7', 'bior3.9', 'bior4.4', 'bior5.5', 'bior6.8'), etc.

● idwt:
  this function performs a single-level signal reconstruction. The command

```
x = idwt(cA,cD,wname)
```

returns the single-level-reconstructed vector x based on approximation and detail
coefficients vectors cA and cD, and using the wavelet wname.

● wavedec:
  this function performs a multi-level wavelet decomposition (Daubechies 1992;
  Mallat 1989; Meyer 1993). *The function wavedec supports only orthogonal
  biorthogonal wavelets*. The command

```
[C,L] = wavedec(x,lev,wname)
```

returns the wavelet decomposition of the signal x at level lev, using the wavelet wname. The parameter lev must be a strictly positive integer. The output decomposition structure contains the wavelet decomposition vector C and the bookkeeping vector L: all the coefficients of a decomposition at level lev (that is, the lev-th approximation coefficients and lev levels of detail coefficients) are returned concatenated into one vector, C, while vector L gives the lengths of each component (see Sect. 14.5.2). To extract approximation and detail coefficients from C, the functions appcoef and detcoef are available. To reconstruct approximation and detail signals on the basis of the signal decomposition, use the function wrcoef.

- waverec:
  this function performs a multi-level wavelet reconstruction using a specific wavelet (wname). *The function waverec supports only orthogonal or biorthogonal wavelets.* The command

```
x = waverec(C,L,wname)
```

reconstructs the signal x based on the multi-level wavelet decomposition structure [C,L] and wavelet wname.

- appcoef:
  this function extracts the approximation coefficients from a signal decomposition. The command

```
cA = appcoef(C,L,wname,j)
```

returns the approximation coefficients at level j using the wavelet decomposition structure [C,L] provided by wavedec. The value of j must be an integer such that $0 \leq j \leq \text{length}(L) - 2$; the command

```
cA = appcoef(C,L,wname)
```

extracts the approximation coefficients at the last level, i.e., $\text{length}(L) - 2$.

- detcoef:
  this function extracts the detail coefficients of a signal, and works like appcoef. For instance, the command

```
cD = detcoef(C,L,j)
```

extracts the detail coefficients at level j from the wavelet decomposition structure [C,L].

- wrcoef:
  this function reconstructs approximations or details, given a wavelet decomposition structure (C and L) and a specified wavelet.

```
y = wrcoef('type',C,L,wname,j)
```

Argument 'type' determines whether level-j approximation ('type' = 'a') or detail ('type' = 'd') signals are reconstructed. When 'type' = 'a', j can be 0; otherwise, a strictly positive integer j is required, with $j \leq \text{length}(L) - 2$. The command

```
y = wrcoef('type',C,L,wname)
```

reconstructs y for the maximum level, lev = length(L) − 2.
- wenergy:
for a wavelet decomposition [C,L] produced by wavedec, the command

```
[Ea,Ed] = wenergy(C,L)
```

returns Ea, which is the percentage of signal energy corresponding to the approximation, and Ed, which is the vector containing the percentages of energy corresponding to all the details.
- wfilters:
this function gives the impulse responses of the four filters (Sect. 14.3) associated with the *orthogonal or biorthogonal wavelet* given in the string wname (Mallat 1989; Daubechies 1992). The command

```
[Lo_D,Hi_D,Lo_R,Hi_R] = wfilters(wname)
```

gives in output

- Lo_D, the decomposition low-pass filter,
- Hi_D, the decomposition high-pass filter,
- Lo_R, the reconstruction low-pass filter,
- Hi_R, the reconstruction high-pass filter;

the command

```
[F1,F2] = wfilters(wname,'type')
```

returns

- Lo_D and Hi_D (decomposition filters) if 'type' = 'd',
- Lo_R and Hi_R (reconstruction filters) if 'type' = 'r',
- Lo_D and Lo_R (lowpass filters) if 'type' = 'l',
- Hi_D and Hi_R (highpass filters) if 'type' = 'h'.

- wvarchg:
This function finds variance change points in a signal x, which must be zero-mean (Lavielle 1999). The command

```
[pts_opt,Kopt,t_est] = wvarchg(x,K,d)
```

computes the estimated change points of the variance in the signal x. The integer number K specifies the maximum number of expected change points; default is 6. The integer number d sets the minimum delay in adimensional discrete-time units between two change points; default is 10. In output, the integer Kopt is the number of detected change points, with $0 \leq Kopt \leq K$; the vector pts_opt contains the discrete times of the detected change points. The variable t_est is such that for $1 \leq k \leq K$, t_est(k + 1,1:k) contains the k time values of the variance change points; therefore

- if Kopt > 0, pts_opt = t_est(Kopt + 1,1:Kopt);
- if Kopt = 0, pts_opt = [] (empty vector).

K and d must be integers such that $1 < K \ll$ length(x) and $1 \leq d \ll$ length(x).

- wnoise:
this function generates noisy wavelet test data, according to the following syntax:

```
1   x = wnoise(fun,nu)
```

The function wnoise returns values of the test signal identified by fun, over 2^{nu} equally-spaced time points in [0,1]. The six available fun types (see Donoho and Johnstone 1994a, 1995) are:

- fun = 1 or 'blocks',
- fun = 2 or 'bumps',
- fun = 3 or 'heavy sine',
- fun = 4 or 'doppler',
- fun = 5 or 'quadchirp',
- fun = 6 or 'mishmash'.

```
1   [x,xn] = wnoise(fun,nu,sqrt_snr)
```

returns a test vector x as above, rescaled such that std(x) = sqrt_snr. The returned vector xn contains the same test vector corrupted by additive Gaussian white noise with zero mean and unit variance. Then, xn has a nominal signal-to-noise ratio of $SNR = (sqrt\_snr)^2$. This can be checked centering both signals thus obtaining two vectors, xc and xnc respectively, and then computing a SNR_check value as

```
1   SNR_check = sum(xc.^2)./sum((xnc−xc).^2)
```

This should be equal to the nominal SNR. Note that by varying sqrt_snr we vary the standard deviation of the reference signal and not the standard deviation of the noise, which remains 1 in all cases.

```
1   [x,xn] = wnoise(fun,nu,sqrt_snr,init)
```

returns the previously mentioned vectors x and xn, but the generator seed for noise is set to init. This option serves for obtaining reproducible results.
Now we will experiment with these functions.

1. First we learn how to compute the impulse responses (and the frequency responses) of the four filters associated with each wavelet. The following script (filter_set) does this: run it for different orthogonal wavelets to check that in any case the four filters lengths are all equal, and are always even. Try also biorthogonal wavelets.

```
1   % filter_set
2   % Compute filter set for given wavelet.
3
4   prompt='wname = (e.g."db5")';
5   wname= input(prompt);
6
7   % Compute the impulse responses of the four filters
```

```
 8    % associated with the wavelet
 9
10    [Lo_D,Hi_D,Lo_R,Hi_R] = wfilters(wname);
11    M2=length(Lo_D);% filter length
12
13    figure(4000)
14    subplot(2,2,1)
15    stem(0:M2−1,Lo_D)
16    set(gca,'XLim',[−1 M2])
17    set(gca,'YLim',[−1 1])
18    hlabelx=get(gca,'Xlabel');
19    set(hlabelx,'String','$n$','Interpreter','Latex','FontSize',16)
20    hlabely=get(gca,'Ylabel');
21    set(hlabely,'String','$h_{\mathrm{LD}}$','Interpreter','Latex','FontSize',16)
22    text(1,0.76,wname,'FontSize',12)
23    title('Decomposition low−pass filter');
24    subplot(2,2,2)
25    stem(0:M2−1,Hi_D)
26    set(gca,'XLim',[−1 M2])
27    set(gca,'YLim',[−1 1])
28    hlabelx=get(gca,'Xlabel');
29    set(hlabelx,'String','$n$','Interpreter','Latex','FontSize',16)
30    hlabely=get(gca,'Ylabel');
31    set(hlabely,'String','$h_{\mathrm{HD}}$','Interpreter','Latex','FontSize',16)
32    title('Decomposition high−pass filter');
33    subplot(2,2,3)
34    stem(0:M2−1,Lo_R)
35    set(gca,'XLim',[−1 M2])
36    set(gca,'YLim',[−1 1])
37    hlabelx=get(gca,'Xlabel');
38    set(hlabelx,'String','$n$','Interpreter','Latex','FontSize',16)
39    hlabely=get(gca,'Ylabel');
40    set(hlabely,'String','$h_{\mathrm{LR}}$','Interpreter','Latex','FontSize',16)
41    title('Reconstruction low−pass filter');
42    subplot(2,2,4)
43    stem(0:M2−1,Hi_R)
44    set(gca,'XLim',[−1 M2])
45    set(gca,'YLim',[−1 1])
46    hlabelx=get(gca,'Xlabel');
47    set(hlabelx,'String','$n$','Interpreter','Latex','FontSize',16)
48    hlabely=get(gca,'Ylabel');
49    set(hlabely,'String','$h_{\mathrm{HR}}$','Interpreter','Latex','FontSize',16)
50    title('Reconstruction high−pass filter');
51
52    % Compute the frequency responses and plot
53    figure(4001)
54    N_f=1024;
55    [H,w]=freqz(Lo_D,1,N_f);
56    plot(w/pi,20*log10(abs(H)),'LineWidth',4)
57    hold on
58    [H,w]=freqz(Hi_D,1,N_f);
59    plot(w/pi,20*log10(abs(H)),'r','LineWidth',4)
60    [H,w]=freqz(Lo_R,1,N_f);
```

```
61  plot(w/pi,20*log10(abs(H)),'g')
62  [H,w]=freqz(Hi_R,1,N_f);
63  plot(w/pi,20*log10(abs(H)),'k')
64  text(0.45,-20,wname,'FontSize',12)
65  legend('LD','HD','LR','HR','Location','South')
66  hold off
67  set(gca,'YLim',[-60 5])
68  hlabelx=get(gca,'Xlabel');
69  set(hlabelx,'String','$\omega/\pi$','Interpreter','Latex','FontSize',16)
70  hlabely=get(gca,'Ylabel');
71  set(hlabely,'String','$20\log10|H|$','Interpreter','Latex','FontSize',16)
```

2. Next we try a single-level decomposition of three different signals:

- the first signal is a stationary signal (a noisy sinusoid). We generate it, then perform a level-1 decomposition using the function dwt and the db8 wavelet and plot the signal and its DWT coefficients;

- the second signal is a sinusoid with a discontinuity introduced on purpose. It is a 3-sample discontinuity starting at sample 302. The example illustrates the possibility of discontinuity detection by DWT: the discontinuity is detected in the 1st level detail coefficients at sample 151, because of the downsampling by a factor of 2 associated with each DWT level;

- the third signal is a non-stationary signal: a sinusoid with changing frequency. The detail coefficients reveal the frequency transitions in the signal;

- the fourth signal is made of noisy blocks with two noise-variance change points. It is generated using wnoise with fun = 1 ('blocks'). A single-level DWT decomposition followed by the reconstruction of the detail D_1 allows us to use the function wvarchg to estimate with good accuracy the position of the variance change points. Indeed, the reconstructed detail at level 1 is mainly signal-free, and captures the features of the noise from a change-points-detection viewpoint, if the interesting part of the signal has a sparse wavelet representation. In the test signal, two change points and three intervals with different variances are present. Since the variances of the noise in these three intervals are very different among themselves, the optimization program detects easily the variance structure, and the estimated change points are fairly close to the true change points. To help wvarchg locating the change points more accurately, we then further replace 2 % of the biggest detail values by the mean, in order to remove almost all of the signal. The estimates of the change points actually improve.

The script (singlelev_decomp) for this exercise is reported below. Note that we leave the extension mode to default ('sym', i.e., symmetric-padding with half-point symmetry). We do not worry too much about the extension mode, since the latter is expected to affect mainly the edges of the processed signal, and also because in our introductory approach to the DWT we did not discuss extension modes in depth.

```matlab
1   % singlelev_decomp
2   % Single-level decompositions of four different signals
3
4   clear
5
6   %1
7   % Single level decomposition
8   % of a stationary signal (noisy sinusoid)
9
10  % Generate sequence
11  f=100;%Hz
12  T=1/f;%s -- 10ms
13  A=4;
14  dur=40;%ms
15  f_s=10000;%Hz
16  T_s=1/f_s;%s
17  T_sm=T_s*1000;%ms
18  N=dur/T_sm;
19  t=0:T_s:(N-1)*T_s;%s
20  x=A*sin(2*pi*f*t);%sinusoid
21  xn=x+0.5*randn(size(x)); % noisy sinusoid
22
23  % Single level decomposition with Daubechies 8 wavelet
24  wname='db8';
25  [cA,cD]=dwt(xn,wname);
26
27  % Plot signal and DWT coefficients
28  figure(4100)
29  subplot(3,1,1)
30  plot(t*1000,xn,'LineWidth',1.5)% express time in ms
31  set(gca,'XLim',[0 40])
32  set(gca,'YLim',[-7.5 7.5])
33  hlabelx=get(gca,'Xlabel');
34  set(hlabelx,'String','$t$ (ms)','Interpreter','Latex','
        FontSize',16)
35  hlabely=get(gca,'Ylabel');
36  set(hlabely,'String','$x$','Interpreter','Latex','FontSize'
        ,16)
37  title('Noisy Sinusoid')
38  subplot(3,1,2)
39  plot(cA,'LineWidth',1.5)
40  set(gca,'XLim',[0 207])
41  set(gca,'YLim',[-7.5 7.5])
42  hlabely=get(gca,'Ylabel');
43  set(hlabely,'String','$C_a$','Interpreter','Latex','FontSize'
        ,16)
44  title('First Level Approx. Coeffs. - db8')
45  subplot(3,1,3)
46  plot(cD,'LineWidth',1.5)
47  set(gca,'XLim',[0 207])
48  hlabelx=get(gca,'Xlabel');
49  set(hlabelx,'String','$k$','Interpreter','Latex','FontSize'
        ,16)
```

```
50  hlabely=get(gca,'Ylabel');
51  set(hlabely,'String','$C_d$','Interpreter','Latex','FontSize'
        ,16)
52  title('First Level Detail Coeffs. - db8')
53
54  % 2
55  % Single level decomposition of a stationary signal
56  % (a sinusoid with a discontinuity):
57  % discontinuity detection
58
59  % Take clean sinusoidal signal generated above
60  % and introduce a 3-sample discontinuity starting at sample
        302
61  x(302:305)=.25;
62
63  % Single level decomposition with Daubechies 8 wavelet
64  wname='db8';
65  [cA,cD]=dwt(x,wname);
66
67  % Plot signal and DWT coefficients.
68  figure(4200)
69  subplot(3,1,1)
70  plot(t*1000,x,'LineWidth',1.5)
71  set(gca,'XLim',[0 40])
72  set(gca,'YLim',[-7.5 7.5])
73  hlabelx=get(gca,'Xlabel');
74  set(hlabelx,'String','$t$ (ms)','Interpreter','Latex','
        FontSize',16)
75  hlabely=get(gca,'Ylabel');
76  set(hlabely,'String','$x$','Interpreter','Latex','FontSize'
        ,16)
77  title('Sinusoid with a discontinuity')
78  subplot(3,1,2)
79  plot(cA,'LineWidth',1.5)
80  set(gca,'XLim',[0 207])
81  set(gca,'YLim',[-7.5 7.5])
82  hlabely=get(gca,'Ylabel');
83  set(hlabely,'String','$C_a$','Interpreter','Latex','FontSize'
        ,16)
84  title('First Level Approx. Coeffs. - db8')
85  subplot(3,1,3)
86  plot(cD,'LineWidth',1.5)
87  set(gca,'XLim',[0 207])
88  set(gca,'YLim',[-0.4 0.2])
89  hlabelx=get(gca,'Xlabel');
90  set(hlabelx,'String','$k$','Interpreter','Latex','FontSize'
        ,16)
91  hlabely=get(gca,'Ylabel');
92  set(hlabely,'String','$C_d$','Interpreter','Latex','FontSize'
        ,16)
93  title('First Level Detail Coeffs. - db8')
94
95  % 3
```

```
96   % Single level Decomposition of a non-stationary signal
97   % (a sinusoid with changing frequency).
98
99   % Generate sequence
100  dur=100;%ms
101  f_s =2500;%Hz
102  T_s =1/f_s ;%s
103  T_sm=T_s*1000;%ms
104  N=dur/T_sm;
105  %
106  f =50;%Hz
107  T=1/f ;%s --- 20ms
108  A=0.8;
109  t =0:T_s:(N-1)*T_s ;%s
110  x1=A*sin(2*pi*f*t);% 5 cycles
111  %
112  f =100;%Hz
113  T=1/f ;%s --- 10ms
114  A=0.18;
115  x2=A*sin(2*pi*f*t);% 10 cycles
116  %
117  f =200;%Hz
118  T=1/f ;%s --- 5ms
119  A=1;
120  x3=A*sin(2*pi*f*t);% 20 cycles
121  %
122  y=cat(2,x1,x2,x3);  % Concatenate the signals
123  %  Concatenate the time vectors
124  t_y =[t  t+dur/1000  t+2*dur/1000];
125
126  % Single level decomposition with Daubechies 8 wavelet
127  wname='db8 ';
128  [cAa,cDa]=dwt(y,wname);
129
130  % Plot signal and DWT coefficients.
131  figure(4300)
132  subplot(3,1,1)
133  plot(t_y*1000,y,'LineWidth',1.5)
134  set(gca,'XLim',[0 300])
135  set(gca,'YLim',[-1.7 1.7])
136  hlabelx=get(gca,'Xlabel');
137  set(hlabelx,'String','$t$ (ms)','Interpreter','Latex','FontSize',16)
138  hlabely=get(gca,'Ylabel');
139  set(hlabely,'String','$x$','Interpreter','Latex','FontSize',16)
140  title('Sinusoid with Changing Frequency')
141  subplot(3,1,2)
142  plot(cAa,'LineWidth',1.5)
143  set(gca,'XLim',[0 382])
144  set(gca,'YLim',[-1.7 1.7])
145  hlabely=get(gca,'Ylabel');
```

```
146   set(hlabely, 'String', '$C_a$', 'Interpreter', 'Latex', 'FontSize'
             ,16)
147   title('First Level Approx. Coeffs. -  db8')
148   subplot(3,1,3)
149   plot(cDa, 'LineWidth',1.5)
150   set(gca, 'XLim',[0  382])
151   set(gca, 'XMinorTick', 'on')
152   set(gca, 'YLim',[-0.03  0.03])
153   hlabelx=get(gca, 'Xlabel');
154   set(hlabelx, 'String', '$k$', 'Interpreter', 'Latex', 'FontSize'
             ,16)
155   hlabely=get(gca, 'Ylabel');
156   set(hlabely, 'String', '$C_d$', 'Interpreter', 'Latex', 'FontSize'
             ,16)
157   title('First Level Detail Coeffs. - db8')
158
159   % 4
160   % Recover variance change points from a signal
161   % (noisy blocks with two noise-variance change points).
162   % by single-level DWT decomposition
163   % and use of wvarchg on detail at level 1.
164
165   % Generate block signal
166   fun='blocks'
167   nu=10;%N=2^nu
168   x = wnoise(fun,nu); %Block signal
169
170   % Make signal noisy, with noise-variance change points
171   % at discrete times 180 and 600
172   rng default;
173   bb = 1.5*randn(1,length(x));
174   cp1 = 180; cp2 = 600;
175   x = x + [bb(1:cp1),bb(cp1+1:cp2)/4,bb(cp2+1:end)];
176   N=length(x);
177
178   % Single level decomposition with Daubechies 3 wavelet
179   wname='db3'
180   [cA,cD]=dwt(x,wname);
181
182   % Reconstruct detail at level 1
183   D1 = idwt([],cD,wname);
184
185   % Use the wvarchg function to estimate the change points
186   % with the following parameters:
187   % minimum delay between two change points = 10 (default,
             omitted);
188   % maximum number of change points = 5 (default is 6).
189
190   % The input signal should be zero mean, so center D1.
191   D1_a=D1-mean(D1);
192
193   [pts_opt,Kopt,t_est] = wvarchg(D1_a,5);
194   str=sprintf('The estimated change points are %d and %d\n',
             pts_opt)
```

```matlab
195
196   % Plots results (signal and detail).
197   figure(4400)
198   subplot(3,1,1)
199   plot([cp1 cp1],1.05*[min(x) max(x)],'r','LineWidth',1.5)
200   hold on
201   plot([cp2 cp2],1.05*[min(x) max(x)],'r','LineWidth',1.5)
202   plot(x)
203   hold off
204   set(gca,'XLim',[0 N])
205   set(gca,'XMinorTick','on')
206   set(gca,'YLim',1.05*[min(x) max(x)])
207   hlabely=get(gca,'Ylabel');
208   set(hlabely,'String','$x[n]$','Interpreter','Latex','FontSize'
            ,16)
209   title('Noisy Block Signal with Noise-Variance Change Points')
210   subplot(3,1,2)
211   plot([cp1 cp1],1.05*[min(D1) max(D1)],'r','LineWidth',1.5)
212   hold on
213   plot([cp2 cp2],1.05*[min(D1) max(D1)],'r','LineWidth',1.5)
214   plot([pts_opt(1) pts_opt(1)],1.05*[min(D1) max(D1)],'g—','
            LineWidth',1.5)
215   plot([pts_opt(2) pts_opt(2)],1.05*[min(D1) max(D1)],'g—','
            LineWidth',1.5)
216   plot(D1)
217   hold off
218   set(gca,'XLim',[0 N])
219   set(gca,'XMinorTick','on')
220   set(gca,'YLim',1.05*[min(D1) max(D1)])
221   hlabely=get(gca,'Ylabel');
222   set(hlabely,'String','$D_1$','Interpreter','Latex','FontSize'
            ,16)
223   text(300,-3,wname,'FontSize',12)
224
225   % To help wvarchg locating the change points more accurately,
226   % remove almost all the signal by replacing 2% of biggest
            values by the mean.
227   y = sort(abs(D1));
228   v2p100 = y(fix(length(y)*0.98));
229   ind = find(abs(D1)>v2p100); % locate 2% of biggest values
230   D1_b=D1;
231   D1_b(ind) = mean(D1); % replace them by the mean
232   D1_b=D1_b-mean(D1_b); % remove mean
233
234   % Use again the wvarchg function to estimate the change points
235   [pts_opt,Kopt,t_est] = wvarchg(D1_b,5);
236   str=sprintf('The estimated change points are %d and %d\n',
            pts_opt)
237
238   % Plot result
239   subplot(3,1,3)
240   plot([cp1 cp1],1.05*[min(D1_b) max(D1_b)],'r','LineWidth',1.5)
```

```
241  hold on
242  plot([cp2 cp2],1.05*[min(D1_b) max(D1_b)],'r','LineWidth',1.5)
243  plot([pts_opt(1) pts_opt(1)],1.05*[min(D1_b) max(D1_b)],'g—',
         'LineWidth',1.5)
244  plot([pts_opt(2) pts_opt(2)],1.05*[min(D1_b) max(D1_b)],'g—',
         'LineWidth',1.5)
245  plot(D1_b)
246  hold off
247  set(gca,'XLim',[0 N])
248  set(gca,'XMinorTick','on')
249  set(gca,'YLim',1.05*[min(D1_b) max(D1_b)])
250  hlabelx=get(gca,'Xlabel');
251  set(hlabelx,'String','$n$','Interpreter','Latex','FontSize'
         ,16)
252  hlabely=get(gca,'Ylabel');
253  set(hlabely,'String','Processed $D_1$','Interpreter','Latex','
         FontSize',16)
```

3. We now load a test signal included in the Matlab distribution (sumsin.mat), containing three sinusoids with different frequencies—two low-frequency components and one high-frequency component. We perform signal decomposition at level 3 using the db1 wavelet. We plot the original signal and the vector C of the DWT coefficients (including the coefficients for the approximation at level 3 and those for the details at levels 1–3) to see how the wavelet decomposition structure looks like, i.e., how these coefficients are stored in memory. The script for this exercise (multilev_decomp_coefstruct) is given below.

```
1   % multilevel_decomp_coefstruct
2   % Check wavelet decomposition structure.
3
4   clear
5
6   % Load signal.
7   load sumsin % test signal with three sinusoids
8   x = sumsin;
9   N=length(x)
10
11  % Perform decomposition at level 3 of s using db1.
12  lev=3
13  wname='db1'
14  [C,L] = wavedec(x,lev,wname);
15  N_DWT=length(C)
16
17  % Plot original signal,
18  % coeffs. for approx. at level 3 and coeffs. for details at levels 1--3
19  % to observe wavelet decomposition structure
20
21  figure(4500)
22  subplot(2,1,1)
23  plot(x)
24  set(gca,'YLim',[-5 5])
25  set(gca,'YTick',[-4:2:4])
26  hlabelx=get(gca,'Xlabel');
```

```
27   set(hlabelx,'String','$n$','Interpreter','Latex','FontSize',16)
28   hlabely=get(gca,'Ylabel');
29   set(hlabely,'String','$x[n]$', 'Interpreter','Latex','FontSize',16)
30   title('Original signal − Three Sinusoids')
31   subplot(2,1,2)
32   plot(C)
33   set(gca,'YLim',[−5 5])
34   set(gca,'YTick',[−4:2:4])
35   hlabelx=get(gca,'Xlabel');
36   set(hlabelx,'String','$k$','Interpreter','Latex','FontSize',16)
37   hlabely=get(gca,'Ylabel');
38   set(hlabely,'String','$C[k]$','Interpreter','Latex','FontSize',16)
39   title('Wavelet decomposition structure − db1, level 3')
```

4. Finally, as the last DWT-decomposition example, we perform a multi-level analysis of a noisy sinusoid up to level 4, using the db8 wavelet. We see how to reconstruct the approximations and the details at various levels. Plotting the signal and its various approximations we can observe that the level-4 approximation constitutes a sort of crude de-noising of the signal. We then reconstruct the original signal from the level 4 decomposition, and plot the original and reconstructed signals together. This test visually proves that a perfect reconstruction has been achieved, but to quantify the quality of the reconstruction, we also compute the maximum absolute deviation between the reconstructed and the original vector, and the root of the sum of squares of deviations (norm of the difference between the two vectors). We also calculate the percentage of the signal energy carried by the approximation and by each detail; the sum gives 100 %. For this exercise, run the script (multilev_decomp) reported below.

```
1    % multilev_decomp
2    % Multi−level wavelet decomposition
3
4    clear
5
6    % Generate signal (noisy sinusoid)
7    f=100;%Hz
8    T=1/f;%s −− 10ms
9    A=4;
10   dur=40;%ms
11   f_s=10000;%Hz
12   T_s=1/f_s;%s
13   T_sm=T_s*1000;%ms
14   N=dur/T_sm
15   t=0:T_s:(N−1)*T_s;%s
16   x=A*sin(2*pi*f*t);
17   xn=x+0.5*randn(size(x));
18
19   % Do a multilevel analysis to level 4 using Daubechies−8 wavelet
20   lev=4
```

```
21   wname='db8'
22   [C,L] = wavedec(xn,lev,wname);
23
24   % Reconstruct the approximations at various levels
25   A1 = wrcoef('a',C,L,wname,1);
26   A2 = wrcoef('a',C,L,wname,2);
27   A3 = wrcoef('a',C,L,wname,3);
28   A4 = wrcoef('a',C,L,wname,4);
29
30   % Reconstruct the details at various levels
31   D1 = wrcoef('d',C,L,wname,1);
32   D2 = wrcoef('d',C,L,wname,2);
33   D3 = wrcoef('d',C,L,wname,3);
34   D4 = wrcoef('d',C,L,wname,4);
35
36   % Plot signal and approximations
37   figure(4600)
38   subplot(5,1,1)
39   plot(t*1000,xn,'LineWidth',1.5)
40   set(gca,'XLim',[0 40])
41   set(gca,'YLim',[−6 6])
42   hlabely=get(gca,'Ylabel');
43   set(hlabely,'String','$x$','Interpreter','Latex','FontSize',16)
44   title('Noisy Sinusoid and Approximations (db8)')
45   subplot(5,1,2)
46   plot(t*1000,A1,'LineWidth',1.5)
47   set(gca,'XLim',[0 40])
48   set(gca,'YLim',[−6 6])
49   hlabely=get(gca,'Ylabel');
50   set(hlabely,'String','$A_1$','Interpreter','Latex','FontSize',16)
51   subplot(5,1,3)
52   plot(t*1000,A2,'LineWidth',1.5)
53   set(gca,'XLim',[0 40])
54   set(gca,'YLim',[−6 6])
55   hlabely=get(gca,'Ylabel');
56   set(hlabely,'String','$A_2$','Interpreter','Latex','FontSize',16)
57   subplot(5,1,4)
58   plot(t*1000,A3,'LineWidth',1.5)
59   set(gca,'XLim',[0 40])
60   set(gca,'YLim',[−6 6])
61   hlabely=get(gca,'Ylabel');
62   set(hlabely,'String','$A_3$','Interpreter','Latex','FontSize',16)
63   subplot(5,1,5)
64   plot(t*1000,A4,'LineWidth',1.5)
65   set(gca,'XLim',[0 40])
```

```
66   set(gca,'YLim',[−6 6])
67   hlabelx=get(gca,'Xlabel');
68   set(hlabelx,'String','$t$ (ms)','Interpreter','Latex','FontSize',16)
69   hlabely=get(gca,'Ylabel');
70   set(hlabely,'String','$A_4$','Interpreter','Latex','FontSize',16)
71
72   % Display the results of the multilevel decomposition at level 4
73   figure(4700)
74   subplot(5,1,1)
75   plot(t*1000,A4,'LineWidth',1.5);
76   set(gca,'XLim',[0 40])
77   set(gca,'YLim',[−6 6])
78   hlabely=get(gca,'Ylabel');
79   set(hlabely,'String','$A_4$','Interpreter','Latex','FontSize',16)
80   title('Decomposition of Noisy Sinusoid at level 4, using db8')
81   subplot(5,1,2)
82   plot(t*1000,D1);
83   set(gca,'XLim',[0 40])
84   hlabely=get(gca,'Ylabel');
85   set(hlabely,'String','$D_1$','Interpreter','Latex','FontSize',16)
86   subplot(5,1,3)
87   plot(t*1000,D2,'LineWidth',1.5);
88   set(gca,'XLim',[0 40])
89   hlabely=get(gca,'Ylabel');
90   set(hlabely,'String','$D_2$', 'Interpreter','Latex','FontSize',16)
91   subplot(5,1,4)
92   plot(t*1000,D3,'LineWidth',1.5);
93   hlabely=get(gca,'Ylabel');
94   set(hlabely,'String','$D_3$','Interpreter','Latex','FontSize',16)
95   subplot(5,1,5)
96   plot(t*1000,D3,'LineWidth',1.5);
97   set(gca,'XLim',[0 40])
98   hlabelx=get(gca,'Xlabel');
99   set(hlabelx,'String','$t$ (ms)','Interpreter','Latex','FontSize',16)
100  hlabely=get(gca,'Ylabel');
101  set(hlabely,'String','$D_4$','Interpreter','Latex','FontSize',16)
102
103  % Reconstruct the original signal from the Level 4 decomposition.
104  A0 = waverec(C,L,wname);
105  N_rec=length(A0)
106
107  % Plot reconstructed signal
108  figure(4800)
109  plot(t*1000,xn,'LineWidth',3)
110  hold on
```

```
111    plot(t*1000,A0,'g','LineWidth',1)
112    hold off
113    set(gca,'XLim',[0 40])
114    hlabelx=get(gca,'Xlabel');
115    set(hlabelx,'String','$t$ (ms)','Interpreter','Latex','FontSize',16)
116    hlabely=get(gca,'Ylabel');
117    set(hlabely,'String','$x$','Interpreter','Latex','FontSize',16)
118    legend('Original','Reconstructed')
119    title('Noisy Sinusoid − Reconstruction from level 4 decomposition using
           db8')
120
121    % Check for perfect reconstruction.
122    errs=xn−A0;  % deviations
123    nor = norm(errs) % norm of deviations
124    errmax = max(abs(errs)) % maximum absolute deviation
125
126    [Ea,Ed] = wenergy(C,L)
127    total_energy=Ea+sum(Ed)
```

Note how significant de-noising occurs with the level-4 approximation. Using wavelets to remove noise from a signal requires identifying which component or components contain the noise, and then reconstructing the signal without those components. In this example, we note that successive approximations become less and less noisy as more and more high-frequency information is filtered out of the signal. The level-4 approximation is quite clean, as can be seen from the comparison between it and the original signal. Of course, in discarding all the high-frequency information, we have also lost many of the original signal's sharpest features. Optimal de-noising requires a more subtle approach, namely, thresholding (Chap. 15).

16.8.2 De-noising and Compression

Here we will examine the most important de-noising- and compression-oriented Matlab functions. For the related theoretical treatment, see Chap. 15. These functions mainly rely on the use of thresholds of the Donoho-Johstone family. The fundamental functions are wdencmp and wden:

- wden performs automatic de-noising, and cannot be used for compression;
- wdencmp is both for de-noising and compression. With respect to wden, wdencmp allows more flexibility, and you can implement your own de-noising or compression strategy.

Automatic De-noising Using wden

The function wden performs an automatic de-noising process of a signal using
wavelets (Donoho 1993, 1995; Donoho and Johnstone 1994a; Donoho et al. 1995;
Antoniadis and Oppenheim 1995). In input to wden, we can provide the signal, or
directly its decomposition structure obtained by a previous call to wavedec. Thus the
syntax can be

```
[xd,Cxd,Lxd] = wden(x,tptr,sorh,scal,lev,wname)
```

or

```
[xd,Cxd,Lxd] = wden(C,L,tptr,sorh,scal,lev,wname)
```

where [C,L] is the wavelet decomposition structure of the signal x obtained using
wavedec up to level lev, in association with the wavelet wname.
The function wden automatically finds the threshold value on the basis of the parameters tptr and scal.

- the tptr string contains the threshold selection rule (Donoho-Johstone family):

  - 'rigrsure' uses the principle of Stein's Unbiased Risk;
  - 'heursure' is an heuristic variant of the first option;
  - 'sqtwolog' uses the so-called universal threshold by Donoho and Johnstone
    (Square2log);
  - 'minimaxi' uses minimax thresholding.

- scal defines the multiplicative threshold rescaling:

  - 'one' for no rescaling (useful for simulations in which noise has unit variance);
  - 'sln' for rescaling using a single estimation of the noise level, based on the
    first-level coefficients of the decomposition structure [C,L] computed using the
    wavelet specified in wname;
  - 'mln' for rescaling done using a level-dependent estimation of the noise level
    (useful for dealing with non-white noise).

  To estimate the noise level, wden calls the function wnoisest: the command

```
sigma = wnoisest(C,L,S)
```

returns robust estimates of the detail coefficients' standard deviation for the levels
contained in the input vector S. The noise-level estimator given in Sect. 15.2.2 is
indeed a robust estimator of the standard deviation of the level-1 detail coefficients
in the case of white noise, or of the level-1, level-2, level-3 etc. in the case of non-white
noise. Thus, for a single estimation of the noise level the command is wnoisest(C,L,1);
for a level-dependent estimation the command is wnoisest(C,L,S), with S containing
the levels for which the noise level must be estimated, e.g., S = [1 2 3 ... lev]. The
outputs of wden are the de-noised signal xd and its wavelet decomposition structure
[Cxd,Lxd].

De-noising or Compression Using wdencmp with Donoho-Johstone Threshold Selection Criteria

The function wdencmp must receive in input the pre-fixed values of four parameter, defining the strategy for DWT-based de-noising or compression of the signal x:

- a scalar opt, declaring if a global threshold is desired, or if a level-dependent threshold is preferred;
- a scalar threshold thr (one threshold for all levels; global threshold) or vector threshold thr (level-dependent threshold), according to the value of opt;
- a scalar sorh, declaring if a soft or hard thresholding must be performed;
- a scalar keepapp, stating if the approximation coefficients must be kept untouched.

To define these input parameters, we can use default values, which are provided by the function ddencmp, or make our own decisions. In the latter case, the selection of the threshold value(s) is the most crucial. A threshold can be established on the basis on some criterion, or calculated exploiting some other Matlab function.

- The function ddencmp returns default values for DWT-based de-noising or compression (Donoho 1995; Donoho and Johnstone 1994a, b) of an input signal x. The syntax is

```
1   [thr,sorh,keepapp] = ddencmp(in1,in2,x)
```

In input, in1 is 'den' for de-noising or 'cmp' for compression; in2 is 'wv' for wavelets.[5] In output, the threshold thr is always computed as the Square2log (sqtwolog) threshold, properly rescaled by the noise-level estimate(s) for the examined signal; the thresholding rule sorh is 's' (soft) for de-noising and 'h' (hard) for compression; keepapp is always 1, i.e., the approximation coefficients are always left untouched. More precisely, the threshold defaults are the following:

- if in1 = 'den' (de-noising),
  the nominal unscaled threshold is thr = sqrt(2*log(N_DWT)), where N_DWT is the length of C. Then the threshold is rescaled according to the value of scal:

```
1   [C,L] = wavedec(x,1,'db1')
2   cD = C(L(1)+1:end)
3   scalingfact = median(abs(cD))
4   thr = thr* scalingfact /0.6745
```

Note the use of the db1 wavelet, independently of the wavelet wname that will be specified for use in wdencmp. The median of the absolute value of level-1 detail coefficients divided by 0.6745 is a robust estimator of the noise level, and this ensures an estimate which is nor contaminated by DWT end effects, which are pure artifacts due to computations on the edges, nor by those possibly signal-related few level-1 detail coefficients that may be present;

---

[5]The function can be used also for wavelets packets.

– if in1 = 'cmp' (compression),
the nominal threshold is set to thr = 1. The rescaled threshold is then computed
as follows:

```
1  [C,L] = wavedec(x,1,'db1')
2  cD = C(L(1)+1:end)
3  medianc_db1 = median(abs(cD))
4  % If medianc_db1=0, simply kill the lowest coefficients
5  if medianc_db1 == 0
6      medianc_db1 = 0.05*max(abs(cd))
7  end
8  thr = thr *medianc_db1
```

- If we do not use ddencmp, how can we choose a threshold? We can use the function
thselect:

```
1  thr = thselect(x,tptr)
```

returns the x-adapted threshold value using the threshold selection rule defined by
the string tptr. Available options belong to the Donoho-Johnston family, and are
the same described when discussing wden input variables. Dealing with non-white
noise can then be handled by rescaling the output threshold thr by a level-dependent
estimation of the noise level.

The wdencmp function performs wavelet coefficients thresholding for both de-
noising and compression purposes (Donoho 1993, 1995; Donoho and Johnstone
1994a, b; Donoho et al. 1995). Its syntax is

```
1  [xd,Cxd,Lxd] = wdencmp(opt,x,wname,lev,thr,sorh,keepapp)
```

for de-noising, and

```
1  [xc,Cxc,Lxc,perf0,perfl2] = wdencmp(opt,x,wname,lev,thr,sorh,keepapp)
```

for compression, where

- x is the signal to be de-noised or compressed using the wavelet wname at level
lev;
- opt = 'gbl' and thr is a positive real number for a global threshold;
- opt = 'lvd' and thr is a vector for a level-dependent threshold; the length of thr
must be equal to lev;
- sorh = 's' (soft) or 'h' (hard) (thresholding rule);
- keepapp = 1 to keep approximation coefficients, and keepapp = 0 to allow for
approximation coefficients thresholding (this option is seldom used).

The function returns a de-noised (xd) or compressed (xc) version of the input signal
x and the corresponding wavelet decomposition structure ([Cxd,Lxd] or [Cxc,Lxc],
respectively). Additional output arguments used in compression are perfl2 and perf0:
the ℓ^2-norm recovery and the compression score in percentage, respectively.

perfl2 = 100*(vector-norm of Cxc/vector-norm of C)^2, where [C,L] denotes the
wavelet decomposition structure of x. If wname is an orthogonal wavelet, perfl2

reduces to $100*|xc|^2/|x|^2$. Other performance parameters we may want to compute for each threshold value are

```
1   retained = 100 − perf0 % Retained coefficients in percentage
2   RMSdev= sqrt(sum((x−xco).^2)/N) % Neglected signal energy
3   RMSdev_check = sqrt(sum((c2−cxc).^2)/N) % Energy in neglected coefficients
```

The function wdencmp can also be called using directly the wavelet decomposition structure [C,L] of the signal to be de-noised or compressed, previously computed at level lev using the wavelet wname: for example,

```
1   [xc,Cxc,Lxc,perf0,perfl2] = wdencmp('gbl',C,L,wname,lev,thr,sorh,keepapp)
```

This is useful if we have already decomposed the signal x:

```
1   [C,L] = wavedec(x,lev,wname)
```

Inside wdencmp (and wden), the actual thresholding is operated by a function called wthresh. This function performs soft or hard thresholding. The command

```
1   y = wthresh(x,sorh,thr)
```

returns the soft (if sorh = 's') or hard (if sorh = 'h') thresholding of the input vector x using the threshold thr.

De-noising or Compression Using wdencmp with Birgé–Massart Threshold Selection Criteria

So far, we described approaches to denoising or compression that rely on thresholds belonging to the Donoho-Johnstone family. Other Matlab functions implement different threshold selection criteria for de-noising or compression. For example, wdcbm uses the Birgé–Massart threshold, while wbmpen uses the Birgé–Massart penalized threshold.

After performing a wavelet decomposition of the signal x at level lev with the wavelet wname,

```
1   [C,L] = wavedec(x,lev,wname)
```

we can

- use wdcbm for selecting level-dependent thresholds for signal de-noising or compression:

```
1   [thr,nkeep] = wdcbm(C,L,alpha,M)
```

Typically, alpha = 1.5 for compression and alpha = 3 for de-noising. A default value for M is M = L(1), the number of the coarsest approximation coefficients. Recommended values for M are from L(1) to 2L(1).

The function returns the level-dependent thresholds thr and numbers of coefficients to be kept, nkeep, for de-noising or compression (Birgé and Massart 1997). The parameters alpha and M must be greater than 1. In output, thr is a vector of length lev; thr(j) contains the threshold for level j, with j = [1,lev]. The other output, nkeep, is a vector of length lev; nkeep(j) contains the number of coefficients to be kept at level j;

- use the wbmpen function for selecting a global threshold for signal de-noising (no compression), which is called Birgé–Massart penalized threshold:

```
thr = wbmpen(C,L,sigma,alpha)
```

Here thr is obtained by a wavelet coefficients selection rule using the penalization method provided by Birgé and Massart (1997). The parameter sigma is the standard deviation of the zero-mean Gaussian white noise in de-noising model (see wnoisest for more information); alpha is a *tuning parameter for the penalty term* included in the Birgé and Massart (1997) method. It must be a real number greater than 1. The sparsity of the wavelet representation of the de-noised signal grows with alpha; typically, alpha = 2.

Once the necessary parameters have been determined, wdencmp can be called. We will now try different shades of de-noising.

1. First we use ddencmp to obtain the default global threshold for wavelet de-noising. In this way we verify that the threshold is equal to the universal threshold of Donoho and Johnstone, scaled by a robust estimate of the noise standard deviation. For this exercise, run the script check_default_glbthresh.

```
1   % check_default_glbthresh
2   % Use ddencmp to obtain the default global threshold for wavelet denoising.
3   % Demonstrate that the threshold is equal to the universal threshold
4   % of Donoho and Johnstone, scaled by a robust estimate of the variance.
5
6   clear
7
8   % Generate white noise signal
9   % Set the random number generator to the default initial settings
10  % for reproducible results.
11  rng default;
12  x = randn(512,1);
13
14  % Compute default global threshold by ddencmp
15  in1= 'den'
16  in2='wv'
17  [thr,sorh,keepapp] = ddencmp(in1,in2,x)
18
19  % Decompose the signal at level 1 by db1
20  wname='db1'
21  [cA,cD] = dwt(x,wname);
22  NA=length(cA);
23  ND=length(cD);
24  N_DWT=NA+ND
25
26  % Compute noise level
27  noiselev = median(abs(cD))/0.6745;
28
29  % Compute universal threshold of Donoho and Johnstone
```

```
30   uthr = sqrt(2*log(N_DWT));
31   thr_check = uthr*noiselev
```

2. As a second step, we try the function wden for automatic de-noising on the "heavy sine" test signal used by Donoho and Johnstone (1994a). We set the SNR and the seed for noise generation (so that results are reproducible). These parameters serve as inputs to the wnoise function that generates the reference signal xref and a noisy version of it, x, constructed adding Gaussian white noise with unit variance to xref, with some prescribed SNR. Four de-noising strategies are then applied to the noisy signal:

   a. x is de-noised using soft Square2log thresholding, working on detail coefficients obtained from the decomposition of x at level 5 by the sym8 wavelet. No multiplicative threshold rescaling is performed, since in this case we know a priori that the standard deviation of noise, which is 1;
   b. x is de-noised using soft SURE thresholding;
   c. x is de-noised using soft heurSURE thresholding;
   d. x is de-noised using soft minimax thresholding.

The de-noised signals are plotted and compared with the reference signal and among themselves.

Automatic de-noising by soft heuristic SURE thresholding is then repeated on another test signal: a noisy block signal with some prescribed SNR. Since only a small number of large coefficients characterize the original signal, the method performs very well. The script for this exercise is automatic_de_noising (see below).

```
1    % automatic_de_noising
2    % Try different automatic de−noising procedures
3    % on noisy heavy sine signal by Donoho and Johnstone:
4    % 1) soft Square2log thresholding,
5    % 2) soft SURE thresholding,
6    % 3) soft heuristic SURE thresholding,
7    % 4) soft minimax thresholding.
8
9    % Repeat soft heuristic SURE thresholding
10   % on noisy blocks signal by Donoho and Johnstone.
11
12   clear
13
14   % Set number of samples
15   nu=11 %N=2^nu
16
17   % Set signal to noise ratio
18   snr = 9;
19   sqrtsnr=sqrt(snr)
```

```
20
21   % Set rand seed for signal generation
22   init = 2055615866;
23
24   % Generate reference heavy sine signal (xref) and noisy version of it (x)
25   % containing additive Gaussian white noise with std=1
26   fun='heavy sine'
27   [xref,x] = wnoise(fun,nu,sqrtsnr,init);
28   %std(xref) = sqrtsnr
29   N = length(x);
30   time=0:N−1;
31
32   % Test noise std − supposed to be 1
33   noise = x−xref;
34   varnoise = var(noise);
35   stdnoise = sqrt(varnoise)
36
37   % De−noise noisy signal using Square2log thresholding of detail
           coefficients
38   % obtained from decomposition of x at level 5 by sym8 wavelet
39   tptr='sqtwolog'
40   sorh='s'
41   scal='one'% noise has sigma=1
42   lev = 5
43   wname = 'sym8'
44   xd = wden(x,tptr,sorh,scal,lev,wname);
45
46   % Plot original and de−noised signals
47   figure(5000)
48   subplot(3,2,1)
49   plot(time,xref)
50   set(gca,'XLim',[−1 N])
51   hlabely=get(gca,'Ylabel');
52   set(hlabely,'String','$x_{\mathrm{ref}}$','Interpreter','Latex','FontSize',16)
53   title('Heavy sine reference signal');
54   subplot(3,2,2)
55   plot(time,x)
56   set(gca,'XLim',[−1 N])
57   hlabely=get(gca,'Ylabel');
58   set(hlabely,'String','$x[n]$','Interpreter','Latex','FontSize',16)
59   title(['Noisy signal − SNR =' num2str(fix(snr))]);
60   subplot(3,2,3)
61   plot(time,xd)
62   set(gca,'XLim',[−1 N])
63   hlabely=get(gca,'Ylabel');
```

```
64  set(hlabely,'String','$x_d$','Interpreter','Latex','FontSize',16)
65  title('De−noised − Square2log − sym8, lev. 5');
66
67  % De−noise noisy signal using soft SURE thresholding of detail
         coefficients
68  % obtained from the decomposition of x at level 5 using sym8 wavelet
69  tptr='rigrsure'
70  xd = wden(x,tptr,sorh,scal,lev,wname);
71
72  % Plot de−noised signal
73  subplot(3,2,4)
74  plot(time,xd)
75  set(gca,'XLim',[−1 N])
76  hlabely=get(gca,'Ylabel');
77  set(hlabely,'String','$x_d$','Interpreter','Latex','FontSize',16)
78  title('De−noised − SURE − sym8, lev. 5');
79
80  % De−noise noisy signal
81  % using soft heuristic SURE thresholding on detail coefficients
82  % obtained from the decomposition of x at level 5 using sym8 wavelet
83  tptr='heursure'
84  xd = wden(x,tptr,sorh,scal,lev,wname);
85
86  % Plot de−noised signal
87  subplot(3,2,5)
88  plot(time,xd)
89  set(gca,'XLim',[−1 N])
90  hlabelx=get(gca,'Xlabel');
91  set(hlabelx,'String','$n$','Interpreter','Latex','FontSize',16)
92  hlabely=get(gca,'Ylabel');
93  set(hlabely,'String','$x_d$','Interpreter','Latex','FontSize',16)
94  title('De−noised − HeurSURE − sym8, lev. 5');
95
96  % De−noise noisy signal using minimax thresholding of detail coefficients
97  % obtained from the decomposition of x at level 5 using sym8 wavelet.
98  tptr='minimaxi'
99  xd = wden(x,tptr,sorh,scal,lev,wname);
100
101 % Plot de−noised signal
102 subplot(3,2,6)
103 plot(time,xd)
104 set(gca,'XLim',[−1 N])
105 hlabelx=get(gca,'Xlabel');
106 set(hlabelx,'String','$n$','Interpreter','Latex','FontSize',16)
107 hlabely=get(gca,'Ylabel');
```

```
108  set(hlabely,'String','$x_d$','Interpreter','Latex','FontSize',16)
109  title('De−noised − Minimax − sym8, lev. 5');
110
111  %
112  % Repeat soft heuristic SURE thresholding on another test signal
113  %
114
115  % Set signal to noise ratio and set rand seed.
116  snr = 16;
117  sqrtsnr = sqrt(snr)
118
119  % Set rand seed for signal generation
120  init = 2055615866;
121
122  % Set number of samples
123  nu=11%N=2^nu
124
125  % Generate reference blocks signal xref and a noisy version of it (x)
126  % adding standard Gaussian white noise with std=1
127  fun='blocks'
128  [xref,x] = wnoise(fun,nu,sqrtsnr,init);
129  N= length(x);
130  time=0:N−1;
131
132  % Test noise std− supposed to be 1
133  noise=x−xref;
134  varnoise=var(noise);
135  stdnoise=sqrt(varnoise)
136
137  % De−noise noisy signal
138  % using soft heuristic SURE thresholding of detail coefficients
139  % obtained from decomposition of x at level 3 using sym8 wavelet
140  tptr='sqtwolog'
141  sorh='s'
142  scal='one'% noise has sigma=1
143  lev = 3
144  wname = 'sym8'
145  xd = wden(x,tptr,sorh,scal,lev,wname);
146
147  % Plot original and de−noised signals
148  figure(5100)
149  subplot(3,1,1)
150  plot(time,xref)
151  set(gca,'XLim',[−1 N])
152  set(gca,'YLim',[−7 17])
```

```
153   hlabely=get(gca,'Ylabel');
154   set(hlabely,'String','$x_{\mathrm{ref}}$','Interpreter','Latex','FontSize',16)
155   title('Blocks reference signal')
156   subplot(3,1,2)
157   plot(time,x)
158   set(gca,'XLim',[-1 N])
159   set(gca,'YLim',[-7 17])
160   hlabely=get(gca,'Ylabel');
161   set(hlabely,'String','$x$','Interpreter','Latex','FontSize',16)
162   title(['Noisy signal - SNR =' num2str(fix(snr))]);
163   subplot(3,1,3)
164   plot(time,xd)
165   set(gca,'YLim',[-7 17])
166   set(gca,'XLim',[-1 N])
167   hlabelx=get(gca,'Xlabel');
168   set(hlabelx,'String','$n$','Interpreter','Latex','FontSize',16)
169   hlabely=get(gca,'Ylabel');
170   set(hlabely,'String','$x_d$','Interpreter','Latex','FontSize',16)
171   title('De-noised - heuristic SURE - sym8, lev. 3');
```

3. Next, we de-noise a noisy bumps signal that is available in the Matlab distribution as noisbump.mat, using the function wdencmp and adopting a Birgé–Massart threshold selection approach (wbmpen). We load the signal and perform a wavelet decomposition of the signal at level 5 using sym6. We then estimate the noise standard deviation from the detail coefficients at level 1 of this decomposition, using wnoisest, and use wbmpen for selecting a global threshold for de-noising, with the tuning parameter alpha = 2. Then we apply wdencmp for de-noising the signal using the above threshold, with soft thresholding and approximation kept. We plot original and de-noised signals. The script reported below (de_noising) performs this task.

```
1    % de_noising
2    % De-noising using wdencmp with Birge-Massart penalized threshold
3
4    clear
5
6    % Load noisy bumps signal.
7    load noisbump;
8    x = noisbump;
9    N=length(x);
10   time=0:N-1;
11
12   % Perform a wavelet decomposition of the signal at level 5 using sym6
13   lev = 5
14   wname = 'sym6'
15   [C,L] = wavedec(x,lev,wname);
16
17   % Estimate the noise standard deviation from the detail coefficients at level 1
```

```
18   sigma = wnoisest(C,L,1)
19
20   % Use wbmpen for selecting global threshold
21   % for signal de−noising, using the tuning parameter
22   alpha = 2;
23   thr = wbmpen(C,L,sigma,alpha)
24
25   % Use wdencmp for de−noising the signal
26   % using the above threshold with soft thresholding and approximation kept
27   o='gbl'
28   sorh='s'
29   keepapp = 1
30   xd = wdencmp(o,C,L,wname,lev,thr,sorh,keepapp);
31
32   % Plot original and de−noised signals
33   figure(5200)
34
35   subplot(2,1,1)
36   plot(time,x)
37   set(gca,'XLim',[−1 N])
38   hlabely=get(gca,'Ylabel');
39   set(hlabely,'String','$x$','Interpreter','Latex','FontSize',16)
40   title('Noisy bump signal')
41   subplot(2,1,2)
42   plot(time,xd)
43   set(gca,'XLim',[−1 N])
44   hlabelx=get(gca,'Xlabel');
45   set(hlabelx,'String','$n$','Interpreter','Latex','FontSize',16)
46   hlabely=get(gca,'Ylabel');
47   set(hlabely,'String','$x_d$','Interpreter','Latex','FontSize',16)
48   title('De−noised − Birge−Massart penalized threshold − sym6, lev. 5')
```

4. As the last exercise on de-noising, we tackle the case of a real-world signal, the same electrical load signal used in Chap. 14 that represents electrical consumption measured over the course of 3 days. This signal is particularly interesting because of noise introduced when a defect developed in the monitoring equipment as the measurements were being made. Wavelet analysis effectively removes the noise. This signal is included in the Matlab distribution as leleccum.mat.
For this example, proceed as follows.

   • Load the signal and select a portion for wavelet analysis.
   • Perform a multi-level wavelet decomposition of the signal. Use the db1 wavelet and level 3. If we look at the details from the decomposition at level 3 we notice that noise in this latter part of the signal causes the details show a great activity. So we can manipulate each of the vectors cD1, cD2, and cD3, setting each vector's element to some fraction of the vectors' peak or average value. Then we can reconstruct new detail signals D1, D2, and D3 from the thresholded coefficients.
   • To de-noise the signal according to these lines, use the ddencmp function specifying the global thresholding option 'gbl', to calculate default parameters; then use the wdencmp command to perform the actual de-noising, specifying

the results of the previous signal's decomposition (C and L) and the wavelet (db1) that was used to perform the decomposition. Display the original and de-noised signal. Notice the noise has been removed without compromising the sharp detail of the original signal. This is a strength of wavelet analysis.

- Repeat the de-noising of the same signal segment using the wavelet db3: this time decompose the signal first at level 2, and then at level 5. Observe the differences in the result with respect to the previous case.
- In this real-world case, the noise might actually be non-white. To deal with the composite nature of noise, thresholds must be rescaled by a level-dependent estimation of the noise level. The idea is to define the rescaled threshold level by level, thus increasing the capability of the de-noising strategies.[6] Try the method on a highly perturbed part of the electrical signal (samples 2000–3450). Memorize the first signal value, deb = x(1), and then de-noise x-deb instead of x, to avoid edge effects. Use wden with the option tptr = 'sqtwolog' (scalar universal threshold) and soft thresholding; use again the db3 wavelet and decompose at level 3. Use a level-dependent estimation of the noise level by setting scal = 'mln' in wden.

Use the script given below (de_noise_leleccum). The result is quite good in spite of the time heterogeneity of the nature of the noise after and before the beginning of the sensor failure around time 2450.

```
1   % de_noise_leleccum.m
2   % Signal de-noising
3
4   clear
5
6   % Load the signal
7   load leleccum;
8   Ntot=length(leleccum)
9   T_s=1 %min
10  t = 0:T_s:(Ntot-1)*T_s; % (3 days)
11
12  % Select a part for de-noising
13  indx = 2000:3920;
14  x = leleccum(indx);
15  N=length(x)
16  time=t(indx);
17
18  % Decompose at level 3 by db1
19  lev = 3;
```

---

[6]In the most general case, the threshold or the set of level-dependent thresholds could be made time-dependent, to handle non-stationary variance noise models. In that case, the signal's model would still be $x[n] = s[n] + \sigma_e e[n]$, but the noise standard deviation σ_e should be allowed to vary with time, because there are several different variance values on several time intervals. The values as well as the intervals could then be found by wvarchg. We will not go deeper into these more advanced de-noising techniques.

```matlab
20   wname = 'db1'
21   [C,L] = wavedec(x,lev,wname);
22
23   % Threshold de-noising
24
25   % Calculate default parameters for de-noising
26   in1='den'
27   in2='wv'
28   [thr,sorh,keepapp] = ddencmp(in1,in2,x)
29
30   % Perform de-noising
31   o='gbl'
32   xd = wdencmp(o,C,L,wname,lev,thr,sorh,keepapp);
33   Nden=length(xd)
34
35   % Display both the original and de-noised signals
36   figure(5300)
37   subplot(4,1,1)
38   plot(time,x)
39   set(gca,'XLim',[time(1) time(end)]);
40   hlabely=get(gca,'Ylabel');
41   set(hlabely,'String','$x$','Interpreter','Latex','FontSize',16)
42   title('Threshold de-noising with default parameters of electrical load
            signal')
43
44   subplot(4,1,2)
45   plot(time,xd)
46   set(gca,'XLim',[time(1) time(end)]);
47   hlabely=get(gca,'Ylabel');
48   set(hlabely,'String','$x_d$','Interpreter','Latex','FontSize',16)
49   text(2100,200,'db1, lev=3','FontSize',12)
50
51   % Decompose at level 2 by db3
52   clear C L
53   lev = 2;
54   wname = 'db3';
55   [C,L] = wavedec(x,lev,wname);
56
57   % Calculate default parameters for de-noising
58   in1='den'
59   in2='wv'
60   [thr,sorh,keepapp] = ddencmp(in1,in2,x)
61
62   % Perform de-noising
63   o='gbl'
```

```
64   xd = wdencmp(o,C,L,wname,lev,thr,sorh,keepapp);
65   Nden=length(xd)
66
67   % Plot results
68   subplot(4,1,3)
69   plot(time,xd);
70   set(gca,'XLim',[time(1) time(end)]);
71   hlabely=get(gca,'Ylabel');
72   set(hlabely,'String','$x_d$','Interpreter','Latex','FontSize',16)
73   text(2100,200,'db3, lev=2','FontSize',12)
74
75   % Decompose at level 5 by db3
76   clear C L
77   lev = 5;
78   wname = 'db3';
79   [C,L] = wavedec(x,lev,wname);
80
81   % Calculate default parameters for de-noising
82   in1='den'
83   in2='wv'
84   [thr,sorh,keepapp] = ddencmp(in1,in2,x)
85
86   % Perform de-noising
87   o='gbl'
88   xd = wdencmp(o,C,L,wname,lev,thr,sorh,keepapp);
89   Nden=length(xd)
90
91   % Plot results
92   subplot(4,1,4)
93   plot(time,xd);
94   set(gca,'XLim',[time(1) time(end)]);
95   hlabelx=get(gca,'Xlabel');
96   set(hlabelx,'String','$t$ (min)','Interpreter','Latex','Fontsize',16)
97   hlabely=get(gca,'Ylabel');
98   set(hlabely,'String','$x_d$','Interpreter','Latex','FontSize',16')
99   text(2100,200,'db3, lev=5','FontSize',12)
100
101  % To deal with the composite noise nature,
102  % try a level-dependent noise size estimation
103
104  % Select a highly perturbed part of the signal
105  clear x time
106  indx = 2000:3450;
107  x = leleccum(indx);
108  N=length(x)
```

```
109   time=t(indx);
110
111   % Find first value in order to avoid edge effects
112   deb = x(1);
113
114   % De−noise signal using a level−dependent estimation of noise level
115   % Use db3 at level 3
116   tptr='sqtwolog'
117   sorh='s'
118   scal = 'mln';
119   lev=3;
120   wname='db3';
121   xd = wden(x−deb,tptr,sorh,scal,lev,wname)+deb;
122   Nden=length(xd)
123
124   figure(5400)
125   subplot(2,1,1)
126   plot(time,x)
127   set(gca,'XLim',[time(1) time(end)]);
128   hlabely=get(gca,'Ylabel');
129   set(hlabely,'String','$x$', 'Interpreter','Latex','FontSize',16)
130   title('De−noising of electrical load signal with level−dependent
          estimation of noise level ')
131
132   subplot(2,1,2)
133   plot(time,xd)
134   set(gca,'XLim',[time(1) time(end)]);
135   hlabelx=get(gca,'Xlabel');
136   set(hlabelx,'String','$t$ (min)','Interpreter','Latex','Fontsize',16)
137   hlabely=get(gca,'Ylabel');
138   set(hlabely,'String','$x_d$','Interpreter','Latex','FontSize',16)
139   text(2100,100,'db3, lev. 3 − Square2log','FontSize',12)
```

Finally, we can experiment with compression. Over a selected part of the electrical load signal, proceed as follows.

• Perform compression by hard thresholding, using wdencmp with a fixed threshold of 35 (see Sect. 15.3). In this compression, about 83 % of the coefficients are set to zero, but 99 % of the energy in the signal is retained. Thus the procedure results in effective signal compression.

• As a last step, perform a wavelet decomposition at level 5 using db3, and use wdcbm for selecting level-dependent thresholds according to the Birgé–Massart scarce-high approach for signal compression, with alpha = 1.5 and M = L(1). Use wdencmp for compressing the signal using the above thresholds with hard thresholding. Observe the original and compressed signals and examine the compression performances of the adopted procedure. Use the following script (compress_leleccum).

```
1    % compress_leleccum.m
2    % Signal compression
3
4    % 1. Use Donoho approach (ddencmp and wdencmp )
5
6    clear
7
8    % Load signal
9    load leleccum;
10   Ntot=length(leleccum)
11   T_s=1 %min
12   t= 0:T_s:(Ntot-1)*T_s; % (3 days)
13
14   % Select a part of load electrical signal
15
16   indx = 2300:3500;
17   x = leleccum(indx);
18   time=t(indx); % time is shorter than t
19   tmin=min(time)
20   tmax=max(time)
21   xmin=min(x)
22   xmax=max(x)
23   N=length(x)
24
25   % Perform wavelet decomposition at level 3 by db3
26   lev=3
27   wname='db3'
28   [C,L] = wavedec(x,lev,wname);
29   N_DWT=length(C)
30   N_DWT_check=sum(L(1:end-1))
31   Nd=sum(L(2:end-1))
32   Na=L(1)
33
34   % Find default parameter values for compression
35   in1='cmp'
36   in2='wv'
37   [thr,sorh,keepapp] = ddencmp(in1,in2,x);
38   thr_default=thr
39
40   % Compress with threshold of 35 (about ten times the default value)
41   thre=35
42   o='gbl'
43   [xc,Cxc,Lxc,perf0,perfl2] = wdencmp(o,C,L,wname,lev,thre,sorh,keepapp);
44   Ncomp=length(xc)
45   perf0_DJ=perf0
46   perfl2_DJ=perfl2
47
48   % Compute more performance parameters for each threshold value
49   retained=100-perf0
50   RMSdev= sqrt(sum((x-xc).^2)/N)
```

```
51
52   % Reconstruct approximation
53   typ='a';
54   A = wrcoef(typ,C,L,wname,lev);
55
56   % Plot results – compare approximation A_3 with compression (threshold=35)
57
58   figure(5500)
59
60   subplot(4,1,1)
61   plot(time,x);
62   set(gca,'XLim',[tmin tmax])
63   hlabely=get(gca,'Ylabel');
64   set(hlabely,'String','$x$','Interpreter','Latex','FontSize',16)
65   title('Compression of electrical load signal according to different criteria')
66
67   subplot(4,1,2)
68   plot(time,A);
69   set(gca,'XLim',[tmin tmax])
70   hlabely=get(gca,'Ylabel');
71   set(hlabely,'String','$A_3$','Interpreter','Latex','FontSize',16)
72   text(2400,200,'db3, lev. 3','FontSize',12)
73
74   subplot(4,1,3)
75   plot(time,xc);
76   set(gca,'XLim',[tmin tmax])
77   hlabely=get(gca,'Ylabel');
78   set(hlabely,'String','$x_c$','Interpreter','Latex','FontSize',16)
79   text(2400,200,'db3, lev. 3, fixed threshold of 35','FontSize',12)
80
81   % 2. Use Birg – Massat approach (wdcbm)
82
83   % Perform decomposition at level 5 using db3.
84   clear C L
85   lev = 5
86   wname = 'db3'
87   [C,L] = wavedec(x,lev,wname);
88
89   % Use wdcbm for selecting level–dependent thresholds
90   % for signal compression using the adviced parameters
91   alpha = 1.5
92   M = L(1)
93   [thr,nkeep] = wdcbm(C,L,alpha,M)
94
95   % Compres the signal using the above thresholds with hard thresholding.
96   o='lvd'
97   sorh='h'
98   [xc,Cxd,Lxd,perf0,perfl2] = wdencmp(o,C,L,wname,lev,thr,sorh);
99   Ncomp=length(xc)
100  perfl0_BM=perf0
101  perfl2_BM=perfl2
102
103  % Plot original and compressed signals
```

```
104 | subplot(4,1,4)
105 | plot(time,xc)
106 | set(gca,'XLim',[tmin tmax])
107 | hlabelx=get(gca,'Xlabel');
108 | set(hlabelx,'String','$t$ (min)','Interpreter','Latex','Fontsize',16)
109 | hlabely=get(gca,'Ylabel');
110 | set(hlabely,'String','$x_c$','Interpreter','Latex','FontSize',16)
111 | text(2400,200,'db3, lev. 5, Birge−Massart','FontSize',12)
```

References

Antoniadis, A., Oppenheim, G. (eds.): Wavelets and Statistics. Lecture Notes in Statistics, vol. 103. Springer, Berlin (1995)

Birgé, L., Massart, P.: From Model Selection to Adaptive Estimation. In: Pollard, D. (ed.) Festschrift for Lucien Le Cam: Research Papers in Probability and Statistics. Springer, New York (1997)

Daubechies, I.: Ten Lectures on Wavelets. CBMS-NSF Regional Conference Series in Applied Mathematics, Society for Industrial and Applied Mathematics (SIAM), Philadelphia (1992)

Donoho, D.L.: Wavelet shrinkage and W.V.D. A ten-minute tour. In: Meyer, Y., Roques, S. (eds.) Progress in Wavelet Analysis and Applications, pp. 109–128. Editions Frontières, Gif-sur-Yvette (Paris) (1993)

Donoho, D.L.: De-noising by soft-thresholding. IEEE Trans. Inf. Theory **41**(3), 613–627 (1995)

Donoho, D.L., Johnstone, I.M.: Ideal spatial adaptation by wavelet shrinkage. Biometrika **81**, 425–455 (1994a)

Donoho, D.L., Johnstone, I.M.: Ideal de-noising in an orthonormal basis chosen from a library of bases. Comptes Rendus de l'Académie des Sciences (CRAS)—Series I, vol. 319, pp. 1317–1322. Paris (1994b)

Donoho, D.L., Johnstone, I.M.: Adapting to unknown smoothness via wavelet shrinkage. J. Am. Stat. Assoc. (JASA) **90**(432), 1200–1224 (1995)

Donoho, D.L., Johnstone, I.M., Kerkyacharian, G., Picard, D.: Wavelet shrinkage: asymptopia. J. Roy. Stat. Soc. Ser. B **57**(2), 301–369 (1995)

Lavielle, M.: Detection of multiple changes in a sequence of dependent variables. Stoch. Proc. Appl. **83**(2), 79–102 (1999)

Mallat, S.: A theory for multiresolution signal decomposition: the wavelet representation. IEEE Trans. Pattern Anal. Mach Intell. **11**(7), 674–693 (1989)

Meyer, Y.: Wavelets and Operators. Cambridge University Press, Cambridge (1993)

Moler, C.B.: Numerical Computing with MATLAB. Society for Industrial and Applied Mathematics (SIAM), Philadelphia (2004)

Oppenheim, A.V., Schafer, R.W.: Discrete-Time Signal Processing. Prentice Hall, Englewood Cliffs (2009)

Orfanidis, S.J.: Introduction to Signal Processing. Prentice Hall, Upper Saddle River (1996)

Smith, C.A., Sardeshmukh, P.: The effect of ENSO on the intraseasonal variance of surface temperature in winter. Int. J. Climatol. **20**, 1543–1557 (2000)

Strang, G., Nguyen, T.: Wavelets and Filter Banks. Wellesley-Cambridge Press, Wellesley (1996)

The MathWorks: Signal Processing Toolbox User's Guide (R2014b). The MathWorks, Inc., Natick (2014). http://www.mathworks.com/help/pdf_doc/signal/signal_tb.pdf. Accessed 22 Nov 2014

The MathWorks: Wavelet Toolbox User's Guide (R2014b). The MathWorks, Inc., Natick (2014). http://www.mathworks.com/help/pdf_doc/wavelet/wavelet_ug.pdf. Accessed 22 Nov 2014

Torrence, C., Compo, G.P.: A practical guide to wavelet analysis. B. Am. Meteor. Soc. **79**(1), 61–78 (1998)

Index

© Springer International Publishing Switzerland 2016
S.M. Alessio, *Digital Signal Processing and Spectral
Analysis for Scientists*, Signals and Communication Technology,
DOI 10.1007/978-3-319-25468-5

Printed in the United States
By Bookmasters